Synthese Library

Studies in Epistemology, Logic, Methodology, and Philosophy of Science

Volume 467

The aim of *Synthese Library* is to provide a forum for the best current work in the methodology and philosophy of science and in epistemology, all broadly understood. A wide variety of different approaches have traditionally been represented in the Library, and every effort is made to maintain this variety, not for its own sake, but because we believe that there are many fruitful and illuminating approaches to the philosophy of science and related disciplines.

Special attention is paid to methodological studies which illustrate the interplay of empirical and philosophical viewpoints and to contributions to the formal (logical, set-theoretical, mathematical, information-theoretical, decision-theoretical, etc.) methodology of empirical sciences. Likewise, the applications of logical methods to epistemology as well as philosophically and methodologically relevant studies in logic are strongly encouraged. The emphasis on logic will be tempered by interest in the psychological, historical, and sociological aspects of science. In addition to monographs *Synthese Library* publishes thematically unified anthologies and edited volumes with a well-defined topical focus inside the aim and scope of the book series. The contributions in the volumes are expected to be focused and structurally organized in accordance with the central theme(s), and should be tied together by an extensive editorial introduction or set of introductions if the volume is divided into parts. An extensive bibliography and index are mandatory.

Ralph Jenkins

A Protocol-theoretic Framework for the Logic of Epistemic Norms

Springer

Ralph Jenkins
CUNY City College of New York
New York, NY, USA

ISSN 0166-6991 ISSN 2542-8292 (electronic)
Synthese Library
ISBN 978-3-031-08596-3 ISBN 978-3-031-08597-0 (eBook)
https://doi.org/10.1007/978-3-031-08597-0

This Springer imprint is published by the registered company Springer Nature Switzerland AG
The registered company address is: Gewerbestrasse 11, 6330 Cham, Switzerland

For Louie, whose pure warmth, attentiveness, and constant companionship have been intertwined with this project from its inception, and whose absence is an unrecoverable loss.

Preface

Epistemic rationality is fundamentally a regulative concept; the point of it is guiding the deliberate control of epistemic action. This means that epistemic norms share an underlying procedural structure with plans, programs, or algorithms, and the logical investigation of epistemic norms should reflect this procedural structure. This, in short, is the set of ideas that motivate the project of which this book is a part.

This book defines a logical system called the Protocol-theoretic Logic of Epistemic Norms (PLEN), it develops PLEN into a formal framework for representing and reasoning about epistemic norms, and it shows that PLEN is theoretically interesting and useful with regard to the aims of such a framework. In order to motivate the project, I defend an account of epistemic norms called *epistemic proceduralism*. This is the view that, in virtue of their indispensable, regulative role in cognitive life, epistemic norms are closely intertwined with—if not reducible to—procedural rules that restrict epistemic actions, procedures, and processes. Epistemic proceduralism is an abstract (or formal) and comprehensive theory; it's a theory of the structure (or content), regulative function, and logic of epistemic norms, without commitment to any specific list of norms or account of the structure of epistemic states (i.e., full belief, acceptance, partial belief, ...). In other words, it doesn't try to tell us what the norms for (e.g.) belief change are; it provides a framework for reasoning about the kinds of things that norms for belief change are and how they work. Nonetheless, the defense of the abstract theory does appeal to substantive observations about such norms, as well as their roles in cognitive life, and it does have consequences for the content of substantive theories of epistemic rationality.

The main technical project of the book is developing PLEN, which is essentially a novel variant of propositional dynamic logic (PDL) distinguished by more or less elaborate revisions of PDL's syntax and semantics. The syntax encodes the procedural content of epistemic norms by means of the well-known protocol or program constructions of dynamic and epistemic logics. It then provides a novel language of operators on protocols, including a range of unique protocol equivalence relations, syntactic operations on protocols, and various procedural relations among protocols in addition to the standard dynamic (modal) operators of PDL. The semantics of the system then interprets protocol expressions and expressions embedding protocols over a class of directed multigraph-like structures rather than the standard labeled transition systems or modal frames. The intent of the system is to better represent epistemic dynamics, build a logic of protocols atop it, and then show that the resulting logic of protocols is useful as a logical framework for epistemic norms. The resulting theory of epistemic norms centers on notions of norm equivalence derived from theories of process equivalence familiar from the study of dynamic and modal logics. The canonical account of protocol equivalence in PLEN turns out to possess a number of interesting formal features, including satisfaction of important conditions on hyperintensional equivalence, a matter of recently recognized importance in the logic of norms, generally.

Given those foci, this book is primarily for philosophers of logic, philosophical logicians, and epistemologists interested in the formal representation of epistemic dynamics, foundational topics concerning epistemic normativity, or the analysis of normative and procedural components of methodologies for belief revision and theory change. More technically inclined readers interested in applications of modal logics (especially dynamic logics) and process theories to the analysis of epistemic rationality may find something of interest or use in this book. Such readers may be disappointed with the level of technical development, however, as I've focused on developing the technical properties of the formal framework only with specific philosophical projects in mind, leaving many of the standard metatheorems as plausible conjectures, and I've deliberately set out to do as much as possible with elementary tools.

This book is a revision and refinement of the project of my doctoral dissertation, *The Philosophical Foundations of PLEN: a Protocol-theoretic Logic of Epistemic Norms* (City University of New York, Graduate Center). As such, some of the material here appears there, though none of it appears unchanged, nor does any of it appear elsewhere. Chapters 1, 2, and 3 of this book refine and expand the arguments and stage-setting of Chaps. 1 and 2 of the dissertation. Chapters 4, 5, 6, and 7 of this book provide a marked refinement of the eponymous system, containing some material from Chap. 3 of the dissertation. Chapters 8, 9, and 10 of this book refine some of the material from Chaps. 4 and 5 of my dissertation, and move past it considerably. There is little direct overlap, but the project here is a direct continuation of that inaugurated in the dissertation.

New York, NY, USA Ralph Jenkins

Acknowledgments

I owe much to many. Tanya Manthey, without whom I would have been lost many times. My small circle of family and close friends, for whom much the same is true. The philosophers, logicians, and computer scientists working in dynamic, epistemic, deontic, and non-classical logics; these fields are driven by sharper minds than mine, and they house many of the sources of technical invention and philosophical insight that have informed this project, and to which I owe a deep debt of gratitude.

This project couldn't have seen completion if not for the patience and guidance of the faculty at the Graduate Center (CUNY), especially the supervisor of my dissertation: Graham Priest. Aside from directly providing virtually endless technical guidance, philosophical insight, and a continual stream of letters of recommendation, Graham's continued defense and development of dialetheism over the years in ever more diverse and surprising ways is an exemplar of the "hard core" research strategy. It's an example that has fundamentally re-shaped my philosophical goals and professional aspirations. Sergei Artemov and David Papineau deserve special consideration as patient and supportive dissertation committee members.

In virtue of this project's genesis from my dissertation research and my reluctance to show work-in-progress, the set of philosophers I owe gratitude to for talking these ideas through are those I associated or worked with closely during grad school—some from my cohort, others from the rotating cast of Graham's dissertation supervision seminars. Nonetheless, there are too many helpful colleagues to name, and here is a frankly poor attempt: Daniel Weissglass, David Nagy, Daniel Boyd, Yuval Abrams, Sophie Bishop, Erik Stei, Tommy Kivatinos, Thomas Ferguson, Amanda Bryant, Hitoshi Omori, Filippo Casati, and Federico Matias Palos, and more besides! If I've omitted anyone, it's a failure of memory on my part rather than a failure of important contribution on theirs.

There are very many ways in which this project could have failed, whether from technical or philosophical error or from the numerous ways in which extra-academic factors get in the way. Somehow, I haven't quite run into them face-first, yet. The explanation lies partly in the combination of guidance and freedom from constraint provided by Graham and the Graduate Center. Another part of the explanation is Rutgers-Newark. Where the Graduate Center guided the path of the education and research that lead to this book, it was Rutgers-Newark's small, excellent, and deeply encouraging philosophy department—and Jeff Buechner, in particular—that grounded the root state and the initial steps of that path.

Special thanks to: John MacFarlane for graciously letting me cite an unpublished paper, an anonymous reviewer that harshly (and rightly) rejected an earlier version of this book, and another anonymous reviewer that checked my technical work and provided encouraging and helpful suggestions.

Contents

List of Abbreviations

Results: Lemmata		Results: Lemmata	
EQID	EQuivalent protocols, IDentical protocol graph sets	GSRTL	Graph-theoretic Syntactic Replacement Theorem Lemma
PGER	Protocol Graph Equivalence Reflexivity	Ix	Implementation rule no. x
PGES	Protocol Graph Equivalence Symmetry	ISOP	ISOmorphism Preserves protocol graphs
PGET	Protocol Graph Equivalence Transitivity	ISOPL	ISOmorphism Preservation Lemma
PRT$x(y)$	Protocol Relation Transmission x, Clause y	NSTL	No-Smaller-Than Lemma
		NSTL Lemma	No-Smaller-Than-Lemma Lemma
		PGIIRC	Protocol Graph Isomorphism Invariance of Regular Constructions

Results: Theorems	
GSRT	Graph-theoretic Syntactic Replacement Theorem
GRT	Graph-theoretic Replacement Theorem
IHT	Immediate Subprotocol Hierarchy Theorem
PFT	Protocol Function Theorem
SUT	Structural Uniqueness Theorem
RRT	Restriction Representation Theorem
RGSRT	Restricted Graph-theoretic Syntactic Replacement Theorem

System Abbreviations		Abbreviated Terms	
AGM	Belief revision theory of Alchourron, Gardenfors, and Makinson	EDG	Epistemic Directed Graph
DEL	Dynamic Epistemic Logic	EXC	Exclusivity Condition
PDL	Propositional Dynamic Logic	LTS	Labeled Transition System
PDeL	Propositional Dynamic Deontic Logic	NST	No-Smaller-Than condition
PLEN	Protocol-theoretic Logic of Epistemic Norms	RC	Recursive Condition
PLED	Protocol-theoretic Logic of Epistemic Deontics	SD$n(m)$	Structural Display relation n, Clause m
PLEDKNA	Protocol-theoretic Logic of Epistemic Deontics, Procedural Knowledge, and Norm Application	(Dn)	Numbered principle codifying a claim of epistemic proceduralism
SDL	Standard Deontic Logic	(DnCm)	Cluster of corollaries of principle codifying claims of epistemic proceduralism

List of Figures

List of Tables

Part I
Epistemic Proceduralism

Chapter 1
Epistemic Proceduralism Stated

Abstract This chapter formulates the conceptual background of the rest of the book by introducing some of the basic formal machinery, providing a first iteration of the articulation of the core ideas of epistemic proceduralism, and outlining the overall argumentative plan that the rest of the book follows. A basic model of epistemic dynamics is spelled out, using the familiar labeled transition systems of the logical dynamics framework. From there, the content of epistemic rationality norms is regimented by two constructions. The first construction is the familiar "norm-kernel", introduced by von Wright and adopted widely since. This construction brings in its train some logical tools, including the deontic operators. The second construction is a novel one. The content of epistemic rationality norms is regimented by means of elementary set-theoretic constructions called "restrictions on epistemic dynamics". This construction harnesses some of the existing formal machinery for deontic logic and logical dynamics in order to provide an adaptable structure for analyzing epistemic norms and developing the protocol-theoretic logic of epistemic norms. From there, a number of interesting terminological disputes is raised and set aside as orthogonal to the project in the rest of the book.

1.1 Introduction

1.1.1 On Epistemic Rationality Norms

Epistemic rationality norms, here conceived of as norms of epistemic regulation or control, are among the central features of philosophical reasoning and ordinary living. We are reasoning and deliberating agents, after all. We form, revise, and sustain systems of beliefs, credences, and other epistemic judgments or attitudes, often with elaborately structured theories as their objects. More, we do the same for systems of goals, values, and norms themselves, then forming and deliberating about (more or less elaborately structured) plans for achieving them or according with them.

R. Jenkins, *A Protocol-theoretic Framework for the Logic of Epistemic Norms*, Synthese Library 467, https://doi.org/10.1007/978-3-031-08597-0_1

Our greatest intellectual and practical projects consist in large proportion of these cognitive activities, and their outcomes frequently depend on whether these activities are executed correctly or well. Deliberate regulation of epistemic action with epistemic rationality norms is thus integral to our epistemic and practical lives, and, as a consequence, so is reasoning with and about the epistemic norms that are used to regulate epistemic processes, structure their dynamics, and control their outcomes.[1] Indeed, many projects in philosophy—or inquiry more broadly—can be construed plausibly as a search for methods, norms, or standards for deciding intellectual disputes about topics for which there aren't antecedently endorsed regulative norms.

1.1.2 On the Logic of Epistemic Norms or Lack Thereof

Nonetheless, in comparison to ethical and legal norms—or even norms as a general class—the logical analysis of epistemic norms has been sparse and narrow. Ethical and legal norms are often studied from the most abstract perspective, with fully general logical frameworks developed for the purpose of representing and reasoning about their structure and content, codifying their logical properties, and validating the distinctive forms of inference they're involved in. Formal frameworks that are devised with the intent to formalize *the logic of norms* (broadly construed) proliferate: standard and dynamic deontic logics, imperative logics, logics of normative propositions, and formal models of normative systems all contend for this title.[2] Applications and modifications of these systems and others to the formalization of legal reasoning and jurisprudential concepts also proliferate.[3]

There have been, to my knowledge, no comparably abstract or general logics of epistemic rationality norms. The nearest systems are those studied in logical dynamics, dynamic epistemic logics, the analysis of multi-agent systems, belief revision theories, deontic logics, and the analysis of the logical relations among epistemic norms in some parts of formal epistemology and philosophy of logic.[4] While nearby systems are plentiful, none of them actually does the trick. These rich and important programmes have not, to my knowledge, produced any system that is either correct or (philosophically) complete as a formal framework for representing

[1] See Chaps. 2 and 3.

[2] See [36], especially the introduction, for an accounting of this proliferation. See also [72].

[3] See again [36] as well as Alchourron's seminal work on normative systems [3, 4]. See also a programmatic application of deontic logic to legal norms in [76].

[4] For some representative examples of the systems I have in mind, see [109, 112–114] (logical dynamics), [10, 116, 117] (dynamic epistemic logics), [26] (multi-agent systems), [5, 37, 38, 46] (belief revision theories), [36, 119, 120] (deontic logics), [19, 29, 30, 85, 87, 103] (formal epistemology), and [12, 31, 32, 98] (philosophy of logic). These examples are chosen explicitly because they are the closest projects to what I have in mind, they provided tools used to work on this project, or they played some role in the initiation or development of this project.

and reasoning about the abstract structure of epistemic norms as a class, their regulative function, or the distinctive forms of inference with or about epistemic norms. Rather, to the extent that the foregoing frameworks can be used for thinking about epistemic normativity at all, they are all designed for entirely different projects with little or nothing to say about epistemic norms, or (inclusively) they have very specific and often dubious assumptions about the logic of epistemic norms or the structure of epistemic states and epistemic dynamics built into them from the ground up.

Aside from this very project, the systems studied in logical dynamics have not been (again, so far as I know) applied to the analysis of epistemic norms. As we will see, there are very rich, un-tapped resources for doing so. Dynamic epistemic logics and multi-agent systems analyze the interaction of the logic of knowledge expressions and parts of epistemic dynamics, but not the logic of the epistemic norms that, given a justification, warrant, or rationality requirement on knowledge, are essential to actually acquiring knowledge by regulating or guiding epistemic actions and processes.[5] Deontic logics (including dynamic deontic logics) are, generally, not designed for thinking about epistemic norms, and, moreover, most systems are explicitly norm-agnostic, providing only analyses of the logic of deontic operators independent of any considerations about (e.g.) the differences in deontic accessibility between distinct systems of norms or the logical distinctions among different kinds (static, dynamic, or procedural) of epistemic norms.

Belief revision theories provide a model of the dynamics of belief systems as they evolve under specific repertoires of norms of rationality, e.g., the AGM rationality postulates, or norms of epistemic entrenchment, or norms of defeasible reasoning. But belief revision theories don't directly address the nature, structure, or logic of epistemic rationality norms at any stage but the pre-theoretical assumptions that shape the formal details. Bayesian systems are no different in this regard, however different they may be in others. Bayesian frameworks provide models of the dynamics of systems of credences as they unfold under specific systems of (e.g.) update and coherence norms, and they have little to say about the processes and forms of reasoning with or about norms that allow them to play a role in the regulation of credence.

Moreover, all of the foregoing systems share a problem that is developed in this book;[6] they're all built atop a general model of dynamics that analyzes actions, processes, and procedures by means of transition relations, functions, or operations

[5] Epistemic logics also tend to analyze a rather different concept of knowledge than that which animates most epistemology, so this is perhaps less an oversight than a product of divergent focus. Justification logics are, naturally, exceptions to this focus away from justification. See [7] for an overview. Justification logic does, of course, offer a formal logic of justifications. Nonetheless, the complaints about the systems in the main text apply as well to justification logics; there may be logical observations about epistemic norms extractable from them in some way, but justification logics won't be logics of epistemic norms because they are not designed to be, and, to my knowledge, they are explicitly norm-agnostic.

[6] See Chaps. 2, 3, and 4.

on states. While this sort of underlying model of dynamics clearly gets much right, it also generates cascades of distinctive errors when building a logic of epistemic norms.

Formal epistemologists and philosophers of logic often explore and defend particular suites of norms of inference, particular rules of justification, or specific conceptions of rationality defined by particular suites of norms, but they seldom give us an account of what kinds of epistemic norms there are, how they function in epistemic practice,[7] or what it even means for two epistemic norms to be equivalent.[8] The abstract structure and logic of epistemic norms is seldom the target of analysis; particular norms imposing particular requirements on particular types of epistemic dynamics are programmatically the exclusive targets of analysis.[9]

Even where epistemic norms as a general class draw focus [16, 68, 102], providing a logical framework for representing and reasoning about them is not the project. Rather, it seems that philosophical work on epistemic norms operates under one or another umbrella of assumptions about the logical framework for epistemic norms, and it is only challenges to or consequences of these assumptions (and the details of the attendant disputes) that typically occupy epistemology.

1.1.3 Toward a Logic of Epistemic Norms

This book is an attempt to fill the gap where the logic of epistemic rationality norms should fit, albeit in a self-consciously schematic and iterative way. In particular, this book develops and defends a novel formal framework for representing and reasoning about epistemic rationality norms (i.e., norms for epistemic guidance, regulation, or control) that is abstract and comprehensive, as well as non-committal about what list of norms defines epistemic rationality for different kinds of epistemic states. It

[7] There are some notable exceptions [16, 30, 90] that motivate this project.

[8] Again, with some motivating exceptions, including [19, 28, 30, 39, 87, 118].

[9] It's worth mentioning systematic projects that do attempt analyses of broad swathes of epistemic norms. I have in mind Carnap's continuum of inductive methods [19], formal learning theory [53, 104], and epistemic utility theory [85–89, 103]. The former two systems can be interpreted as frameworks for formulating and deriving logical relations among inductive norms. The third framework has the distinction of being explicitly a framework for formulating and deriving logical relations among epistemic norms in general (static and dynamic, inductive and deductive). All three provide, in some sense, rigorous theories of classes of epistemic norms, and can be seen as establishing technical results that provide justifications for epistemic norms in the form of derivations from more fundamental norms. Nonetheless, even these remarkable systems are ultimately not logical accounts of epistemic norms of the kind being sought. The former two frameworks are focused only on inductive rules, and so not general in the sense required by this project. Moreover, none of these three frameworks is designed with an eye toward analyzing the regulative function of norms, i.e., how we adopt and use norms, the complex dynamics of applying epistemic norms, or, indeed, the logic of the distinctive forms of reasoning about epistemic norms that underly their most interesting practical and theoretical uses.

attempts to codify the nature, structure, and logic of epistemic rationality norms, including the distinctive forms of reasoning with and about these norms that are integral to their regulative role in epistemic life, and it develops a novel theory of epistemic norm equivalence. The formal framework developed is a *protocol-theoretic framework*, based on an elaboration of propositional dynamic logic (PDL), and its conceptual underpinnings consist of a *protocol-theoretic account* of epistemic rationality norms. These are potentially unfamiliar terms, so let me spell them out.

Protocols are structures built out of procedural operations and action terms that, like plans, programs, or algorithms, conditionally and sometimes categorically dictate or otherwise determine the configuration of the sequences, choices, and combinations of actions that must or may be carried out by systems (or agents) that would execute them.[10] Indeed, protocols are, in some sense, formal representations of the shared, underlying structure of plans, programs, and algorithms; protocols are useful tools for representing the abstract procedural structures by which basic actions and processes can be composed into complexes. The determination of the structure of paths of action, in turn, determines the required or permissible processes, changes of state, and even the admissible or reachable space of states for systems (or agents) that would execute a protocol.

As guides for regulating or controlling epistemic action, *epistemic rationality norms* restrict the space of possible epistemic factors to the correct (i.e., rational, epistemically permissible, epistemically obligatory, etc.) proper subset, allowing would-be epistemically rational systems or agents to reason about what those norms require, permit, and forbid (or what their paths of action and state change must or may be, if rational), and deliberate about whether and which courses of epistemic action accord with them, in order to choose courses of epistemic action. Epistemic rationality norms are thus the rules, principles, and policies that discriminate rational, justified, warranted, or otherwise correct epistemic factors (e.g., correct

[10] The terms "protocol", "plan", "program", and "algorithm" all have dual senses. For each, the term can be used to denote both structured procedures over some kind of action space or formalizations of such procedural structures. One can refer to plans in the abstract or plans as objects in a formal theory (i.e., Bratman's plans [17] vs Gibbard's [30, 39] or Pollock's plans [91]). One can refer to programs as procedures for task execution or as formal objects, i.e., specific chunks of code that carry out specific tasks in a specific coding language, or syntactic objects in propositional dynamic logic that encode an abstract procedural structure. One can refer to algorithms with or without the more rigorous elements added by the theory of computation (e.g., finiteness, well-definedness, effectiveness, correctness). I employ the term "protocols" where, in propositional dynamic logic, one would use "programs" to denote formalized procedural constructions over action libraries. This follows the conventions of some of the germinal works for this project [26, 51, 81, 121]. The primary reason for this terminological convention is that the term "protocol" has normative uses that the other terms lack; protocols are both regulative and evaluative. The term is also more neutral about domain. Plans are procedural operations that have to do specifically with practical reasoning tasks, while programs and algorithms are procedural operations that have to do with computational tasks. Protocols are defined over the actions of individuals and collectives or agents and subpersonal systems, and they can be defined over computations, operations on the world, communications, and even collective behaviors.

epistemic states, actions, processes, and procedures) from their complements (irrational, unjustified, unwarranted, or incorrect epistemic factors). Epistemic norms can and, arguably, typically achieve this by conditionally attributing epistemic deontic statuses (e.g., rational obligation, permission, and prohibition) to their objects.

The *protocol-theoretic account of epistemic rationality norms*, then, can be summed by the following thesis:

(T0) Epistemic rationality norms are protocols for epistemic planning and control.

The core theses of this book, then, are the following, which resolve the main elements of T0:

(T1) There are norms of epistemic rationality that are procedural; epistemic rationality is partially defined by program-like procedural norms that restrict the possible ways in which epistemic actions and processes can be combined by means of control operators including (inter alia) sequential composition, parallel execution, non-deterministic choice, and iteration.

(T2) Procedural epistemic rationality norms are not reducible to strictly static or merely dynamic norms, and there are theoretically central clusters of principles—denoted later in this book with the terms D1, D2, and D3, each accompanied by sub-clauses and corollaries—that characterize the regulative function (epistemic planning and control), structure, and logic of procedural epistemic norms; moreover, these clusters of principles stand as theoretical criteria or tests of adequacy for systems that would be logical frameworks for representing and reasoning about epistemic norms.

(T3) Procedural epistemic norms can be formally represented and reasoned about by means of a *protocol-theoretic logical framework*—a logical framework that deploys *protocols* to directly encode the procedural structure or content of procedural epistemic norms; more, the protocol-theoretic framework I develop and defend in this book can be shown to be theoretically interesting and useful, in that each of the theoretically central principle clusters characterizing procedural norms (D1–D3 and all sub-clauses and corollaries) can be formally analyzed in a plausible way and then derived from the protocol logic of the framework.

Thesis T1 forms the backbone of what might best be called *epistemic proceduralism*, and thesis T2 (when fully spelled out) articulates one spelled-out account of this philosophical position. Proving T3 is the project of this book. In what follows, I present the Protocol-theoretic Logic of Epistemic Norms (PLEN) framework and develop its properties with respect to analyzing the account of epistemic proceduralism codified by D1–D3 (and their sub-clauses and corollaries).

Of course, this shouldn't be done without some reflection on T1 and T2, so the book opens with a trio of chapters (1, 2, and 3) that articulates D1–D3 (et al) and presents the main arguments, as I see them, for taking D1–D3 (et al) to be theoretically central to the analysis of epistemic norms. This is Part I of the book, and while far from a full exploration of the relevant philosophical questions (which could occupy separate volumes) it's a starting point for introducing PLEN. As such, Part I can be skipped and revisited at a later time for readers that appreciate the main conceptual thrust of the system and are eager to see the logic. Part II develops and refines the PLEN Framework. Chapter 4 defines PLEN, walks through some of the obligatory metatheorems, and discusses its conceptual underpinnings. Chapters 5 and 6 revise the protocol logic of PLEN, elaborating on the basic system and deriving its most distinctive and interesting properties, with focus on the theory of protocol equivalence in PLEN. Chapter 7 builds a logic of protocols around the theory of protocol equivalence, and puts it to work formalizing the logic of procedural epistemic norms. The remainder of the book comprises Part III, which elaborates further on PLEN in order to derive D1–D3 (et al). Chapter 8 builds a protocol-theoretic epistemic deontic logic and a logic of procedural knowledge and norm application into PLEN. These systems are operative in representation theorems that closely bind the syntax of protocols to the content of epistemic norms and in the formalization of D1–D3 (et al). Chapter 9 articulates, derives, and briefly discusses the aforementioned representation theorems. Finally, Chap. 10 derives (or outlines the derivation of) the formal counterparts of D1–D3 (et al), thereby bookending the presentation of PLEN; Part I presents a motivating philosophical task, and the final chapter of Part III completes that task by applying the system developed in between.

1.1.4 Philosophical Landscape

If all of the moving parts of this project work, where would a protocol-theoretic logic of epistemic norms sit in the philosophical landscape? There are two answers. The first is an academic one. PLEN intertwines three prominent lines of recent research in philosophical logic and epistemology:

- The Dynamic Turn in Logic, Belief Revision, and Methodology
- The Hyperintensionality Revolution
- The Foundations of Formal Epistemology

The dynamic turn is, roughly, the collective effort to reappropriate formal tools for modeling dynamics and dynamic systems (e.g., dynamic logic and game theory) for traditional logical tasks, such as modeling (principles of) inference, intelligent agency, rational belief, or the unfolding of scientific theories under methodological constraints.[11] The intensionality revolution of the latter half of the twentieth century

[11] The dynamic turn in these fields is, arguably, most directly and summarily encapsulated by a sampling of van Benthem's research [109–113, 115]. I have yet to find any earlier uses of the phrase "the dynamic turn in logic" outside of van Benthem's work, but the roots of it are present at least as far back as Pratt [94] and other early work on dynamic logics. All of this work, in turn, owes to the dynamic developments in semantics. Gochet [43] traces the dynamic turn in semantics

involved the collective realization that possible world semantics could be developed and deployed as a useful and unified technical framework for the analysis for fundamental philosophical concepts like information, knowledge, belief, deontics, explanation, and meaning. (To say nothing of its use in metaphysics.) The dynamic turn itself was a significant part of this revolution.[12] The hyperintensionality revolution,[13] roughly, is the collective realization[14] that the analysis of these very concepts (inter alia) requires a technical framework that can distinguish its objects at a finer grain than necessary equivalence. There is not yet a unified framework, but the framework seems to be emerging in a piecemeal fashion. Formal epistemology, roughly, is the collective effort of philosophers to avail themselves of any formal tools that might work for analyzing doxastic and credal states, knowledge, epistemic state change and reasoning, epistemic values, and the principles or norms of reasoning. Frequently enough, this dovetails with the intensional and dynamic turns.[15] The roots of formal epistemology draw on the intertwining of threads from Bayes to Boole to Hintikka.

If all goes right, PLEN is a hyperintensional framework for analyzing epistemic states, epistemic dynamics, and epistemic norms. It is a product of the dynamic turn, clearly enough, given its use of dynamic logical tools, but it goes beyond them to (as Chap. 6 argues) give hyperintensional criteria for epistemic norm equivalence and a partly hyperintensional logic of epistemic norms. Given its amenability to simple modifications (e.g., changing out the model of epistemic states for credal states, amending its model of dynamics with probability or preference functions, exchanging its core propositional logic for a non-classical logic), it provides a unified foundational framework for formal epistemology; it captures aspects that are common to the central formal models of (e.g.) rational belief and credence update, and solves problems inherent to them.

The second answer is a somewhat more mundane or traditional philosophical project: an abiding interest in finite, flawed epistemic agents, their goals, and the tools they use to achieve them. Some epistemic agents, for instance, aim to achieve the goals of inquiry—things like truth, knowledge, and rationally justified belief. But every observable instance of such an agent is prone to performance error, subject to bias, and limited in cognitive resources. In order to meet their epistemic goals, such reasoners must find ways to detect and recover from errors, to detect bias and course correct beforehand, and to overcome their natural epistemic inclinations or

to work by Hintikka on game-theoretic semantics even earlier, and beyond that to Pierce. See also [84].

[12] See, for a few instances among many others, including the references in prior footnotes: [5, 10, 11, 14, 50].

[13] The phrase is, I think, due to Nolan [78]. See [15] for an overview and a brief history.

[14] See, for one compelling line of argument, [27]. For a core research program driving the hyperintensionality revolution, see [7].

[15] See, for just a tiny sampling: [2, 5, 10, 11, 45, 46, 49, 54, 89, 113].

limitations. In other words, the question is: how can beings like us overcome our frailties to meet the goals of inquiry? The answer, I think, is *epistemic methodology*.

Epistemic methodology consists of (1) systems of norms that we endorse and deploy to evaluate and regulate our epistemic behavior and (2) our reflection thereon and deliberation thereover. We don't (always) blindly endorse and deploy procedures and standards for carrying out our epistemic projects; we reflect on their properties (e.g., the effects of following them, the restrictions they impose, the commitments they entail), and deliberate over which of them to keep endorsing and deploying. We even devise and adopt new and better epistemic norms from time to time. After all, the formal techniques of logic and probability theory are indispensable to reasoning through complexity and uncertainty, and extensive, exact, and deliberate use of these things is hardly native endowment.[16]

This project is part of the analysis of epistemic methodology or, to borrow terminology, procedural epistemology.[17] The focus of the project is not the analysis of fundamental elements of epistemic value. Rather, it focuses on the kinds of normative structures that are needed in epistemic regulation or epistemic advice. These, it turns out, must focus on the distinctive features of procedures and processes involved in changes of mind in order to be useful, as must our deliberation about which of them to adopt. Ultimately, this book is designed to contribute something to the reflective and deliberative activities involved in epistemic methodology; it's an attempt to contribute to the collective project of thinking more carefully about how we think and how we ought to think in order that we might do better.

1.2 Epistemic Proceduralism Stated

Epistemic proceduralism is the account of the structure, cognitive role, and logic of epistemic rationality norms that grounds and motivates the protocol-theoretic framework for the logic of epistemic norms. The core of epistemic proceduralism consists of the following clusters[18] of principles:

(D1) Epistemic rationality norms define restrictions on epistemic dynamics distinct from constraints of capacity and feasibility, including restrictions on

[16] See the rather large empirical literature on what can, with some controversy [40–42], be called "irrationality", starting from the seminal [52] and continuing through all of [2, 25, 100, 107, 108] and beyond.

[17] Arguably, this puts the project in line with the true mainstream of epistemology, in the long run. For influential modern and recent examples, see: [8, 9, 24, 73, 90, 92, 93]. The terminology is Pollock's [92].

[18] Epistemic proceduralism is not wholly original. See (inter alia) [30, 31, 58, 62] for defenses—or at least articulations—of some of its core claims. Indeed, it's frequently assumed as background for other epistemological projects. Nonetheless, the formulation of epistemic proceduralism explicitly by means of clusters D1 through D3 (including sub-clauses and corollaries) is, to the author's knowledge, original.

the content of epistemic states, restrictions on transitions and trajectories of evolution of epistemic state, and restrictions on procedures and processes of epistemic state change.

(D1C1) Epistemic rationality is dynamic; there are epistemic restrictions on transitions among epistemic states, not just states or contents of states.

(D1C2) Epistemic rationality is ineliminably procedural; there are epistemic restrictions on epistemic actions, processes, and procedures, not just states or transitions among states, and restrictions on epistemic procedures and processes provide an irreducible unifying structure for interpreting and reasoning about even apparently non-procedural (epistemic) evaluations and regulative (epistemic) normative verdicts, and such restrictions are practically indispensable in epistemic cognition.

(D2) Procedural epistemic rationality norms define restrictions on the construction of procedures; they determine how basic epistemic actions are to be sequenced, iterated, combined, chosen among, conditioned on tests, avoided, reversed, and cautiously executed.

(D2C1) Procedural norms of epistemic rationality have the internal structure of programs or plans, conditioning actions and procedures on tests, and dictating procedures built with complex operations on action like conditional choice and guarded iteration, which are composed out of more basic procedural operations.

(D3) Epistemic rationality norms can be reasoned about explicitly; terms denoting epistemic rationality norms and statements about the properties of epistemic norms can be embedded in valid arguments.

(D3C1) Where N_i is a term denoting an arbitrary norm, A_i is a formula embedding a norm term (letting $A(N_i, x)$ be a special case encoding N_i's verdict on an arbitrary epistemic factor, x), C_i is an arbitrary formula stating some condition, and $\equiv_x$ is an arbitrary form of normative equivalence, the following table summarizes schematic inference rules for the foregoing kinds of reasoning:[19]

[19] The forms of reasoning articulated schematically in this table are more fully explicated in Sect. 3.1 of Chap. 3.

Basic forms	Equivalence	(In)Dependence, sufficiency
$\dfrac{\begin{array}{c}A_1(N_1)\\ C\end{array}}{A_2(N_2)}$	$\dfrac{\begin{array}{c}N_1 \equiv_x N_2\\ A_1(N_1)\end{array}}{A_2(N_2)}$	$\dfrac{\begin{array}{c}N_1 \supset_x N_2\\ A_1(N_1)\end{array}}{A_2(N_2)}$
$\dfrac{\begin{array}{c}A_1(N_1)\\ C_1\end{array}}{C_2}$	$\dfrac{\begin{array}{c}A_1(N_1)\\ A_2(N_2)\\ \dots\end{array}}{N_1 \equiv_x N_2}$	$\dfrac{\begin{array}{c}\neg(N_1 \supset_x N_2) \wedge \neg(N_2 \supset_x N_1)\\ A_1(N_1)\end{array}}{A_2(N_2)}$
$\dfrac{\begin{array}{c}C_1\\ C_2\\ \dots\end{array}}{A(N_i)}, \dfrac{\begin{array}{c}C_1\\ C_2\\ \dots\end{array}}{A(N_i, x)}$	$\dfrac{N_1 \equiv_x N_2}{N_1 \equiv_y N_2}$	$\dfrac{\begin{array}{c}A_1(N_1)\\ A_2(N_2)\\ \dots\end{array}}{N_1 \supset_x N_2}$
$\dfrac{\begin{array}{c}A(N_1, x)\\ C\end{array}}{A(N_2, y)}, \dfrac{\begin{array}{c}A(N_1, x)\\ C\end{array}}{A(N_2, x)}$		$\dfrac{\begin{array}{c}A_1(N_1)\\ A_2(N_2)\\ \dots\end{array}}{\neg(N_1 \supset_x N_2) \wedge \neg(N_2 \supset_x N_1)}$

(D3C2) Epistemic rationality norms assign deontic properties to epistemic factors including states, transitions, actions, processes, and complex procedures.

(D3C3) Epistemic rationality norms can be deliberately applied (as opposed to merely complied with).

There is considerable resistance to some of these clusters. There is no consensus in epistemology as to whether there are even dynamic norms governing (e.g.) rational changes of mind in addition to those static norms governing rational states of mind, let alone that there are even finer grained distinctions to be drawn between the processes that result in changes of mind and the patterns of changes of mind that result. Indeed, there are research programs in formal and traditional epistemology built around the assumption that epistemic states are the only proper objects of epistemic normativity.[20]

Defending an account of epistemic proceduralism is therefore paramount to building the protocol-theoretic logic of epistemic norms. This chapter prepares the ground for a systematic defense of D1 through D3C3 by setting out some of the terminology invoked in developing and defending epistemic proceduralism. From there, I turn to setting aside some terminological matters, and planning the course for pursuing the primary objectives of the first part of the book. The next two chapters

[20] For some debate about the matter, see [16, 18, 57, 58]. For examples of epistemological projects committed to this kind of "epistemic staticism", see [33, 48, 75, 122]. Really, there's not even consensus about whether there are epistemic norms governing states of mind, either! See [83] for a defense of the claim that there are no properly epistemic norms of belief.

devote most of their spans to articulating and systematically defending, respectively, the pair of D1 and D2 (and all sub-clauses and corollaries), and D3 through D3C3.

1.3 Terminological and Conceptual Groundwork

1.3.1 Epistemic Dynamics and Actions

Call any system that deliberately reasons, learns, or acquires and processes information an *epistemic agent* or *epistemic system*.[21] Epistemic systems are usefully analyzed in terms of two factors: *epistemic states* and *epistemic dynamics*.

Let *epistemic states* be an epistemic system's representational states, the models of the world that the agent constructs, manipulates, and transforms in order—among other things—to guide its actions. At a minimum, epistemic states are composed of *epistemic judgments* and their *contents*. Contents are usefully encoded with sentences of a language, natural or formal. Standard taxonomies of epistemic judgments include full and partial belief (credence), acceptance, and other cognates or alternatives. So, for instance, one's epistemic states can be partially described by a set of believed sentences or by a credence function on a set of sentences. See Table 1.1 for a simple array[22] of these kinds of partial epistemic state description with some common (and uncommon[23]) bits of notation for them and some references in which one might see them in action.

---

[21] The term "deliberately" is not accidental, here. There are worries—e.g., Alston's classic argument in [6]—that discussions of epistemic normativity—at least deontic kinds of normativity—are moot because epistemic actions are not voluntary. At this stage, I'm simply declaring commitment to the claim that this worry is mistaken. In Sect. 1.4, I defend the claim that much epistemic action—and most epistemic action of interest in this project—is voluntary. My response to the classic involuntarist arguments is rather similar to that of [71].

[22] Taking on different categories of epistemic judgment as sort of a founding conceptual assumption is common in various fields. Full or all-or-nothing belief is prominent in, e.g., Epistemic Logic [49], Doxastic Logic [64, 67, 105], Belief Revision [37, 38, 46], and in epistemology, generally [1, 6, 12, 21, 23, 33, 44, 47, 68, 69, 73, 90–92, 101]. See, especially, Staffel's [106] critical discussion of its prominence and Leitgeb's work relating full belief to credences [63]. Credences are prominent, of course, in Bayesian epistemology [20, 61, 63, 79] and also in certain branches of belief revision [34, 60] and doxastic logic [11]. Acceptance and rejection are discussed independently by Cohen [22] and Restall [13], with roots in philosophy of science, re: theoretical commitments [96]. Supposition is a medieval logical notion [97] given an interesting analysis in an inductive context by Levi [65]. Perhaps the list of epistemic judgments is even more expansive. Some potential add-ons: predicting P, suspecting P, guessing P, having the hunch that P, and so on. This is one of many points that suggests future work in analyzing epistemic dynamics.

[23] The notation $b(P)$ is mine, and will be used freely throughout. Additionally, representation of acceptance and rejection as membership in a set of formulae is used to very interesting effect in [98]. It is also the norm in syntactic epistemic logics [59], contrary to the norm in doxastic and epistemic logic, where belief is often encoded with a modal operator. In any case, I omit a full taxonomy of notations.

Table 1.1 Epistemic judgments

Judgment	Content	Notation	References
(Full) Belief	$P, Q, R, \ldots$	$b(P), b(\neg P), P \in K$ :: P (resp. $\neg P$) is in one's belief set K, P is fully believed	[38, 47]
(Full) Disbelief	$P, Q, R, \ldots$	$d(P), d(\neg P), P \notin K$, $\neg P \in K$ :: P (resp. $\neg P$) is in one's disbelief set (the complement of K), P is fully disbelieved	[38, 47]
Credence (Partial (Dis)Belief)	$P, Q, R, \ldots$	$Pr(P), Pr(\neg P)$, $Pr(P \mid Q)$ :: P (resp. $\neg Q$) has subjective probability Pr (given Q)	[20, 63, 79, 118]
Acceptance	$P, Q, R, \ldots$	$P \in X, \neg P \in X, P \notin X$ :: P is accepted, p is in one's acceptance set X	[22, 98]
Rejection	$P, Q, R, \ldots$	$P \in Y, \neg P \in Y, P \notin Y$ :: P is rejected, p is in one's rejection set Y	[22, 98]
Supposition	$P, Q, R, \ldots$	$P \in T(K, P)$—with $T(K, P)$ the suppositional corpus supported by belief set K :: P is a supposition supported by one's belief set, K $P > Q$—conditional belief on the basis of supposition :: Supposing P, Q rationally follows (were one to fully believe P, one would be committed to fully believing Q)	[65, 66]

Epistemic states change over time as a result of various processes. *Doxastic* or belief-related states, for instance, can be changed by processes such as reasoning, perception, memory, and testimony. Full beliefs are expanded, contracted, revised, or sustained, and credences are updated by reasoning processes, processes of observation, processes of learning, and so on. Take a toy example:[24]

[24] Any noticeable lack of realism is probably due in part to the distorting effects of simplification for the sake of illustration, implausibly convenient stipulations about the epistemic considerations of the situation, and, perhaps, to my having adapted and augmented a scenario from a comic book, [80], to produce it. Nonetheless, it serves as an example of complex epistemic dynamics with interactions among information generating and inferential actions.

The Researcher Dr. Dubrovsky arrives on-site to study Phenomenon A, an anomalous causal process of some interest in their field. They immediately begin development of a novel theory about the target phenomenon. This epistemic process is temporally extended and consists of steps that trace a path through possible belief or theory revision states:

- First, they devise a set of possible qualitative theories of the origin, structure, and predicted developmental trajectory of Phenomenon A: $T_1, \ldots, T_n$. They then rank each T_i in terms of prior probability or plausibility on the basis of a priori theoretical constraints on the space of possible theories. Let $T_1 > \ldots > T_n$ be the (unusually linear) ranking.
- Second, they consider the set of applicable mathematical frameworks for formalizing theories $T_1, \ldots, T_n$, which, conveniently, satisfy a linear ordering of elegance and rigor, $M_1 > \ldots > M_n$ and map $1-1$ to $T_1, \ldots, T_n$.
- Third, they receive a mixed data set from the researchers already on-site. It consists of first-hand observational reports from on-the-ground sources. These reports vary with respect to multiple axes of quality, including a priori credibility of sources, and they partition into inconsistent, non-convergent subsets with respect to their support for $T_1, \ldots, T_n$. That is, if a data set supports T_i, then it doesn't also support any inequivalent T_j, and the sets of reports supporting T_i are not consistent with those supporting T_j.
- Fourth, they carry out a set of interventions (experiments) that allow them to provide confirmatory evidence for a number of counterfactuals derived—uniquely—from theory T_1 by means of framework M_1.

There are multiple convergent and divergent processes of theory selection or belief formation, revision, and retention possible for Dr. Dubrovsky. Here is a non-exhaustive selection:

- Processes that converge on T_1:
  - P_1 : Adopt T_1 with formalization M_1 at the first step, on the basis of a priori plausibility.
  - P_2 : Adopt T_1 with formalization M_1 at the second step, on the basis of a theory selection procedure that prioritizes rigor, elegance, and prior plausibility over adequacy to data.
  - P_3 : Adopt T_1 with formalization M_1 at the fourth step, on the basis of a theory selection procedure that equally prioritizes rigor, elegance, and prior plausibility with adequacy to data.
  - P_4 : Adopt T_1 with formalization M_1 at any step, on the basis of a Tarot card reading.
- Processes that diverge from T_1:
  - P_5 : Adopt any T_i, where $i \neq 1$, with formalization M_i, at the third step, on the basis of a theory selection procedure that prioritizes adequacy to data over all else, and selectively privileges some subset, D_i, of first-hand reports that support T_i to the exclusion of the rest.

- P_6 : Adopt T_j, where $j \neq 1$, with formalization M_j, at the third step, on the basis of a theory selection procedure that prioritizes adequacy to data over all else, and selectively privileges some subset, D_j, of first-hand reports.

In the end, let us stipulate, Dr. Dubrovsky carries out process P_3.

The researcher's epistemic state is partially described by their beliefs (or, in this case, at least the theories they accept and their beliefs about them). The content differences (non-identity of doxastic state) between the researcher's epistemic states at some pair of indices, s_i and s_{i+1}, imply the distinctness of the epistemic states at s_i and s_{i+1}. The additions of belief at successor states are thus processes that result in changes from one epistemic state to another over time.

Let changes of epistemic state be *epistemic transitions*. Then the space of (in some sense) formally possible epistemic transitions is defined by the set of all pairs of possible epistemic states. A subset of these formally possible transitions is within the agent's *capacity*; these are the transitions such that there is some epistemic process that could actually bring about the transition. Call these the possible transitions for the agent. The transitions from s to another state, s', that are within an agent's capacity are those that are *feasible* at s.

There are multiple kinds of paths through a space of epistemic states. First, there are sequences of epistemic transitions or *trajectories*. There are also *epistemic actions*, which are actions that result in epistemic transitions (though not necessarily any change in the content of epistemic judgments, i.e., not all epistemic actions result in (e.g.) doxastic state change). An epistemic action may be *deterministic*, connecting any given root state to one terminal state, or *indeterministic*, connecting any given root state to two or more terminal states. Chains of actions together with the resulting trajectories are, for our immediate purposes, *epistemic processes*.

The actual, feasible, and possible epistemic transitions, and the actual, feasible, and possible processes of epistemic transition resulting from (actual, feasible, or possible) sequences and choices among actions, and the properties of and relations among the foregoing items constitute a system's *epistemic dynamics*.[25]

Epistemic factors will be the catch-all term for epistemic states and all of the various parts that compose epistemic dynamics, including epistemic actions and processes as well as the procedural patterns or structures of complex epistemic process construction.[26]

The Researcher case illustrates a number of important features of epistemic dynamics. First, it illustrates the way that complex processes can be constructed

---

[25] The phrase "epistemic dynamics" can also denote the study or theory of these epistemic changes. I'll let context disambiguate. I won't distinguish between *epistemic kinematics* (the "laws of motion" taking states to other states given inputs or forces of change) and epistemic dynamics as some kind of more comprehensive analysis of the epistemological features of inputs as well as the kinematics. Both [38, 118] discuss the distinction usefully.

[26] This analysis of epistemic factors will remind readers of the *epistemological theories* of [37, 38, 118]. The primary difference being that we are not interested in epistemic commitment functions or the like, at this stage.

out of more basic actions. The transition from the first belief state to the final theory adoption state is effected by sequencing multiple, more basic processes, often making choices among alternatives. Second, it illustrates why it's important for epistemological analyses to focus on the alternative, unactualized trajectories as well as their relations with each other and with the actual one.

Presumably, the researcher's beliefs evolved from earlier states to those with which the illustration begins and will unfold past those with which it terminates, but even if Dr. Dubrovsky was born with innate knowledge at Step 1 and dies after Step 4, the actual sequence of doxastic changes is merely a part of Dubrovsky's epistemic dynamics that matters for epistemological purposes. Given the stipulation that they could have based their theory adoption on experimental data, a priori theoretical considerations, or Tarot readings, there are other possible trajectories through the space of possible doxastic states than the one they actually realized. There are properties of and relations among these other possible processes and trajectories worth noting.

Some processes are more complex or more difficult to carry out than others. Some distinct pairs of processes effect identical changes of state. Some processes are normatively correct, and others are not. These properties might interact; the complexity or difficulty of processes might, for instance, impact their normative statuses. For instance, some processes *enable* other processes. The formal work at Step 2 makes process P_2 possible. The evidence acquisition at Steps 3 and 4 make process P_3 as well as processes P_5 and P_6 possible. Processes P_2 and P_3 are more complicated than P_1; P_3 in particular may take months, years, or longer to carry out after numerous experimental failures and refinements. This obviously matters for practical or cognitive control purposes, and it also matters if, for instance, epistemic ought implies epistemic can. Given such an ought implies can principle, it can't be the case that the researcher ought to carry out an evidentially grounded theory selection procedure if her prior actions did not enable it, and it can't be the case that the researcher ought to carry out such procedures if they are not feasible. Of course, such a norm may not be given, but that's beside the point.

Third, the case illustrates that procedurally, methodologically, and normatively distinct processes can converge on the same theory adoptions, and that procedurally, methodologically, and normatively similar (though still distinct) processes can diverge in theory adoptions. The Tarot card based theory selection procedure produces, in this case, the same theory selection as the evidence-oriented and formally oriented theory selection procedures. The selection procedures that selectively weigh data sets are all, in some sense, empirical theory selection procedures (deficient as they may be), and the relatively fine-grained differences among them result in very different theory adoptions. It will be essential, in the second part of the chapter, to note that distinct processes can result in the same epistemic changes and yet still be normatively distinct. Several of the researcher's empirical theory adoption procedures give the same conclusion as accepting the outputs of the Tarot reading consultation, despite their manifest epistemological differences.[27] This observation carries some epistemological weight, as I will argue at length in Chap. 2.

1.3.2 Labeled Transition Systems

It's useful to think of epistemic dynamics in terms of *labeled transition systems* (LTSs) or (polymodal) *Kripke structures*, though slightly more complicated structures will be introduced to model epistemic dynamics in future chapters, as the next three chapters raise some serious problems for certain ways of applying LTSs as models of epistemic dynamics. A labeled transition system, $\langle S, \rightarrow_{a\in A}\rangle$, is a non-empty set, S, with a set of relations, $\rightarrow_{a\in A}$, on S. Each element of $\rightarrow_{a\in A}$ is a binary relation on S, $\rightarrow_a$, that is indexed to a member, a, of the set A. These are the *transition relations*. Each ordered pair of elements of S, $\langle s, s'\rangle$, is a *transition*, which represents a change from the left element, s, to the right element, s'. The relations, $\rightarrow_a$, are *labeled transition relations*; each is indexed with an element of A. In a transition from state s to state s', s is the *root state* and s' is the *terminal state*. As a matter of notation, transitions $\langle s, s'\rangle$ and $\langle s, s''\rangle$ can both be *executed* from state s or both transitions are *available* at s.

Now, labeled transition systems are standardly defined in terms of both labeled binary transition relations with indices in A, $\rightarrow_{a\in A}\subseteq S \times S$, and also in terms of labeled ternary transition relations, $\rightarrow_{a\in A}\subseteq S \times A \times S$. Note, however, that "bare" transitions, $\langle s, s'\rangle$, in a labeled transition relation are distinct from what I will call *named transitions* or *edges*, which are triples, $\langle s, a, s'\rangle$. Consider two labels in A, a and b. Let $\langle s, s'\rangle \in\rightarrow_a$ and $\langle s, s'\rangle \in\rightarrow_b$. Of course, $\langle s, s'\rangle = \langle s, s'\rangle$. Indeed, it's possible that $\rightarrow_a=\rightarrow_b$ where $\rightarrow_a$ and $\rightarrow_b$ are labeled binary transition relations. The labeled ternary transitions, $\langle s, a, s'\rangle$ and $\langle s, b, s'\rangle$, however, are clearly distinct, despite connecting s and s' in the unique transition $\langle s, s'\rangle$. Thus, where labeled ternary transition relations are defined in such a way that they are labeled with their middle terms (i.e., $\rightarrow_a\subseteq S \times a \times S$), it's not possible for two ternary transition relations, $\rightarrow_a$ and $\rightarrow_b$, to be identical where $a \neq b$. Thus, ternary transition relations can be used to distinguish *trajectories* in LTSs (sequences of states, $\langle s_1, \ldots, s_n\rangle$) from *paths* formed out of edges, $\langle s_1, a_1, \ldots, a_n, s_{n+1}\rangle$.

[27] I take it for granted that the former group is rationally acceptable in ways that the latter isn't, and the former group is itself epistemologically diverse. There may be much more to think about with respect to this toy example. For instance: Is Dubrovsky's procedure a rational one? Can it generate rational belief in T_1 in this context (i.e., it was stipulated that T_1 makes unique empirical predictions and that these predictions were confirmed, but nothing was stipulated about the predictions made by the rest of the theories; it's plausible that an adequate theory selection procedure wouldn't discount empirical adequacy enough to immediately prefer T_1 given this lacuna)? These are interesting questions that would take us far afield. I won't be digging into this case further than consideration of how to represent certain aspects of its dynamics.

Labeled transition systems are useful for representing and reasoning about dynamic systems in general, and can be applied quite naturally as tools for thinking about epistemic dynamics. The structures of labeled transitions systems can be mapped onto epistemic factors. For the purposes of this exposition, let any given LTS, $\langle S, \rightarrow_{a \in A} \rangle$, be a model of a single epistemic agent's epistemic dynamics. Let S be a set of possible epistemic states (letting the content of epistemic states be a replaceable parameter), let A be the set or repertoire of basic epistemic actions, and let $S \times S$ be the set of all formally possible epistemic transitions. From there, let T_A be the set of all possible epistemic transitions that are elements of any labeled binary transition relation (or where those states are the first and last terms of a labeled ternary transition relation). This contains the set of feasible epistemic transitions[28]—the transitions that can actually be carried out from any state in S. Let $S*$ be the set of all finite sequences of elements of S. Let T_A* be the set of all finite sequences, $\langle s_1, \ldots, s_n \rangle$, of elements of S such that, for every s_i and s_{i+1}, there is some $a \in A$ such that $\langle s_i, s_{i+1} \rangle \in \rightarrow_a$. Let T_A* be the set of all epistemic trajectories, the set of all sequences of epistemic states that a system represented by $\langle S, \rightarrow_{a \in A} \rangle$ can traverse by any sequence of actions in its repertoire of basic actions, A.

An LTS then displays a space of possible trajectories through the space of epistemic states and the paths of action, if any, that can take a system through each one. The LTS additionally identifies the structure of epistemic processes and procedures by several means. One means of identifying the procedural structure of epistemic processes is to let paths represent epistemic processes and then to identify the choices and sequences of epistemic actions in the tree-like structure of an LTS (i.e., identify sequences of edges and branches among paths). Another means is by identifying equivalence classes among the parts of LTSs. One well-known way of doing so is by identifying bisimulations among LTSs or parts of LTSs, and positing that such equivalence classes identify processes (or process types), of which individual paths and trajectories are instances. By way of illustration, let states in an LTS be associated with sets of schematic formulae (say, via an interpretation function or just a map from states to sets of formulae) representing the accepted propositions of the single agent modeled by an LTS. The Researcher's epistemic dynamics can, for example, be represented with the LTS in Fig. 1.1, granting some abuse of notation.

[28] I consider feasibility and capacity more closely in Chap. 4, by using PLEN. Informally, think of the feasibility of a dynamic or procedural object (e.g., epistemic transition, action, process, or procedure) as the possibility of executing it under the local (and typically contingent) conditions for a system—its current state, s, and the current state of the world. Think of the capacity of a dynamic or procedural object (less awkwardly, think of a dynamic or procedural object being within an agent's capacity) as the possibility of executing it under some conditions that the system could be in, whether local or non-local. Locality can be cashed out in terms of nearby possible worlds or some other mechanism.

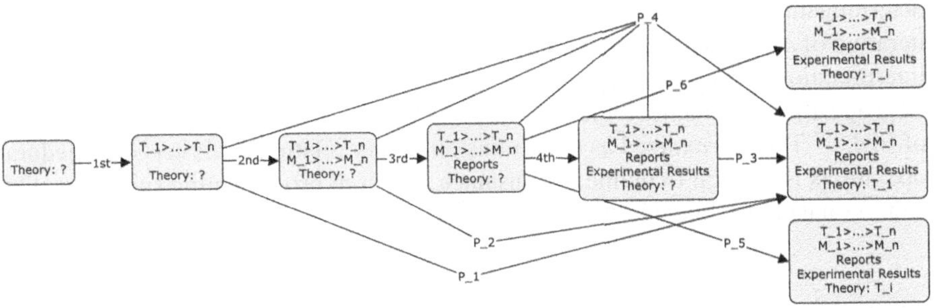

Fig. 1.1 The Researcher's LTS: The researcher, Dr. Dubrovsky, starts at the leftmost state and moves right over time. The labeled edges connecting states from left to right represent the possible transitions for the researcher that result from various procedures, inferences, or processes. These omit very much structural information—as in the complexity of temporal extension of (e.g.) experimentation. After the first step, Dubrovsky has an idea of the space of possible theories and their relative plausibility. After the second, they know how to formalize them as well as the respective rigor and elegance of these formalizations. The processes, P_1 through P_6, represent inferential transitions on the basis of Dubrovsky's epistemic state at a given time

1.3.3 Epistemic Procedures

Complex epistemic actions and processes can be broken down into more basic actions and processes organized by sequential compositions, combinations, choices, and other *procedural operations*. Conversely, basic epistemic actions and processes can be combined into more complex procedures by means of procedural operations.

Procedural operations are operations that take sets of actions and string them together into sequences, string iterations of the same action into a sequence, combine or interleave actions into parallel processes, and bind actions non deterministically into conditionally branching processes.[29] The standard repertoire of procedural operations includes sequential compositions, finite iterations, parallel executions, and non-deterministic choices. These procedural operations define the repertoire of possible *procedural constructions* or ways of building procedures out of a set of basic actions.

Relational algebra provides a framework for analyzing procedural operations by means of the general model of dynamics provided by LTSs. In LTSs, actions are represented with transition relations. Thus, letting the semi-colon, ";" stand in for the operation of sequential compositions, and letting LTSs provide our model of dynamics, the sequential composition of actions a and b, $a; b$, can be (and standardly is) modeled by the composition function on $\rightarrow_a$ and $\rightarrow_b$, the non-

[29] This is not an exhaustive list of procedural operations, though it contains the standard core. As we will see, there are several other basic combinatorial instructions that are useful in specifying the procedural content of norms. Some other procedural operations of note: avoidance of specific actions or procedures, reversal of an action or procedure, and execution of an action with minimal change in the world.

deterministic choice, $a \cup b$, is modeled by the union of $\rightarrow_a$ and $\rightarrow_b$, and the iteration of a, $a*$, is modeled by the reflexive, transitive closure of $\rightarrow_a$. Indeed, this interpretation of relational operations over LTSs is the core of the standard semantics of propositional dynamic logic (PDL), and so lies at the core of the program logic of PDL.[30] Thus formalized as operations on binary relations, this set of procedural operations is enough to define, for instance, the core programming structures, like conditional choice and guarded iteration.[31]

1.3.4 Epistemic Rationality Norms and Norm-Kernels

Epistemic rationality imposes various kinds of requirements on reasoners and their epistemic factors. Recall that the set of epistemic factors encompass the *epistemic states* (doxastic states, acceptance states, propositional attitudes, preferences and intentions) and *epistemic dynamics* (ways of changing epistemic states, possible state transitions, actions, processes, even complex procedures in an agent's repertoire) of an epistemic agent or system. Then, in order to be epistemically rational, the epistemic factors of an epistemic agent must meet certain qualitative conditions, have quantitative properties that fall within certain ranges, or have certain properties. If a factor fails to meet those conditions, it falls outside the bounds of rationality; it is, in some sense or to some degree, irrational.

Let epistemic rationality define *restrictions on epistemic factors* in the sense that the criteria of epistemic rationality (whatever they are) divide epistemic factors into correct and incorrect sets, determine which epistemic factors are permissible, required, or prohibited, or determine that some epistemic factors are greater than others with respect to some epistemic normative status. Restrictions on epistemic dynamics, for instance, may partition epistemic transitions or processes into rational or irrational sets or into permissible or impermissible sets, or determine a preference ranking over them (epistemic transition t is more rational than transition t', process p is rational where p' is not). The factors that meet the requirements are in the restricted or designated set defined by the norm (e.g., the rational set, the permitted or required set, the preferred set), and those that don't fall outside of it.

Call the rules, policies, or principles that define these restrictions on epistemic factors the *norms of epistemic rationality* or, interchangeably, *epistemic rationality*

[30] For the impatient reader: Yes, this set of procedural operations also lies at the core of the protocol logic of PLEN in virtue of being derived from PDL. In PLEN, however, the semantics of expressions like $a; b$ and $a \cup b$ are more complicated. PLEN interprets these operations not just over binary transition relations but over ternary transition relations and paths of edges as well. This explicitly builds subsystems that, in due course, will be seen to formalize different approaches to modeling epistemic dynamics within the overarching system of PLEN.

[31] See [109, 110, 112, 115] for far more thorough explication of the uses of LTSs and relational algebra for thinking about dynamic systems in general.

norms. Epistemic rationality norms are encoded in natural or regimented languages via norm-kernels:[32]

(Norm-Kernel) If C, then $R(x)$.

The condition C is commonly called the *condition of application* of the norm; it gives us a condition under which the normative claim, $R(x)$, holds. The variable x stands in for a term denoting something in the domain of the norm, something to which the norm applies. In other words, x is the *object* of the norm. The domains of norms are the sets of objects that they govern. The consequent, $R(x)$, is a normative claim of some kind, often a deontic proposition, though not necessarily. $R(x)$ states what is sometimes called the *valence* of the epistemic appraisal. The object of the norm may, for instance, be declared (by the norm), under condition C, to be rational or irrational, rationally obligatory or permissible, more or less rational relative to a standard or an alternative object, and so on. To be more exact, let $R(x)$ state the *normative verdict* of a norm-kernel—the assignment of rationality, irrationality, rational permissibility, epistemic deontics to epistemic factors by norm-kernels under their conditions of application—and let the *valence of the verdict* be the specific normative predicate or deontic operator assigned by the verdict.

Conditions of application and normative verdicts should not be identified strictly as the antecedent and consequent, respectively, of conditionals that state sufficient conditions for a monadic normative verdict. As suggested by prior discussion, there is room for norm-kernels of different and more complicated forms, including necessary conditions, conjunctive verdicts, and n-ary relational verdicts:

Norm-kernels	Sufficient condition	Necessary condition
Monadic	If C, then $R(x)$	If $R(x)$, then C
n-adic	If C, then $R(x_1, \ldots, x_n)$	If $R(x_1, \ldots, x_n)$, then C
Conjunctive monadic	If C, then $R(x)$ and $R'(y)$	If $R(x)$ and $R'(y)$, then C
Conjunctive n-adic	If C, then $R(x_1, \ldots, x_n)$ and $R(y_1, \ldots, y_n)$	If $R(x_1, \ldots, x_n)$ and $R(y, \ldots, y_n)$, then C

These rule-like expressions are a focal point of analysis [68, 69, 120] because they plausibly articulate the essential structure and properties of norms; it is this form of conditional rule that enables norm-kernels to define restrictions on epistemic factors by stating necessary and sufficient conditions for the satisfaction of deontic or other normative properties.[33]

[32] Following von Wright's initial analysis of norms [119, 120] and its later adoption in studying epistemic norms specifically, e.g., [68]. There may be other constructions that encode epistemic rationality norms (conditional imperatives come to mind), but norm-kernels are an uncontroversial and familiar starting point for the analysis of the content of epistemic rationality norms.

[33] Exploring the logic of these conditional rules (i.e., whether the logic of the conditional is material, intensional, non-classical, relevant, etc.) is largely beyond the scope of this book. As such, norm-kernels are treated as material conditionals. Chapter 7, Sect. 7.5 very briefly explores

In sum, epistemic rationality delivers normative verdicts about what states, transitions, and processes are rational (inter alia), thereby defining restrictions on epistemic factors. The norms of epistemic rationality are the rules that structure these verdicts by specifying conditions under which their objects possess some valence or another. Norm-kernels can explicitly state these rules and thereby define restrictions on epistemic factors. Thus, if one wants to achieve or execute epistemically rational factors, then one's factors must lie within the bounds set by the norms of epistemic rationality, as characterized by norm-kernels.

For convenience, say that any predicate that takes epistemic states as their arguments is a *static* predicate while those that take state transitions or sequences thereof are *dynamic* predicates, and apply the same convention to properties. That is, a normative property (like permissibility) is *dynamic* or there is a dynamic variant of that property if the objects that bear the property are changes of states or sequences of state changes (or sets thereof). Just so, a normative property is *static* if the objects that bear the property are epistemic states (or sets thereof). The *synchronic properties* of epistemic states are the properties that a state, s, has at any particular time, t, or that relate s to other things (e.g., the parts of s, the contents of s, other epistemic states, s') only at t. *Diachronic* and *polychronic* properties are those that involve change over time. *Procedural* predicates are the (typically intensional or even hyperintensional) predicates that apply to procedures.

I will say that epistemic rationality or its norms *distinguish* objects of epistemic appraisal x and y when x and y differ in the verdicts assigned to them by some norm-kernel, and I'll specify epistemic distinction with respect to individual norms, e.g., "norm N distinguishes x and y" when N assigns a verdict to x that it does not assign to y. I will also say, equivalently, that x and y are *rationally distinct* or *rationally distinguishable* (with respect to N). I will say that rationality (or specific norm N) *distinguishes x and y on the basis of C* when x and y are rationally distinct and C is what makes the difference in valence.

Finally, let there be distinct kinds of epistemic restriction for the distinct kinds of epistemic factors (restrictions on epistemic states, restrictions on epistemic dynamics, restrictions on epistemic processes, restrictions on epistemic procedures, restrictions on epistemic actions,...), and, accordingly, distinct kinds of epistemic norms and norm-kernels. *Static norms and norm-kernels* take epistemic states (or sets thereof) as their objects. *Dynamic norms and norm-kernels* take transitions or trajectories (or sets thereof) as their objects. *Procedural norms and procedural norm-kernels* take epistemic processes or sets thereof as their objects.

---

the possibility of importing a relevant logic with intensional conditional into PLEN, and the effects on the logic of norm-kernels are left implicit. Further exploration of the logic of norm-kernels is left to future work. It's tempting, but a garden path, to continue considering complications. I'll settle for one: relevance constraints on norm-kernels. It seems plausible that the conditions of application ought to state properties of the objects of the normative verdict; else, we are assigning normative verdicts on the basis of the properties of things that are irrelevant to the epistemic factor under consideration. These constraints are not built in for the sake of generality.

1.3.5 Restrictions on Epistemic Dynamics

As LTSs provide elementary, set-theoretic models of epistemic dynamics, and epistemic norms deliver verdicts that restrict epistemic factors (presumably, including dynamic factors), let me try to codify the content of epistemic norms by means of elementary set-theoretic constructions. The intuitive idea of a *restriction on epistemic dynamics* is a codification of the verdicts of an epistemic norm. So, what's needed is a construction that divides epistemic factors by normative statuses, whether categorical, deontic, or comparative. For that purpose, let's use the following definition:

Definition 1 (Restrictions on Epistemic Dynamics (over LTSs)) Let M be an LTS, $\langle S, \rightarrow_{a \in A}\rangle$. Let X be a finite, possibly empty set of elements of M, i.e., some set of states in S, transitions in $S \times S$ or T_A, paths in $S \times A \times \ldots \times A \times S$ or T_A*, or some specified subset thereof. Let $P(X)$ be the powerset of X. Let $\preceq$ be an arbitrary partial order on the elements, P_i, of $P(X)$. Now, let $R_X, T_X, I_X \in P(X)$. Then:

(Restrictions on Epistemic Dynamics) Every structure defined over X of the form

$$R_{\preceq}(X) = (P(X), \preceq, T_X, R_X, I_X)$$

is a *restriction on epistemic dynamics* (defined over LTSs) iff:

(1)

$$\forall P_i \in P(X) \forall x \in X \cap P_i (P_i \preceq T_X \equiv x \in I_X)$$

(2)

$$\forall P_i \in P(X) \forall x \in X \cap P_i (P_i \succ T_X \equiv x \in R_X)$$

(Restriction Conventions 1)

(1) Let R_X and I_X be the *verdict sets* of $R_{\preceq}(X)$.
   (a) R_X is the rational set.
   (b) I_X is the irrational set.
(2) Let T_X be the *threshold set* of $R_{\preceq}(X)$.
(3) Let $x \in R_X$, $x \in I_X$, and statements of the form $x \in P_i \succeq P_j$ be the *verdicts* of $R_{\preceq}(X)$. Respectively, they say:
   (a) $x \in R_X$ :: x is rational (rationally permissible).
   (b) $x \in I_X$ :: x is not rational (rationally impermissible).
   (c) $x \in P_i \succeq P_j$ :: x is more rational than any y in P_j.
   (d) $x \succeq y$ :: x is more rational than y (i.e., $x \in P_i$, $y \in P_j$, and $P_i \succeq P_j$)
(4) Denote arbitrary restrictions on X with $R_{\preceq}(X)$ or, where the set of epistemic factors is irrelevant or determined by context, $R_{\preceq}$.

(a) Say that $R_{\preceq}(X)$ is defined over X, and, that any restriction defined over any set of elements of epistemic dynamics is a restriction on epistemic dynamics.
(b) Introduce indexing of restrictions on the same set, $R^i_{\preceq}(X)$, where necessary.

The basic formal idea of restrictions on epistemic dynamics is that the structures of these elementary set-theoretic constructions correspond to the important features of rational distinction or the assignment of rational verdicts conditional on the basis of the properties of epistemic factors. Technically, each restriction comprises an exclusive (though not necessarily exhaustive) sorting of epistemic factors (i.e., the components of LTSs that represent the components of epistemic dynamics) into R_X and I_X, and defines an ordering over the powerset of this set of components of epistemic dynamics. The powerset of X provides a naive account of the extensions of properties of the elements of X. The thought is that the verdict sets, R_X and I_X, represent the way that categorical norms distinguish the objects to which they apply, with T_X being the threshold of irrationality with respect to the ordering $\preceq$. R_X is the set of rational elements of X according to $R_{\preceq}(X)$, and I_X is the irrational set of elements of X. By setting R_X (for instance) to some element, C, of the powerset over X, the restriction on epistemic dynamics codifies the idea that C is necessary and sufficient for whatever normative status R_X represents (e.g., rationality, rational permissibility, ...). Of course, C may be the union, $C_1 \cup \ldots \cup C_n$, of the extension of many properties, and so each C_i is represented by such a restriction as sufficient but not necessary for rationality (et al). Where C is the intersection of some set of $C_1, \ldots, C_n$, each property, C_i, is represented as necessary but not sufficient for rationality (et al) by such a restriction. It may be useful to think of restrictions as boundaries separating different classes of epistemic factors. See the Fig. 1.2 for diagrammatic help.

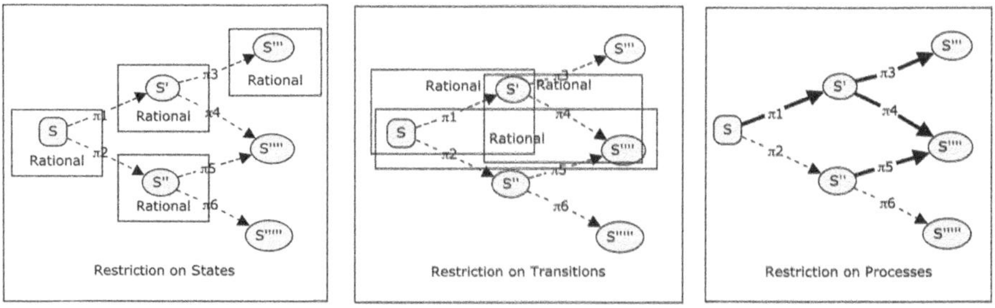

Fig. 1.2 Restrictions on Epistemic Dynamics: If the restriction is on epistemic states, the boundary encircles the rational states, placing the irrational states out of bounds, not to be trespassed upon. If the restriction is on transitions, the boundary encircles the rational transitions, placing the irrational transitions out of bounds, and cutting across any paths to their terminal states, blocking access. If the restriction is on processes, the boundary encircles the rational processes (which is harder to draw, and so represented with bold arrows); these are the processes that are to be executed, and even if a transition to a process' terminal state is rational, the boundary may place some ways of carrying out the transition out of bounds. With the subsets R_X and I_X, restrictions codify this kind of binding structure

The core of the construction codifies the conditional assignment of categorical normative verdicts. Norms are sometimes simple in that they conditionally assign components of epistemic dynamics exclusive, categorical statuses like "rational" and "irrational", "permissible" and "impermissible" on the basis of the properties of those components. Restrictions codify the content of these categorical norms by means of the designation of verdict sets among the elements of the powerset of X; each $P_i \in P(X)$ represents the extension of a (possibly logically complex) property of the epistemic factors in X, and the verdict sets capture the verdicts assigned to those factors on the basis of their properties. For example, the set of epistemic states can be partitioned into those with inconsistent belief sets and those without. Formally, let $P(S)$ partition the set of epistemic states, S, into just P_C and P_{IC}, where P_C is the set of consistent belief sets and P_{IC} is the set of inconsistent belief sets. Let $P_{IC} \prec P_C$, $P_{IC} = T_S = I_S$, and $R_S = P_C$. Then the following structure is a restriction on epistemic states: $R_{\preceq}(S) = \langle P(S), \preceq, P_{IC}, P_C, P_{IC} \rangle$. A restriction like the foregoing completely captures the content of a full-stop consistency norm such as:

(Full-Stop Consistency (FSC)) For any belief state, s, s is rational iff there is no sentence A such that both A and $\neg A$ are believed in s.[34]

It's easy to see that this norm defines a class of restrictions exactly like $R_{\preceq}(S)$, though perhaps varying in the properties of their ordering relations. FSC quantifies over all belief states and partitions them into rational and irrational subsets on the basis of inconsistency. The ordering, $\prec$, of a restriction need only satisfy the condition that $P_{IC} \prec P_C$. Full-Stop Consistency (FSC) tells us that the set of all and only the consistent states is R_S, and the set of inconsistent states is to be identified with both I_S and T_S. Every other set of belief states may be incomparable in the ordering; the norm says nothing about states on the basis of their other properties.

The ordering over powersets of elements of epistemic dynamics provides a representation of more nuanced normative verdicts. Continuing with the boundaries metaphor, the ordering over the elements of the powerset of components works like a concentric series of boundaries; this enables the representation of comparative or relational normative statuses (x is outside fewer boundaries than y) and thresholds (some of the boundaries are fine to cross as long as one doesn't go too far or pass a cutoff point). The ranking may be so defined that it equates the possession of some property with a normative verdict, e.g., associating consistency or degree of entrenchment with greater rationality but not a categorical rational status. After all, epistemic norms, expressed via norm-kernels, typically tell us that entire classes of objects (e.g., states, transitions, or processes) possess some normative valence if they satisfy C, the condition of application. Given that the ordering is partial, there may be nothing else to say about the other elements of the power set with respect to that restriction—the ordering may leave them all incomparable. This is a virtue

[34] This norm is, for familiar reasons [47, 69], both implausibly permissive and implausibly strict. Still, it serves as a suitable formal example.

in terms of the representation of norms; a consistency norm like FSC may tell us nothing about a mixed set of consistent and inconsistent belief states other than that the consistent subset is a set of rational states.

More complex restrictions can be defined by more subtle norms. A strictly negative consistency norm, for instance, might declare inconsistent belief states irrational but not declare consistent belief states automatically rational. A more complex division of (arbitrary) X and ordering thereon might then be defined by the norm. Whatever the level of nuance, each epistemic rationality norm defines some class of restrictions on epistemic dynamics of some level of complexity. For a second example, note that the threshold element, T_X, can be modified in a number of ways. The definition of restrictions builds in the notion of a threshold element, T_X, that does intuitive work: epistemic rationality norms are frequently categorical, as FSC is. The threshold element just matches the ordering to the categorical verdict that results from dividing states into R_S and I_S. However, for more nuanced norms that issue no all-or-nothing verdicts with respect to their objects, the threshold T_X may be safely ignored while the ordering is employed; for any restriction, there is always a restriction with an identical rational subset and ranking but that $T_X = I_X = \emptyset$. These ranking relations then can represent preference relations on epistemic factors. Thus, subtler epistemic norms than FSC that merely recommend some epistemic factors over others can be captured by restrictions. Of course, T_X might also represent a threshold with respect to the ranking by letting $I_X \prec T_X$ that doesn't partition $P(X)$ into rational and irrational subsets.

Norms that deliver deontic verdicts can also be modeled with restrictions on epistemic dynamics. Deontics can be encoded with restrictions by taking I_X to be the prohibited factors, R_X to be the set of all permissible factors, and some $\prec$ maximal subset of R_X to be the obligatory or required factors. An alternative and simpler encoding of obligation on procedural epistemic factors reduces it to permission and structural relations: let any factor, a, that is a part of every epistemic factor in R_X or otherwise accompanies every permissible factor be obligatory. The prohibited factors are thus all of those factors that preclude the execution of a. To illustrate: if one's norms permit all and only the courses of action p_1, p_2, and p_3, and there is a fourth course of action, q, that must be carried out in order to carry out any of the foregoing p_i, then q is obligated by those norms. This will be the case if q is a subprocess of each p_i, if q must be carried out prior to carrying out any p_i, or if one must do q concurrently with any permissible p_i. In other words, the obligatory actions are simply those actions such that either (i) they are the uniquely permissible actions under some condition or (ii) they must be executed in addition to any other permissible action. All one needs to analyze obligation in terms of restrictions is machinery for thinking about these kinds of features of action. With obligation translated into the permissions defined by restrictions on epistemic dynamics, prohibition comes along for free by the obvious intertranslations.

As for terminological conventions, the idea of restrictions on epistemic dynamics is a construction that codifies the verdicts of epistemic norms. Thus, when a norm or norm-kernel, N, assigns normative verdicts to the elements of a set of objects, X, and those verdicts are codified by restriction on X, $R_{\preceq}(X)$, I'll say that N

restricts X or that *N defines restrictions on X* or *over X*. In virtue of the intended correlation between the structures of restrictions and the verdicts of norms and norm-kernels, I'll also write interchangeably of the *verdicts* or *verdict sets* of norm-kernels, restrictions on epistemic dynamics, and norms, when there are no technical or philosophical considerations that hinge on the distinction.[35]

1.4 Terminological Disputes Set Aside

Now, let me set aside some disputes I won't be taking up any further in this chapter or book, but that may be worth litigating further in future work.

1.4.1 Epistemic Rationality vs Other Terms of Epistemic Appraisal

I deploy the term "epistemic rationality" as a catch-all term for epistemic correctness. The core idea is that epistemic rationality is a normative concept the central function of which is cognitive control and evaluation.[36] Epistemic rationality functions primarily to distinguish some epistemic factors as those to be executed, arrived at, brought about, or sustained and others as those not to be executed, to be avoided, or to be revised or ended. In taking rationality as a catch-all term, I may draw the criticism of failing to draw proper distinctions among terms of epistemic assessment. For instance, one might think that epistemic justification is right standing with respect to a rule, and that rationality is a matter of bearing some proper relation or another to reasons or evidence, and there may be serious questions about how these notions hang together or whether there are reductions in the neighborhood. Nothing in this project, as far as I can see, hangs on the outcomes of these questions or the contours of these finer distinctions, so my terminological choice treats these concepts as species of the same regulative kind.

1.4.2 Epistemic Factors vs Epistemic Agents

Further, it might be thought that it is reasoners themselves that are the unique appropriate objects of epistemic appraisal—particularly where rationality is concerned. This can't be right. It's a commonplace feature of the normative deployment of the

[35] One context in which the distinction matters: when justifying the claim that restrictions on epistemic dynamics represent the verdict assignment of norm-kernels, as in Chap. 9.

[36] See Chap. 2 for the main arguments for this claim.

concept of rationality that we distinguish the rationality of a view that agents hold from the rationality of the agents themselves. One may have some epistemically irrational views or even systematically irrational views about some domain while being an epistemically rational person or being free from epistemic blame. While it is plausible that appraisals of rationality apply to agents themselves in addition to their epistemic factors, I ignore this complication in all of what follows. There should be no difficulty extending the work here to agents. As such, in all of what follows, I focus on the epistemic rationality—broadly construed—of epistemic factors rather than agents.

1.4.3 Epistemic vs Non-epistemic Norms

There is a great variety of proposed epistemic rationality norms across the fields of epistemology and philosophy of logic. The debate about these often centers on whether the norms are themselves rational—or, if one prefers, plausible or defensible—as norms for controlling epistemic state change. See, for instance, the debates about (e.g.) deductive closure, conditionalization, or even logical consistency [2, 20, 47, 95]. I will not be directly weighing in on these debates. Rather, I'll be pulling examples of epistemic norms liberally from the literature (as well as the aether) in order to illustrate concepts or articulate arguments. While it will sometimes matter whether these examples are cases of rational epistemic rationality norms, it will be assumed that they are all norms of epistemic rationality as opposed to norms of some other kind.

As for what distinguishes epistemic norms from non-epistemic norms, I wish to remain ultimately non-committal. The primary reason for this commitment to being non-committal is that doing so is unnecessary; nothing hangs on the matter. This project will develop a formal framework for thinking about norms that govern epistemic dynamics, i.e., norms articulated by norm-kernels that take elements of epistemic dynamics as their objects, that deliver verdicts about elements of epistemic dynamics, or that define restrictions on epistemic dynamics. In this chapter, I will argue that norms governing epistemic dynamics are both irreducible and indispensable to epistemic life, and that they can provide reductions of norms governing epistemic states. Thus, if it turns out that the norms that are the target of analysis in this project are not really epistemic norms in some specified sense, so much the worse for that specified sense.[37]

Nevertheless, it's worthwhile to point out my working conception of epistemic norms in contrast to non-epistemic norms. I employ a fairly permissive criterion:

[37] I expand on this point immediately below in Sect. 1.4.5.

(Epistemic Norms (vs Non-epistemic Norms)) A norm-kernel, *N*, articulates an epistemic norm iff *N*'s objects are epistemic factors, and the valences of *N* are conditioned on:

(i) logical, evidential, or alethic properties of its objects,
(ii) the structures of epistemic procedures related to its objects, or
(iii) the logical, evidential, or alethic properties of epistemic procedures related to its objects.

From this, it follows that, if the valence that *N* assigns its objects is conditioned on (only) practical or moral properties of epistemic factors, then, roughly speaking, it's a non-epistemic norm. This rough and ready rule will serve to delineate the class of epistemic norms that are the targets of analysis, although it's a matter of irrelevance to this project whether this is the final statement of the necessary and sufficient conditions for "epistemic-ness".

1.4.4 Epistemic Involuntarism

I deploy a capacious notion of epistemic action in concert with the permissive conception of epistemic norms above. The set of epistemic actions is the set of actions that generate, transform, reveal, or obscure information in any ways relevant to epistemic cognition. This includes actions as diverse as those overt actions involved in the construction and manipulation of external representations (e.g., carrying out calculations or derivations on a blackboard, crafting a physical model and manipulating it[38]), methodical evidence acquisition or generation (i.e., experimentation or physical manipulations of the environment that otherwise generate information[39]), as well as those actions involved in more internally oriented but still voluntary cognitive behavior like consciously attempting to think of alternative explanations, directing attention to particular bodies of evidence, drawing inferences consciously, or the actions involved in deliberate self-stimulation of inference or supposition [30, 56, 71, 77].

---

[38] This usage is especially consilient with lines of empirical research on epistemic action, e.g., [56] and [70], that show that manipulation of models can improve performance on cognitive tasks. For instance, the manipulation of physical models of arrangements of blocks improves performance on later abstract block arrangement tasks, specifically as exhibited in playing the game Tetris [55]. There is also much philosophical research on the construction and manipulation of models in philosophy of science [35, 74, 99]. This research tends to focus on what epistemic gains are made by manipulating models despite their admitted distance from the phenomena they model. However, that such gains are made coupled with the fact that model manipulation is a volitional activity implies the point being made: that the sorts of non-inferential or non-cognitive actions involved in manipulation of models is epistemically useful, perhaps because they trigger inferences or other cognitive processes.

[39] For instance, turning over cards, looking under cups, reading letters, or any of the other standard actions in the DEL literature, re: Ch. 5 of [117].

As a result of this permissive conception of epistemic actions, I'll be setting aside any worries about doxastic involuntarism and related issues. One might argue, as Alston [6] does, that epistemic norms cannot require or impose normative requirements on things like inference or belief formation because normative requirements can only apply to things that agents have control over, and inference and belief formation are not among these things. However, even if we accept the claim that inference and belief formation are not actions over which agents wield direct control, an indirect form of control over inference and belief change can be exerted by performing actions that trigger these involuntary cognitive processes [71]. For instance, the deliberate direction of attention to logical relations among believed propositions can trigger expansion or contraction of beliefs. More generally, the deliberate direction of attention to specific features of argument structure, evidence, environments, or our models can function as self-stimulating actions that trigger inference and belief change. The deliberate construction, modification, and manipulation of models can do the same—even if only because deliberately building, tinkering with, and using models can direct attention to logical relations, evidence, etc.

By these means one can *see to it that* one infers a certain way or forms a belief, as long as there are regular connections between actions that one can directly control and those cognitive processes. That is, one can wield control over otherwise involuntary cognitive processes by carrying out epistemic actions in the broad sense above. Moreover, one can exert full voluntary control over some forms of epistemic state, as with acceptance.[40] So, in place of a full engagement with doxastic voluntarism and involuntarism, I'll just stipulate that epistemic actions subsume both the kinds of actions described above as well as the kinds of action that reliably stimulate inference or belief. So, where I assert that norms require inference or formation of belief under some conditions, the doxastic involuntarist can re-phrase this as saying that norms really require that one see to it that one infers or forms beliefs under those conditions by carrying out the requisite self-stimulation actions.

This point extends to complex cognitive processes involving involuntary and even unconscious subprocesses. Epistemic actions can be strung together to form epistemic processes. The epistemic actions out of which a process is formed are basic with respect to the processes they comprise. Complex processes can be strung together to form more complex processes. For instance, in proving a theorem, a logician may combine and sequence many different kinds of processes, from motor control processes of manipulation of the environment (writing on a chalkboard or in a notebook) to immediate inferences to reading and exchange of information with collaborators. These combinatorial assumptions coupled with the capacious notion of epistemic action—perhaps controversially—run together processes of epistemic cognition and processes formed out of sequenced epistemic actions, as defined above. Such epistemic processes might involve the interleaving of cognitive processes (i.e., less controversially epistemic processes) and actions on

[40] In the sense of Cohen [22].

the environment. One might wish to draw some sharp line between processes that affect the outside world and those that don't. However, I doubt that this boundary is much of one for the reasons that, as argued above, (i) actions that affect the world generate, reveal, or conceal information for cognitive processes to use, and (ii) they also stimulate or trigger such processes, inputing that information in a usable form or bringing the salient information to awareness. A defense of this line of thought is in [71]. The boundary is thoroughly questioned independent of my worries [56, 77].

1.4.5 *Procedural Epistemology*

Epistemic Proceduralism is the view that the requirements of epistemic rationality are defined by procedural norms that rule in and rule out processes formed out of more basic processes in certain ways. As I'll argue in Chap. 2, these procedural norms are integral to epistemic rationality. The attention paid to these procedural norms might seem alien to some perspectives in epistemology. In what follows, I frequently defend putative procedural requirements by appeal to means-ends considerations rather than what might be thought of as "fundamental" epistemological considerations, and this might suggest that my target isn't properly epistemic norms at all.

For instance, I argue that epistemic rationality distinguishes between epistemic processes even where they effect identical transitions among epistemic states because the rational process is built in some specified way out of more basic epistemic actions and processes by means of procedural operations of sequencing and choice (inter alia). Some of these arguments are based on the observation that procedural features (like the sequencing of subprocesses) of our epistemic processes impact whether and how well we can attain epistemic goals.

To this general approach, it might be objected that the "epistemic" norms that I focus on at such length are merely practical rules for achieving properly epistemic ends or complying with properly epistemic norms; they are not themselves fundamental nor properly epistemological. Consider, for instance, the monistic view that truth is the sole, fundamental epistemic norm; it's the only standard against which epistemic factors can be evaluated. On this view, beliefs can be evaluated on the basis of evidential support and coherence only in virtue of the (putative) relations between these features of belief systems and truth, and reasoning processes can be evaluated only with respect to their reliability for producing true beliefs (or perhaps indirectly on the basis of their production of evidential support or coherence). On this view, there might be rules that pick out some processes from others in virtue of having telltale features associated with reliability, but these rules will not be properly epistemological. They'll be strictly practical (non-epistemic) rules the normative force of which reduces to conditional imperatives of the form:[41] "If you want to have true beliefs, then you ought to embody such-and-such mental processes, since they are reliable for producing truths." [82] (p.xi)

My response to this objection is twofold. First, it's not really an objection. My project is compatible with the view above; I just think that the derivative and "not fundamental and properly epistemological" procedural or pragmatic norms in question are extremely important for any theorist that wants to understand epistemic dynamics and methodology, and I use the phrase "epistemic norms" to talk about them. The basic epistemological perspective in this book is that of procedural epistemology [92]. As Papineau [82] puts it, I think of epistemology "...as a practical enterprise, concerned to ensure that the processes we embody are reliable for truth, as opposed to a purely theoretical enterprise within the realm of belief." (p.xviii) In other words, I'm concerned with epistemic methodology—the kinds of rules, principles, and concepts that limited epistemic agents can use to achieve the aims of inquiry. To that end, I pay special attention to considerations about deliberate epistemic self-regulation, which requires norms governing procedural construction. If these regulative and methodological concerns are not properly epistemological, proper epistemology is poorer for omitting them.[42]

Second, if I'm right in my characterization of epistemic proceduralism, there is a clear sense in which the "not properly epistemological norms" are indispensable to epistemology for reasons that go beyond the analysis of self-regulation. I show (eventually[43]) that all of the important normative information that is carried by any "static" epistemic norm[44] can be read off of some procedural norm and that the converse fails. If that's all correct, whether we call procedural norms "properly epistemic" or not is of little relevance, as a model of epistemic rationality that rigorously analyzes procedural norms will get a good analysis of the "fundamental and properly epistemic" norms for free.

1.5 Toward the Defense

With those disputes set aside, a reader may wonder what arguments this book will take up. In Part I, I want to explicitly spell out and systematically defend the account of epistemic proceduralism driving this project. In the next chapter,

[41] While [82] is the source of a clear statement of the view described, Papineau isn't a proponent of it. On Papineau's view, one fundamental epistemic norm is one too many. True belief is desirable because it's a prerequisite for successful action; there are no fundamental epistemic norms [82, 83]. I don't think there is a serious conflict between the epistemological approach taken here and Papineau's view.

[42] Indeed, as Pollock points out in [92], "proper epistemology" diverges from the main lines of modern epistemology as embodied in Descartes' regulae [24]. It does so to its detriment.

[43] The impatient reader will have to wait until Chap. 9.

[44] Static norms being norms that concern only the contents of epistemic states at a time.

I articulate and defend clusters of principles, D1 and D2, along with clusters of corollaries thereof, that characterize the structure, content, and regulative function of epistemic rationality norms, as well as exploring some of their logical interactions. In particular, I devote much of Chap. 2 to the analysis of correspondence and reduction relations among prominent classes of epistemic norms (static, dynamic, and procedural). Each clause of each cluster is argued for in turn, often utilizing multiple lines of reasoning. Chapter 3 will defend a third cluster, D3, by exploring the logic of epistemic rationality norms, as it appears in the distinctive forms of reasoning with and about epistemic norms in their regulative role and in philosophical study. Parts II and III turn, respectively, to developing the protocol-theoretic logical framework for epistemic norms, and to applying it to the analysis of the structure and logic of epistemic rationality norms, i.e., to the formalization of the core elements of epistemic proceduralism.

References

1. Adler, J.E. 2009. Another argument for the knowledge norm. *Analysis* 69(3):407–411.
2. Adler, J.E., and L.J. Rips. (eds.). 2008. *Reasoning: Studies of human inference and its foundations*. Cambridge University Press.
3. Alchourrón, C.E. 1991. Conflicts of norms and the revision of normative systems. *Law and Philosophy* 10(4):413–425.
4. Alchourrón, C., and E. Bulygin. 1971. Normative systems. *Crítica: Revista Hispanoamericana de Filosofía* 5(14):120–123.
5. Alchourrón, C., P. Gardenfors, and D. Makinson. 1985. On the logic of theory change: Partial meet contraction and revision functions. *The Journal of Symbolic Logic* 50(2):510–530.
6. Alston, W. 1989. *Epistemic justification*. Cornell University Press.
7. Artemov, S., and M. Fitting. 2021. Justification logic. In *The Stanford Encyclopedia of Philosophy* (Spring 2021 Edition), ed. E.N. Zalta. https://plato.stanford.edu/archives/spr2021/entries/logic-justification/.
8. Bacon, F. 1851. *The advancement of learning*. Modern Library.
9. Bacon, F. 1960. *The new organon, and related writings*. New York: Liberal Arts Press.
10. Baltag, A., and L.S. Moss. 2004. Logics for epistemic programs. *Synthese* 139:165–224.
11. Baltag, A., and S. Smets. 2008. Probabilistic dynamic belief revision. *Synthese* 165(2):179–202.
12. Beall, J.c. 2015. Free of detachment: Logic, rationality, and gluts. *Noûs* 49(2):410–423.
13. Beall, J.c., and G. Restall. 2005. *Logical pluralism*. Oxford University Press.
14. Belnap, N., M. Perloff, and M. Xu. 2001. *Facing the future*. Oxford: Oxford University Press.
15. Berto, F., and D. Nolan. 2021. Hyperintensionality. In *The Stanford Encyclopedia of Philosophy (Summer 2021 Edition)*, ed. E.N. Zalta. https://plato.stanford.edu/archives/sum2021/entries/hyperintensionality/.
16. Boghossian, P. 2008. Epistemic rules. *Journal of Philosophy* 105(9):472–500.
17. Bratman, M. 1998. *Intention, plans, and practical reason*. Center for the Study of Language and Information.
18. Broome, J. 2000. Normative requirements. In *Normativity*, ed. J. Dancy. Oxford: Blackwell.
19. Carnap, R. 1951. *The continuum of inductive methods*. University of Chicago.
20. Christensen, D. 2004. *Putting logic in its place*. Oxford: Oxford University Press.
21. Clifford, W.K. 1999. *The ethics of belief and other essays*. Prometheus Books.
22. Cohen, L.J. 1992. *An essay on belief and acceptance*. Oxford: Oxford University Press.

23. Conee, E.B. 2004. *Evidentialism: essays in epistemology*. Oxford University Press.
24. Descartes, R. 1964. *Philosophical essays: Discourse on method; Meditations; Rules for the direction of the mind*. Indianapolis: Bobbs-Merrill.
25. Evans, J., and K.E. Stanovich. (2013). Dual-process theories of higher cognition advancing the debate. *Perspectives on Psychological Science* 8(3):223–241.
26. Fagin, R., J. Halpern, Y. Moses, and M. Vardi. 2004. *Reasoning about knowledge*. MIT Press.
27. Faroldi, F. 2019. *Hyperintensionality and normativity*. Springer.
28. Field, H. 2001. *Truth and the absence of fact*. Oxford: Oxford University.
29. Field, H. 2008. *Saving truth from paradox*. Oxford: Oxford University Press.
30. Field, H. 2009a. Epistemology Without Metaphysics. *Philosophical Studies*. Volume 143, Issue 2, pp 249–290.
31. Field, H. 2009b. Pluralism in logic. *Review of Symbolic Logic* 2(2):342–359.
32. Field, H. 2009c. What is the normative role of logic? *Aristotelian Society Supplementary Volume* 83(1):251–268.
33. Foley, R. 1993. *Working without a net*. Oxford University Press.
34. Forrest, P. 1986. *The dynamics of belief*. London: Blackwell.
35. Frigg, R., and S. Hartmann. 2017. Models in science. In *The Stanford Encyclopedia of Philosophy* (Spring 2017 Edition), ed. E.N. Zalta. https://plato.stanford.edu/archives/spr2017/entries/models-science/.
36. Gabby, D., J. Horty, and X. Parent, (eds.). 2013. *Handbook of deontic logic and normative systems*. College Publications.
37. Gardenfors, P. 1992. *Belief revision*. Cambridge: Cambridge University Press.
38. Gardenfors, P. 2008. *Knowledge in flux*. Reprint. Original: 1988. London: College Publications.
39. Gibbard, A. 2003. *Thinking how to live*. Harvard University Press.
40. Gigerenzer, G. 2007. *Gut feelings: The intelligence of the unconscious*. London, England: Penguin.
41. Gigerenzer, G. 2011. Personal reflections on theory and psychology. *Theory & Psychology* 20:733–743.
42. Gigerenzer, G., and T. Regier. 1996. How do we tell an association from a rule? Comment on Sloman (1996). *Psychological Bulletin* 119:23–26.
43. Gochet, P. 2002. The dynamic turn in twentieth century logic. *Synthese* 130(2):175–184.
44. Goldman, A. 1979. What is Justified Belief? In *Justification and knowledge*, 1–25, ed. G. Pappas. Boston: D. Reidel.
45. Hansson, S.O. 2000. Formalization in philosophy. *Bulletin of Symbolic Logic* 6(2):162–175.
46. Hansson, S.O. 2017. Logic of belief revision. In *The Stanford Encyclopedia of Philosophy* (Winter 2017 Edition), ed. E.N. Zalta, forthcoming. https://plato.stanford.edu/archives/win2017/entries/logic-belief-revision/.
47. Harman, G. 1986. *Change in view*. Cambridge: MIT Press.
48. Hedden, B. 2015. Time-slice rationality. *Mind* 124(494):449–491.
49. Hintikka, J. 1962. *Knowledge and belief*. Ithaca: Cornell University Press.
50. Horty, J. 2001. *Agency and deontic logic*. Oxford: Oxford University Press.
51. Hoshi, T. 2009. *Epistemic dynamics and protocol information*. ILLC Dissertation Series
52. Kahneman, D., P. Slovic, and A. Tversky, (eds.). 1982. *Judgment under uncertainty: Heuristics and biases*. Cambridge University Press.
53. Kelly, K.T. 1988. Formal learning theory and the philosophy of science. *PSA: Proceedings of the Biennial Meeting of the Philosophy of Science Association* 1988:413–423.
54. Kelly, K.T., O. Schulte, and V. Hendricks. 1995. Reliable belief revision. *Logic and scientific methods: Volume one of the tenth international congress of logic, methodology and philosophy of science, Florence*, 383–398.
55. Kirsh, D., and P. Maglio. 1992. Some epistemic benefits of action-tetris, a case study. In *Proceedings of the 14th annual conference of the cognitive science society*.
56. Kirsh, D., and P. Maglio. 1994. On distinguishing epistemic from pragmatic action. *Cognitive Science* 18(4):513–549.

57. Kolodny, N. 2005. Why be rational? *Mind* 114(455):509–563.
58. Kolodny, N. 2007. State or process requirements? *Mind* 116(462):371–385.
59. Konolige, K. 1986. *A deduction model of belief*. Morgan Kaufmann Publishers.
60. Kooi, B.P. 2003. Probabilistic dynamic epistemic logic. *Journal of Logic, Language and Information* 12(4):381–408.
61. Kyburg, H. 1983. *Epistemology and inference*. Univ. of Minnesota Press.
62. Lam, B. Unpublished. The Dynamic Foundations of Epistemic Rationality. Ph.D Dissertation: Vassar College.
63. Leitgeb, H. 2017. *The stability of belief: How rational belief coheres with probability*. Oxford University Press.
64. Leitgeb, H., and K. Segerberg. 2007. Dynamic doxastic logic: why, how, and where to? *Synthese* 155(2):167–190.
65. Levi, I. 1996. *For the sake of the argument: Ramsey test conditionals, inductive inference, and nonmonotonic reasoning*. Cambridge University Press.
66. Levi, I. 2004. *Mild contraction: Evaluating loss of information due to loss of belief*. Oxford University Press.
67. Lindström, S., and W. Rabinowicz. 1999. DDL unlimited: Dynamic doxastic logic for introspective agents. *Erkenntnis* 50(2-3):353–385.
68. Littlejohn, C., and J. Turri. *Epistemic norms: New essays on action, belief, and assertion*. Oxford: Oxford University Press.
69. MacFarlane, J. 2004. In What Sense (If Any) Is Logic Normative for Thought? Unpublished. (Presented at the 2004 Central Division APA symposium on the normativity of logic.) Accessed online: http://johnmacfarlane.net/normativity_of_logic.pdf.
70. Maglio, P.P., M.J. Wenger, and A.M. Copeland. 2003. The benefits of epistemic action outweigh the costs. In *Proceedings of the Annual Meeting of the Cognitive Science Society*, vol. 25. Permalink: https://escholarship.org/uc/item/5bg5p6bv.
71. Mandelbaum, E., and N. Levy. 2014. The Powers that bind: doxastic voluntarism and epistemic obligation. In *The ethics of belief*, 12–33, ed. J. Matheson. New York: Oxford University Press.
72. McNamara, P. 2019. Deontic logic. In *The Stanford Encyclopedia of Philosophy* (Summer 2019 Edition), ed. E.N. Zalta, forthcoming. https://plato.stanford.edu/archives/sum2019/entries/logic-deontic/.
73. Mill, J.S. 1843. *A system of logic, ratiocinative and inductive being a connected view of the principles of evidence and the methods of scientific investigation*. Longmans, Green, Reader, and Dyer.
74. Morgan, M. 1999. Learning from Models. In *Models as mediators: Perspectives on natural and social science. Ideas in context*, 347–388, ed. Morgan, M.S., and M. Morrison, (eds.). Cambridge, UK: Cambridge University Press.
75. Moss, S. 2015. Time-slice epistemology and action under indeterminacy. In *Oxford studies in epistemology*, 172–94, ed. T.S. Gendler, and J. Hawthorne. Oxford University Press.
76. Navarro, P.E., and J.L. Rodríguez. 2014. *Deontic logic and legal systems*. Cambridge University Press.
77. Neth, H., and T. Müller. 2008. Thinking by doing and doing by thinking: a taxonomy of actions. In *30th Annual conference of the cognitive science society*, 993–998, ed. V. Sloutsky, B. Love, and K. McRae. Austin, TX: Cognitive Science Society.
78. Nolan, D. 2014. Hyperintensional metaphysics. *Philosophical Studies* 171(1):149–160. https://doi.org/10.1007/s11098-013-0251-2
79. Oaksford, M., and N. Chater. 2007. *Bayesian rationality: The probabilistic approach to human reasoning*. Oxford University Press.
80. Ōtomo, K. 2011. *Akira*, vol. 5. New York: Kodansha Comics.
81. Pacuit, E., and S. Simon. 2010. Reasoning with protocols under imperfect information. *Review of Symbolic Logic* 4(3):412–444.
82. Papineau, D. 1987. *Reality and representation*. Blackwell.

83. Papineau, D. 2013. There are no norms of belief. In *The aim of belief*, ed. T. Chan. Oxford University Press.
84. Peregrin, J. (ed.) 2003. *Meaning: The dynamic turn*. Elsevier Science.
85. Pettigrew, R. 2011. An improper introduction to epistemic utility theory. In *Proceedings of EPSA: Amsterdam '09*, 287–301, ed. H. de Regt, S. Okasha, and S. Hartmann. Springer.
86. Pettigrew, R. 2013a. Epistemic utility and norms for credences. *Philosophy Compass* 8(10):897–908.
87. Pettigrew, R. 2013b. A new epistemic utility argument for the principal principle. *Episteme* 10(1):19–35.
88. Pettigrew, R. 2016. Epistemic utility arguments for probabilism. In *The Stanford Encyclopedia of Philosophy* (Spring 2016 Edition), ed. E.N. Zalta. https://plato.stanford.edu/archives/spr2016/entries/epistemic-utility/.
89. Pettigrew, R. 2017. Epistemic utility and the normativity of logic. *Logos and Episteme* 8(4):455–492.
90. Pollock, J.L. 1987. Epistemic norms. *Synthese* 71(1):61–95.
91. Pollock, J.L. 1995. *Cognitive carpentry*. MIT Press.
92. Pollock, J.L. 1998. Procedural epistemology. In *The digital phoenix: How computers are changing philosophy*, 17, ed. T.W. Bynum, and J. Moor. Blackwell.
93. Pollock, J.L. 2009. A recursive semantics for defeasible reasoning. In *Argumentation in artificial intelligence,* 173–197, ed. G. Simari, and I. Rahwan. Boston, MA: Springer.
94. Pratt, V.R. 1980. Application of modal logic to programming. *Studia Logica* 34(2/3):257–274.
95. Priest, G. 2006. *Doubt truth to be a liar.* Oxford: Oxford University.
96. Quine, W.V. 1969. *Ontological relativity and other essays*. Columbia University Press.
97. Read, S. 2015. Medieval theories: Properties of terms. In *The Stanford Encyclopedia of Philosophy* (Spring 2015 Edition), E.N. Zalta. https://plato.stanford.edu/archives/spr2015/entries/medieval-terms/.
98. Restall, G. 2005. Multiple conclusions. *Logic, methodology and philosophy of science: Proceedings of the twelfth international congress.*
99. Reutlinger, A., D. Hangleiter, and S. Hartmann. 2016 (Preprint). Understanding (with) toy models. *British Journal for the Philosophy of Science*.
100. Rips, L.J. 1994. *The psychology of proof: Deductive reasoning in human thinking*. MIT Press.
101. Schafer, K. 2014. Doxastic planning and epistemic internalism. *Synthese* 191(12):2571–2591.
102. Schmechtig, P., and M. Grajner, (eds.). 2016. *Epistemic reasons, norms, and goals*. De Gruyter.
103. Schulte, O. 1999. Means-ends epistemology. *British Journal for the Philosophy of Science* 50(1):1–31.
104. Schulte, O. 2017. Formal learning theory. In *The Stanford Encyclopedia of Philosophy* (Spring 2017 Edition), E.N. Zalta. https://plato.stanford.edu/archives/spr2017/entries/learning-formal/.
105. Segerberg, K. 1995. AGM as a dynamic, doxastic logic. *Bulletin of Symbolic Logic* 1.
106. Staffel, J. 2013. Can there be reasoning with degrees of belief? *Synthese* 190(16):3535–3551.
107. Stanovich, K.E. 2005. *The robot's rebellion: Finding meaning in the age of Darwin*. University of Chicago Press.
108. Stenning, K., and M. van Lambalgen. 2008. *Human reasoning and cognitive science*. Boston, USA: MIT Press.
109. van Benthem, J. 1990. General dynamics. *ITLI prepublication series for logic, semantics, and philosophy of language*. University of Amsterdam.
110. van Benthem, J.F.A.K. 1994. Programming operations that are safe for bisimulation. *Studia Logica* 60(2):311–330.
111. van Benthem, J.F.A.K. 1995. *Language in action categories, lambdas and dynamic logic*.
112. van Benthem, J.F.A.K. 1996. *Exploring logical dynamics*. Stanford: CSLI Publications.
113. van Benthem, J.F.A.K. 2007. Dynamic logic for belief revision. *Journal of Applied Non-Classical Logics* 17(2):129–155.

114. van Benthem, J.F.A.K. 2010. *Modal logic for open minds*. Center for the Study of Language and Information.
115. van Benthem, J.F.A.K. 2014. *Logical dynamics of information and interaction*. Cambridge: Cambridge University Press.
116. van Ditmarsch, H., J. Halpern, W. Van Der Hoek, and B. Kooi. 2015. *Handbook of epistemic logic*. College Publications.
117. van Ditmarsch, H., J. Halpern, W. Van Der Hoek, and B. Kooi. 2007. *Dynamic epistemic logic*. Springer.
118. van Fraassen, B. 1980. Rational belief and probability kinematics. *Philosophy of Science* 47(2):165–187.
119. von Wright, G.H. 1951. Deontic logic. *Mind* 60:1–15.
120. von Wright, G.H. 1960. *Norm and action*. Routledge.
121. Wang, Y. 2010. Epistemic modeling and protocol dynamics. Ph.D. Dissertation.
122. Williamson, T. 2000. *Knowledge and its limits*. Oxford University Press.

Chapter 2
Epistemic Proceduralism Defended I

Abstract Epistemic proceduralism is spelled out by clusters of principles characterizing the structure, function, and logic of procedural epistemic norms, norms that define restrictions on epistemic actions, processes, and procedures. With the content of epistemic norms regimented by means of norm-kernels and restrictions on epistemic dynamics, this chapter articulates and defends clusters of principles, D1 and D2. Roughly, these clusters of principles state that the content of procedural norms is program-like, dictating complex procedural operations on epistemic action, and that this program-like structure entails their irreducibility to static and dynamic norms. One of the core lines of argument in this chapter links the regulative function of epistemic norms to the existence and irreducibility of procedural norms. In the context of the book, the clusters of principles articulating these results stand as theoretical criteria or tests applicable to formal frameworks for representing and reasoning about epistemic norms.

2.1 Epistemic Proceduralism Defended

With the preliminaries of Chap. 1 completed, I want to explicitly spell out and systematically defend the account of epistemic proceduralism driving this project. In this chapter, I articulate clusters of principles, D1 and D2, along with clusters of corollaries thereof, that partially characterize the structure, content, epistemic role, and logical relations among epistemic norms. Each clause of each cluster is argued for in turn, often utilizing multiple lines of reasoning.

R. Jenkins, *A Protocol-theoretic Framework for the Logic of Epistemic Norms*, Synthese Library 467, https://doi.org/10.1007/978-3-031-08597-0_2

2.2 The First Cluster of Principles (D1)

2.2.1 *The Core of the Matter*

Claim 2 (D1) Epistemic rationality norms define restrictions on epistemic dynamics distinct from constraints of capacity and feasibility, including:

(1) restrictions on the content of epistemic states,
(2) restrictions on transitions and trajectories of evolution of epistemic state, and
(3) restrictions on procedures and processes of epistemic state change.

That there are restrictions on the content of epistemic states is trivial to defend; it's one of the core assumptions of essentially all projects in normative epistemology (whether traditional or formal) as well as many projects in philosophical logic. Recall the formal or informal notions of restrictions on epistemic dynamics: restrictions on epistemic dynamics are divisions of the elements of epistemic dynamics (states, transitions, processes, actions, procedures) into correct and incorrect sets or preference rankings on the basis of their (extensional) properties.[1] Restrictions on epistemic dynamics, for instance, may partition epistemic states into rational or irrational sets or into permissible or impermissible sets on the basis of some static property. Thus, if there are any defensible norms of consistency, coherence, evidential support, or justification of belief, for instance, and epistemic agents can violate those norms, then D1.1 is plainly true. That a norm of coherence (et al) can be violated entails that there are epistemic states that agents can be in (i.e., that are feasible for the agent to be in or within the agent's capacity to be in) that are not coherent (inter alia), and thus, in some sense, epistemically wrong (according to that epistemic norm). Thus, the verdicts delivered by norms of coherence (et al) define a division of epistemic states by their coherence (et al) into rational or irrational, better or worse, sets, and this is distinct from any divisions of feasibility or capacity.

The argument thus far only supports D1.1. The existence of restrictions on epistemic states is immediate and uncontroversial for most philosophers that recognize epistemic norms at all. Restrictions on epistemic transitions, trajectories, actions, processes, or procedural constructions, on the other hand, is more controversial. To establish D1.2 and D1.3, it's useful to break each of them down into its own cluster of claims, and to directly argue for them in turn. The rest of this chapter is prosecuted

[1] Restrictions on epistemic dynamics divide epistemic factors extensionally—the elements of the powerset of X are, as I suggest above, to be taken to be the extensions of properties of elements of X. That restrictions on epistemic dynamics divide epistemic factors on the basis of their extensional properties is a limitation on the range of epistemic norms that can be represented by them. There are compelling reasons to think that normative contexts in general, and normative epistemic contexts in particular are intensional or even hyperintensional. Nonetheless, the current extensional analysis of the properties of epistemic factors implicit in the construction suffices to account for the content of a wide range of epistemic norms; indeed, the range contains the most important kinds of epistemic norms. The reader may supply a device for dividing sets of epistemic factors by hyperintensional properties. The rest of the construction should suffice to do the basic work of restrictions—dividing these factors by their properties into classes or sets that represent the verdicts of norms.

accordingly in a step by step, abstract and no-frills manner. The reader is encouraged to supply examples from any work of formal epistemology that explicitly defends norms for full or partial belief revision [2, 3, 16, 18, 29, 33, 56, 63–68, 70, 71] or any work of philosophical logic [4, 7, 8, 22–26, 57, 58] that defends norms of inference. These should map straightfowardly onto the bones of epistemic dynamics articulated above.[2]

2.2.2 The Existence and Irreducibility of Dynamic Norms

Claim 3 (D1C1) Epistemic rationality is dynamic; there are epistemic restrictions on transitions among epistemic states, not just states or contents of states, and there are characteristic relations among restrictions on states and restrictions on transitions. Parsing this claim, we get:

(1) There are norms of epistemic rationality that take epistemic transitions as their objects.[3]
    (a) (D1C1.1C) Norms that restrict transitions can be combined to construct norms that restrict trajectories.
(2) There are norms of epistemic rationality that take properties of epistemic transitions as their conditions of application.
(3) (StoT) Norms that restrict epistemic states can be used to generate norms that restrict epistemic transitions in the following sense: there are norms that restrict epistemic states, N_S, such that a corresponding dynamic norm, N_T, matching N_S's verdicts can be defined on the basis of N_S's verdicts on states and the relations between states and transitions.

---

[2] Map conditions on beliefs and credences to classes of epistemic states, inferences to classes of transitions, and the principles articulated in these works to restrictions on epistemic dynamics of the requisite types.

[3] I'll write of norms, norm-kernels, and restrictions on epistemic dynamics indifferently in stating D1–D3 and their subclauses and corollaries. Thus, D1C1.1 could read, alternately:

(1') There are norm-kernels of epistemic rationality that take epistemic transitions as their objects.
(1") There are norms of epistemic rationality that define restrictions on epistemic transitions.

These are equivalent. Norm-kernels that take epistemic transitions as objects assign verdicts to them, which divides transitions into verdict sets that, by the reasoning in the main text for encoding deontic verdicts with restrictions on dynamics, can be formalized as restrictions on epistemic transitions. Restrictions on epistemic transitions can be specified by means of norm-kernels that condition membership in R_{T_A} (resp. T_{T_A}, I_{T_A}) on some condition shared by all elements (perhaps a tautological or vacuous one). More, both of these constructions have been offered as regimentations or tools for representing or codifying the content of epistemic rationality norms, so this terminological choice should do no violence to the statement of these clusters of principles or the arguments for them.

(4) (TtoS) Norms that restrict epistemic transitions can be used to generate norms that restrict epistemic states in the following sense: there are norms that restrict epistemic transitions, N_T, such that a corresponding static norm, N_S, matching N_T's verdicts can be defined on the basis of N_T's verdicts on transitions and the relations between states and transitions.

(5) (TtoS Irreducibility) There are dynamic norms, N_T, such that there is no corresponding static norm, N_S, that can be defined on the basis of N_T's verdicts on transitions and the relations between states and transitions.

(6) (StoT Reducibility) If N_S is a norm that restricts epistemic states, then a corresponding dynamic norm, N_T, matching N_S's verdicts can be defined on the basis of N_S's verdicts on states and the relations between states and transitions.

This cluster goes further than is needed, as D1C1.1 suffices to establish D1.2, and D1C1.2 through D1C1.6 articulate important relations of dependence and independence between dynamic and static norms.

2.2.2.1 (D1C1.1) Argument from Cases

There are classes of pairs of transitions, $\langle s, s' \rangle$ and $\langle s, s'' \rangle$, that are epistemically rationally distinct. If epistemic rationality distinguishes classes of epistemic factors, then there are epistemic rationality norm-kernels that take those epistemic factors as their objects. Thus, there are epistemic rationality norm-kernels that take transitions as their objects.[4] Thus, given that all rules that conditionally assign normative (epistemic) verdicts to epistemic factors are constructions that capture the content of some epistemic norm or another, there are epistemic rationality norms that take transitions as their objects.

The first premise relies on antecedently plausible schematic cases that display pairs of rationally distinct epistemic transitions. Each of these cases, if the normative verdict is plausible, articulates a pattern or regularity of rational distinction among epistemic transitions; the schematic cases show that there are classes of epistemic transitions that are rationally distinct. These can be interpreted as statements of schematic epistemic norms that take epistemic transitions as their objects. The schematic cases[5] each present a pair of schematic transitions among epistemic states

---

[4] The point is made for pairs of transitions sharing root states, and the cases below that illustrate it are formulated accordingly. The point for classes of pairs of transitions sharing terminal states can be made with minor modifications of the following cases.

[5] These are examples of kinds of epistemic norm, not items from the final theory of epistemic rationality. The argument stands if any norms like these (i.e., norms with dynamic objects and conditions of application), however elaborated, are plausible. Only a complete reduction of all norms like these would suffice to undermine the point. See Sect. 2.2.2.7. Moreover, in all of these cases, take "$\vdash$" to symbolize the logical validity relation of the reader's preference. That is, whichever logical consequence relation, proof-theoretic or model-theoretic, makes the example most plausible to the reader.

that are rationally distinct (the verdict) where the terminal states equally satisfy the usual suite of static norms:

Defeasible Closure 1 (DCl1) Let a belief set, s, contain $B_1, \ldots, B_n$, the true belief that $B_1, \ldots, B_n \vdash A$, and let it contain no counterevidence for A. Let s' be identical to s but that it contains A. Let s'' be identical to s but that it contains $\neg A$.

(Verdict) If these conditions hold, it is rational for an agent to transition from s to s', and it is not rational to transition from s to s''.

Demand for Evidence 1 (DE1) Let a belief set, s, contain only $B_1, \ldots, B_n$, none of which individually or collectively bears logical, probabilistic, explanatory, or any other evidential relations to A. Let s' be identical to s but that it contains A. Let s'' be identical to s but that it does not contain A (or, if your account of epistemic states can handle it, it contains explicit suspension of judgment on A).

(Verdict) If these conditions hold, it is not rational for an agent to transition from s to s', and it is rational to transition to s''.

Minimal Mutilation (MM) Let a belief set, s, contain $B_1, \ldots, B_n$ and be logically consistent. Let s' be some belief set with an empty intersection with s. Let s'' be any belief set that satisfies every condition of minimal mutilation, conservatism, or preservation of entrenchment relative to s defended in the epistemological literature (or some preferred, consistent set thereof).

(Verdict) If these conditions hold, it is not rational for an agent to transition from s to s', and it is rational to transition to s''.

If the verdicts proposed above are correct (or near enough), then there are classes of pairs of transitions that are rationally distinct. Thus, there are norms that rationally distinguish transitions rather than states and condition their valences, irrespective of object, on the properties of transitions rather than merely the properties of states. Let me belabor this presently.

The second premise is almost immediate from the assumption that there are regularities governing the rational distinguishability of epistemic factors.[6] Let such

[6] Really, it can be supported in a number of ways. One alternative: Let such regularities be thought of in the broadest sense as patterns of correlation between conditions satisfied by epistemic factors and the valences of normative verdicts. Let x and y be rationally distinct; assume $R(x)$ and $R'(y)$ where R and R' are not identical. Either x and y jointly satisfy some condition, C, or they do not. In the former case, there is, trivially, some pair of true material conditionals, $C \rightarrow R(x)$ and $C \rightarrow R'(y)$, linking the satisfaction of C to the satisfaction of different normative verdicts, R and R'. If they jointly satisfy no condition, then some pair of true material conditionals, $C \rightarrow R(x)$ and $C' \rightarrow R'(y)$, link the satisfaction of C and C' to the satisfaction of different normative verdicts, R and R', respectively. Indeed, *material* conditionals of these forms follow immediately from the assumption without the intermediary reasoning. In either case, the resulting conditionals are paradigmatic norm-kernels. But let's not trust the properties of the material conditional for this result.

regularities be thought of in the broadest sense as patterns of correlation between conditions satisfied by classes of epistemic factors and the valences of normative verdicts satisfied by (the elements of) those classes of epistemic factors. For instance, where x and y are rationally distinct, there are normative valences, R and R', that are distinct, and where this difference is an instance of a regularity, there is some correlation between the satisfaction of some condition C (or, more generally, conditions $C_1, \ldots, C_n$) and the truth of $R(x)$ and $R'(y)$ (respectively). Assume that there are such regularities. Correlations can be stated as conditionals of some kind. Conditionals linking the satisfaction of conditions (by the relevant epistemic factors or not) to normative verdicts on epistemic factors are epistemic norm-kernels taking those factors as objects, by definition. Thus, any such regularity can be stated as a rule that takes the form of a norm-kernel. Indeed, there are no other forms that a regularity of rational distinguishability could take than that of a rule that connects some conditions (even vacuous or tautological conditions) to normative verdicts with some logical force; where there are no conditional rules (of any kind) correlating the conditions satisfied by epistemic factors to the valences of normative verdicts, there are no regularities of rational distinction. In that case, there simply would be no patterns of rational distinction among epistemic factors at all. Thus, if there are regularities of rational distinguishability over a class of epistemic factors, then there are norm-kernels capturing them that take those factors as objects. Thus, if the foregoing cases (or even radically different cases) capture genuine regularities governing the rational distinguishability of epistemic transitions, then there are norm-kernels taking epistemic transitions as their objects.

Now, the final step of the inference moves from the existence of epistemic norm-kernels to the existence of epistemic norms. This is also unproblematic. The inference does not rely on the claim that there is a defensible or plausible norm or that there is a norm that is otherwise worth following for every norm-kernel. Rather, the thought is that every norm-kernel states some possible norm, and, in the schematic cases given above, the norm-kernel articulated captures a plausible regularity in the verdicts of epistemic rationality. The latter conjunct is what entails that there is an epistemic rationality norm answering to the norm-kernels articulated. The final step of the inference, then, relies on the plausibility of the regularities that we start with. Thus, given that DCl1, DE1, and MM are plausible, we have DC1.1, as these cases seem clearly to state (broadly) plausible norms that distinguish epistemic transitions.

2.2.2.2 (D1C1.2) Argument from Cases

Some of the pairs of transitions, $\langle s, s' \rangle$ and $\langle s, s'' \rangle$, identified in DCl1, DE1, and MM, differ only in the properties of the transitions themselves. Thus, $\langle s, s' \rangle$ and $\langle s, s'' \rangle$ are distinguished on the basis of the properties of transitions. If rationality distinguishes x and y on the basis of C, then there is a norm-kernel (hence, norm) in which C is the condition of application. Thus, the conditions of application of some epistemic rationality norms concern properties of transitions. Again, the conditional underlying this argument is straightforward: if there are conditionals that codify the distinguishability of $\langle s, s' \rangle$ and $\langle s, s'' \rangle$, then there are norm-kernels (hence, norms) that distinguish these transitions on the basis of their properties (i.e., that take their properties as conditions of application).

2.2.2.3 Definitions for Arguments by Construction

For any two restrictions on epistemic dynamics, $R_{\preceq}(X)$ and $R'_{\preceq}(Y)$, say that:

Definition 4 (Restriction Correspondence) $R_{\preceq}(X)$ and $R'_{\preceq}(Y)$ *correspond* iff for any $x \in X$ there is a $y \in Y$ such that the verdict of $R_{\preceq}(X)$ for x (rational, irrational, obligatory, permissible, placement in the ranking,...) is the same as the verdict of $R'_{\preceq}(Y)$ for y, and the converse holds.

Similarly, for any two norm-kernels, N_1 and N_2, that take epistemic factors as their objects, say that:

Definition 5 (Norm-kernel Correspondence) N_1 and N_2 *correspond* iff for any x in the domain of N_1 there is a y in domain of N_2 such that the verdict of N_1 for x (rational, irrational, obligatory, permissible, placement in the ranking, ...) is the same as the verdict of N_2 for y, and the converse holds.

A final terminological note:

Definition 6 (Reduction of Norm-kernels or Restrictions) If, between two classes of norm-kernels or restrictions on epistemic dynamics, C_1 and C_2, there is a rule that generates, for every norm-kernel/restriction in C_1, a corresponding norm-kernel/restriction in C_2, say that the class C_2 *reduces* the class C_1 or that C_1 is *reducible to* C_2.

So, for instance, if, for every dynamic norm-kernel, there is a corresponding static norm-kernel, the static norm-kernels reduce the dynamic norm-kernels or the dynamic norm-kernels are reducible to the static norm-kernels.

2.2.2.4 (D1C1.1C) Argument by Construction

Let a norm-kernel, N, restrict the transition set, T_A, in some LTS, $\langle S, \rightarrow_{a \in A} \rangle$, and let N have a domain of transitions that do not share any trajectories (i.e., there are no sequences of states that share first and last elements). Let N_{T_A} be a norm-kernel such that, N assigns $\langle s, s' \rangle$ to the rational (irrational, threshold) set iff N_{T_A} assigns any trajectory, $\langle s, \ldots, s' \rangle$, in T_A* to the rational (irrational, threshold) set. This rule immediately maps norm-kernels restricting transitions to corresponding norm-kernels restricting trajectories. Let N_1 and N_2 be norm-kernels that restrict T_A (with the same conditions on sequence sharing as above). Let $N_{1;2}$ be a norm-kernel such that, N_1 assigns $\langle s, s' \rangle$ to the rational (irrational, threshold) set, and N_2 assigns $\langle s', s'' \rangle$ to the rational (irrational, threshold) set iff $N_{1;2}$ assigns $\langle s, s', s'' \rangle$ to the rational (irrational, threshold) set. This rule immediately maps norm-kernels restricting T_A to a norm-kernel restricting trajectories of three states in a compositional way. There's no upper limit on compositionality, here, so a similar rule can be written that maps norm-kernels restricting T_A to norm-kernels restricting T_A*, generally (i.e., restricting arbitrarily long trajectories). Later sections argue that the restriction to transitions that don't share trajectories is necessary to the generation of corresponding norm-kernels that restrict trajectories.

2.2.2.5 (D1C1.3) Argument by Construction

Let N_C be a coherence norm-kernel; i.e., a norm-kernel conditioning normative verdicts on epistemic states on their coherence. For a simple case, let N_C state that if a state, s, is such that the epistemic judgments that define it are coherent in some proprietary sense, then s is rational. Generalizing the following reasoning to other forms of norm-kernel is straightforward.

Define a norm-kernel, $N_{C,T}$, such that: the domain of $N_{C,T}$ is transitions that do not share terminal states, and for any transition, t, s is rational according to N_C and $t = \langle x, s \rangle$ iff t is rational. The rule specifying $N_{C,T}$ is itself a norm-kernel that corresponds to N_C, and it also immediately maps norm-kernels restricting states to further corresponding norm-kernels restricting transitions by a cut-like move; e.g., the terminal state of t is coherent (in the aforementioned proprietary sense) iff t is rational.

Fully generalize $N_{C,T}$ by defining a dynamic norm-kernel, $N_{S,T}$, over arbitrary static norm-kernels, N_S, and filling in the slot for verdicts as necessary: the domain of $N_{S,T}$ is transitions that do not share terminal states, for any transition, t, s is rational (irrational, higher in the ranking than some other element, s', etc.) according to N_S and $t = \langle x, s \rangle$ iff t is rational (irrational, higher in the ranking than some other element, t', etc.). The rule specifying $N_{S,T}$ is itself a norm-kernel that corresponds to N_S, and it also immediately maps norm-kernels restricting states to further corresponding norm-kernels restricting transitions by a cut-like move; e.g., the terminal state of t is coherent (in the aforementioned proprietary sense) iff t is rational.

The same procedure works for dynamic norm-kernels that condition their verdicts on the verdicts of N_S on root states of transitions. This reasoning sets a template for the other arguments by construction in this and the remaining sections of Chap. 2. Note the restriction to transitions that do not share terminal states. Call these converging transitions. Similarly, call transitions that share root states branching transitions.

2.2.2.6 (D1C1.4) Argument by Construction

Let N_T be an arbitrary norm-kernel that restricts transitions such that no two transitions in N_T's domain converge in their terminal states. Let N_S be a norm-kernel such that: for any state, s, t is rational (irrational, higher in the ranking than some other element, etc.) according to N_T, and $t = \langle x, s \rangle$ iff s is rational (irrational, higher in the ranking than some other element, etc.). This rule immediately maps norm-kernels restricting (non-convergent) transitions to corresponding norm-kernels restricting states, including the rule specifying N_S itself, as in the prior argument. The procedure works the same for restrictions on non-branching transitions. As the following two sections argue, the restriction to non-converging (or non-branching) transitions is not a matter of indifference.

2.2.2.7 (D1C1.5) Argument from the Failure of Reduction Strategies

Counterexamples that suffice to establish this proposition are given in Sects. 2.2.3.2 and 2.2.3.3. Here, I argue against putative "reduction strategies", which would falsify TtoS Irreducibility.

D1C1.5 (and the argument for D1C1.1) might be thought to fail if there were successful strategies for reducing each example of a dynamic norm to some more fundamental static norm. The thought would be that, if we accept the claim that the only norms that are real[7] are the fundamental norms (i.e., those norms in the domain not reducible to others in the same domain), then a successful reduction strategy straightforwardly shows that the dynamic norms above aren't real, and so can't plausibly impose any conditions on the adequacy of formal reconstructions of epistemic normativity and epistemic dynamics.

This is a mistake, though. First, there is no reason at all to accept the claim above about the reality of derivative norms. This line of thought is more fully spelled out in Sect. 2.2.3.13 with respect to procedural norms, but the reasoning applies to dynamic norms as well. Second, and more important, there are no available, successful reduction strategies for dynamic and procedural norms. There are counterexamples to extant reduction strategies, and there are general philosophical considerations against reduction strategies. I'll build those here. In Sect. 2.2.2.8, I give a very simple formal argument against any reduction of restrictions on transitions to restrictions on states.

---

[7] "Real" is used in a non-committal way to capture the idea that dynamic norms, if reduction strategies are successful, are not theoretically worthwhile in modeling in a formal theory of epistemic rationality. Perhaps "legitimate" or "acceptable" would be appropriate cognates.

Several philosophers[8] have defended principles to the rough effect that:

(Reduction 1)

(i) Epistemic states are rational iff they satisfy the requirements of whatever canon of synchronic rationality one prefers (e.g., probabilistic coherence).
(ii) A change of epistemic state from s to s' is rational iff s' is rational.

The idea is that the rationality of the change from s to s' depends only on the rationality of s'. Now, obviously, this is itself a dynamic norm, as it defines restrictions on epistemic transitions, and so a strict staticist couldn't coherently endorse it. Setting that aside, if it were supposed that this is the sole requirement of epistemic rationality on changes of epistemic state, this would offer a formula for determining whether any particular change of states is rational, and it would do so on the basis of the rationality of epistemic states. However, this can't be the sole requirement on changes of epistemic state; Reduction 1 is clearly mistaken. Consider:

Non-Arbitrariness (NA) Let a belief set, s, contain $B_1, \ldots, B_n$ and an inconsistency, $A \wedge \neg A$. Let $B_i, \ldots, B_j$ where $i \geq 1$ and $j \leq n$ be counterevidence for A. Let s' be identical to s but that it does not contain A or $A \wedge \neg A$, and it satisfies the canons of synchronic rationality as fully as possible. Let s'' be a non-empty belief set satisfying the synchronic canons of rationality as fully as possible that also has an empty intersection with s.

(Verdict) If these conditions hold, it is not rational for an agent to transition from s to s'', and it is rational to transition to s'.

Even where s'' satisfies the conditions of synchronic rationality, it's still clearly irrational to move, willy nilly, to just any random statically rational state or another with no consideration of the evidence one has already acquired. This is not a rational change of belief. Thus, because these two transitions are indistinguishable in terms of the rational distinguishability of the relevant states, it must be the properties of the transitions that makes the difference.

The strongest response[9] to counterexamples like this is to deny that there are any cases where the relevant terminal states are rationally indistinguishable. This takes the discussion down one of two paths: one takes us into the territory of minimal mutilation or epistemic conservatism (of the Quinean variety) of epistemic states

[8] Moss [60] takes this approach explicitly, as do Foley [28] and Levi [56]. This is also essentially the strategy behind the rationality postulates of the belief revision paradigm [29, 30] and probabilism [16].

[9] One could, as Levi [56] does, just bite the bullet and accept the bizarreness of taking arbitrary transitions among internally coherent but utterly disconnected epistemic states to be rational. One could even argue that this is the best available theoretical option given the virtues of staticist models, and so our intuitions simply need re-training. This is a live option, but I think the rest of this section makes that bullet too bitter to bite, whatever else an all-things-considered argument might motivate.

and the other takes us into the territory of *epistemic basing relations* [51, 52, 87]. Let's take these paths in turn.

A tempting response to the examples above is the replacement of Reduction 1 with considerations of conservatism or minimal mutilation:

(Reduction 2)

(i) An epistemic state, s', is rational iff it satisfies the canons of synchronic rationality and it represents a minimal mutilation or a maximal conservation of the state that precedes it in the course of the evolution of a reasoner's epistemic states.

(ii) A change of epistemic state from s to s' is rational iff s' is rational.

Reduction 2 is schematic, and we can let minimal mutilation be cashed out in terms of some preferred kind of comparison between the content of s and s' with respect to, e.g., entrenchment, conservation of central elements in a web of belief, etc.. There are competing accounts of minimal mutilation in the epistemological literature, and nothing hangs on the details of these. The idea is that, as long as minimal mutilation or conservatism rules out all states with an empty intersection with its predecessor, then the distinctions that Non-Arbitrariness appears to make can be captured by Reduction 2.

This reduction strategy doesn't work. First, if correct, it supports D1C1.1 and D1C1.2 directly—it doesn't reduce (e.g.) MM above to static notions only—and it posits an irreducible (or at least unreduced) dynamic norm in clause (ii), thereby supporting D1C1.5. This is decisive enough, but there are also counterexamples. Note that Reduction 2 fails to handle the following case:

Defeasible Consistency 1 (DCo1) Let a belief set, s, contain $B_1, \ldots, B_n$, the true belief that $B_1, \ldots, B_n \vdash A$, and counterevidence for A, $C_1, \ldots, C_n$. Let s' be identical to s but that it contains A and $\neg A$. Let a distinct belief set, s'' be identical to s but that doesn't contain A, it contains $\neg A$, does not contain some B_i, and contains defeaters for B_i.

(Verdict) If both of these transitions are feasible for a reasoner, then it is rational to transition from s to s'', and it is not rational to transition from s to s'. If the transition $\langle s, s'' \rangle$ is not feasible for a reasoner, it is rational to transition to s'.

This is a schema for non-paradoxical Preface-like cases.[10] Let A be the conjunction of all B_i. Let the counterevidence for A, $C_1, \ldots, C_n$, amount to all available evidence for fallibilism about A—that any suitably large class of beliefs I have is overwhelmingly likely to contain a false belief and that $B_1, \ldots, B_n$ satisfies the size requirement. Standard Preface cases show that, arguably, there are cases where it is more rational to maintain the inconsistency that results from adding A despite the counterevidence for it (the evidence for fallibilism about A) than it is to randomly or

[10] For the inaugural discussion of Preface Paradox, see [59].

arbitrarily reject some of the particular conjuncts for which I have strong evidence.[11] It is crucial to these cases that rejection of any B_i would be random or arbitrary. The reasoner in the Preface case cannot find a way to rationally select any of her B_i for revision; else, there isn't much of a problem at all. If acquiring sufficiently strong counterevidence to some B_i is feasible for an agent, then there is no serious conflict. In any Preface case, if we could undermine the evidence for one or other of the belief sets in tension, there would be no reason to accept the inconsistency.

Now, despite the fact that the terminal states differ across these transitions, it's clear that the static properties are not doing the work of differentiating the transitions, (s, s') and (s, s''). Which transition is rational depends on which transitions are available for execution, not merely the static states involved. Thus, if DCo1 is plausible as a constraint on epistemic transitions, there is at least one epistemic norm restricting transitions that can't be reduced to the properties of epistemic states.

Now, onto basing relations. The usual thought [51, 52] is that epistemic states inherently carry information about the evidential bases of beliefs. The epistemic basing relation is the relation that holds between a set of beliefs $B_1, \ldots, B_n$ and some other belief A just when the belief A is based on $B_1, \ldots, B_n$. According to the basing theorist, it is only relative to the beliefs on which a belief is based that it is rational or justified. Only states with appropriate basing relations are rational, and basing relations track (inter alia) inferential relationships. Consider a schematic reduction strategy in terms of basing:

(Reduction 3)

(i) An epistemic state is rational only if the *epistemic basing relation* of that state satisfies (for my purposes, arbitrary) condition C.
(ii) A change of epistemic state from s to s' is rational iff s' is rational.

The thought of Reduction 3 is that the terminal states of rationally distinct transitions are not actually identical or even rationally indistinguishable. The terminal states that result from the processes in these cases fail to have identical basing relations, and this results in rational distinguishability between them. On the basis of this difference, the transitions are rationally distinguishable. Deploying the basing concept, Reduction 3 appears to effectively reduce Non-Arbitrariness 1. Willy nilly changes to synchronically rational belief states are prohibited by Reduction 3—the willy nilliness of the transition would be recorded in inappropriate basing relations.

This strategy fails. There are three broad families of accounts of the epistemic basing relation, causal accounts, counterfactual causal accounts, and doxastic

[11] Alternately, Preface cases may be cases that justify refusing to accept some logical consequences of one's beliefs or even cases that show that classical conjunction rules are not valid in some sense. Whether one takes the hard Preface cases to be of one or the other kind is irrelevant to the argument in the main text; if there is no serious conflict because there is some identifiably rejectable conjunct, then none of the more radical conclusions can be sustained.

accounts. The details of these views are irrelevant to the thrust of argument, here.[12] The doxastic basing relation is the only version of the epistemic basing relation that could work for Reduction 3, and this won't do.

Suppose epistemic basing is a matter of the causal relations between states, actual or counterfactual, whatever preferred account one gives along these lines. Reduction 3 would then tell us that a transition is rational iff the terminal state is such that it is linked to the root state by an appropriate causal process. It is transparent that Reduction 3, spelled out in terms of causal basing, directly supports D1C1.1 and D1C1.2, and also offers no further reduction of the dynamic norm it posits, thereby supporting D1C1.5. Indeed, if the basing relation depends on differences between causal processes of epistemic transition, it actually supports D1C2 rather than any putative reduction to static norms. Defined in terms of doxastic basing, however, Reduction 3 fails to handle DCo1. There, the norm distinguished transitions on the basis of the availability of other transitions, not on anything reducible to doxastic basing; the feasibility of epistemically or rationally preferable epistemic processes is not a function of epistemic basing.

More important, the basic assumption underlying Reduction 3, defined in terms of doxastic basing, is that no two distinct processes produce identical epistemic states with respect to doxastic basing relations. This is false on its face. There are ubiquitous processes that obscure basing relations, and so produce identical terminal states without instantiating the right transitions, according to Reduction 3. Bounded agents forget their reasoning; doxastic basing relations fade over time. Epistemic processes are frequently inscrutable, producing judgments and leaving no traces whatever in their resultant doxastic states. See, for instance, the general inscrutability of many heuristic, subpersonal, and even person-level processes [21, 34, 45, 80, 84], as well as Lam's considerations [54]. I illustrate this point in more detail in Sect. 4.3.3.2.

2.2.2.8 (D1C1.5) Formal Argument

There are technical grounds for the irreducibility of restrictions on transitions to restrictions on states, as well; the claim is essentially a trivial consequence of the fact that some rationally distinct transitions converge in their root or terminal states. Explicitly, there are sets, S, of epistemic states, such that there are sets of pairs of distinct transitions that converge on elements of S or that branch out from elements of S. Thus, if restrictions on states are analyzed in terms of sets (sets of rational states, sets of obligatory states, etc.), then there are pairs of transitions that cannot be distinguished by the verdict sets that their terminal (second) states are contained in and there are pairs of transitions that cannot be distinguished by the verdict sets that their root (first) states are contained in. In other words, for these transitions, there can be no (consistent) rule that directly maps the verdicts of corresponding

[12] For these irrelevant details, see: [51, 52, 87].

restrictions on root (resp. terminal) states to the verdicts of restrictions on transitions that distinguish them. More formally:

Proposition 7 *If* (S, T_S) *is a simple transition system where* S *is a finite set of states and* $T_S \subseteq S \times S$ *is a transition relation over* S*,* $Part(S)$ *is a partitioning of* S *into disjoint subsets,* $Rational(S)$ *and* $Irrational(S)$*, and* $Part(T_S)$ *is a partitioning of* T_S *into disjoint subsets,* $Rational(T_S)$ *and* $Irrational(T_S)$*, then there can be no function,* $f : S \longrightarrow T_S$*, such that, for all* s *and* s'*,* $s \in Rational(S)$ *iff* $\langle s', s \rangle \in Rational(T_S)$*.*

Proof By contraposition. Let $s, s', s'' \in S$. Let $\langle s, s' \rangle, \langle s'', s' \rangle \in T_S$. Let $Part(T_S)$ be such that $\langle s, s' \rangle \in Rational(T_S)$ and $\langle s'', s' \rangle \in Irrational(T_S)$. Suppose there is a function from S to T_S as described. Then, immediately, $s' \in Rational(S)$ iff $\langle s, s' \rangle \in Rational(T_S)$ and $s' \in Irrational(S)$ iff $\langle s'', s' \rangle \in Irrational(T_S)$. Thus, $s' \in Rational(S)$ and $s' \in Irrational(S)$. Thus, $Rational(S)$ and $Irrational(S)$ are not disjoint. □

Moreover, if it's possible that such pairs of transitions have rationally indistinguishable terminal (resp. root) states, then restrictions on transitions that distinguish them cannot be mapped to a corresponding restriction on states at all. Consider transitions $t = \langle s, s' \rangle$ and $t' = \langle s'', s' \rangle$ where transition from s'' to s' minimally mutilates s'' but transition from s to s' maximally mutilates s but s and s'' are, themselves, rationally indistinguishable with respect to any consistent static restriction. It's straightforward that no consistent rule can distinguish t and t' on the basis of their terminal states. By stipulation, there can be no consistent rule that can distinguish t and t' on the basis of their root states. Thus, for a restriction on transitions that distinguishes t and t' (on the basis, in this case, of minimal mutilation), there can be no corresponding restriction on states. Minimal mutilation is inessential to this argument, of course. Thus, the norms captured by such restrictions on transitions cannot be reduced to norms captured by restrictions on states.

The point is amplified given the foregoing examples of norm-kernels that condition their verdicts on transitions on the availability of other transitions. This technical fact plus the counterexamples and the failure of reduction strategies provide nearly decisive reasons for accepting D1C1.5.

Moreover, it might be thought that there are mappings from the verdicts of pairs (or arbitrarily large sets) of restrictions on root states and terminal states to the verdicts of restrictions on transitions. Let arbitrary t and t' share root states but diverge in terminal states. Let t be rational according to arbitrary restriction, $R_{\preceq}(T)$, for the set of transitions, T, in some LTS, M. Let t' be in the irrational set. Let $R_{\preceq}(S)$ and $R'_{\preceq}(S)$ be restrictions on the state set, S, of M, such that $f(t) = f(t')$ is rational according to both, but $l(t)$ is rational according to $R_{\preceq}(S)$ and $l(t')$ is not, and vice versa for $R'_{\preceq}(S)$. Define a restriction on transitions, $R'_{\preceq}(T)$, such that t is in the rational set iff if $t = \langle s, s' \rangle$, then s is rational (or irrational) according to both $R_{\preceq}(S)$ and $R'_{\preceq}(S)$, and s' is rational according to only one of $R_{\preceq}(S)$ and $R'_{\preceq}(S)$. Such restrictions will distinguish transitions that have shared root or terminal states (and, of course, transitions that share both are identical). Since $R_{\preceq}(T)$ was arbitrary, we have something of a reduction of restrictions on transitions to sets of (at least two) restrictions on states. Run the same argument transposing terminal states and root states for the case where arbitrary t and t' share terminal states.

This does not falsify D1C1.5, however. First, note that these constructions are consistent with the argument above, which shows that there is no 1–1 reduction from restrictions on transitions to restrictions on states, and this is enough to establish an asymmetry between static and dynamic restrictions. Second, procedural norms will be shown below to define restrictions on processes and restrictions on transitions. Processes can, in principle, be distinct and yet transition equivalent (over all LTSs). Thus, restrictions defined over processes can distinguish necessarily transition-equivalent processes, and it follows that no restriction on transitions (or even sets of transitions) can reduce such a restriction on processes, and therefore that no set of restrictions on states can reduce such a restriction on processes. Thus, there are procedural norms (captured by the foregoing restrictions) that restrict epistemic transitions that cannot be reduced to norms that restrict epistemic states.

2.2.2.9 (D1C1.6) Construction Rule

Every partitioning of states into rational and irrational classes can be generated by the rules in Sects. 2.2.2.3 through 2.2.2.6. This is easy to see, so I'll show only that it works for terminal states. Assume there is some N_S that dictates that s is rational (etc.). If s is the terminal state of any transition, t, in T_A, then there is some partitioning, N_T, of transitions into rational and irrational sets such that t is rational according to N_T and converse; just define N_T as in Sect. 2.2.2.5. The result is a corresponding norm-kernel that restricts transitions on the basis of the verdicts of N_S and the relations between states and transitions. Now, work the other direction by starting with N_T, and applying the reasoning Sect. 2.2.2.6 to define $N_{S'}$. $N_{S'}$ will match N_S's verdicts. This reasoning applies most readily to the case where s is connected to other states by epistemic actions in A, but in the odd case where s is detached from any other state (completely unreachable by any epistemic process), the restriction on transitions need only operate on looping transitions, e.g., $\langle s, s\rangle$.

2.2.3 *The Existence and Irreducibility of Procedural Norms*

Claim 8 (D1C2) Epistemic rationality is ineliminably procedural; there are epistemic restrictions on epistemic actions, processes, and procedures, not just states or transitions among states, and restrictions on epistemic procedures and processes provide an irreducible unifying structure for interpreting and reasoning about even apparently non-procedural (epistemic) evaluations and regulative (epistemic) normative verdicts, and such restrictions are practically indispensable in epistemic cognition. Parsing this claim, we get:

(1) There are norms of epistemic rationality that take epistemic actions, processes, and procedures as their objects.
(2) The norms of epistemic rationality take properties of epistemic actions, processes, and procedures as their conditions of application.
(3) Procedural epistemic norms define restrictions on epistemic procedures, which in turn define restrictions on epistemic factors of other kinds.
(4) The procedural norms of epistemic rationality that define restrictions on actions, processes, and procedures are indispensable for the regulative role of epistemic norms.
(5) (TtoA) Norms that restrict epistemic transitions can be used to generate norms in the following sense: there are norms that restrict epistemic transitions, N_T, such that a corresponding norm that restricts actions, N_A, matching N_T's verdicts can be defined on the basis of N_T's verdicts on transitions and the relations between actions and transitions.
(6) (TtoP) Norms that restrict epistemic transitions can be used to generate norms in the following sense: there are norms that restrict epistemic transitions, N_T, such that a corresponding norm that restricts processes, N_P, matching N_T's verdicts can be defined on the basis of N_T's verdicts on transitions and the relations between processes and transitions.
(7) (TtoPr) Norms that restrict epistemic transitions can be used to generate norms in the following sense: there are norms that restrict epistemic transitions, N_T, such that a corresponding norm that restricts procedures, N_{Pr}, matching N_T's verdicts can be defined on the basis of N_T's verdicts on transitions and the relations between procedures and transitions.
(8) (APPrtoT) Norms that restrict epistemic actions, processes, or procedures can be used to generate norms in the following sense: there are norms that restrict epistemic actions, processes, or procedures (respectively), N_A, N_P, or N_{Pr} (respectively), such that a corresponding norm that restricts transitions, N_T, matching N_A's verdicts (resp. N_P's verdicts or N_{Pr}'s verdicts) can be defined on the basis of N_A's verdicts (resp. N_P or N_{Pr}) on actions (resp. processes, procedures) and the relations between actions (resp. processes, procedures) and transitions.
(9) (APPrtoS Irreducibility) Not all norms restricting epistemic actions, processes, and procedures can be generated from norms restricting epistemic states on the basis of the verdicts on states and the relations between states and actions/processes/procedures: There are procedural norms, N_A (resp. N_P, N_{Pr}), such that there is no corresponding static norm, N_S, that can be defined on the basis of N_A's verdicts (resp. N_P or N_{Pr}) on actions (resp. processes, procedures) and the relations between states and actions (resp. processes, procedures).

(10) (APPrtoT Irreducibility) Not all norms restricting epistemic actions, processes, and procedures can be generated from norms restricting epistemic transitions. on the basis of the verdicts on transitions and the relations between actions/processes/procedures and transitions: There are procedural norms, N_A (resp. N_P, N_{Pr}), such that there is no corresponding dynamic norm, N_T, that can be defined on the basis of N_A's verdicts (resp. N_P or N_{Pr}) on actions (resp. processes, procedures) and the relations between transitions and actions (resp. processes, procedures).

This cluster goes further than is needed, as D1C2.1 suffices to establish D1.3, and D1C2.2 through D1C2.8 articulate important relations of dependence and independence between static, dynamic, and procedural norms.

2.2.3.1 (D1C2 Core) Argument from the Regulative Function of Epistemic Norms

There are two obvious roles for epistemic norms to play in cognition, only one of which is essential to the notion of norms:

(Descriptive) Epistemic norms could be regularities that show up in the best (idealized) models of an agent's cognition. They "govern" ideal epistemic dynamics the way the laws of physics govern the evolution of physical systems.

(Normative) Epistemic norms could be objects that play a role in evaluating and regulating cognition, not just in describing regularities. They govern evaluations of epistemic dynamics the way that laws govern legal evaluations of behavior, and they guide epistemic dynamics the way that professional methodologies guide professional behavior.

Descriptive uses of norms are interesting on their own and plausibly inform any normative use of a given set of rules or standards, but the proper functions of norms are evaluation and regulation. The description of the regulative role of epistemic norms suggests two kinds[13] of regulative role for norms to play:

(Post Hoc Evaluation) Assessment epistemology: evaluating an epistemic state, change of epistemic state, or process of epistemic state change after it's occurred. Epistemic norms can be used as standards or normative ideals that a system can realize, fail to realize, or approximate to some greater or lesser degree. That is, norms can be used to determine, post hoc, that a system is in the wrong state or that it got to its current state by the wrong means.

(Ante Hoc Regulation) Directive epistemology: deployment of norms to select epistemic actions, procedures, or processes. Epistemic norms can be used as standards or normative ideals that a system's current or future states and potential courses of action can realize, fail to realize, or approximate to some greater or

[13] The assessment of credit and culpability may be a third, distinct use of epistemic norms, re: [83]. The forthcoming arguments do not hinge on acknowledging this.

lesser degree. The relations that current and future states and courses of action bear to these standards can be used in selecting future courses of action.

To illustrate post hoc evaluation, we can consider the normal uses of various kinds of standards. For instance, we can use the dictionary to look up "comprise" and "compose" in order to determine whether a sentence with one of these terms that has already been spoken was grammatically correct. Additionally, we often want to tell whether someone has followed an experimental protocol or proper methods of data analysis in presenting us with a new finding or whether someone's already constructed proof actually works. Post hoc evaluation of reasoning processes is also a standard practice. Knowing that I'm prone to wishful thinking and recognizing that wishful thinking is a reasoning process that frequently guarantees error, I can ask whether I got to my conclusion by merely being wishful.

Ante hoc regulation is about projecting or otherwise figuring out what norms require of future actions, and trying to ensure that our actions accord with those requirements. In other words, ante hoc epistemic regulation resolves into processes of epistemic planning and control. For instance, we can look up "comprise" and "compose" in order to determine how to construct a sentence with one of these terms (rather than the other). With some conception of how the sentence ought to come out, one can use the dictionary to guide the process of constructing the sentence in a way that matches grammatical usage. For further parallel examples, empirical scientists employ experimental protocols to guide their information generating actions in the lab, statisticians use established methods of data analysis to guide the way they draw conclusions from data sets, and mathematicians use methods of structural and transfinite induction or various other techniques or structural templates to guide their construction of proofs. Again, ante hoc guidance of reasoning processes is also standard practice. Knowing that I'm prone to wishful thinking and recognizing that wishful thinking is a reasoning process that frequently guarantees error, I can suppress this process and substitute better ways of reasoning.

Naturally, a useful framework for thinking about ante hoc regulation is epistemic planning,[14] as it's understood in artificial intelligence research [11, 31, 39, 70]. Essentially, epistemic planning is just a special case of planning. Planning is conceived of in terms of planning problems and solutions to planning problems.

---

[14] The epistemic planning problems I'm focusing on are a bit different from the norm, re: [6, 31, 70]. There, epistemic planning problems are problems of verifying epistemic formulae that encode features of knowledge or belief; goal states concern only whether a system ends up with certain kinds of epistemic or doxastic states, and say nothing about executing actions that comply with epistemic norms. There may or may not be epistemic formulae in a given framework that encode this latter kind of epistemic goal. In other words, epistemic planning is typically focused on a proper subset of the broader space of epistemic planning problems that are considered in the main text. Of course, if the logic in which one is formalizing planning problems has formulae that encode things like "Procedure P complies with norm N", then this worry disappears. The PLEN Framework, introduced and developed over the course of Chaps. 4 through 7, does contain such formulae. For now, these technical considerations won't be material in making the argument of the main text.

The basic background is the by now familiar conception of dynamic systems. There is a set of possible states of the relevant system—whether the system is a robot, set of interacting agents, or single agent in an environment—and a set of actions that the system can perform to transition from state to state. An *initial state* is an arbitrary state of the system. A *goal state* is a state of the system that has some specified set of properties that encode a goal for the system. An *action library* is a subset of the set of all possible actions. *Planning problems* are triples of initial state, goal state, and action library.

Planning problems are selection problems. The fundamental challenge is figuring out out how to get from the initial state, by some sequence of actions in the action library, to the goal state. A *plan* is a sequence or other procedural structure built out of procedural operations over an action library. *Solutions* to planning problems are thus plans that suffice to transition the system from the initial state to the goal state (these plans may involve arbitrarily many choice points or parallel processes).

Epistemic planning problems are planning problems defined over epistemic states (initial, goal) and epistemic action libraries that build epistemic normativity into the problem in two ways. On the one hand, an epistemic planning problem may build epistemic normativity into the goal state by specifying that the state have certain normative features (or that it satisfy certain formulae). On the other, an epistemic planning problem may build normativity into the space of planning solutions by imposing restrictions on the space of plans defined over the epistemic action library. For instance, the goal state may be any state that satisfies some condition like consistency, deductive closure, or proper basing relations, and the action library may involve epistemic actions involving proof construction and (seeing to it that) inference (occurs), where only these are permissible plans. Generally, a planning problem might impose restrictions on the kinds of plans that can be used to reach the goal state; a norm might require that certain epistemic actions be carried out, that epistemic actions be sequenced in certain ways, or that the result process omit certain kinds of action.[15] One could potentially collapse these in the formalization of planning problems; the goal state might specify that only a certain kind of procedure was used to reach it.

Solving an epistemic planning problem, then, requires reasoning about, first, what one's epistemic norms require. This has two dimensions. Epistemic norms may impose restrictions on the space of possible goal states given an initial state or even determine specific goal states with respect to specific initial states. Epistemic norms may also impose restrictions on the space of plans defined over one's action library. From there, a plan must be constructed (in line with the prior reasoning) and shown to satisfy those requirements prior to adoption and use in epistemic control.

[15] This point is explicitly defended in multiple places in this chapter, as well as directly expanded below.

With the epistemic planning framework in mind, the following argument directly establishes the core of D1C2:

(1) If epistemic rationality norms can be deployed or applied in deliberate, ante hoc regulation or epistemic planning, then they must define restrictions on epistemic dynamics including:

  (a) restrictions on epistemic states,
  (b) restrictions epistemic state transitions and trajectories,
  (c) restrictions on epistemic actions, procedures, and processes, and
  (d) such restrictions are distinct from constraints of feasibility and capacity.

(2) Epistemic rationality norms can—and, if they are to have any role in cognition other than description of regularities in practice, must—be deployed or applied in deliberate, ante hoc regulation and epistemic planning.

(3) ∴ Epistemic rationality norms define restrictions on epistemic dynamics including restrictions on epistemic states, restrictions on epistemic state transitions and trajectories, restrictions on epistemic actions, procedures, and processes, and such restrictions are distinct from constraints of feasibility and capacity.

More exactly, the argument over-provides; the inclusion of clause c in the conclusion establishes the core of D1C2.

Premise 1 can be derived from direct analysis of deliberate, ante hoc regulation or (inclusive) by thinking more explicitly in terms of epistemic planning problems. Suppose that an agent endorses a set of norms, $\{N_1, \ldots, N_n\}$, and sets out to follow them. At any initial state, s, that the agent can be in, there is a set of possible epistemic states, $\{s_1, \ldots, s_m\}$, that the agent can reach by applying some set of epistemic processes, $\{p_1, \ldots, p_k\}$, formed out of the actions in the agent's action library.[16] According to $\{N_1, \ldots, N_n\}$, some of these epistemic states are rational, others irrational, some transitions from s to s_i are rational, others not, and some processes are rational, others not. The goal state can involve arriving at a normative state, carrying out a normative state change, or carrying out a normative action or process. Deliberate ante hoc regulation is solving the planning problem: reasoning about which of the p_i satisfy the conditions imposed by $\{N_1, \ldots, N_n\}$ and also suffice to meet the conditions imposed by $\{N_1, \ldots, N_n\}$ on the goal state, and then selecting and carrying out the p_i that best does so (or perhaps arbitrarily selecting one of the p_i that does so, if many do equally well). To do this, the agent must discover (at least) the array of features of $\{N_1, \ldots, N_n\}$ laid out in Table 2.1.

The necessity of these questions in deliberate ante hoc regulation is somewhat self-explanatory. If one is trying to deliberately guide one's belief formation toward a rational state, one must figure out which rational states one can reach by chaining actions in the action library. To do this, one must figure out which states are rational,

[16] Note that it's possible that $n = k = m$ and that $n \neq k \neq m$, as processes may diverge and converge on their terminal states, and norms may converge or diverge on their verdicts with respect to processes and states. If, for instance, $k < m$, then some processes are nondeterministic—they lead to multiple terminal states. If $m < k$, then some processes lead to the same state.

Table 2.1 Solutions to epistemic planning problems

States	Transitions	Actions, processes, procedures
Whether a possible epistemic state, s, is rational or not with respect to N_i • What the static indicators of rationality and irrationality are with respect to N_i • Whether the static indicators of rationality and irrationality with respect to N_i hold for s	Whether a possible epistemic transition, t, is rational or not with respect to N_i • Whether the current state is rational or not with respect to N_i. • What the dynamic indicators of rationality and irrationality are with respect to N_i • Whether the dynamic indicators of rationality and irrationality with respect to N_i hold for t • Whether t starts or terminates with a rational state w.r.t. N_i	Whether a possible epistemic action, a, process, p, or procedure, P, is rational or irrational with respect to N_i • What the procedural indicators of rationality and irrationality are with respect to N_i – Whether a, p, or P guarantee or permit rational (or irrational) states or transitions w.r.t. N_i – Whether p or P have impermissible steps, sequences, or subprocedures w.r.t. N_i • Whether the procedural indicators of rationality and irrationality with respect to N_i hold for a, p, or P
Whether the current state is rational or not with respect to N_i	Whether a transition that is rational with respect to N_i is feasible (available or executable)	Whether a process that is rational with respect to N_i is feasible (available or executable)
Which feasible (available or executable) transitions and processes lead to rational states, and which lead to irrational states	What the consequences of executing a rational transition with respect to N_i will be (e.g., what state one ends up at)	What the consequences of executing a rational process with respect to N_i will be (e.g., what state one ends up at, what transition is effected)

which in turn requires being able to recognize the indicators of the normative verdicts of states. More, one will need to discover whether one even needs to carry out a revision or whether one can stay where one is and still be rational. In order to deliberately select a normative action or process, one must be able to figure out what feasible actions and processes are ruled in by the norms one endorses. If one cannot distinguish the normative courses of action from those that are merely feasible, the use of norms in action guidance will be frustrated. One would figure out what one's norms required, confuse those with the merely feasible actions, and be no better equipped to act normatively. Follow-up questions about indicators and feasibility (etc.) similar to those concerning states accompany these tasks, and some of these are matters of the relations between procedural structure and the resulting states and transitions.

Thus, if epistemic norms are useful in deliberate, ante hoc regulation, then they must somehow provide answers to questions like those in Table 2.1. More, in addition to dividing epistemic factors into sets by normative valence, norms useful for ante hoc regulation must provide information that distinguishes the merely feasible actions from the rational ones.

From the definition of restrictions on epistemic dynamics, it follows that any epistemic norm that provides answers to these questions specifies restrictions on epistemic dynamics of the specified forms—including restrictions on epistemic procedures (thereby implying D1C2.4). More precisely, if these questions, with respect to norm N (one of the $N_i \in \{N_1, \ldots, N_n\}$ but with subscripts omitted as unnecessary), even have answers, then N divides epistemic factors including epistemic procedures into distinct subsets and identifies them as rational or irrational, more rational or less, etc.. Thus, the definition of restrictions on dynamics immediately or mechanically implies that there is a restriction on epistemic dynamics that agrees with N. For any norm N, define an order, $\preceq$, on the powerset of N's domain, $Dom(N)$, such that for all $x, y \in Dom(N)$, and all P_i and P_j in $P(Dom(N))$, $x \in P_i$, $y \in P_j$, and $P_i \succ P_j$ iff either (i) x is rational according to N, and y is irrational according to N, or (ii) x is more rational according to N than y. Then define a restriction on N's domain, $R_{\preceq}(Dom(N)) = (P(Dom(N)), \preceq, T_{Dom(N)}, R_{Dom(N)}, I_{Dom(N)})$, such that (iii) for all such P_i and P_j, $P_i \subseteq R_{Dom(N)}$ and $P_j \subseteq I_{Dom(N)}$, or (iv) the ordering over the powerset of the domain is identical to the ordering in the restriction. There may be many structures that satisfy these conditions with respect to N, but the important point is that, by definition, all of them will be restrictions on epistemic dynamics defined over N's domain. Thus, the answers to Table 2.1s questions imply the content of some restriction(s) on epistemic dynamics defined over the relevant class of epistemic factors. Moreover, such restrictions will be usable for answering Table 2.1s questions.[17]

In sum, to deploy a norm N in deliberate, ante hoc regulation, one must answer the questions above. Answering the questions above is straightforwardly limning restrictions on epistemic dynamics including restrictions on procedural construction (i.e., some norm-kernels assign verdicts to actions, sequences of actions, or procedures). Thus, the usefulness of epistemic norms in ante hoc regulation implies that they define restrictions on epistemic dynamics of types (a)–(c). The restrictions outlined by answering the deliberation questions must be distinct from the constraints of capacity and feasibility if it's to be useful, so usefulness in ante hoc regulation implies that the restrictions defined by epistemic norms meet the condition (d) of the consequent of premise 1. Thus, we have premise 1: the usefulness of epistemic norms in ante hoc epistemic regulation implies that they define restrictions on epistemic dynamics including restrictions on actions,

[17] Strictly speaking, if N is a static norm, then we would also need to consult the model of epistemic dynamics over which the restriction is defined (i.e., the LTS over which $R_{\preceq}(Dom(N))$ is defined) to answer all of the questions in Table 2.1. Working from actions, processes, and procedures is easier, as they are richer objects; a set of actions, processes, or procedures—analyzed by means of the structures of EDGs as in earlier sections—can already be used to figure out (e.g.) the relevant properties of the states, transitions, and trajectories associated with them.

processes, and procedures. This gives us the basic idea of D1C2 conditional on the antecedent of premise 1.

The second premise, which gives the required antecedent, is easy to see. It's formulated conjunctively—epistemic rationality norms are usable in deliberate, ante hoc regulation, and they must be if epistemic norms are not entirely inert in cognitive life. The former conjunct suffices to detach the conclusion of the argument. To see that epistemic norms can be deliberately applied, all that's necessary is (1) noting that empirically supported models of cognition and cognitive behavior presuppose deliberate application of epistemic norms, and (2) direct observation of epistemic behavior reveals deliberate application of epistemic norms.

First, note the vast empirical literature[18] on the two systems (a.k.a., dual processes, a.k.a. default interventionist) model of reasoning, as well as the related research on cognitive biases and de-biasing strategies [2, 19–21, 45, 80]. The thought that runs through this research is that our innate cognitive endowments are "fast and frugal" but lead us into error in complicated, abstract, or novel situations. In order to override these error-prone strategies, cognizers must halt their cognitive processes and enact a normative alternative. These normative alternatives can be supplied by explicit knowledge of logical and statistical methods. That many epistemically damaging biases can be counteracted by training in various formal methods establishes the prevalence and usefulness of effortful, reflective, and otherwise deliberate ante hoc regulation [19, 20, 61]. Indeed, the influential (though not uncontroversial, e.g. [32, 53]) two systems approach to the cognitive psychology of reasoning is entirely built around this observation, and codifies it in empirically robust models [19–21, 81, 84].

Second, the sketch of deliberate application of epistemic norms above is, in many respects, notably similar to the entirely independent[19] EMIL-A model of the cognitive dynamics of norm compliance [17]. The EMIL-A model directly provides a place for restrictions on dynamics to fit; basically, the array of features that need to be discovered in the procedure above correspond to beliefs that must be developed in the EMIL-A model.

[18] How does this empirical literature hold up under the replication crisis in empirical psychology? I don't know, and I'm unsure of the ramifications of the crisis and even its direness. Fully investigating this question is far beyond the scope of this book, though plausibly within the scope of the ultimate project of which this book is a part. As such, it must remain for future work. A tentative, initial judgment, however: I am unaware of any detailed reviews of this paradigm in light of the replication crisis, but many of the core phenomena underlying the dual processes/dual systems/default interventionist model appear to be robust in the sense of [94]; they have been rediscovered by many, independent lines of research, and so bear a sort of replicability. See [20, 21] for this line of defense.

[19] It may be interesting to note that the sketch of deliberate application was outlined independently—indeed, entirely ignorant of—EMIL-A, and that there are a number of other independent sketches of norm adoption that basically capture the contours of the model. See, for instance, the synoptic power of EMIL-A in [17]. This might suggest that there is something robust to the main ideas of the sketch.

The EMIL-A model is broken into an epistemic and a pragmatic component. The epistemic component—comprising the norm recognition module—transforms information inputs into normative beliefs. Normative beliefs come in three varieties: main normative beliefs, norm pertinence beliefs, and norm enforcement beliefs. Main normative beliefs essentially comprise the information contained in norm-kernels: beliefs about the verdicts of norms under their conditions of application, beliefs about whether the conditions of application hold, about what specific actions in the agent's repertoire are assigned what deontic operators under the present conditions, and other beliefs, including those about the authority or sources of the norm. Norm pertinence beliefs are about whether the norm binds the agent. Norm enforcement beliefs are about the sanctions underwriting the norm.

Normative beliefs are transformed by the norm-adoption module in EMIL-A into normative goals by the pragmatic component. Normative goals are then sent to a planning system that solves planning problems in which the goal states are shaped by normative goals, which generates normative intentions and, finally, actions. I overlook alternative models of norm compliance here, for the reason that EMIL-A synthesizes the insights of available alternative accounts of norm compliance.[20]

The parallels between the ante hoc regulation procedure and the EMIL-A model are clear: both models require figuring out the features of norms outlined in the table above as part of synthesizing a normative plan of action. The main normative beliefs generated by the norm recognition module are explicit answers to the questions articulated in that table. Restrictions on epistemic dynamics are exactly the kind of construction from which those answers can be derived. The EMIL-A model thus requires the use of representations that carry the information that can be derived from restrictions on epistemic dynamics.

Third, recall the above examples of deliberate, ante hoc regulation. Empirical scientists employ experimental protocols to reflectively guide their information generating actions in the lab. Statisticians use established methods of data analysis to guide the way they draw conclusions from data sets. Mathematicians use methods of structural and transfinite induction to guide their construction of proofs. Philosophers deploy all of the above from time to time as well as established argumentative methods to approach theses proposed by colleagues and forerunners. More homely examples abound. Knowing that I'm prone to wishful thinking and recognizing that wishful thinking is a reasoning process that frequently guarantees error, I can suppress this process and substitute better ways of reasoning. All of these examples involve both pre-emptive commitment to certain courses of action as well as self-correction or the halting of natural or intuitive processes followed by the execution of normative courses of action. These kinds of occurrence of deliberate, ante hoc regulation have not only been empirically catalogued but can be readily

[20] This is argued at some length in [17].

witnessed. The second premise of the argument is secure, and thus the conclusion follows.[21]

The second conjunct is less important for the argument, but worth lingering on. Suppose that epistemic norms play no regulative role. Under this supposition, epistemic norms might still be useful to post hoc evaluation; they could be the standards by which we make post hoc evaluative judgments. But what could possibly be the point of post hoc evaluations in the absence of ante hoc regulation? If post hoc evaluations could be used to deliberately guide future epistemic behavior, it's unclear how this wouldn't just be deliberate ante hoc regulation. Suppose that post hoc evaluation shows that state s_{i-1} was irrational according to norm N. An agent might bypass directly reasoning about the verdicts of N by reasoning about future states by analogy with s_{i-1}. Future state s_{i+1} might have exactly the features of s_{i-1} that made it irrational according to N, and so a reasoner might seek to avoid s_{i+1}. But this is just a circuitous way of answering the questions in the table above, and so just a circuitous form of ante hoc regulation. If post hoc evaluations couldn't be used in at least this way, epistemic norms could be nothing other than tools for pointless score keeping.[22]

The pointlessness of such scorekeeping is worth contemplating. If there were only post hoc evaluation according to norms, errors in reasoning would be entrenched and incorrigible. To suppose that there are no means of ante hoc regulation is to suppose that there is no way to drive the negative post hoc evaluations into future corrections. Any such correction requires some way of guiding future action according to the verdicts of our epistemic norms. Suppose

[21] One caveat may be important to address, at this point. I am not defending what Pollock [68, 72] calls the "intellectualist model" of normative regulation. The intellectualist model represents the channeling of epistemic behavior along the restrictions defined by norms (i.e., the projection of the verdicts of norms and guidance of action in accordance with them) as a process that essentially involves explicit reasoning and epistemic judgment. Pollock argues that this cannot be the case on pain of regress. Grappling with that argument in detail is beyond the scope of this book. I have argued only that (i) there is normative regulation of the deliberate, ante hoc variety, but I will argue in further sections that (ii) there must be ante hoc regulation (deliberate or otherwise) if epistemic norms are not useless, and that (iii) deliberate or "intellectualist" forms of normative regulation and explicit processes of reasoning about epistemic norms are important in many of our epistemic and epistemological projects. I don't argue that such explicit processes are essential or unique forms of ante hoc regulation. What is defended here is compatible with at least (and at most) some of the processes of ante hoc regulation falling in line with Pollock's planning model of normative regulation, which posits that epistemic norms are acquired plans (i.e., non-propositional procedural knowledge) that, once internalized, automatically channel the behavior of cognitive mechanisms into the restrictions the norms define. The intellectualist regress may ground out in automatic normative regulation, but this doesn't gainsay the manifest reality and importance of deliberate normative regulation. Additionally, note that Pollock's model still presupposes that epistemic norms define restrictions on epistemic dynamics by means of their plan-like structure.

[22] More, even post hoc evaluation is a dynamic process that can be done correctly or incorrectly. I may, after all, find out that I misapplied an evaluative norm. The norms of epistemic rationality must be used to guide this process, or else all we can do is keep score about our score keeping behavior.

that post hoc assessment of my prior reasoning reveals that I drew conclusions on insufficient evidence. If there were no way to employ norms in deliberate, ante hoc regulation, then there is no way for me to deliberately ensure that my future belief states accord with norms of sufficient evidence. Deliberately doing so would require answering questions in the table above with respect to such evidence norms and then, somehow, selecting future courses of inference in accordance with those answers. To assume there is no way to engage in ante hoc regulation at all (deliberate or otherwise) is to assume that there is no way to channel my epistemic actions into the restriction defined by my epistemic norms. Thus, if there were nothing that counts as ante hoc regulation, then there would simply be no way to guide my future epistemic actions.

Put another way, in the absence of some way of suppressing my erroneous dispositions to act and supplying a normative course of action, I would, at best, only be able to recognize that I have made errors while being utterly powerless to do anything about them. In particular, without the ability to project the verdicts of my norms to future epistemic actions and to then guide my future courses of epistemic action in accordance with those verdicts, I would be sentenced to a kind of epistemic hell—a future of entrenched errors each of which I can recognize and perhaps even anticipate but that I am powerless to correct or revise.[23] Thus, ante hoc epistemic regulation is axiologically central to our epistemic lives. This goes further than is necessary. It suffices for the argument to note that this kind of epistemic fatalism isn't merely troubling, it's demonstrably not our epistemic situation.

[23] Perhaps this is too dire. It's possible, after all, that we can drive normative requirements into behavior by means of (e.g.) corrective training or Pollock-style automatic plan acquisition without deliberately figuring out any solution to planning problems by using some kind of conditioning regimen based on post hoc evaluation. I consider this in a footnote because it is, with respect to the main thrust of argument, an aside. The value or importance of epistemic regulation is, strictly speaking, a garden path with respect to the argument. The main thrust of the premise in question is that epistemic norms *can be* applied in deliberate ante hoc regulation, and that this implies that epistemic norms define restrictions on procedures. Nonetheless, there are two responses to the possibility of such non-deliberative corrective training or plan acquisition. First, the planning problem above can be thought of as a rational reconstruction of the process that one carries out implicitly through corrective training or norm internalization. One has to answer these questions at least implicitly in order to actually select the right course of action with respect to a norm. The training and automatic plan acquisition just answer them implicitly by conditioning and internalization processes. Thus, this kind of corrective training would amount to ante hoc regulation, even if not of the deliberative kind. Second, it's not clear that it's possible to be trained in a norm without some deliberate application of it. Training to use a normative or methodological rule system typically inculcates a sense of (i.e., at least implicit beliefs about) the verdicts and restrictions defined by it, and, in situations of novelty, uncertainty, or a multitude of options, one consciously, explicitly deliberates about what those norms require and tries to apply them in the process of training. So it seems plausible that such non-deliberative means of ante hoc regulation (at t) require deliberative regulation elsewhere (at some prior t').

2.2.3.2 (D1C2.1) Argument from Cases

The basic argument schemata for D1C1 can be modified and applied to get D1C2. There are classes of pairs of processes, $\langle s, a, s' \rangle$ and $\langle s, b, s'' \rangle$, that are rationally distinct. If rationality distinguishes classes of epistemic factors, then there is a norm-kernel (hence, norm) that takes those factors as objects. Thus, epistemic processes are the objects of some epistemic rationality norm(-kernel)s. The key premise is given by examples with a shared form:

Bad Reasoning (BR) Let a belief set at t be $s = \{B, A \rightarrow B, C \rightarrow A, C\}$ where $\rightarrow$ is a validly detachable conditional, and A is consistent with B and C. Let the belief set at $t+1$ be $s' = \{B, A \rightarrow B, C \rightarrow A, C, A\}$. Consider two possible process types for transitioning from s to s':

(Process 1) Detach A from the conditional $C \rightarrow A$ by affirming the antecedent[24] C.

(Process 2) Detach A from the conditional $A \rightarrow B$ by deliberately affirming the consequent, B, and *not* performing any sort of explanatory or confirmatory inference to the same conclusion.

(Verdict) If these conditions hold, Process 2 is bad reasoning; an agent carrying out Process 2 in lieu of Process 1 is not executing a rational epistemic process.

Non Reasoning (NR) Let a belief set at t be $s = \{B, A \rightarrow B, C \rightarrow A, C\}$ where $\rightarrow$ is a validly detachable conditional, and A is consistent with B and C. Let the belief set at $t+1$ be $s' = \{B, A \rightarrow B, C \rightarrow A, C, A\}$. Consider two possible process types for transitioning from s to s':

(Process 1) Detach A from the conditional $C \rightarrow A$ by affirming the antecedent C.

(Process 2) Hold an umbrella out during a lightning storm in the hope that it causes belief that A and, by chance, get struck by lightning and end up with the belief that A.

(Verdict) If these conditions hold, Process 2 is not even reasoning; an agent carrying out Process 2 in lieu of Process 1 is not executing a rational epistemic process.

Non-Arbitrary Reasoning 1 (NAR1) Let a belief set, s, contain $B_1, \ldots, B_n$ and be logically consistent. Consider two possible processes:

(Process 1) Subtract every B_i such that i is a multiple of 3, and restore consistency by removing any element of s that implies any such B_i. Let the resulting state be s'.

---

[24] Perhaps more plausible, suppose the process comprises explicitly formally representing my belief set, plugging my beliefs into the valid argument schema of conditional detachment and thereby recognizing the validity of such an argument, then expanding my beliefs with A automatically by an internalized process of deductive closure.

(Process 2) Transition to s'', which is an arbitrary, consistent expansion of s with a logical consequence, A, of $B_1, \ldots, B_m$.

(Verdict) Process 1 is an utterly arbitrary change of belief, as opposed to Process 2, which is partially arbitrary, but driven by a logically cogent reason. If these conditions hold, then an agent carrying out Process 1 in lieu of Process 2 is not executing a rational epistemic process.

Non-Arbitrary Reasoning 2 (NAR2) Let a belief set, s, contain $B_1, \ldots, B_n$ and be logically consistent. Let s' be a belief set with an empty intersection with s that satisfies any static conditions of rationality one wishes. Let s'' be any belief set that satisfies every condition of minimal mutilation, conservatism, or preservation of entrenchment relative to s defended in the epistemological literature (or some preferred, consistent set thereof). Consider two possible processes:

(Process 1) Subtract every B_i such that i is a multiple of 3, and restore consistency by removing any element of s that implies any such B_i until arriving at s''.

(Process 2) Expand s with an observation, C_i that contradicts some B_i, then remove the B_i and anything implying it to restore consistency. Iterate this process until reaching s'.

(Verdict) Process 1 is a process of chaining utterly arbitrary changes of belief while every step of Process 2 is a rational belief change. If these conditions hold, then an agent carrying out Process 1 in lieu of Process 2 is not executing a rational epistemic process.

Non-Arbitrary Reasoning 3 (NAR3) Let a belief set, s, contain $B_1, \ldots, B_n$ and be logically consistent. Let s' be a belief set that satisfies any static conditions of rationality one wishes, as well as every condition of minimal mutilation, conservatism, or preservation of entrenchment relative to s defended in the epistemological literature (or some preferred, consistent set thereof). Consider two possible processes for reaching s':

(Process 1) Subtract every B_i such that i is a multiple of 3, and restore consistency by removing any element of s that implies any such B_i until arriving at s'.

(Process 2) Expand s with an observation, C_i that contradicts some B_i, then remove the B_i and anything implying it to restore consistency. Iterate this process until reaching s'.

(Verdict) Process 1 is a process of chaining utterly arbitrary changes of belief while every step of Process 2 is a rational belief change. If these conditions hold, then an agent carrying out Process 1 in lieu of Process 2 is not executing a rational epistemic process.

Each example introduces a pair of process types or process tokens that satisfy a set of conditions; in particular, each of the processes involved equally satisfies some specified set of dynamic restrictions. There is then a normative verdict; it is claimed that one or the other process type is, under the conditions articulated, rational or irrational, more or less rational than the alternative. The logic of the argument is then straightforward; if the verdict claim of an example is correct, then the example suffices to establish the first premise of the argument. There are, then, token processes that equally satisfy the sorts of dynamic restrictions typically defended that are nonetheless rationally distinct. Readers are free to convince themselves of the premise with the example they find most plausible.

As for the cases themselves, each of these is an example of a case that illustrates a putative epistemic norm that implicitly defines a restriction on epistemic dynamics that is finer grained than any restriction on epistemic transitions. In particular, Bad Reasoning and Non-Reasoning indicate putative norms that distinguish process types that are transition-convergent. That is, both of the process types in each case converge in some transitions, and, if the normative verdict introduced at the end of the case is correct, the tokens of these types are nonetheless (according to these norms) rationally distinct even where they converge in the epistemic transitions that bookend them. That, at least, is the apparent correct verdict in these cases. Thus, the first premise of the argument is true, if either of the verdicts in BR and NR are correct.

The Non-Arbitrary Reasoning variants strengthen the case for the first premise of the argument, as each imposes a restriction on epistemic procedures and processes that distinguishes process types that equally satisfy dynamic restrictions of minimal mutilation, entrenchment, etc.. If the normative verdict introduced at the end of any of NAR1, NAR2, or NAR3 is correct, then there are rationally distinct process types that equally well satisfy the most common forms of dynamic restriction. Thus, if any of these verdicts is correct, the first premise is true.

(D1C2.1) Second Argument from Cases

See also Table 2.3 for an array of types of methodological norms that are procedural, and the fields of research in which they are studied.

2.2.3.3 (D1C2.2) Argument from Cases

Some of the pairs of processes, $\langle s, a, s' \rangle$ and $\langle s, b, s'' \rangle$, in the illustrations of Sect. 2.2.2.1 differ only in the properties of the transitions they result in or the processes themselves. Thus, there are processes, (s, a, s') and (s, b, s''), that are distinguished on the basis of the properties of transitions or processes. If rationality distinguishes x and y on the basis of C, then there is a norm in which C is the condition of application. Thus, the conditions of application of some epistemic rationality norms concern properties of transitions or processes.

2.2.3.4 (D1C2.3) Argument by Construction

First, epistemic rationality norms define the requirements of epistemic rationality. I take it that this is built into the concept of such norms given that epistemic rationality is not merely a free for all. That is, if there are requirements of epistemic rationality that one could fail to satisfy under some condition or another, and there are any regularities that govern or even describe these conditional requirements, then it is epistemic norms that define them. That is what at least a subset of norms do; they are the rules or policies that specify what conditions an epistemic factor is to meet in order to be rational, what conditions suffice for irrationality, and so on.

Second, the requirements of rationality divide states, changes of state, actions, processes, and procedures into rational and irrational classes or perhaps into more complex divisions involving rankings or preferences among these epistemic factors. I take it that, given D1C2.1, this is simply built into the concept of normative requirements; for any given requirement, the set of relevant objects is divided into those that meet it and those that don't, those that are assigned some particular verdict and those that aren't. Given D1C1 and D1C2.1, the premise is trivial. Now, sometimes requirements are looser or more nuanced; they might divide objects into a ranking according to how many requirements are met or how heavily weighted the met requirements are. Nonetheless, such requirements divide the objects of norms into various categories, though perhaps not exclusively or exhaustively. Thus, the norm-kernels that specify requirements of rationality divide the epistemic factors that they take as objects into rational and irrational subsets or they effect some other more complex division, such as a ranking.

Finally, any division of epistemic factors into rational and irrational classes or ranking of epistemic factors (with or without a threshold that assigns the locations in the ranking lower than it to the irrational class) defines restrictions on epistemic dynamics. For any norm-kernel N, define an order on the powerset of N's domain (set of objects), $\preceq$, such that for all $x, y \in Dom(N)$, and all P_i and P_j in $P(Dom(N))$, $x \in P_i$, $y \in P_j$, and $P_i \succ P_j$ iff either (i) x is rational according to N, and y is irrational according to N, or (ii) x is more rational according to N than y. Then define a restriction on N's domain, $R_{\preceq}(Dom(N)) = (P(Dom(N)), \preceq, T_{Dom(N)}, R_{Dom(N)}, I_{Dom(N)})$, and (iii) for all such P_i and P_j, $P_i \subseteq R_{Dom(N)}$ and $P_j \subseteq I_{Dom(N)}$, or (iv) the ordering over the powerset of the domain with respect to the norm is identical to the ordering in the restriction. For more complex divisions, restrictions with more complicated structures can be given, as outlined in Sect. 1.3.5. Thus, every norm-kernel (hence, norm) that divides its objects defines a restriction on epistemic dynamics over its objects.

It follows from the general result that all norms define restrictions on epistemic dynamics defined over their objects, that procedural epistemic norms define restrictions on epistemic procedures. In particular, procedural epistemic norms that distinguish epistemic procedures on the basis of features of their procedural construction (see section on D2) define restrictions on the class of procedural constructions. That is, a procedural epistemic norm that, for instance, sets procedural constructions with feature F to the rational set and those with G to the irrational set, parses the class of procedural constructions on the basis of their properties, distinguishing elements of the powerset of procedural constructions. Thus, the norms in Table 2.3, the norms presented as cases for the argument for D1C2.1, and any procedural epistemic norms take procedural constructions or ways of building epistemic procedures out of repertoires of basic epistemic actions as their objects, all define restrictions on epistemic dynamics.[25]

From there, the remaining clause of the principle is derived from the observation that there are systematic and differential relations between the structures of procedures and the states that they bring about, the transitions they effect, or the processes they compose (this latter clause is implied by the clauses D2.4–D2.6, below). In any plausible model of epistemic dynamics for any plausible epistemic agent or system, there are causal constraints that systematically correlate procedural constructions and the terminal states that can be reached by them, the transitions that can be achieved by them, and the epistemic processes produced or triggered by them. In general, a procedure may guarantee or permit the bringing about of some epistemic states and not others (or guarantee that some states are not brought about), a procedure may guarantee or permit some epistemic transitions and not others, a procedure may define one set of processes or causally trigger some set of processes and not others. One formal proof construction procedure determines a different range of processes of thought than another. Some procedures draw our attention to information that causes a sudden insight, and others don't. And so on. Thus, restrictions on states, transitions, and processes can be derived from restrictions on procedural construction; the restriction on procedures means that whatever states, transitions, or processes are incompatible with the rational set of procedures are implicitly ruled out.

2.2.3.5 (D1C2.4) Derivation from D1C2.9 and D1C2.10

If D1C2.9 and D1C2.10 are true, then there are procedural epistemic norms that cannot be reduced to static or dynamic epistemic norms. The argument presupposes a methodological norm: if there are no reductions of a class of epistemic norms to another class, then, insofar as that class of norms finds important uses in epistemic regulation or in epistemological theory, then the irreducible class is indispensable to epistemic regulation and to epistemological theory. The next two sections argue that there are indispensable uses of procedural epistemic norms in epistemic regulation. The following six sections explore the possibilities for reduction.

[25] Section 9.2.2 spells out the notions of procedural constructions and restrictions on procedural construction more explicitly.

2.2.3.6 (D1C2.4) Argument from the Regulative Function of Epistemic Norms and Epistemic Planning

There are two ways to derive D1C2.4 from the foregoing considerations about the regulative function of epistemic norms. First, to deploy a norm N in ante hoc regulation, one must answer the sorts of questions in Table 2.1. Answering the questions in Table 2.1 is straightforwardly limning a restriction on epistemic procedures. Given that deliberate, ante hoc regulation is indispensable, the clause results. Making the same point from another direction, solutions to epistemic planning problems directly impose restrictions on epistemic plans, which just are restrictions on epistemic procedures:

(1) Ante hoc regulation requires solving epistemic planning problems.
(2) Solutions to epistemic planning problems are restrictions on (epistemic) procedural construction.
(3) ∴ Ante hoc regulation requires restrictions on (epistemic) procedural construction.

Ante hoc regulation is the process of solving epistemic planning problems. The solution to an epistemic planning problem is a set of epistemic plans that each suffice to achieve the goal state determined by some set of epistemic norms. A set of epistemic plans that achieve a goal state just is a restriction on (epistemic) procedural construction because epistemic plans are themselves sets of instructions built out of procedural operations of conditional choice, sequencing, etc. over a set of elements of the epistemic action library. Given a restriction on permissible plans over epistemic actions, there is a restriction on permissible sets of instructions for building epistemic procedures. Thus, insofar as ante hoc regulation requires solutions to epistemic planning problems, ante hoc regulation requires that there be restrictions on procedural construction. These restrictions will be at least those derivable from the norms that define the goal state. In other words, if there are any norms from which to derive restrictions on goal state, there will be derivative norms that restrict procedural construction.

Typically, only a subset of possible plans constructible from the action library will reach the goal state. This is often due to the way the plans are constructed, the ordering of the subtasks in the plan, the proper conditionalizing of choices in the plan, and so on. For instance, whether a semantic tableaux procedure gives a correct answer about the validity of an argument depends on how one orders one's choice of rules to apply. The requirement that one get right answers entails a requirement that one use procedures constructed with certain sequential compositions of rule applications and not others. This point applies to nearly all of the examples in the table of procedural requirements above. So, the goal state, in concert with the features of the constructible plans, divides the set of procedures into those that are effective for reaching the goal state and those not. Insofar as the goal state is defined by the requirements of some set of norms, those norms indirectly restrict procedural constructions. This brings us to a second argument.

Second, in cases where epistemic outcomes depend on procedures being constructed in certain ways (as in inductive rules, proof and model construction procedures, numerous other formal methods involving algorithms for the construction and manipulation of models, and de-biasing strategies), norms requiring those epistemic outcomes derivatively require those procedural constructions. In other words, if epistemic rationality defined no restrictions on procedures, then it would define no restrictions on states, transitions, or processes, for restrictions on states, transitions, and processes, if they are to play any regulative role, must imply restrictions on procedures.

This is immediate from the fact that procedures differentially bring about epistemic states, produce epistemic transitions, and structure or trigger epistemic processes. If one is, for instance, required by a norm N to end up at state s (or a state with property F), and some epistemic procedures guarantee termination at s (or termination at a state with property F) while others don't, then N at least implicitly requires that one execute one of the former procedures and not one of the latter. If one is required, by norm N, to carry out transition from s to s' and some epistemic procedures guarantee this transition and others don't (or others even actively prevent such a transition), then one is required by N to execute a procedure that effects that transition, and not one that fails to. If N dictates a specific epistemic process, p, (or some process in a set, P) and some procedures compose p (or an element of P) out of epistemic actions in one's repertoire or bring about a sufficient trigger for process p (or one in P) while others lack these properties, then N dictates, at least derivatively, that one is to execute such a procedure (and not others). In each case, we can interpret the situation as there being a derivative norm, N', that takes procedures as its objects that is logically related to the original norm, N.

Metaphorically, the idea is that if, contrary to popular opinion, some roads lead to Rome and others don't, and you are required to get to Rome, then you are implicitly required to travel some roads and not others. Thus, there must be procedural epistemic norms given that epistemic norms play a regulative role; i.e., we can't be obligated not to be in state s if we are not also (at least implicitly) obligated not to do what guarantees that we end up in state s.

2.2.3.7 (D1C2.4) Argument from Restrictions on Actions and Processes

There is a straightforward derivation of this clause from the putative cases of restrictions on actions and processes.

(1) Every restriction on processes can be used to derive a set of restrictions on procedural construction.
(2) Every epistemic rationality norm can be used to derive a set of restrictions on processes.
(3) $\therefore$ Every epistemic rationality norm can be used to derive a set of restrictions on procedural construction.

The first premise is direct. Processes have procedural structures, so sets of processes capture patterns of procedural structure. Restrictions on processes divide processes into verdict sets, so restrictions on processes divide patterns of procedural structure into corresponding verdict sets. From the prior arguments concerning the construction of restrictions on epistemic dynamics, restrictions on procedural structures result.

Consider an arbitrary restriction on epistemic processes, $R_{\preceq}(Pr)$. Consider the rational set of processes according to this restriction, R_{Pr}. Note that epistemic processes generally break down into basic actions and sequences of basic actions. At the very least, epistemic processes can be built from sequences of more basic actions or processes, and so all but indivisible processes break down this way. All sets of processes instantiate some pattern of procedural structure expressible in terms of procedural operations. All of the processes, p, in the set might contain particular actions a or subprocesses q in a specific sequence, or they might be so formed that, at a given point in a sequence, every process contains one element, p_i, of a particular set of subprocesses, $\{p_1, \ldots, p_n\}$. Some pattern of sequential compositions and names for processes and subprocesses will suffice to denote these sets of processes. In the first case, processes without a particular sequencing of subprocesses are ruled out. In the latter, any processes without a choice among $\{p_1, \ldots, p_n\}$ are ruled out. In the absence of any such procedural patterns, there is a general procedural construction defined by any set of processes: choice over the set of token processes in the set.

Thus, for any set of rational processes in $R_{\preceq}(Pr)$, R_{Pr}, there is a corresponding division of ways of building procedures out of actions and processes into rational and irrational sets. Thus, by the definition of restrictions, if we can identify sets, X, of patterns of procedural constructions (e.g., expressions in relational algebra with variables that range over processes or, more to the point, expressions in the program language of PDL), then every restriction on processes defines a unique set of restrictions on procedural constructions—the set of procedural constructions such that the processes in R_{Pr} match those patterns of construction that can be extracted from R_{Pr}.

Identify sets of procedural constructions with sequences of procedural operations on actions: sequencing, choices, iterations, etc.. A process, p, matches a procedural construction iff there is a set of subprocesses, $\{p_1, \ldots, p_n\}$, that, when interspersed with the sequence of procedural operations, is identical to p. That is, a process matches a procedural construction iff one can find a way of building p out of a set of subprocesses using the exact sequence of procedural operations. This definition yields numerous sets of restrictions on procedural constructions per restriction on processes. For a developed example of methods of associating procedural operations on (structures representing) actions and processes with the relational algebraic operators, see the semantics of PLEN in the next Part or the semantics of PDL [38].

The second premise is easier given D1C2.1. Epistemic norms that take processes as objects straightforwardly define restrictions on processes. Moreover, the construction rules from the arguments for the remaining clauses of D1C2 show how restrictions on processes can be generated from restrictions on states or transitions (though, of course, not all restriction on processes can be so generated). Thus, those construction rules show how every norm that restricts any epistemic factor is associated with a set of restrictions on processes (and that not every restriction on processes can be so associated).

Finally, the modal "can be used to derive" is transitive. Suppose that there is some procedure for reading off sets of restrictions on processes from a given epistemic rationality norm and that there is some procedure for reading a set of restrictions on procedural construction from a given restriction on processes. Then chaining those procedures forms a procedure for reading a set of restrictions on procedural construction off of a given epistemic norm. From this fact, and the prior premises, every norm can be seen to be associated with a set of restrictions on procedural construction.[26] Thus, every restriction on states or transitions can be used to derive a restriction on processes, which can then be used to derive a restriction on procedural constructions. Again, that there are restrictions on epistemic dynamics seems to imply that there are restrictions on procedural construction.

2.2.3.8 (D1C2.5) Construction Rule

Let N_T be a norm-kernel that restricts transitions that are exclusive with respect to their action labels (i.e., no two transitions share the same action labels). Let N_A be a norm-kernel such that: for any action, a, t is rational (irrational, higher in the ranking than some other element, etc.) according to N_T, $t = \langle s, s' \rangle$, and $\langle s, s' \rangle \in \rightarrow_a$ iff a is rational (irrational, higher in the ranking than some other element, etc.). This rule immediately maps norm-kernels restricting transitions to norm-kernels restricting actions, including the rule specifying N_A itself, as in the arguments in Sect. 2.2.2.4. Sections 2.2.3.12 and 2.2.3.13 show that the restriction to domains of transitions that do not converge in their action labels is not a matter of indifference.

2.2.3.9 (D1C2.6) Construction Rule

Let N_T be a norm-kernel that restricts transitions that are exclusive with respect to their action labels (i.e., no two transitions share the same action labels). Let N_P be a norm-kernel restricting processes (i.e., paths) such that: for any process, $p = \langle s, a_1, \ldots, a_n, s' \rangle$, t is rational (irrational, higher in the ranking than some other element, etc.) according to N_T, $t = \langle s, s' \rangle$, and $\langle s, a_1, \ldots, s_n, s' \rangle \in T_A*$ iff p is rational (irrational, higher in the ranking than some other element, etc.). This rule immediately maps norm-kernels restricting transitions to corresponding

[26] The logic of the argument may not be immediately apparent. A more precise form of the inference is given by the Reduction Lemma in Chap. 9. Indeed, it's a virtue of PLEN that it provides a formal analysis of this intuitive reasoning.

norm-kernels restricting processes, including the rule specifying N_P itself, as in the arguments in Sect. 2.2.2.4. Sections 2.2.3.12 and 2.2.3.13 show that the restriction to domains of transitions that do not converge in their action labels is not a matter of indifference.

2.2.3.10 (D1C2.7) Construction Rule

Let N_T be a norm-kernel that restricts transitions that are exclusive with respect to their action labels (i.e., no two transitions share the same action labels). Let N_{Pr} be a norm-kernel restricting processes (i.e., sets of paths) such that: for any set of processes, $Pr = \langle p_1, \ldots, p_n \rangle$, t is rational (irrational, higher in the ranking than some other element, etc.) according to N_T, $t = \langle s, s' \rangle$, and, for each p_i, $p_i = \langle s, a_1, \ldots, s_n, s' \rangle \in T_A*$ iff p_i is rational (irrational, higher in the ranking than some other element, etc.). This rule immediately maps norm-kernels restricting transitions to corresponding norm-kernels restricting sets of paths (i.e., procedures), including the rule specifying N_{Pr} itself, as in the arguments in Sect. 2.2.2.4. Sections 2.2.3.12 and 2.2.3.13 show that the restriction to domains of transitions that do not converge in their action labels is not a matter of indifference.

2.2.3.11 (D1C2.8) Construction Rule

Let N_A be a norm-kernel that restricts actions. Let N_T be a norm-kernel that restricts transitions that are exclusive with respect to their action labels such that: for any transition, t, a is rational (irrational, higher in the ranking than some other element, etc.) according to N_A, $t = \langle s, s' \rangle$, and $\langle s, s' \rangle \in \rightarrow_a$ iff t is rational (irrational, higher in the ranking than some other element, etc.).

Let N_P be a norm-kernel that restricts processes (paths). Let N_T be a norm-kernel that restricts transitions that are exclusive with respect to their process labels such that: for any transition, t, $p = \langle s, a_1, \ldots, a_n, s' \rangle$ is rational (irrational, higher in the ranking than some other element, etc.) according to N_P, and $t = \langle s, s' \rangle$ iff t is rational (irrational, higher in the ranking than some other element, etc.).

Let N_{Pr} be a norm-kernel that restricts sets of processes. Let N_T be a norm-kernel that restricts transitions that are exclusive with respect to their sets of process labels such that: for any transition, t, $Pr = \langle p_1, \ldots, p_n \rangle$ is rational (irrational, higher in the ranking than some other element, etc.) according to N_P, $t = \langle s, s' \rangle$, and, for each p_i, $p_i = \langle s, a_1, \ldots, s_n, s' \rangle \in T_A*$ iff t is rational.

These rules immediately map norm-kernels restricting transitions to corresponding norm-kernels restricting sets of actions, processes, and procedures, as in the arguments in Sect. 2.2.2.4. This result is only a "generation" rule. It does not show that all norm-kernels restricting transitions can be reduced to norm-kernels restricting actions, processes, or procedures. Readers should notice, however, that the following two sections leave open a real question of logical asymmetry between static and dynamic norm-kernels on the one hand and procedural norm-kernels on the other. If the next sections are right, and it can be shown that procedural norm-kernels can reduce all static and dynamic norm-kernels, there is a powerful argument for the logical primacy of procedural norm-kernels. This question is left open until Chap. 9, where it is closed by the Reduction Lemma for Representation Theorem 1, showing that all static and dynamic restrictions can, indeed, be reduced to procedural restrictions while the converse fails, and this extends also to norm-kernels.

2.2.3.12 (D1C2.9) Direct Argument

The formal argument for D1C1.5, coupled with the argument for D1C2.10 directly imply this result. Processes can, in principle, be distinct and yet transition equivalent (over all LTSs). Thus, restrictions defined over processes can distinguish necessarily transition-equivalent processes, and it follows that no restriction on transitions (or even sets of transitions) can reduce such a restriction on processes, and therefore that, even if (sets of) restrictions on states reduce restrictions on transitions, no (set of) restrictions on states can reduce such a restriction on processes. Thus, there are procedural norms (captured by the foregoing restrictions) that (restrict epistemic transitions that) cannot be reduced to norms that restrict epistemic states.

2.2.3.13 (D1C2.10) Direct Argument

It's possible for epistemic actions and processes to be associated with identical binary transition relations despite being distinct in epistemically relevant ways. Thus, D1C2.10 follows by reasoning similar to that in Sect. 2.2.2.1. Consider the following case.

(Honest Twin) Suppose that I know methods for discerning the validity of arguments in first-order logic that are infallible if carried out correctly. Suppose that I have a twin that knows all of the same methods I know but is infallibly honest and diligent when asked about validity; so much so that he walks me through every proof construction if I but ask. In any situation, for any argument, if I were to successfully carry out a method for discerning validity, the execution of the methods I know would tell me exactly what my twin would tell me, were I to ask, and vice versa. In all possible cases, we converge on identical proofs. In cases where there are multiple proofs, my twin always carries out the one that I would have, had I carried out the process. Let a represent the process of testing arguments using my deductive and model-checking methods. Let b represent the process of asking my twin about the validity of first order arguments and coming to believe what he says. These two processes are certainly equivalent in a sense; given an argument, they return identical verdicts if carried out correctly; they even give the same proofs. Thus, they terminate with identical epistemic states from any initial state. Let T_a be the transition relations associated with a; the set of all pairs of initial states and output states of carrying out a. Let T_b be defined in a parallel fashion. Then $T_a = T_b$.

In this case, $a \neq b$, and they are distinct in senses more substantive than mere non-identity. Let $t \in T_a$. Thus, of course, $t \in T_b$. For an arbitrary norm, N, call N's verdict on a with respect to a situation, s, be $N(a, s)$. Let $f(t)$ be the first state in the transition, t. There are norms, N_i, such that $N_i(a, f(t)) \neq N_i(b, f(t))$ and even $\neg(N_i(a, f(t)) \equiv N_i(b, f(t)))$. This difference in verdicts may be due to numerous properties of a and b, including their respective: reliability, implementability, efficiency, complexity, speed, power, the internal accessibility of their properties, or even their non-doxastic features (e.g., whether they serve to better ingrain some cognitive skill).

Generally, transition equivalent actions like a and b may differ with respect to these properties despite their transition equivalence. The following intuitive (putative, plausible) rules of epistemic responsibility may suffice to distinguish a and b on the basis of (a subset of) such properties:

- Always use a reliable process even if an unreliable process suffices in this case for true belief.
- It's irrational to use a process of unknown reliability, speed, complexity, power, etc. instead of a process of known reliability, speed, complexity, power, etc. even if the unknown process suffices in this case for true belief or is, in fact, equal with respect to the relevant property.
- It cannot be equally obligatory to use an unimplementable process as it is to use an implementable process.
- It cannot be equally rational to use an unimplementable/unknown process as it is to use an implementable/known process.

Since the processes, a and b, are transition equivalent but epistemically inequivalent, there must be a norm that distinguishes them (by reasoning like that in Sect. 2.2.2.1). It immediately follows that there is no way to map a norm distinguishing transitions to one distinguishing processes that will properly distinguish a and b, since a and b have identical transition relations; there can be no mapping from conditions of application to verdicts that distinguish the transitions of a and b, so there can be no norm from which a mapping from conditions of application to verdicts that distinguish a and b can be derived.

A formal argument similar to that for D1C1.5 can be given for this claim:

Proposition 9 *If* (S, T_S, A_T) *is a simple transition system where* S *is a finite set of states,* $T_S \subseteq S \times S$ *is a transition relation over* S, *and* A_T *is a set of actions associated with transition relations that are subsets of* T_S, $Part(S)$ *is a partitioning of* S *into disjoint subsets,* $Rational(S)$ *and* $Irrational(S)$, $Part(T_S)$ *is a partitioning of* S *into disjoint subsets,* $Rational(T_S)$ *and* $Irrational(T_S)$, *and $Part(A_T)$ is a partitioning of A_T into disjoint subsets $Rational(A_T)$ and $Irrational(A_T)$, then:*

(1) There can be no function, $f : T_S \longrightarrow A_T$, such that, for all s and s', $a \in Rational(A_T)$ iff $\langle s, s' \rangle \in Rational(T_S)$ and $\langle s, s' \rangle$ is associated with a.
(2) There can be no function, $f : S \longrightarrow A_T$, such that, for all s and s', $s' \in Rational(S)$ iff $a \in Rational(A_T)$ and $\langle s, s' \rangle$ is associated with a.

Proof By contraposition.

(1) Let a and b be nonidentical actions associated with one transition, $\langle s, s' \rangle$ (i.e., let $\langle s, s' \rangle$ be in the transition relations associated with a and b). Let $Part(A_T)$ be such that $a \in Rational(A_T)$ and $b \in Irrational(A_T)$. Suppose there is a function from T_S to A_T as described. Then, immediately, $\langle s, s' \rangle \in Rational(T_S)$ iff $a \in Rational(A_T)$ and $\langle s, s' \rangle \in Irrational(T_S)$ iff $b \in Irrational(A_T)$. Thus, $\langle s, s' \rangle \in Rational(T_S)$ and $\langle s, s' \rangle \in Irrational(T_S)$. Thus, $Rational(T_S)$ and $Irrational(T_S)$ are not disjoint.
(2) Let $\langle s, s' \rangle, \langle s'', s' \rangle \in T_S$. Let a and b be nonidentical actions such that $\langle s, s' \rangle$ is associated with a and $\langle s'', s' \rangle$ is associated with b. Let $Part(A_T)$ be such that $a \in Rational(A_T)$ and $b \in Irrational(A_T)$. Suppose there is a function from S to A_T as described. Then, immediately, $s' \in Rational(S)$ iff $a \in Rational(A_T)$ and $s' \in Irrational(S)$ iff $b \in Irrational(A_T)$. Thus, $s' \in Rational(S)$ and $s' \in Irrational(S)$. Thus, $Rational(S)$ and $Irrational(S)$ are not disjoint.

□

Thus, if restrictions on transitions are analyzed in terms of disjoint sets (sets of rational—as opposed to irrational—transitions, sets of permissible—as opposed to impermissible—transitions, etc.), then there are pairs of processes that cannot be rationally distinguished by the verdict sets that the transitions comprising their root (first) and terminal (last) states are contained in. In other words, for these processes, there can be no rule that directly maps the verdicts of restrictions on transitions to the verdicts of restrictions on processes that satisfies the conditions of correspondence. Moreover, this is not just a formal point. There can be no rule that maps the verdicts of sets of restrictions on transitions (or even trajectories) to the verdicts of restrictions on processes that distinguish such transition equivalent processes as appear in (e.g.) the Honest Twin case or even such transition convergent processes as appear in earlier cases. Thus, the norms captured by restrictions on processes cannot in general be reduced to norms captured by corresponding restrictions on transitions even in cases of plausible norms.[27]

[27] Additionally, the failure of reducibility between restrictions on transitions and restrictions on actions is not symmetrical, though that's not yet relevant to the main line of argument. To be clear, there is a symmetry in the neighborhood: there is no function, $f : A_T \longrightarrow T_S$, satisfying the conditions of the antecedent of Proposition 9 such that for all transitions, $\langle s, s' \rangle$, $a \in Rational(A_T)$ iff $\langle s, s' \rangle \in Rational(T_S)$ and $\langle s, s' \rangle$ is associated with a. The counterexample is any action, $a \in A_T$, associated with distinct transitions, t and t' such that $t \in Rational(T_S)$ and

2.3 The Second Cluster of Principles (D2)

Where D1C2 stated that there are procedural epistemic norms, and that these are indispensable to the regulative role of epistemic norms, the second cluster of principles spells out the program-like structure of procedural epistemic norms, and the contents of the restrictions on dynamics that they define.

Claim 10 (D2) Epistemic rationality norms define restrictions on the construction of procedures; they determine how basic epistemic actions are to be sequenced, iterated, combined, chosen among, conditioned on tests, avoided, reversed, and cautiously executed. A further parsing:

(1) Epistemic rationality requires the execution of sequential and iterative procedures under some conditions.
(2) Epistemic rationality requires parallel execution of procedures under some conditions.
(3) Epistemic rationality requires nondeterministic choice among procedures under some conditions.
(4) Sequential and iterative procedures determine sequentially composed paths through epistemic state space.
(5) Parallel execution procedures determine converging and corresponding paths through epistemic state space.
(6) Procedures instructing nondeterministic choice determine branching paths through epistemic state space.
(7) (PrCtoP) Restrictions on procedural construction can be used to generate restrictions on processes in the following sense: there are restrictions on procedural construction, R_{Pr}, such that a corresponding restriction on processes, R_P, matching R_{Pr}'s verdicts can be defined on the basis of R_{Pr}'s verdicts on procedural constructions and the relations between procedural constructions and processes.

2.3.1 (D2 Core) Argument from Cases

The fundamental assumption of D2 is that epistemic processes can be decomposed into actions (or executions of actions) and sequences, choices, or other combinations

$t' \in Irrational(T_S)$. This symmetrical result is a product of treating actions only as action *labels* associated with transitions. If we use finer-grained tools to represent actions (that also suffice to associate action labels with transitions), the symmetry disappears. This is shown in Chap. 8 by means of the Reduction Lemma which states, among other things, that if epistemic actions and processes are identified not with mere labels associated with binary relations but with edges (or concatenated paths formed out of edges) in multi-graphs, then, with respect to an arbitrary set of epistemic states and the restrictions on transitions and actions/processes defined over it, (1) it is not possible to reduce all restrictions on actions/processes to corresponding restrictions on transitions, and (2) it is possible to reduce all restrictions on transitions to corresponding restrictions on actions/processes.

Table 2.2 Procedural operations

Instruction	Procedural operation	Relational structure
Do a, then do b!	Sequential composition "$a; b$"	Relational Composition
Do a some finite n times!	Kleene Iteration "$a*$"	Reflexive, Transitive Closure
Comply with both the instructions for a and b!	Parallel Execution "$a \cap b$"	Intersection
Select either a or b (inclusive)!	Indeterministic Choice "$a \cup b$"	Union
Avoid a, do anything other than a!	Complement "$-a$"	Complement
Reverse the effects of a!	Reversal "$\curvearrowleft a$"	Inverse of Relation
Do a with minimal effects!	Fixed Point "∇a"	Fixed Point

thereof by means of some kind of action operator or, from the other direction, basic actions, including epistemic actions, can be combined, sequenced, or otherwise connected via procedural operations. There is a well-studied set of procedural operations in relational algebra [88, 89]. The procedural operations of relational algebra include sequential composition, iteration, parallel execution, choice, test, complement, reversal, and fixed point.

Perhaps the most familiar procedural operation is sequential composition, which is an operation that orders actions in a linear sequence. In general, there are graph-theoretic and set-theoretic structures associated to each procedural operation. We will see more on this in later chapters. For now, procedural operations can be thought of as instructions. Sequential composition instructs one to carry out an action a and then, come what may, to carry out an action b. Other familiar procedural operations are associated with familiar instructions: do a some finite number of times, do a while doing b, choose among a and b, do a conditional on some condition C holding, avoid a, undo the effects of a, do a extra carefully, etc.. These are associated in Table 2.2.

Procedural operations define ways of constructing epistemic procedures. Procedural epistemic norms, then, are epistemic norms that restrict the ways of constructing epistemic procedures. More precisely, let $\{a_1, \ldots, a_n\}$ be a library of relatively basic epistemic actions. This is not to say that $a_1, \ldots, a_n$ are atomic in any absolute sense; simply let the a_i be basic with respect to more complex procedures defined over this set of actions. Procedural operations form complex procedures out of these basic actions in straightforward ways. The foregoing set of procedural operations, $Op = \{; , *, \cap, \cup, -, \curvearrowleft, \nabla\}$, defines a class of possible procedural constructions: the union of all well-formed[28] sequences of procedural operations and basic actions. Call this set $Proc$, where $Proc \subseteq (Op \cup \{a_1, \ldots, a_n\})*$. Procedural epistemic norms define restrictions on the class of well-formed procedural

[28] Well-formed with respect to some recursive definition of well-formedness, of course. See the syntax of PLEN in Chaps. 4 and 5 for an illustration.

constructions, parsing them into rational and irrational (inter alia) sets and rankings. A restriction on procedural constructions is, accordingly, a restriction on epistemic dynamics defined over the set of procedural constructions: $R_{\preceq}(Proc)$.

That procedural epistemic norms exist can be seen by numerous means. First, one can simply point to both accepted and proposed normative and methodological standards that dictate or require particular procedural operations on epistemic actions, and show that these have at least rudimentary program-like structure. Table 2.3 outlines some procedural operations and some classes of norms that frequently dictate these kinds of procedural operations.

Second, one can recall the arguments for D1C2, which (a) purported to establish both the existence and irreducibility of procedural epistemic norms by deriving them from restrictions on actions and processes and from solutions to planning problems, which are indispensable to epistemic regulation, as well as (b) encompassed a cluster of explicitly spelled out direct arguments from cases.

The fact that epistemic norms restrict ways of constructing procedures directly implies that epistemic norms convey program-like information. Methods or ways of constructing a procedure, as analyzed above, are formed out of the standard procedural operations out of which programs are built—sequential composition, choice, conditioning on tests, etc.. In short, every program is a way of constructing procedures and every way of constructing procedures is a program. This is just to say that every procedural epistemic norm directly conveys program-like information by instructing the sequential, conditional, or structured choice of actions.

To explicate just a handful of these examples, first, consider "argumentation methods", which consist of epistemic procedures that should be at least marginally familiar to any reader of this book. By this, I just mean, generally, the procedures that are explicated in logic and critical thinking texts for analyzing and evaluating arguments. In standard textbook presentations,[29] the following processes typically have chapters or subchapters devoted to them: initial presentation of argument (excising it from its original spoken or written context; denote this with π_p), search for enthymemes (π_e), mapping out putative evidential or support relations (π_m), formalizing the argument in some logical language (e.g., translating the argument into first-order logic; π_f), testing putative support relations using formal and informal methods (e.g., deploying semantic tableau methods to prove the arguments valid or invalid; π_{ft}), and evaluating the truth of premises (π_{pr}). The chapters usually involve an array of algorithms, informal procedures, or rules of thumb for structuring these processes.

The norms that govern the overall process of analyzing and evaluating arguments dictate particular sequences of these subprocesses. For instance, the requirement that one interpret arguments charitably requires that one not rest one's case on the evaluation of the strength of the argument without having completed the search for enthymemes. That is, one can't complete the evaluation of argument well without searching for enthymemes. At the very least, this suggests a ranking of possible

[29] Among many others: [5, 36, 46, 50, 55, 75, 92].

Table 2.3 Epistemic norms requiring regular procedural operation

Procedural operation	Epistemic norm/procedure/method	Field of study	Illustrative reference
Sequential Composition	• Argumentation Methods • Hypothesis Testing Strategies • Proof Construction Algorithms • Model Checking Algorithms • Algorithms (misc.) • Test Enabling and Prevention Procedures • Inductive Strategies	• Informal Logic • Formal Learning Theory (FLT) • Formal Logic • Formal Logic • Formal Methods (field relative) • DEL, Laboratory Protocols • Philosophy of Science (Statistical Methods)	• [5, 91–93] • [47, 49, 77] • [50, 75] • [50, 75] • [13, 35, 37, 44, 69] • [48, 90] • [43, 44]
Iteration	• Argumentation Methods • Hypothesis Testing Strategies • Proof Construction Algorithms • Model Checking Algorithms • Algorithms (misc.) • Iterated Belief Revision • Iterated Credence Update	• Informal Logic • Formal Learning Theory (FLT) • Formal Logic • Formal Logic • Formal Methods (field relative) • Belief Revision • Philosophy of Science (Bayes)	• [5, 91–93] • [47, 49, 77] • [50, 75] • [50, 75] • [13, 35, 37, 44, 69] • [3, 29, 30, 40, 79, 85] • [14, 44, 62]
Parallel Execution	• Every system in which multiple norms must be followed at once.	• Traditional Epistemology, Ethics of Belief	• [1, 2, 10, 15, 78]
Choice	• Argumentation Methods • Logical Constraints • Proof Construction and Model Checking Techniques • Belief Revision	• Informal Logic • Philosophy of Logic • Formal Logic • Belief Revision	• [5, 91–93] • [7–9, 12, 23–25, 74, 76, 82] • [50, 75] • [3, 29, 30, 40, 79, 85]

From left to right, the first column lists the procedural operation required by norm systems or methods listed in the second column that are studied or devised or employed in the fields listed in the third column. The fourth column lists references from the fields in the third column or else references that provide explicit illustrations of the requisite sort from those fields. These are not to be taken to be comprehensive sets of citations; they are merely illustrative.

procedures formed out of the foregoing subprocesses that places the search for enthymemes prior to completion of testing support relations. That is, the principle of charity requires that the process of analysis and evaluation of argument take some procedural form[30] that matches the sequence: $\ldots; \pi_e; \ldots; \pi_{ft}; \ldots$. The program that produces such sequences will be an iteration of complicated conditional choices.[31] Thus, the norm of charity divides these ways of constructing epistemic processes into relatively acceptable and unacceptable subsets. Insofar as charity is a norm of epistemic rationality, the point is made. But perhaps this is too capacious a sense of epistemic norms. More direct examples are available.

Second, the usual model-checking and proof construction procedures [27, 50, 73] involved in the formal testing phase of argument evaluation themselves require fairly straightforward sequential orderings of substeps; indeed, these are formulated in terms of explicit algorithms! For a concrete instance, see the verification procedure for PDL devised in [38] (Chap. 8) in order to establish the decidability of PDL. This is not an unusual or idiosyncratic procedure. For something closer to the concerns of epistemic methodology, note that one must apply the non-branching rules to a semantic tableaux prior to applying the branching rules; else, in systems that countenance infinite trees, one's attempt at proving the validity of an argument will be vulnerable to erroneous failure to close the tree [27].

Semantic verification algorithms are a commonplace, as are proof construction algorithms. There wouldn't be automated theorem proving software if not for the program-likeness of these epistemic procedures. Insofar as accurate verdicts on the validity of arguments shape requirements of rationality—either because accuracy is itself a norm of epistemic rationality or because epistemic rationality conditions rationality of belief and inference on logical validity—epistemic rationality divides ways of ordering steps of proof construction into relatively acceptable and unacceptable subsets. That is, well-known methods for applying logic in rational belief revision defines program-like restrictions on procedural construction.

2.3.2 (D2 Core) Arguments from Epistemic Regulation and Planning

D2 has independent support from consideration of ante hoc regulation and epistemic planning in light of the examples of norms that restrict procedural construction. I reiterate and expand on some of the arguments for D1C2.4 for emphasis.

---

[30] Really, an algorithm for doing this would probably involve looping from formalization and testing procedures $\pi_f; \pi_{ft}$ back to π_e; a good test for whether or not one has charitably interpreted an argument is how obvious it seems that the argument is invalid. If it's trivially invalid, and your interlocutor is smart, you're probably (but not necessarily) being uncharitable!

[31] See D2C1 for more on this sort of procedural construction.

First, note that the consideration of argumentation methods and formal proof construction techniques in the prior section make one important presupposition: the results of an epistemic process can depend on the ordering of epistemic actions and other procedural features of the construction of the process. Thus, the norms that restrict these outcomes provide more than a merely incidental restriction on the ordering (or other procedural features) of epistemic actions.

This is a familiar fact of epistemic life that is well-recognized in iterated belief revision and DEL. It's well-known that some ways of iterating belief revision functions are utterly untenable[32] in virtue of their sequencing. There are also dynamic paradoxes of belief revision, as Harman [41, 42] and others [40, 79] point out. For instance, given a limited set of available actions, it may be impossible to revise one's belief states out of an inconsistent state once one ends up at one (given common belief revision constraints).

One of the sets of driving observations for dynamic epistemic logic (DEL) is that some actions can obscure or reveal information, and their procedural placement is thus of significant epistemic relevance. One needn't consider complex subterfuge to see this, as the revelation of information is readily apparent in the basic public announcement logic (PAL) example of the muddy children and many other standard DEL cases (e.g., hidden cards). Though those cases were multi-agent cases, one can easily see the observation in single agent cases, as in The Researcher case.

The two phenomena of import, here, are illumination and obfuscation. Performing some actions, for instance, makes information available (illumination) or unavailable (obfuscation) for later processes of inference. Research, as a general class of actions, is typically aimed at illumination. In the toy example of The Researcher, we see how formalization and experimentation produce information that is necessary for enabling later processes of inference and theory selection. The formalization step made an execution of an oddly calibrated theory selection procedure feasible, as a procedure that prioritizes formal elegance can't be carried out without the proper information about formal elegance. The experimentation made a more reasonably calibrated theory selection procedure feasible, as a procedure that hinges on empirical evidence can't be carried out without that evidence.

Obfuscation is ubiquitous. One can routinely perform actions that prevent later epistemic actions from being feasible. If the researcher, for instance, had insulted the Tarot reader or destroyed the laboratory equipment, the relevant processes would have become infeasible. These are epistemic actions precisely because of their effects on the feasibility of later epistemic processes. Scientific practice reveals the ubiquity of this sort of thing. In order to identify a substance, one might need to destroy it by an identification process (as is common in forensic sciences). This prevents other such processes from being carried out. Processes of concept acquisition and revision may illuminate and so make new epistemic processes possible, and they may also obfuscate, ruling out other processes [86]. The explicit ordering of epistemic actions can be used as a form of obfuscation,

[32] Here, I have in mind Tennant's Degeneracy results from [85].

and this is sometimes normative. Consider de-biasing strategies. There are cases in which, in order to suppress a heuristic process from triggering, one simply has to prevent oneself from accessing some information (as in blind review) or one has to sequence the ordering of tasks to prevent other processes from being accidentally or intentionally executed (as in the discrete separating of data gathering from data analysis processes in order to avoid p-hacking [43]).

In sum, the foregoing table of norms and methodological rules in Table 2.3 and the subsequent discussion shows, if I'm right, that the procedural construction of epistemic processes is not a matter of indifference. Insofar as epistemic rationality norms divide outcomes into rational and irrational subsets (et al), the general observation that procedural structure regularly determines epistemic outcomes implies that epistemic rationality divides procedural constructions into rational and irrational subsets. Let me belabor this point.

First, some of the norms articulated explicitly divide processes into rational and irrational sets on the basis of the procedural operations out of which they are built. Thus, from the definition of restrictions on epistemic dynamics, there are restrictions on procedural construction.[33] Second, even if the static outcomes of processes are the only features of normative relevance, the static outcomes often depend on the procedural structure of processes. Thus, in such cases of dependence, the existence of restrictions on static outcomes of processes implies the existence of restrictions on ways of constructing processes out of actions. More, processes constructed in certain ways may be related to normatively relevant outcomes in certain modal ways. As in the cases above, particular ways of (e.g.) procedurally operating on actions may be necessary for producing (or avoiding) an outcome, or they may be reliable for doing so, even if there are other feasible processes that also generate that outcome. Norms frequently dictate the execution of the reliable processes over unreliable processes that converge, in some cases, on outcomes with the reliable process. Thus, even in the absence of norms that directly specify restrictions on procedural construction, the norms that specify restrictions on static outcomes indirectly specify restrictions on procedural construction, and the structural restriction is not incidental. The restriction on procedural construction is based on a dependence relation between the procedural structure and the epistemic outcomes of the procedure.

From these observations, we can derive two direct arguments for the core of D2. First, restrictions on procedural construction are necessary for ante hoc regulation. From the foregoing considerations, it follows that, in cases where epistemic outcomes depend on procedures being constructed in certain ways (as in inductive rules, proof and model construction procedures, numerous other formal methods involving algorithms for the construction and manipulation of models, and de-biasing strategies), norms requiring those epistemic outcomes derivatively require those procedural constructions. For instance, whether a semantic tableaux procedure gives a correct answer about the validity of an argument depends on how

[33] The next section spells this out.

one orders one's choice of rules to apply. The requirement that one get right answers entails a requirement that one use procedures constructed with certain sequential compositions of rule applications and not others. This point applies to nearly all of the examples in the table of procedural requirements above.

Second, connecting the foregoing observations about procedural structure to epistemic planning further strengthens the inference to the D2.

(1) Ante hoc regulation requires solving epistemic planning problems.
(2) Solutions to planning problems are restrictions on (epistemic) procedural construction.
(3) ∴ Ante hoc regulation requires restrictions on (epistemic) procedural construction.

Ante hoc regulation is the process of solving epistemic planning problems. The solution to an epistemic planning problem is a set of epistemic plans that each suffice to achieve the goal state determined by some set of epistemic norms. A set of epistemic plans that achieve a goal state just is a restriction on (epistemic) procedural construction because epistemic plans are themselves sets of instructions built out of procedural operations of conditional choice, sequencing, etc. over a set of elements of the epistemic action library. Thus, insofar as ante hoc regulation requires solutions to epistemic planning problems, ante hoc regulation requires that there be restrictions on procedural construction. These restrictions will be specified by the norms that define the goal state. Typically, only a subset of possible plans constructible from the action library will reach the goal state. As noted above, this is often due to the way the plans are constructed, the ordering of the subtasks in the plan, the proper conditionalizing of choices in the plan, and so on. So, the goal state, in concert with the features of the constructible plans, divides the set of procedures into those that are effective for reaching the goal state and those not. Insofar as the goal state is defined by the requirements of some set of norms, those norms indirectly restrict procedural constructions.

2.3.3 (D2.1)–(D2.3) Arguments from Cases

See Table 2.3 for classes of examples of norms requiring procedural constructions of each kind. See also the examples in prior sections and the attendant discussion. See also the consideration of the procedural operations in Table 2.2 and their correspondence to operations on structures (in this case, binary relations) representing actions. These cases, coupled with the general considerations about procedural structure, analyzed by means of the program constructions enumerated in D2.1–D2.3, suffice for the program-likeness of procedural epistemic norms, generally, which suffices to imply that there are norms requiring each kind of program construction.

2.3.4 (D2.4)–(D2.6) Argument from Correlation of Procedural Operations with Path Structures

This claim follows immediately from the correlation of procedural operations with operations on relations, as shown in Table 2.2. Sequential composition of actions, for instance, corresponds to composition of the binary relations representing actions. Now, given that epistemic norms plausibly restrict the space of procedural structures, D2.4–D2.6 immediately follow.

Of course, the philosophical importance or plausibility of this feature of epistemic norms is tempered by how plausible binary relations are as models of actions or processes. The prior arguments about the importance of procedural norms implies that binary relations are not particularly plausible as models of actions or processes. Nonetheless, operations on more plausible structures can just as well be associated with the procedural operations discussed above (i.e., paths), and this correspondence will directly imply D2.4–D2.6.

2.3.5 (D2.7) Construction Rule

This follows from D2.5–D2.6. Let N_{Prc} be a norm-kernel that restricts procedural constructions. Let N_P be a norm-kernel that restricts processes such that: a procedural specification, $(\circ_1, \ldots, \circ_n)$, on a set of actions, $(a_1, \ldots, a_m)$, is rational (etc.), and $(t_1, \ldots, t_m)$ is a set of transitions that correspond 1–1 to each a_i iff the path formed by taking the operations, $(\circ_1, \ldots, \circ_n)$, over the transitions, $\langle t_1 \circ_1 \ldots \circ_n t_m \rangle$, is rational (etc.). This rule immediately generates a corresponding restriction on processes. Indeed, this rule is a norm-kernel that restricts processes, and one that does so on the basis of the procedural structure of the process.

2.3.6 Epistemic Norms as Programs or Plans

Claim 11 (D2C1) Procedural norms of epistemic rationality have the internal structure of programs or plans, conditioning actions and procedures on tests, and dictating procedures built with complex operations on action like conditional choice and guarded iteration, which are composed out of more basic procedural operations.

(1) Epistemic rationality norms dictate or instruct the execution of epistemic actions and procedures.
(2) Epistemic rationality norms specify the conditions under which epistemic actions and procedures are to be executed.
(3) Epistemic rationality norms dictate the construction of complex epistemic procedures formed with relatively basic epistemic actions (and procedures) and

the procedural operations of relational algebra, including (inter alia) sequential composition, nondeterministic choice, tests.

(4) The foregoing procedural operations can be combined to form more complex operations like conditional choice and guarded iteration.

2.3.7 (D2C1.1–D2C1.4) Argument from Cases

The program-likeness of the information conveyed by epistemic norms falls out of observations about the generality of the procedural operations introduced in Table 2.2. First, one can simply point to both accepted and proposed normative and methodological standards that dictate or require particular procedural operations on epistemic actions, and show that these have at least rudimentary program-like structure.

Programs are fundamentally constructed out of procedural operations or with decision structures. The most basic program instructs the sequential execution of a set of actions. That is, the most basic program is a linear sequence of steps. More complex programs are built out of more complex instructions that enable a program to unfold via alternative sequences of steps. Generally, boolean constructions enable the decision structure of programs to branch into complex trees. Conditional constructions formed from test, sequential composition, and choice, in particular, allow one to skip and select out of the original queue of actions. Choice structures allow a branching of sequences of program steps. This set of procedural operations defines a space of possible procedural constructions that contains the central program structures.

Some useful (indeed, ubiquitous) and slightly more complex basic program constructions are conditional choice (or guarded choice) and guarded iteration:

Conditional Choice If A, then do a; (or) if $\neg A$, then do b. (Equivalently (given bivalence): If A, do a; else, do b.)

Guarded Iteration While A, do a; else, don't do a.

Examples of epistemic norms that have these kinds of structures are fairly numerous. Consider the defeasible closure rule:

(Defeasible Closure) If A implies B, and there are no defeaters for B, then see to it that you don't believe A without believing B!

This rule dictates a conditional choice structure that can be encoded by the "pseudocode" program in Fig. 2.1, which highlights a "program" for complying with Defeasible Closure.[34]

[34] The displayed pseudocode is just one way—and probably not the best way—to implement the requirements of this defeasible closure rule. Two obvious improvements: adding conditional choice clauses for addressing the relative strength of A against the strength of the defeaters for B as well

```
If: A implies B;
    Then:
       If: A is believed;
          Then:
             If: B is not believed;
               Then:
                 If: There are no defeaters for B that are believed;
                   Then: Believe B;
                   Else: Revise away belief in A.
             If: B is believed;
               Then:
                 If: There are no defeaters for B that are believed;
                   Then: Halt;
                   Else: Revise away belief in B; Revise away belief in A.
          Else: Halt.
    Else: Halt.
```

Fig. 2.1 Pseudocode for defeasible closure

As is shown in Chap. 4, this program can be encoded with the procedural operations, including only sequential composition, choice, and test. The foregoing example additionally shows how the dictation of epistemic action under specified conditions even creates complex procedural structure; conditioning the execution of some actions on one condition and other actions on distinct conditions creates a conditional choice.

It's clear that the defeasible closure program above is a conditional choice containing nested conditional choices. Of course, the conditional form of norm-kernels suggests that epistemic norms convey exactly these kinds of instructions, generally. Consider the following norm:

(Defeasible Consistency 2) If a belief set, s, contains A and a non-empty subset, $\{B_1, \ldots, B_n\}$, such that each B_i is inconsistent with A, and, for each B_i, s contains no evidence for B_i comparable in strength to the evidence for A, then s is irrational.

Defeasible Consistency 2 instructs a rational agent to reinstate consistency or revise one's evidence if the conditions of application hold. To put it in more program-like terms, Defeasible Consistency 2 gives the following instruction: If s contains A and $B_1, \ldots, B_n$ such that each B_i is inconsistent with A, and the evidence favors A over each B_i, then, if evidence can be found for some B_i that favors it, then reject A; else, reject each B_i.

Examples of translation of norms into program-like constructions can be carried out at will. Table 2.3 provides a table outlining some procedural operations and some classes of norms that frequently dictate these kinds of procedural operations.

as clauses that address the possibility of defeaters for A. The optimality of this bit of pseudocode is, of course, a garden path with respect to the argument.

In general, any norm that can be used to solve planning problems or to answer the questions in the ante hoc regulation table will specify norms with program-like structure. Thus, procedural epistemic norms will, in general, have program-like structures satisfying D2C1.1–D2C1.4; all program constructions dictate the execution of actions, conditional choices condition actions on detectable features of the world, the foregoing constructions are built from more basic actions and operations, and there is no upper limit on the complexity of these constructions (other than the heat death of the universe).

As for the decomposition of conditional choice and guarded iteration into more basic procedural operations, this much is well-known. See the following table for a translation into the relational procedural operations of PDL.

Name	English w/ metavariables	PDL
Conditional Choice	If A, do π_1; else, do π_2!	$((A)?;\pi_1) \cup ((-(A)?;\pi_2)$
Generalization of Conditional Choice	If A, do π_1; if B, do π_2; ...!	$((A)?;\pi_1) \cup ((B)?;\pi_2) \cup \ldots$
Guarded Iteration	While A, do π	$((A)?;\pi)*;(\neg A)?$
Guarded Repetition	Repeat π until A is true.	$(\pi;(\neg A)?)*;(A)?$

2.4 Toward Further Defense

The regulative function arguments we've considered thus far show, if correct, that the fundamental use of epistemic rationality norms in epistemic regulation requires or entails the existence of procedural epistemic rationality norms partially in virtue of the resolution of regulation into processes of epistemic planning and control. Examination of the prospects for reduction of procedural epistemic norms to static and dynamic epistemic norms, in turn, shows that procedural epistemic rationality norms are irreducible to these others kinds of epistemic norms, and so procedural epistemic norms are indispensable for any would-be theory of epistemic rationality norms, let alone a proper logic of epistemic norms. A key step in the analysis of these indispensable procedural norms shows that they are program-like in structure; they operate on the procedural structures of complex action.

The next chapter completes the defense of epistemic proceduralism by examining the forms of reasoning with and about epistemic rationality norms that are central to their first-order use in ante hoc regulation as well as their second-order role in epistemology and philosophical logic. Together, clusters D1–D3 fully spell out T2 and, if reasonably well established, suffice to establish T1 (reasonably well). The next chapter should, if all goes right, thereby complete the articulation and defense of two of the three core theses of the protocol-theoretic account of epistemic rationality norms, leaving the main project of the book: proving T3.

References

1. Adler, J. 2002. *Belief's own ethics*. MIT Press.
2. Adler, J.E., and L.J. Rips (eds.). 2008. *Reasoning: Studies of human inference and its foundations*. Cambridge University Press.
3. Alchourron, C., P. Gardenfors, and D. Makinson. 1985. On the logic of theory change: Partial meet contraction and revision functions. *The Journal of Symbolic Logic* 50(2):510–530.
4. Anderson, A.R., and N. Belnap. 1975. *Entailment: The logic of relevance and necessity*, Vol. I. Princeton: Princeton University Press.
5. Angell, R.B. 1964. *Reasoning and logic*. New York: Appleton-Century-Crofts.
6. Aucher, G. 2012. DEL-sequents for regression and epistemic planning. *Journal of Applied Non-Classical Logics* 22(4):337–367.
7. Beall, J. 2011. Multiple-conclusion LP and default classicality. *Review of Symbolic Logic* 4(2):326–336.
8. Beall, J. 2015. Free of detachment: Logic, rationality, and gluts. *Noûs* 49(2):410–423.
9. Beall, J.c., and G. Restall. 2005. *Logical pluralism*. Oxford University Press.
10. Boghossian, P. 2008. Epistemic rules. *Journal of Philosophy* 105(9):472–500.
11. Bolander, T., and M.B. Andersen. 2011. Epistemic planning for single- and multi-agent systems. *Journal of Applied Non-Classical Logics* 21(1):9–34.
12. Brown, B., and G. Priest. 2004. Chunk and permeate, a paraconsistent inference strategy. Part I: The infinitesimal calculus. *Journal of Philosophical Logic* 33(4):379–388.
13. Carlson, E. 2010. Formal methods in ethics. In *The routledge companion to ethics*, ed. J. Skorupski. Routledge.
14. Carnap, R. 1951. *The continuum of inductive methods*. University of Chicago.
15. Chisolm, R.M. 1981. Epistemic principles. *Dialectica*. 35(3):341–345.
16. Christensen, D. 2004. *Putting logic in its place*. Oxford: Oxford University Press.
17. Conte, R., G. Andrighetto, and M. Campennl (eds.). 2013. *Minding norms: Mechanisms and dynamics of social order in agent societies*. Oxford University Press.
18. Ellis, B. 1979. *Rational belief systems*. London: Blackwell.
19. Evans, J.St.B.T., S. Newstead, J. Allen, and P. Pollard. 1994. Debiasing by instruction: The case of belief bias. *European Journal of Cognitive Psychology* 6:263–285.
20. Evans, J.St.B.T., and D.E. Over. 1996. *Rationality and reasoning*. Hove, England: Psychology Press.
21. Evans, J., and K.E. Stanovich. 2013. Dual-process theories of higher cognition advancing the debate. *Perspectives on Psychological Science* 8(3):223–241.
22. Field, H. 2001. *Truth and the absence of fact*. Oxford: Oxford University.
23. Field, H. 2009a. Epistemology without metaphysics. *Philosophical Studies* 143(2):249–290.
24. Field, H. 2009b. Pluralism in logic. *Review of Symbolic Logic* 2(2):342–359.
25. Field, H. 2009c. What is the normative role of logic? *Aristotelian Society Supplementary Volume* 83(1):251–268.
26. Field, H. 2015. What is logical validity? In *Foundations of logical consequence*, ed. C. Caret, and O. Hjortland. Oxford: Oxford University Press.
27. Fitting, M., R. Mendelsohn, and R.A. Girle. 2002. First-order modal logic. *Bulletin of Symbolic Logic* 8(3):429–430.
28. Foley, R. 1993. *Working without a net*. Oxford University Press.
29. Gardenfors, P. 2008. *Knowledge in flux*. Reprint. Original: 1988. London: College Publications.
30. Gardenfors, P. 1992. *Belief revision*. Cambridge: Cambridge University Press.
31. Georgeff, M. (ed.). 1987. *Reasoning about actions and plans*. Kaufmann.
32. Gigerenzer, G. 2007. *Gut feelings: The intelligence of the unconscious*. London, England: Penguin.
33. Glymour, C. 1980. *Theory and evidence*. Princeton: Princeton University Press.
34. Glymour, C. 1986. *Epistemology and cognition*. Harvard University Press.
35. Gupta, A. 1992. Formal hardware verification methods: A survey. *Formal Methods in System Design* 1:151–238. Springer.

36. Guttenplan, S.D. 1997. *The languages of logic: An introduction to formal logic*. Blackwell.
37. Hansson, S.O. 2000. Formalization in philosophy. *Bulletin of Symbolic Logic* 6(2):162–175.
38. Harel, D., D. Kozen, and J. Tiuryn. 2000. *Dynamic logic*. MIT Press.
39. Haddawy, P.F. 1994. *Representing plans under uncertainty: A logic of time, chance, and action*. Springer-Verlag.
40. Hansson, S.O. 2017. Logic of belief revision. *The Stanford Encyclopedia of Philosophy* (Winter 2017 Edition), ed. E.N. Zalta, forthcoming. https://plato.stanford.edu/archives/win2017/entries/logic-belief-revision/.
41. Harman, G. 1983. Logic and probability theory versus canons of rationality. *Behavioral and Brain Sciences* 6(2):251.
42. Harman, G. 1986. *Change in view*. Cambridge: MIT Press.
43. Head, M.L., L. Holman, R. Lanfear, A.T. Kahn, and M.D. Jennions. 2015. The extent and consequences of P-hacking in science. *PLoS Biology* 13(3):e1002106.
44. Horsten, L., and I. Douven. 2008. Formal methods in the philosophy of science. *Studia Logica* 89(2):151–162.
45. Kahneman, D., P. Slovic, and A. Tversky. (eds.). 1982. *Judgment under uncertainty: Heuristics and biases*. Cambridge University Press.
46. Kalish, D., R. Montague, and G. Mar. 1964. *Logic: Techniques of formal reasoning*. Oxford University Press USA.
47. Kelly, K.T. 1988. Formal learning theory and the philosophy of science. *PSA: Proceedings of the Biennial Meeting of the Philosophy of Science Association* 1988:413–423.
48. Kelly, M.A., and P.L. Haddix. 2015. *The fundamentals of scientific research*. John Wiley & Sons, Incorporated.
49. Kelly, K.T., O. Schulte, and V. Hendricks. 1995. Reliable belief revision. *Logic and scientific methods*: *Volume One of the Tenth International Congress of Logic, Methodology and Philosophy of Science, Florence*, 383–398.
50. Kleene, S.C. 1967. *Mathematical logic*. Dover Publications.
51. Korcz, K. 1997. Recent work on the basing relation. *American Philosophical Quarterly* 34(2):171–191.
52. Korcz, K.A. 2015. The epistemic basing relation. *The Stanford Encyclopedia of Philosophy* (Fall 2015 Edition), ed. E.N. Zalta. https://plato.stanford.edu/archives/fall2015/entries/basing-epistemic/.
53. Kruglanski, A.W., and G. Gigerenzer. 2011. Intuitive and deliberative judgements are based on common principles. *Psychological Review* 118:97–109.
54. Lam, B. Unpublished. The Dynamic Foundations of Epistemic Rationality. Ph.D Dissertation: Vassar College.
55. Lepore, E. 2000. *Meaning and argument: An introduction to logic through language*. Wiley-Blackwell.
56. Levi, I. 1997. *The covenant of reason: Rationality and the commitments of thought*. Cambridge University Press.
57. MacFarlane, J. 2004. In What Sense (If Any) Is Logic Normative for Thought? Unpublished. (Presented at the 2004 Central Division APA symposium on the normativity of logic.) Accessed online: http://johnmacfarlane.net/normativity_of_logic.pdf.
58. MacFarlane, J. 2017. Is Logic a Normative Discipline? Unpublished. (Conference on the Normativity of Logic University of Bergen) Accessed online: https://johnmacfarlane.net/normative.pdf
59. Makinson, D.C. 1965. The paradox of the preface. *Analysis* 25(6):205–207.
60. Moss, S. 2015. Time-slice epistemology and action under indeterminacy. In *Oxford studies in epistemology*, 172–194, ed. T.S. Gendler, and J. Hawthorne. Oxford University Press.
61. Moutier, S., N. Angeard, and O. Houde. 2002. Deductive reasoning and matching-bias inhibition training: Evidence from a debiasing paradigm. *Thinking and Reasoning* 8:205–224.
62. Oaksford, M., and N. Chater. 2007. *Bayesian rationality: The probabilistic approach to human reasoning*. Oxford University Press.
63. Pettigrew, R. 2011. An improper introduction to epistemic utility theory. In *Proceedings of EPSA: Amsterdam '09*, 287–301, ed. H. de Regt, S. Okasha & S. Hartmann. Springer.

64. Pettigrew, R. 2013a. Epistemic utility and norms for credences. *Philosophy Compass* 8(10):897–908.
65. Pettigrew, R. 2013b. A new epistemic utility argument for the principal principle. *Episteme* 10(1):19–35.
66. Pettigrew, R. 2016. Epistemic utility arguments for probabilism. In *The Stanford Encyclopedia of Philosophy* (Spring 2016 Edition), ed. E.N. Zalta. https://plato.stanford.edu/archives/spr2016/entries/epistemic-utility/.
67. Pettigrew, R. 2017. Epistemic utility and the normativity of logic. *Logos and Episteme* 8(4):455–492.
68. Pollock, J.L. 1987. Epistemic norms. *Synthese* 71(1):61–95.
69. Pollock, J.L. 1990. *Technical methods in philosophy*. Westview Press.
70. Pollock, J.L. 1995. *Cognitive carpentry*. MIT Press.
71. Pollock, J.L. 2009. A recursive semantics for defeasible reasoning. In *Argumentation in Artificial Intelligence*, 173–197, ed. G. Simari, and I. Rahwan. Boston, MA: Springer.
72. Pollock, J., and J. Cruz. 1999. *Contemporary theories of knowledge*, 2nd Edn. Rowman & Littlefield.
73. Priest, G. 2008. *If to is: An introduction to non-classical logics*. Cambridge: Cambridge University.
74. Priest, G. 2016. Logical disputes and the A priori. *Logique et Analyse* 236:347–366.
75. Quine, W.V. 1950. *Methods of logic*. Harvard University Press.
76. Restall, G. 2005. Multiple conclusions. *Logic, Methodology and Philosophy of Science: Proceedings of the Twelfth International Congress*.
77. Schulte, O. 2017. Formal learning theory. In *The Stanford Encyclopedia of Philosophy* (Spring 2017 Edition), ed. E.N. Zalta. https://plato.stanford.edu/archives/spr2017/entries/learning-formal/.
78. Stalnaker, R. 2002. Epistemic consequentialism: Robert Stalnaker. *Aristotelian Society Supplementary* 76(1):153–168.
79. Stalnaker, R. 2009. Iterated belief revision. *Erkenntnis* 70(2):189–209.
80. Stanovich, K.E. 2005. *The Robot's rebellion: Finding meaning in the age of Darwin*. University of Chicago Press.
81. Stanovich, K.E. 2008. Individual differences in reasoning and the algorithmic/intentional level distinction in cognitive science. In *Reasoning: Studies of human inference and its foundations*, 414–436, ed. L. Rips, and J. Adler. Cambridge: Cambridge University Press.
82. Steinberger, F. 2017. Consequence and normative guidance. *Philosophy and Phenomenological Research* 95(1).
83. Steinberger, F. 2019. Three ways in which logic might be normative. *Journal of Philosophy* 116(1):5–31.
84. Stenning, K., and M. van Lambalgen. 2008. *Human reasoning and cognitive science*. Boston, USA: MIT Press.
85. Tennant, N. 2006. On the degeneracy of the full AGM-theory of theory-revision. *Journal of Symbolic Logic* 71(2):661–676.
86. Thagard, P.R. 1990. Concepts and conceptual change. *Synthese* 82(2):255–74.
87. Turri, J. 2011. Believing for a reason. *Erkenntnis* 74(3):383–397.
88. van Benthem, J. 1990. General dynamics. *ITLI prepublication series for logic, semantics, and philosophy of language*. University of Amsterdam.
89. van Benthem, J. 1996. *Exploring logical dynamics*. Stanford: CSLI Publications.
90. van Ditmarsch, H., J. Halpern, W. Van Der Hoek, and B. Kooi. 2007. *Dynamic epistemic logic*. Springer.
91. Walton, D. 1992. Rules for plausible reasoning. *Informal Logic* 14(1):33–51.
92. Walton, D. 2013. *Methods of argumentation*. Cambridge University Press.
93. Walton, D., C. Tindale, and T.Gordon. 2014. Applying recent argumentation methods to some ancient examples of plausible reasoning. *Argumentation* 28(1):85–119.
94. Wimsatt, W. 2007. *Re-engineering philosophy for limited beings: Piecewise approximations to reality*. Harvard University Press.

Chapter 3
Epistemic Proceduralism Defended II

Abstract Epistemic proceduralism is spelled out by clusters of principles characterizing the structure, function, and logic of procedural epistemic norms, norms that define restrictions on epistemic actions, processes, and procedures. With the content of epistemic norms regimented by means of norm-kernels and restrictions on epistemic dynamics, this chapter completes the defense of epistemic proceduralism. It articulates and defends the cluster, D3, along with its subclauses and corollaries, D3C1, D3C2, and D3C3. Roughly, this cluster of principles identifies the core forms of reasoning with and about epistemic rationality norms required by their first-order use in deliberate, ante hoc regulation (i.e., epistemic planning and control) and by their second-order use as objects of epistemological and logical analysis. Reasoning about norm equivalence, dependence, and independence relations as well as interactions between these logical properties and their normative verdicts prove to be of central importance, as does reasoning about their connections to deontic operators and the logical decomposition of ante hoc regulation. In the context of the book, the clusters of principles articulating these results stand as theoretical criteria or tests applicable to formal frameworks for representing and reasoning about epistemic norms. The chapter closes with two exercises helpful for framing the rest of the book. The first is a pair of brief arguments for the claim that the clusters of principles, D1–D3C3, are criteria on formal frameworks for the logic of epistemic norms. The second is an application of D1-D3C3 to other formal frameworks, which serves to highlight the main ideas of PLEN by contrast.

3.1 The Logic of Epistemic Rationality Norms

The prior clusters of principles articulated features of the structure and regulative function of procedural epistemic norms. The final cluster articulates principles or valid forms of reasoning with and about epistemic norms that are paramount in both their regulative adoption and application as well as the study of epistemic norms. The thought is that these should be the targets of codification in a logic of epistemic norms.

R. Jenkins, *A Protocol-theoretic Framework for the Logic of Epistemic Norms*, Synthese Library 467, https://doi.org/10.1007/978-3-031-08597-0_3

3.1.1 Reasoning with and About Epistemic Norms

Claim 12 (D3) Epistemic rationality norms can be reasoned about explicitly; terms denoting epistemic rationality norms and statements about the properties of epistemic norms can be embedded in valid arguments. Parsing this claim:

(1) Epistemic rationality norms can be equivalent in multiple, distinct ways; reasoning about epistemic rationality must at least sometimes involve reasoning about these equivalence relations.
(2) The essential forms of reasoning about and with epistemic rationality norms include:
    (a) Arguments establishing norm equivalence and deriving facts about norms from premises about norm equivalence.
        (i) Arguments from the features of norms, norm-kernels, or restrictions to norm equivalence.
        (ii) Arguments from norm equivalence to propositions about norms, norm-kernels, or restrictions on epistemic dynamics.
        (iii) Arguments replacing embedded norms with equivalents in normative, procedural, deontic, modal, and propositional contexts.
    (b) Arguments deriving normative verdicts (valences, deontic statuses, procedural requirements) from conditions of application, procedural structure, sentential or propositional contexts (including modal and deontic propositions).
    (c) Arguments establishing dependence, sufficiency, and independence relations among norms.

Claim 13 (D3C1) Where N_i is a term denoting an arbitrary norm, A_i is a formula embedding a norm term (letting $A(N_i, x)$ be a special case encoding N_i's verdict on an arbitrary epistemic factor, x), C_i is an arbitrary formula stating some condition, and $\equiv_x$ is an arbitrary form of normative equivalence, the D3C1 Table summarizes and explicates schematic inference rules for the foregoing kinds of reasoning (Table 3.1).

Remark 14 In stating D3 and D3C1, I implicitly draw the distinction between *reasoning with* an epistemic norm and *reasoning about* an epistemic norm. By the former, I mean using an epistemic norm to guide one's reasoning, as in ante hoc regulation of belief formation. By the latter, I mean reasoning about epistemic norms as the objects of one's thought (perhaps while reasoning with the very norms one is reasoning about), as in thinking about whether deductive closure conditions should be maintained in preface cases [14] or whether conditionalization can be derived from other, more fundamental veritistic norms [50].

Table 3.1 D3C1 table

Basic forms	Equivalence	(In)Dependence, sufficiency
$\begin{array}{c} A_1(N_1) \\ C \\ \hline A_2(N_2) \end{array}$	$\begin{array}{c} N_1 \equiv_x N_2 \\ A_1(N_1) \\ \hline A_2(N_2) \end{array}$	$\begin{array}{c} N_1 \supset_x N_2 \\ A_1(N_1) \\ \hline A_2(N_2) \end{array}$
$\begin{array}{c} A_1(N_1) \\ C_1 \\ \hline C_2 \end{array}$	$\begin{array}{c} A_1(N_1) \\ A_2(N_2) \\ \dots \\ \hline N_1 \equiv_x N_2 \end{array}$	$\begin{array}{c} \neg(N_1 \supset_x N_2) \wedge \neg(N_2 \supset_x N_1) \\ A_1(N_1) \\ \hline A_2(N_2) \end{array}$
$\begin{array}{c} C_1 \\ C_2 \\ \dots \\ \hline A(N_i) \end{array}, \begin{array}{c} C_1 \\ C_2 \\ \dots \\ \hline A(N_i, x) \end{array}$	$\begin{array}{c} N_1 \equiv_x N_2 \\ \hline N_1 \equiv_y N_2 \end{array}$	$\begin{array}{c} A_1(N_1) \\ A_2(N_2) \\ \dots \\ \hline N_1 \supset_x N_2 \end{array}$
$\begin{array}{c} A(N_1, x) \\ C \\ \hline A(N_2, y) \end{array}, \begin{array}{c} A(N_1, x) \\ C \\ \hline A(N_2, x) \end{array}$		$\begin{array}{c} A_1(N_1) \\ A_2(N_2) \\ \dots \\ \hline \neg(N_1 \supset_x N_2) \wedge \neg(N_2 \supset_x N_1) \end{array}$

Interpreting the D3C1 Table: Each rule is a schematic argument form encoding the logical properties of sentential contexts embedding norm terms that encode the properties of norms

(1) Column 1: General Logical Properties

(a) Row 1: Argument form encoding how arbitrary sentential contexts for norms combine with arbitrary conditions to imply sentential contexts for norms

(b) Row 2: Argument form encoding how arbitrary sentential contexts for norms combine with arbitrary conditions to imply other arbitrary conditions (not embedding norms)

(c) Row 3: Argument form encoding how arbitrary conditions may be sufficient conditions for a given sentential context for norms, with the special case of the sufficient conditions for sentential contexts encoding normative verdicts

(d) Row 4: Argument form encoding how the sentential contexts for norms encoding normative verdicts combine with arbitrary conditions to imply other sentential contexts for norms encoding normative verdicts

(2) Column 2: Logic of Norm Equivalence

(a) Row 1: Argument form encoding how sentential contexts for norms that encode different types of norm equivalence combine with arbitrary sentential contexts for norms to imply other sentential contexts for norms

(b) Row 2: Argument form encoding how arbitrary sentential contexts for norms encode sufficient conditions for protocol equivalences of different types

(c) Row 3: Argument form encoding logical relations among sentential contexts for norms that encode different types of norm equivalence

(3) Column 3: Logic of Norm Dependence, Sufficiency, and Independence

(a) Row 1: Argument form encoding how sentential contexts for norms that encode different types of norm implication/sufficiency/dependence combine with arbitrary sentential contexts for norms to imply other sentential contexts for norms

(b) Row 2: Argument form encoding how sentential contexts for norms that encode different types of norm implication failure/independence/insufficiency combine with arbitrary sentential contexts for norms to imply other sentential contexts for norms

(c) Row 3: Argument form encoding how arbitrary sentential contexts for norms may be sufficient for different types of norm implication/sufficiency/dependence

(d) Row 4: Argument form encoding how arbitrary sentential contexts for norms may be sufficient for different types of norm implication failure/insufficiency/independence

3.1.1.1 (D3.1) Direct Argument

Let N, N', and N'' be epistemic rationality norm-kernels. One obvious form of equivalence between N and N' is identity of verdicts about objects in the domain of both N and N' given the same conditions of application. This means that, for any state s, transition t, or process p in the domains of both N and N', s (resp. t or p) is assigned valence R by N under C iff s is assigned valence R by N' under C. Trivially, the relation of identity of verdicts with respect to conditions of application is reflexive, symmetric, and transitive;[1] if N delivers the same verdicts as N' and N' delivers the same verdicts as N'', N delivers the same verdicts as N'' wherever $Dom(N) = Dom(N')$ and $Dom(N') = Dom(N'')$ or, alternately, transitivity holds for $Dom(N) \cap Dom(N') \cap Dom(N'')$. All of the relations defined in this section are reflexive, symmetric, and transitive. I omit the arguments establishing this because it should be obvious in each case.

Now, the foregoing notion of equivalence doesn't imply that N and N' share the same domains or the same conditions of application. It means only that, where there is overlap in domains and conditions of application, N and N' coincide in verdicts. That is, there may be plenty of conditions under which N and N' part ways with respect to object x, due to x not being in the intersection of their domains, and there may be plenty of objects for which N and N' part ways with respect to condition C due to C not being a shared condition of application. Really, this form of equivalence runs together with mere convergence in verdicts. Stricter forms of equivalence can be defined for norms that have logically equivalent conditions of application, that have identical sets of objects, or both; the final case will be equivalence of verdicts for all objects and conditions of application. By "stricter", I mean that fewer pairs of norms will bear that equivalence relation than more "permissive" forms of equivalence, which will have a larger extension. Equivalently, say that stricter equivalences are stronger, and more permissive equivalences are weaker.

Thinking in terms of the rough, LTS based model of epistemic dynamics, it's easy to see that norm-kernels define restrictions on epistemic dynamics. Where the object, x, of a norm-kernel, N, is a state, transition, action, process, or procedure, all that is required is that the valence R that N assigns to x under C be the sort of thing amenable to representation with a ranking on the powerset of elements to which x belongs. All normative valences are so amenable, at least with qualifications.[2] So,

[1] I'll leave this bit of proof off of further discussion. It should be clear that the equivalence relations I define—or even merely gesture at—below are genuine equivalence relations. This is easy to see from the definitions, which typically employ the identity of some set of objects. Identity is, of course, reflexive, symmetric, and transitive.

[2] For instance, as argued above, deontics are fairly well representable with ranked sets of epistemic factors, and this is quite natural where deontics are deployed to shape choices among actions. There may be some complications. For one, the usual assumption is that obligation and forbiddance/prohibition are duals. If doing a is obligatory, then not doing a is forbidden. The translation sketched in Sect. 1.3.5 doesn't capture this without machinery for omissions of action.

if R is full-stop rationality, then let R_X be the set of objects that N determines is rational. If R is a deontic, then set the ranking to reflect this. To build conditions of application into a restriction on dynamics, note that conditions of application typically encode properties of the objects of norms. So, simply let the rational subset (resp. appropriate deontic subset) in the restriction be the union of the sets of objects of N that satisfy C. For instance, for a closure norm that assigns "rationally obligatory" to deductively closed belief states, let the restriction be such that R_X is the set of all closed belief states, and the ranking relation places R_X above all other comparable sets of belief states. If N tells us that action a is rational under C, then define the restriction so that R_X is the set of actions that satisfy C. And so on. Norm-kernels can thusly be mapped to restrictions on epistemic dynamics that accurately represent their verdicts.

The strictest form of equivalence among norm-kernels (short of identity) that can be defined based on restrictions is this: N and N' are equivalent iff every restriction on dynamics that N defines is identical to a restriction on dynamics that N' defines and the converse also holds. If that's right, then for any x in the domain of N, then x is in the domain of N', and N's verdict on x will be N''s verdict on x. Identity of restrictions extends to the rank ordering of restrictions. Identity of restrictions on sets of epistemic factors preserve the rankings in those restrictions. So, if p is ranked higher than q by N, identity of restrictions on processes implies that p is ranked higher than q by N' and the converse also holds. This kind of equivalence between N and N' thus tells us that any process that complies with N (or the preferences defined by N) also complies with (that of) N' and the converse also holds.

Generally, defining equivalence in terms of restrictions on particular sets of epistemic factors defines a strength ordering of equivalence relations. To illustrate, a relatively strict[3] form of norm equivalence can be defined by identity of the restrictions on processes. Suppose that norm N tells us that process p is rational if it satisfies C. Suppose that N' also restricts processes. Under the given equivalence, the verdict of N on p is identical to N''s verdict on p. Generally, it may be the case that norm-kernels that are equivalent in terms of transitions are not equivalent in terms of processes, that norm-kernels equivalent in terms of states are not equivalent in terms of transitions, and so on. This claim has to be clarified. After all, two possibilities suggest that norm-kernels equivalent in restricting one set of epistemic factors might just define radically different restrictions on other epistemic factors. First, norm-kernels that restrict (e.g.) processes, need not say anything explicit about transitions or states. Second, it's possible that a norm-kernel takes a disjunctive form over an inclusive domain: "If C, then $R(p)$ or $R'(t)$ or $R''(s)$". However, identical verdicts on subsets of other epistemic factors can be derived from the generation

There are options. One can capture not doing a with $-a$ (interpreted as complementation of the executions of a in an LTS), and define a restriction such that, if $a \in P_A \succ R_A$ where R_A is read as the permitted actions, then $-a \in I_A$. This isn't obviously incoherent, nor is it obviously adequate.

[3] It's less strict than straightforward identity among restrictions and stricter than the other relations to come in this section.

rules described in prior sections, and this defines a strength ordering. For instance, consider:

(**GenAPPRtoT**) If a (resp. p, P_π) is rational according to N and performing a (resp. completing p, implementing P_π) at s results in a transition from s to s', then $t = \langle s, s' \rangle$ is rational.

This rule connects the restriction on actions/processes/procedures to a restriction on transitions. From there, if N and N' are "process restriction equivalent", then for the subset of transitions that bookend the processes in either N's restriction on processes or N''s restriction on processes, the verdicts of N on transitions will be identical to the verdicts of N' and the converse also holds. The same reasoning can be applied to states.

Call the restrictions defined in virtue of application of generation rules "generated restrictions". Call the sets of epistemic factors that meet the conditions of generation rules generated sets of epistemic factors. Equivalence defined in terms of restrictions on procedures thus implies equivalence of generated restrictions on processes, which implies equivalence of generated restrictions on transitions, which implies equivalence of generated restrictions on states. Weaker (more permissive) equivalences can be defined by identity between restrictions on transitions or restrictions on states. That these are weaker is easy to see. Suppose N and N' define identical restrictions on states. Then N and N' may deliver identical verdicts on all states, but still disagree on the generated transitions unless their sets of rational transitions converge on the same terminal states. Suppose they define identical restrictions on transitions. Then N and N' may deliver identical verdicts on transitions, but still disagree on processes if their sets of rational transitions bookend different processes. Thus, restrictions on epistemic factors define strength orderings at least among the generated restrictions.

From there, the generation rules for translating restrictions on one set of epistemic factors to other sets enable further equivalence relations among norms. We can get more permissive forms of equivalence by plugging a process restricting N into (GenAPPRtoT), which gives us:

(**Gen*N*toT**) If p is rational according to N and completing p at s results in a transition from s to s', then $t = \langle s, s' \rangle$ is rational.

Define a restriction on transitions that meets the conditions implied by this rule. Do the same for N'. Then a somewhat permissive form of norm equivalence can be defined by appealing to identity of the derivative restrictions on transitions only. This kind of equivalence between N and N' does not tell us that any process that complies with N also complies with N' (or converse), but it does tell us that any transition that complies with Gen$N$toT also complies with GenN'toT and the converse also holds.

Finally, still weaker equivalence relations can be defined on parts of restrictions on dynamics. For instance, one can define equivalence of N and N' in terms of identity of rational—and not irrational—sets of epistemic factors, identity of rankings but different thresholds of irrationality, or convergence of rankings above some particular point. In these cases, N and N' might only agree on positive verdicts or comparative verdicts or subsets of comparative verdicts.

3.1.1.2 (D3.2) Arguments from the Regulative Function of Epistemic Norms and Epistemic Deliberation

The regulative function of epistemic norms requires ante hoc regulation, which, in turn, requires epistemic planning and epistemic deliberation. These forms of reasoning require precisely the forms of explicit reasoning about epistemic norms articulated.

The first argument, simply put, is this:

(1) Deliberate application of norms requires deliberation about how to follow or apply norms or deliberation about which norms to apply.
(2) Deliberation about how to apply norms requires explicit reasoning about them.
(3) Deliberation about which norms to apply requires explicit reasoning about them.
(4) ∴ Deliberate application of norms requires explicit reasoning about norms.

The first premise is, if you will, analytic. By application of norms, I mean following norms, carrying out their dictates, or meeting the requirements they specify under their conditions of application. One applies a norm that requires consistency of belief when, upon discovering inconsistency, one revises away the inconsistency. For instance:

(Defeasible Consistency 2) If a belief set, s, contains A and a non-empty subset, $\{B_1, \ldots, B_n\}$, such that each B_i is inconsistent with A, and, for each B_i, s contains no evidence for B_i comparable in strength to the evidence for A, then s is irrational.

The core requirement specified by this norm is to not maintain inconsistent beliefs in the face of an uneven balance of evidence. There are several ways to meet the requirements of rationality according to this norm: one option is to reinstate consistency by rejecting each B_i, another is to do so by shifting the balance of evidence away from A and then rejecting A, a third way might be to shift the balance in favor of some B_i while retaining the inconsistency. Application of this norm is carrying out one of these options upon discovery of the inconsistency or the discovery of the relative lack of evidence for the B_is that contradict A.

Deliberation is comparative reasoning about each of an array of options in order to select one. Typically, one deliberates about actions. One has a repertoire of actions, $\{a_1, \ldots, a_n\}$, that one can carry out, and one considers which of them to actually do on the basis of familiar practical and ethical considerations, though practical and ethical considerations need not exhaust the kinds of consideration that are relevant.

There are thus two ways in which deliberate application of norms is possible. First, deliberate application of a norm may take the form of the construction of plans or the selection of courses of action that meet the requirements of the norm. If one realizes that inconsistency in belief is the condition of application of a consistency norm, and, seeing multiple ways to successfully revise the inconsistency away, one deliberates about which is best and does it, one deliberately applies the norm. In AI terminology, deliberate application of a norm, N, is solving a planning problem in which the initial condition satisfies the condition of application of N and the goal state is defined by N. Defeasible Consistency 2 sets the goal state as either a consistent belief state or a state in which all inconsistencies are backed up by comparably strong, conflicting lines of evidence. Applying it involves constructing a plan to achieve this goal state.

The second way to deliberately apply norms is opened up by the possibility of norm revision. There are cases in which (a) there are multiple, distinct norms none of which we are forced by any logical, metaphysical, physical, or less general natural laws to apply, and (b) we can voluntarily apply each of the distinct norms. In those cases, we can engage in comparative reasoning about the rival norms, determine which we should endorse, and then apply it. In those cases, we can decide which norms to endorse and (from that point onward) apply, either for all conditions of application or just under the current conditions. From there, there may be more than one way to apply the norm selected, and so it can be deliberately applied in the first way as well.

The remaining premises follow from consideration of both forms of deliberation about norms. In solving the planning problem imposed by a norm N, one considers a set of plans, $\{p_1, \ldots, p_n\}$, and, at a minimum,[4] which of the p_i achieve the goal state of N or meet N's requirements under the current epistemic circumstances. Comparative reasoning about whether p_i meets the requirements of N involves determining what N's requirements are given the current conditions. Suppose, arbitrarily, that I want to deliberately apply Defeasible Consistency 2. Table 3.2 presents an array of plans[5] for applying it.

There is no way for me to figure out which of these plans achieve the goal state of Defeasible Consistency 2 without figuring out, at a minimum, what the rule requires. Generally, these plans all revise s to either falsify the conditions of application by either removing the inconsistency or revising the balance of evidence; some are "biased" toward retaining A, others toward retaining the B_i. More, some of them

[4] Matters of optimization with respect to considerations of effectiveness, time to completion, costliness, etc., can complicate plan construction.

[5] One might object that these are not legitimate plans because accepting and rejecting propositions are not voluntary actions. However, on a Cohen style account of acceptance [16], it's quite plausible that acceptance and rejection are voluntary because the methodological and research oriented actions Cohen describes are clearly voluntary or at least semi-voluntary actions. However, if one thinks that acceptance and rejection are more like belief, it's not so plausible. But in this case, there may be self-stimulatory actions that are voluntary that reliably trigger cognitive effects like belief update [32, 33, 40]. The usual suite of such actions includes directing attention to one's evidence, carefully constructing and evaluating arguments, thinking up new explanations for accepted data, etc.. These plans are specified at a very abstract level. In Bratman's terms, they're partial plans rather than complete plans [10].

Table 3.2 Norm application plans

Bias	Consistency	Evidence
A-bias	**A-bias Consistency 1** Under the conditions of application, reject every B_i. **A-bias Consistency 2** Under the conditions of application, reject every B_i and every C_i such that C_i implies some B_i.	**A-bias Evidential 1** Under the conditions of application, acquire evidence that favors A over all B_i; reject every B_i. **A-bias Evidential 2** Under the conditions of application, acquire evidence that favors A over all B_i; reject every B_i and every C_i such that C_i implies some B_i.
B-bias	**B-bias Consistency 1** Under the conditions of application, reject A. **B-bias Consistency 2** Under the conditions of application, reject A and every C_i such that C_i implies some A .	**B-bias Evidential 1** Under the conditions of application, acquire evidence comparable to that for A for some B_i; retain s. **B-bias Evidential 2** Under the conditions of application, acquire evidence comparable to that for A for some B_i; reject A. **B-bias Evidential 3** Under the conditions of application, acquire evidence comparable to that for A for some B_i; reject A and every C_i that implies A.
Neutral	**Neutral Consistency 1** Under the conditions of application, reject either A or (inclusively) every B_i. **Neutral Consistency 2** Under the conditions of application, reject either A or every B_i (and every C_i such that C_i implies some B_i). **Neutral Consistency 3** Under the conditions of application, reject either A (and every C_i such that C_i implies A) or every B_i. **Neutral Consistency 4** Under the conditions of application, reject either A (and every C_i such that C_i implies A) or every B_i (and every C_j such that C_j implies some B_i).	**Neutral Evidential 1** Under the conditions of application, acquire evidence neutrally; (i) if the evidence favors A over all B_i, then reject all B_i, (ii) if, for some B_i, the evidence is comparable to that for A, then retain s. **Neutral Evidential 2** Under the conditions of application, acquire evidence neutrally; (i) if the evidence favors A over all B_i, then reject all B_i, (ii) if, for some B_i, the evidence is comparable to that for A, then reject A. **Neutral Evidential 3** Under the conditions of application, acquire evidence neutrally; (i) if the evidence favors A over all B_i, then reject all B_i, and every C_j such that C_j implies some B_i, (ii) if, for some B_i, the evidence is comparable to that for A, then reject A. **Neutral Evidential 4** Under the conditions of application, acquire evidence neutrally; (i) if the evidence favors A, then reject each B_i, (ii) if the evidence is comparable to that for A for some B_i, then reject A and every C_i such that C_i implies A. **Neutral Evidential 5** Under the conditions of application, acquire evidence neutrally; (i) if the evidence favors A over all B_i, then reject all B_i, and every C_j such that C_j implies some B_i, (ii) if, for some B_i, the evidence is comparable to that for A, then reject A and every C_i such that C_i implies A.

go beyond the requirements to ensure compliance with deductive closure without inconsistency, and some of them fail to do so.[6] The rule does not require that I reject A nor does it require that I reject any B_i; it allows somewhat free choice among these options. If one is to retain A, one must find evidence that favors A over all B_i; else, the conditions of application hold, and s is irrational according to the rule. However, rejection of A suffices to falsify the conditions of application, and so the resulting state may be rational. But figuring out which of these plans meets the requirements requires reasoning about the conditions under which the conditions of application hold, reasoning about what objects of the norm satisfy these conditions, and, of course, reasoning about which plan brings the object of the norm into alignment with it. Thus, in order to deliberate over this set of plans, one must reason explicitly about what Defeasible Consistency 2 requires.

Now, the second form of deliberate application of norms is much easier to see requiring explicit reasoning. If one deliberates over which of $N_1, \ldots, N_n$ to endorse and apply, one must reason about what they require and what they permit in order (inter alia) to project the static, dynamic, and procedural consequences of applying them. Distinct epistemic norms may have distinct properties concerning (e.g.) the difficulty of applying them in addition to the outcomes of following them. The point about outcomes has been argued for above. The point concerning the difficulty of application relies on the *function-algorithm* distinction [42]. Pavese [48] provides an exceptionally clear example by giving a recursive method for calculating Fibonacci numbers and comparing it to the functionally equivalent Binet algorithm. These two procedures are functionally identical; abstractly, from any input state, they return the same output state. Nonetheless, Binet's algorithm is far more efficient than the recursive method. Additionally, distinct norms may result in violation of or compliance with other norms. For another example, one can conceive of alternative consistency rules that absolutely require the elimination of inconsistency, or that merely strongly recommend it. Defeasible Consistency 2, after all, seems to say that finding evidence for some B_i and leaving s inconsistent is just as rational (ceteris paribus) as removing the inconsistency. Perhaps this speaks in its favor or against it. For another example, debates over alternatives to simple conditionalization involve explicit arguments about the practical properties of such credence updating rules [24, 50]. Some updating rules enable convergence of multiple agents from distinct priors in fewer iterations than others. Some updating rules fail to make sense of common counterexamples. And so on.

Less formal examples abound. Suppose one endorses a demanding Humean [28] or Cliffordian [15] norm like this:

(Humeford) Belief in proposition A is rational iff the conditional probability of A given total evidence is above threshold T.

[6] Every plan that doesn't also reject the propositions that imply A or B_i will, under closure, just reinstate the inconsistency, and so reinstate the conditions of application of Defeasible Consistency 2, resulting in an irrational state s' after all is said and done.

Let T be specified however one wants—just keep it higher than .75. The rule requires that all individual propositions believed are supported by suitably strong evidence. Belief that falls below the threshold T is not rational belief. Then, in the face of conclusive defeat of arguments against A, even the utter absence of counterevidence to A, it does not follow that belief in A is rational. However, from a more permissive view[7] that requires that only beliefs with suitably strong counterevidence be rejected, such apologetics may well be sufficient for rational belief. Consider:

(Anti-Humeford) Belief in proposition A is irrational iff the conditional probability of A given total evidence is below threshold T'.

Given that $T \neq T'$, these are distinct norms and have different consequences for agents that follow them. At high T, the Humeford rule is extremely demanding; rational belief may well be nonexistent. At low T', the Anti-Humeford rule may license forms of extreme dogmatism. The conditions on rational belief defined by Anti-Humeford may well be so expansive that the basic task of dispute resolution is failed utterly. These are among the practical considerations that are relevant to adopting and applying these norms

In sum, there are two ways to deliberately apply norms: deliberate about how to apply the norms one follows and then apply them or deliberate about which norms to apply. Deliberation about how to apply norms cannot be carried out without reasoning explicitly about norms. The same holds for deliberation about which norms to apply. In particular, solving the planning problem for epistemic norms involves inferences about what the norms require, e.g., what properties of the objects of the norms constitute goal states, what the conditions of application imply, and what plans actually accomplish the goal states of the norm. Deliberation about which norms to apply involves inferences about what the norms require as well as the outcomes of applying them. That is, deliberate application requires reasoning about the verdicts of epistemic rationality norms (e.g., is a particular action required/obligated by the norm or is it merely permitted because it suffices to restore rationality?), what the procedural content of a norm implies about how to solve the planning problem (e.g., if a norm requires specific procedural operations directly, this will rule out some plans; if the goal state can only be reached by some chaining of procedural operations, the norm rules out some plans), and about how modal and propositional facts interact with the norm (e.g., whether the planning problem can be solved depends, for instance, on what plans are feasible and what outcomes the plans have). Thus, deliberate application of norms requires explicit reasoning about norms of the articulated forms. Given the arguments of preceding sections, no complete account of the cognitive function of epistemic rationality norms can omit analysis of how we reason about them.

---

[7] Arguably, like the forms of epistemic conservatism in any of [27, 38, 39], for instance.

3.1.1.3 (D3.2) Arguments from Epistemic Planning

Finally, D3.2 also has independent support from consideration of ante hoc regulation and epistemic planning in light of the examples of norms that restrict procedural construction. The same argument applies:

(1) Ante hoc regulation requires solving epistemic planning problems.
(2) Solutions to planning problems are restrictions on procedural construction.
(3) ∴ Ante hoc regulation requires restrictions on procedural construction.

But solving these problems requires directly reasoning about the verdicts of epistemic norms. It requires reasoning about the verdicts with respect to the states, transitions, and processes determined by epistemic plans, and it requires reasoning about the verdicts with respect to the (procedural construction of the) plans themselves (especially where this is relevant to the differential outcomes of the plans). These forms of reasoning will make use of norm equivalences, and such reasoning will be impacted by dependence and independence results. Norms equivalent with respect to ranges of verdicts (over ranges of objects) will be intersubstitutable over some contexts of such reasoning. Norm dependence will intertwine the verdicts of norms. Norm independence will sever them.

3.1.1.4 (D3.2) Arguments from Formal Epistemology

Turning to epistemology and philosophical logic provides an abundance of examples of explicit, deductively structured argumentation about epistemic rationality norms. Work in formal epistemology, traditional epistemology, and seminal work in philosophical logic implies D3.2, as each of these fields is rife with analyses of epistemic norms and complex argument construction out of propositions about them. There are (seemingly) logically valid arguments in these fields that are about (e.g.) epistemic norms, their contents, and the outcomes of implementing or following them. This reasoning takes the form of: arguments establishing equivalence among norms, deriving results from assumptions about the procedural structure or conditions of application of norms, or establishing dependence relations among norms. Such arguments appear in virtually any paper about the preface paradox, about constraints on rational belief (full or partial), about logical normativity, about the nature or properties of evidence, about conditionalization, about probabilistic coherence, about deductive or inductive strategies or methods, about the norm of assertion, about the norm of belief, about justification or warrant transmission, and so on.

Explicitly, the argument is this:

(1) Formal epistemological (et al) reasoning requires valid argumentation about the general logical properties (logic of normative verdicts, etc.) of epistemic rationality norms, as well as the properties of equivalence, dependence, independence relations among epistemic norms.

(2) Valid argumentation about the general logical properties (logic of normative verdicts, etc.) of epistemic rationality norms, as well as the properties of equivalence, dependence, independence relations among epistemic norms requires the logical analysis of the validity of the argument forms articulated by D3C1.
(3) ∴ Formal epistemological (et al) reasoning requires the logical analysis of the validity of the arguments forms articulated by D3C1.

Thus, any logic of epistemic norms adequate to these philosophical projects must analyze the logical principles governing epistemic norms that underly these forms of argument. Such principles must encompass the rules governing (e.g.) epistemic norm equivalence, dependence, independence, and the derivation of normative verdicts.

The two premises are clear from engagement with formal epistemology, epistemology, or philosophical logic. The former field is nearly entirely composed of arguments about the properties, structure, content, and nature of epistemic norms (often under the guise of projection rules, methods, or procedures), and work in the other fields rather typically take the form of complex arguments built out of propositions about epistemic norms and their properties. These arguments tend to belong to one of three kinds:

- Arguments deriving verdicts (valences, deontic statuses, procedural requirements) from conditions of application, procedural structure, or other contexts.
- Arguments establishing equivalence between norms or deriving conclusions from equivalence.
- Arguments establishing dependence, sufficiency, and independence relations among norms.

Consider this small, representative set of examples:

Inductive Rule Let j and k be fixed real numbers, $0 < j < k$. Consider the rule, discussed in Field's [17]: "If in the past you observed n ravens, and m of them have been black, you should believe to degree $(m + j)/(n + k)$ of any raven not yet observed that it is black." (p. 375)
Concerning this rule, Field formulates the following explicit arguments, among others (my emphasis):

> The rule is to be generalized to apply to other pairs of predicates besides 'raven' and 'black', though as the 'grue' paradox makes vivid, using the rule for some pairs requires not using it for certain others. *The rule given is in effect an instance of Carnap's* λ*−continuum* (at least when k/j is an integer: it results rom taking $\lambda = k$ and the number of kinds as k/j. (275, fn 23)
>
> [A]s observations of raven color accumulate, that *initial degree of belief gradually becomes swamped by the observed frequency.* (This happens slowly if k is large, quickly if k is small.) This is of course a thoroughly implausible rule: among other defects, (i) it takes no account of any evidence other than observations of raven color; (ii) it takes no account of any regularities in the ordering of black raven color; (iii) it takes no account of how ravens were selected for observation. (375)
>
> Suppose our initial bias was $j = 1$, $k = 10$, so that the degree of belief assigned to a given raven being black was only .1. And suppose we observe 20 ravens, of which 19 are black. Then the rule tells us to believe to degree .667 (20/30) of an unobserved

> raven that it is black. The rule has in a sense told us to modify our biases. In another sense, though, the biases go unchanged: the initial bias, represented by the pair $\langle j, k\rangle$ from which we started, is still there producing the new degree of belief. Of course, the initial bias produces that new degree of belief only with the accumulated evidence, and *the effect of using the initial bias with the accumulated evidence is in a sense equivalent to altering the bias: it is equivalent to altering the initial bias to* $j = 20$ *and* $k = 30$, *and then dropping the observation of the first 20 ravens from our evidence.* [...] In a sense then our rule is 'self-correcting': with accumulating evidence the old rule behaves like an altered rule would without the added evidence. Because it is 'self-correcting' in this way, there is no need for it to be revised when evidence of its unreliability in previous applications is accumulated. (376)

Preface Paradox The preface paradox [41] poses a serious problem for models of epistemic rationality that require naive or indefeasible forms of consistency and deductive closure and basic forms of evidence proportioning. Let s be an all or nothing belief set. Let rejection of A be encoded by s's containing $\neg A$. The following set of norms runs afoul of preface cases:

Consistency If s contains A and $\neg A$ or any $\{B_1, \ldots, B_n\}$ that implies both A and $\neg A$, then s is irrational.

Closure If $A_1, \ldots, A_n \vdash B$, then if s contains $A_1, \ldots, A_n$ and does not contain B, s is irrational.

Evidence If $Pr(A_i \mid E) > t$, then if s does not contain A_i and does believe E, then s is irrational.

Let $A_1, \ldots, A_n$ be a set of beliefs in an agent's belief state s such that for each A_i, $Pr(A_i \mid E)$ and $E \subseteq s$. More, let the agent also have decisive evidence that she is fallible, and thus that she's wrong about some A_i, which is to say that she believes $\neg A_1 \vee \ldots \vee \neg A_n$ in order to be rational (let it be the case that either $Pr(\neg A_1 \vee \ldots \vee \neg A_n \mid E)$ or $Pr(Fallibility \mid E)$ and Fallibility implies $\neg A_1 \vee \ldots \vee \neg A_n$). But $A_1, \ldots, A_n \vdash A_1 \wedge \ldots \wedge A_n$, of course, so by closure, the agent must believe $A_1 \wedge \ldots \wedge A_n$ in order to be rational. Thus, *in order to satisfy closure, she must violate Consistency. If she were to reject the conjunction or the disjunction, she would violate Closure*. The only way to avoid violation of one or the other is to reject some A_i, but which is she to pick? *Given the evidence she has for each* A_i, *the agent will have to throw out some belief for which she has evidence exceeding the threshold* t, *and so she violates Evidence.*

Formal Learning Theory Formal learning theory [63] is a mathematical framework for modeling different methods or norms by which epistemic agents come to conclusions on the basis of observation. Often conceived of as epistemology done from the engineering or means-ends perspective [62], the basic approach is to outline a number of epistemic goals and to design and evaluate procedures or rules for taking an agent from observations to conclusions. Basically, it's about formalizing inductive rules and evaluating their procedural structure, features, and consequences. The standard consideration raised about epistemic norms in this context is the reliability of a method of inference.

A basic example [63]: Let $grue(t)$ be the predicate such that, x is $grue(t)$ iff if x is observed prior to time t, it will be green, and if x is observed at or after

t, it will be blue. Consider two rules for accepting a universal hypothesis about emeralds:

(Green) If all emeralds are observed to be green prior to t, then accept hypothesis: "All emeralds are green", and if there is some emerald observed to be blue at t, then accept hypothesis: "All emeralds are $grue(t)$".

(Grue) If all emeralds are observed to be green prior t, then accept hypothesis: "All emeralds are $grue(t)$", and if there is some emerald observed to be blue at t, then accept hypothesis: "All emeralds are $grue(t+1)$".

Now, consider two sequences of observations of emeralds:

$$\langle g, g, g, g, g, g, g, g, g \rangle$$

and

$$\langle g, g, g, g, g, g, g, g, b, b, b, b, b, b \rangle$$

The Green rule will accept the correct generalization (with suitably restricted quantification) about either of these sets of emeralds. This can be seen by straightforward conditional reasoning. After the first observation in the first sequence, the Green rule has the agent accept "All emeralds are green". The same result happens after each observation of a green emerald in the first or second sequences. In the second sequence, the first observation of a blue emerald at t has Green rule accept "All emeralds are $grue(t)$". *The Grue rule will never accept the correct generalization for the first sequence*, it will always accept the next grue hypothesis that isn't ruled out by the observations. It will eventually settle on the right generalization for the second sequence.

Relevance Standard arguments [3, 55] for Relevant Logic (RL) against Classical Logic (CL) run like so: Logics are theories of the validity relation. There are relevance constraints on reasoning—i.e., there are moves in reasoning that are ruled out by our norms for rational inquiry on the basis of irrelevance. Thus, if a logic does not generate (rules/laws encoding) relevance constraints, the logic is wrong and must be revised (preferably, in the direction of relevant logics). Classical Logic (CL) undergenerates with respect to relevance constraints on reasoning and overgenerates with respect to other constraints on reasoning. That is, CL is wrong because CL licenses irrelevant reasoning. Consider versions of this argument put forward, respectively, by Meyer [43] and the seminal Anderson and Belnap [3]:

(Meyer) Let the set of accepted propositions for some fallible, person-like agent, a, be X. Let it be the case that (1) X contains inconsistencies. (1) will almost certainly be the case for anything like a. Consider: (2) if a is committed to every proposition in X, then a is committed to the logical closure of X. Contradictions imply triviality in CL, so *(1) and (2) entail that, wherever a is committed to all of its beliefs, a must be committed to absolutely every expressible proposition*. That's an intuitively absurd ("odd, silly, and ridiculous" (p. 814)) constraint on any rational agent's beliefs. So, CL must be wrong; the inference from $A \wedge \neg A$ to arbitrary B must not be valid because it mischaracterizes acceptable reasoning.

(Anderson and Belnap) Let T be some theorem of some branch of mathematics. Then T is necessarily true if it's true at all. Classically, $A \vdash T$ for any A. *It's intuitively absurd to argue, infer, or reason from just any arbitrary A to T in numerous contexts*. For example, if you tried to publish an article claiming you'd found a novel proof of T from A where A is "Dogs sometimes get hungry", (Assume A, then, by CL, T!), your reviewers would correctly reject your article. *This is straightforwardly not acceptable reasoning due to the irrelevance of A to T*, so $A \vdash T$ is invalid because it doesn't correctly characterize acceptable reasoning.

Epistemic Utility Theory I'll let Pettigrew explain the epistemic utility theory project in his own words [49]:

> One of the central projects of formal epistemology concerns the formulation and justification of epistemic norms. The project has three stages: First, the formal epistemologist produces a mathematical model of an agent's epistemic states—call this the descriptive stage. Next, she formulates, in terms of this model, putative norms that she claims govern these states—call this the normative stage. Finally, she provides a justification for these norms—call this the justificatory stage. It is one of the great virtues of formal epistemology that the final justificatory stage can be made mathematically precise. The strategy is this: the formal epistemologist states an epistemic norm that is taken to be more general and fundamental than those given in the normative stage; this norm is then formulated in terms of the mathematical model provided by the descriptive stage; and *the norms posited in the normative stage are derived from this fundamental norm by means of a mathematical theorem*. [...]
>
> In epistemic utility theory, the descriptive and normative stages are not novel. [...] Thus, the novelty of epistemic utility lies in the justificatory stage. (p.287)

If it is not obvious that these passages contain arguments of the forms designated by D3C1, let me belabor that point a bit.

First, Field explicitly articulates an epistemic norm for governing degrees of belief and builds arguments of each kind delineated above. The first quote reports his result that the rule articulated above, when suitably generalized, is equivalent to one of the norms in Carnap's continuum of c-functions. The second explicitly argues for the self-correcting property of the rule by explicitly stipulating some conditions of application and deriving verdicts from them. By stipulating the initial bias, the rule delivers a verdict about what one's posterior credences should be in each prediction of the blackness of an unobserved raven. The argument then goes on to argue for a kind of equivalence. Given that initial bias individuates rules, the verdicts of the rule above, under certain preconditions, are equivalent to those of a distinct rule (one with different initial bias). The third quote makes procedural observations about the rule, noting that it lacks a number of procedural requirements that could be epistemologically relevant. For instance, it's plausible—at least Field takes it as plausible—that observations of things other than ravens (e.g., other kinds of birds) should be able to provide some kind of evidence for inferences about the blackness of ravens.

Second, assessing preface cases requires straightforward reasoning about the verdicts of epistemic norms under varying conditions of application. One directly reasons from the conditions in the preface case to violations of the requirements of Consistency, Closure, and Evidence. In order to figure out that these norms are violated, one has to understand what they require under these conditions (given some stipulations). Generally speaking, to even engage in reasoning about the preface paradox involves straightforward argumentation about the verdicts of norms.

Third, formal learning theory is an explicit, formal framework for articulating epistemic rules (inductive rules, in particular), and this framework supports all manner of argumentation about the verdicts, equivalence relations, procedural structures, and outcomes of applying rules under varying conditions. Standard results include arguments establishing convergence among different norms or methods (i.e., a form of equivalence), deriving explicit verdicts, as above, and deriving other features of norms including (centrally) reliability, power, and speed [52, 63].

Fourth, the example from defenses of relevant logic is a straightforward example of arguing from assumptions about epistemic rationality to conclusions about validity. Both the Meyer and Anderson and Belnap arguments assume that some form of naive closure is a constraint on reasoning. In fact, the arguments they provide must take a rather straightforward schematic form: If $X \vdash A$ is valid, then reasoning process p is rationally acceptable.[8] The arguments then articulate cases in which an argument is valid according to CL though the process that corresponds to that argument is not rationally acceptable. Whatever norms are assumed in the background, the instances that are appealed to must conflict with the verdicts of epistemic rationality, so these examples straightforwardly employ some of the forms of argument outlined above; we must be able to derive comparative verdicts about the same objects. If that's right, then these examples require further complex reasoning about epistemic norms.

Fifth, Pettigrew explicitly articulates the main point of the argument. Formal epistemology, insofar as it is concerned with the justification of epistemic norms, must explicitly characterize them (with mathematical precision) and then derive them from some more fundamental norm. That is, the justificatory project is one of establishing dependence, independence, sufficiency, and equivalence relations among norms. One can potentially undermine a justificatory project by showing that putative norms are independent of any more fundamental norm (imagine if probabilism were entirely independent of preference consistencies, long-term convergence on the truth, or non-dutchbookability). Justification of norms can be undermined or established by identifying equivalences among putative epistemic norms.

[8] I will resist the impulse to digress on what kind of conditional this is. Pick your favorite contraposition validating conditional, else, the Meyer and Anderson/Belnap arguments straightforwardly fail!

More, it's not uncommon outside of the foregoing contexts to see arguments establishing dependence, sufficiency, and independence relations. For a quick instance, there are entire debates around whether or not dependence relations hold among epistemic norms or among epistemic and non-epistemic norms. See, for instance Papineau's contention that all norms of belief reduce to pragmatic or moral norms [47]. Arguing for dependence relations among norms that run the other direction, Montminy [46] argues that the norms of assertion and practical reasoning reduce to the same epistemic norms. Gerken [23] takes up the argument. For a final quick instance, see the debate between Broome and Kolodny over the scope of rationality norms for plenty of examples of deriving verdicts from rationality norms and arguments reducing norms to others [11, 34, 35].

Finally, the resolution of normative conflicts [1], which is a goal in the formal analysis of norms in general, requires complex reasoning about norms. This consideration also applies to epistemic norms, insofar as they can be used to regulate epistemic action. Normative conflict arises primarily in the context of deliberate application of norms. In figuring out how to follow one's norms one may come across a situation in which two norms, N and N', specify incompatible requirements. In order to determine that there is a conflict, one must be able to derive the verdicts of those norms under the relevant conditions. The resolution of these conflicts will typically involve *balancing* and *specification* of these norms [8]. That is, one must determine whether N overrides N' under the given conditions (balancing them) or whether there is a way to reduce the indeterminateness of N and N' in a way that shows that there was actually no conflict in the precise requirements of N and N'. Given that epistemic norms can be deliberately applied, balancing and specification, which require the forms of argument above, will be needed in the context of epistemic rationality.

This is a very small sample that can be drawn from a large pool of work in philosophical logic and epistemology (traditional or formal) that is built out of arguments of the designated forms.[9] These examples transparently display the argument structures characterized in premise 2. The conclusion follows trivially.[10] Now, the kinds of properties of norms that accounts of valid reasoning about norms

[9] See, for instance, this proper superset of references: [2, 4, 6, 7, 9, 12, 17–19, 22, 23, 31, 37, 46, 49–54, 56–62, 64, 67–70, 76]. It might be thought that the ubiquity of arguments of these forms, concerning the logical properties of epistemic norms, somehow gainsays my earlier insistence upon the neglect of the logic of epistemic norms. This would be a mistake. These are arguments attempting to establish particular logical properties or relations among particular epistemic norms or kinds of epistemic norm, under various umbrellas of assumptions and theoretical commitments, rather than attempts to build general or comprehensive logical frameworks for representing and reasoning about epistemic norms. These arguments presuppose a logical framework. Indeed, many of these cases were explicitly referenced in making the putatively gainsaid point.

[10] In addition to the forms of reasoning pointed out in formal epistemology and philosophical logic above, there is an interesting interaction between the arguments for D1 and the arguments for D3. Arguments for reduction of (e.g.) procedural epistemic norms to static norms themselves require valid reasoning about the relations (equivalence or at least sufficiency and dependence) among verdicts of epistemic norms. This presents a bit of a trap for would-be opponents of epistemic

characterize are the central targets of analysis for any formal framework. The rules and laws of valid argumentation about some object, x, are just those rules that characterize which inferences from propositions about x preserve truth. So, these rules will, among other things, characterize which properties of x imply other properties, and which relations among x and other objects imply other relations. So, given the foregoing conclusion, it's plausible that formal frameworks for epistemic norms must seek to explicitly codify valid reasoning (of the forms articulated) about norms of the forms articulated.

3.1.1.5 (D3C1) Implication from Prior Arguments

The prior arguments for D3.1 and D3.2 enumerated uses of the kinds of reasoning in the table defining D3C1.

3.1.2 The Logic of Epistemic Deontics

The remaining corollaries of D3 are almost immediate from the foregoing arguments in this chapter and the prior chapter.

Claim 15 (D3C2) Epistemic rationality norms assign deontic properties to epistemic factors including states, transitions, actions, processes, and complex procedures:

(1) Deontic concepts of obligation, permission, forbiddance, and reason can be translated into the procedural concepts of execution, dictation, and compatibility of actions and processes with procedures.
(2) Deontic operators are defined relative to sets of norms.
(3) Epistemic rationality norms assign deontic statuses to epistemic states, transitions, processes, and propositions both directly and indirectly, and these can be conceptualized as normative goal states or postconditions.

3.1.2.1 (D3C2 Core)

The analysis of epistemic norms in terms of norm-kernels presupposes the core of this principle, and D2 immediately implies the range of objects that these norm-kernels take.

proceduralism—one can weaken the irreducibility claims of D1 only by strengthening the claims of D3.

3.1.2.2 (D3C2.1)

The first clause of the principle must be the case given the regulative function of epistemic norms; the deontic verdicts of epistemic norm-kernels must, at some point, be translatable into instructions for (epistemic) action on pain of their inertness. Additionally, many of the paradoxes of standard deontic logic (SDL) disappear when deontics are translated into procedural concepts.[11] This produces some consilience with the well-known oddness of SDL; deontic operators (and normative language in general) apply most comfortably to actions, not to states of affairs, facts, or propositions. This, of course, is consilient with the regulative function of epistemic norms.

3.1.2.3 (D3C2.2)

The second clause is immediate; different epistemic norms define different obligations and permissions. Norms of logical consistency, famously, impose obligations of closure over the conjuncts of conjunctions that norms of probabilistic consistency do not. No logic of epistemic norms can afford to ignore this. This doesn't, of course, mean that the logic of deontic operators must countenance systems of norms. The point is rather that deontic logics must do so if deontic logic is to be the logic of norms (rather than the logic of the general, norm-agnostic notions of obligation and permission). There is also a methodological justification for defining the deontic operators of a logic of epistemic norms in a norm-relative fashion; doing so results in a system that is general enough to countenance different meta-normative views. Defining epistemic deontics relative to different systems of epistemic norms makes room for reasoning about the specific obligations and permissions defined by different sets of norms, without presuming that any one system defines epistemically rational obligation and permission. This leaves open the possibility that there is a single, correct normative system without prejudging the matter at the level of design of the logical system.

3.1.2.4 (D3C2.3)

The first conjunct of the clause is immediately implied by D1 and D2, granting that epistemic norms do take states as their objects in addition to the dynamic and procedural objects defended above. The second conjunct follows from the regulative function arguments. Suppose that epistemic verdicts on states have a regulative role. How can they function? By specifying goal states in planning problems, which are then to be conceived of postconditions of plans.

[11] See [44, 45] for one approach to this. See [65, 66] for another. See Chap. 8, Sect. 8.2 of this book for yet another.

3.1.3 The Logic of Norm Application

Claim 16 (D3C3) Epistemic rationality norms can be deliberately applied and adhered to as opposed to merely followed or complied with.

(1) Epistemic actions can be procedurally operated upon in knowing compliance with epistemic rationality norms, and this involves determining divergences in actual and possible practice from the rational case.
(2) Natural, heuristic, unconscious epistemic processes can be halted by effortful cognition.
(3) Epistemic processes can be deliberately re-directed in accordance with the restrictions defined by epistemic norms.
(4) Deliberate application of a norm requires knowing adherence to it.
(5) Knowing adherence of a norm is distinct from merely following a norm.
(6) Deliberate application of norms requires reasoning about epistemic rationality norms.
(7) Deliberate construction of procedures out of procedural operations on epistemic action in compliance with norms requires reasoning about procedural content.

3.1.3.1 (D3C3 Core)

The arguments from the regulative function of epistemic norms and from epistemic planning directly support the core of this principle. The instances used to illustrate the regulative role of epistemic norms require that there be epistemic deliberation. Indeed, the argument from the importance of the regulative role suggests that epistemic deliberation is one of the centerpieces of epistemic life.

3.1.3.2 (D3C3.1)

Table 2.3 presents numerous cases of norms that require specific procedural constructions—sequential compositions, choices, etc.. Criterion D2 spells these procedural requirements out further. The possibility of deliberately applying such norms implies the first conjunct of D3C3.1. The process of epistemic deliberation, as outlined in Sects. 2.2.3 and 3.1.1 supports the second conjunct of D3C3.1. One must model one's own practices and their divergences and convergences with the restrictions defined by a norm to apply it.

3.1.3.3 (D3C3.2-3)

The models of epistemic regulation, epistemic deliberation, and epistemic planning, as well as the litany of example cases in Sects. 2.2.2 and 2.2.3 imply these clauses.

This is, moreover, the essence of the empirically well-supported two systems model of reasoning.

3.1.3.4 (D3C3.4)

Let an agent knowingly adhere to an epistemic norm iff the agent knows that a process, P, complies with a norm, knows that they can execute other processes, $P_1,\ldots,P_n$, that comply with other norms, $N_1,\ldots,N_n$, and executes P anyway. Conceptually, deliberate adherence requires knowing adherence precisely because the necessary and sufficient conditions for knowing adherence are all necessary for deliberate application of the norm, at least with respect to all of the models of deliberation articulated above. One can't deliberately apply a norm, N, for instance, if one doesn't know whether a process, P, within one's repertoire of courses of action complies with N. One can't deliberately apply N if one has no choice in the matter, either; there's no deliberation to carry out if one must execute P in compliance with N (though one can go through the motions of deliberation, as it were).

3.1.3.5 (D3C3.5)

This is an immediate conceptual observation; one can accidentally or blindly act in accordance with a norm. Generally, let someone follow a norm iff they act in accordance with it. In other words, one (merely) follows an epistemic norm when one carries out, effects, or arrives at an epistemic factor in the restriction on epistemic dynamics defined by it. This conceptual distinction accounts for the observation with which we started.

3.1.3.6 (D3C3.6)

This clause is immediate from the account of epistemic deliberation above. Deliberation about how to apply an epistemic norm requires reasoning about what that norm itself requires, whether it's feasible, and what the outcomes of applying it will be, and deliberation about what norms to apply require reasoning about the properties of the norms on the table. A fortiori, deliberate application of epistemic rationality norms requires reasoning about epistemic rationality norms.

3.1.3.7 (D3C3.7)

Again, the models of epistemic regulation, epistemic deliberation, and epistemic planning imply this clause. In order to regulate one's epistemic dynamics with a procedural norm, one must figure out what restrictions on procedural construction that norm defines, and, in any cases where these norms are not internalized to the point of tacitness, this requires explicitly reasoning about the requirements of the norm.

3.2 Epistemic Proceduralism as Criteria on the Logic of Epistemic Norms

The prior section concludes the preliminary defense of epistemic proceduralism, as specified by clusters of principles D1–D3, and their corollaries D1C1–D3C3.3. If the arguments throughout this part of the book are sound, then the formal analysis and derivation of each of these clusters of principles defining epistemic proceduralism is a criterion of adequacy on any putative logic of epistemic norms. This, for two reasons.

First, clusters D1 and D2 capture fundamental aspects of the nature, structure, and function of epistemic norms, and D3 outlines some epistemologically central forms of reasoning with and about them. As such, the arguments of this densely packed and perhaps simultaneously too cursory Part I concern a number of fundamental debates in epistemology. Among these are debates about the basic taxonomy of epistemic norms (i.e., epistemic staticism vs. alternative accounts) and the fundamental role of epistemic norms in epistemic life (i.e., ante hoc regulation vs. post hoc evaluation). A logic of epistemic norms ought to be compatible with the best (or at least most defensible) positions on these matters and provide formalizations of the logical properties of epistemic norms entailed by them.

Second, the cluster D3 outlines the shape of the logic of epistemic norms. It provides a schematic codification of some forms of reasoning that underlie the most important and ubiquitous functions of epistemic norms as well as interesting and rigorous work in relatively recent epistemology and philosophy of logic. Both forms of inquiry were argued to rely centrally on reasoning about norm equivalence, dependence, independence, and the derivation of normative verdicts. The logic of epistemic norms should aim at codifying the central and reliable forms of epistemological reasoning, as exemplified in these arenas.

3.3 Applying the Criteria of Epistemic Proceduralism

In preparation for developing the protocol-theoretic framework of the book, it might be helpful to draw some preliminary contrasts between existing formal frameworks and the system to be developed.[12] The criteria defended up to this point provide a straightforward way of doing this. Here, I will show how the base or core

[12] My thanks to an anonymous reviewer for suggesting this.

systems of propositional dynamic logic (PDL), dynamic epistemic logic (DEL), and AGM belief revision theory (AGM) fail to satisfy some of the criteria, D1–D3C3. This should not, of course, be taken to be comprehensive criticism of the research programs that build upon these core systems, nor all of the systems that are developed under their aegis.[13] Rather, the intention of this section is to display the features of PLEN's nearest kin that fail the criteria of epistemic proceduralism so that the revisions that define PLEN are more easily grasped when the system is defined in Part II. I show in Part III how PLEN satisfies all of these criteria.

3.3.1 Propositional Dynamic Logic

First, define (the basic form of) PDL. PDL has two languages, a language of programs, P_{PDL}, and a language of program-theoretic formulae, L_{PDL}. Start from two sets of atomics: PR and AT, where PR is a set of atomic programs and AT is a set of atomic formulae.

Definition 17 (Infix Notation Program Language (P_{PDL})) Where $a_i \in PR$, π_i is a metavariable for an arbitrary program with no occurrences of self-reference, and A is a metavariable for an arbitrary protocol-theoretic formula with no occurrences of self-reference, P_{PDL} can be defined by the following grammar:

$$(a_i) \mid (\pi_1; \pi_2) \mid (\pi_1 \cup \pi_2) \mid (\pi_1 *) \mid (A)?$$

Definition 18 (Infix Notation Program-theoretic Language (L_{PDL})) Where $P_i \in AT$, π_i is a metavariable for an arbitrary program with no occurrences of self-reference, and A, B, ... are metavariables for arbitrary program-theoretic formulae with no occurrences of self-reference, L_{PDL} can be defined by the following grammar:

$$(P_i) \mid (A \wedge B) \mid (A \vee B) \mid (A \rightarrow B) \mid (A \equiv B) \mid (\neg A) \mid ([\pi_i] A) \mid (\langle \pi_i \rangle A)$$

[13] Indeed, this section is not even meant as criticism of these other systems at all. That would require that they are incomplete or inaccurate for the purposes for which they were developed. The arguments here merely show that they are incomplete or inaccurate for the purposes of developing a logic of epistemic norms, a project for which none of them were initially devised. Strictly, the problems raised for these systems merely show that they are not themselves adequate for the purposes of devising a logic of epistemic norms, and so motivate revisions and additions to them, which is what produced PLEN. For a somewhat more detailed development of the material in this section, see Chapter 5 of [30]. In particular, that chapter shows that the problems articulated here can be extended to many of the systems derived from these core systems. Now, it may not be the case that any of the problems articulated are impossible to resolve within the toolsets of these core systems (i.e., without appeal to the revisions of the tools that define PLEN). Nonetheless, the arguments there do the work of highlighting the main ideas of PLEN while also providing the rudiments of a theory choice argument favoring PLEN over its rivals, which I take to be an adequate justification for pursuing the development of PLEN.

Definition 19 The program language of PDL is interpreted over labeled transition systems, $\langle S, \rightarrow_{a \in PR} \rangle$. The idea is that atomic programs, a, in PR are mapped to binary relations, $\rightarrow_a$, defined over S. Execution rules extending the binary relations to complex programs are defined on this basis:
(PDL Execution Rules)

(1) (Base) For all $a \in PR$: $T_a = \rightarrow_a$
(2) (;) $T_{\pi_1;\pi_2} = T_{\pi_1} \circ T_{\pi_2}$
(3) (*) $T_{\pi *} = RFT(T_\pi)$ where $RFT(T_\pi)$ is the reflective, transitive closure of T_π.
(4) (∪) $T_{\pi_1 \cup \pi_2} = T_{\pi_1} \cup T_{\pi_2}$
(5) ((-)?) $T_{(A)?} = \{\langle s, s' \rangle \in T_g \mid s \Vdash A, s' \Vdash A\}$ where T_g is the union of all of binary relations in $\rightarrow_{a \in PR}$.

(Truth Rules) Valuations, $v_s(A)$, assign semantic values to atomic program-theoretic formulae at states. Extend verification for atomics to complex formulae by the following rules:

(1) (Base) $s \Vdash A$ iff $v_s(A) = 1$.
(2) (Conn) The standard classical rules for the truth-functions, $\vee, \wedge, \neg, \rightarrow$.
(3) ($\langle \pi \rangle A$) $s \Vdash \langle \pi \rangle A$ iff $\exists t \in T_\pi$ $(s = first(t), last(t) \Vdash A)$.
(4) ($[\pi]A$) $s \Vdash [\pi]A$ iff $\forall t \in T_\pi$ (if $s = first(p)$, then $last(t) \Vdash A$).

The typical definition of PDL defines LTSs over binary relations each mapped or labeled with atomic actions. It can, alternatively, be defined in terms of ternary transition relations; see Ch. 4 of [72]. This is, with respect to the language of PDL, equivalent to the "textbook", binary relation definition of PDL [25] given above, as the execution rules for complex programs are defined in such a way as to take into account only the binary relations. However, with more expressive languages, the equivalence disappears. Suppose that we define the structures (the modal frames over which PDL is defined) over ternary relations, i.e., each atomic program is mapped to a ternary relation and then the execution rules map each complex program to a ternary relation on the basis of the ternary relations of the atomics. Then, we will have both sets of edges (triples), $\langle s, \pi, s' \rangle$, and sets of binary relations, $\langle s, s' \rangle$ assigned to each program (the binary relations can just be those extracted from the sets of edges).

Define two program equivalence relations: Rp and Rr such that: (1) $Rp(\pi, \pi')$ is true at s iff for all edges, $\langle s, \pi, s' \rangle$, $\langle s, \pi, s' \rangle$ is in the set mapped to π iff it is also in the set mapped to π' and vice versa, and (2) $Rr(\pi, \pi')$ is true at s iff for all transitions, $\langle s, s' \rangle$, $\langle s, s' \rangle$ is in the set mapped to π iff it is also in the set mapped to π' and vice versa. Then $Rp(\pi, \pi')'$ can fail at states at which $Rr(\pi, \pi')$ holds. Consider a ternary or labeled transition system with only states s and s' and labeled transitions $\langle s, \pi, s' \rangle$ and $\langle s, \pi', s' \rangle$. Since $\langle s, \pi, s' \rangle$ and $\langle s, \pi', s' \rangle$ are distinct triples, $\langle s, \pi, s' \rangle$ may be in the set of π edges but not in the set of π' edges, and so $Rp(\pi, \pi')$ fails at s, despite the fact that $Rr(\pi, \pi')$ holds at s ($\{\langle s, s' \rangle\}$ is both the total set of transitions for both π and π'). So, shifting from binary transition relations to ternary transition relations isn't an insignificant move; though the language of PDL can't distinguish these structural nuances.

Now, it should be fairly clear where PDL fails our criteria: D1C2 and D3. First, note that the promise of PDL as a model of epistemic dynamics and epistemic norms is one of the primary motivators of this entire project. In examining PDL, interpret the states in PDL's model theory as epistemic states, the propositional language as one that contains epistemic or doxastic expressions, and the program expressions as encoding the content of epistemic norms. Keep this adaptation of PDL in mind in what follows.

PDL does well with respect to criterion D1, at first glance. Being a centerpiece of the dynamic turn in logic, it recognizes the importance of dynamics and, of course, if programs analyze the structure of epistemic norms, it provides an analysis of dynamic epistemic norms. It does well with D2 generally, as it explicitly provides an analysis of epistemic procedures and, in virtue of that, the content of epistemic norms by means of the syntax of programs. So far, so good.

Now, consider D1C2, which, in its sub-clauses explicitly distinguish procedures, actions, and processes from the epistemic transitions they result in. Moreover, the sub-clauses explicitly state the irreducibility of restrictions on epistemic processes (et al) to restrictions on epistemic transitions. For instance, Proposition 9 shows that restrictions on processes (et al) defined over LTSs are not reducible to restrictions on transitions defined over LTSs.

The model of epistemic dynamics provided by PDL gets this exactly wrong. Let the binary transition relations mapped to π and π' in all PDL structures be identical. Thus, π and π' are equivalent in all contexts in the language of PDL; they are perfectly intersubstitutable. Formally, if the binary transition relation, T_π, mapped to π is identical to that mapped to π' ($T_{\pi'}$) for every structure of PDL, then $\models_{PDL} A(\pi) \equiv A(\pi')$ where $A(x)$ is any formula of PDL containing x. This follows from substitution of identicals in the definition of the truth conditions for $A(\pi)$ and $A(\pi')$. If we take programs to represent procedures, processes, or actions, the perfect equivalence of π and π' violates D1C2 by rendering procedures, processes, and actions indistinguishable from transitions. If we take programs to represent norms that restrict procedures, processes, or actions, then restrictions on procedures, processes, and actions are indistinguishable from those on transitions. Summarily, there is nothing in the language of PDL that allows us to say that π and π' are distinct, despite sharing transitions.

Thus, PDL fails D1C2. As was argued extensively in Chap. 2 (re: Proposition 9), there are classes of assertions that may be true of π that are false of π' even if π and π' are mapped to equivalent transition relations. Moreover, the analysis of executions of programs purely in terms of binary relations either wrongly implies that restrictions on transitions are indistinguishable from restrictions on actions and processes, or it is simply incomplete, saying nothing about restrictions on actions and processes. In either case, PDL fails to live up to D1C2. Of course, if we define the semantics of PDL in terms of ternary relations, then there is semantic machinery for distinguishing processes (action, et al) from transitions; we could represent process (action) types with ternary relations and process (action) tokens with triples. For instance, we can interpret the triples, $\langle s, \pi, s' \rangle$ and $\langle s, \pi', s' \rangle$, as constructions that represent distinct processes. Nonetheless, the semantics of PDL, as they stand, will not enable us to distinguish these triples or the programs they execute.

As for D3, PDL cannot adequately formalize the forms of argument articulated above (Sects. 3.1, 3.2) for the simple reason that PDL cannot analyze restrictions on processes and actions in a way that distinguishes them from restrictions on transitions. It cannot, for instance, accurately validate argument forms leading from premises about procedural structure or procedural equivalence to other properties if those equivalences are finer grained than transitional equivalence and vice versa. As noted above, for any π and π', if $T_\pi = T_{\pi'}$, it follows that $\models_{PDL} A(\pi) \equiv A(\pi')$. This was argued in Chaps. 2 and 3 (and further in Chap. 4) to be quite incorrect. However, it's possible that, for any π and π', if some (hyperintensional) equivalence holds between π and π', then it follows that $\models_{PDL} A(\pi) \equiv A(\pi')$. PDL can't be used to formulate any candidates for this sort of equivalence. Conversely, suppose that $\models_{PDL} A(\pi) \equiv A(\pi')$. It follows that $T_\pi = T_{\pi'}$. Does it follow that some (hyperintensional) equivalence holds between π and π'? We can't even formulate the question in PDL. What could be the sufficient conditions for equivalence relations stronger than transitional equivalence? Similarly, PDL cannot tell us what follows from various forms of norm equivalence, dependence, or independence finer grained than transitional equivalence or what other conditions might suffice to entail them. Thus, the forms of reasoning that PDL can formalize are limited to arguments about properties that can be analyzed in terms of transitions only. PDL cannot, for instance, formalize an argument establishing that, in the Honest Twin case, a norm that dictates constructing proofs myself is inequivalent in any important way to a norm that dictates either asking my honest twin or doing the proofs myself.[14]

Further, consider a feature of PDL's expressiveness. Denote bisimulation with $\sim$. Formula A is bisimulation invariant iff for any graphs, g and g' such that $s \in S$ and $s' \in S'$, if $s \sim s'$, then $s \Vdash A$ iff $s' \Vdash A$. In other words, A is bisimulation variant iff it's not bisimulation invariant. It is well-known that the control operators of PDL are bisimulation invariant. As a result, the language of PDL contains no expressions that are bisimulation variant.[15] Bisimulation invariance is a kind of intensional invariance. If a finer grained analysis of norm equivalence is needed, PDL will not do the trick.

The foregoing arguments make no reference to how epistemic states are represented. All of the foregoing complaints have to do with the program logic that PDL employs. Fundamentally, PDL is constructed by attaching a bit of relational algebra to a modal logic. Actions are all interpreted with binary transition relations, and

[14] Formally, one norm dictates carrying out a transition in transition relation T_d—deducing for myself. The other dictates carrying out a transition in $T_t \cup T_d$—the union of the deduction and twin-asking transition relations. Of course, if $T_t = T_d$, then $T_t \cup T_d = T_t = T_d$. The two norms have the same requirements under PDL. Nonetheless, the Honest Twin case shows that they are pre-theoretically distinct. One can comply with the latter norm by carrying out courses of action that don't comply with the other. This is a product of the granularity of the analysis: binary transition relations ignore relevant levels of grain.

[15] Both of these technical results are proved and explored in [71–73].

so are programs. It's this feature that generates the problems for PDL. This is an important point because many formalisms that are plausible rivals of PLEN simply build the action and program logic of PDL on top of some innovation at the level of states or logic or they offer mild modifications of the action and program logic of PDL.

These arguments are further spelled out in Sect. 4.3.3 by means of the explication of the "processes problem". There, I articulate some forms of reasoning that PDL fails to formalize. That material reinforces the rest of this section. PLEN is designed to address all of these concerns by building in a robust array of distinct structures of representing process and transition distinctly, building constructions in its languages that require finer grained semantics than PDL, and analyzing the logic of an array of program equivalence relations, including hyperintensional protocol equivalences. This is addressed throughout Part II, culimating in Part III, in which it is shown that PLEN satisfies all of D1–D3C3. In the remaining parts of this section, I articulate some of the features of the base systems of DEL and AGM that run against the criteria defended up to this point. I leave the drawing of explicit contrasts with PLEN as an exercise.

3.3.2 Dynamic Epistemic Logic

Dynamic epistemic logic (DEL) has the best chances of being a logic of epistemic norms out of the rival formalisms of PLEN. Basic forms of DEL can be modified to incorporate action logics and even programs in addition to their well known and fruitful structural representations of epistemic states and the dynamics thereof. As such, such modifications of DEL seem plausible as formalisms with structures for representing both epistemic dynamics and epistemic norms.

First, let's define the core system of DEL.[16] First, the arguments we consider here won't depend on any details of the language, so let us say that the language of DEL is L_{DEL}. We start with epistemic models, $M = \langle S, R, V \rangle$ where S is a set of possible worlds, R is a set of binary relations each mapped to an agent representing uncertainty relations (i.e., $R_i(s, s')$ tells us that agent i can't tell the difference between s and s'), and V is a valuation function on the atomics of L_{DEL}. Epistemic models represent epistemic uncertainty relations (per agent), and these are used to read off properties and contents of epistemic states. That is, for agent i, R_i in model M tells us everything we need to know about what i knows (or, in a doxastic logic, what i believes), e.g., what's true at the worlds connected by R_i in M gives a snapshot of the contents of i's knowledge and beliefs. We build toward a dynamic logic by defining event models, $E = \langle \Sigma, R_\Sigma, Pre \rangle$, where Σ is a set of events, R_Σ is a set of uncertainty relations between events, and Pre is a function

[16] Following only the textbook presentation in [74] and, to a lesser extent, the overall contours of DEL outlined in [5].

from Σ into $P(L_{DEL})$. Event models add to the content of an agent's epistemic state by representing uncertainty relations among events as well as some "external" information about the preconditions ($Pre(e)$) of events.

Product updates represent transitions from one set of uncertainty relations to another that are effected by particular events. Formally, product updates are just functions from pairs, (M, E), of epistemic and event models to epistemic models that combine the uncertainty relations among states in M and events in E while preserving valuations for atomics at all states. That is, epistemic dynamics are conceptualized in terms of transitions among epistemic states associated with events. The properties of epistemic dynamics can be read off of these updates.

Call a dynamic epistemic logic with structures analyzing actions an Epistemic Action Logic (EAL).[17] To analyze the epistemic features of actions, DEL deploys pointed models. First, we consider pointed epistemic models (M, s) where s is one of the states in M. Second, we consider pointed action models, (E, e), where E is an event model and e is a particular event in the set of events in E. Product update generates a pointed epistemic model $(M', (s, e))$ in a straightforward way that restricts or, in more fully generalized versions (re: [5, 74]), expands the uncertainty relation. The thought is that action is a species of event, uncertainty relations encode the properties and content of epistemic states, and to see the epistemic effects of an action, one must see how the action updates the uncertainty relations at the current world.

An action algebra can be added to DEL with EAL to generate a Logic of Epistemic Programs (LEP) [5]. Here are two ways to do this. First, simply take the foregoing machinery and add program control operators to the language so that the basic actions can be combined via procedural operations. At a minimum, consider choice $\cup$ and sequential composition ;. Then define functions on product updates in order to assign product updates to epistemic programs. For an action a, we know that there is an action model (E, a) and a product update function that identifies all transitions among epistemic models for a. The same, mutatis mutandis, works for b. For the program $a; b$, then, the product update of the program is just the composition of the product updates of a and b. The choice is the union of them. A second way to do this is to assign to each program a program model, (E, π), which is an action model such that π is a designated set of actions in E. Then the update for (E, π) is the update for E restricted to the designated actions, i.e., we update any (M, s) to $(M', (s, e))$ where $e \in \pi$. In either case, there is a way in which complex procedures built out of epistemic actions can be analyzed, and the underlying model of epistemic dynamics moves beyond input assimilation to tree-like representations of possible unfoldings of a system.

LEP is a dynamic epistemic logic with a program logic on top. As such, it provides a prima facie plausible underlying model of epistemic dynamics and a plausible looking model of epistemic normativity in just the same way that PDL or PLEN do. Let the programs in LEP represent procedural epistemic norms

[17] See [74] for technical details.

that dictate operations on epistemic actions. The product updates represent the transitions that result from executing the instructions of these programs and so represent the rational courses of action according to the norms the programs represent.

Now, the fundamental problems of DEL (generalized with EAL) are two-fold. First, the use of Kripke models as the basis for structures that represent epistemic states generates the well-known problem of logical omniscience and closure of knowledge [29, 36]. This is a result of defining the semantics of knowledge and belief operators in terms of transition relations among worlds. I won't comment on this more, except to note that syntactic epistemic logics do not have this bug, and hyperintensional analyses might provide another way out.

Second, and the more serious for the goal of pre-emptively contrasting DEL to PLEN, DEL does not satisfy criterion D1C2, and thus cannot satisfy D3. In contrast to PDL, however, DEL conflates processes and transitions not by omitting machinery for representing processes, but by omitting machinery for representing transitions *as distinct from processes*. No two distinct actions represented in DEL can be transition equivalent. The epistemic features of actions in DEL are represented by event models, but the dynamic features of actions are defined by product updates. Consider actions a and b. It's tempting to take the phrase "action model" on its face, and think that a can be represented with pointed event models. But this is a mistake; actions are encoded syntactically, as in PLEN, by means of the symbols in action models, and their dynamic features are modeled indirectly by the mapping provided by product updates. Recall that product update is a function from pairs of pointed epistemic models and pointed action models, $((M, s), (E, a))$, to pointed epistemic models, (M', s'), where the states in the resulting epistemic model are pairs, $s' \in S \times E$. The transition relations that define execution of actions in DEL, then, are defined like so: $(M, s) \mapsto_{(E,a)} (M', s')$ iff the preconditions of a are all satisfied at s, M' is the update product of $((M, s), (E, a))$, and $s' = (s, a)$.

This rules out transition equivalent yet distinct actions. Consider the following action models. $(E, a) = \langle E, \rightarrow_i, pre\rangle$ where $a \in E$, and $(E', b) = \langle E', \rightarrow'_i, pre'\rangle$ where $b \in E'$. Suppose that a and b define identical product updates. Thus, for any pointed epistemic model $(M, s) = \langle S, R, V, s\rangle$, the product update from $((M, s), (E, a)) = (M', (s, a)) = ((M, s), (E, b))$. But this just means that $a = b$, that $\longrightarrow_i = \longrightarrow'_i$, and that $pre = pre'$. Thus, $(E, a) = (E, b)$. Thus, two distinct actions, a and b, cannot be transition equivalent; the terminal states in transitions assigned to actions via the product update carry a "record" of which action brought about the change, and so no two actions, DEL, can possibly be associated with equivalent transition relations.

The foregoing property is a serious failure to accurately model epistemic dynamics as well as the content of epistemic rationality norms. First, as argued in Chaps. 2 and 3, it's plausible that there really are transition equivalent epistemic actions. This means that DEL straightforwardly mischaracterizes epistemic actions by ruling out an entire class of potentially—and, if Chaps. 2 and 3 are right—epistemologically relevant actions by technical fiat. Second, the argument above shows that there cannot even be transition convergent actions in DEL. For no pair of actions, a and b, is it possible for there to be any (M, s) such that (E, a) and (E, b) to lead to the same terminal state from (M, s) while $a \neq b$. Explicitly, the execution rules of DEL rule this out. $(M, s) \mapsto_{(E,a)} (M', s')$ iff the preconditions of a are all satisfied at s, M' is the update product of $((M, s), (E, a))$, and $s' = (s, a)$, and $(M, s) \mapsto_{(E,b)} (M', s')$ iff the preconditions of b are all satisfied at s, M' is the update product of $((M, s), (E, b))$, and $s' = (s, b)$. If a and b converged, for even one transition, $((M, s), (M', s')$, then $s' = (s, a) = (s, b)$, and so $a = b$. As a and b are arbitrary, this is a general feature of this basic form DEL.[18]

The problem is that the distinction between processes is built directly into the machinery for representing transitions. What is necessary is machinery that can represent transitions and processes separately while also capturing their relations. By building machinery that distinguishes processes at a fine grain (the syntactic level) into the terminal states of transitions, this assimilates all of the properties of transitions to those of processes rather than, as PDL does, the other way around. But, again, it's quite clear that there are distinct epistemic actions and complex epistemic processes that are transition convergent. Thus, DEL gets epistemic dynamics wrong.

All of this adds up to DEL failing to adequately formalize D1C2 because epistemic rationality can distinguish transition equivalent and transition convergent actions and processes (as well as restrictions on epistemic dynamics that distinguish them), and nothing in the base form of DEL can approximate such rational distinctions because DEL cannot even recognize these categories of actions (et al). As a result, D3 can't be formalized in DEL (extended to LEP). The restrictions on epistemic dynamics defined by procedural norms cannot be accurately represented (though dynamic restrictions might be well represented) or reasoned about, especially if we want to reason about the equivalence, dependence, and independence relations among transitionally equivalent norms. If we wanted to think, for instance, about whether procedurally inequivalent restrictions on dynamics do not reduce to transition equivalent restrictions, we are out of luck; there is no machinery for doing so in the base form of DEL. The forms of reasoning we need that distinguish epistemic norms that are equivalent in one sense and inequivalent in others will not be available without some modification or addition to DEL (extended to LEP).

3.3.3 AGM Belief Revision Theory

In many ways, AGM is the most natural model of epistemic dynamics. Epistemic states are represented simply and intuitively with sets of sentences of a propositional language. The core constructions of AGM are belief sets, deductively closed sets of sentences K (i.e., $K = Cn(K)$). This gives them rather transparent appeal as

---

18 It does, of course, not necessarily apply to all formalisms derived from the main ideas of DEL. There are intensional protocol variants of DEL, for instance, that seem not to have this problem [75].

models of the content of beliefs. This modeling decision also makes it easy to apply classical logic to the analysis.

There are numerous belief revision theories based on the AGM framework or, if you will, versions of AGM that can be built by modifying the basic formalism [20, 26, 76]. The core problem among AGM theories is that epistemic dynamics reduces to transitions among epistemic states. Thus, AGM theories suffer at least the processes problem, and, as with PDL, this spreads to the failure to formalize D1C2 and D3.

Explicitly, dynamics in AGM theories is defined by a set of epistemic commitment functions that take belief sets and inputs (particular sentences in the base formalism) to belief sets. For instance, letting $K!$ be the set of all belief sets and L a logical language, an epistemic commitment function can be defined encoding the operation of belief set expansion: $K{+}_A : K! \times L \longrightarrow K!$. From there, we can identify the set of epistemic commitment functions that satisfy rationality postulates encoding various sorts of epistemic norms. Indeed, the closure condition already encodes a normative assumption.

As Gardenfors shows [22], these functions are equivalent to relations on belief sets. Hence, the basic actions of belief change in AGM theories are analyzed with transition relations among belief sets. This is as natural a model of epistemic dynamics as the model of epistemic dynamics; changes of epistemic state consist of transitions among representations of the contents of belief. Epistemic norms, then, are specified by rationality or entrenchment postulates, and the restrictions on epistemic dynamics that they define are the epistemic commitment functions that satisfy those postulates. Unfortunately, this generates a couple serious problems with respect to the goal of providing an underlying model of dynamics for an account of epistemic norms.

AGM belief revision theories fail to satisfy D1C2, D2, and D3. AGM inherits the problems for PDL and others besides. Epistemic dynamics, even if we translate into the iterated case, entirely reduces to transitions among belief sets, however those are defined.[19] In other words, no AGM theory can represent the basic distinction between processes and transitions; every process of epistemic change in AGM theories just is a sequence of transitions. AGM theories simply provide a model of the space of all formally possible transitions among belief sets and the subset that are defined by the epistemic commitment functions. Thus, there is no distinction between processes and epistemic transitions in AGM theories, and this leaves AGM theories powerless to represent the numerous distinctions and properties of epistemic dynamics articulated in Chaps. 2 and 3.

As a result of these problems, no AGM theory can accurately represent properties like methodological inclusion, procedural information, or epistemic distinctions between processes. See the processes problem in Sect. 4.3 for a full development. For now, it can be seen intuitively. Procedural information is entirely invisible in

[19] This needn't be stepped through for each case; just look at the explicit definitions of the formalisms, re: [13, 21, 22, 57, 59, 76].

AGM theories except where it corresponds to differences in belief set transitions. In AGM theories, if two processes (e.g., those in the Honest Twin case) define identical transition relations (epistemic commitment functions), they are identical in AGM. Expanding with *A* by inference is entirely indistinguishable from expanding by observation. Expanding with *A* by inferring in accordance with modus ponens is indistinguishable from inferring *A* by flipping a coin. Revising as a result of applying some inconsistency or consequence detecting algorithm is indistinguishable from revising as a result of paying undue credence to psychic testimony. And so on. This straightforwardly implies that AGM theories can say nothing of the procedural differences in cases like the Honest Twin, and so fail with respect to criterion D1C2.

Normative information is also obscured or neglected. Despite the features of AGM theories that represent epistemic norms (the rationality postulates, entrenchment conditions, etc.) and compliance with them (epistemic commitment functions), nothing in them represents all or even any of the same features of procedural epistemic norms that the arguments of Chaps. 2 and 3 argue are important. The rational distinctions between epistemic processes in the Researcher case (the deduction vs the psychic testimony), Bad Reasoning case, or the Honest Twin case are invisible in AGM. Thus, AGM theories cannot adequately represent epistemic dynamics nor can they adequately represent the verdicts of procedural epistemic rationality norms or, especially, can it provide a rigorous framework for reasoning about their divergences, dependences, equivalences, and so on.

Moreover, AGM implicitly reduces all restrictions on epistemic dynamics to restrictions on states and transitions. This straightforwardly violates Proposition 9 and so fails D1C2. As a result of all of this, AGM theories uniformly and straightforwardly fail to formalize the forms of reasoning articulated by D3. AGM theories don't recognize procedural norms at all, let alone the restrictions that procedural norms specify or the logical relations among them that we want to reason about.

3.4 From Epistemic Proceduralism to the Protocol-Theoretic Framework

The primary objective of this chapter was outlining the logic of epistemic norms—or at least some of the features required by an adequate logic of epistemic norms—by articulating clusters of forms of reasoning that are central to the regulative use of epistemic norms. If the arguments of Chaps. 2 and 3 are correct, this shows that D1 through D3C3 are criteria on putative formal frameworks for the logic of epistemic norms.

The secondary objective was to apply these criteria to highlight the main ideas of PLEN in advance by showing how at least the basic forms of the formalisms from which PLEN derives fail to satisfy criteria D1–D3C3. From here, the rest of the book presents an extended argument for the claim that the protocol-theoretic logic of epistemic norms (PLEN) is a logical framework that satisfies these criteria. The next part of the book defines and develops PLEN for this purpose.

References

1. Alchourrón, C.E. 1991. Conflicts of norms and the revision of normative systems. *Law and Philosophy* 10(4):413–425.
2. Alechina, N., M. Jago, and B. Logan. 2008. Preference-based belief revision for rule-based agents. *Synthese* 165(2):159–177.
3. Anderson, A.R. and N. Belnap. 1975. *Entailment: The logic of relevance and necessity*, vol. I. Princeton: Princeton University Press.
4. Arieli, O. and A. Avron. 2000. General patterns for nonmonotonic reasoning: from basic entailments to plausible relations. *Logic Journal of the IGPL* 8(2):119–148.
5. Baltag, A. and L.S. Moss. 2004. Logics for epistemic programs. *Synthese* 139:165–224.
6. Beall, J. 2011. Multiple-conclusion LP and Default Classicality. *Review of Symbolic Logic* 4(2):326–336.
7. Beall, J. 2015. Free of Detachment: logic, rationality, and gluts. *Noûs* 49(2):410–423.
8. Beauchamp, T.L. and J. Childress. 1994. *Principles of biomedical ethics*. Oxford University Press.
9. Boghossian, P. 2008. Epistemic rules. *Journal of Philosophy* 105(9):472–500.
10. Bratman, M. 1998. *Intention, plans, and practical reason*. Center for the Study of Language and Information.
11. Broome, J. 2000. Normative requirements. In *Normativity*, ed. J. Dancy. Oxford: Blackwell.
12. Cherniak, C. 1986. *Minimal rationality*. MIT Press.
13. Chopra, S. and Parikh, R. 2000. Relevance sensitive belief structures. *Annals of Mathematics and Artificial Intelligence*, 259–285.
14. Christensen, D. 2004. *Putting logic in its place*. Oxford: Oxford University Press.
15. Clifford, W.K. 1999. *The ethics of belief and other essays*. Prometheus Books.
16. Cohen, L.J. 1992. *An essay on belief and acceptance*. Oxford: Oxford University Press
17. Field, H. 2001. *Truth and the absence of fact*. Oxford: Oxford University.
18. Field, H. 2009a. Epistemology without metaphysics. *Philosophical Studies* 143(2):249–290.
19. Field, H. 2009b. Pluralism in logic. *Review of Symbolic Logic* 2(2):342–359.
20. Gabbay, D., O. Rodrigues, and A. Russo. 2008. Belief revision in non-classical logics. *Review of Symbolic Logic* 1(3):267–304.
21. Gardenfors, P. 1992. *Belief revision*. Cambridge: Cambridge University Press.
22. Gardenfors, P. 2008. *Knowledge in flux*. Reprint. Original: 1988. London: College Publications.
23. Gerken, M. 2014. Same, same but different: the epistemic norms of assertion, action and practical reasoning. *Philosophical Studies* 168(3):725–744.
24. Glymour, C. 1980. *Theory and evidence*. Princeton: Princeton University Press.
25. Harel, D., D. Kozen, and J. Tiuryn. 2000. *Dynamic logic*. MIT Press.
26. Hansson, S.O. 2017. Logic of belief revision. In *The Stanford Encyclopedia of Philosophy* (Winter 2017 Edition), ed. E.N. Zalta, forthcoming. https://plato.stanford.edu/archives/win2017/entries/logic-belief-revision/.
27. Harman, G. 1986 *Change in view*. Cambridge: MIT Press.
28. Hume, D. 1998. *An enquiry concerning human understanding: A critical edition*. Oxford University Press.
29. Jago, M. 2009. Epistemic logic for rule-based agents. *Journal of Logic, Language and Information* 18(1):131–158.

30. Jenkins, R. 2018. The Philosophical Foundations of PLEN: A Protocol-theoretic Logic of Epistemic Norms. Ph.D Dissertation: CUNY Graduate Center.
31. Kelly, K.T., O. Schulte, and V. Hendricks. 1995. Reliable belief revision. *Logic and Scientific Methods: Volume One of the Tenth International Congress of Logic, Methodology and Philosophy of Science, Florence*, 383–398.
32. Kirsh, D. and P. Maglio. 1992. Some epistemic benefits of action-tetris, a case study. In *Proceedings of the 14th annual conference of the cognitive science society.*
33. Kirsh, D. and P. Maglio. 1994. On distinguishing epistemic from pragmatic action. *Cognitive Science* 18(4):513–549.
34. Kolodny, N. 2005. Why be rational? *Mind* 114(455):509–563.
35. Kolodny, N. 2007. State or process requirements? *Mind* 116(462):371–385.
36. Konolige, K. 1986. *A deduction model of belief.* Morgan Kaufmann Publishers.
37. Leitgeb, H. and R. Pettigrew. 2010. An objective justification of Bayesianism I: Measuring inaccuracy. *Philosophy of Science* 77(2):201–235.
38. Levi, I. 1996. *For the sake of the argument: Ramsey test conditionals, inductive inference, and nonmonotonic reasoning*. Cambridge University Press.
39. Levi, I. 2004. *Mild contraction: Evaluating loss of information due to loss of belief.* Oxford University Press.
40. Maglio, P.P., M.J. Wenger, A.M. Copeland. 2003. The benefits of epistemic action outweigh the costs. In *Proceedings of the Annual Meeting of the Cognitive Science Society,* vol. 25. Permalink: https://escholarship.org/uc/item/5bg5p6bv
41. Makinson, D.C. 1965. The paradox of the preface. *Analysis* 25(6):205–207.
42. Marr, D. 1982. *Vision: A computational investigation into the human representation and processing of visual information*. New York, NY, USA: Henry Holt.
43. Meyer, R.K. 1971. Entailment. *Journal of Philosophy* 68(21):808–818.
44. Meyer, J.-J. 1987. A different approach to deontic logic: deontic logic viewed as a variant of dynamic logic. *Notre Dame Journal of Formal Logic* 29(1):109–136.
45. Meyer, J.-J. 2000. Dynamic logic for reasoning about actions and agents. In *Logic-based artificial intelligence,* ed. J. Minker. Kluwer International Series in Engineering and Computer Science, vol. 597, 281–311. Norwell, MA, USA: Kluwer Academic Publishers.
46. Montminy, M. 2013. Why assertion and practical reasoning must be governed by the same epistemic norm. *Pacific Philosophical Quarterly* 94(1):57–68.
47. Papineau, D. 2013. There are no norms of belief. In *The aim of belief,* ed. T. Chan. Oxford University Press.
48. Pavese, C. 2015. Practical senses. *Philosophers' imprint* 15.
49. Pettigrew, R. 2011. An improper introduction to epistemic utility theory. In *Proceedings of EPSA: Amsterdam '09,* 287–301, ed. H. de Regt, S. Okasha, S. Hartmann. Springer.
50. Pettigrew, R. 2013a. Epistemic utility and norms for credences. *Philosophy Compass* 8(10):897–908.
51. Pettigrew, R. 2013b. A new epistemic utility argument for the principal principle. *Episteme* 10(1):19–35.
52. Pettigrew, R. 2016. Epistemic utility arguments for probabilism. In *The Stanford Encyclopedia of Philosophy* (Spring 2016 Edition), ed. E.N. Zalta. https://plato.stanford.edu/archives/spr2016/entries/epistemic-utility/.
53. Pettigrew, R. 2017. Epistemic utility and the normativity of logic. *Logos and Episteme* 8(4):455–492.
54. Priest, G. 2006. *Doubt truth to be a liar.* Oxford: Oxford University.
55. Read, S. 2010. *Relevant logic: A philosophical examination of inference*. Originally Published by Basil Blackwell 1988. Revised Edition (c) Stephen Read 2010.
56. Restall, G. 2005. Multiple conclusions. *Logic, methodology and philosophy of science: Proceedings of the twelfth international congress.*
57. Restall, G. and J. Slaney. 1995. Realistic belief revision. In *Proceedings of the second world conference on the fundamentals of artificial intelligence*, 367–378, ed. M. De Glas (Paris, July 1995)

58. Rott, H. and M. Pagnucco. 1999. Severe withdrawal (and recovery). *Journal of Philosophical Logic* 28(5):501–547.
59. Ryan, M. 1996. Belief revision and ordered theory presentations. In *Logic, action, information*, ed. A. Fuhrmann, H. Rott, 129–151. Berlin: De Gruyter.
60. Schafer, K. 2014. Doxastic planning and epistemic internalism. *Synthese* 191(12):2571–2591.
61. Schmechtig, P. and M. Grajner. (eds.) 2016. *Epistemic reasons, norms, and goals*. De Gruyter.
62. Schulte, O. 1999. Means-ends epistemology. *British Journal for the Philosophy of Science* 50(1):1–31.
63. Schulte, O. 2017. Formal learning theory. In *The Stanford Encyclopedia of Philosophy* (Spring 2017 Edition), ed. E.N. Zalta. https://plato.stanford.edu/archives/spr2017/entries/learning-formal/.
64. Schurz, G. 2011. Truth-conduciveness as the primary epistemic justification of normative systems of reasoning. *Behavioral and Brain Sciences* 34(5):266–267.
65. Schurz, G. 2011. Routines. *Synthese* 65(2):185–210.
66. Segerberg, K. 2012. $D\triangle L$: Dynamic deontic logic. *Synthese* 185(S1):1–17.
67. Spohn, W. 2012. *The laws of belief: ranking theory and its philosophical applications*. Oxford: Oxford University Press.
68. Staffel, J. 2009. Iterated belief revision. *Erkenntnis* 70(2):189–209.
69. Staffel, J. 2013. Can there be reasoning with degrees of belief? *Synthese* 190(16):3535–3551.
70. Tennant, N. 2012. *Changes of mind: An essay on rational belief revision*. Oxford: Oxford University Press.
71. van Benthem, J.F.A.K. 1994. Programming operations that are safe for bisimulation. *Studia Logica* 60(2):311–330.
72. van Benthem, J.F.A.K. 1996. *Exploring logical dynamics*. Stanford: CSLI Publications.
73. van Benthem, J.F.A.K. 1998. Program constructions that are safe for bisimulation. *Studia Logica* 60(2):311–330.
74. van Ditmarsch, H., J. Halpern, W. Van Der Hoek, and B. Kooi. 2007. *Dynamic epistemic logic*. Springer.
75. van Lee, H.S., R.K. Rendsvig, and S. van Wijk. 2019. Intensional protocols for dynamic epistemic logic. *Journal of Philosophical Logic* 48(6):1077–1118.
76. Wassermann, R. 2011. On AGM for non-classical logics. *Journal of Philosophical Logic* 40(2):271–294.

Part II
Developing PLEN

Chapter 4
PLEN: A Protocol-Theoretic Logic of Epistemic Norms

Abstract In order to motivate the protocol-theoretic framework for the logic of epistemic norms, the prior chapters articulated and defended several clusters of principles characterizing the structure, regulative role, and logic of epistemic norms. This chapter develops the core of the protocol-theoretic framework: the protocol-theoretic logic of epistemic norms (PLEN). PLEN is an elaborate revision of propositional dynamic logic (PDL). It encodes the content of epistemic norms with protocols, syntactic expressions that represent complex procedural operations on actions, and it models epistemic dynamics with process models that distinguish epistemic transitions from the actions, processes, and procedures that result in them while also characterizing their interrelations. The main elaborations involve formulae that encode properties of protocols that are invisible in the multi-modal logic of PDL. The main work of the chapter is defining the framework, and exploring it mainly from a conceptual perspective, interpreting the logic of protocols in PLEN as a logic of procedural epistemic norms defined over a model of single-agent epistemic dynamics, with minimal discussion of the standard metatheory. As far as the standard metatheory goes, it is shown that the deductive system of PLEN is sound with respect to PLEN's semantics, and it is conjectured that the standard proofs of completeness and decidability for PDL can be extended to PLEN. This main work lays the initial groundwork for codifying epistemic proceduralism in PLEN.

4.1 Introduction

4.1.1 The Protocol-Theoretic Logic of Epistemic Norms (PLEN) and the PLEN Framework

The protocol-theoretic logic of epistemic norms (PLEN) is a formal system with which to represent and reason about epistemic norms. The protocols of PLEN are, syntactically, the program constructions of propositional dynamic logic (PDL) or dynamic epistemic logic (DEL). These familiar syntactic objects are interpreted in PLEN over more elaborate structures—for the purposes of representing epistemic

R. Jenkins, *A Protocol-theoretic Framework for the Logic of Epistemic Norms*, Synthese Library 467, https://doi.org/10.1007/978-3-031-08597-0_4

actions and norms—than in PDL or DEL, resulting in significant differences in the resulting logic of protocols.[1]

The design objective of PLEN is to devise a framework in which distinct sublogics of protocols can be built and compared with regard, inter alia, to their theories of protocol equivalence. It would be useful, for instance, to have a system in which to rigorously draw the consequences of different conceptions of protocol equivalence, especially with respect to using the resulting logic for reasoning about epistemic norms.[2] The idea is to develop the PLEN Framework, a full framework for building and analyzing protocol logics, out of PLEN. This project is presented iteratively. This chapter presents the core system of PLEN with the goal in mind of getting a handle on how the formal machinery works and how it is intended to be used to model epistemic norms and epistemic dynamics. The next two chapters elaborate on this system with the goal in mind of providing a theory of protocol equivalence and a powerful logic of protocols with which to, ultimately, formalize epistemic proceduralism. The formalization of epistemic proceduralism is presented in stages over Chaps. 7–10.

Accordingly, the chapter is broken into two parts. First, I define the system in a relatively no-frills way, remarking on some minimal exploration of the system's metalogical properties. Then, in the second half of the chapter, I present the main ideas of the philosophical interpretation of PLEN as a logic of epistemic norms,

[1] Protocols have been well-analyzed from perspectives that diverge quite a bit from the interests animating PLEN. See [37] for an analysis of protocols as abstract, procedural constructions governing human social behavior (of a piece with Parikh's social software research programme [26]). There, Wang deploys a number of upgrades of PDL and the action models of DEL to analyze his target semantic phenomena. In particular, see [18] for the use of the notion of protocols to bind DEL and temporal logic (TL) into a novel system of temporal dynamic epistemic logic (TDEL). See [25] for a slightly more elaborate analysis of protocols, and the embedding of protocols into models of reasoning under imperfect information. These three projects were germinal for PLEN, so there are some convergences with the ideas presented there, however different their philosophical and technical foci. The protocols of [25] can be seen as the prototypes of the epistemic directed graphs introduced below and the protocol graphs introduced in Chap. 5, which compose a slightly more abstract class of constructions. For further relevant analysis of protocols, see: [19, 36] within the PDL and DEL framework. For another germinal take on protocols, from the mult-agent systems or interpreted systems perspective, see [8]. This is a small sampling of work on protocols. See the references of the cited works for a more complete picture of the literature on protocols. The point of this brief excursion is that the logic of protocols is not itself novel, and the development of PLEN is not the most advanced technical exploration of possible frameworks for analyzing the logic of protocols. The protocol logic of PLEN is, however, more elaborate—at least with respect to the philosophical distinctions driving PLEN—than that of most forms of PDL, DEL, or multi-agent systems. Moreover, the focused application of this (perhaps parochially) more elaborate logic of protocols to the analysis of procedural epistemic norms (guided by the undergirding philosophical programme) is novel, as far as I know. There is, for instance, a marked contrast between PLEN's approach to the procedural norms impacting epistemic state change and recent dynamic doxastic logics or logics of belief revision (typically built in the PDL or DEL framework), e.g., [3, 23, 24, 29, 30, 35]. These systems are not focused on analyzing the same phenomena.

[2] Recall, of course, the distinctive forms of reasoning about and with epistemic norms articulated in Chap. 3, Sect. 3.1.

illustrate how the system works by putting PLEN to work on some sample cases, and then I explicate the basic philosophical underpinnings of the system that distinguish it from near neighbors in the space of formalisms.

4.2 Technical Development

There are some notational conventions to note prior to getting on with the chapter. First, in all of what follows, I deploy a less than optimized and often ad hoc notation that may deploy familiar symbols in novel ways. Where this is the case, I explicitly define the novel use, though context should disambiguate on pain of category error. For instance, I use $*$ both as the iteration operator on protocols and as the finite sequence operator on sets, i.e., where S is a set, $S*$ is the set of all finite sequences of elements of S. For another instance, for sets, X and Y, I will use X/Y and $X - Y$ interchangeably to denote relative complement (set difference). This is for readability.

Second, I will define the system using first-order language where it is more or less clearly unproblematic, and I will use abbreviated quantificational expressions in ways that should be transparent. For instance, $\exists x \in S(F(x))$ is equivalent to $\exists x(x \in S \wedge F(x))$, as usual. Third, despite my abbreviations, I sometimes formulate complex definitions or conditions with first order language expressions that are quite long. To save horizontal space, I present these in vertically tiered or arrayed form. These should be intuitively readable, e.g., expressions of the form $\forall x \in S(F(x) \wedge ... \wedge G(x))$ are, if they are quite long, often presented in the form $\forall x \in S\left(\bigwedge\begin{pmatrix} F(x) \\ ... \\ G(x) \end{pmatrix}\right)$ or $\forall x \in S\left[\bigwedge\begin{bmatrix} F(x) \\ ... \\ G(x) \end{bmatrix}\right]$. I may omit conjunction in lieu of commas between entries, though disjunctions will always be explicit. I will also present conditional expressions in a rule like form, e.g., $\forall x \in S(F(x) \rightarrow G(x))$ may be presented as $\forall x \in S\left(\frac{F(x)}{G(x)}\right)$. The conditional interpretation of these expressions will apply in the formulation of definitions, while they denote a metalevel consequence relation when rules outside the context of a definition are formulated this way, unless explicitly noted. These notational conventions are in force throughout the rest of the book.

4.2.1 Syntactic System

As a logic, PLEN is a formal system comprising three subsystems. These systems are presented in Sects. 4.2.1, 4.2.2, and 4.2.3 in turn.

The first subsystem is a two-sorted, ω-typed syntactic system. The first sort of syntactic object, *protocols*, is defined by the protocol language P_{PLEN}. The second sort of syntactic object, *protocol-theoretic formulae*, is defined by the protocol-theoretic language, L_{PLEN}. P_{PLEN} and L_{PLEN} are defined by simultaneous recursion, resulting in an ascending hierarchy of types of syntactic objects with protocols of type 0 and formulae of type 0 at the base, protocols of higher types defined over both protocols and formulae of lower types, and formulae of higher types defined over both protocols and formulae of lower types. The protocol-theoretic formulae defined in PLEN codify logical or procedural relations among protocols (including multiple kinds of what will eventually be classed as local, execution-theoretic protocol equivalence relations), the modal or dynamic properties of protocols (via modal operators that embed each protocol), and pre and postconditions of protocols. It also, somewhat uniquely, has a number of formulae that encode syntactic features of protocols. These are commented upon after introduction in Sect. 4.2.1 and also after their "semantics" are introduced in Sect. 4.2.2.

Remark 20 The syntax is defined in both a typed string-algebraic form and in an untyped infix notation with explicit prohibition of self-reference. For the intents of this project, they are equivalent languages. The typed string-algebraic form is for the rigorous definition of construction procedures and for the proof of core results. The infix notation form is for ease of use in devising examples and formulating claims.

Definition 21 (String Algebra)

Let x be a string of finite length n over set X iff x is a function from $\langle 1, ..., n\rangle$ $(n \in \mathbb{N})$ into X; $x = f_x : \langle 1, ..., n\rangle \longrightarrow X$.

Let $String(X)$ be the set of all strings of finite length defined over X.

Let $\frown$ be a string concatenation function such that: If $a = f_a : \langle 1, ..., n\rangle \longrightarrow X$ and $b = f_b : \langle 1, ..., m\rangle \longrightarrow X$, then $a \frown b = f_{a\frown b} : \langle 1, ..., n+m\rangle \longrightarrow X$ where, for any input, $l \in \langle 1, ..., n+m\rangle$, $a \frown b(l) = \begin{cases} f_a(l) & \text{if } l \leq n \\ f_b(l-n) & \text{if } l > n \end{cases}$.

(String Algebra) $String(X)$ is a *string algebra* iff:

(1) $X \subseteq String(X)$ (Base),
(2) $a \frown (b \frown c) = (a \frown b) \frown c$ (Associativity),
(3) $\forall a \in X \neg\exists b, c \in String(X)[a = b \frown c]$ (Atomicity),
(4) $\forall a \in String(X)\exists b_1, ..., b_n \in X[b_1 \frown ... \frown b_n = a]$ where n is finite (Finite Generativity),
(5) If $a = x \frown b$, $a = y \frown c$, and $x, y \in X$, then $x = y$ and $b = c$ (Unique Decomposition/Readability).

Remark 22 The syntactic system, comprising both P_{PLEN} and L_{PLEN}, is defined by simultaneous recursion, though, of course, I have no choice but to present the definitions in sequence; first, I present the protocol language in string-algebraic and then infix form, and then the protocol-theoretic language in string-algebraic and then infix form.

Definition 23 (String Algebraic Protocols of Type i (P^i_{PLEN})) Let $\Pi_0 = \{a_1, ..., a_n, ...\}$ be a countable, non-empty set of atomic protocols.

Let $Prot = \{; , \cap, \cup, *\}$ and $Prot+ = \{(-)?, -, \curvearrowleft, \nabla\}$ be the *regular* and *irregular control operators*, respectively.

The arity of each control operator is denoted $a(\circ)$ and explicitly specified for each control operator: if $\circ \in Prot \cup Prot + / \{*, (A)?, -, \curvearrowleft, \nabla\}$, then $a(\circ) = 2$; else, $a(\circ) = 1$ and $\circ \frown y_1 \frown y_{a(\circ)}$ is elliptical for $\circ \frown y_1$.

(Protocols of Type i) The set of protocols of type i, P^i_{PLEN}, is the smallest subset of $String(\Pi_0 \cup Prot \cup Prot + \cup \bigcup_{j=1}^{i-1} L^j_{PLEN})$ closed under the following rules:

(1) If $x \in \Pi_0$, then $x \in P^0_{PLEN}$.
(2) If $x = \circ \frown y_1 \frown y_{a(\circ)}$, where each $y_j = \langle x_1 \frown ... \frown x_n \rangle$, each $y_j \in \bigcup_{j=1}^{i-1} P^j_{PLEN}$, some $y_j \in P^{i-1}_{PLEN}$, and $\circ \in Prot \cup Prot+$, then $x \in P^i_{PLEN}$.
   (a) For any protocol, x, $x = \circ \frown y_1 \frown y_{a(\circ)}$ or $x = \circ \frown y_{a(\circ)}$ iff $\circ$ is the *main control operator* of x.
      (i) iff $MC(x) = \circ$.
      (ii) iff $\circ$ is the first symbol in the string.
(3) If A is a metavariable for a formula of L^{i-1}_{PLEN}, then $(A)? \in P^i_{PLEN}$.

Definition 24 (String Algebraic Protocol Language (P_{PLEN}))

$$P_{PLEN} = \bigcup_{i=1}^{\omega} P^i_{PLEN}$$

Definition 25 (Infix Notation Protocol Language (P_{PLEN})) Where $a_i \in \Pi_0$, π_i is a metavariable for an arbitrary protocol with no occurrences of self-reference, and A is a metavariable for an arbitrary protocol-theoretic formula with no occurrences of self-reference, P_{PLEN} can be defined by the following grammar:

$$(a_i) \mid (\pi_1; \pi_2) \mid (\pi_1 \cap \pi_2) \mid (\pi_1 \cup \pi_2) \mid (\pi_1 *) \mid (A)? \mid (-\pi_1) \mid (\curvearrowleft \pi_1) \mid (\nabla \pi_1)$$

Definition 26 (Protocol-Theoretic Formulae of Type i (L^i_{PLEN})) Let $AForm = \{P_i, Q_i, ..., \top, ...\}$ be a countable, non-empty set of atomic protocol-theoretic formulae, and define the following sets of operators for building complex expressions and embedding protocol expressions, where the variable $x \in \{p, r, t, c\}$ and π_i is a meta-variable for an arbitrary protocol:

$$Syn = \{\in Sub, \in PSub, \in ISub, \rho(/)() =, M\}$$

$$Op = \left\{[\pi_i]_x, \langle \pi_i \rangle_x, \Box, b, d\right\}$$

$$Conn = \{\wedge, \vee, \rightarrow, \equiv, \neg\}$$

$$Pop = \{\leftrightarrow_x, \curlywedge_x, \curlyvee_x\}$$

The arity of each control operator is denoted $a(\circ)$ and explicitly specified for each operator and connective:

(1) If $\circ \in Syn - \{\rho(/)() =, M\}$, then $a(\circ) = 2$; else,

  (a) $\circ = M, a(\circ) = 1$, and $\circ \frown y_1 \frown y_{a(\circ)}$ is elliptical for $\circ \frown y_1$ OR
  (b) $a(\circ) = 4$ and $\circ \frown y_1 \frown y_{a(\circ)}$ is elliptical for $\circ \frown y_1 \frown y_2 \frown y_3 \frown y_4$.

(2) If $\circ \in Op \cup Conn/\{\neg, \Box, b, d\}$, then $a(\circ) = 2$; else, $a(\circ) = 1$ and $\circ \frown y_1 \frown y_{a(\circ)}$ is elliptical for $\circ \frown y_1$,
(3) If $\circ \in Pop$, then $a(\circ) = 2$.

(Protocol-theoretic Formulae of Type i) The set of protocol-theoretic formulae of type i, L^i_{PLEN}, is the smallest subset of

$$String(AForm \cup Conn \cup Op \cup Pop \cup Syn \cup \bigcup_{j=1}^{i-1} P^j_{PLEN})$$

closed under the following rules:

(1) If $x \in AForm$, then $x \in L^0_{PLEN}$.
(2) If $x = \circ \frown y_1 \frown y_{a(\circ)}$, where each $y_j = \langle x_1 \frown ... \frown x_n \rangle$, each $y_j \in \bigcup_{j=1}^{i-1} L^j_{PLEN}$, some $y_j \in L^{i-1}_{PLEN}$, and $\circ \in Conn$, then $x \in L^i_{PLEN}$.
(3) If $x = \circ \frown y_1 \frown y_{a(\circ)}$, where each $y_j = \langle x_1 \frown ... \frown x_n \rangle$, $y_1 \in P^{i-1}_{PLEN}$, $y_{a(\circ)} \in \bigcup_{j=1}^{i-1} L^j_{PLEN}$, and $\circ \in Op/\{\Box, b, d\}$, then $x \in L^i_{PLEN}$.
(4) If $x = \circ \frown y_1 \frown y_{a(\circ)}$, where each $y_j = \langle x_1 \frown ... \frown x_n \rangle$, each $y_j \in \bigcup_{j=1}^{i-1} L^j_{PLEN}$, some $y_j \in L^{i-1}_{PLEN}$, and $\circ \in \{\Box, b, d\}$, then $x \in L^i_{PLEN}$.
(5) If $x = \circ \frown y_1 \frown ... \frown y_{a(\circ)}$, where each $y_j = \langle x_1 \frown ... \frown x_n \rangle$, each $y_j \in \bigcup_{j=1}^{i-1} P^j_{PLEN}$, some $y_j \in P^{i-1}_{PLEN}$, and $\circ \in Pop \cup Syn$, then $x \in L^i_{PLEN}$.

  (a) If $a(\circ) = 1$, then $x = \circ \frown y_{a(\circ)}$.
  (b) For $\langle x_1 \frown ... \frown x_n \rangle$, if $n = 1$, then $\langle x_1 \frown ... \frown x_n \rangle = \langle x_1 \rangle$.

(6) If $x = \circ \frown y_1 \frown y_{a(\circ)}$ or $x = \circ \frown y_{a(\circ)}$, then $\circ$ is the *main operator* of x.

Definition 27 (String Algebraic Protocol-Theoretic Language (L_{PLEN}))

$$L_{PLEN} = \bigcup_{i=1}^{\omega} L^i_{PLEN}$$

Definition 28 (Infix Notation Protocol Language (L_{PLEN})) Where $P_i \in AForm$, π_i is a metavariable for an arbitrary protocol with no occurrences of self-reference,[3] A, B, ... are metavariables for arbitrary protocol-theoretic formulae with no occurrences of self-reference, and filling in subscripts $y \in \{p, r, t, c\}$ produces p, r, t, and c variants of those formulae, L_{PLEN} can be defined by the following grammar:

$$(\pi_1 \in Sub(\pi_2)) \mid (\pi_1 \in PSub(\pi_2)) \mid (\pi_1 \in ISub(\pi_2)) \mid (\rho(\pi_1/\pi_2)(X_{\pi_1}) = X_{\pi_2}) \mid (M(\pi))$$

$$(\pi_1 \leftrightarrow_y \pi_2) \mid (\pi_1 \curlywedge_y \pi_2) \mid (\pi_1 \curlyvee_y \pi_2) \mid ([\pi_i]_y A) \mid (\langle \pi_i \rangle_y A)$$

$$(P_i) \mid (A \wedge B) \mid (A \vee B) \mid (A \rightarrow B) \mid (A \equiv B) \mid (\neg A)$$

$$(\Box A) \mid (b(A)) \mid (d(A))$$

Remark 29 Note that the string-algebraic language is unambiguous, and so omits any use of parentheses. Without parentheses, infix notation protocols of the form, $\pi_1\ o_1\ \pi_2\ o_2\ \pi_3$, are ambiguous; $(\pi_1\ o_1\ \pi_2)\ o_2\ \pi_3$ and $\pi_1\ o_1\ (\pi_2\ o_2\ \pi_3)$ are syntactically distinct in the language and, as we will see, may be semantically distinct. Hence, such ambiguous protocols are ruled out as ungrammatical. The string-algebraic syntax distinguishes between protocols of these forms immediately, the corresponding forms being $o_2 \frown o_1 \frown \pi_1 \frown \pi_2 \frown \pi_3$ and $o_1 \frown \pi_1 \frown o_2 \frown \pi_2 \frown \pi_3$, respectively. Similarly, for the four-place $\rho(/)() =, \mapsto$, $\rho(\pi/\pi')(\pi'') = \pi'''$ will be $\rho \frown \pi \frown \pi' \frown \pi'' \frown \pi'''$. For the protocol-theoretic dynamic operators in Op, the string-algebraic syntax works by taking elements of both languages, for example: $[x]_p \frown \pi \frown A$. Formulae formed from metalevel operators (Syn) aren't assigned truth conditions that vary by state or EDG. Pop is the set of object level operators; these are assigned truth conditions that vary by state and by EDG. For the protocol-theoretic dynamic operators in Op, the string-algebraic syntax works by treating them just like protocol control operators, but taking elements of both languages. For example: $[x]_p \frown \pi \frown A$. The metalevel operators work the same. For the four-place $\rho(/)() =, \mapsto$, $\rho(\pi/\pi')(\pi'') = \pi'''$ will be $\rho \frown \pi \frown \pi' \frown \pi'' \frown \pi'''$.

Claim 30 (Simultaneous Recursion for PLEN's Languages) The syntactic system of PLEN is defined so that there is a pair of interwoven ascending hierarchies of pairs of protocol and protocol-theoretic languages:

[3] The typed syntactic system of PLEN rules out liar formulae $L = \neg(\langle (L)? \rangle \top)$ and liar protocols $\pi = x_1 \frown ... \frown -\pi \frown ... \frown x_n$ as not well-formed. These would be paradoxical given the semantics in 4.2.2, so they're explicitly ill-formed in the infix notation.

Protocol Type	Protocols are strings in:
0	Π_0
1	$String(\Pi_0 \cup Prot \cup Prot + \cup L^0_{PLEN})$
2	$String(\Pi_0 \cup Prot \cup Prot + \cup L^0_{PLEN} \cup L^1_{PLEN})$
i	$String(\Pi_0 \cup Prot \cup Prot + \cup \bigcup_{j=1}^{i-1} L^j_{PLEN})$
$i+1$	$String(\Pi_0 \cup Prot \cup Prot + \cup \bigcup_{j=1}^{i} L^j_{PLEN})$

Formula Type	Protocol-theoretic Formulae are strings in:
0	$AForm$
1	$String(AForm \cup Conn \cup Op \cup Pop \cup Syn \cup P^0_{PLEN})$
2	$String(AForm \cup Conn \cup Op \cup Pop \cup Syn \cup P^1_{PLEN} \cup P^0_{PLEN})$
i	$String(AForm \cup Conn \cup Op \cup Pop \cup Syn \cup \bigcup_{j=1}^{i-1} P^j_{PLEN})$
$i+1$	$String(AForm \cup Conn \cup Op \cup Pop \cup Syn \cup \bigcup_{j=1}^{i} P^j_{PLEN})$

4.2.1.1 Syntactic Constructions

Let me set some terminological and notational conventions for discussing PLEN.

Definition 31 (Terminological and Notational Conventions 1)

(1) For arbitrary set or class, X, let $P(X)$ denote the powerset of X.
(2) For arbitrary set or class, X, let $X*$ denote the set of all finite length sequences of elements of X.
(3) For any string or sequence, x, let $len(x)$ be the length of the string.

  (a) For regular protocols, $\pi = x_1 \frown ... \frown x_n$, where each x_i is an atomic protocol or control operator, $len(\pi) = n$. Equivalently, the length of a protocol is determined by the length of the sequence of natural numbers in the domain of the protocol, e.g., $len(\langle 1, ..., 100\rangle) = 100$, thus, $len(\pi) = 100$ if

$$f_\pi : \langle 1, ..., 100\rangle \longrightarrow \Pi_0 \cup Prot \cup Prot + \cup \bigcup_{j=1}^{i-1} L^j_{PLEN}$$

  (b) For $\langle x_1 \frown ... \frown x_n\rangle$, if $n = 1$, then $\langle x_1 \frown ... \frown x_n\rangle$ is elliptical for $\langle x_1\rangle$.

(4) Let $x = x_1 \frown ... \frown x_n$ be a string over some set X such that $X_x = \{x_1, ..., x_n\}$. Let $y = x_j \frown ... \frown x_k$ be a string over X_x. Then: y is a *substring* of x iff $j \geq 1$, $k \leq n$, and, if $j \leq i < i+1 \leq k$, $x_i \frown x_{i+1}$ appears in y only if it appears in x.

(5) For arbitrary protocol, π, let $At(\pi)$ be the set of all atomic substrings used to form π and let $\circ(\pi)$ be the set of control operators used to form π.

(6) *Protocol meta-variables* are variables for protocols in the metalanguage of PLEN. Explicitly, the syntactic system is formulated using protocol and formulae meta-variables. Common meta-variables for protocols will be be π, π', π'', ..., π_1, ..., π_n. Common meta-variables for protocol-theoretic formulae will follow that pattern.

(7) I will frequently employ notation that formulates protocol meta-variables that identify protocols of specified syntactic forms. More precisely:

   (a) If a protocol meta-variable doesn't specify a syntactic form, then it takes all protocols as admissible values. Just so for formulae meta-variables.
   (b) Special meta-variables that specify important or commonly used syntactic forms:
      (i) Let X_π be a protocol meta-variable for any protocol such that π is subprotocol of X_π.
      (ii) Let $\pi_1 \circ \pi_{a(\circ)}$ (in infix notation) or, in string-algebraic notation, $\circ \frown \pi_1 \frown \pi_{a(\circ)}$, be a protocol meta-variable for any protocol such that $MC(\pi) = \circ$. For instance:
         (A) Let $\pi_1; \pi_2$ and $; \frown \pi_1 \frown \pi_{a(\circ)}$ be meta-variables for protocols with main control operator ;.
         (B) Let $\pi_1 \cap \pi_2$ and $\cap \frown \pi_1 \frown \pi_{a(\circ)}$ be meta-variables for protocols with main control operator $\cap$.
         (C) Let $\pi_1 \cup \pi_2$ and $\cup \frown \pi_1 \frown \pi_{a(\circ)}$ be meta-variables for protocols with main control operator $\cup$.
         (D) Let π_1* and $* \frown \pi_1 \frown \pi_{a(\circ)}$ $(* \frown \pi_1)$ be meta-variables for protocols with main control operator $*$.

(8) Formula meta-variables designate contexts for protocols or contexts for subformulae. For instance, $A_{(\pi)}$ is a formula meta-variable for any formula that embeds any value of the protocol meta-variable π in some protocol-theoretic operator, and $A_{(B)}$ is a meta-variable for any formula that embeds (a value of) B as a subformula.

   (a) Meta-variables can embed freely. Syntactic forms for embedded protocols or formulae can be specified by the above notation. For instance, $A_{(\pi_1;\pi_2)}$ can only embed sequential compositions, and $A_{(X_\pi)}$ can only embed protocols that contain protocols specified by π as subprotocols.

(9) To formalize the notion of a protocol meta-variable, let every protocol meta-variable, π, be associated with an operation or mapping of the form, f_π : $(P(\Pi_0*), (Prot \cup Prot+)*, I(x)) \longrightarrow P(P_{PLEN})$.

   (a) For a given meta-variable operation, all sequences of atomics in its domain are equal length, n.
   (b) Where the length of the sequence of control operators is m, $I(x)$ is an indexing function that assigns exactly one number in a finite sequence,

$\langle x_1, ..., x_{n+m} \rangle$, of natural numbers to each element of each sequence of atomic protocols and control operators such that:

(i) For all sequences, $\langle a_1, ..., a_n \rangle$ and $\langle b_1, ..., b_n \rangle$, $I(a_i) = I(b_i)$.

(ii) There is a protocol in the codomain of f_π in which the sequence of control operators and atomics is identical to the output of replacing each x_k in $\langle x_1, ..., x_{n+m} \rangle$ with a_i or $\circ_j$ wherever $I(a_i) = x_k$ or $I(\circ_j) = x_k$.

(c) The indexing function is optional, and introduces a level of precision not necessary to employ protocol meta-variables to refer to classes of protocols.

(i) This definition rules out the possibility of a protocol meta-variable mapping the same pair of sequence of atomics and sequence of control operators to sets of non-identical protocols.

(ii) The simpler formulation of protocol meta-variables, $f_\pi : (P(\Pi_0*), (Prot \cup Prot+)*) \longrightarrow P(P_{PLEN})$, permits this, and is equally useful in articulating and developing the PLEN Framework.

Claim 32 Protocol meta-variables map pairs of sets of sequences of atomics and sequences of control operators (i.e., structures encoding syntactic forms) into sets of protocols of certain syntactic forms, with the limit case of a meta-variable identifying a singleton set (when $At(\pi)*$ contains only a singleton set of sequences):

$$f_\pi : (At(\pi)*, \circ(\pi)*, I(x)) = \{\circ_1 \frown ... \frown a_{n+m}\}$$

The form of the protocol meta-variable operation provides a general form for partial or total specifications of such protocol meta-variables. Not all operations of this form are sensible meta-variables, of course. A sequence of n many control operators over $n - m$ length sequences of atomics will not pick out any grammatical protocols. However, all sensible protocol meta-variables are captured by operations of this form.

4.2.1.2 Important Syntactic Constructions I

With these notations in hand, we can set up consideration of some of the more important syntactic constructions. First, substrings of protocols or subprotocols will be essential to stating and proving most of the important results of PLEN, and there are several types. The kinds of subformulae of protocol-theoretic formulae are defined in a parallel manner. I omit their explicit definition.

Definition 33 (Immediate Subprotocols) For any protocol, π, the set of *immediate subprotocols of* π, $ISub(\pi)$, is the subset of P_{PLEN} such that: if $\pi = \circ \frown x_1 \frown x_{a(\circ)}$, $\circ \in Prot \cup Prot+$, and $x_1, x_{a(\circ)} \in P_{PLEN}$, then $ISub(\pi) = \{\pi, x_1, x_{a(\circ)}\}$.

(1) Equivalently:

$$ISub(\pi) = \{\pi' \in P_{PLEN} \mid \pi = \circ \frown \pi_1 \frown \pi_{a(\circ)} \rightarrow (\pi' = \pi_1 \vee \pi' = \pi_{a(\circ)} \vee \pi' = \pi)\}$$

(a) As a relation: $\pi' \in ISub(\pi)$ iff $\pi = \pi' \circ \pi_i$, $\pi = \pi_i \circ \pi'$, $\pi = \circ\pi'$, or $\pi = \pi'$.

Definition 34 (Subprotocols) For any protocol, π, the set of *subprotocols of* π, $Sub(\pi)$, is the set of protocols that are substrings of π:

(1) Equivalently:

$$Sub(\pi) = \{\pi' \in P_{PLEN} \mid \pi = \circ \frown x_1 \frown \dots \frown x_n,$$
$$\exists x_i \frown \dots \frown x_j((i \geq 1 \wedge j \leq n) \wedge \pi' = x_i \frown \dots \frown x_j)\}$$

(a) In other words, $Sub(\pi)$ is the subset of P_{PLEN} such that:
  (i) $\pi \in Sub(\pi)$,
  (ii) If a string, x, is such that $x \in P_{PLEN}$ and $x \in ISub(\pi)$, then $x \in Sub(\pi)$,
  (iii) If $\pi_i \in Sub(\pi)$ and $\pi_j \in ISub(\pi_i)$, then $\pi_j \in Sub(\pi)$.
(b) As a relation: $\pi' \in Sub(\pi)$ iff $\pi = \pi'$, $\pi' \in ISub(\pi)$, or there is a $\pi'' \in Sub(\pi)$ such that $\pi' \in ISub(\pi'')$.

Remark 35 For any protocol, π, the set of *subprotocols of* π, $Sub(\pi)$, is the smallest subset of P_{PLEN} closed under the immediate subprotocol relation.

Definition 36 (Proper Subprotocols) Let $PSub(\pi)$ denote the *proper subprotocols of* π. Then: $PSub(\pi) = Sub(\pi) - \{\pi\}$

Definition 37 (Atomic Subprotocols) Denote the set of *atomic subprotocols* of a protocol, π, with $At(\pi)$. Then: $At(\pi) = Sub(\pi) \cap \Pi_0$.

4.2.1.3 Important Syntactic Constructions II

The three most central constructions that emerge in the syntactic system are the class of regular protocols, the proper immediate subprotocol relation, and syntactic replacement operations.

Regular protocols are those formed only by means of the regular control operators: sequential composition, ";", finite, non-deterministic iteration, "$*$", parallel execution, "$\cap$", and non-deterministic choice, "$\cup$". The regular control operators are those that are safe for protocol graph isomorphism,[4] and this property logically undergirds the theory of protocol equivalence in PLEN. Accordingly, the core results of the PLEN Framework are restricted to quantification over only regular protocols.

Definition 38 (Regular Protocols) The set of all *regular protocols* is $String(\Pi_0 \cup Prot) \cap P_{PLEN}$.

[4] See Chap. 6.

The *proper immediate subprotocol relation* undergirds the central method of metalevel proof/derivation in PLEN as it is a well-founded relation over regular protocols and over the subprotocols of any regular protocol. Thus, structural induction over the proper immediate subprotocols of arbitrary regular protocols is valid, as it is a form of well-founded induction, and it is deployed in the derivation of the core results of the system.

Definition 39 (Proper Immediate Subprotocols) For any protocol, π, the set of *proper immediate subprotocols of* π, $PISub(\pi)$, is the subset of P_{PLEN} such that: if $\pi = \circ \frown x_1 \frown x_{a(\circ)}$, $\circ \in Prot \cup Prot+$, and $x_1, x_{a(\circ)} \in P_{PLEN}$, then $PISub(\pi) = \{x_1, x_{a(\circ)}\}$.

(1) Equivalently:

$$PISub(\pi) = PSub(\pi) \cap ISub(\pi)$$

(2) Equivalently:

$$PISub(\pi) = \{\pi' \in P_{PLEN} \mid \pi = \circ \frown \pi_1 \frown \pi_{a(\circ)} \rightarrow (\pi' = \pi_1 \vee \pi' = \pi_{a(\circ)})\}$$

(3) String-algebraic definition:

$$\begin{aligned} PISub(\pi) = \{\pi' \in P_{PLEN} \mid \pi &= \circ \frown x_1 \frown \ldots \frown x_n \\ &= \circ \frown s_1 \frown s_{a(\circ)}, (\pi' = s_1 \vee \pi' = s_2)\} \end{aligned}$$

where:

(a) $s_1 = x_1 \frown \ldots \frown x_i$ where $(i \geq 1 \wedge i \leq n)$
(b) if $i = n$, then $\pi = \circ \frown s_1$
(c) if $i \neq n$, then $s_2 = x_{i+1} \frown \ldots \frown x_n$

The *syntactic replacement operation on protocols* frames the theory of protocol equivalence developed in PLEN; *syntactic replacement* in the context of protocols and protocol-theoretic formulae is the core syntactic criterion of all protocol equivalence relations. Generally, equivalence relations are governed by semantic value preserving replacement or intersubstitution in various contexts; equivalent objects must be equivalent with respect to some set of properties. Thus, the valid replacement rules that substitute equivalent protocols across contexts (that codify the properties of protocols) codify the power of equivalence relations to preserve protocol-theoretic properties and determine the strictness, fine-grainedness, and informativeness of equivalence relations. These properties determine the equivalence relation's quality as an individuation criterion. The analysis of systems of equivalence relations thus proceeds by enumerating the properties preserved by various classes of instances of syntactic replacement. Protocol equivalences can be formalized by the property preserving syntactic replacement rules, i.e., the rules governing syntactic replacements that preserve: execution conditions, properties and relations of protocols (mappings to specialized constructions and properties of specialized constructions), and semantic values of classes of formulae.

Definition 40 (Syntactic Replacement Operations) Syntactic replacement operations are restricted to syntactic sorts; the syntactic replacement operation on protocols is distinct from the syntactic replacement operation on protocol-theoretic formulae, and protocols cannot be replaced within formulae, nor formulae within protocols. To effect the approximation of replacement of protocols in formulae, sequence syntactic replacement of protocols with syntactic replacement of the formulae embedding the resulting protocols.

(Syntactic Replacement On Protocols) For each pair of protocols in P_{PLEN}, π and π', define an operation on protocols, $\rho(\pi/\pi') : P_{PLEN} \longrightarrow P_{PLEN}$, such that $\rho(\pi/\pi')(X_\pi) = X_{\pi'}$ iff $(\lambda\pi[X_\pi])\pi' \rhd X_\pi[\pi := \pi'] = X_{\pi'}$ iff for concatenated strings, X_π and $X_{\pi'}$:

(1) $X_\pi = \langle x_1 \frown \ldots \frown x_n \rangle$
(2) $X_{\pi'} = \langle y_1 \frown \ldots \frown y_m \rangle$
(3) $\pi = \big\langle x_i \frown \ldots \frown x_j \big\rangle$
(4) $\pi' = \big\langle y_{i'} \frown \ldots \frown y_{j'} \big\rangle$
(5) $\forall x_k, y_{k'} \big[((k \notin [i, j], k' \notin [i', j'], d(k) = d(k')) \rightarrow (x_k = y_{k'})\big]$

$$\text{(a) where } d(i) = \begin{cases} i & if\, n \neq i < l \\ (i - l') + l & if\, n \neq i > l' \end{cases}$$

$$\text{(b) where } d(i') = \begin{cases} i' & if\, m \neq i' < h \\ (i' - h') + h & if\, m \neq i' > h' \end{cases}$$

(6) $n - l(\pi) = m - l(\pi'), i - 1 = i' - 1, n - i = m - j'$

(Syntactic Replacement On Protocol-theoretic Formulae) For each pair of protocol-theoretic formulae in L_{PLEN}, A and A', define an operation, $\rho(A/A') : L_{PLEN} \longrightarrow L_{PLEN}$, such that $\rho(A/A')(B(A)) = B(A')$ iff $(\lambda A[B(A)])A' \rhd B(A)[A := A'] = B(A')$ iff for concatenated strings, $B(A)$ and $B(A')$:

(1) $B(A) = \langle x_1 \frown \ldots \frown x_n \rangle$
(2) $B(A') = \langle y_1 \frown \ldots \frown y_m \rangle$
(3) $A = \big\langle x_i \frown \ldots \frown x_j \big\rangle$
(4) $A' = \big\langle y_{i'} \frown \ldots \frown y_{j'} \big\rangle$
(5) $\forall x_k, y_{k'} \big[((k \notin [i, j], k' \notin [i', j'], d(k) = d(k')) \rightarrow (x_k = y_{k'})\big]$

$$\text{(a) where } d(i) = \begin{cases} i & if\, n \neq i < l \\ (i - l') + l & if\, n \neq i > l' \end{cases}$$

$$\text{(b) where } d(i') = \begin{cases} i' & if\, m \neq i' < h \\ (i' - h') + h & if\, m \neq i' > h' \end{cases}$$

(6) $n - l(\pi) = m - l(\pi'), i - 1 = i' - 1, n - i = m - j'$

Example 41 Syntactic replacement of protocols:

$$\rho(\pi/\pi')(\circ \frown \pi \frown \pi_{a(\circ)}) = \circ \frown \pi' \frown \pi_{a(\circ)}$$

Example 42 Syntactic replacement of formulae:

$$\rho(A(\pi)/A(\pi'))(\circ \frown A(\pi) \frown B_{a(\circ)}) = \circ \frown A(\pi') \frown B_{a(\circ)}$$

Example 43 The derivation rules of PLEN telescope applications of both syntactic replacement operations. For instance, consider the following rule (RGSRT), which we will see in Chap. 6:

$$\left[\frac{\pi \equiv_G \pi',\ \rho(\pi/\pi')(X_\pi) = X_{\pi'},\ A_{(X\pi)}}{A_{(X\pi')}}\right]$$

This rule is elliptical for:

$$\left[\frac{\pi \equiv_G \pi',\ \rho(\pi/\pi')(X_\pi) = X_{\pi'},\ A_{(X\pi)}}{\rho(B(X_\pi)/B(X_{\pi'}))(A_{(X\pi)})}\right]$$

Remark 44 (Syntactic Formula Meta-Variables) The foregoing syntactic constructions are meta-level constructions, but, in much the same way that protocol meta-variables are used to identify classes of protocols of certain syntactic forms, I'll use the notation for these constructions within the context of syntactic protocol-theoretic formulae to construct meta-variables for formulae. This is the function of the protocol operators in Syn, that were used to define P_{PLEN}. The resulting formula meta-variables pick out classes of protocols that satisfy the meta-level syntactic properties above. So, for instance, the foregoing examples of rules really only contain two premises that have semantic values in PLEN, the middle premise just specifies the subset of protocols that can be plugged into those premises of the rule, while the conclusion specifies the protocol-theoretic formulae that can be plugged into the conclusion of the rule. In the foregoing example, the rule is schematic, saying that, if protocols, π and π', are procedurally equivalent,[5] and A holds of a protocol, X_π, that contains π, and $X_{\pi'}$ is the result of syntactic replacement of π with π' in X_π, then A holds of $X_{\pi'}$ as well. The following rule is an instance of a similar rule:

$$\left[\frac{(a;(b\cup c)) \leftrightarrow_p ((a;b)\cup(a;c)),\ A_{((a;(b\cup c))*)}}{A_{(((a;b)\cup(a;c))*)}}\right]$$

The syntactic formula meta-variable, $\rho(\pi/\pi')(X_\pi) = X_{\pi'}$, identifies $(a;(b\cup c))$, $((a;b)\cup(a;c))$, $(a;(b\cup c))*$, and $(((a;b)\cup(a;c))*$ as protocols that can slot

[5] See the next chapter.

into the rest of the rule. Subprotocol, immediate subprotocol, and proper immediate subprotocol formula meta-variables will be used similarly. For instance:

$$\left[\frac{\pi_1 \leftrightarrow_p \pi_2,\ \pi_1 \in PISub(\pi'),\ \pi' \leftrightarrow_p \pi''}{\rho(\pi_1/\pi_2)(\pi') \leftrightarrow_p \pi''}\right]$$

This schematic rule identifies a class of instances, including:

$$\left[\frac{(a;(b \cup c)) \leftrightarrow_p ((a;b) \cup (a;c)),\ ((a;(b \cup c)) \cap d) \leftrightarrow_p (d \cup e)}{(((a;b) \cup (a;c)) \cap d) \leftrightarrow_p (d \cup e)}\right]$$

The formula meta-variable, $\pi_1 \in PISub(\pi')$, identifies $(a;(b \cup c))$, $((a;b) \cup (a;c))$, $((a;(b \cup c)) \cap d)$, and $(((a;b) \cup (a;c)) \cap d)$ as protocols that can slot into the rest of the rule. Finally, note that syntactic replacement, as defined, applies only to one occurrence of its target protocol (π) in the protocol context (X_π) to which it is applied. To replace n occurrences of π, simply iterate the syntactic replacement operation n times over the occurrences of π in the string-algebraic form of X_π starting from left to right.

4.2.2 Semantic System

The second subsystem is a semantic system defined over the class of directed, multigraph-like structures called *epistemic directed graphs* and over systems of specialized constructions. *Execution rules* map substructures of EDGs to every protocol, binding the syntactic form of protocols to the graph-theoretic properties of EDGs. Sequential composition is mapped to structure concatenation, parallel execution is mapped to structure correspondence, non-deterministic choice is mapped to structure branching, and finite iteration is mapped to looping, branching, and concatenation (roughly, reflexive, transitive, closure). A unique feature of PLEN, however, is that each protocol is mapped to three distinct sets of execution structures in each graph; each protocol is mapped to an execution set defined in terms of paths, a set defined in terms of ternary relations, and a set defined in terms of binary relations. This supports PLEN's ability to embed different protocol logics, each with a semantics based on one of the types of execution sets.

Definition 45 (Directed Multigraph-Like Structures (DMSs))

Let S be a countable, non-empty set (of epistemic states), $S = \{s_1, ...s_n, ...\}$.

Let $\Pi_g \subseteq \Pi_0$. For each $a_i \in \Pi_g$, map a ternary relation (a set of edges), R_{a_i}, to it: $R_{a_i} \subseteq S \times \{a_i\} \times S$. The *atomic action set* in g is the set of all ternary relations mapped to atomic protocols in Π_g: $R = \{R_\pi \mid \exists\pi(\pi \in \Pi_g \wedge R_\pi \subseteq S \times \{\pi\} \times S)\}$ (formulated with meta-variable, π). Define the *atomic edge set* R_0 as the union of each set of edges mapped to each π in Π_g: $R_0 = R_{\pi_1} \cup ... \cup R_{\pi_n}$ for each $\pi_i \in \Pi_g$.

Let $\Pi_g^+ \subseteq P_{PLEN}/\Pi_0$. For each $\pi \in \Pi_g^+$, map a distinct ternary relation (set of edges), $R{+}_\pi \subseteq S \times P_{PLEN} \times S$, to it such that $R{+}_\pi \subseteq S \times \{\pi\} \times S$. The *action set* in g is the set of ternary relations mapped to each complex protocol in Π_g^+: $R{+} = \{R{+}_\pi \mid \exists\pi(\pi \in P_{PLEN}/\Pi_0 \wedge R{+}_\pi \subseteq S \times \{\pi\} \times S)\}$. Let the union of the ternary relations mapped to complex protocols in Π_g^+ be $R{+}_0 = R{+}_{\pi_1} \cup ... \cup R{+}_{\pi_m}$; call this the *edge set* in g.

(Directed Multigraph-like Structures)

(1) For any set-theoretic structure, g, *g is a directed, multigraph-like structure* ($g \in DMS$) *iff*

$$g = \left\langle S, \Pi_g, R, R_0, \Pi_g^+, R{+}, R{+}_0 \right\rangle$$

(2) Let the class of all directed, multigraph-like structures be DMS.

   (a) $|DMS| > \aleph_0$: Π_0 and $P_{PLEN} - \Pi_0$ are countable sets. The class of all DMSs must be built from the powersets of these sets, which are uncountable. Even setting a fixed set, S, of epistemic states, the class of all structurally distinct DMSs (i.e., DMSs with distinct edge sets) is uncountable:

$$DMS \subseteq S \times P(\Pi_0) \times P(S \times \Pi_0 \times S) \times P(P_{PLEN} - \Pi_0) \times P(S \times P_{PLEN} - \Pi_0 \times S)$$

The semantics of PLEN are defined over a restricted subset of DMSs, in particular, those that satisfy the following restriction rules:

Definition 46 (Restriction Rules (*Rr*)) A DMS, g, *satisfies the Restriction Rules*, $Rr(g)$, iff, for all protocols:

(1) If there are edges, $\langle s, \pi_1, s' \rangle$ and $\langle s', \pi_2, s'' \rangle$, such that $\langle s, \pi_1, s' \rangle \in R{+}_{\pi_1}$ and $\langle s', \pi_2, s'' \rangle \in R{+}_{\pi_2}$, then there is an edge, $\langle s, \pi_1; \pi_2, s'' \rangle \in R{+}_{\pi_1;\pi_2}$.

(2) If there are sequences, $\langle s_1, ..., s_n \rangle$ and $\langle p_1, ..., p_m \rangle$, such that $p_i \in P_\pi$ for every $i \leq m$, $s_1 = f(p_1)$, and $s_n = l(p_m)$, then $\langle s_1, \pi*, s_n \rangle \in R{+}_{\pi*}$.

   (a) Equivalently: If there are sequences, $\langle s_1, ..., s_n \rangle$, such that $\langle s_i, s_{i+1} \rangle \in T_\pi$ for every $i \leq n-1$, then $\langle s_i, \pi*, s_n \rangle \in R{+}_{\pi*}$.

(3) If $\pi \in P_{PLEN} \cap String(\Pi_g \cup Prot \cup Prot{+})$, and there are no executions of π at s, then $\exists \langle s, \pi*, s \rangle \in R{+}_{\pi*}$.

(4) If there are edges, $\langle s, \pi_1, s' \rangle$ and $\langle s, \pi_2, s' \rangle$, such that $\langle s, \pi_1, s' \rangle \in R{+}_{\pi_1}$ and $\langle s, \pi_2, s' \rangle \in R{+}_{\pi_2}$, then there are edges:

   (a) $\langle s, \pi_1 \cap \pi_2, s' \rangle \in R{+}_{\pi_1 \cap \pi_2}$ and $\langle s, \pi_1 \cup \pi_2, s' \rangle \in R{+}_{\pi_1 \cup \pi_2}$.
   (b) $\langle s, \pi_2 \cap \pi_1, s' \rangle \in R{+}_{\pi_2 \cap \pi_1}$ and $\langle s, \pi_2 \cup \pi_1, s' \rangle \in R{+}_{\pi_2 \cup \pi_1}$.

(5) If there are edges, $\langle s, \pi_1, s' \rangle$ and $\langle s, \pi_2, s'' \rangle$, such that $\langle s, \pi_1, s' \rangle \in R{+}_{\pi_1}$ and $\langle s, \pi_2, s'' \rangle \in R{+}_{\pi_2}$, then there are edges:

   (a) $\langle s, \pi_1 \cup \pi_2, s' \rangle \in R{+}_{\pi_1 \cup \pi_2}$ and $\langle s, \pi_2 \cup \pi_1, s' \rangle \in R{+}_{\pi_2 \cup \pi_1}$.
   (b) $\langle s, \pi_1 \cup \pi_2, s'' \rangle \in R{+}_{\pi_1 \cup \pi_2}$ and $\langle s, \pi_2 \cup \pi_1, s'' \rangle \in R{+}_{\pi_2 \cup \pi_1}$.

Definition 47 (Epistemic Directed Graphs (EDGs))
Let g_{dms} be an arbitrary DMS:

$$g_{dms} = \left\langle S, \Pi_g, R, R_0, \Pi_g^+, R+, R+_0 \right\rangle$$

Let v be a valuation function, $v : S \times L^0_{PLEN} \longrightarrow \{0, 1\}$.
Let c be a *content function*, $c : S \longrightarrow P(L_{PLEN}) \times P(L_{PLEN}) \times P(P_{PLEN})$. Call the triples in $P(L_{PLEN}) \times P(L_{PLEN}) \times P(P_{PLEN})$ *content states.*
(Epistemic Directed Graphs)

(1) Structure, g, *is an epistemic directed graph* ($EDG(g)$ *or* $g \in EDG$) *iff*

$$g = \left\langle S, \Pi_g, R, R_0, \Pi_g^+, R+, R+_0, c, v \right\rangle$$

and the Restriction Rules hold for g ($Rr(g)$).

(a) Throughout this book, the terms g, g', and indexed variants thereof denote arbitrary epistemic directed graphs rather than merely set-theoretic structures or DMSs unless further or other structure is specified.

(2) Equivalently, g *is an epistemic directed graph* ($EDG(g)$ *or* $g \in EDG$) *iff* $g = \langle g_{dms}, c, v \rangle$ *and* $Rr(g_{dms})$.
(3) The class of all EDGs, EDG, is the set of all DMSs, paired with content states and valuations, that satisfy the restriction rules.
(4) That is:

$$EDG = \left\{ g \mid g = \left\langle S, \Pi_g, R, R_0, \Pi_g^+, R+, R+_0, c, v \right\rangle, Rr(g) \right\}$$

(5) Equivalently:

$$EDG = \{ g \mid \exists g_{dms} (g = \langle g_{dms}, c, v \rangle , Rr(g)) \}$$

(6) Use the notation $[X : Y : \Pi_M]$ for content states, where $X, Y \in P(L_{PLEN})$ and $\Pi_M \in P(P_{PLEN})$.

Remark 48 To define the distinctive execution rules for protocols, numerous graph-theoretic features or substructures of EDGs must be identified.

Definition 49 (Procedural Paths) Let g be an EDG. Finite sequences of atomic edges in g, $\langle e_1, ..., e_n \rangle$, such that, for all e_i where $i \leq n$, $last(e_i) = first(e_{i+1})$ are *procedural paths*. Equivalently, procedural paths are finite, alternating sequences of epistemic states and protocols, $p = \langle s_1, \pi_1, ..., \pi_n, s_{n+1} \rangle$, comprising finite sequences of edges.

(1) (Procedural Path Sets) For any EDG, g, let P_g be the set of all *procedural paths* in g.

(2) (Root and Terminal States) For any edge or path, the *root state* is $first(...)$, and the *terminal state* is $last(...)$.

(a) For any edge, $e = \langle s, \pi, s' \rangle$, $first(e) = s$, and $last(e) = s'$.
(b) For any path, $p = \langle s_1, \pi_1, ..., \pi_n, s_{n+1} \rangle$, $first(p) = s_1$, and $last(p) = s_{n+1}$.
(c) When confusion does not threaten, abbreviate $first(...)$ to $f(...)$ and $last(...)$ to $l(...)$.

Definition 50 ((Epistemic) Transitions) Every pair of epistemic states, $\langle s, s' \rangle$, is an *epistemic transition* or, where confusion does not threaten, interchangeably, a *transition*.

(1) (Procedural Transition) Every epistemic transition, $\langle s, s' \rangle$, such that $s = first(p)$ and $s' = last(p)$ for some $p \in P_g$, is a procedural transition.
(2) (Procedural Transition Set) The set of procedural transitions in arbitrary EDG, g, is T_g.

Definition 51 (Subgraphs)

(1) (Subgraphs) For any EDGs, g and g', g is a *subgraph* of g', $g \sqsubseteq g'$, iff, conjunctively: $S \subseteq S'$,$R \subseteq R'$, and $P_g \subseteq P_{g'}$.
(2) (Proper Subgraphs) g is a *proper subgraph* of g', $g \sqsubset g'$, iff $g \sqsubseteq g'$ and, disjunctively: $S \subset S'$, $R \subset R'$, and $P_g \subset P_{g'}$.
(3) (Subpath) All procedural paths are graphs. Thus, p is a subpath of p' iff $p \sqsubseteq p'$.
(4) (Proper Subpaths) Procedural path, p, is a proper subpath of p' iff $p \sqsubset p'$.
(5) (Sets of Subgraphs) Denote the set of subgraphs of g (resp. p) with $Sub(g)$ (resp. $Sub(p)$) and the proper subgraphs with $PSub(g)$ (resp. $PSub(p)$). These apply as well to sets of graphs; that is, $Sub(P_g)$ or $Sub(P_\pi)$ denote the set of all subpaths of paths in g and the set of all subpaths of executions of π (in an unspecified g).

Definition 52 (Graph-Theoretic Structures)

(1) (Procedural Path Operations) Let $p = \langle e_1, ..., e_n \rangle$ and $q = \langle e'_1, ..., e'_m \rangle$. Use the notation "$p \oplus q$" to denote an arbitrary operation on procedural paths, p and q. Procedural path operations, $p \oplus q$: $P_g \times P_g \longrightarrow Sub(g)$, map pairs of paths in an EDG to a unique subgraph of that EDG.

(a) The operation $\oplus$ is schematic for the following operations on paths, which are central to understanding the execution rules and many of the core results of PLEN.

(2) (Path Concatenation) Let $p = \langle e_1, ..., e_n \rangle$ and $q = \langle e'_1, ..., e'_m \rangle$. Then $pq = r$ iff $r = \langle e_1, ..., e_n, e'_1, ..., e'_m \rangle$ or, equivalently, $p \sqsubseteq r$, $q \sqsubseteq r$, and $last(p) = first(q)$.

(a) The result of path concatenation on p and q is pq. Say that the procedural path, pq, is a path concatenation of procedural paths p and q or that path p concatenates with q or q concatenates on p. Since edges are paths, the foregoing notation for paths, $p = \langle e_1, ..., e_n \rangle$, using angled brackets and commas can be simplified: $p = e_1...e_n$.

(b) If a path, p, is identical to a concatenation of paths, q and r, then say that $p = qr$

(c) (Concatenated Path Length) If $len(p) = n$ and $len(q) = m$, then $len(pq) = n + m$.

(3) (Iterated Concatenation) Consider a set of paths, $\{p_1, ..., p_n\}$. Let $p^n_{i \leq n} = p_1...p_n$ be the iterated concatenation of paths, $p_1,...,p_n$, in order with each p_j and p_{j+1} ($j \leq n$) such that $l(p_j) = f(p_{j+1})$.

(a) Explicitly, where $p_i = \left\langle e^i_1, ..., e^i_{m(i)} \right\rangle$ with i the numerical subscript of p_i and $m(i)$ the last edge in p_i, the iterated concatentation of all $p_i \in \{p_1, ..., p_n\}$ in order is $p^n_{i \leq n} = \left\langle e^1_1, ..., e^1_{m(1)}, ..., e^n_1, ..., e^n_{m(n)} \right\rangle$ where:

(i) for each e^i_j and e^i_{j+1} where $j, j+1 \leq m(i)$, $l(e^i_j) = f(e^i_{j+1})$, and

(ii) for each $e^i_{m(i)}$ and $e^{i+1}_{m(i+1)}$ in that sequence of edges, $l(e^i_{m(i)}) = f(e^{i+1}_{m(i+1)})$.

(b) Say that $p_1 p_2$ is the result of p_2 concatenating on p_1. Say that $p_1 p_2 p_3$ is the result of p_3 concatenating on $p_1 p_2$. And so on.

(4) (Branching Iteration) Let $p_1, ..., p_m$ be an arbitrary set of m many paths. The branching path iteration of $p_1, ..., p_m$ is $p^m_{i \leq m}* = p^m_{i \leq m} \cup p^{m-1}_{i \leq m-1} \cup ... \cup p^1_{i \leq 1}$ - a branching of concatenated iterations of paths in the specified set of paths in the specified order of descending length.

(5) (Execution Iteration) Let $p_1, ..., p_m$ be an arbitrary set of m many executions of a protocol, π. The execution iteration of $p_1, ..., p_m$ is $p_\pi* = p^{n(\pi)}_{i \leq n} \cup p^{n-1(\pi)}_{i \leq n-1} \cup ... \cup p^{1(\pi)}_{i \leq 1}$ - a branching of concatenated iterations of arbitrary paths in the specified set of executions of π up to n times.

(a) Note: Say that $P_\pi* = p_\pi* = p^{n(m)}_{i \leq n} \cup p^{n-1(m)}_{i \leq n-1} \cup ... \cup p^{1(m)}_{i \leq 1}$ iff $P_\pi = \{p_1, ..., p_m\}$.

(6) (Root and Terminal Correspondence) Let $p = \langle e_1, ..., e_n \rangle$ and $q = \left\langle e'_1, ..., e'_m \right\rangle$. Then $p \rightleftharpoons q = r$ iff $first(e_1) = first(e'_1) \wedge last(e_n) = last(e'_n)$ and $r = p$ or $r = q$.

(a) Alternately, say that path p *corresponds* with q, $(p \rightleftharpoons q)$, iff $(first(e_1) = first(e'_1) \wedge last(e_n) = last(e'_n))$.

(7) (Full Path Correspondence) Let $p = \langle e_1, ..., e_n \rangle$ and $q = \left\langle e'_1, ..., e'_m \right\rangle$. Then $p \rightleftarrows q = r$ iff $\forall i \leq n(first(e_i) = first(e'_i) \wedge last(e_i) = last(e'_i))$ (equivalently, $s(p) = s(q)$) and $r = p$ or $r = q$.

(a) Path p *fully corresponds* with q, $(p \rightleftarrows q)$, iff $n = m$, and $\forall i \leq n(first(e_i) = first(e'_i) \wedge last(e_i) = last(e'_i))$ (equivalently, $s(p) = s(q)$).

(8) (Path Branching) Let $p = \langle e_1, ..., e_n \rangle$ and $q = \langle e'_1, ..., e'_m \rangle$. . Then $p \cup q = r$ iff $first(e_1) = first(e'_1)$ and $r = p$ or $r = q$.

  (a) Say that paths p and q branch, $(p \cup q)$, iff $f(p) = f(q)$.

(9) (Path Iteration) Let $p = \langle s_1, \pi_1, ..., \pi_n, s_{n+1} \rangle$. Let $p_1, ..., p_m$ be a set of paths such that, for each p_i where $i \leq m$, $p_i = \langle s'_1, \pi_1, ..., \pi_n, s'_{n+1} \rangle$ and s'_j is an arbitrary state; that is, each p_i is a path of equal length, formed from concatenated edges for identical sequences of atomic protocols. The finite, non-deterministic path iteration of p, $p*$, is a graph such that for some $n \geq 1$, $p* = p^m_{i \leq m}* = p^m_{i \leq m} \cup p^{m-1}_{i \leq m-1} \cup ... \cup p^1_{i \leq 1})$ - the branching of all concatenations of p with paths built from identical sequences of atomics to itself up to n times.

  (a) Each p_i is isomorphic to p, in the sense defined in Sect. 5.4.1.

(10) (Path Looping) For any edge, e, or path, p, e loops iff $first(e) = last(e)$, and p loops iff there are some paths, q and r, such that $p = qr$ and either $first(q) = last(q)$ or $first(r) = last(r)$.

(11) (State-Sequences) Finite sequences of transitions, $\langle t_1, ..., t_n \rangle$, such that, for all t_i where $i \leq n$, $last(t_i) = first(t_{i+1})$ are *state-sequences*.

  (a) Let $s(p)$ denote the state sequence of p.
  (b) Let $seq(p)$ denote the set of all state sequences in p.

(12) (Inverting Functions) Let P_g be the set of all procedural paths in an EDG, g. Let $p \in P_g$. Let $\overline{e_i} = \langle s, \pi, s' \rangle$ iff $e_i = \langle s', \pi, s \rangle$. Define an *inverting function* $f_I : P_g \longrightarrow P_g$ such that $f_I(p) = \langle e_1, ..., e_n \rangle$ iff $p = \langle \overline{e_n}, ..., \overline{e_1} \rangle$.

(13) (Path Length) Denote the length of procedural path, $p = \langle e_1, ..., e_n \rangle$ is n; $len(p) = n$.

(14) (Maximal Paths) Procedural path, p, is *maximal* iff:

  (a) there is no p' such that $p \sqsubset p'$ and
  (b) there are no looping edges, e, such that $e \sqsubseteq p$.
  (c) (Maximal Path Sets) The set of maximal paths with respect to a graph, g, is defined:

$$maxpath(g) = \{p \in P_g \mid [\neg\exists q(q \in P_g \wedge p \sqsubset q) \wedge \neg\exists e(f(e) = l(e) \wedge e \sqsubseteq p)]\}$$

(15) (Closure of a Set of Paths) Denote the closure of a set of paths in a graph, g, P_g, under concatenation (the set of all paths formed out of concatenation of paths in P) with $Cl \frown P_g$. Extend this to all other sets of graphs, including execution sets or maximal paths (defined below).

Remark 53 If $len(p)$ is defined in terms of states or alternating sequences of states and protocols, then path concatenation isn't additive in the way that string concatenation is. In that case, if $len(p) = n$ and $len(q) = m$, then $len(pq) = (n + m) - 1$.

4.2.2.1 Execution Rules

As PLEN is ultimately to serve as a system in which to build and compare distinct sublogics of protocols, there are three kinds of executions for each protocol, each kind presenting a different level of logical granularity. This is a feature of PLEN that comes into more focus in Chaps. 5 and 6. The following sets of execution rules map protocols to sets of paths (n-ary relations), binary relations, or ternary relations.

Definition 54 (Procedural Execution Rules)

(1) (Base) For all $\pi \in \Pi_0$: $P_\pi = R_\pi$
(2) (;) $P_{\pi_1;\pi_2} =$

$$\{p \in P_g \mid p = qr \wedge q \in P_{\pi_1} \wedge r \in P_{\pi_2}\}$$

(3) (*) $P_{\pi *} =$

$$\left\{p \in P_g \mid \exists n \left(\bigvee \begin{bmatrix} \exists \langle q_1, \ldots, q_n \rangle (p = q_1 \ldots q_n, \forall i \leq n(q_i \in P_\pi)) \\ p = \langle s, \pi *, s \rangle \end{bmatrix} \right)\right\}$$

(4) ($\cap$) $P_{\pi_1 \cap \pi_2} =$

$$\left\{p \in P_g \mid \bigvee \begin{bmatrix} (p \in P_{\pi_1} \wedge \exists p' \in P_{\pi_2}(p \rightleftharpoons q)) \\ (p \in P_{\pi_2} \wedge \exists p' \in P_{\pi_1}(p \rightleftharpoons q)) \end{bmatrix}\right\}$$

(5) ($\cup$) $P_{\pi_1 \cup \pi_2} =$

$$\left\{p \in P_g \mid f(p) \neq l(p), \bigvee \begin{bmatrix} (p \in P_{\pi_1} \wedge \exists p' \in P_{\pi_2}(f(p) = f(p'))) \\ (p \in P_{\pi_2} \wedge \exists p' \in P_{\pi_1}(f(p) = f(p'))) \end{bmatrix}\right\}$$

(6) ((−)?) $P_{(A)?} =$

$$\{p \in P_g \mid first(p) \Vdash A, last(p) \Vdash A \wedge b(A)\}$$

(7) (−) $P_{-\pi} =$

$$\{p \in P_g \mid p \notin P_\pi \wedge \neg \exists q \in P_\pi (q \sqsubset p \vee q \leftrightharpoons p)\}$$

(8) ($\curvearrowleft$) $P_{\curvearrowleft \pi} =$

$$\left\{p \in P_g \mid \left[\exists q \bigwedge \begin{bmatrix} q \in P_\pi, \\ p = f_I(q) \end{bmatrix} \right]\right\}$$

(9) (∇) $P_{\nabla\pi} =$

$$\left\{p \in P_g \mid \bigwedge \begin{bmatrix} p \in P_\pi, \\ first(p) = last(p) \end{bmatrix}\right\}$$

Definition 55 (Binary Execution Rules)

(1) (Base) For all $\pi \in \Pi_g$:

$$T_\pi = \left\{\langle s, s' \rangle \in S \times S \mid \exists e(e \in R_0, s = first(e), s' = last(e))\right\}$$

(2) (;) $T_{\pi_1;\pi_2} = T_{\pi_1} \circ T_{\pi_2} =$

$$\left\{\langle s, s' \rangle \in T_g \mid \exists x \in S \bigwedge \begin{bmatrix} \langle s, x \rangle \in T_{\pi_1} \\ \langle x, s' \rangle \in T_{\pi_2} \end{bmatrix}\right\}$$

(3) ($*$) $T_{\pi*} =$

$$\left\{\langle s, s' \rangle \in T_g \mid \exists n \geq 1 \begin{bmatrix} \exists x_1 ... x_n (\forall i \leq n (x_i \in S)) \\ \bigwedge \begin{bmatrix} s = x_1, s' = x_n \\ \forall i \leq n-1 \, \langle x_i, x_{i+1} \rangle \in T_\pi \end{bmatrix} \end{bmatrix}\right\}$$

(4) ($\cap$) $T_{\pi_1 \cap \pi_2} =$

$$\left\{t \in T_g \mid \exists p \in P_{\pi_1 \cap \pi_2}(t = \langle first(p), last(p) \rangle)\right\}$$

(5) ($\cup$) $T_{\pi_1 \cup \pi_2} =$

$$\left\{t \in T_g \mid f(t) \neq l(t), \bigvee \begin{bmatrix} (t \in T_{\pi_1} \wedge \exists t' \in T_{\pi_2}(f(t) = f(t'))) \\ (t \in t_{\pi_2} \wedge \exists t' \in T_{\pi_1}(f(t) = f(t'))) \end{bmatrix}\right\}$$

(6) ((-)?) $T_{(A)?} =$

$$\left\{\langle s, s' \rangle \in T_g \mid s \Vdash A, s' \Vdash A \wedge b(A)\right\}$$

(7) ($-$) $T_{-\pi} =$

$$\left\{\langle s, s' \rangle \in T_g \mid \exists \langle s, s' \rangle \notin T_\pi\right\} = T_g - T_\pi$$

(8) ($\curvearrowleft$) $T_{\curvearrowleft\pi} =$

$$\left\{\langle s, s' \rangle \in T_g \mid \exists \langle s', s \rangle \in T_\pi\right\}$$

(9) (∇) $T_{\nabla\pi} =$

$$\{\langle s, s\rangle \in T_g \mid \exists \langle s, s\rangle \in T_\pi\}$$

Definition 56 (Ternary Relational Execution Rules)

(1) (Base) For all $\pi \in \Pi_g$:

$$R_\pi = R{+}_\pi = \{\langle s, \pi, s'\rangle \mid \langle s, \pi, s'\rangle \in S \times \{\pi\} \times S\}$$

(2) (;) $R{+}_{\pi_1;\pi_2} =$

$$\left\{\langle s, \pi_1; \pi_2, s'\rangle \in R{+}_0 \mid \; \exists x \in S \bigwedge \begin{bmatrix} \langle s, \pi_1, x\rangle \in R_{\pi_1} \\ \langle x, \pi_2, s'\rangle \in R_{\pi_2} \end{bmatrix} \right\}$$

(3) ($*$) $R{+}_{\pi*} =$

$$\left\{\langle s, \pi*, s'\rangle \in R{+}_0 \mid \exists n \geq 1 \begin{bmatrix} \exists x_1 ... x_n (\forall i \leq n (x_i \in S)) \\ \bigwedge \begin{bmatrix} s = x_1, s' = x_n, \\ \forall i \leq n-1 \; \langle x_i, \pi, x_{i+1}\rangle \in R{+}_\pi \end{bmatrix} \end{bmatrix} \right\}$$

(4) ($\cap$) $R{+}_{\pi_1 \cap \pi_2} =$

$$\left\{\langle s, \pi_1 \cap \pi_2, s'\rangle \in R{+}_0 \mid \bigwedge \begin{bmatrix} \exists \langle s, \pi_1, s'\rangle \in R{+}_{\pi_1} \\ \exists \langle s, \pi_2, s'\rangle \in R{+}_{\pi_2} \\ \exists p \in P_{\pi_1 \cap \pi_2} (s = f(p) \wedge s' = l(p)) \end{bmatrix} \right\}$$

(5) ($\cup$) $R{+}_{\pi_1 \cup \pi_2} =$

$$\left\{ \langle s, \pi_1 \cup \pi_2, s'\rangle \right.$$

$$\left. \in R{+}_0 \mid s \neq s', \bigvee \begin{bmatrix} (\exists e \in R+_{\pi_1} \exists e' \in R+_{\pi_2} (f(e) = f(e') = s \wedge l(e) = s')) \\ (\exists e \in R+_{\pi_1} \exists e' \in R+_{\pi_2} (f(e) = f(e') = s \wedge l(e) = s')) \end{bmatrix} \right\}$$

(6) ((-)?) $R{+}_{(A)?} =$

$$\{\langle s, (A)?, s'\rangle \in R{+}_0 \mid s \Vdash A, s' \Vdash A \wedge b(A)\}$$

(7) ($-$) $R{+}_{-\pi} =$

$$\{\langle s, -\pi, s'\rangle \in R{+}_0 \mid \neg\exists e (e \in R+_\pi \wedge e \leftrightharpoons \langle s, -\pi, s'\rangle)\}$$

(8) ($\curvearrowleft$) $R+_{\curvearrowleft\pi} =$

$$\{\langle s, \curvearrowleft\pi, s'\rangle \in R+_0 \mid \exists \langle s', \pi, s\rangle \in R+_\pi\}$$

(9) (∇) $R+_{\nabla\pi} =$

$$\{\langle s, \nabla\pi, s\rangle \in R+_0 \mid \exists \langle s, \pi, s\rangle \in R+_\pi\}$$

Remark 57 Protocols are mapped to classes of execution sets in a way that binds the graph-theoretic structure of the substructures of EDGs containing non-empty execution sets closely but not uniquely to the syntactic form of protocols. Sequential composition is bound to path concatenation, choice is bound to path branching, parallel execution is bound to path correspondence, and iteration is bound to path iteration (branching and concatenation). The execution sets mapped to each protocol are unique in the sense that only protocols equivalent in most contexts definable in PLEN can have execution sets that are pairwise isomorphic. Thus, the syntactic form of each protocol determines a somewhat unique, abstract, graph-theoretic structure determined by the execution rules, as pictured below.

Induced Path Structure

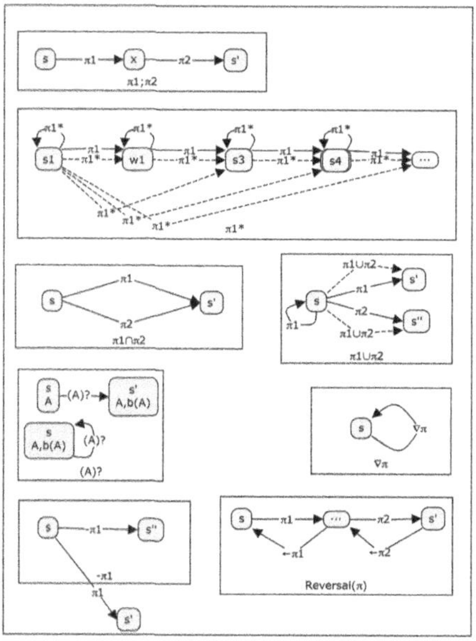

With the exception of the iteration diagram, these graphs ignore R_0+.

Induced Edge Structure

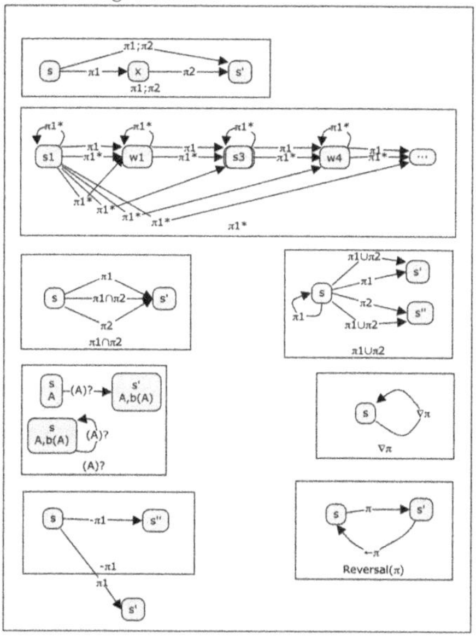

With the exception of the iteration diagram, these graphs ignore some of the edges in R_0+.

4.2.2.2 Truth Rules

On the basis of the interpretation of protocols by means of the execution rules, EDGs can be used as models of the protocol-theoretic language. Truth conditions for protocol-theoretic formulae, including those embedding relations of protocol equivalence and modal operators with embedded protocols, are defined in terms of the properties of the substructures and specialized constructions of EDGs identified by the execution rules or mapped to protocols on the basis of their syntax. Protocol equivalences are defined in terms of the properties of the substructures of EDGs mapped to protocols by the execution rules; each set of executions thus produces a distinctive protocol equivalence relation.

Definition 58 Terminological and Notational Conventions 2

(1) For arbitrary π, let any t in T_π be a *transition-theoretic* or *binary execution of* π.
(2) For arbitrary π, let any e, $R{+}_\pi$ be an *edge-theoretic* or *ternary execution of* π.
(3) For arbitrary π, let any p in P_π be a *path-theoretic* or *procedural execution of* π.

Definition 59 (Verification/Satisfaction) An EDG, g, *verifies/satisfies* atomic protocol-theoretic formula, A, *by* s, $g, s \Vdash A$, iff $s \in S$ and $v_s(A) = 1$.

Definition 60 (Basic Truth Rules) Valuations, $v_s(A)$, assign semantic values to atomic protocol-theoretic formulae at states. Extend verification for atomics to complex formulae by the following rules:

(1) (Base) $s \Vdash A$ iff $v_s(A) = 1$.

  (a) Let $v_s(\top) = 1$ for all s in all EDGs.

(2) (Conn) The standard classical[6] rules for the truth-functions, $\vee, \wedge, \neg, \rightarrow$.
(3) ($\Box$) $s \Vdash \Box A$ iff $\forall s' \in S(s \Vdash A)$.
(4) (b) $s \Vdash b(A)$ iff $c(s) = [X : Y : \Pi_M]$, and $A \in X$.
(5) (d) $s \Vdash d(A)$ iff $c(s) = [X : Y : \Pi_M]$, and $A \in Y$.
(6) ($M(\pi)$) $s \Vdash M(\pi)$ iff $c(s) = [X : Y : \Pi_M]$, and $\pi \in \Pi_M$.

Definition 61 (Procedural Truth Rules) Where A is any complex formula of L_{PLEN}, extend the valuations to define verification for procedural variants of protocol-theoretic formulae:

(1) ($\langle\pi\rangle_p A$) $s \Vdash \langle\pi\rangle_p A$ iff $\exists p$ ($s = first(p)$, $p \in P_\pi$, and $s \Vdash A$).
(2) ($[\pi]_p A$) $s \Vdash [\pi]_p A$ iff $\forall p$ (if $s = first(p)$, and $p \in P_\pi$, then $s \Vdash A$).
(3) ($\pi \leftrightarrow_p \pi'$) $s \Vdash \pi \leftrightarrow_p \pi'$ iff $\forall p$ (if $s = first(p)$, then $p \in P_\pi$ iff $p \in P_{\pi'}$).
(4) ($\pi \curlywedge_p \pi'$) $s \Vdash \pi \curlywedge_p \pi'$ iff $\exists p$ ($s = first(p)$, $p \in P_\pi$, and $p \in P_{\pi'}$).

---

[6] I'm partial to adding intensional conditionals or replacing classical material conditionals with intensional conditionals: $v_s(A \rightarrow B)$ iff for all s', if $v_{s'}(A) = 1$, then $v_{s'}(B) = 1$. Nothing in the main results of PLEN hangs on the conditional being classical and material.

(5) $(\pi \curlyvee_p \pi')\ s \Vdash \pi \curlyvee_p \pi'$ iff $\exists p\ (s = first(p)$, and either $p \in P_\pi - P_{\pi'}$ or $p \in P_{\pi'} - P_\pi)$.

Definition 62 (Binary Truth Rules) Where A is any complex formula of L_{PLEN}, extend the valuations to define verification for binary variants of protocol-theoretic formulae:

(1) $(\langle\pi\rangle_t A)\ s \Vdash \langle\pi\rangle_t A$ iff $\exists s'\ (\langle s, s'\rangle \in T_\pi$, and $s' \Vdash A)$.
(2) $([\pi]_t A)\ s \Vdash [\pi]_t A$ iff $\forall s'$ (if $\langle s, s'\rangle \in T_\pi$, then $s' \Vdash A)$.
(3) $(\pi \leftrightarrow_t \pi')\ s \Vdash \pi \leftrightarrow_t \pi'$ iff $\forall s'\ (\langle s, s'\rangle \in T_\pi$ iff $\langle s, s'\rangle \in T_{\pi'})$.
(4) $(\pi \curlywedge_t \pi')\ s \Vdash \pi \curlywedge_t \pi'$ iff $\exists s'\ (\langle s, s'\rangle \in T_\pi$, and $\langle s, s'\rangle \in T_{\pi'})$.
(5) $(\pi \curlyvee_t \pi')\ s \Vdash \pi \curlyvee_t \pi'$ iff $\exists s'\ (\langle s, s'\rangle \in T_\pi$ and $\langle s, s'\rangle \notin T_{\pi'}$, or $\langle s, s'\rangle \in R_\pi$ and $\langle s, s'\rangle \notin T_{\pi'})$.

Definition 63 (Ternary Truth Rules) Where A is any complex formula of L_{PLEN}, extend the valuations to define verification for ternary variants of protocol-theoretic formulae:

(1) $(\langle\pi\rangle_r A)\ s \Vdash \langle\pi\rangle_r A$ iff $\exists s'\ (\langle s, \pi, s'\rangle \in R{+}_\pi$, and $s' \Vdash A)$.
(2) $([\pi]_r A)\ s \Vdash [\pi]_r A$ iff $\forall s'$ if $(\langle s, \pi, s'\rangle \in R{+}_\pi$, then $s' \Vdash A)$.
(3) $(\pi \leftrightarrow_r \pi')\ s \Vdash \pi \leftrightarrow_r \pi'$ iff $\forall e\ (e \in R{+}_\pi$ iff $e \in R{+}_{\pi'})$.
(4) $(\pi \curlywedge_r \pi')\ s \Vdash \pi \curlywedge_r \pi'$ iff $\exists e\ (e \in R{+}_\pi$, and $e \in R{+}_{\pi'})$.
(5) $(\pi \curlyvee_r \pi')\ s \Vdash \pi \curlyvee_r \pi'$ iff $\exists e\ (e \in R{+}_\pi$ and $e \notin R{+}_{\pi'}$, or $e \in R{+}_\pi$ and $e \notin R{+}_{\pi'})$.

Definition 64 (Correspondence Truth Rules) Where A is any complex formula of L_{PLEN}, extend the valuations to define verification for correspondence variants of protocol-theoretic formulae:

(1) $(\langle\pi\rangle_c A)\ s \Vdash \langle\pi\rangle_c A$ iff $\exists p, \langle s, s_1, ..., s_n\rangle\ (\langle s, s_1, ..., s_n\rangle = s(p)$, $p \in P_\pi$, and $s \Vdash A)$.
(2) $([\pi]_c A)\ s \Vdash [\pi]_c A$ iff $\forall p, \langle s, s_1, ..., s_n\rangle$ (if $\langle s, s_1, ..., s_n\rangle = s(p)$ and $p \in P_\pi$, then $s \Vdash A)$.
(3) $(\pi \leftrightarrow_c \pi')\ s \Vdash \pi \leftrightarrow_c \pi'$ iff $\forall p, \langle s, s_1, ..., s_n\rangle\ (\langle s, s_1, ..., s_n\rangle = s(p)$ and $p \in P_\pi$ iff $\exists p', \langle s, s_1, ..., s_n\rangle$ such that $\langle s, s_1, ..., s_n\rangle = s(p'))$ and $p' \in P_{\pi'})$ and vice versa.
(4) $(\pi \curlywedge_c \pi')\ s \Vdash \pi \curlywedge_c \pi'$ iff $\exists p, \langle s, s_1, ..., s_n\rangle\ (\langle s, s_1, ..., s_n\rangle = s(p)$, $p \in P_\pi$, and $p \in P_{\pi'})$.
(5) $(\pi \curlyvee_c \pi')\ s \Vdash \pi \curlyvee_c \pi'$ iff $\exists p, \langle s, s_1, ..., s_n\rangle\ (\langle s, s_1, ..., s_n\rangle = s(p)$, and either $p \in P_\pi - P_{\pi'}$ or $p \in P_{\pi'} - P_\pi)$.

Definition 65 (Truth Rules for Syntactic Formulae) The following formulae are "syntactic" in that they are not interpreted over the graph-theoretic properties of EDGs or the properties of valuations but rather in terms of the syntactic structures of protocols.

(1) $(\pi \in Sub(\pi'))\ s \Vdash \pi \in Sub(\pi')$ iff π is a subprotocol of π'.
(2) $(\pi \in PSub(\pi'))\ s \Vdash \pi \in PSub(\pi')$ iff π is a proper subprotocol of π'.
(3) $(\pi \in ISub(\pi'))\ s \Vdash \pi \in ISub(\pi')$ iff π is an immediate subprotocol of π'.

(4) $(\rho(\pi/\pi')(X_\pi) = X_{\pi'})$ $s \Vdash \rho(\pi/\pi')(X_\pi) = X_{\pi'}$ iff $(\lambda\pi[X_\pi])\pi' \triangleright X_\pi[\pi := \pi'] = X_{\pi'}$.

Remark 66 In line with Remark 44, the "truth conditions" for the variants of the subprotocol and syntactic replacement formulae should be interpreted as rules for identifying instances of these formulae rather than assigning truth at states; every such formula is necessarily true in every EDG if it's true in any, as it's a meta-level property of syntactic objects. Their inclusion in the object language of PLEN is designed to streamline the presentations of the derivation system presented in this chapter and the revised version in Chap. 5. There, the derivation systems are presented uniformly using the meta-level syntactic replacement operations and subprotocol relations. The syntactic replacement and subprotocol "formulae" in the premise slots of such derivation rules (whether meta-level or object-level) should be interpreted as shorthand specifications of classes of instances of rule schemata without them. For instance, the rule schema $\frac{\pi \leftrightarrow_P \pi',\ \rho(\pi/\pi')(X_\pi) = X_{\pi'}}{X_\pi \leftrightarrow_P X_{\pi'}}$ should be interpreted as specifying the class of rules of the form, $\frac{\pi \leftrightarrow_P \pi'}{X_\pi \leftrightarrow_P X_{\pi'}}$, where $\rho(\pi/\pi')(X_\pi) = X_{\pi'}$ holds for the protocol meta-variables. For an instance to be introduced in Chap. 5, the rule schema $\frac{\pi \equiv_G \pi',\ \rho(\pi/\pi')(X_\pi) = X_{\pi'},\ \pi \in ISub(X_\pi)}{X_\pi \equiv_G X_{\pi'}}$ should be interpreted as specifying the class of rules of the form, $\frac{\pi \equiv_G \pi'}{X_\pi \equiv_G X_{\pi'}}$, where $\rho(\pi/\pi')(X_\pi) = X_{\pi'}$ and $\pi \in ISub(X_\pi)$ hold for the protocol meta-variables. For the semantics, there is no harm done in treating these as rules identifying instances of necessities at all states.

4.2.2.3 Validity (Semantic)

Remark 67 The analysis of valid reasoning in the semantics is standard.

Definition 68 (Logical Truth/Tautology) For any $A \in L_{PLEN}$, A is a *logical truth/tautology* in PLEN ($\models_{PLEN} A$) iff A is verified by all states in all EDGs, i.e., every state, s, in every EDG, g, is such that $s \Vdash A$.

Definition 69 (Logical Consequence) For any $X \subseteq L_{PLEN}$ and $A \in L_{PLEN}$, A is a *logical consequence in PLEN* of X ($X \models_{PLEN} A$) iff for every g and every s in S, if $\forall B_i \subseteq X(s \Vdash B_i)$, then $s \Vdash A$.

4.2.3 Derivation System

The third subsystem is a *derivation system* that extends the syntax from well-formed protocol-theoretic formulae (strings) to structures (sequences, trees, and partial orders) built out of protocol-theoretic formulae (strings). The derivation system comprises *object-level rules and axioms* governing the object-level counterparts of the syntactic and graph-theoretic properties of protocols, the dynamic properties of protocols, and the logic of modality and propositions in PLEN (via a Fitch-style natural deduction system). The core of PLEN's logic of protocols comprises the *theory of protocol equivalence* and the *theory of protocol dynamics* built into it. The theory of protocol equivalence is the system of rules and axioms governing the properties of the protocol equivalence relations in PLEN. The theory of protocol dynamics is the system of rules and axioms governing the interaction of the pre and postconditions of protocols and the modal operators that embed protocols. The derivation system is sound and plausibly complete, but it is intended to be a *basic* system in the sense that it is a starting point for the protocol-theoretic logic of epistemic norms. As this basic derivation system is developed iteratively, the initial derivation system of PLEN developed immediately below goes unnamed or referred to simply as the PLEN derivation system, and the revised derivation system introduced in Chap. 5 is accordingly called the *R-Derivation system.*

4.2.3.1 Execution-Theoretic Rules and Axioms

Definition 70 (Protocol Equivalence Rules (Execution-Theoretic))

(1) For $NSEP$, the formula meta-variable, $A(\pi)$, cannot embed π directly in schematic formulae of the form: $M(\pi)$, $\pi \leftrightarrow_r ...$, $\pi \curlywedge_r ...$, or any of the purely syntactic expressions (Sub, $PSub$, $ISub$, $\rho()() =$).
(2) For SEP, the formulae meta-variable, $A(\pi)$, cannot embed π directly in schematic formulae of the form: $\Box ...$, $[...]_p A(\pi)$, $[...]_t A(\pi)$, $[...]_r A(\pi)$, $\langle ... \rangle_p A(\pi)$, $\langle ... \rangle_t A(\pi)$, $\langle ... \rangle_r A(\pi)$, $M(\pi)$, $\pi \leftrightarrow_r ...$, $\pi \curlywedge_r ...$, or any of the purely syntactic expressions (Sub, $PSub$, $ISub$, $\rho()() =$).
(3) Where $\cap \notin \circ(\pi) \cup \circ(\pi') \cup \circ(\pi'')$, $A(\pi)$ can embed π directly in $\pi \curlyvee_r \pi''$ in both rules.
(4) For $NSEP_t$, the formula meta-variable, $A(\pi)$, cannot embed π directly in schematic formulae of the form: $M(\pi)$, $\pi \leftrightarrow_r ...$, $\pi \curlywedge_r ...$, $\pi \leftrightarrow_p ...$, $\pi \curlywedge_p ...$, $\pi \curlyvee_p ...$, $\pi \curlyvee_r ...$ or any of the purely syntactic expressions (Sub, $PSub$, $ISub$, $\rho()() =$).
(5) For SEP_t, the formula meta-variable, $A(\pi)$, cannot embed π directly in schematic formulae of the form: $\Box ...$, $[...]_p A(\pi)$, $[...]_t A(\pi)$, $[...]_r A(\pi)$, $\langle ... \rangle_p A(\pi)$, $\langle ... \rangle_t A(\pi)$, $\langle ... \rangle_r A(\pi)$, $\pi \leftrightarrow_p ...$, $\pi \curlywedge_p ...$, $M(\pi)$, $\pi \leftrightarrow_r ...$, $\pi \curlywedge_r ...$, $\pi \curlyvee_p ...$, $\pi \curlyvee_r ...$, or any of the purely syntactic expressions (Sub, $PSub$, $ISub$, $\rho()() =$).
(6) For $NSEP_r$ and SEP_r, the formula meta-variable, $A(\pi)$, has no restrictions.

Path-theoretic (n-ary)	Transition-theoretic (Binary)	Edge-theoretic (Ternary)
$\left[\frac{\Box(\pi \leftrightarrow_p \pi'),\ \rho(\pi/\pi')(X_\pi) = X_{\pi'}}{\Box(X_\pi \leftrightarrow_p X_{\pi'})} \quad \frac{\Box(\pi \leftrightarrow_p \pi'),\ \Box(\pi' \leftrightarrow_p \pi'')}{\Box(\pi \leftrightarrow_p \pi'')} \quad \frac{}{\Box(\pi \leftrightarrow_p \pi)} \quad \frac{\Box(\pi \leftrightarrow_p \pi')}{\Box(\pi' \leftrightarrow_p \pi)} \quad \frac{\Box(\pi \leftrightarrow_p \pi')}{\Box(\pi \leftrightarrow_t \pi')}\right]$	$\left[\frac{\Box(\pi \leftrightarrow_t \pi'),\ \rho(\pi/\pi')(X_\pi) = X_{\pi'}}{\Box(X_\pi \leftrightarrow_t X_{\pi'})} \quad \frac{\Box(\pi \leftrightarrow_t \pi',\ \Box(\pi' \leftrightarrow_t \pi'')}{\Box(\pi \leftrightarrow_t \pi'')} \quad \frac{}{\Box(\pi \leftrightarrow_t \pi)} \quad \frac{\Box(\pi \leftrightarrow_t \pi')}{\Box(\pi' \leftrightarrow_t \pi)}\right]$	$\left[\frac{\Box(\pi \leftrightarrow_r \pi'),\ \rho(\pi/\pi')(X_\pi) = X_{\pi'}}{X_\pi \leftrightarrow_r X_{\pi'})} \quad \frac{\Box(\pi \leftrightarrow_r \pi'),\ \Box(\pi' \leftrightarrow_r \pi'')}{\Box(\pi \leftrightarrow_r \pi'')} \quad \frac{}{\Box(\pi \leftrightarrow_r \pi)} \quad \frac{\Box(\pi \leftrightarrow_r \pi')}{\Box(\pi' \leftrightarrow_r \pi)} \quad \frac{\Box(\pi \leftrightarrow_r \pi')}{\pi = \pi'}\right]$
$\left[\frac{\pi \leftrightarrow_p \pi',\ \rho(\pi/\pi')(X_\pi) = X_{\pi'}}{X_\pi \leftrightarrow_p X_{\pi'}} \quad \frac{\pi \leftrightarrow_p \pi',\ \pi' \leftrightarrow_p \pi''}{\pi \leftrightarrow_p \pi''} \quad \frac{}{\pi \leftrightarrow_p \pi} \quad \frac{\pi \leftrightarrow_p \pi'}{\pi' \leftrightarrow_p \pi} \quad \frac{\pi \leftrightarrow_p \pi'}{\pi \leftrightarrow_t \pi'}\right]$	$\left[\frac{\pi \leftrightarrow_t \pi',\ \rho(\pi/\pi')(X_\pi) = X_{\pi'}}{X_\pi \leftrightarrow_t X_{\pi'}} \quad \frac{\pi \leftrightarrow_t \pi',\ \pi' \leftrightarrow_t \pi''}{\pi \leftrightarrow_t \pi''} \quad \frac{}{\pi \leftrightarrow_t \pi} \quad \frac{\pi \leftrightarrow_t \pi'}{\pi' \leftrightarrow_t \pi}\right]$	$\left[\frac{\pi \leftrightarrow_r \pi',\ \rho(\pi/\pi')(X_\pi) = X_{\pi'}}{X_\pi \leftrightarrow_r X_{\pi'}} \quad \frac{\pi \leftrightarrow_r \pi',\ \pi' \leftrightarrow_r \pi''}{\pi \leftrightarrow_r \pi''} \quad \frac{}{\pi \leftrightarrow_r \pi} \quad \frac{\pi \leftrightarrow_r \pi'}{\pi' \leftrightarrow_r \pi} \quad \frac{\pi \leftrightarrow_r \pi'}{\pi = \pi'}\right]$
$\left[\frac{\Box(\pi \leftrightarrow_p \pi'),\ A(\pi)}{A(\pi')}(NSEP) \quad \frac{(\pi \leftrightarrow_p \pi'),\ A(\pi)}{A(\pi')}(SEP) \quad \frac{(\pi \curlywedge_p \pi'),\ [\pi]_p A}{\langle\pi'\rangle_p A}(CON) \quad \frac{(\pi \curlyvee_p \pi')}{\neg(\pi \leftrightarrow_p \pi')}(DIV)\right]$	$\left[\frac{\Box(\pi \leftrightarrow_t \pi'),\ A(\pi)}{A(\pi')}(NSEP_t) \quad \frac{(\pi \leftrightarrow_t \pi'),\ A(\pi)}{A(\pi')}(SEP_t) \quad \frac{(\pi \curlywedge_t \pi'),\ [\pi]_t A}{\langle\pi'\rangle_t A}(CON_t) \quad \frac{(\pi \curlyvee_t \pi')}{\neg(\pi \leftrightarrow_t \pi')}(DIV_t)\right]$	$\left[\frac{\Box(\pi \leftrightarrow_r \pi'),\ A(\pi)}{A(\pi')}(NSEP_r) \quad \frac{(\pi \leftrightarrow_r \pi'),\ A(\pi)}{A(\pi')}(SEP_r) \quad \frac{(\pi \curlywedge_r \pi'),\ [\pi]_r A}{\langle\pi'\rangle_r A}(CON_r) \quad \frac{(\pi \curlyvee_r \pi')}{\neg(\pi \leftrightarrow_r \pi')}(DIV_r)\right]$

Definition 71 (Protocol Dynamics Rules)

Path-theoretic (n-ary)	Transition-theoretic (Binary)	Edge-theoretic (Ternary)
$\left[\dfrac{\langle\pi\rangle_p A}{\langle\pi\rangle_r A}(DP1)\quad \dfrac{\langle\pi\rangle_p A}{\langle\pi\rangle_t A}(DP2)\right]$	$\left[\dfrac{\langle\pi\rangle_t A}{\langle\pi\rangle_p A}(DT1)\quad \dfrac{\langle\pi\rangle_t A}{\langle\pi\rangle_r A}(DT2)\right]$	$\left[\dfrac{\langle\pi\rangle_r A}{\langle\pi\rangle_p A}(DR1)\quad \dfrac{\langle\pi\rangle_r A}{\langle\pi\rangle_t A}(DR2)\right]$
$\dfrac{[\pi]_p A,\ \langle\pi\rangle_p \top}{\langle\pi\rangle_p A}(MP1)$ $\dfrac{\langle\pi\rangle_p A}{\langle\pi\rangle_p \top}(MP2)$ $\dfrac{A \vdash_{PLEN} B,\ [\pi]_p A}{[\pi]_p B}(MP3)$ $\dfrac{A \vdash_{PLEN} B,\ \langle\pi\rangle_p A}{\langle\pi\rangle_p B}(MP5)$ $\dfrac{[\pi]_p(A \to B),\ [\pi]_p A}{[\pi]_p B}(MP6)$ $\dfrac{[\pi]_p(A \to B),\ \langle\pi\rangle_p A}{\langle\pi\rangle_p B}(MP7)$ $\dfrac{\langle\pi\rangle_p (A \to B),\ [\pi]_p A}{\langle\pi\rangle_p B}(MP8)$	$\dfrac{[\pi]_t A,\ \langle\pi\rangle_t \top}{\langle\pi\rangle_t A}(MT1)$ $\dfrac{\langle\pi\rangle_t A}{\langle\pi\rangle_t \top}(MT2)$ $\dfrac{A \vdash_{PLEN} B,\ [\pi]_t A}{[\pi]_t B}(MT3)$ $\dfrac{A \vdash_{PLEN} B,\ \langle\pi\rangle_t A}{\langle\pi\rangle_t B}(MT5)$ $\dfrac{[\pi]_t(A \to B),\ [\pi]_t A}{[\pi]_t B}(MT6)$ $\dfrac{[\pi]_t(A \to B),\ \langle\pi\rangle_t A}{\langle\pi\rangle_t B}(MT7)$ $\dfrac{\langle\pi\rangle_t (A \to B),\ [\pi]_t A}{\langle\pi\rangle_t B}(MT8)$	$\dfrac{[\pi]_r A,\ \langle\pi\rangle_r \top}{\langle\pi\rangle_r A}(MR1)$ $\dfrac{\langle\pi\rangle_r A}{\langle\pi\rangle_r \top}(M2)$ $\dfrac{A \vdash_{PLEN} B,\ [\pi]_r A}{[\pi]_r B}(MR3)$ $\dfrac{A \vdash_{PLEN} B,\ \langle\pi\rangle_r A}{\langle\pi\rangle_r B}(MR5)$ $\dfrac{[\pi]_r(A \to B),\ [\pi]_r A}{[\pi]_r B}(MR6)$ $\dfrac{[\pi]_r(A \to B),\ \langle\pi\rangle_r A}{\langle\pi\rangle_r B}(MR7)$ $\dfrac{\langle\pi\rangle_r (A \to B),\ [\pi]_r A}{\langle\pi\rangle_r B}(MR8)$

Definition 72 (Execution-Theoretic Axioms)

Edge-theoretic (Ternary)	Path-theoretic (n-ary)	Transition-theoretic (Binary)
$[\pi_1]_r[\pi_2]_r A \equiv [\pi_1; \pi_2]_r A$	$[\pi_1; \pi_2]_p A \equiv [\pi_1]_p[\pi_2]_p A$	$[\pi_1; \pi_2]_t A \equiv [\pi_1]_t[\pi_2]_t A$
$\langle\pi_1; \pi_2\rangle_r A \equiv \langle\pi_1\rangle_r \langle\pi_2\rangle_r A$	$\langle\pi_1; \pi_2\rangle_p A \equiv \langle\pi_1\rangle_p \langle\pi_2\rangle_p A$	$\langle\pi_1; \pi_2\rangle_t A \equiv \langle\pi_1\rangle_t \langle\pi_2\rangle_t A$
$\langle\pi_1 \cap \pi_2\rangle_r A \to (\langle\pi_1\rangle_r A \wedge \langle\pi_2\rangle_r A)$	$\langle\pi_1 \cap \pi_2\rangle_p A \to (\langle\pi_1\rangle_p A \wedge \langle\pi_2\rangle_p A)$	$\langle\pi_1 \cap \pi_2\rangle_t A \to (\langle\pi_1\rangle_t A \wedge \langle\pi_2\rangle_t A)$
$([\pi_1]_r A \wedge [\pi_2]_r B) \to [\pi_1 \cap \pi_2]_r(A \wedge B)$	$([\pi_1]_p A \wedge [\pi_2]_p B) \to [\pi_1 \cap \pi_2]_p(A \wedge B)$	$([\pi_1]_t A \wedge [\pi_2]_t B) \to [\pi_1 \cap \pi_2]_t(A \wedge B)$
$([\pi_1]_r A \wedge [\pi_2]_r A) \to [\pi_1 \cup \pi_2]_r A$	$([\pi_1]_p A \wedge [\pi_2]_p A) \to [\pi_1 \cup \pi_2]_p A$	$([\pi_1]_t A \wedge [\pi_2]_t A) \to [\pi_1 \cup \pi_2]_t A$
$\langle\pi_1 \cup \pi_2\rangle_r A \to (\langle\pi_1\rangle_r A \vee \langle\pi_2\rangle_r A)$	$\langle\pi_1 \cup \pi_2\rangle_p A \to (\langle\pi_1\rangle_p A \vee \langle\pi_2\rangle_p A)$	$\langle\pi_1 \cup \pi_2\rangle_t A \to (\langle\pi_1\rangle_t A \vee \langle\pi_2\rangle_t A)$
$(\langle\pi_1\rangle_r A \wedge \langle\pi_2\rangle_r A) \to \langle\pi_1 \cup \pi_2\rangle_r A$	$(\langle\pi_1\rangle_p A \wedge \langle\pi_2\rangle_p A) \to \langle\pi_1 \cup \pi_2\rangle_p A$	$(\langle\pi_1\rangle_t A \wedge \langle\pi_2\rangle_t A) \to \langle\pi_1 \cup \pi_2\rangle_t A$
$([\pi]_r[\pi*]_r A) \wedge A \equiv [\pi*]_r A$	$([\pi]_p[\pi*]_p A) \wedge A \equiv [\pi*]_p A$	$([\pi]_t[\pi*]_t A) \wedge A \equiv [\pi*]_t A$

Edge-theoretic (Ternary)	Path-theoretic (n-ary)	Transition-theoretic (Binary)
$\langle\pi*\rangle_r A \rightarrow ((\langle\pi\rangle_r \langle\pi*\rangle_r A) \vee A$	$\langle\pi*\rangle_p A \rightarrow ((\langle\pi\rangle_p \langle\pi*\rangle_p A) \vee A$	$\langle\pi*\rangle_t A \rightarrow ((\langle\pi\rangle_t \langle\pi*\rangle_t A) \vee A$
$(A \wedge [\pi*]_r(A \rightarrow [\pi]_r A)) \rightarrow [\pi*]_r A$	$(A \wedge [\pi*]_p(A \rightarrow [\pi]_p A)) \rightarrow [\pi*]_p A$	$(A \wedge [\pi*]_t(A \rightarrow [\pi]_t A)) \rightarrow [\pi*]_t A$
$\langle(A)?\rangle_r B \rightarrow (A \wedge B)$	$\langle(A)?\rangle_p B \rightarrow (A \wedge B)$	$\langle(A)?\rangle_t B \rightarrow (A \wedge B)$
$\langle(A)?\rangle_r \top \rightarrow A$	$\langle(A)?\rangle_p \top \rightarrow A$	$\langle(A)?\rangle_t \top \rightarrow A$
$A \equiv [\nabla\pi]_r A$	$A \equiv [\nabla\pi]_p A$	$A \equiv [\nabla\pi]_t A$
$\Box(A \equiv \langle\pi\rangle_r \top) \rightarrow \Box[\curvearrowleft \pi]_r A$	$\Box(A \equiv \langle\pi\rangle_p \top) \rightarrow \Box[\curvearrowleft \pi]_p A$	$\Box(A \equiv \langle\pi\rangle_t \top) \rightarrow \Box[\curvearrowleft \pi]_t A$
$([\pi]_r A \wedge \langle\pi\rangle_r \top) \rightarrow \langle\pi\rangle_r A$	$([\pi]_p A \wedge \langle\pi\rangle_p \top) \rightarrow \langle\pi\rangle_p A$	$([\pi]_t A \wedge \langle\pi\rangle_t \top) \rightarrow \langle\pi\rangle_t A$
$\langle\pi\rangle_r A \rightarrow \langle\pi\rangle_r \top$	$\langle\pi\rangle_p A \rightarrow \langle\pi\rangle_p \top$	$\langle\pi\rangle_t A \rightarrow \langle\pi\rangle_t \top$
$[\pi]_r(A \wedge B) \equiv ([\pi]_r A) \wedge [\pi]_r B)$	$[\pi]_p(A \wedge B) \equiv ([\pi]_p A) \wedge [\pi]_p B)$	$[\pi]_t(A \wedge B) \equiv ([\pi]_t A) \wedge [\pi]_t B)$
$([\pi]_r(A \rightarrow B) \wedge [\pi]_r A) \rightarrow [\pi]_r B$	$([\pi]_p(A \rightarrow B) \wedge [\pi]_p A) \rightarrow [\pi]_p B$	$([\pi]_t(A \rightarrow B) \wedge [\pi]_t A) \rightarrow [\pi]_t B$
$([\pi]_r(A \rightarrow B) \wedge \langle\pi\rangle_r A) \rightarrow \langle\pi\rangle_r B$	$([\pi]_p(A \rightarrow B) \wedge \langle\pi\rangle_p A) \rightarrow \langle\pi\rangle_p B$	$([\pi]_t(A \rightarrow B) \wedge \langle\pi\rangle_t A) \rightarrow \langle\pi\rangle_t B$
$(\Box(\pi \leftrightarrow_r \pi') \wedge A(\pi)) \rightarrow A(\pi')$	$(\Box(\pi \leftrightarrow_p \pi') \wedge A(\pi)) \rightarrow A(\pi')$	$(\Box(\pi \leftrightarrow_t \pi') \wedge A(\pi)) \rightarrow A(\pi')$
For $(\Box(\pi \leftrightarrow_r \pi') \wedge A(\pi)) \rightarrow A(\pi')$, there are no syntactic restrictions on $A(\pi)$.	For $(\Box(\pi \leftrightarrow_p \pi') \wedge A(\pi)) \rightarrow A(\pi')$, $A(\pi)$ may not embed π in any of the subprotocol formulae, in formulae that imply protocol identity, in graph-theoretic equivalences, in implementation formulae, nor in methodological inclusion formulae.	For $(\Box(\pi \leftrightarrow_t \pi') \wedge A(\pi)) \rightarrow A(\pi')$, $A(\pi)$ may not embed π in any of the subprotocol formulae, in formulae that imply protocol identity, in graph-theoretic equivalences or procedural equivalences, in implementation formulae, in procedural convergence formulae, nor in methodological inclusion formulae.
$((\pi \leftrightarrow_r \pi') \wedge A(\pi)) \rightarrow A(\pi')$	$((\pi \leftrightarrow_p \pi') \wedge A(\pi)) \rightarrow A(\pi')$	$((\pi \leftrightarrow_t \pi') \wedge A(\pi)) \rightarrow A(\pi')$

Edge-theoretic (Ternary)	Path-theoretic (n-ary)	Transition-theoretic (Binary)
For $((\pi \leftrightarrow_r \pi') \wedge A(\pi)) \rightarrow A(\pi')$, there are no syntactic restrictions on $A(\pi)$.	For $((\pi \leftrightarrow_p \pi') \wedge A(\pi)) \rightarrow A(\pi')$, $A(\pi)$ may not embed π in any of the subprotocol formulae, in formulae that imply protocol identity, in graph-theoretic equivalences, in implementation formulae, in formulae in the scope of modals, nor in methodological inclusion formulae.	For $((\pi \leftrightarrow_t \pi') \wedge A(\pi)) \rightarrow A(\pi')$, $A(\pi)$ may not embed π in any of the subprotocol formulae, in formulae that imply protocol identity, in graph-theoretic equivalences or procedural equivalences, in implementation formulae, in procedural convergence formulae, in formulae in the scope of modals, nor in methodological inclusion formulae.
$((\pi \curlywedge_r \pi') \wedge [\pi]_r A) \rightarrow \langle \pi' \rangle_r A$	$((\pi \curlywedge_p \pi') \wedge [\pi]_p A) \rightarrow \langle \pi' \rangle_p A$	$((\pi \curlywedge_t \pi') \wedge [\pi]_t A) \rightarrow \langle \pi' \rangle_t A$
$(\pi \curlyvee_r \pi') \equiv \neg(\pi \leftrightarrow_r \pi') \equiv \pi \neq \pi'$	$(\pi \curlyvee_p \pi') \equiv \neg(\pi \leftrightarrow_p \pi') \equiv \pi \neq \pi'$	$(\pi \curlyvee_t \pi') \equiv \neg(\pi \leftrightarrow_t \pi') \equiv \pi \neq \pi'$
$\pi \curlywedge_r \pi' \rightarrow \pi \leftrightarrow_r \pi'$		
$\Box A \rightarrow [\pi]_r A$	$\Box A \rightarrow [\pi]_p A$	$\Box A \rightarrow [\pi]_t A$

Example 73 (Derivative Laws) GRT[7] and the graph-theoretic equivalence axioms entail (assuming the aforementioned restrictions):

$$\begin{bmatrix} \Box(A(\pi_1 \cap \pi_2) \equiv A(\pi_2 \cap \pi_1)) \\ \Box(A(\pi_1 \cup \pi_2) \equiv A(\pi_2 \cup \pi_1)) \end{bmatrix}$$

Example 74 (Derivative Laws) Deduction conjoined with RP and PR (or, metatheoretically, the restriction rules) entail:

$$\begin{bmatrix} [\pi]_r A \equiv [\pi]_p A \\ \langle \pi \rangle_r A \equiv \langle \pi \rangle_p A \end{bmatrix}$$

Definition 75 Identity Rules

(1) $(A = A') \rightarrow (B \equiv \rho(A/A')(B))$
(2) $(\pi = \pi') \rightarrow (A(\pi) = A(\pi'))$

4.2.3.2 Propositional and Modal Logic

Definition 76 (Logic in PLEN) For the basic system, assume any standard suite of propositional and modal rules or axioms. Here is a system of rules that will be of use in the book: Let $\Sigma = \{A_1, ..., A_n\}$ and $\Sigma^\wedge = A_1 \wedge (...(... \wedge A_n))$.

[7] See Chap. 6.

Introduction

$$\frac{\Sigma \vdash_{PLEN} B}{\Sigma^{\wedge} \rightarrow B}(\rightarrow I)$$

$$\frac{\neg A \vee B}{A \rightarrow B}(Imp - I)$$

$$\frac{A \rightarrow B}{\neg B \rightarrow \neg A}(Contra - I)$$

$$\frac{A}{A \vee B}(\vee Il)$$

$$\frac{B}{A \vee B}(\vee Ir)$$

$$\frac{A, B}{A \wedge B}(\wedge I)$$

Elimination

$$\frac{A, A \rightarrow B}{B}(\rightarrow E)$$

$$\frac{A \rightarrow B}{\neg A \vee B}(Imp - E)$$

$$\frac{\neg B \rightarrow \neg A}{A \rightarrow B}(Contra - E)$$

$$\frac{A \rightarrow C, B \rightarrow C}{(A \vee B) \rightarrow C}(\vee E)$$

$$\frac{A \wedge B}{A}(\wedge El)$$

$$\frac{A \wedge B}{B}(\wedge Er)$$

Exchange

$$\frac{A \wedge B}{B \wedge A}(\wedge Comm) \quad \frac{A \vee B}{B \vee A}(\vee Comm) \quad \frac{A \wedge A}{A}(\wedge Simp)$$

$$\frac{A \vee A}{A}(\vee Simp)$$

$$\frac{(A \wedge B) \wedge C}{A \wedge (B \wedge C)}(L \wedge Assoc) \quad \frac{(A \vee B) \vee C}{A \vee (B \vee C)}(L \vee Assoc)$$

$$\frac{A \wedge (B \wedge C)}{(A \wedge B) \wedge C}(R \wedge Assoc) \quad \frac{A \vee (B \vee C)}{(A \vee B) \vee C}(R \vee Assoc)$$

$$\frac{\Sigma \vdash_{PLEN} \perp}{\neg(\Sigma^{\wedge})}(\neg I) \qquad \frac{A, \neg A}{\perp}(\neg E)$$

$$\frac{\perp}{A}(EFQ)$$

$$\frac{A}{\neg\neg A}(DNI) \qquad \frac{A}{\neg\neg A}(DNE)$$

$$\frac{\Box(A \rightarrow B)}{(\Box A \rightarrow \Box B)}(\Box Dist - U) \qquad \frac{(\Box A \rightarrow \Box B)}{\Box(A \rightarrow B)}(\Box Dist - D)$$

$$\frac{\vdash_{PLEN} A}{\Box A}(\Box I) \qquad \frac{\Box A}{A}(\Box E)$$

$$\frac{\Diamond A}{\Box\Diamond A}(\Box\Diamond I) \qquad \frac{\Box\Diamond A}{\Diamond A}(\Box\Diamond E)$$

Claim 77 PLEN builds in a classical propositional logic and an unusual, "universal box" modal logic, but PLEN is interestingly modular. The protocol logic of PLEN, presented in the rules and axioms above, is largely distinct from the propositional and modal logics. As such, it can sustain non-classical replacements of its propositional and modal logics, without change to most of its protocol logic. See Chap. 7, Sect. 7.5 for more on this.

4.2.3.3 Validity (Syntactic)

Remark 78 The analysis of valid reasoning in the derivation system is standard, or close to it, though the definition of derivations is somewhat ad hoc.

Definition 79 (Rule-Based Ordering) For any $X \subseteq L_{PLEN}$, R is a Rule-based Ordering over some set of protocol-theoretic formulae, X, with respect to the derivation rules of PLEN, iff R is a partial order over X such that:

(1) xRx' iff $x, x' \in X$ and, disjunctively:

(a) x' is an axiom of PLEN,

(b) there is some derivation rule of PLEN, Dr, such that $\frac{x}{x'}(Dr)$,

(c) there is an $x'' \in X$, such that $\frac{x, x''}{x'}(Dr)$ for some derivation rule of PLEN, Dr,

(d) there are x'', x''' in X with xRx'' and xRx''' such that $\frac{x'', x'''}{x'}(Dr)$ for some derivation rule of PLEN, Dr,

(e) there is an x'' in X such that xRx'' and $x''Rx'$,

(f) x' is introduced by conditional introduction or negation introduction either as an assumption or as a consequence of a conditional or negation introduction rule.

(i) Assumptions introduce branching in the R relation: if Dr is $\frac{[A] \vdash B}{A \rightarrow B}(\rightarrow I)$ or $\frac{[A] \models \bot}{\neg A}(\neg I)$, then $\exists\psi(x' = \neg\psi)$ or $\exists\phi, \psi(x' = \phi \rightarrow \psi)$, and there is a z such that $x' \neq z$ on a separate branch (x' and z are incomparable with respect to R), and whatever is ultimately derived from z and all of its R ancestors lies on a separate branch from that built by adding x' (all descendants of z are incomparable to x'). Say that z is an assumption introduced by the rule on the separate branch.

(2) For any subset, X', of X, say that $X'Rx'$ iff there is some element, $x'' \in X'$, such that $x''Rx'$.

Definition 80 (Derivation) For any $X \subseteq L_{PLEN}$ and $B \in L_{PLEN}$, a *derivation* of protocol-theoretic formula B from a set of premises (protocol-theoretic formulae), $X = \{A_1, ..., A_n\}$, in the basic system is any tree, $\langle L, X, R, B\rangle$, such that:

(1) $(L \cup X \cup \{B\}) \subseteq L_{PLEN}$ is the set of nodes.
(2) The elements of X are in arbitrary, linear order.
(3) R is a Rule-based Ordering over $L \cup \{B\}$ with respect to the derivation rules of PLEN, and for all $x \in L \cup \{B\}$, XRx.
(4) If $x \in L \cup \{B\} - X$, then for all $A_i \in X$ it holds that A_iRx.
(5) The sequence of nodes in X is the *trunk* of the tree, and each sequence of R-ordered nodes is a *branch*. Each branch extends from the trunk, but

distinguish a *core branch* composed only of the trunk, axiom nodes, and nodes added by derivation rules (including conditional and negation introduction; no assumptions and no formulae derived from assumptions can be on the core branch).

(6) B is the terminal node of the core branch (for all x on the core branch such that $x \neq B, xRB$.

Definition 81 (Derivability) Let $A \in L_{PLEN}, X \subseteq L_{PLEN}$, and $\{\emptyset, B_1, ..., B_n\} = X$. Then: A is *derivable* from X in PLEN ($X \vdash_{PLEN} A$) iff there is a derivation of A from X, $\langle L, X = \{\emptyset, B_1, ..., B_n\}, R, A\rangle$, and R is characterized by the derivation rules and axioms of the derivation system.

Definition 82 (Theorem) For any $A \in L_{PLEN}$, A is a *theorem* in PLEN ($\vdash_{PLEN} A$) iff A is an instance of an axiom or A can be derived from the axioms of PLEN (i.e., there is a derivation of A of the form $\langle L, X = \{\emptyset\}, R, A\rangle$).

4.2.4 Metalogic

I want to briefly comment on the standard metalogical properties of PLEN. Contrary to the norm in development of philosophical logics, I am interested primarily in developing PLEN with respect to the nonstandard features of its protocol logic, e.g., its capacity for representing or codifying the most theoretically central properties of epistemic norms and its capacity for supporting interesting theories of protocol equivalence. The only standard metatheorem of import for these uses of PLEN is its soundness; I aim to show only that (i) the semantic system of PLEN provides plausible and useful models of epistemic dynamics and features of epistemic normativity, and (ii) that the protocol logic of PLEN contains rules that characterize the logic of epistemic norms that are sound with respect to this semantics (i.e., sound with respect to these plausible models of epistemic dynamics). Establishing these claims suffices to show that PLEN is interesting and useful for representing and reasoning about epistemic norms.

With that programmatic caveat in place, I show in Appendix 1 that the derivation system of PLEN is sound with respect to its semantics. The usual definitions and forms of reasoning are easily modified for PLEN:

Definition 83 (Soundness) The PLEN derivation system is sound with respect to PLEN's semantics iff if $X \vdash_{PLEN} A$, then $X \models_{PLEN} A$.

Theorem 84 (Soundness Theorem) *If $X \vdash_{PLEN} A$, then $X \models_{PLEN} A$.*

The proof proceeds by cases for the rules and axiom schemata of PLEN's derivation system; each rule is shown to preserve verification over arbitrary states in arbitrary EDGs, and each axiom schema is shown to be verifiable at arbitrary states in arbitrary EDGs. The most interesting and novel steps are for the protocol equivalence rules, in virtue of the centrality that protocol equivalence comes to play in the system. Let me sketch the reasoning for the procedural protocol equivalence rules, reserving sketches for the rest of the equivalence rules for Appendix 1:

Example 85 (Execution-Theoretic Protocol Equivalence Rules) The following clusters of rules are sound.

$$\left[\begin{array}{l} \dfrac{\Box(\pi \leftrightarrow_p \pi'),\ \rho(\pi/\pi')(X_\pi) = X_{\pi'}}{\Box(X_\pi \leftrightarrow_p X_{\pi'})} \\ \dfrac{\Box(\pi \leftrightarrow_p \pi'),\ \Box(\pi' \leftrightarrow_p \pi'')}{\Box(\pi \leftrightarrow_p \pi'')} \\ \dfrac{}{\Box(\pi \leftrightarrow_p \pi)} \\ \dfrac{\Box(\pi \leftrightarrow_p \pi')}{\Box(\pi' \leftrightarrow_p \pi)} \\ \dfrac{\Box(\pi \leftrightarrow_p \pi')}{\Box(\pi \leftrightarrow_t \pi')} \end{array}\right] \left[\begin{array}{l} \dfrac{\pi \leftrightarrow_p \pi',\ \rho(\pi/\pi')(X_\pi) = X_{\pi'}}{X_\pi \leftrightarrow_p X_{\pi'}} \\ \dfrac{\pi \leftrightarrow_p \pi',\ \pi' \leftrightarrow_p \pi''}{\pi \leftrightarrow_p \pi''} \\ \dfrac{}{\pi \leftrightarrow_p \pi} \\ \dfrac{\pi \leftrightarrow_p \pi'}{\pi' \leftrightarrow_p \pi} \\ \dfrac{\pi \leftrightarrow_p \pi'}{\pi \leftrightarrow_t \pi'} \end{array}\right]$$

***Proof** (Sketch)* The first rules of both clusters can be proven in one move: induction over proper immediate subprotocols. I prove that the first rule of the righthand cluster is sound over arbitrary s in arbitrary EDG, g, which gives the boxed version immediately:[8]

(1) (Base) Assume that $\pi \in PISub(X_\pi)$. Let $p \in P_{X_\pi}$ where $s = f(p)$. By syntactic replacement, $X_{\pi'} = \pi' \circ \pi_i$. Let P^s_π for any π be the set of executions of π in g that start at s. Let:

$$P^s_{X_\pi} = \left\{p \mid s = f(p),\ p = q \oplus^\pi_\circ r, q \in P^s_\pi, r \in P_{\pi_i}\right\}$$

where $\oplus^\pi_\circ$ corresponds to the graph-theoretic relations among executions of the immediate subprotocols of $\pi' \circ \pi_i$ required by the procedural execution rules. For instance, sequential composition corresponds to path concatenation, parallel execution corresponds to path correspondence, and choice corresponds to branching from a shared root state. By assumption, for any p in g, $p \in P^s_\pi$ iff $p \in P^s_{\pi'}$. Thus, $P^s_\pi = P^s_{\pi'}$. Thus, by substitution of identical sets, $P^s_{X_\pi} = P^s_{X_{\pi'}}$. The converse follows by transposition.

(2) (Induction) If the result holds for all proper immediate subprotocols of X_π, then it holds for X_π.

 (a) From the application of the prior reasoning to the induction assumption, the result immediately follows.

(3) Given results proven in Chaps. 5 and 6 (the immediate subprotocol hierarchy theorem (proven in Sect. 5.7.2) and the graph-theoretic syntactic replacement

[8] The reasoning showing the unboxed version of these protocol equivalence rules valid can just be iterated at every state to show that the boxed versions of these rules are valid.

theorem lemma (proven in Sect. 6.4.2), the foregoing reasoning generalizes to $\pi \in Sub(X_\pi)$ case. This suffices for the first rule in the righthand column.

Elementary set theory supplies the transitivity, reflexivity, and symmetry rules immediately, which suffices for the rest of the rules in the righthand column save the last. The foregoing reasoning was applied to arbitrary s in g, and so can be be applied all states in g, validating the boxed version of the rules in the lefthand column.

The last rule in the righthand column is immediate from the definition of path-theoretic executions: if π and π' share all path executions, then, given that there are no detached transitional executions of any protocols (transitions that are not the root or terminal states of any paths in the EDG in question), they share all transitional executions. □

The rest of the protocol equivalence rules can be shown mechanically by adapting the foregoing reasoning to binary and ternary executions. Most of the execution-theoretic axiom schemata can be adapted with some modification from proofs of soundness for PDL, and they are derivable by mechanical model-checking, anyway. The axiom schemata governing the operators that are new to PLEN are novel, but they are ultimately mechanically derivable, and there are quite a few of them, so I reserve explicit model-checking of illustrative cases for each new operator for Appendix 1. Some of the distinctive steps are made easier by means of results from Chaps. 5 and 6. I leave those also for Appendices 1–3. I make only a few comments on the propositional and modal logics below, as these are mostly familiar.

The reader may have noticed one unusual feature of PLEN's derivation system, given that PLEN is a dynamic logic. It supports conditional proof and proof by contradiction. This is somewhat unusual in dynamic (indeed, in modal) logics, given the variety of logical consequence notions definable therein. Conditional proof would, for instance, fail to be sound in the (textbook) semantics of PDL where logical consequence is defined as preservation of logical truth in models [15]. This is due to formulae of the form A and $[\pi]A$. Similarly, if logical consequence were so defined in PLEN, conditional proof would be unsound (due to formulae of the form A and $[\pi]_x A$). Suppose that the logical truth of A implies the logical truth of $[\pi]_x A$, which, of course, it does: if A is true at all states in arbitrary EDG, g, then $[\pi]_x A$ is true at all states in g. However, it doesn't follow that $A \to [\pi]_x A$ is true at all states in g. Suppose that, compatible with the logical truth preservation above, A is not true at all states in g but is true at one state, s, and that A fails at s' where s' is the terminal state of some path, transition, and edge execution of π that is rooted at s. Then (each variant of) $[\pi]_x A$ fails at s, and so does the conditional.

Of course, logical consequence in PLEN is defined as preservation of truth or verification at any state over models, so such formulae do not generate counterexamples for the conditional proof rule of PLEN. The soundness of the conditional proof rule in Definition 76 is immediate from the definition of logical consequence:

Proof Suppose that there is a sound derivation of B from A using rules excluding conditional proof. If s verifies A, then B is verified at s by the assumption of the soundness of the derivation, and s thus verifies $A \to B$. Else, s verifies $A \to B$ by its truth rule (material conditional). For derivations using nested Conditional Proofs, induction on the depth of the nesting suffices where the base case is the reasoning just given, and the next level up in the nesting is immediately sound by the induction hypothesis. □

This rule facilitates the proof of the LtoR half of the Deduction Theorem for PLEN:

Theorem 86 (Deduction Theorem for PLEN) $X \cup \{B\} \vdash_{PLEN} A$ *iff* $X \vdash_{PLEN} B \rightarrow A$.

Proof (LtoR) Let $X \cup \{B\} \vdash_{PLEN} A$. Induction on the length of core branches of proofs: If the result holds for any premises and any formulae derived by steps that precede or are equal to step n, then the result holds for the premises and any formula derived by step $n + 1$. Suppose the result holds for any premises and any formulae derived by any step up to step n.

(1) Consider the last step of a derivation of A from the premises $X \cup \{B\}$; the last formula derived, A, was derived by step $n + 1$.
(2) Either $A \in X$, $A = B$, A is an instance of an axiom schema, or A is an instance of the consequent of a derivation rule in PLEN such that instances of all of the premises of the rule are derivable from $X \cup \{B\}$ by steps up to step n. For the first three cases:
    (a) Let $A \in X$. Then A is derivable from X by one step: $X \vdash_{PLEN} A$. But then $X \vdash_{PLEN} A \rightarrow (B \rightarrow A)$ by propositional logic. Thus, by conditional elimination, $X \vdash_{PLEN} (B \rightarrow A)$.
    (b) Let $A = B$. Then $X \vdash_{PLEN} B \rightarrow B$ by propositional logic. Thus, $X \vdash_{PLEN} A \rightarrow A$.
    (c) Let A be an instance of an axiom schema. Then A is derivable from X by one step: $X \vdash_{PLEN} A$. But then $X \vdash_{PLEN} A \rightarrow (B \rightarrow A)$ by propositional logic. Thus, by conditional elimination, $X \vdash_{PLEN} (B \rightarrow A)$ by conditional elimination.

(3) Let A be an instance of the consequent of a derivation rule in PLEN, (Dr), such that there are instances of all of the premises of the rule that are derivable from $X \cup \{B\}$.
    (a) There are two cases: (Dr) is either a one-premise rule or a two-premise rule. The two-premise rule case is shown explicitly, and the one-premise rule case simplifies it.
    (b) If A is derivable from $X \cup \{B\}$ by any two-premise rule, (Dr), then there are some formulae, C_1 and C_2 of formulae such that:
        (i) $X \cup \{B\} \vdash_{PLEN} C_1$ and $X \cup \{B\} \vdash_{PLEN} C_2$ by steps of the proof that precede or are equal to step n.
        (ii) C_1 and C_2 are instances of the premises of (Dr).
        (iii) A is an instance of the consequent of (Dr).
        (iv) Thus, $X \vdash_{PLEN} B \rightarrow C_1$ by the induction hypothesis and i.
        (v) Thus, $X \vdash_{PLEN} B \rightarrow C_2$ by the induction hypothesis and i.

(vi) Trivially, if A follows by means of rule (Dr) from C_1 and C_2, then for any set of premises Y, $Y \cup \{C_1\} \cup \{C_2\} \vdash_{PLEN} A$ by monotonicity.
(vii) Thus, $\emptyset \cup \{C_1\} \cup \{C_2\} \vdash_{PLEN} A$ from vi.
(viii) Thus, $\emptyset \vdash_{PLEN} (C_1 \wedge C_2) \rightarrow A$ from vii by Conditional Proof.
(ix) Thus, $X \vdash_{PLEN} (C_1 \wedge C_2) \rightarrow A$ from viii and monotonicity, since $X = X \cup \emptyset$.
(x) Thus, $X \vdash_{PLEN} B \rightarrow (C_1 \wedge C_2)$ from iv and v by propositional logic.
(xi) Hypothetical Syllogism (HS) is a derived law in the R-Derivation System (Conditional Proof establishes it):
(A) $\emptyset \vdash_{PLEN} (A \rightarrow B) \rightarrow ((B \rightarrow C) \rightarrow (A \rightarrow C))$
(B) The relevant instance: $\emptyset \vdash_{PLEN} (B \rightarrow (C_1 \wedge C_2)) \rightarrow (((C_1 \wedge C_2) \rightarrow A) \rightarrow (B \rightarrow A))$.
(xii) Thus, $X \vdash_{PLEN} B \rightarrow A$ from x and xi by monotonicity and two iterations of Conditional Elimination (Modus Ponens).

The RtoL direction is immediate by detachment of the conditional. □

The PLEN derivation system is not shown to be complete, as it is formulated here. Of course, completeness is a standard technical goal in designing a logical system, but the completeness of the system is orthogonal to the main argument of the project. The purpose of PLEN is to provide a useful formal framework for representing and reasoning about procedural epistemic norms and epistemic dynamics. Soundness with respect to the underlying models of dynamics is essential; completeness is merely nice but strictly unnecessary.

Nonetheless, it's plausible that the completeness theorem of PDL extends to the fragment of PLEN defined over regular protocols without syntactic formulae.[9] PLEN is, after all, just a complicated variant of PDL. Consider these PLEN-specific modifications of the textbook definitions for and proofs of completeness for PDL presented in [15]:

Definition 87 (Fischer-Ladner Closure (FLC)) To define the Fischer-Ladner Closure for the regular protocol fragment of PLEN, let FL and FL_π, respectively, be simultaneously defined functions such that:

(1) (FL)

$$L_{PLEN} \rightarrow P(L_{PLEN})$$

(a) For all atomic formulae, P: $FL(P) = \{P\}$
(b) $FL(\neg A) = \{\neg A\} \cup \{A\}$
(c) $FL(A \wedge B) = \{A \wedge B\} \cup \{A\} \cup \{B\}$

[9] Trivially, there are no rules governing syntactic formulae, so there are no derivations to or from syntactic formulae, even if they are semantically valid. Note also that all inferences to true syntactic formulae are valid, as truth for syntactic formulae is defined independent of any features of EDGs.

(d) $FL(A \vee B) = \{A \vee B\} \cup \{A\} \cup \{B\}$
(e) $FL(A \rightarrow B) = \{A \rightarrow B\} \cup \{A\} \cup \{B\}$
(f) $FL(\top) = \{\top\}$
(g) $FL(\bot) = \{\bot\}$
(h) $FL([\pi]_x A) = FL_\pi([\pi]_x A) \cup FL(A)$
(i) For any protocols, $\pi_1,...,\pi_4$:
(i)

$$FL(R_x(\pi_1, \pi_2)) = \{R_x(\pi_1, \pi_2)\} \cup FL_\pi([\pi_1]_x \top) \cup FL_\pi([\pi_2]_x \top)$$

(ii)

$$FL(R_x(\pi_1, \pi_2, \pi_3, \pi_4)) = \{R_x(\pi_1, \pi_2, \pi_3, \pi_4)\} \cup FL_\pi([\pi_1]_x \top) \\ \times \cup FL_\pi([\pi_2]_x \top) \cup FL_\pi([\pi_3]_x \top) \cup FL_\pi([\pi_4]_x \top)$$

(2) (FL_π)

$$\{[\pi]_x A \mid \pi \in P_{PLEN}, A \in L_{PLEN}\} \rightarrow P(L_{PLEN})$$

(a) For all atomic protocols, a: $FL_\pi([a]_x A) = \{[a]_x A\}$
(b) $FL_\pi([\pi_1 \cup \pi_2]_x A) = \{[\pi_1 \cup \pi_2]_x A\} \cup FL_\pi([\pi_1]_x A) \cup FL_\pi([\pi_2]_x A)$
(c) $FL_\pi([\pi_1 \cap \pi_2]_x A) = \{[\pi_1 \cap \pi_2]_x A\} \cup FL_\pi([\pi_1]_x A) \cup FL_\pi([\pi_2]_x A)$
(d) $FL_\pi([\pi_1; \pi_2]_x A) = \{[\pi_1; \pi_2]_x A\} \cup FL_\pi([\pi_1]_x [\pi_2]_x A) \cup FL_\pi([\pi_2]_x A)$
(e) $FL_\pi([\pi_1 *]_x A) = \{[\pi_1 *]_x A\} \cup FL_\pi([\pi_1]_x [\pi_1 *]_x A)$
(f) $FL_\pi([(A)?]_x B) = \{[(A)?]_x B\} \cup FL(A)$

Remark 88 The syntax of PLEN works identically for all r and p variants of the operators, $[\pi]_x$ and $\langle\pi\rangle_x$ in defining FLC.

Remark 89 The extension of filtrations to PLEN is very lengthy but only somewhat complicated. The basic idea is that a filtration of EDG, g, with respect to $FL(A)$ where A embeds protocols, perfectly replicates the binary, ternary, and procedural structure of g with respect to the protocols in A.

Definition 90 (EDG Filtration) Let g be an EDG. Then the filtration of g by $FL(A)$ is an EDG of the form:

$$g_{FL(A)} = \left\langle S_{FL(A)}, \Pi_g, R, R_0, \Pi_g^+, R+, R+_0, T_{\pi/FL(A)}, c, v_{FL(A)} \right\rangle$$

such that:

(1) For any states in g, s and s', $s \Leftrightarrow s'$ iff:

(a) $\forall A \in FL(B)(v_s(A) = 1 \equiv v_{s'}(A) = 1)$
(b) $c(s) = [X : Y : \Pi_M]$, $c(s') = [X' : Y' : \Pi'_M]$, and:

(i)

$$\forall A \in FL(B)((A \in X \text{ iff } A \in X) \wedge (A \in Y \text{ iff } A \in Y))$$

(ii)

$$\forall \pi' \in FL_\pi(B)(\pi' \in \Pi_M \text{ iff } \pi' \in \Pi'_M)$$

(2) $S_{FL(A)} = \{[s] \mid s \in S\}$ where $[s] = \{s' \mid s \Leftrightarrow s'\}$.
(3) For atomic formula, A: $v_{s/FL(A)} = \{[s] \mid v_s(A) = 1\}$
(4) The valuations and content functions for complex formulae are extended in the obvious fashion.

   (a) Valuations are extended by the truth rules.
   (b) Content functions are simply reproduced, i.e., $b(A)$ is verified at s iff $A \in X$ for $c([s])$.

(5) For atomic protocol, a:

   (a) $T_{a/FL(A)} = \{([s],[t]) \mid (s,t) \in T_a\}$
      (a) Define $T_{g/FL(A)}$ accordingly.
   (b) $R+_{a/FL(A)} = \{([s], a, [t]) \mid (s,t) \in T_a\}$
      (i) Define $R_{g/FL(A)}$ and $R{+}_{0/FL(A)}$ accordingly.
      (ii) Collapse all atomic edges, (s, a, s') and (t, a, t'), such that $[s] = [t]$ and $[s'] = [t']$.
   (c) $P_{a/FL(A)} = R_{\pi/FL(A)}$
      (i) Define $P_{g/FL(A)}$ and $P_{0/FL(A)}$ accordingly.

(6) The structure of $g_{FL(B)}$ is defined by the following sets of rules for complex protocols:

   (a) (Transition Sets)
      (i) (;) $T_{\pi_1;\pi_2/FL(A)} = T_{\pi_1/FL(A)} \circ T_{\pi_2/FL(A)} =$

$$\left\{([s],[t]) \in T_{g/FL(A)} \mid \exists [x] \in S_{FL(A)} \bigwedge \begin{bmatrix} ([s],[x]) \in T_{\pi_1/FL(A)} \\ ([s],[x]) \in T_{\pi_2/FL(A)} \end{bmatrix}\right\}$$

      (ii) ($*$) $T_{\pi*/FL(A)} =$

$$\left\{([s],[t]) \in T_{g/FL(A)} \mid \exists n \geq 1 \begin{bmatrix} \exists [x_1]...[x_n](\forall i \leq n([x_i] \in S_{FL(A)})) \\ \bigwedge \begin{bmatrix} [s] = [x_1], [s'] = [x_n] \\ \forall i \leq n-1 \, \langle [x_i], [x_{i+1}] \rangle \in T_{\pi/FL(A)} \end{bmatrix} \end{bmatrix}\right\}$$

      (iii) ($\cap$) $T_{\pi_1 \cap \pi_2/FL(A)} =$

$$\{([s],[t]) \in T_{g/FL(A)} \mid \exists p \in P_{\pi_1 \cap \pi_2/FL(A)}(([s],[t]) = \langle first(p), last(p) \rangle)\}$$

(iv) ($\cup$) $T_{\pi_1 \cup \pi_2/FL(A)} =$

$$\left\{ ([s],[t]) \in T_{g/FL(A)} \mid [s] \neq [t], \bigvee \begin{bmatrix} (([s],[t]) \in T_{\pi_1/FL(A)} \wedge \exists([s'],[t']) \in T_{\pi_2/FL(A)}([s]=[s'])) \\ (([s],[t]) \in t_{\pi_2/FL(A)} \wedge \exists([s'],[t']) \in T_{\pi_1/FL(A)}([s]=[s'])) \end{bmatrix} \right\}$$

(v) ((-)?)

$$T_{(A)?/FL(A)} = \left\{ ([s],[s']) \in T_{g/FL(A)} \mid [s] \Vdash A \wedge b(A), [s'] \Vdash A \wedge b(A) \right\}$$

(vi) ($-$)

$$T_{-\pi/FL(A)} = \left\{ ([s],[t]) \in T_{g/FL(A)} \mid \exists([s],[t]) \notin T_\pi \right\} = T_{g/FL(A)} - T_{\pi/FL(A)}$$

(vii) ($\curvearrowleft$)

$$T_{\curvearrowleft\pi/FL(A)} = \left\{ ([s],[t]) \in T_{g/FL(A)} \mid \exists([t],[s]) \in T_{\pi/FL(A)} \right\}$$

(viii) (∇)

$$T_{\nabla\pi/FL(A)} = \left\{ ([s],[s]) \in T_{g/FL(A)} \mid \exists([s],[s]) \in T_{\pi/FL(A)} \right\}$$

(b) (Edge Sets)

(i) (;) $R+_{\pi_1;\pi_2/FL(A)} =$

$$\left\{ \langle [s], \pi_1; \pi_2, [s'] \rangle \in R+_{0/FL(A)} \mid \exists]x] \in S_{FL(A)} \bigwedge \begin{bmatrix} \langle [s], \pi_1, [x] \rangle \in R_{\pi_1/FL(A)} \\ \langle [x], \pi_2, [s'] \rangle \in R_{\pi_{2/}/FL(A)} \end{bmatrix} \right\}$$

(ii) ($*$) $R+_{\pi*/FL(A)} =$

$$\left\{ \langle [s], \pi*, [s'] \rangle \in R+_{0/FL(A)} \mid \exists n \geq 1 \begin{bmatrix} \exists [x_1]...[x_n](\forall i \leq n(x_i \in S_{FL(A)})) \\ \bigwedge \begin{bmatrix} [s]=[x_1],[s'=x_n], \\ \forall i \leq n-1 \, \langle [x_i], \pi, [x_{i+1}] \rangle \in R+_{\pi/FL(A)} \end{bmatrix} \end{bmatrix} \right\}$$

(iii) ($\cap$) $R+_{\pi_1 \cap \pi_2 / FL(A)} =$

$$\left\{ \langle [s], \pi_1 \cap \pi_2, [s'] \rangle \in R+_{0/FL(A)} \mid \right.$$

$$\left. \bigwedge \begin{bmatrix} \langle [s], \pi_1, [s'] \rangle \in R+_{\pi_1/FL(A)} \\ \langle [s], \pi_2, [s'] \rangle \in R+_{\pi_2/FL(A)} \\ \exists p \in P_{\pi_1 \cap \pi_2}([s] = f(p) \wedge [s'] = l(p)) \end{bmatrix} \right\}$$

(iv) ($\cup$) $R+_{\pi_1 \cup \pi_2 / FL(A)} =$

$$\left\{ \langle [s], \pi_1 \cup \pi_2, [s'] \rangle \in R+_0 \mid [s] \neq [s'], \right.$$

$$\left. \bigvee \begin{bmatrix} (e \in R+_{\pi_1} \wedge \exists e' \in R+_{\pi_2} (f(e) = f(e') = [s] \wedge l(e) = [s'])) \\ (e \in R+_{\pi_1} \wedge \exists e' \in R+_{\pi_2} (f(e) = f(e') = [s] \wedge l(e) = [s'])) \end{bmatrix} \right\}$$

(v) ((-)?) $R+_{(A)?/FL(A)} =$

$$\{ \langle [s], (A)?, [s'] \rangle \in R+_{0/FL(A)} \mid [s] \Vdash A \wedge b(A), [s'] \Vdash A \wedge b(A) \}$$

(vi) ($-$) $R+_{-\pi/FL(A)} =$

$$\{ \langle [s], -\pi, [s'] \rangle \in R+_{0/FL(A)} \mid \neg \exists e (e \in R+_{\pi/FL(A)} \wedge e \leftrightarrows \langle s, -\pi, s' \rangle) \}$$

(vii) ($\curvearrowleft$) $R+_{\curvearrowleft \pi/FL(A)} =$

$$\{ \langle [s], \curvearrowleft \pi, [s'] \rangle \in R+_{0/FL(A)} \mid \exists \langle s', \pi, s \rangle \in R+_{\pi/FL(A)} \}$$

(viii) (∇) $R+_{\nabla \pi/FL(A)} =$

$$\{ \langle [s], \nabla \pi, [s] \rangle \in R+_0 \mid \exists \langle s, \pi, s \rangle \in R+_{\pi/FL(A)} \}$$

(c) (Path Sets) For all $p \in P_g$, there is a $q \in P_{g/FL(A)}$ such that $p \simeq q$ and $first(p)$ is generated graph isomorphic[10] with $first(q)$, and vice versa.

(i) (;) $P_{\pi_1;\pi_2/FL(A)} =$

$$\{ p \in P_{g/FL(A)} \mid p = qr \wedge q \in P_{\pi_1/FL(A)} \wedge r \in P_{\pi_2/FL(A)} \}$$

[10] This is a well-known relation from logical dynamics [33, Ch. 4] and is defined explicitly in this volume, Chap. 5. For the impatient reader: generated graph isomorphism means that every EDG with $first(p)$ as a root state is isomorphic to every EDG with $first(q)$ as a root state.

(ii) $(*)$ $P_{\pi*/FL(A)} =$

$$\left\{p \in P_{g/FL(A)} \mid \exists n \left(\bigvee\begin{bmatrix} \exists \langle q_1, ..., q_n\rangle \,(p = q_1 ... q_n, \forall i \leq n(q_i \in P_{\pi/FL(A)})) \\ p = \langle s, \pi*, s\rangle \end{bmatrix}\right)\right\}$$

(iii) $(\cap)$ $P_{\pi_1 \cap \pi_2/FL(A)} =$

$$\left\{p \in P_{g/FL(A)} \mid \bigvee\begin{bmatrix} (p \in P_{\pi_1/FL(A)} \wedge \exists p' \in P_{\pi_2/FL(A)}(p \rightleftharpoons q)) \\ (p \in P_{\pi_2/FL(A)} \wedge \exists p' \in P_{\pi_1/FL(A)}(p \rightleftharpoons q)) \end{bmatrix}\right\}$$

(iv) $(\cup)$ $P_{\pi_1 \cup \pi_2/FL(A)} =$

$$\left\{p \in P_g \mid f(p) \neq l(p), \bigvee\begin{bmatrix} (p \in P_{\pi_1/FL(A)} \wedge \exists p' \in P_{\pi_2/FL(A)}(first(p) = first(p'))) \\ (p \in P_{\pi_2/FL(A)} \wedge \exists p' \in P_{\pi_1/FL(A)}(first(p) = first(p'))) \end{bmatrix}\right\}$$

(v) $((-)?)$ $P_{(A)?/FL(A)} =$

$$\left\{p \in P_{g/FL(A)} \mid \bigwedge\begin{bmatrix} [first(p)] \Vdash A \wedge b(A), \\ [last(p)] \Vdash A \wedge b(A) \end{bmatrix}\right\}$$

(vi) $(-)$ $P_{-\pi/FL(A)} =$

$$\{p \in P_{g/FL(A)} \mid p \notin P_{\pi/FL(A)} \wedge \neg\exists q \in P_\pi (q \sqsubset p \vee q \leftrightharpoons p)\}$$

(vii) $(\curvearrowleft)$ $P_{\curvearrowleft\pi/FL(A)} =$

$$\left\{p \in P_{g/FL(A)} \mid \left[\exists q \bigwedge\begin{bmatrix} q \in P_{\pi/FL(A)}, \\ p = f_I(q) \end{bmatrix}\right]\right\}$$

(viii) (∇) $P_{\nabla\pi/FL(A)} =$

$$\left\{p \in P_{g/FL(A)} \mid \bigwedge\begin{bmatrix} p \in P_{\pi/FL(A)}, \\ first(p) = last(p) \end{bmatrix}\right\}$$

Remark 91 Given these definitions, the following two conjectures seem plausible and, together, would yield a small model property for PLEN and the requisite results to prove completeness:

Conjecture 92 (PLEN Fischer-Ladner Lemma (PFLL)) The following properties of the Fischer-Ladner Closure on PLEN hold:

(1) (FLL1)

(a) If $A \in FL(B)$, then $FL(A) \subseteq FL(B)$.
(b) If $A \in FL_\pi([\pi]_x B)$, then $FL(A) \subseteq FL_\pi([\pi]_x B) \cup FL(B)$.
(c) If $A \in FL(R_x(\pi_1, \pi_2))$, then

$$FL(A) \subseteq \{R_x(\pi_1, \pi_2)\} \cup FL_\pi([\pi_1]_x \top) \cup FL_\pi([\pi_2]_x \top)$$

(d) If $A \in FL(R_x(\pi_1, \pi_2, \pi_3, \pi_4))$, then

$$FL(A) \subseteq \{R_x(\pi_1, \pi_2, \pi_3, \pi_4)\} \cup FL_\pi([\pi_1]_x \top)$$
$$\cup\ FL_\pi([\pi_2]_x \top \cup FL_\pi([\pi_3]_x \top) \cup FL_\pi([\pi_4]_x \top$$

(2) (FLL2)

(a) If $[\pi]_x A \in FL(B)$, then $A \in FL(B)$.
(b) If $[(C)?]_x A \in FL(B)$, then $C \in FL(B)$.
(c) If $[\pi_1 \cup \pi_2]_x A \in FL(B)$, then $[\pi_1]_x A \in FL(B)$ and $[\pi_2]_x A \in FL(B)$.
(d) If $[\pi_1 \cap \pi_2]_x A \in FL(B)$, then $[\pi_1]_x A \in FL(B)$ and $[\pi_2]_x A \in FL(B)$.
(e) If $[\pi_1; \pi_2]_x A \in FL(B)$, then $[\pi_1]_x [\pi_2]_x A \in FL(B)$ and $[\pi_2]_x A \in FL(B)$.
(f) If $[\pi_1 *]_x A \in FL(B)$, then $[\pi_1]_x [\pi_1 *]_x A \in FL(B)$.
(g) If $R_x(\pi_1, \pi_2) \in FL(B)$, then

$$\{R_x(\pi_1, \pi_2)\} \cup FL_\pi([\pi_1]_x \top) \cup FL_\pi([\pi_2]_x \top) \subseteq FL(B)$$

(h) If $R_x(\pi_1, \pi_2, \pi_3, \pi_4) \in FL(B)$, then

$$\{R_x(\pi_1, \pi_2, \pi_3, \pi_4)\} \cup FL_\pi([\pi_1]_x \top) \cup FL_\pi([\pi_2]_x \top) \cup FL_\pi([\pi_3]_x \top)$$
$$\cup\ FL_\pi([\pi_4]_x \top) \subseteq FL(B)$$

(3) (FLL3)

(a) For any formula, A, $|FL(A)| \leq len(A)$.
(b) For any formula, $[\pi]_x A$, $|FL_\pi([\pi]_x A)| \leq len(\pi)$.
(c) For any formula, $R_x(\pi_1, \pi_2)$,

$$|FL(R_x(\pi_1, \pi_2))| \leq (len(\pi_1) + len(\pi_2) + |FL_\pi([\pi_1]_x \top)| + |FL_\pi([\pi_2]_x \top)|)$$

(d) For any formula, $R_x(\pi_1, \pi_2, \pi_3, \pi_4)$,

$$|FL(R_x(\pi_1, \pi_2, \pi_3, \pi_4))| \leq (len(\pi_1) + ... + len(\pi_2) + |FL_\pi([\pi_1]_x \top)| \\ +... + |FL_\pi([\pi_4]_x \top)|)$$

Conjecture 93 (PLEN Filtration Lemma (PFL)) Let g be an EDG, and let s and t be states in S. Then $g_{FL(B)}$ is an EDG such that:

(1) For all $A \in FL(B)$, $v_s(A) = 1$ iff $v_{s/FL(B)}(A) = 1$.
(2) For all $[\pi]_x A \in FL(B)$:

  (a) if $(s, s') \in T_\pi$, then $([s], [s']) \in T_{\pi/FL(B)}$
  (b) if $([s], [s']) \in T_{\pi/FL(B)}$ and $v_s([\pi]_x A) = 1$, then $v_{s'}(A) = 1$.

(3) For all $R_x(\pi_1, \pi_2) \in FL(B)$,

  (a) for each π_i,
    (i) if $t \in T_{\pi_i}$ with $s = f(t)$ and $s' = l(t)$, then $t_{FL(B)} \in T_{\pi_i/FL(B)}$ and $[s] = f(t_{FL(B)})$ and $[s'] = l(t_{FL(B)})$.
    (ii) if $e \in R+_{\pi_i}$ with $s = f(e)$ and $s' = l(e)$, then $e_{FL(B)} \in R+_{\pi_i/FL(B)}$ and $[s] = f(e_{FL(B)})$ and $[s'] = l(e_{FL(B)})$.
    (iii) if $p \in P_{\pi_i}$ with $s = f(p)$ and $s' = l(p)$, then $p_{FL(B)} \in P_{\pi_i/FL(B)}$ and $[s] = f(p_{FL(B)})$ and $[s'] = l(p_{FL(B)})$.
  (b) for each π_i, the following conditionals and their converses hold:
    (i) if $t_{FL(B)} \in T_{\pi_i/FL(B)}$ and $[s] = f(t_{FL(B)})$ and $[s'] = l(t_{FL(B)})$, then:
      (A) if $t_{FL(B)} \in T_{\pi_i/FL(B)}$ and $f(t_{FL(B)}) = [f(t)]$, then $t \in T_{\pi_i}$.
      (B) if $t_{FL(B)} \in T_{\pi_1/FL(B)}$ iff $t_{FL(B)} \in T_{\pi_2/FL(B)}$ and $f(t_{FL(B)}) = [f(t)]$, then $t \in T_{\pi_1}$ iff $t \in T_{\pi_2}$.
      (C) if $t_{FL(B)} \in T_{\pi_1/FL(B)} \cap T_{\pi_2/FL(B)}$ and $f(t_{FL(B)}) = [f(t)]$, then $t \in T_{\pi_1} \cap T_{\pi_2}$.
      (D) if $t_{FL(B)} \in T_{\pi_1/FL(B)} - T_{\pi_2/FL(B)}$ and $f(t_{FL(B)}) = [f(t)]$, then $t \in T_{\pi_1} - T_{\pi_2}$.
    (ii) if $e_{FL(B)} \in R+_{\pi_i/FL(B)}$ and $[s] = f(e_{FL(B)})$ and $[s'] = l(e_{FL(B)})$, then:
      (A) if $e_{FL(B)} \in R+_{\pi_i/FL(B)}$ and $f(e_{FL(B)}) = [f(e)]$, then $e \in R+_{\pi_i}$.
      (B) if $e_{FL(B)} \in R+_{\pi_1/FL(B)}$ iff $e_{FL(B)} \in R+_{\pi_2/FL(B)}$ and $f(e_{FL(B)}) = [f(e)]$, then $e \in R+_{\pi_1}$ iff $e \in R+_{\pi_2}$.
      (C) if $e_{FL(B)} \in R+_{\pi_1/FL(B)} \cap R+_{\pi_2/FL(B)}$ and $f(e_{FL(B)}) = [f(e)]$, then $e \in R+_{\pi_1} \cap R+_{\pi_2}$.
      (D) if $e_{FL(B)} \in R+_{\pi_1/FL(B)} - R+_{\pi_2/FL(B)}$ and $f(e_{FL(B)}) = [f(e)]$, then $e \in R+_{\pi_1} - R+_{\pi_2}$.
    (iii) if $p_{FL(B)} \in P_{\pi_i/FL(B)}$ and $[s] = f(p_{FL(B)})$ and $[s'] = l(p_{FL(B)})$, then:
      (A) if $p_{FL(B)} \in P_{\pi_i/FL(B)}$ and $f(p_{FL(B)}) = [f(p)]$, then $p \in P_{\pi_i}$.
      (B) if $p_{FL(B)} \in P_{\pi_1/FL(B)}$ iff $p_{FL(B)} \in P_{\pi_2/FL(B)}$ and $f(p_{FL(B)}) = [f(p)]$, then $p \in P_{\pi_1}$ iff $p \in P_{\pi_2}$.

(C) if $p_{FL(B)} \in P_{\pi_1/FL(B)} \cap P_{\pi_2/FL(B)}$ and $f(p_{FL(B)}) = [f(p)]$, then $p \in P_{\pi_1} \cap P_{\pi_2}$.

(D) if $p_{FL(B)} \in P_{\pi_1/FL(B)} - P_{\pi_2/FL(B)}$ and $f(p_{FL(B)}) = [f(p)]$, then $p \in P_{\pi_1} - P_{\pi_2}$.

(4) For all $R_x(\pi_1, \pi_2, \pi_3, \pi_4) \in FL(B)$, $v_s(R_x(\pi_1, \pi_2, \pi_3, \pi_4)) = 1$ iff $v_{s/FL(B)}(R_x(\pi_1, \pi_2, \pi_3, \pi_4)) = 1$.

Remark 94 As the completeness of PLEN is new but not central to the main argument, I leave the proof sketches, which are perhaps mildly interesting, to Appendix 1. If these lemmata are correct, though, the verifiability lemma of PDL plausibly extends to PLEN, and so, given all of the foregoing, it seems that we arrive at weak completeness:

Conjecture 95 $\models_{PLEN} A$ iff $\vdash_{PLEN} A$

The deduction theorem then allows the generalization from weak completeness to strong completeness:

Conjecture 96 If $X \models_{PLEN} A$, then $X \vdash_{PLEN} A$.

So much for the plausible strong completeness of PLEN. As for decidability and complexity, it's at least plausible that PLEN, in its current shape, is decidable. Consider the following small model property:

Conjecture 97 If A is a verifiable formula of L_{PLEN}, then A is verifiable in an EDG, g, such that

$$|S| \leq 2^{(len(\pi_1)+...+len(\pi_4)+|FL_\pi([\pi_1]_x\top)|+...+|FL_\pi([\pi_4]_x\top)|)}$$

This property is also sketchily derived from the foregoing revision of the textbook proofs in Appendix 1. If this is correct, then it's straightforward to design a naive effective algorithm for deciding the verifiability problem of PLEN. Given that the language of PLEN contains that of PDL, and its semantics is considerably more complicated, PLEN is likely as least as complex as PDL (at least EXP-TIME complete) and non-compact. Again, these are left open, if plausible, in this work, in favor of pursuing more interesting, if off the beaten path, results.

4.3 Philosophical Development

4.3.1 Philosophical Interpretation

Protocols are most naturally interpreted as programs or procedural instructions. Given epistemic proceduralism, this makes protocols natural constructions for representing epistemic norms, especially procedural epistemic norms. However, one might want to draw a strong distinction between epistemic norms, on the one hand, and the rules that define epistemic procedures (the constitutive norms for epistemic procedures), on the other. After all, it seems that epistemic norms might dictate epistemic procedures, and the rules that dictate them are distinct from those that define the procedures (i.e., that determine which courses of action correctly execute the procedure and which don't). The difference seems to be one between high level regulative norms—norms concerning right epistemic action—and low level constitutive norms—norms defining what courses of action count as performing a procedure.

For instance, naive belief revision rules like "See to it that you don't believe *A* and the negations of the logical consequences of *A* at once" seem to be high-level regulative norms. They certainly dictate epistemic action (they require that one see to it that one's beliefs have a closure property), but they do not specify specific courses of action or seem to define any particular procedures. The rule does not specify, for instance, what inferential procedures to carry out in order to find out what the consequences of *A* are. Similarly, Beall's multiple conclusion theory revision rule [4, 5] offers only high level constraints on what theories to accept given the contours of a multiple-conclusion logical consequence relation. What specific belief revisions, inferences, proof constructions, or theory adoption procedures one carries out to make one's beliefs comply with those rules is left as a matter of free choice. Just the same, rules in the family of "Given the defeasibility of the arguments in favor of theory T_i, default to a comparative theory selection procedure with respect to the alternatives on the table, $T_1, \ldots, T_n$", give high level regulative guidance about what to do in the face of non-conclusive reasons, but leave a great deal of freedom in actual procedure. In each case, the idea is that high level regulative norms are distinct from what we might think of as constitutive norms with respect to some precise procedure.

The distinction between these high level norms and the more restrictive constitutive norms might relate to a more philosophical distinction. It seems, prima facie, that a belief revision rule is normative in a way that a proof rule for a logical system is not. One is forming, retaining, or revising beliefs wrongly in the case of violation of the former, but simply failing to carry out a procedure correctly in the latter.

These distinctions make no difference for the projects of PLEN. All regulative norms are constitutive procedural norms, and all constitutive procedural norms are epistemically normative. Both the high level regulative rules and the more exacting (constitutive) procedural norms are equally procedural epistemic norms. Certainly, they have formal properties that distinguish them. High level regulative norms, are formed from large non-deterministic choices of many, many exact procedures. "Constitutive" procedural norms are more fine-grained or restrictive. Both types of norms define procedures; they just define procedures with different structures, which, in turn prescribe epistemic action at differing levels of granularity. Moreover, the secondary difference above doesn't imply a categorical difference between high level norms and constitutive procedural norms; it just shows that there are epistemic norms that are constitutive for different tasks. Proof construction is an epistemic process, and one can do it more or less correctly. It may not be as important to us, all things considered, as the higher level processes of belief revision in which this subprocess is involved, but that doesn't entail that the higher level processes are not constitutive norms nor that the lower level procedural norms are somehow "less normative". High level regulative norms of (e.g.) belief revision are constitutive norms with respect to the procedure of rational belief revision. Low level constitutive norms are normative with respect to specific courses of epistemic action.

The thrust, then, is that, though protocols can equally well be thought of as constitutive rules for defining types of epistemic action, process, or procedure, this does no harm to the use of protocols to represent epistemic norms precisely because epistemic norms are constitutive rules for defining types of epistemic action, processes, or procedure. Thus, the intended interpretation of PLEN begins with the stipulation that procedural epistemic norms are protocols.

4.3.1.1 Procedural Epistemic Norms Are Protocols, and Protocols in PLEN Are Procedural Epistemic Norms

Procedural epistemic norms are program-like constructions that define restrictions on procedural construction (i.e., restrictions on how epistemic actions are to be combined or the sequences of choices of epistemic action that one is to face), which then, in turn, define restrictions on epistemic dynamics defined over other epistemic factors.[11] Protocols are the formal objects in PLEN that encode the program-like structure of the content of procedural epistemic norms.[12] From here on, I will freely refer to protocols as procedural epistemic norms, and denote procedural epistemic

[11] This is one of the main results of Chaps. 2 and 3. An independent argument is suggested by recent analyses of procedural knowledge [27]. (1) Operational semantics can be used to give an analysis of practical senses as programs. (2) Practical senses are the core parts of the content of procedural knowledge. (3) Procedural knowledge analysis of norm content solves the Intellectualist Regress and accounts for the cognitive and regulative roles of norms. From these premises, it follows plausibly that the content of norms is fruitfully analyzed in terms of programs. As this line of thinking was a late discovery in the preparation of this book, I leave more detailed development of the argument to future work.

[12] A technical note: The protocol control operators—sequential composition, iteration, choice, etc.—are syntactic operators; they operate on syntactic objects only. So, strictly speaking, these should not be confused with the relational operations of the same names discussed in Chap. 1 as analyzans for program structure, which are operations on binary relations. Nonetheless, the semantics of PLEN maps the syntactic control operators to relational and graph-theoretic operations. First, atomic protocols are mapped to relations (binary and ternary) in EDGs, and then the control operators are mapped to relational or graph-theoretic operations on the corresponding relations. It is in this way that protocols analyze program or program-like structures.

norms with protocol metavariables, π_i, π_j, X_π, while also interpreting protocol-theoretic formulae as expressing claims about procedural epistemic norms.

Procedural epistemic norms are thus directly embedded in the protocol-theoretic formulae in PLEN. These formulae are thus expressions that encode language for reasoning directly about epistemic norms. The various equivalence relations in L_{PLEN} enable the construction of formulae that encode statements about the equivalence of procedural epistemic norms. The variants of modal operators in which protocols are embedded enable the construction of formulae that encode statements about the deontic accessibility relations associated with different epistemic norms. That is, different epistemic norms, π_1 and π_2, may state that some courses of epistemic action are obligatory or permissible, and other epistemic norms may specify otherwise. The deontic accessibility relations of these norms, T_{π_1} and T_{π_2}, define notions of obligation and permission for π_1 and π_2, respectively. These operators, $[\pi_i]_x$... and $\langle\pi_i\rangle_x$ ... thus tell us what results from compliance with epistemic norms, along with the modality of these postconditions. The formula, $[\pi_i]_p A$, for instance, tells us that A always results from the execution of any epistemic process complying with π_i (that unfolds from the current state), while $\langle\pi_i\rangle_t A$ tells us that we can get to a state at which A holds without violating π_i. These are the most elementary schemata, but they serve to illustrate the intended interpretation of PLEN.

The syntactic system of PLEN thus comprises a language of procedural epistemic norms and a language of epistemic normative propositions. The semantics of PLEN models the content of (i.e., restrictions on epistemic dynamics defined by) procedural epistemic norms by means of execution rules, and it provides truth rules for interpreting formulae about epistemic norms. These are all defined over PLEN's models of epistemic dynamics, which, as argued in Sect. 2.3.3, are relatively fine-grained models of epistemic dynamics, in the sense that they distinguish epistemic actions, processes, and procedures from the transitions and sequences of transitions that result from the execution of them.

4.3.1.2 EDGs Are Single-Agent Models of Epistemic Dynamics

Epistemic directed graphs are taken, in PLEN, to be abstract models of single-agent epistemic dynamics, with each (informal) element of epistemic dynamics mapped to a feature of epistemic directed graphs with the intent that the mapping produces a kind of structural resemblance:

Epistemic States: Judgment and Content	*Represented By* $\Longrightarrow$	States: $S = \{s_1, ..., s_n\}$, Content States: $\langle X : Y : \Pi_M \rangle$, Content functions: $c(s)$
Epistemic State Transitions	*Represented By* $\Longrightarrow$	Transitions: $t = \langle s, s' \rangle$
Epistemic Action Tokens	*Represented By* $\Longrightarrow$	Edges: $e = \langle s, a, s' \rangle$
Epistemic Action Types	*Represented By* $\Longrightarrow$	Execution Sets for Atomic Protocols: $R_a = \{e \in S \times \Pi_0 \times S \mid e \in S \times a \times S\}$ Edge Sets for Complex Protocols: $R+_\pi = \{e \in S \times \Pi_C \times S \mid e \in S \times \pi \times S\}$
Epistemic Process Tokens	*Represented By* $\Longrightarrow$	Procedural Paths: $p = \langle e_1, ..., e_n \rangle$ where for each e_i and e_{i+1}, $last(e_i) = first(e_{i+1})$ and there are π_i and π_{i+1} such that $e_i \in R_{\pi i}$ and $e_{i+1} \in R_{\pi i+1}$
Epistemic Process Types (Epistemic Procedures)	*Represented By* $\Longrightarrow$	Execution Sets for Complex Protocols: P_π such that $p \in P_\pi$ iff p satisfies the relevant execution rules with respect to π
Feasibility of a Transition under C	*Represented By* $\Longrightarrow$	Let t be the transition. Then t is feasible under C iff $first(t) \Vdash C$.
Feasibility of an Action under C	*Represented By* $\Longrightarrow$	Let e be the action token. Then e is feasible under C iff $first(e) \Vdash C$.
Feasibility of a Process under C	*Represented By* $\Longrightarrow$	Let p be the process token. Then p is feasible under C iff $first(p) \Vdash C$.
Capability of a Transition	*Represented By* $\Longrightarrow$	Let t be the transition. Then t is within the capacities of agent i iff g represents i's epistemic dynamics and there is some s such that $s = first(t)$.
Capability of an Action	*Represented By* $\Longrightarrow$	Let e be the action token. Then e is within the capacities of agent i iff g represents i's epistemic dynamics and there is some s such that $s = first(e)$ (iff $e \in R$).
Capability of a Process	*Represented By* $\Longrightarrow$	Let p be the process token. Then p is within the capacities of agent i iff g represents i's epistemic dynamics and there is some s such that $s = first(p)$ (iff $e \in P_g$).

4.3.2 Detailed Examples

4.3.2.1 Programming Constructions

The syntax of PLEN is an elaboration on the syntax of PDL, which is already a powerful language for expressing program constructions. The intended interpretation of protocols as structures representing the program-like content of procedural epistemic norms is quite natural, as a result. This section comments a bit on the basic constructions in P_{PLEN} and illustrates their use in analyzing program-like information.

The protocol $\pi_1; \pi_2$ is the *sequential composition* of π_1 and π_2; the procedure that requires doing π_1 and then immediately doing π_2 from the state reached by π_1. The protocol $\pi_1 \cap \pi_2$ is the *parallel execution* of π_1 and π_2; the procedure that requires doing π_1 while doing π_2. The protocol $\pi_1 \cup \pi_2$ is the *nondeterministic choice* of π_1 and π_2; the procedure that requires nondeterministic selection from among π_1 and π_2 (including the parallel execution of both). The protocol $\pi*$ is the finite *iteration* of π; the procedure that requires doing π any n times sequentially.

The protocol $(A)?$ is a test for A; a unique protocol that checks the current state for the truth of A and, in PLEN, produces the belief or acceptance of A. The protocol $-\pi$ is the *complement* of π; the procedure comprising execution of the set of protocols other than π incompatible with doing π. The protocol $\curvearrowleft \pi$ is the *reversal* of π; the procedure that reverses the effects that are brought about by doing π. The protocol $\nabla\pi$ is the *fixed point of* π; the version of π that makes no changes to the world or to any epistemic states.

Sequential composition, parallel execution, and choice are binary, the rest are unary, and P_{PLEN} is recursively enumerable. Complex protocols can be composed to create more complex protocols. In the infix notation, parentheses can be used, as normal, to clarify structure. For example, the following are well-formed protocols:

- $(\pi_1; \pi_2); (\pi_3; \pi_4)$
- $(\pi_1; \pi_2*); (\pi_3; \pi_4)*$
- $\pi_1; (\pi_2; (\pi_3; \pi_4))$
- $((\pi_1; \pi_2); (\pi_3; \pi_4)) \cup (\pi_4 \cap (\pi_6 \cup \pi_1))$
- $-((\pi_1; \pi_2); (\pi_3; \pi_4))$
- $(A)?; ((\pi_1; \pi_2); (\pi_3; \pi_4))$
- $\curvearrowleft ((\pi_1; \pi_2); (\pi_3; \pi_4))$
- $-(\pi_1; \pi_2); (\pi_3; \pi_4)$
- $((A)?; (\pi_1; \pi_2)); (\pi_3; \pi_4)$
- $\curvearrowleft (\pi_1; \pi_2); (\pi_3; \pi_4)$

Without parentheses, the protocols, $\pi_1; \pi_2; \pi_3; \pi_4$, $\pi_1 \cup \pi_2 \cup \pi_3$, and $\pi_1 \cap \pi_2 \cup \pi_3$ are each not well-formed, but for any complex protocol formed from multiple iterations of a single control operator, this doesn't matter semantically. For example, $(\pi_1; \pi_2); (\pi_3; \pi_4)$ and $\pi_1; (\pi_2; (\pi_3; \pi_4))$, while syntactically distinct protocols, are equivalent (in ways to be explicated), as are $(\pi_1 \cup \pi_2) \cup (\pi_3 \cup \pi_4)$ and $\pi_1 \cup (\pi_2 \cup (\pi_3 \cup \pi_4))$, and $(\pi_1 \cap \pi_2) \cap (\pi_3 \cap \pi_4)$ and $\pi_1 \cap (\pi_2 \cap (\pi_3 \cap \pi_4))$. In general, from a semantic perspective, each of the binary control operators is associative, while parallel execution and choice are commutative, as well. Permutations on the order of constituent protocols thus make no semantic difference for choice and parallel execution, though, obviously, they do for sequential composition.

The string algebraic protocol language is less user-friendly, but is unambiguous, and so makes no use of parentheses. The infix notation protocols, $(\pi_1; \pi_2); (\pi_3; \pi_4)$ and $\pi_1; (\pi_2; (\pi_3; \pi_4))$, are not likely to be confused in the string algebraic syntax.

They translate to $; \frown; \frown \pi_1 \frown \pi_2 \frown; \frown \pi_3 \frown \pi_4$ and $; \frown \pi_1 \frown; \frown \pi_2 \frown; \frown \pi_3 \frown \pi_4$, respectively.

As for intuitively reading off what each protocol expression says, protocols can be read as programs, recipes, or instructions for carrying out complex procedures that are formed out of basic actions. If an atomic protocol expression, π, is an instruction of the form "Do π!", then:

- $\pi_1; \pi_2$ reads "Do π_1, then do π_2!"
- $\pi_1 \cap \pi_2$ reads "Do π_1 concurrently with π_2!" or perhaps "Do π_1 while doing π_2!"
- $\pi_1 \cup \pi_2$ reads "Do π_1 or do π_2!"
- $\pi*$ reads "Do π any $n \geq 0$ times!" or "Do π zero or more times!"
- $(A)?$ reads "Test for A!" or "Determine whether A!"
- $-\pi$ reads "Do anything other than π, and don't do π!"
- $\curvearrowleft \pi$ reads "Undo the changes wrought by π!"
- $\nabla\pi$ reads "Do π in a way that doesn't change anything!"

Consequently, protocols formalize programs. In particular, the most important programming and decision structures can be expressed in PLEN. Here is a small sampling:[13]

[13] In PDL (inter alia), conditional choice is formulated more simply: $(((A)?); \pi_1) \cup (((\neg A)?); \pi_2)$. However, this doesn't work properly with the execution rules for nondeterministic choice defined in PLEN, as the choice requires both proper immediate subprotocols to be executable. This is motivated by the observation that nondeterministic choice should obey something like the principle of alternative possibilities. Just as it makes no sense to say "You are free to choose between options A and B but you can't choose option B", it makes no sense to say that you can execute protocol $\pi_1 \cup \pi_2$ but you can't execute π_2. However, this formal reflection of a philosophical observation comes at the cost of making conditional choice behave somewhat oddly in PLEN; conditional choice is now an iterative protocol.

If A is true, then, under the PDL formulation, the test on the right hand side of the choice (as presented) is simply not executable, so the execution of the conditional choice implies forces the execution of the other branch of the protocol. In PLEN, the execution conditions of the conditional choice require $((-(A)? \cap (\neg A)?); \pi_2)*$ to be executable. Thus, if A is true, then $((-(A)? \cap (\neg A)?); \pi_2)*$ must be executable by a single, looping edge, thereby implying that the only execution of the conditional choice is the path that executes $((A)? \cap -(\neg A)?); \pi_1)*$. The oddness of this can be removed by considering the notion of bounded iterations in Chap. 7, and replacing the iteration, $*$, here with the bounded iteration $*_1$. This will give something that functions essentially like the PDL formulation.

It's a matter for future work to determine whether this is the best formalization of conditional choice in PLEN or, indeed, whether the theoretical benefits of the current nondeterministic choice execution rules are worth bearing given this odd law. See Sect. 8.2.5 for some accounting of the benefits with respect to PLEN's resources for resolving the paradoxes of deontic logic. One can always revise the execution rules for choice with rules closer to that of PDL, and then let conditional choice be formulated more typically.

Name	English w/ metavariables	PLEN
Conditional Choice	If A, do π_1; else, do π_2!	$(((A)? \cap -(\neg A)?); \pi_1) * \cup((-(A)? \cap (\neg A)?); \pi_2)*$
Generalization of Conditional Choice	If A, do π_1; if B, do π_2; ...!	$(((A)? \cap -(B)?); \pi_1) * \cup(((B)? \cap -(A)?); \pi_2)* \cup ...$
Guarded Iteration	While A, do π	$((A)?; \pi)*; (\neg A)?$
Guarded Repetition	Repeat π until A is true.	$(\pi; (\neg A)?)*; (A)?$

The point can be grasped (in one way, at least) by considering the notion of *pseudocode*. Pseudocode is a term in programming instruction and practice that denotes the result of laying out the structure and behavior of a program in natural language using decision structures. The idea is to understand a program in a way that doesn't depend on the specific functions and syntactic conventions of a programming language.

For example, the following pseudocode states, in a regimented but not fully formal (i.e., written in a specific programming language) language, a program that asks for the user's age and their preferred retirement age and then returns either the news that they can retire now or the year in which the user will be able to retire and the number of years from the current date.

```
Get user's current age;
Get user's preferred retirement age;
Get current date and time;
If the user can't retire right now:

  tell how long it will be and what year it will be;

    else: tell them they can retire now
```

The following is the actual code of a program in the language Python that (more or less) does the trick:

```
age1 = input(" What is your current age? " )
age2 = input(" At what age would you like to retire? " )
def retirement1(age1, age2):

  return(int(age2) - int(age1))

from datetime import datetime
if retirement1(age1, age2) > 0:

  print(" You have " + str(retirement1(age1, age2)) +

    "  years left until you retire, because you'll be able to
       retire in "
    + str(int(datetime.now().year) + int(age2) - int(age1)) + "
       ." );
    else: print(" Congratulations! You can retire now
       (in principle)!" )
```

For another example, the following pseudocode could be found in a video game.[14] The goal is to find a program that causes an object in the game—in this case, a computer controlled character called an NPC—to initiate a default conversation with the player character despite being hostile to them as soon as the NPC perceives the player character. This might be used in a game to initiate a sense of narrative tension—as, in the context of the game, the outcome of the conversation may resolve the hostility or provoke it. A pseudocode rendition of such a program might be:

```
set a default conversation
if the npc recognizes a valid object:

  if

    the valid object is the last thing the npc has seen

      and

    the npc is hostile to that object:
    initiate default conversation;

      else: ignore and return to routine
```

The following is the actual code of a program in the language C with numerous custom functions designed that (more or less) does the trick, given that the default conversation has already been defined:

```
void   main()
{
object oNoticed = GetLastPerceived();

  if(GetIsObjectValid(oNoticed))
  {

    if (GetLastPerceptionSeen() && GetIsEnemy(oNoticed))
    {

      ClearAllActions();
      ActionStartConversation(oNoticed);
      }

  }
}
```

These programs are constructed by means of sequential composition and conditional choice. Vastly more complex programs can be built out of these fundamental decision structures over libraries of basic actions (i.e., functions or variable assignments). Of course, one can then build complex programs out of complex programs. The atomic protocol set represents potential libraries of actions. Protocol syntax encodes the fundamental decision structures by means of control operators, and so,

---

[14] Strictly speaking, a computer role playing game that more or less formalizes the third edition of *Dungeons and Dragons* called *Neverwinter Nights*.

in virtue of its recursiveness, encodes more complex programming constructions built out of them.

The program-likeness of the information conveyed by epistemic norms falls out of observations about the generality of these procedural operations. One can simply point to both accepted and proposed normative and methodological standards that dictate or require particular procedural operations on epistemic actions, and show that these have at least rudimentary program-like structure.

Recall the Defeasible Closure Rule and the attending pseudocode for complying with it (Fig. 2.1 from Chap. 2):

(Defeasible Closure): If A implies B, and there are no defeaters for B, then see to it that you don't believe A without believing B!

Given the right repertoire of basic actions, the protocol constructions of PLEN can be straightforwardly used to encode this rule:

Example 98 Let $D(B)$ be elliptical for a conjunction of propositions that are epistemic defeaters[15] for B. Let π_B be an arbitrary protocol such that $\models_{PLEN} [\pi_B]_p b(B)$. Let π_{-A} be an arbitrary protocol such that $\models_{PLEN} [\pi_{-A}]_p(\neg b(A) \wedge d(A))$. Let $\pi_\perp$ be a halt protocol. For our purposes, it's any protocol; it should be read as going on with one's life. Then, the following infix notation protocol formalizes the Defeasible Closure program:

$$[(((A \rightarrow B)? \cap -(\neg(A \rightarrow B))?); [((((b(A))? \cap -(\neg b(A))?); [((((\neg b(B))? \cap -(b(B))?); ((((\neg D(B))? \cap -(D(B))?); \pi_B) * \cup(((D(B))? \cap -(\neg D(B))?); \pi_{-A}) * \cup((b(B))? \cap (-(\neg b(B))?); (((\neg D(B))? \cap -(D(B))?); \pi_\perp) * \cup(((D(B))? \cap -(\neg D(B))?); \pi_{-B}) *] * \cup(((\neg b(A))? \cap -(b(A))?); \pi_\perp) *] \cup [((\neg(A \rightarrow B))? \cap -(A \rightarrow B)?); \pi_\perp) *]$$

4.3.2.2 The Researcher Case

Now, to return to some of the motivating examples and apply PLEN to their formalization. Recall the Researcher Case from Chap. 1:

Example 99 Let $b(T_i)$ stand in for accepting the entirety of one of the theories on the table for Dubrovsky. Let $D(T_i)$ be elliptical for a conjunction of propositions that are epistemic defeaters for T_i; this conjunction, in other words, is the counterevidence for T_i. Let π_i denote the data gathering steps outlined (themselves, likely to decompose into more complex protocols). Let π_{P_i} be an arbitrary protocol

[15] See [28] for an account of epistemic defeaters. Given the protocol-theoretic analysis of defeasible reasons in Chap. 8, one can, alternately, take $D(B)$ to stand in for some conjunction of defeasible reasons for $\neg B$.

such that, for $P_i \in \{P_1, \ldots, P_6\}$ in the case, $\models_{PLEN} [\pi_{P_i}]_p b(T_1)$ for $i \leq 4$, and $\models_{PLEN} [\pi_{P_j}]_p b(T_k)$ for $j > 4$ and $k \neq 1$. Then:

The following infix notation protocol permits all and only Dubrovsky's listed choices:

$$[\pi_1; ((\pi_{P_1} \cup \pi_{P_4})) \cup (\pi_2; (\pi_{P_2} \cup \pi_{P_4})) \cup (\pi_2; (\pi_3; (\pi_4 \cup (\pi_{P_4} \cup (\pi_{P_5} \cup \pi_{P_6})))) \\ \cup (\pi_2; (\pi_3; (\pi_4; (\pi_{P_4} \cup \pi_{P_3}))))]$$

The following more restrictive protocols restrict Dubrovsky's choices to omit the theory selection based on Tarot reading, and so seem to encode epistemic rationality norms applicable to the Researcher Case more plausible than the fully permissible protocol above:

- $\pi_1; (\pi_2; (\pi_3; (\pi_4 \cup (\pi_{P_5} \cup \pi_{P_6}))))$
- $\pi_1; (\pi_2; (\pi_3; (\pi_4; \pi_{P_3})))$

4.3.2.3 The Honest Twin Case

Recall the Honest Twin Case from Chap. 2:

Example 100 First, let π_a and π_b be protocols standing in for the procedures involved in carrying out the proof and asking my twin, respectively. I use protocol metavariables here so as not to create the impression that these are unanalyzable epistemic actions.[16] The following protocol-theoretic formulae capture many of the features of this case:

- Deriving the result myself, π_a, is in a methodology, M, but asking my twin, π_b, is not, despite their transition equivalence:

$$(M(\pi_a) \wedge \neg M(\pi_b)) \wedge (\pi_a \leftrightarrow_t \pi_b)$$

- Deriving the result myself, π_a, has procedural features that asking my twin, π_b does not despite their transition equivalence:
  - For some substep of formal proof, π_f, the executability of π_a implies that π_f is possible to execute, while the executability of π_b implies no such thing, despite their transition equivalence:

$$((\langle \pi_a \rangle_p \top \rightarrow \Diamond(\langle \pi_f \rangle_p \top) \wedge \neg(\langle \pi_b \rangle_p \top \rightarrow \Diamond \langle \pi_f \rangle_p \top)) \wedge (\pi_a \leftrightarrow_t \pi_b)$$

[16] The notation for a and b in the description of the case might misleadingly suggest this, given that a and b are atomic protocols in the syntax of PLEN.

- Where A encodes the possession of some theorem and a proof thereof, π_a and π_b equally produce possession of the theorem, and, despite their transition equivalence, π_a and π_b diverge in the epistemic processes:

$$((\langle\pi_a\rangle_p A \equiv \langle\pi_b\rangle_p A) \wedge (\pi_a \curlyvee_p \pi_b)) \wedge (\pi_a \leftrightarrow_t \pi_b)$$

- Transition equivalence fails to imply procedural equivalence:

$$\neg((\pi_a \leftrightarrow_t \pi_b) \rightarrow (\pi_a \leftrightarrow_p \pi_b))$$

- Deriving the result myself, π_a, requires executing formal substep, π_f, at some stage, but asking my twin, π_b, does not, despite their transition equivalence:

$$[\pi_a]_p(\langle\pi_f\rangle_p \top \vee \Diamond\langle\pi_f\rangle_p \top \vee \Diamond\Diamond\langle\pi_f\rangle_p \top \vee \ldots \vee \Diamond\ldots\Diamond\langle\pi_f\rangle_p \top$$

- The foregoing formulae are all satisfiable at some states in some EDGs.
- At the meta-level, the normative differences between π_a and π_b are visible, as there may be protocols (hence, procedural norms) that dictate the execution of π_a and prohibit the execution of π_b: let there be a π such that, for all g:

$$P_\pi = \{p \in P_g \mid \exists q \in P_{\pi_a}(q \sqsubseteq p \vee p \rightleftharpoons q) \wedge \forall r \in P_{\pi_b}(r \in P_g - P_\pi\}$$

- Features and their formal counterparts can be multiplied at will.

4.3.3 *Philosophical Distinctions and Rival Systems*

4.3.3.1 Norm Exogeny and Protocol Information

One of the primary virtues of PLEN is that it is a norm exogenous logic. Let me spell that out. Exogenous action logics are those logics of action that contain expressions that explicitly name actions and that build the truth conditions of complex propositions in their languages out of some interpretation of the action expressions. The standard example of an exogenous logic is PDL. To see exactly what exogeny comes to, compare PDL to the standard example of endogenous logic: temporal logic.[17]

PDL is exogenous because its syntax contains a sublanguage of expressions that explicitly name actions and analyze program constructions built out of them; there is a syntax of programs built out of complex actions. The semantics of PDL assigns structures to the basic actions and give a recipe for assigning structures to complex programs which are then used to determine the truth values of the basic

[17] See [15, p. 157] for more on this distinction.

propositional expressions. From there, formulae that embed programs in modal operators can be used to reason about the accessibility relations of the programs. The properties of these accessibility relations are then characterized in part by the syntactic structures of programs. For instance, the program logic of PDL comprises axioms about the structure (i.e., relational composition) of accessibility relations like $[a; b]A \leftrightarrow [a][b]A$.

Temporal logics, in contrast, contain no language of actions, but rather just temporal operators that are interpreted over structures that represent the evolution of a world or system state over time (sets of histories or paths formed from sequences of words or states). The program that determines how the system unfolds is built into the structures over which the language is interpreted—the properties of the transition relations among worlds, in particular, determine the temporal "program". That is, what changes can be made to the current state of the world are encoded in the structure of a model of temporal logic, and there are no expressions explicitly encoding their properties. Temporal logics thus don't contain anything that encodes valid reasoning about the programs themselves.

Thus, PDL is action exogenous because its syntax makes explicit room for terms that name (or at least represent) actions, its semantics interprets those terms and builds truth conditions for complex formulae embedding them, and its logic provides a system for reasoning about those formulae.

A logic of norms is *norm exogenous,* then, if it contains syntactic expressions that name or represent norms with a syntactic item, interprets those items and builds the truth conditions of complex formulae out of the interpretation of norm expressions, and provides a logic of those formulae. Now, why is norm exogeny a virtue?

There are two straightforward, if abstract, arguments for thinking that the logic of epistemic norms ought to be norm exogenous. The first argument is simple: given D3 and D3C1, the logic of epistemic norms ought to formalize direct reasoning about epistemic norms, and this requires norm exogeny. By "direct reasoning about norms", I mean reasoning of the forms articulated by D3C1. Roughly, a logic of norms should provide a logical framework for analyzing and validating (or invalidating) reasoning of forms like "Norms N_1, N_2, ..., N_n bear relation, F; therefore, ...". This directly entails norm exogeny, because these arguments are built out of statements that directly embed norms.

The second argument is more methodological:

(1) The equivalence, dependence, and verdict relations among epistemic norms (and, in the third case, objects of norms) are relations among epistemic norms that partially determine the logical relations among epistemic normative propositions.
(2) The logic of epistemic normative propositions must directly formalize reasoning about the relations among epistemic norms that even partially determines the logical relations among epistemic normative propositions.
(3) If a logic is to formalize reasoning about equivalence, dependence, and verdict relations among epistemic norms, then it must solve the embedding problem with respect to epistemic norms.

(4) If a logic is to solve the embedding problem with respect to epistemic norms, then it must be norm exogenous.
(5) ∴ The logic of epistemic normative propositions must be norm exogenous.

The first premise is defended directly in Chap. 3, Sect. 3.1. There, it is argued that epistemic norms bear the sorts of relations in question, and some of their properties are discussed. In particular, epistemic norms bear equivalence relations with each other of varying kinds, and these are genuine; the relations in question are reflexive, symmetric, and transitive. Dependence relations are reflexive, asymmetric, and transitive. Verdict relations pick out a class of relations of implication among claims about the verdicts of norms with respect to an object. Now, perhaps these aren't really "logical" relations in some strong sense.[18] Nonetheless, they are general relations that support numerous forms of valid inference. Whether the protocol equivalence relations themselves are properly logical relations is a matter of indifference to the main thrust of the argument; they are relations that a logic of norms must analyze that are abstract, general, and formal, insofar as they hold between protocols in virtue of interactions between protocol syntax and execution rules irrespective of the semantics of formulae. Moreover, there are necessary and logically protocol equivalences, which are more easily defended as logical relations among protocols.

These kinds of relations partially determine the logical relations among normative propositions. For each equivalence relation defended in Sect. 3.1, if N and N' are equivalent, then they deliver identical verdicts for some range of objects. More, the fundamental notions of equivalence require that N and N' are intersubstitutable in some range of contexts. Dependence relations deliver implication relations about verdicts. And verdict relations deliver implication relations among postconditions or the outcomes of following norms. Thus, the logical behavior (implications, equivalences, etc.) of normative propositions[19] like "x is irrational with respect to N", and "N dictates the performance of x under C" (etc.) depends on the logical relations that hold among norms.

In other words, there is a sort of embedding problem that only exogenous logics can solve (i.e., premise 3). First, a logic of normative propositions must provide systematic, analytic, and recursive recipes for analyzing the logical relations among normative propositions [9, 12]. But, given the argument here, the logical relations among epistemic normative propositions are systematically determined by the logical relations among epistemic norms. Thus, a logic of normative propositions needs to be able to analyze—or at least characterize—the relations

[18] Because, for instance, they aren't properly formal or normative [6], perhaps because they don't hold between truth-evaluable items [20].

[19] Note that one could get away with having norm-relative deontic or epistemic operators in the language of the logic of norms, as in SDL. So, in the absence of an argument like the current one, one needn't have a norm exogenous logic in order to have a deontic or epistemic logic. In such a logic, however, norms won't be first-class citizens. They'll operate behind the scenes of the interpretation of the operators. I argue below (chapters 5 and 8) that this is a problematic limitation.

among epistemic norms and show how they are built into the truth conditions for normative propositions. In a slogan: the logic of epistemic normative propositions must be built atop the logic of epistemic norms.

This transparently requires formulae or other expressions that can reasonably taken to represent such relations (or the content of fragments of language that assert them). Obviously, to solve the embedding problem for such expressions, there must be terms or expressions for norms and interpretations of them that can be composed into truth conditions for the complex expressions embedding the norm expressions. Thus, such a system requires norm exogeny, giving us premise 4.

From another direction, a logic can't do the foregoing work without being able to tell us what follows from "N is equivalent to N'", "N is dependent on N'", and so on. Of course, logics only tell us what follows from what by means of its validity relation and its logical truths, which are characterized by meta-level rules and object-level laws formulated in terms of the system's object language. Thus, the logic of norms requires norm expressions and systematic recipes for interpreting them in order to fully analyze the logical relations among normative propositions.

So much for premises 1, 3, and 4. A basic desire for comprehensiveness motivates premise 2. Given that epistemic norms truly bear the kinds of relations in question, the logic of norms must have something to say about these relations, their properties, and how they fit into the logic of epistemic normative propositions. Thus, a non-exogenous (modal or multi-modal) approach to the logic of epistemic norms won't do.

The protocol-theoretic logic of epistemic norms uses protocol information to formalize procedural epistemic norms, thereby solving the foregoing problems. That is, protocols shape accessibility relations in dynamic logics, and PLEN codifies information about the protocols themselves in addition to information about the accessibility relations (i.e., mere dynamics) that protocols shape. PLEN allows us to reason directly about protocols themselves and, indeed, even to reason about what one protocol allows or enables an agent to think or discover about other protocols. Given the intended interpretation of PLEN, this means that PLEN is an exogenous logic of procedural epistemic norms, formalizing direct reasoning about such norms and even formalizing reasoning about what one system of norms allows or enables an agent to think or discover about other systems of norms.

The basic thrust of the solution is that the protocol logic of PLEN solves the embedding problem for procedural epistemic norms.[20] In PLEN, protocols are

[20] The embedding problem with respect to epistemic norms is solved by the basic formalism of PLEN. The language of protocols provided expressions that, in the meta-language, can be quantified over and reasoned about explicitly, and, at the object-level, can be embedded in expressions that codify the application of predicates to protocols and the construction of propositions out of these predicating clauses. In other words, systematic recipes were provided for assigning semantic values to propositions about protocols on the basis of the assignment of "semantic" values (execution sets) to protocols. More, logical relations among protocols—i.e., equivalence—were connected by systematic recipes to the semantic values of meta-language and object-language expressions. Finally, systems were provided for constructing valid complex

embedded in contexts of protocol equivalence (and, derivatively, forms of protocol implication, dependence, sufficiency, and independence), as well as modal operators that characterize accessibility (i.e., dynamics) with respect to a protocol (these can be thought of as outcomes or postconditions of executing the protocols). This allows us to reason directly about protocols and to model direct reasoning about protocols. For instance, derivations of properties of protocols using valid rules like $\frac{\pi \leftrightarrow_p \pi', \pi' \leftrightarrow_t \pi''}{\pi \leftrightarrow_t \pi''}$ can be rather easily constructed and validated in PLEN. Thus, the protocol-theoretic formulae in which protocols are embedded allow us to say things directly about protocols and to characterize valid reasoning formulated with such statements. Procedural epistemic norms are encoded in PLEN by means of protocols. Thus, given the intended interpretation, this provides a logic for direct reasoning about procedural epistemic norms.

Further, while the PLEN framework imposes no conditions on content states (i.e., no constraints on formula membership in acceptance or rejection sets), it's straightforward to see how sets of axioms or premises characterizing acceptance and rejection rules (e.g., $\Box\neg(b(A) \wedge d(A))$ or $b(A) \rightarrow \neg d(A)$) or relations among protocols and acceptance and rejection (e.g., $[\pi_1]_p((b(A) \wedge b(\neg A \vee B)) \rightarrow \neg d(B)))$ can be built in PLEN. From there, the protocol logic can be used derive further properties of doxastic states and the relations among protocols and doxastic states. These formulae thus allow us to say things about what the protocol one is following allows one to discover about another protocol. We can, for instance, figure out what follows from claims like $\langle\pi_1\rangle_p b(\pi_2 \leftrightarrow_t \pi_3)$, $[\pi_1]_p\neg b(\pi_2 \leftrightarrow_t \pi_3)$, or, more generally, for classes of formulae of forms like $\langle\pi_1\rangle_p b(...\pi_2...)$ and $[\pi_1]_p\neg b(...\pi_2...)$. These are extremely elementary, schematic examples, but they suffice to show that the system allows statements about epistemic norms and their regulative effects on doxastic states of arbitrary (finite) complexity, and this includes doxastic states involving beliefs about epistemic norms. It also validates chains of inferences of arbitrary complexity formulated with these statements. Thus, PLEN provides a logical framework for direct reasoning about epistemic norms and, as promised, for reasoning about what different epistemic norms allow or enable us to think about other epistemic norms.

This is a virtue of PLEN not shared with traditional (static) epistemic logics, temporal doxastic logics, belief revision theories, deontic logics, or probabilistic theories of epistemic dynamics. These formal frameworks hide the protocol information that would enable direct reasoning about epistemic norms (at the object level) or that would facilitate meta-level reasoning about epistemic norms. Of course, one can, using these systems, reason about some suites of epistemic norms (e.g., the rationality or entrenchment postulates of AGM-style belief revision

arguments and derivations out of meta-language and object-language expressions. This could only be the case, given that the statements comprising the inferential steps were built out of expressions that embed names for protocols (hence, epistemic rationality norms) and predications applied to protocols, if the embedding problem were solved. This is all developed further in Chaps. 5 and 6.

theories, the consistency axioms and conditionalization, update, or imaging rules of probabilistic theories). But these are "parochial" epistemological theories, rather than general theories of the nature and logic of epistemic norms. The logic of epistemic programs and dynamic epistemic logics are much better for this purpose, and represent a thoroughly developed alternative to PLEN's approach.

4.3.3.2 The Processes Problem

The second primary virtue of PLEN is that it solves a problem in modeling epistemic processes that familiar classes of systems stumble over. The first part of the problem is the observation that processes are distinct from the transitions they result in, and this is epistemically relevant in at least some contexts. The second part of the problem is that logics of norms built atop models of dynamics that don't recognize the distinction between processes and transitions will inevitably get the normative requirements of epistemic rationality wrong. PLEN builds the distinction between processes and transitions into its semantics from the beginning, and so handily evades this problem.

In the abstract, the processes problem is this:

Problem 101 (Processes Problem) There are pairs of epistemic actions, procedures, or processes that are transition convergent and, more important, there are pairs that are transition equivalent. Some pairs of transition-convergent and transition-equivalent epistemic actions, procedures, or processes are (epistemically) normatively distinct. If we conceive of epistemic normativity in terms only of the properties of the contents of epistemic states, whether synchronic, diachronic, or polychronic, there is no way to make sense of these distinctions.

Let me emphasize how much of a problem this is and, thereby, emphasize how much of a theoretical virtue it is that PLEN can immediately—and perhaps uniquely—be seen to resolve this problem in the next section.

The problem can be seen most readily by considering a case where a token action a and a token action b would, if either is carried out, result in the same transition $t = \langle s, s' \rangle$. Let $N_C(a)$ say that norm N specifies that this particular token of a is rational to carry out under condition C. Assume that such statements must be interpreted over the transitions among epistemic states associated with a and b. If this interpretation is extensional in the sense that we consider, at state s, only the transitions that unfold from s in interpreting $N_C(a)$, then, at s, $N_C(a) \equiv N_C(b)$. But this is false. According to N, a may be rational and b may not be despite the fact that these actions result in the same transition. There are plenty of examples to be had. Consider inference by Modus Ponens from the set of statements $\{A, A \supset B, B \supset C, C\}$ as opposed to inference by Affirming the Consequent on the same set. It's clear that one is a rational inference, and the other (all things considered, typically) not, and this despite the fact that any statement about rationality must, by assumption, be interpreted only over the transitions involved, which, for this initial state, will be the same for both forms of inference. Thus, the failure to distinguish processes from transitions results in a mischaracterization of the requirements of epistemic rationality.

Let me spell this out with slightly more precision. Let actions a and b be such that $a \neq b$, and let $t \in T_a \cap T_b$; a and b are distinct, transition-convergent actions. For an arbitrary norm, N, call N's verdict on a with respect to a situation, s, be $N(a, s)$. Let $f(t)$ be the first state in the transition, t. There are norms, N_i, such that $N_i(a, f(t)) \neq N_i(b, f(t))$ and even $\neg(N_i(a, f(t)) \equiv N_i(b, f(t)))$. Chapter 2 was replete with examples. If any of those norms are actually plausible as rules governing the selection of processes, the argument is complete; transition-based analyses of dynamics will be doomed to get the verdicts of these norms wrong or else get the verdicts of epistemic rationality wrong insofar as they incorporate these rules.

The problem is serious enough for transition convergent actions, processes, and procedures. Transition equivalent actions, processes, and procedures exacerbate it.[21] Assume that methodological, procedural, and normative expressions in a putative logic, L_{EN}, of epistemic norms are to be analyzed in terms only of the properties of epistemic states and transitions. More precisely, let the semantics of such statements in our hypothetical logic of norms be defined in terms of functions on states, binary relations, or n-ary relations over the epistemic state set. Which one is chosen is a matter of irrelevance; the logics of these kinds of expressions will be substantially wrong. Recall the Honest Twin case above. Let a be the procedure of carrying out the derivation myself and let b be the procedure of asking my honest twin.

First, conflating actions (procedures, processes) with transitions (or even n-ary transition relations over states) will get claims about methodological inclusion wrong. Let $a!$ and $b!$ be instructions for carrying out a and b. Let $M_i(a!)$ say that the instruction encoded by $a!$ is part of an explicit methodology that I've learned, M. Now, in a model of epistemic dynamics in which actions and processes are analyzed in terms of transitions only, the statements about actions and processes will be interpreted with the same truth rule schema as any other formula in L_{EN}: $s \Vdash M_i(a!)$ iff (... T_a ...) where the righthand side of the definition specifies some condition that T_a satisfies.[22] Thus, whatever the rule says, $T_a = T_b$ implies that $s \Vdash M_i(a!)$ iff $s \Vdash M_i(b!)$ for any s in any model. Thus, (e.g.) $\models_{L_{EN}} M_i(a!) \equiv M_i(b!)$. Of course, this equivalence fails in the Honest Twin Case; $b!$ may not be in my explicit methodology even if $a!$ is. Additionally, let $\langle M_i \rangle \top$ be a statement to the effect that some procedure in my methodology M has been executed. This will be true if $a! \in M$ and a has been executed. But if a and b are equivalent just if they lead to the same transitions, then executing b suffices for executing a. So, even if b is not in M (let $\{a\} = M$), $\langle M_i \rangle \top$ will be true whenever b is executed.

[21] So, even if transition convergent actions (procedures, processes) can be distinguished in non-extensional ways that provide a means of drawing the necessary distinctions, transition equivalent actions (et al) will still present a processes problem.

[22] The alternative is to assign semantics to such formulae without reference to the parts of the models that interpret action or program expressions. But any such route will fail to solve the embedding problem raised above with respect to epistemic normative terms.

But the deductive methodology I've learned may not include (and could perhaps explicitly prohibit or instruct the avoidance of executing) the instruction to ask my twin, and so asking will not carry out anything in my methodology (and may actually constitute a breach of the methods). As shown, conflating processes with transitions will get this wrong. Thus, the analysis of actions with transitions also gets facts about the feasibility of following our methodologies wrong as well as the foregoing facts about methodological inclusion.

Second, procedural information about a and b is incorrectly equated by the identity of transition relations. If any of the propositions of L_{EN} encode procedural information, e.g., information about the number of steps a procedure requires, how those steps are combined via procedural operations, what facts hold over the course of carrying a procedure out, or information about particular substeps of carrying out a procedure, then the equivalence of binary relations implies false equivalences among these assertions.[23] The action a may require numerous substeps to complete while b is relatively basic. If $Com(a)$ says, in PDL, that a has some procedural complexity (like those identified above), then the above fact that $\models_{L_{EN}} A(a) \equiv A(b)$ has a false instance: $Com(a) \equiv Com(b)$. $Com(b)$ may be false while $Com(a)$ is true for all that identity of transitions establishes. Asking my twin does not, of course, need to be as complex a process as figuring out a proof myself. More, a might require execution of an action c while b does not. If a is actually a complex action, $c; d$, while b is atomic, then it's true that a requires execution of c but b does not.[24] More, the process of asking my twin certainly contains sub-actions that doing the proof doesn't, namely, asking my twin. If there are expressions in L_{EN} for encoding this, like $Com(a = c; d)$ and $Com(b = e; f)$, the semantics of L_{EN}—or any strictly transition-based model of dynamics—will simply get them wrong because, as above, it must interpret them in terms of T_a and T_b. Such an interpretation is already intuitively absurd; of course information about transitions doesn't reveal information about procedures. More directly, though, asking my counterpart and constructing a proof do not involve the same procedural operations on the same sets of basic actions; the equivalence is simply false. Thus, the conflation of processes with transitions fails to accurately capture procedural information. Now, this particular case might be resolved by looking at n-ary transition relations assigned to a and b, but only if c and e are not themselves transition equivalent.

Third, and most important for the purposes of this project, the analysis of actions and processes with transitions (i.e., structures defined over states only) will mischaracterize the verdicts of epistemic norms. Explicitly, let $N_C(a)$ say that norm N specifies that a is a rational action (type) to carry out under condition C. Then,

---

[23] That this is not visible in the usual form of PDL is an expressive limitation; its languages don't have any formulae that can distinguish transitions from procedures. The arguments here essentially examine what happens when you add new expressions that do so.

[24] I'm abusing notation a bit. Here, a and b are being used as variables for the perhaps complex procedures deducing and asking my twin, respectively. They are not meant to be atomic actions.

as above, $T_a = T_b$ implies that $\models_{LEN} N_C(a) \equiv N_C(b)$. This is a false equivalence. In situations like that described in the Honest Twin Case, $N_C(b)$ may be false while $N_C(a)$ is true for all that identity of transitions establishes. The transition equivalent actions, a and b—as opposed to the merely transition-convergent actions in the modus ponens vs affirming the consequent case—are rationally distinct. The rules of responsibility above (Sect. 2.2.3.13) also suffice for this result.

This problem depends on the properties of the norms of epistemic rationality. In particular, it requires that these norms distinguish processes a and b even if a and b are transition equivalent. One could resist this normative problem by noting that the foregoing arguments only outline an abstract form of error that a theory of epistemic rationality norms might make. The argument above, if right, merely showed that *if* there are normative propositions $N_C(a)$ that differ in truth value for transition-equivalent actions, a and b, then the process-transition distinction makes a difference to the logic of such propositions. One could then insist that there are no actual instances of such an error because there are no norms or even normative considerations that distinguish transition-equivalent processes, procedures, or actions, and so a theory of epistemic rationality and its norms that ignores the process-transition distinction is safe with respect to normative verdicts.

This line of objection won't work. Epistemic norms distinguish transition convergent and transition equivalent processes.[25] Under some circumstances, it may be rational to carry out a proof myself and not rational to ask my honest twin. For instance, if I am unaware of my counterpart's inexhaustible honesty or even have strong but defeasible—and ultimately misleading—reason to distrust him but I have extremely powerful reasons for trusting my own reasoning processes, it is more rational to deploy my own proof methods. Similarly, if I have good evidence my twin is infallible at proofs and strong but defeasible—and ultimately misleading—evidence for distrusting my own reasoning, I have a powerful reason to ask my twin rather than rely on my own thinking. In the selection of epistemic actions, there is something epistemically irresponsible from the agent-relative perspective with selecting the action that is an unknown quantity or apparently the worse option [9, 10]. The norms of epistemic responsibility in Sect. 2.2.3.13 clarify and strengthen this point. Thus, there are cases where verdicts about the rationality of processes a and b come apart even if a and b are transition equivalent.

There are, as in the transition convergent case, differences between procedures in Honest Twin-like cases that are of epistemic relevance. Among these are features of procedures like (e.g.) efficiency, problem-solving power, speed, and likelihood of error [14]. Consider likelihood of error. In a noisy environment, asking my twin may be more likely to be erroneously executed than deploying my own proof processes, however faster or more efficient asking might be. As we try to guard against this likelihood of error in a noisy environment, we will encounter a tradeoff with efficiency. More, different actions may have cognitive effects that are not part of the representation of epistemic states. Suppose that total procedural knowledge

[25] See Chaps. 2 and 3.

is not represented in a model's representation of epistemic states. Then, it's possible that a and b have different effects on my procedural knowledge; being walked through a proof by my honest twin may not provide the same sort of procedural knowledge that working through the proof on my own provides (or vice versa). This is epistemically relevant, as it affects what future epistemic actions are available to me, as well as the likelihood of performance errors, efficiency, problem solving power, and speed.

One might object that the arguments that derive these three classes of problems from the processes problem are formulated in a way that depends on the class of "transition equivalent" pairs of actions being non-empty, and this is not a given. It might be argued[26] that the basic premise of these arguments, that there are any actions a and b such that $T_a = T_b$, is false. Specifically, it might be argued that, in any case like the Honest Twin, a and b can't be identified with identical transitions because at the states that a reaches, one believes that one reached the state by a, and at the states reachable by b, one believes that one reached it by b.

This objection won't work. First, note that if a and b are *individuated* in terms of their transition relations over states (however complicated), then the beliefs in question are identical or logically equivalent with each other. In the first case, $T_a = T_b$. In the second case, if epistemic states are closed under equivalence, $T_a = T_b$. If epistemic states aren't so closed, then T_a and T_b may be nonidentical. But, second, in that case or the case in which a and b aren't individuated in terms of their transition relations, the objection still doesn't work. It relies on the implausible pair of assumptions that epistemic processes always deliver traces of their execution and that these traces are accurate. If either of these claims is not true, one can simply construct a modification of the Honest Twin case in which a' and b' either give mixed up beliefs about what process was carried out or they simply fail to give beliefs about what processes were carried out at all. Both modifications suffice to undermine the objection. Let me spell these replies out in an intertwined fashion.

There is no reason to think that all epistemic processes of (e.g.) belief change deliver accurate beliefs about the traces of the processes that have been used in addition to the beliefs they directly add. Many beliefs about the processes that

[26] In the main text, I only consider the objection from direct differences in the contents of the terminal states of a and b. I won't consider objections from some putative differences in, for instance, basing relations between the terminal states. First, the Honest Twin case explicitly builds in the stipulation that a and b provide exactly the same proofs, hence, exactly the same non-causal basing relations. Second, to insist on a difference of causal basing relations will reinstate the very problem of normative verdicts above: differences of process (defined in terms of causal structure) are operative in normative judgments about beliefs. Such insistence is also either misguided or provides a direct reason to favor models that distinguish processes from transitions. If the causal basing relation is defined in terms of the prior states, in which case the causal basing relations just are identical between a and b because those processes apply in all of the same situations, then $T_a = T_b$. If causal basing relations are defined in terms of something more fine-grained, then T_a is not T_b but only because differences in the processes are operative in normative judgments about belief, which returns us to the normative verdicts problem.

generate belief either fade over time, are not scrutable in the first place, or are simply misleading.[27]

To see this, simply consider a modification of the Honest Twin Case that extends the processes over long periods of time. Let a' be the process of carrying out the deduction or model-check and then going on with one's life as normal for ten years, and let b' be the counterpart composed of asking my twin and then going on with life for ten years. Assume that, as normal, I forget the details of how I came to the conclusion about validity that I came to by doing a', but I remember the outcome. In this case, $T_{a'} = T_{b'}$ because in the long term, the two processes converge on exactly the same states. Schematically, let $a' = a; c$ and let $b' = b; c$ where c is a process of forgetting (at most) what processes led to what beliefs, and the original argument goes through. Alternately, many cognitive processes just don't give us the necessary kinds of receipt; instances of the availability heuristic don't come stamped with the information that one conflated a readily available case with a probable case. One can figure out that one has done this, but only on post hoc reflection and only on the basis of hypothetical reasoning about oneself. One needn't modify the original Honest Twin Case to support this possibility. Finally, suppose that I form the correct belief about validity on the basis of a misremembered process while correctly recalling the outcome. The relevant modification of the Honest Twin Case: let $a'' = a; d$ and let $b'' = b; d'$ where d' is a process that results in a simple transposition error in memory retrieval; d results in correctly remembering that one used a to result in a belief change, and d' is the process of misremembering that one used a in place of b. In each case, the objection fails; all of these are empirically plausible—even commonplace—cognitive phenomena. The objection neither logically suffices to undermine the derivation of the foregoing theoretical errors, nor does it square with the empirical facts about epistemic processes.

4.3.3.3 PLEN vs The Processes Problem

It's straightforward, from Sect. 4.3.1, that PLEN solves this problem. First, PLEN conceives of epistemic normativity explicitly in terms of protocols and the restrictions on epistemic dynamics they define in virtue of their execution conditions and relations to other protocols. Second, the models of epistemic dynamics underlying PLEN, EDGs, sharply distinguish epistemic actions, processes, and procedures from epistemic transitions and state-sequences (trajectories through sets of states), though they are structurally interrelated. Thus, there are, trivially, actions, processes, and procedures in EDGs that share root and terminal states and state sequences, and these can be distinguished by protocols.

The protocol-theoretic formula schema, $(\pi_1 \curlyvee_p \pi_2) \wedge (\pi_1 \leftrightarrow_t \pi_2)$, for instance, tells us that protocols π_1 and π_2 are transition equivalent and procedurally

[27] See, for instance, the general inscrutability of many heuristic, subpersonal, and even person-level processes [7, 14, 21, 31, 32], as well as Lam's considerations [22].

inequivalent. Thus, where instances of this formula are verified at a given s, it tells us that the EDG in which s resides is one that distinguishes the transitions executing π_1 and π_2 from the epistemic processes, sequences of epistemic action, and procedures that do so.

From there, the mischaracterizations of methodologies and procedural structure that cascade from the processes problem are blocked. Transition convergence and transition equivalence do not imply convergence or equivalence with respect to methodological inclusion, procedural structure, or normative verdicts. These results are mechanical, given the machinery developed in Chaps. 4 through 7.

4.3.3.4 PLEN vs Rival Systems

The ability to solve the processes problem is a virtue of PLEN not generally shared by belief revision theories, PDL, standard deontic logics (SDL), multi-modal variants of SDL, dynamic deontic logics, or, arguably, dynamic epistemic logic (DEL). Fully litigating this claim is beyond the scope of this particular book, which primarily aims to display PLEN's virtues rather than the vices of its rivals. Nonetheless, it's worthwhile to spell out a point of qualification and a point of support.

First, the qualification. The claim is not that there is no machinery present in these families of systems that could be leveraged to solve these problems (let alone that there is no possible such machinery). It's not even that there are no extant variations of them that can solve these problems. The claim is, rather, that the base systems for these families demonstrably suffer this problem, and that, to my knowledge, no such leveraging of machinery or application of extant variations of these systems to this problem exists.

Second, the support, which simply reiterates the points articulated first in Sect. 3.3 in light of the preceding sections (Sects. 4.3.3.1 and 4.3.3.2). For belief revision theories, standard formulations of PDL, or dynamic deontic logics based on standard PDL, the point is immediate: even where there is a program or action logic in the system (not typical of belief revision systems), PDL based systems define the semantics of programs (and the model of dynamics) in terms of binary relations or state-to-state functions. So, there are no definable formulae comparable to $(\pi_1 \curlyvee_p \pi_2) \wedge (\pi_1 \nleftrightarrow_t \pi_2)$. Novel machinery must be added to these kinds of systems in order to solve the problem.

Standard deontic logic (SDL) and multi-modal approaches to deontic logic will suffer this problem, in addition to the well-known raft of others.[28] The general point is that any system in which deontic accessibility relations defined over binary relations on states (worlds, models) are the only tools that the system can deploy to individuate actions, processes, or procedures will suffer this problem; binary transition relations can't distinguish transition equivalent actions, processes, and

[28] See Sects. 8.1 and 8.2 for further detail on this matter.

procedures. Thus, the deontic notions defined in terms of such binary accessibility relations are incapable of distinguishing transition equivalent actions, processes, and procedures.

The case with the base system of DEL is less clear.[29] The interpretation of epistemic actions in the basic form of DEL is defined in terms of product updates, $\otimes$, on pairs of pointed epistemic models, (M, s), and action models, (E, a), which result in transition relations associated with actions where the terminal state contains a clear trace of the actions that result in that state. Omitting irrelevant detail,[30] for (M, s) and (E, a), $(M, s) \otimes (E, a) = (M', s')$ where $s' = (s, a)$ where the states in the resulting epistemic model are pairs, $s' \in S \times E$. The transition relations that define execution of actions in DEL, then, are mappings, $(M, s) \mapsto_{(E,a)} (M', s')$, where the preconditions of a are all satisfied at s. So, it seems there can't be nonidentical transition equivalent actions in DEL.[31] This doesn't square with what are, if the foregoing reasoning is correct, the well-grounded observations of this chapter and Chaps. 2 and 3.

However, DEL is a very well-developed technical framework, with many branches of development. Some of these branches[32] may have the machinery to solve the problems PLEN is designed to solve as well or better, and this comes as an addition to DEL's more elaborate model of epistemic states. Nonetheless, application of DEL based machinery to the problems outlined here isn't carried out anywhere that I know of, and, moreover, it is nowhere made clear that DEL formalizes the key insights of epistemic proceduralism. As per the main argument of the book, this will be shown to be one of PLEN's most distinctive virtues.

To complete the reflection of Sect. 3.3, consider the failure of norm exogeny in AGM belief revision theory. Both PDL and DEL have the virtue of being norm exogenous if we interpret their program constructions as items that encode the content of epistemic norms. AGM doesn't have this feature. AGM theories or AGM-based theories all fail the narrow condition of norm exogeny. Strictly speaking, AGM is a logic-based analysis of epistemic dynamics and doesn't itself provide a logic of epistemic dynamics. As such, it doesn't have an object language into which

[29] Here, I'm taking the textbook presentation in [34] or the more general, programmatic [2] to present what can be taken as the base system of DEL.

[30] Nothing about this point hangs on the conditions on the indistinguishability relations among states and actions and valuations, the conditions on preconditions, etc.

[31] Consider the following action models. $(E, a) = \langle E, \rightarrow_i, pre\rangle$ where $a \in E$, and $(E', b) = \langle E', \rightarrow'_i, pre'\rangle$ where $b \in E'$. Suppose that (E, a) and (E', b) define identical transition relations: $\mapsto_{(E,a)} = \mapsto_{(E',b)}$. Thus, for any pointed epistemic model $(M, s) = \langle S, R, V, s\rangle$, the product update from $((M, s), (E, a)) = (M', (s, a)) = ((M, s), (E', b))$. But this just means that $a = b$, that $\longrightarrow_i = \longrightarrow'_i$, and that $pre = pre'$. Of course, a and b were arbitrary actions in the signatures of E and E'. Thus, $(E, a) = (E', b)$. Thus, two distinct actions, a and b, and two distinct action models, (E, a) and (E', b) cannot be transition equivalent; the terminal states in transitions assigned to actions via the product update carry a "record" of which action brought about the change, and so no two actions, DEL, can possibly be associated with equivalent transition relations.

[32] Many thanks to a reviewer who suggested this point, and directed me to [13, 19, 36] as possible instances.

to embed norm expressions. Thus, AGM theories fail to provide a logical analysis of valid reasoning about epistemic rationality norms.

There are, however, constructions in AGM theories that arguably represent or encode epistemic rationality norms: the rationality and entrenchment postulates,[33] for instance:

Rationality Postulates for Contraction: For propositional formula A, a belief set K, an an independently defined expansion function $K+_A$, a rational contraction function satisfies the following postulates:

(-1): $K-_A = Cn(K-_A)$
(-2): $K-_A \subseteq K$
(-3): If $A \notin K$, then $K-_A = K$.
(-4): If $A \in K-_A$, then $\vdash A$.
(-5): $K \subseteq (K-_A)+_A$
(-6): If A is equivalent to B, then $K-_A = K-_B$.
(-7): $(K -_A \cap K-_B) \subseteq K-_{A\wedge B}$
(-8): If $A \notin K-_{A\wedge B}$, then $K-_{A\wedge B} \subseteq K-_A$.

These are motivated independently in AGM, but can also be shown to be equivalent to the entrenchment postulates.[34] Let $A \leq B$ say that either both A and B are logical truths, that A is not believed, or that the need to discard one of A or B leads to discarding A. The intuitive though is that $\leq$ encodes entrenchment; it says that B is at least as entrenched as A. Then entrenchment is thought to satisfy these conditions:

Entrenchment Postulates:

(EE1): If $A \leq B$ and $B \leq C$, then $A \leq C$.
(EE2): If $A \vdash B$, then $A \leq B$.
(EE3): $A \leq A \wedge B$ or $B \leq A \wedge B$.
(EE4): If K is consistent, then $A \notin K$ iff $A \leq B$ for all B.
(EE5): If $A \leq B$ for all A, then $\vdash B$.

Given a rational contraction function, the following rule defines epistemic entrenchment:

(CtoE): $A \leq B$ iff $A \notin K-_{A\wedge B}$ or $\vdash A \wedge B$.

Given entrenchment, the following rule defines a rational contraction function:

(EtoC): $K-_A = K \cap \{B \mid A < A \vee B\}$ if $\nvdash A$, and $K-_A = K$ otherwise.

The rationality and entrenchment postulates are the rules in AGM that specify what the requirements of rational belief revision are. This suffices for a broad kind of

[33] See, for instance, [1] and [17]. I follow the presentation in [11] pretty closely.

[34] An epistemic commitment function is a partial meet contraction function iff it satisfies (-1)–(-6).

norm exogeny: entrenchment and rationality do perspicuously correspond to norms of epistemic rationality, and the system provides a formal way to think about them. As a result, there is a simple model of compliance with epistemic rationality norms: the epistemic commitment functions. Dynamics in AGM theories just are models of ideally rational belief change with respect to these norms.

The entrenchment and rationality norms are not, however, really adequate to the task of modeling epistemic rationality norms in any general way. The entrenchment and rationality postulates fail to encode crucial properties of epistemic rationality norms. For one, AGM theories say absolutely nothing about epistemic actions that change the world or epistemic programs that are formed from mixed world-changing and information changing actions. As such, there are no ways in AGM to represent norms that restrict the interaction of these kinds of action or the forms of reasoning D3 requires about them. More generally, AGM theories don't say anything about actions or processes as distinct from transitions, and this leaves AGM theories entirely powerless to represent normative restrictions on action or processes where they come apart from restrictions on transitions. Thus, given the vices introduced by the processes problem and those in Sect. 3.3, the model of norms implicit in AGM really can't satisfy D1C2, D2, or D3.

4.4 From PLEN to the Revised PLEN Framework

The norm exogney and readily available resolution of the processes problem is at least a prima facie powerful reason to explore PLEN as a formal framework for approaching the analysis of epistemic dynamics and the logic of epistemic norms. The next three chapters revise and elaborate on this initial formulation of PLEN in order to devise a protocol logic that provides a theory of protocol equivalence. The resulting protocol logic is, accordingly, the core of the protocol-theoretic logic of epistemic norms.

Appendix 1

Remark 102 This appendix spells out the results discussed in Chap. 4.

Completeness

Soundness

Theorem 103 (Soundness Theorem) *If* $X \vdash_{PLEN} A$*, then* $X \models_{PLEN} A$.

Proof (Proof-Plan) Direct proof by enumeration. From the definition of derivations, it follows that, if there is a derivation of B from X and every rule and axiom schema is valid, then the truth of each element of X necessarily preserves the truth of B in the object-level semantics of PLEN. I omit the proofs of the modal and propositional rules, as these are standard parts of soundness proofs for modal and propositional logics. Generally, I focus on proofs for rules or axioms that are interesting (e.g., in cases where the logics for control operators diverge from familiar dynamic logics or where the operators are new to PLEN). I omit almost all proofs for the binary axioms, as they are easily extractable from the proofs of the other axioms or familiar from PDL. Some of the following proofs are simplified by applying the results from Chap. 6 (or Appendix 3). I point those out explicitly.

(1) (Execution-theoretic Protocol Equivalence Rules)

(a)
$$\left[\begin{array}{l} \dfrac{\Box(\pi \leftrightarrow_p \pi'),\ \rho(\pi/\pi')(X_\pi) = X_{\pi'}}{\Box(X_\pi \leftrightarrow_p X_{\pi'})} \\ \dfrac{\Box(\pi \leftrightarrow_p \pi'),\ \Box(\pi' \leftrightarrow_p \pi'')}{\Box(\pi \leftrightarrow_p \pi'')} \\ \dfrac{}{\Box(\pi \leftrightarrow_p \pi)} \\ \dfrac{\Box(\pi \leftrightarrow_p \pi')}{\Box(\pi' \leftrightarrow_p \pi)} \\ \dfrac{\Box(\pi \leftrightarrow_p \pi')}{\Box(\pi \leftrightarrow_t \pi')} \end{array}\right] \quad \left[\begin{array}{l} \dfrac{\pi \leftrightarrow_p \pi',\ \rho(\pi/\pi')(X_\pi) = X_{\pi'}}{X_\pi \leftrightarrow_p X_{\pi'}} \\ \dfrac{\pi \leftrightarrow_p \pi',\ \pi' \leftrightarrow_p \pi''}{\pi \leftrightarrow_p \pi''} \\ \dfrac{}{\pi \leftrightarrow_p \pi} \\ \dfrac{\pi \leftrightarrow_p \pi'}{\pi' \leftrightarrow_p \pi} \end{array}\right]$$

(a) Presented in Chap. 4.

(b)
$$\left[\begin{array}{l} \dfrac{\Box(\pi \leftrightarrow_t \pi'),\ \rho(\pi/\pi')(X_\pi) = X_{\pi'}}{\Box(X_\pi \leftrightarrow_t X_{\pi'})} \\ \dfrac{\Box(\pi \leftrightarrow_t \pi'),\ \Box(\pi' \leftrightarrow_t \pi'')}{\Box(\pi \leftrightarrow_t \pi'')} \\ \dfrac{}{\Box(\pi \leftrightarrow_t \pi)} \\ \dfrac{\Box(\pi \leftrightarrow_t \pi')}{\Box(\pi' \leftrightarrow_t \pi)} \end{array}\right] \quad \left[\begin{array}{l} \dfrac{\pi \leftrightarrow_t \pi',\ \rho(\pi/\pi')(X_\pi) = X_{\pi'}}{X_\pi \leftrightarrow_t X_{\pi'}} \\ \dfrac{\pi \leftrightarrow_t \pi',\ \pi' \leftrightarrow_t \pi''}{\pi \leftrightarrow_t \pi''} \\ \dfrac{}{\pi \leftrightarrow_t \pi} \\ \dfrac{\pi \leftrightarrow_t \pi'}{\pi' \leftrightarrow_t \pi} \end{array}\right]$$

(i) For the first rule in the lefthand column: Induction over proper immediate subprotocols. For arbitrary $\pi, \pi', X_\pi, X_{\pi'}$, let an arbitrary EDG, g, verify all of the premises at s: $\pi \leftrightarrow_t \pi'$, $\rho(\pi/\pi')(X_\pi) = X_{\pi'}$, and $\pi \in ISub(X_\pi)$.

(A) (Base) Assume that $\pi \in PISub(X_\pi)$. Let $t \in T_{X_\pi}$ where $s = f(t)$. By syntactic replacement, $X_{\pi'} = \pi' \circ \pi_i$. $T^s_{X_\pi} = \{t \mid s = f(t), t = t_1 \oplus^\pi_\circ t_2, t_1 \in T^s_\pi, t_2 \in T_{\pi_i}\}$ where $\oplus^\pi_\circ$ corresponds to the relations among executions of the immediate subprotocols of $\pi' \circ \pi_i$. For instance, sequential composition corresponds to composition of binary relations, parallel execution corresponds to transition correspondence, and choice corresponds to branching from a shared root state. By assumption, for any t in g, $t \in T^s_\pi$ iff $t \in T^s_{\pi'}$. Thus, $T^s_\pi = T^s_{\pi'}$. Indeed, by substitution of identical sets, $T^s_{X_\pi} = T^s_{X_{\pi'}}$. The converse follows by transposition.

(B) (Induction) If the result holds for all proper immediate sub-protocols of X_π, then it holds for X_π. (From the application of the prior reasoning to the induction assumption, the result immediately follows.)

(C) The foregoing reasoning was applied to arbitrary s in g, and so can be be applied all states in g, validating the boxed version of the rule.

(ii) Elementary set theory supplies the transitivity, reflexivity, and symmetry rules immediately.

(c)
$$\left[\begin{array}{c}\dfrac{\Box(\pi \leftrightarrow_r \pi'),\ \rho(\pi/\pi')(X_\pi) = X_{\pi'}}{\Box(X_\pi \leftrightarrow_r X_{\pi'})} \\ \dfrac{\Box(\pi \leftrightarrow_r \pi'),\ \Box(\pi' \leftrightarrow_r \pi'')}{\Box(\pi \leftrightarrow_r \pi'')} \\ \dfrac{}{\Box(\pi \leftrightarrow_r \pi)} \\ \dfrac{\Box(\pi \leftrightarrow_r \pi')}{\Box(\pi' \leftrightarrow_p \pi)} \\ \dfrac{\Box(\pi \leftrightarrow_r \pi')}{\Box(\pi \leftrightarrow_t \pi')} \\ \dfrac{\pi \leftrightarrow_r \pi'}{\pi = \pi'}\end{array}\right] \quad \left[\begin{array}{c}\dfrac{\pi \leftrightarrow_r \pi',\ \rho(\pi/\pi')(X_\pi) = X_{\pi'}}{X_\pi \leftrightarrow_r X_{\pi'}} \\ \dfrac{\pi \leftrightarrow_r \pi',\ \pi' \leftrightarrow_r \pi''}{\pi \leftrightarrow_r \pi''} \\ \dfrac{}{\pi \leftrightarrow_r \pi} \\ \dfrac{\pi \leftrightarrow_r \pi'}{\pi' \leftrightarrow_r \pi} \\ \dfrac{\pi \leftrightarrow_r \pi'}{\pi = \pi'}\end{array}\right]$$

(i) All of the rules follow immediately from the last. Let $s \Vdash \pi \leftrightarrow_r \pi'$ at any s in any g. Then there is an edge, e, with $f(e) = s$ that is an execution of both π and π'. But, by the definition of ternary or edge executions in PLEN, the middle term of any edge that executes any π must itself be π. Thus, since e executes both π and π', $\pi = \pi'$. Substitution of identical protocols immediately gives all of the other rules.

(d) $\dfrac{\Box\pi \leftrightarrow_p \pi',\ A(\pi)}{A(\pi')}(NSEP)$, $\dfrac{\Box\pi \leftrightarrow_r \pi',\ A(\pi)}{A(\pi')}(NSEP_r)$, $\dfrac{\Box\pi \leftrightarrow_t \pi',\ A(\pi)}{A(\pi')}$ $(NSEP_t)$

(i) First, the restrictions: The expression $A(\pi)$ cannot embed π directly in $M(\pi)$, $\pi \leftrightarrow_r$ ..., $\pi \curlywedge_r$ ..., $\pi \curlyvee_r$ ..., or any of the purely syntactic expressions (Sub, $PSub$, $ISub$, $\rho()() =$). This is trivial: Note that $\pi \equiv_G \pi'$ implies $\pi \equiv_p \pi'$ (as per the PRT results) and graph-theoretic equivalence doesn't preserve these expressions (as shown in the GRT Failure results). Thus, path-theoretic equivalence can't preserve them; else, contradiction. Now, the positive argument: The formula $\Box\pi \leftrightarrow_p \pi'$ is basically a graph-relative form of $\pi \equiv_p \pi'$. Within a particular graph, it has the same properties. Thus, from PRT (proven above), $\Box\pi \leftrightarrow_t \pi'$ follows. So, anything that $\Box\pi \leftrightarrow_t \pi'$ implies is implied by $\Box\pi \leftrightarrow_p \pi'$. More, the rest of the expressions are defined in terms of paths or terminal states of paths. Thus, if π and π' share all paths from a given state s, they share all terminal states of all paths from s. Thus, any expression in the remaining subset of expressions, $A(\pi)$, true of π at s is also true of π' at s. Additionally, for the remaining ternary-theoretic expressions, the restriction rules imply that every ternary execution of π shares root and terminal states with a procedural execution of π and converse. Thus, the expressions with truth conditions defined in terms of terminal states are preserved. Finally, for modal expressions, the necessitation rule implies that π and π' are (state-relative) procedurally equivalent at all states in g, so the same arguments above can be used for any other state. The truth functions on these expressions are trivial.

(ii) From the EGI rules, $\pi = \pi'$. Thus, any expression true of π at any state is true of π'.

(iii) First, the restrictions: The expression $A(\pi)$ cannot embed π directly in $M(\pi)$, $\pi \leftrightarrow_r ...$, $\pi \curlywedge_r ...$, $\pi \curlyvee_r ...$, $\pi \leftrightarrow_p ...$, $\pi \curlywedge_p ...$, $\pi \curlyvee_p ...$, $\pi \curlyvee_r ...$, or any of the purely syntactic expressions (Sub, $PSub$, $ISub$, $\rho()() =$). This is because the expression $\Box\pi \leftrightarrow_t \pi'$ is basically a graph-relative form of $\pi \equiv_t \pi'$. Thus, because $\pi \equiv_p \pi'$ implies $\pi \equiv_t \pi'$, if path-theoretic equivalence doesn't preserve these formulae, transition-theoretic equivalence can't either; else, contradiction. Now, the positive argument: Of course, path-theoretic equivalence does preserve the path-theoretic expressions (as shown above). That transition-theoretic equivalence doesn't preserve these can be seen mechanically. See the GRT Failure results for some counterexamples. Generally, from the fact that two protocols share transitions, it doesn't follow that they share paths.

(e) $$\frac{\pi \leftrightarrow_p \pi',\ A(\pi)}{A(\pi')}(SEP),\quad \frac{\pi \leftrightarrow_r \pi',\ A(\pi)}{A(\pi')}(SEP_r),\quad \frac{\pi \leftrightarrow_t \pi',\ A(\pi)}{A(\pi')}(SEP_t)$$

(i) First, the restrictions: The expression $A(\pi)$ cannot embed π directly in $\Box$, $[...]_p A(\pi)$, $[...]_t A(\pi)$, $[...]_r A(\pi)$, $\langle ...\rangle_p A(\pi)$, $\langle ...\rangle_t A(\pi)$, $\langle ...\rangle_r A(\pi)$, $M(\pi)$, $\pi \leftrightarrow_r ...$, $\pi \curlywedge_r ...$, $\pi \curlyvee_r ...$, or any of the purely syntactic expressions (Sub, $PSub$, $ISub$, $\rho()() =$). First, the NSEP proof works for any expressions in the intersection of the restrictions on SEP and NSEP. For the new expressions, assume that the premises are true at s, then note that, from facts about the terminal states of executions of protocols from s, nothing follows about reachable, non-identical states, let alone arbitrary non-identical states. Now, the positive argument: The expression $A(\pi)$ can embed π directly in $[\pi]_p$, $\langle\pi\rangle_p$, $\pi \leftrightarrow_p ...$, $\pi \curlywedge_p ...$, or $\pi \curlyvee_p ...$, the transition-theoretic variants of these, and truth-functions on these. Modulo π'. Assume $s \Vdash \pi \leftrightarrow_p \pi'$ and $s \Vdash [\pi]_p A$. Then since all paths p such that $s = first(p)$ are in P_π iff they are in $P_{\pi'}$, then any $p' \in P_{\pi'}$ is such that $last(p') \Vdash A$. Assume $s \Vdash \pi \leftrightarrow_p \pi'$ and $s \Vdash \langle\pi\rangle_p A$. Then there is some p such that $s = first(p)$, $p \in P_\pi \cap P_{\pi'}$, and $last(p) \Vdash A$. Thus, $s \Vdash \langle\pi'\rangle_p A$. Assume $s \Vdash \pi \leftrightarrow_p \pi'$ and $s \Vdash \pi \leftrightarrow_p \pi''$. Then any p ($s = first(p)$) that executes π executes π'' and vice versa. Then since all paths p such that $s = first(p)$ are in P_π iff they are in $P_{\pi'}$, then any $p' \in P_{\pi'}$ is such that $p' \in P_{\pi''}$ and vice versa. Thus, $s \Vdash \pi' \leftrightarrow_p \pi''$. Assume $s \Vdash \pi \leftrightarrow_p \pi'$ and $s \Vdash \pi \curlywedge_p \pi''$. Then some p ($s = first(p)$) that executes π also executes π''. Then since all paths p such that $s = first(p)$ are in P_π iff they are in $P_{\pi'}$, then p is such that $p \in P_{\pi'}$ and $p \in P_{\pi *}$. Thus, $s \Vdash \pi' \curlywedge_p \pi''$. Assume $s \Vdash \pi \leftrightarrow_p \pi'$ and $s \Vdash \pi \curlyvee_p \pi''$. Then some p ($s = first(p)$) that executes π doesn't execute $\pi *$ or vice versa. Then since all paths p' such that $s = first(p')$ are in P_π iff they are in $P_{\pi'}$, then $p \notin P_{\pi'}$ or $p \notin P_{\pi''}$. Thus, $s \Vdash \pi' \curlyvee_p \pi''$. Again, $\pi \leftrightarrow_p \pi'$ implies $\pi \leftrightarrow_t \pi'$, so the transition-theoretic expressions are preserved. The truth functions are trivial, from this point.

(ii) Trivial from the identity of π and π', which follows from EGI or the PRT results.

(iii) The expression $A(\pi)$ can only embed π directly in $[\pi]_t$, $\langle\pi\rangle_t$, $\pi \leftrightarrow_t \ldots$, $\pi \curlywedge_t \ldots$, or $\pi \curlyvee_t \ldots$, and truth-functions on these. The arguments follow the same format as that for (1), but stated in terms of transitions. First, $\pi \leftrightarrow_p \pi'$ implies $\pi \leftrightarrow_t \pi'$, so transition-theoretic equivalence can't preserve any expressions that path-theoretic equivalence doesn't; else, contradiction. So, that takes care of most of the restriction. That transition-theoretic equivalence doesn't preserve path-theoretic expressions is just as in the argument for $NSEP_t$. Generally, the truth conditions of transition-theoretic expressions are defined in terms of transitions and terminal states of transitions, so identical transition relations assigned to π and π' result in identically satisfied truth conditions. The truth functions are trivial, from this point.

(f) $\dfrac{(\pi \curlywedge_p \pi'),\ [\pi]_p A}{\langle\pi'\rangle_p A}(CON)$, $\dfrac{(\pi \curlywedge_r \pi'),\ [\pi]_r A}{\langle\pi'\rangle_r A}(CON_r)$, $\dfrac{(\pi \curlywedge_t \pi'),\ [\pi]_t A}{\langle\pi'\rangle_t A}$ (CON_t)

(i) Assume $s \Vdash \pi \curlywedge_p \pi'$ and $s \Vdash [\pi]_p A$. Then there is some p such that $s = first(p)$ in $P_\pi \cap P_{\pi'}$ such that $last(p) \Vdash A$. Thus, $s \Vdash \langle\pi'\rangle_p A$.

(ii) Assume $s \Vdash \pi \curlywedge_r \pi'$ and $s \Vdash [\pi]_r A$. Then there is some e such that $s = first(e)$ in $R_\pi \cap R_{\pi'}$ such that $last(e) \Vdash A$. Thus, $s \Vdash \langle\pi'\rangle_r A$.

(iii) Assume $s \Vdash \pi \curlywedge_t \pi'$ and $s \Vdash [\pi]_t A$. Then there is some t such that $s = first(t)$ in $T_\pi \cap T_{\pi'}$ such that $last(t) \Vdash A$. Thus, $s \Vdash \langle\pi'\rangle_t A$.

(g) (DIV Variants Proof)

(h) $\dfrac{(\pi \curlyvee_p \pi')}{\neg(\pi \leftrightarrow_p \pi')}(DIV)$, $\dfrac{(\pi \curlyvee_r \pi')}{\neg(\pi \leftrightarrow_r \pi')}(DIV_r)$, $\dfrac{(\pi \curlyvee_t \pi')}{\neg(\pi \leftrightarrow_t \pi')}(DIV_t)$

(i) (1) Assume $s \Vdash \pi \curlyvee\ _p\pi'$. Then there is some p such that $s = first(p)$ in $P_\pi / P_{\pi'}$. Thus, $s \Vdash \neg(\pi \leftrightarrow_p \pi')$. Or vice versa with identical results.

(ii) (2) Assume $s \Vdash \pi \curlyvee_r \pi'$. Then there is some e such that $s = first(e)$ in $R_\pi / R_{\pi'}$. Thus, $s \Vdash \neg(\pi \leftrightarrow_r \pi')$. Or vice versa with identical results.

(iii) (3) Assume $s \Vdash \pi \curlyvee_t \pi'$. Then there is some p such that $s = first(t)$ in $T_\pi / T_{\pi'}$. Thus, $s \Vdash \neg(\pi \leftrightarrow_t \pi')$. Or vice versa with identical results.

(i) $\Box(A(\pi_1 \cap \pi_2) \equiv A(\pi_2 \cap \pi_1)), \Box(A(\pi_1 \cup \pi_2) \equiv A(\pi_2 \cup \pi_1))$

(i) These are derivable from GSRT, GRT, GR, or combinations of SEP and PRT. Alternately, it suffices to note that all execution rules and, consequently, truth rules are insensitive to the syntactic ordering of subprotocols by these operators.

(j) The (DP), (DT), and (DR) pairs are immediate from the restriction rules.

(2) (Miscellaneous Postcondition Rules)

(a)
$$\left[\begin{array}{ll}
\dfrac{[\pi]_r A,\ \langle\pi\rangle_r \top}{\langle\pi\rangle_r A}(MR1) & \dfrac{[\pi]_r(A \to B),\ [\pi]_r A}{[\pi]_r B}(MR6) \\
\dfrac{\langle\pi\rangle_r A}{\langle\pi\rangle_r \top}(M2) & \dfrac{[\pi]_r(A \to B),\ \langle\pi\rangle_r A}{\langle\pi\rangle_r B}(MR7) \\
\dfrac{A \vdash_{PLEN} B,\ [\pi]_r A}{[\pi]_r B}(MR3) & \dfrac{\langle\pi\rangle_r (A \to B),\ [\pi]_r A}{\langle\pi\rangle_r B}(MR8) \\
\dfrac{A \vdash_{PLEN} B,\ \langle\pi\rangle_r A}{\langle\pi\rangle_r B}(MR5) &
\end{array}\right]$$

(b)
$$\left[\begin{array}{ll}
\dfrac{[\pi]_p A,\ \langle\pi\rangle_p \top}{\langle\pi\rangle_p A}(MP1) & \dfrac{[\pi]_p(A \to B),\ [\pi]_p A}{[\pi]_p B}(MP6) \\
\dfrac{\langle\pi\rangle_p A}{\langle\pi\rangle_p \top}(MP2) & \dfrac{[\pi]_p(A \to B),\ \langle\pi\rangle_p A}{\langle\pi\rangle_p B}(MP7) \\
\dfrac{A \vdash_{PLEN} B,\ [\pi]_p A}{[\pi]_p B}(MP3) & \dfrac{\langle\pi\rangle_p (A \to B),\ [\pi]_p A}{\langle\pi\rangle_p B}(MP8) \\
\dfrac{A \vdash_{PLEN} B,\ \langle\pi\rangle_p A}{\langle\pi\rangle_p B}(MP5) &
\end{array}\right]$$

(c)
$$\left[\begin{array}{ll}
\dfrac{[\pi]_t A,\ \langle\pi\rangle_t \top}{\langle\pi\rangle_t A}(MT1) & \dfrac{[\pi]_t(A \to B),\ [\pi]_t A}{[\pi]_t B}(MT6) \\
\dfrac{\langle\pi\rangle_t A}{\langle\pi\rangle_t \top}(MT2) & \dfrac{[\pi]_t(A \to B),\ \langle\pi\rangle_t A}{\langle\pi\rangle_t B}(MT7) \\
\dfrac{A \vdash_{PLEN} B,\ [\pi]_t A}{[\pi]_t B}(MT3) & \dfrac{\langle\pi\rangle_t (A \to B),\ [\pi]_t A}{\langle\pi\rangle_t B}(MT8) \\
\dfrac{A \vdash_{PLEN} B,\ \langle\pi\rangle_t A}{\langle\pi\rangle_t B}(MT5) &
\end{array}\right]$$

(d) The MR and MP rules are mechanical, and the MT rules follow from the equivalence rules and the MP rules.[34]

[34] Or from the MP rules by the PRT rules proven in Appendix 2.

(3) (Procedural Axioms) I prove a subset of procedural axioms by showing them schematically to be verified at arbitrary states at arbitrary EDGs. These are all essentially mechanical. The ternary variants can be derived by modification of the reasoning presented for the procedural axioms. The binary axioms are essentially axioms of PDL, and so known to be sound.

(a) Note that, for any axiom in the system that involves syntactic replacement, there is an implicit restriction that rules out subprotocol, syntactic identity, modal, methodological, graph-theoretic, and some ternary expressions. For instance, the axiom schema, $((\pi \leftrightarrow_p \pi') \wedge A(\pi)) \rightarrow A(\pi')$, cannot embed π in any expression like (inter alia) $M(\pi)$ or $\pi \curlywedge_r \pi''$. In specifying epistemic directed graphs for proving procedural axioms, I mostly ignore $R+$ and $R+_0$, as nothing in the following results other than iteration axioms is impacted by these constructions.

(b) (1p) $[\pi_1; \pi_2]_p A \equiv [\pi_1]_p[\pi_2]_p A$

(i) LtoR. Let $s \Vdash [\pi_1; \pi_2]_p A$. Then $\forall p[s = first(p) \wedge p \in P_{\pi_1;\pi_2} \rightarrow last(p) \Vdash A]$. Every p in $P_{\pi_1;\pi_2}$ is such that $p = qr$, $last(q) = first(r)$, $q \in P_{\pi_1}$, and $r \in P_{\pi_2}$. Suppose $s \nVdash [\pi_1]_p[\pi_2]_p A$. Then $\exists q[first(q) = s, q \in P_{\pi_1}, last(q) \nVdash [\pi_2]_p A]$. Thus, $\exists r[r \in P_{\pi_2}, first(r) = last(q), last(r) \nVdash A]$. But then there is some $qr \in P_{\pi_1;\pi_2}$ such that $first(qr) = s$ and $last(qr) \nVdash A$, contradicting the assumption.

(ii) RtoL Let $s \Vdash [\pi_1]_p[\pi_2]_p$. Then $\forall q[s = first(q) \wedge q \in P_{\pi_1} \rightarrow last(q) \Vdash [\pi_2]_p A]$. Thus, $\forall r[last(q) = first(r) \wedge r \in P_{\pi_2} \rightarrow last(r) \Vdash A]$. Thus, by the execution rules, all qr are such that if $s = first(qr)$, $q \in P_{\pi_1}$, and $r \in P_{\pi_2}$, then $last(qr) \Vdash A$. Suppose $s \nVdash [\pi_1; \pi_2]_p A$. Then $\exists p[first(p) = s, p = qr, q \in P_{\pi_1}, r \in P_{\pi_2}, last(r) \nVdash A]$, contradicting the assumption.

(c) (3p) $\langle \pi_1 \cap \pi_2 \rangle_p A \rightarrow (\langle \pi_1 \rangle_p A \wedge \langle \pi_2 \rangle_p A)$

(i) Assume $\langle \pi_1 \cap \pi_2 \rangle_p A$. Then there is a $p \in P_{\pi_1}$ or $p \in P_{\pi_2}$ such that if $p \in P_{\pi_1}$, then there is a $q \in P_{\pi_2}$ and $p \rightleftharpoons q$, and if $p \in P_{\pi_2}$, then there is a $q \in P_{\pi_1}$ and $p \rightleftharpoons q$. In either case, $last(p) \Vdash A$. By the definition of correspondence, $last(q) \Vdash A$ in either case. Thus, $\langle \pi_1 \rangle_p A \wedge \langle \pi_2 \rangle_p A$.

(ii) (Converse Failure) Let $g = \langle S = \{s, s', s''\}, ..., R = \{\langle s, \pi_1, s' \rangle, \langle s, \pi_2, s'' \rangle\} \rangle$, $s' \Vdash A$, $s'' \Vdash A$, and $s' \neq s''$.

(d) (4p) $([\pi_1]_p A \wedge [\pi_2]_p B) \rightarrow [\pi_1 \cap \pi_2]_p (A \wedge B)$

(i) All $p \in P_{\pi_1}$ verify A at $last(p)$. All $q \in P_{\pi_2}$ verify B at $last(q)$. Thus, for any such p, q, if $p \rightleftharpoons q$, then $last(p) = last(q)$. Thus, $last(p) \Vdash A \wedge B$ and $last(q) \Vdash A \wedge B$. Thus, $[\pi_1 \cap \pi_2]_p (A \wedge B)$.

(ii) (Converse Failure) Let $g = \langle S = \{s, s', s''\}, ..., R = \{\langle s, \pi_1, s' \rangle, \langle s, \pi_2, s' \rangle, \langle s, \pi_1, s'' \rangle\} \rangle$, $s' \Vdash A$, $s'' \nVdash A$, and $s' \neq s''$.

(e) (5p) $([\pi_1]_p A \wedge [\pi_2]_p A) \rightarrow [\pi_1 \cup \pi_2]_p A$

(i) RtoL Failure. Assume $s \nVdash \langle \pi_1 \cup \pi_2 \rangle_p \top$, $s \Vdash \langle \pi_1 \rangle_p \top$, and $s \nVdash \langle \pi_1 \rangle_p A$. Then $s \Vdash [\pi_1 \cup \pi_2]_p A$ is vacuously true; every path starting at s is such that, if it's an execution of the choice, then it terminates at a state verifying A—there are no such paths. Thus, $s \nVdash [\pi_1]_p A \wedge [\pi_2]_p A$.

(ii) LtoR. Assume $[\pi_1]_p A \wedge [\pi_2]_p A$. Assume that there is some $p \in P_{\pi_1 \cup \pi_2}$ and $last(p) \nVdash A$. Then since $P_{\pi_1} \cup P_{\pi_2} = P_{\pi_1 \cup \pi_2}$, contradiction.

(f) (6p1) $(\langle\pi_1\rangle_p A \vee \langle\pi_2\rangle_p A) \to \langle\pi_1 \cup \pi_2\rangle_p A$

(i) RtoL Failure. Assume $\langle\pi_1\rangle_p A \vee \langle\pi_2\rangle_p A$. Then there is some $p \in P_{\pi_1}$ or $p \in P_{\pi_2}$ such that $last(p) \Vdash A$. By, from the execution rules, neither of these need be executions of the choice, which may be unexecutable at the current state.

(ii) LtoR. Assume $\langle\pi_1 \cup \pi_2\rangle_p A$. Then there is some $p \in P_{\pi_1 \cup \pi_2}$ and $last(p) \Vdash A$. Thus, there is some $p \in P_{\pi_1}$ or $p \in P_{\pi_2}$ such that $last(p) \Vdash A$. Thus, $\langle\pi_1\rangle_p A \vee \langle\pi_2\rangle_p A$.

(g) (6p2) $(\langle\pi_1\rangle_p A \wedge \langle\pi_2\rangle_p A) \to \langle\pi_1 \cup \pi_2\rangle_p A$

(i) LtoR. Assume $\langle\pi_1\rangle_p A \wedge \langle\pi_2\rangle_p A$ at s. Then there are executions, p and p' of both subprotocols that start at s, thereby executing the choice. The rules state that there must be edges labeled with the choice that share end states with both of e and e'. Thus, there is at least one edge labeled with the choice that terminates at a state verifying A.

(ii) RtoL Failure. Assume $\langle\pi_1 \cup \pi_2\rangle_p A$ at s. Then some edge, p, such that p executes the choice is such that it terminates at a state verifying A. Assume that there are no $p' \in R_{\pi_i}$ such that $last(p') \nVdash A$ and e starts at s but there is also an $p'' \in R_{\pi_j}$ starting at s that does verify A at its final state. No contradiction. Thus, the conjunction may fail.

(h) (7p) $[\pi*]_p A \equiv [\pi]_p[\pi*]_p A \wedge A$ for some n.

(i) LtoR. Let $s \Vdash [\pi*]_p A$. Then for every p starting at s such that, for some $q_1, ..., q_n$, $p = \langle q_1, ..., q_n\rangle$ and for every $n \geq 1$, such that each $q_i \in P_\pi$, it follows that $last(q_n) \Vdash A$ and $s \Vdash A$. Suppose that $s \nVdash [\pi]_p[\pi*]_p A$. Then for some p' such that $p' \in P_\pi$, $last(p') \nVdash [\pi*]_p A$. Thus, there must be some path r starting at $last(p')$ that is composed of paths executing π for each $n \geq 1$ such that $last(r) \nVdash A$. But then there is a path starting at s, $p'r$, such that $last(p'r) \nVdash A$ and $p'r$ is a concatenated sequence of π executions. But this contradicts the assumption.

(ii) RtoL. Let $s \Vdash [\pi]_p[\pi*]_p A \wedge A$ for some n. Then $s \Vdash A$ and, for all $p \in P_\pi$ such that $first(p) = s$, $last(p) \Vdash [\pi*]_p A$. Thus, for all p' such that $p' = \langle q_1, ..., q_n\rangle$, each $q_i \in P_\pi$, and $first(p') = last(p)$, it holds that $last(q_n) \Vdash A$. From the first conjunct, $\pi*$ verifies A at the terminal state of its $n = 1$ case execution. From the second conjunct, every p'' starting at s such that $p'' = \langle p, q_i, ..., q_n\rangle$ for some $n \geq 0$ and each q_i executes π is such that $last(p'') \Vdash A$. So, for any n greater than or equal to 1, there is a path of that length starting at s and verifying A at its last state. Thus, $s \Vdash [\pi*]_p A$.

(i) (8p) $\langle\pi*\rangle_p A \equiv \langle\pi\rangle_p \langle\pi*\rangle_p A \vee A$ for some n.

(i) LtoR. Let $s \Vdash \langle\pi*\rangle_P A$. Then for some p of length n for some $n \geq 1$ starting at s is such that $p = \langle q_1, ..., q_n\rangle$, and each $q_i \in P_\pi$, it follows that $last(q_n) \Vdash A$. Suppose that $s \nVdash \langle\pi\rangle_p \langle\pi*\rangle_p A \vee A$. Then $s \nVdash A$ and $s \nVdash \langle\pi\rangle_p \langle\pi*\rangle_p A$. So, for all p' such that $p' \in P_\pi$, $last(p') \nVdash \langle\pi*\rangle_p A$. Thus, every path r starting at $last(p')$ that is composed of paths executing π for each $n \geq 1$ such that $last(r) \nVdash A$. But then every $n \geq 1$ length path starting at s composed of executions of π is such that $last(p'r) \nVdash A$. But this contradicts the assumption.

(ii) RtoL. Let $s \Vdash \langle\pi\rangle_p \langle\pi*\rangle_p A \vee A$. Then, for some $n \geq 1$, there is some $p \in P_\pi$ such that $first(p) = s$, $last(p) \Vdash \langle\pi*\rangle_p A$ or $s \Vdash A$. Suppose the former. Then, for some p' such that $p' = \langle q_1, ..., q_n\rangle$, each $q_i \in P_\pi$, $first(p') = last(p)$ and $last(q_n) \Vdash A$. Thus, trivially, there is some p'' starting at s such that $p'' = pp'$ such that $last(p'') \Vdash A$ and p'' is composed of executions of π. Thus, $p'' \in P_{\pi*}$, so $s \Vdash \langle\pi*\rangle_p A$. Suppose the latter. Then, for the $n = 1$ case of the execution rule, $p' \in P_{\pi*}$ and $last(p) = first(p') = last(p') \Vdash A$. But then there is an execution of $\pi*$ starting at s that verifies A; $s \Vdash \langle\pi*\rangle_p A$.

(j) (9p) $(A \wedge [\pi*]_p(A \rightarrow [\pi]_p A)) \rightarrow [\pi*]_p A$

(i) Let $s \Vdash A \wedge [\pi*]_p(A \rightarrow [\pi]_p A)$. Then $s \Vdash A$, and $s \Vdash [\pi*]_p(A \rightarrow [\pi]_p A)$. Thus, for any path p such that there is a sequence of paths $\langle p_1, ..., p_n\rangle = p$ of any n with each $p_i \in P_\pi$, $p \in P_{\pi*}$ and $last(p) = last(p_n) \Vdash (A \rightarrow [\pi]_p A)$. But s itself is the terminal state of the $n = 1$ case execution of $\pi*$ (for $n = 1$, every path such that all of its $n < 1$ states satisfy the condition (there are no states, so vacuously, all such states satisfy the condition) are executions of $\pi*$—and that's all paths composed of one state). Thus, an execution of $\pi*$ terminates at s. Thus, $s \Vdash (A \rightarrow [\pi]_p A)$. Thus, by conditional elimination, $[\pi]_p A$. Thus, for any s' reachable by π from s, $s' \Vdash A$. But then there is a sequence of paths where each path executes π, so there is a path ending at s' that executes $\pi*$. So from $s \Vdash [\pi*]_p(A \rightarrow [\pi]_p A)$, $s' \Vdash (A \rightarrow [\pi]_p A)$. Thus, by conditional elimination, $s' \Vdash [\pi]_p A$. But then there is a sequence of paths where each path executes π, thus there is an execution of $\pi*$ that ends with s''. So from $s \Vdash [\pi*]_p(A \rightarrow [\pi]_p A)$, $s'' \Vdash (A \rightarrow [\pi]_p A)$. Thus, $s'' \Vdash [\pi]_p A$. ... Iterate this reasoning for any finite $n \geq 1$. Thus, every execution of $\pi*$ terminates at a state that verifies A.

(ii) (Induction) In general, suppose that some sequence, p, of edges of length m such that each edge executes π is such that the terminal state verifies A and $[\pi*]_p(A \rightarrow [\pi]_p A)$. Then $last(p)$ itself is the 1 length sequence of states starting at $last(p)$. So $last(p) \Vdash (A \rightarrow [\pi]_p A)$. Thus, by conditional elimination, $last(p) \Vdash [\pi]_p A$. Thus, for any $next(p)$ reachable by π from $last(p)$, $next(p) \Vdash A$. But then, $next(p) \Vdash (A \rightarrow [\pi]_p A)$ by the truth rules for $[\pi*]_p$, so it follows that $next(p) \Vdash [\pi]_p A$. Thus, a fortiori, all sequences of edges of length $m + 1$, such that each edge executes π is such that the terminal state verifies A. Thus, for any sequence of edges of length n such that the root state verifies A and $[\pi*]_p(A \rightarrow [\pi]_p A)$, the terminal state verifies A. Thus, at the root state, $[\pi*]_p A$.

(k) (10p) $\langle (A)? \rangle_p B \rightarrow A \wedge B$

(i) Assume the antecedent. Then any $p \in R_{(A)?}$, $first(p) = last(p)$, $first(p) \Vdash A$, and $last(p) \Vdash B$. Thus, $last(p) \Vdash A \wedge B$.

(l) (11p) $\langle (A)? \rangle_p \top \rightarrow A$

(i) Assume the antecedent. Then any $p \in R_{(A)?}$, $first(p) = last(p)$, $first(p) \Vdash A$, and $last(p) \Vdash \top$. Thus, $last(p) \Vdash A$.

(m) (12p) $A \equiv [\nabla\pi]_p A$

(i) Let $s \Vdash A$. Then, $\forall p[(p \in R_{\nabla\pi} \wedge s = first(p)) \rightarrow s = last(p)]$. Thus, $last(p) \Vdash A$. Converse is trivial. □

(n) (13p) $\Box(A \equiv \langle\pi\rangle_p \top) \rightarrow \Box[\curvearrowleft \pi]_p A$

(i) Let $s \Vdash \Box(A \equiv \langle\pi\rangle_p \top)$. Then for all $s' \in S$, $s' \Vdash A \equiv \langle\pi\rangle_p \top$. Suppose $s \nVdash \Box[\curvearrowleft \pi]_p A$. Then there is some s'' at which $[\curvearrowleft \pi]_p A$ fails. Thus, there is a $p \in P_{\curvearrowleft\pi}$, $s'' = first(p)$, $last(p) = s'''$, and $s''' \nVdash A$. But $p \in P_{\curvearrowleft\pi}$, $s'' = first(p)$, $last(p) = s'''$, iff there is a $p' \in P_\pi$ such that $s''' = first(p)$, and $last(p) = s''$. Thus, $s''' \Vdash \langle\pi\rangle_p \top$. Thus, $s''' \Vdash A$. Contradiction.

(ii) (Converse Failure) The limit case of paths is edges. Consider an EDG:

$$g = \left\langle \begin{bmatrix} S = \{s, s', s'', s'''\}, \\ R = \left\{ \begin{bmatrix} \langle s, \pi, s' \rangle, & \langle s', \pi, s'' \rangle, \\ \langle s, \pi, s''' \rangle, & \langle s''', \curvearrowleft \pi, s \rangle \end{bmatrix} \right\}, \\ \ldots \end{bmatrix} \right\rangle$$

Let g be such that: $s \Vdash A$, $s' \nVdash A$, $s'' \nVdash A$, $s''' \nVdash A$, and $s' \neq s'' \neq s'''$. At every state at which $\curvearrowleft \pi$ can be executed, it brings about A. However, at s', π is executable but A fails.

(o) (20p)

(i) $(\Box(\pi \leftrightarrow_p \pi') \wedge A(\pi)) \rightarrow A(\pi')$

(i) From NSEP and the Deduction Theorem.

(ii) $((\pi \leftrightarrow_p \pi') \wedge A(\pi)) \rightarrow A(\pi')$

(i) From SEP and the Deduction Theorem.

(p) (21p) $((\pi \curlywedge_p \pi') \wedge [\pi]_p A) \rightarrow \langle\pi'\rangle_p A$

(i) Assume $s \Vdash \pi \curlywedge_p \pi'$ and $s \Vdash [\pi]_p A$. Then there is some p such that $s = first(p)$ in $P_\pi \cap P_{\pi'}$ such that $last(p) \Vdash A$. Thus, $s \Vdash \langle\pi'\rangle_p A$. □

(q) (22p) $(\pi \curlyvee_p \pi') \equiv \neg(\pi \leftrightarrow_p \pi')$

(i) Assume $s \Vdash \pi \curlyvee_p \pi'$. Then there is some p such that $s = first(p)$ in $P_\pi / P_{\pi'}$. Thus, $s \Vdash \neg(\pi \leftrightarrow_p \pi')$. The reasoning transposing with π' generates identical results.

(r) (23p) $\Box A \rightarrow [\pi]_p A$

(i) Assume $s \Vdash \Box A$. Suppose that $s' = last(p)$ for any arbitrary p that executes π. Then $s' \Vdash A$. Thus, all terminal states for all procedural executions of π verify A.

(4) (Binary Axioms)

(a) (5t) $([\pi_1]_t A \wedge [\pi_2]_t A) \rightarrow [\pi_1 \cup \pi_2]_t A$

(i) RtoL Failure. Assume $s \nVdash \langle \pi_1 \cup \pi_2 \rangle_t \top$, $s \Vdash \langle \pi_1 \rangle_t \top$, and $s \nVdash \langle \pi_1 \rangle_t A$. Then $s \Vdash [\pi_1 \cup \pi_2]_t A$ is vacuously true; every transition starting at s is such that, if it's an execution of the choice, then it terminates at a state verifying A—there are no such transitions. Thus, $s \nVdash [\pi_1]_t A \wedge [\pi_2]_t A$.

(ii) LtoR. Assume $[\pi_1]_t A \wedge [\pi_2]_t A$. Assume that there is some $t \in T_{\pi_1 \cup \pi_2}$ and $last(t) \nVdash A$. Then since $T_{\pi_1} \cup T_{\pi_2} = T_{\pi_1 \cup \pi_2}$, contradiction.

(5) (Derived Laws)

(a) $\begin{bmatrix} [\pi]_r A \equiv [\pi]_p A \\ \langle \pi \rangle_r A \equiv \langle \pi \rangle_p A \end{bmatrix}$

(i) The Restriction Rules and the execution rules together imply that $\forall e \in R_\pi^+(\exists p \in P_\pi(f(e) = f(p) \wedge l(e) = l(p)))$ and $\forall p \in P_\pi(\exists e \in R_\pi^+(f(e) = f(p) \wedge l(e) = l(p)))$. The rules follow mechanically from these.

(b) $\begin{bmatrix} [\pi]_p A \equiv [\pi]_t A \\ \langle \pi \rangle_p A \equiv \langle \pi \rangle_t A \end{bmatrix}$

(i) The Restriction Rules and the execution rules together imply that $\forall p \in P_\pi(\exists t \in T_\pi(f(t) = f(p) \wedge l(t) = l(p)))$ and $\forall t \in T_\pi(\exists p \in P_\pi(f(t) = f(p) \wedge l(e) = l(p)))$. The rules follow mechanically from these.

(6) (Identity Rules)

(a) $(A = A') \rightarrow (B \equiv \rho(A/A')(B))$

(b) $(\pi = \pi') \rightarrow (A(\pi) = A(\pi'))$

(i) Both are immediate; there is nothing unusual about substitution of identity in PLEN.

(ii) The latter rule is unrestricted.

(Logic):

(1) I omit proofs for most of the propositional and modal rules; each is easy to check given the definition of logical consequence in PLEN.

(a) At a minimum, standard conjunction and disjunction introduction and elimination, detachment of entailment/valid implication, introduction and detachment of the conditional, and the usual definition of $\equiv$ are built into PLEN.

(b) Let $\Diamond$ be defined as the complement of $\Box$: $\Diamond = \neg\Box\neg$. Then the rule $\dfrac{\Diamond A}{\Box\Diamond A}(\Box\Diamond I)$ is clearly sound. Let s verify $\Diamond A$ in g. Then it's not the case that all states verify the negation of A. So some state, s', verifies A. Thus, at all states in g, it's true that there is some state that verifies A. The universal box' semantics are defined without regard to accessibility relations.

(2) Deduction Elimination is redundant as the basic feature of derivability. Nonetheless it's built into PLEN as a formal rule where it functions like a cut, iteration, or repetition rule.

(3) Deduction Introduction is a version of the Deduction Theorem for the derivation system built directly into the system. The Deduction Theorem is proven below; it is a corollary of the soundness theorem, as the soundness of Conditional Introduction makes the proof mechanical.

   (a) The proof of the soundness of Conditional Introduction is immediate from the definition of logical consequence:

      (i) Suppose that there is a sound derivation of B from A using rules excluding conditional proof. If s verifies A, then B is verified at s by the assumption of the soundness of the derivation, and s thus verifies $A \rightarrow B$. Else, s verifies $A \rightarrow B$ by its truth rule (material conditional). For derivations using nested Conditional Proofs, induction on the depth of the nesting suffices where the base case is the reasoning just given, and the next level up in the nesting is immediately sound by the induction hypothesis.

      (ii) Conditional Proof generalizes Conditional Introduction and is sound given the definition of logical consequence in PLEN. The proof is immediate by the same reasoning as that for Conditional Introduction: the sound derivability of A from Σ by rules excluding Conditional Proof entails that, wherever all of Σ is verified, A is verified. Thus, where $\Sigma^\wedge$ is verified, A is verified and so $\Sigma^\wedge \rightarrow B$ is verified. Else, $\Sigma^\wedge \rightarrow B$ is verified where any conjunct of $\Sigma^\wedge$ fails.

      (iii) Proof by Contradiction generalizes Negation Introduction, and it is also sound given the definition of logical consequence in PLEN. These formal rules are built into the logic of PLEN and each is sound by immediate arguments omitted from the soundness proof.

(4) The box of PLEN is a universal modal operator, and the modal/dynamic operators for each protocol, π, work depending on the properties of T_π, though the restriction rules imply that every $[\pi *]_x$ behaves like the necessity of S4. In a generalized version of PLEN, where EDGs don't satisfy the Restriction Rules, $[\pi *]_x$ might be equivalent to a K box at s, and, at s', it might be equivalent to an S5 box. The detachment and necessitation rules for $\Box$ as a universal modality hold, but necessitation is restricted to derivations from no assumptions (i.e., to axioms and formulae derivable from axioms).[35]

[35] This feature makes the system compatible with additional machinery for reasoning from premises without the ugly—or at least perplexing—results that often accompany such an addition. In particular, this revision makes room for a valid deduction theorem and so makes it possible to have rules for introducing conditionals in the usual natural deduction style. See [16] for discussion of the matter and the machinery for adding a valid deduction theorem to normal modal logics. The machinery that Hakli and Negri construct show that a deduction theorem for PLEN is no issue, nor is conditional proof.

(a) Modal operators provide no barrier to the deduction theorem in PLEN in virtue of the prior property.

The proof of the axiom schemata and the logical rules completes the proof of the soundness of the derivation system. □

Completeness Conjecture

PLEN Fischer-Ladner Lemma

Fact 104 *The definition of executions in PLEN maps all transitional executions to at least one edge execution and one path execution such that the transition comprises the root and terminal states of the edge and path.*

Conjecture 105 (Fischer-Ladner Lemma (FLL)) The following properties of the Fischer-Ladner Closure on PLEN hold:

(1) (FLL1)

(a) If $A \in FL(B)$, then $FL(A) \subseteq FL(B)$.
(b) If $A \in FL_\pi([\pi]_x B)$, then $FL(A) \subseteq FL_\pi([\pi]_x B) \cup FL(B)$.
(c) If $A \in FL(R_x(\pi_1, \pi_2))$, then

$$FL(A) \subseteq \{R_x(\pi_1, \pi_2)\} \cup FL_\pi([\pi_1]_x \top) \cup FL_\pi([\pi_2]_x \top)$$

(d) If $A \in FL(R_x(\pi_1, \pi_2, \pi_3, \pi_4))$, then

$$FL(A) \subseteq \{R_x(\pi_1, \pi_2, \pi_3, \pi_4)\} \cup FL_\pi([\pi_1]_x \top)$$
$$\cup\ FL_\pi([\pi_2]_x \top \cup FL_\pi([\pi_3]_x \top) \cup FL_\pi([\pi_4]_x \top$$

(2) (FLL2)

(a) If $[\pi]_x A \in FL(B)$, then $A \in FL(B)$.
(b) If $[(C)?]_x A \in FL(B)$, then $C \in FL(B)$.
(c) If $[\pi_1 \cup \pi_2]_x A \in FL(B)$, then $[\pi_1]_x A \in FL(B)$ and $[\pi_2]_x A \in FL(B)$.
(d) If $[\pi_1 \cap \pi_2]_x A \in FL(B)$, then $[\pi_1]_x A \in FL(B)$ and $[\pi_2]_x A \in FL(B)$.
(e) If $[\pi_1; \pi_2]_x A \in FL(B)$, then $[\pi_1]_x [\pi_2]_x A \in FL(B)$ and $[\pi_2]_x A \in FL(B)$.
(f) If $[\pi_1 *]_x A \in FL(B)$, then $[\pi_1]_x [\pi_1 *]_x A \in FL(B)$.
(g) If $R_x(\pi_1, \pi_2) \in FL(B)$, then

$$\{R_x(\pi_1, \pi_2)\} \cup FL_\pi([\pi_1]_x \top) \cup FL_\pi([\pi_2]_x \top) \subseteq FL(B)$$

(h) If $R_x(\pi_1, \pi_2, \pi_3, \pi_4) \in FL(B)$, then

$$\{R_x(\pi_1, \pi_2, \pi_3, \pi_4)\} \cup FL_\pi([\pi_1]_x \top) \cup FL_\pi([\pi_2]_x \top) \cup FL_\pi([\pi_3]_x \top) \cup FL_\pi([\pi_4]_x \top) \subseteq FL(B)$$

(3) (FLL3)

(a) For any formula, A, $|FL(A)| \leq len(A)$.
(b) For any formula, $[\pi]_x A$, $|FL_\pi([\pi]_x A)| \leq len(\pi)$.
(c) For any formula, $R_x(\pi_1, \pi_2)$,

$$|FL(R_x(\pi_1, \pi_2))| \leq (len(\pi_1) + len(\pi_2) + \left|FL_\pi([\pi_1]_x \top)\right| + \left|FL_\pi([\pi_2]_x \top)\right|)$$

(d) For any formula, $R_x(\pi_1, \pi_2, \pi_3, \pi_4)$,

$$|FL(R_x(\pi_1, \pi_2, \pi_3, \pi_4))| \leq (len(\pi_1) + ... + len(\pi_2) + |FL_\pi([\pi_1]_x \top)| + ... + |FL_\pi([\pi_4]_x \top)|)$$

Proof (Sketch) The reasoning is by induction over proper immediate subprotocols and proper immediate subformulae, and is mechanical by means of the definition of Fischer-Ladner Closure and the semantics of PLEN; the definition of executions in PLEN maps all transitional executions to at least one edge execution and one path execution such that the transition comprises the root and terminal states of the edge and path. Most of the clauses, here, are present in the textbook proof in [15]. The reasoning here only explicitly works through the some of the novel the clauses for PLEN.

(1) (FLL1)

(a) (Base)
(i) For atomic formula, B, the result is mechanical: if B is atomic, then $FL(B) = \{B\}$, and so $A = B$, $FL(A) = \{B\}$, and $FL(A) \subseteq FL(B)$.
(ii) For atomic protocol, π, and atomic formula, B: $FL_\pi([\pi]_x B) = \{[\pi]_x B\}$, so $FL([\pi]_x B) = \{[\pi]_x B\} \cup FL(B)$. So, if $A \in FL_\pi([\pi]_x B)$, then $A = [\pi]_x B$, and thus $FL(A) \subseteq FL_\pi([\pi]_x B)$.
(iii) For atomic protocols, the novel clauses, c and d, follow mechanically from the definition of FL_π and the prior step, ii.

(b) (Induction) Assume that FLL1a and FLL1b hold for all proper immediate subprotocols and proper immediate subformulae of arbitrary π and A.
(i) Let $A \in FL(B)$. There are six possible syntactic forms for B, and so six cases. Each case is mechanical. I illustrate two cases: If $B = C_1 \rightarrow C_2$, then $FL(B) = \{C_1 \rightarrow C_2\} \cup FL(C_1) \cup FL(C_2)$. Thus, either $A = C_1 \rightarrow C_2$, or $A \in FL(C_1)$ or $A \in FL(C_2)$. In the first case, $FL(A) = FL(C_1 \rightarrow C_2)$, as required. For the remaining cases, the inductive hypothesis implies that $FL(A) \subseteq FL(C_1)$ and $FL(A) \subseteq$

$FL(C_2)$. But $FL(C_1)$ and $FL(C_2)$ are subsets of $FL(C_1 \to C_2)$, by definition. Thus, $FL(A) \subseteq FL(B)$. If $B = [\pi]_x C$, then $A \in FL_\pi([\pi]_x C)$ or $A \in FL(C)$. By the induction hypothesis, if $A \in FL_\pi([\pi]_x C)$, then $FL(A) \subseteq FL_\pi([\pi]_x C) \cup FL(C)$. If $A \in FL(C)$, then by the induction hypothesis, $FL(A) \subseteq FL(C)$, and so $FL(A) \subseteq FL_\pi([\pi]_x C) \cup FL(C)$.

(ii) Let $A \in FL_\pi([\pi] B)$. There are five possible syntactic forms for π cases, and so five cases. Each is mechanical. I illustrate one case: Let $\pi = \pi_1; \pi_2$. Then

$$FL_\pi([\pi_1;\pi_2]_x B) = \{[\pi_1;\pi_2]_x B\} \cup FL_\pi([\pi_1]_x [\pi_2]_x B) \cup FL_\pi([\pi_2]_x B)$$

Thus, $A = [\pi_1;\pi_2]_x B$, or $A \in FL_\pi([\pi_1]_x [\pi_2]_x B)$, or $A \in FL_\pi([\pi_2]_x B)$. In the first case, $FL(A) = FL_\pi([\pi_1;\pi_2]_x B) \cup FL(B)$, as required. In the second case, $FL(A) \subseteq FL_\pi([\pi_1]_x [\pi_2]_x B) \cup FL([\pi_2]_x B)$ by induction hypothesis. But then, by definition of FL_π, $FL(A) \subseteq FL_\pi([\pi_1]_x [\pi_2]_x B) \cup FL_\pi([\pi_2]_x B) \cup FL(B)$. But this immediately implies that $FL(A) \subseteq FL_\pi([\pi_1;\pi_2]_x B) \cup FL(B)$. In the third case, $FL(A) \subseteq FL_\pi([\pi_2]_x B) \cup FL(B)$. So, immediately, $FL(A) \subseteq FL_\pi([\pi_1;\pi_2]_x B) \cup FL(B)$.

(iii) For complex protocols, the novel clauses, c and d, follow mechanically from the definition of FL_π and the prior step, ii.

(2) (FLL2) Each clause is mechanical from the definitions of FL and FL_π.

(a) The novel clauses (i.e., those for the novel operators ($\cup, \cap, *$) of this revised system) follow from the corresponding clauses of the definition of FL_π.

(b) Clauses (g) and (h) follow immediately from clause (i) of the definition of FL.

(3) (FLL3)

(a) (Base)

(i) For atomics, A, $len(A) = 1 = |FL(A)|$.

(ii) For atomics, π, $len(\pi) = 1 = |FL_\pi([\pi]_x A|$

(iii) For any formula, $R_x(\pi_1, \pi_2)$, the definition of FL_π requires that:

$$|FL(R_x(\pi_1,\pi_2))| \leq (len(\pi_1) + len(\pi_2) + |FL_\pi([\pi_1]_x \top)| + |FL_\pi([\pi_2]_x \top)|)$$

where $|FL_\pi([\pi_i]_x \top)|$ is as in the argument for clause ii, immediately above. Thus,

$$|FL(R_x(\pi_1,\pi_2))| \leq (len(\pi_1) + len(\pi_2) + |FL_\pi([\pi_1]_x \top)| + |FL_\pi([\pi_2]_x \top)|) = 1 + 1 + 1 + 1$$

(iv) For any formula, $R_x(\pi_1, \pi_2, \pi_3, \pi_4)$, the definition of FL_π requires that:

$$|FL(R_x(\pi_1, \pi_2, \pi_3, \pi_4))| \leq (len(\pi_1) + ... + len(\pi_4) + \left|FL_\pi([\pi_1]_x \top)\right| + ... + \left|FL_\pi([\pi_4]_x \top)\right|$$

where $|FL_\pi([\pi_i]_x \top)|$ is as in the argument for clause ii, immediately above. Thus,

$$|FL(R_x(\pi_1, \pi_2))| \leq (len(\pi_1) + ... + len(\pi_4) + \left|FL_\pi([\pi_1]_x \top)\right| +... + \left|FL_\pi([\pi_4]_x \top)\right|) = 1 + 1 + 1 + 1 + 1 + 1$$

(b) (Induction)

(i) By definition of $FL(A \circ B)$ for any binary connective, $\circ$, $|FL(A \circ B)| = 1 + |FL(A)| + |FL(B)|$. By definition of $FL(\circ A)$ for any unary connective, $\circ$, $|FL(\circ A)| = 1 + |FL(A)|$. By definition of $FL([\pi] A)$, $|FL([\pi] A)| = |FL_\pi([\pi] A| + |FL(A)|$. By induction hypothesis, then:

(A) $|FL(A \circ B)| = 1 + |FL(A)| + |FL(B)| = 1 + len(A) + len(B) = len(A \circ B$.

(B) $|FL(\circ A)| = 1 + |FL(A)| = 1 + len(A) = len(\circ A$.

(C) $|FL([\pi]_x A)| = |FL_\pi([\pi]_x A| + |FL(A)| = len(\pi) + len(A)$.

(ii) By definition of $FL_\pi([\pi_1 \cup \pi_2]_x A)$ and $FL_\pi([\pi_1 \cap \pi_2]_x A)$, $|FL_\pi([\pi_1 \circ \pi_2] A)| = 1 + |FL_\pi([\pi_1]_x A)| + |FL_\pi([\pi_2]_x A)|$. By definition of $FL_\pi([\pi_1; \pi_2]_x A)$, $|FL_\pi([\pi_1; \pi_2]_x A)| = 1 + |FL_\pi([\pi_1]_x [\pi_2]_x A)| + |FL_\pi([\pi_2]_x A)|$. By definition of $FL_\pi([\pi*]_x A)$, $|FL_\pi([\pi*]_x A)| = 1 + |FL_\pi([\pi]_x [\pi*]_x A)|$. By induction hypothesis, then:

(A) $|FL_\pi([\pi_1 \circ \pi_2]_x A)| = 1 + |FL_\pi([\pi_1]_x A)| + |FL_\pi([\pi_2]_x A)| = 1 + len(\pi_1) + len(\pi_2)$.

(B) $FL_\pi([\pi_1; \pi_2]_x A) = 1 + |FL_\pi([\pi_1]_x [\pi_2]_x A)| + |FL_\pi([\pi_2]_x A)| = 1 + len(\pi_2) + len(\pi_2)$.

(C) $|FL_\pi([\pi*]_x A)| = 1 + |FL_\pi([\pi]_x [\pi*]_x A)| - 1 + len(\pi*)$.

(iii) For any formula, $R_x(\pi_1, \pi_2)$, the definition of FL_π requires that:

$$FL_\pi(R_x(\pi_1, \pi_2)) = \{R_x(\pi_1, \pi_2)\} \cup FL_\pi([\pi_1]_x \top) \cup FL_\pi([\pi_2]_x \top)$$

Thus, by definition,

$$|FL(R_x(\pi_1, \pi_2))| \leq (len(\pi_1) + len(\pi_2) + |FL_\pi([\pi_1]_x \top)| + |FL_\pi([\pi_2]_x \top)|)$$

where $|FL_\pi([\pi_i]_x \top)|$ is as in the argument for clause ii, immediately above.

(iv) For any formula, $R_x(\pi_1, \pi_2, \pi_3, \pi_4)$, the definition of FL_π requires that:

$$FL_\pi(R_x(\pi_1, \pi_2)) = \{R_x(\pi_1, \pi_2, \pi_3, \pi_4)\} \cup FL_\pi([\pi_1]_x \top) \cup ... \cup FL_\pi([\pi_4]_x \top)$$

Thus, by definition,

$$|FL(R_x(\pi_1, \pi_2, \pi_3, \pi_4))| \leq (len(\pi_1) + ... + len(\pi_4)$$
$$+ |FL_\pi([\pi_1]_x \top)| + ... + |FL_\pi([\pi_4]_x \top)|)$$

where $|FL_\pi([\pi_i]_x \top)|$ is as in the argument for clause ii, immediately above.

□

PLEN Filtration Lemma

Conjecture 106 (PLEN Filtration Lemma) Let g be an EDG, and let s and t be states in S. Then $g_{FL(B)}$ is an EDG such that:

(1) For all $A \in FL(B)$, $v_s(A) = 1$ iff $v_{s/FL(B)}(A) = 1$.
(2) For all $[\pi]_x A \in FL(B)$:

  (a) if $(s, s') \in T_\pi$, then $([s], [s']) \in T_{\pi/FL(B)}$
  (b) if $([s], [s']) \in T_{\pi/FL(B)}$ and $v_s([\pi]_x A) = 1$, then $v_{s'}(A) = 1$.

(3) For all $R_x(\pi_1, \pi_2) \in FL(B)$,

  (a) for each π_i,
    (i) if $t \in T_{\pi_i}$ with $s = f(t)$ and $s' = l(t)$, then $t_{FL(B)} \in T_{\pi_i/FL(B)}$ and $[s] = f(t_{FL(B)})$ and $[s'] = l(t_{FL(B)})$.
    (ii) if $e \in R{+}_{\pi_i}$ with $s = f(e)$ and $s' = l(e)$, then $e_{FL(B)} \in R{+}_{\pi_i/FL(B)}$ and $[s] = f(e_{FL(B)})$ and $[s'] = l(e_{FL(B)})$.
    (iii) if $p \in P_{\pi_i}$ with $s = f(p)$ and $s' = l(p)$, then $p_{FL(B)} \in P_{\pi_i/FL(B)}$ and $[s] = f(p_{FL(B)})$ and $[s'] = l(p_{FL(B)})$.
  (b) for each π_i, the following conditionals and their converses hold:
    (i) if $t_{FL(B)} \in T_{\pi_i/FL(B)}$ and $[s] = f(t_{FL(B)})$ and $[s'] = l(t_{FL(B)})$, then:
      (A) if $t_{FL(B)} \in T_{\pi_i/FL(B)}$ and $f(t_{FL(B)}) = [f(t)]$, then $t \in T_{\pi_i}$.
      (B) if $t_{FL(B)} \in T_{\pi_1/FL(B)}$ iff $t_{FL(B)} \in T_{\pi_2/FL(B)}$ and $f(t_{FL(B)}) = [f(t)]$, then $t \in T_{\pi_1}$ iff $t \in T_{\pi_2}$.
      (C) if $t_{FL(B)} \in T_{\pi_1/FL(B)} \cap T_{\pi_2/FL(B)}$ and $f(t_{FL(B)}) = [f(t)]$, then $t \in T_{\pi_1} \cap T_{\pi_2}$.
      (D) if $t_{FL(B)} \in T_{\pi_1/FL(B)} - T_{\pi_2/FL(B)}$ and $f(t_{FL(B)}) = [f(t)]$, then $t \in T_{\pi_1} - T_{\pi_2}$.

(ii) if $e_{FL(B)} \in R{+}_{\pi_i/FL(B)}$ and $[s] = f(e_{FL(B)})$ and $[s'] = l(e_{FL(B)})$, then:

(A) if $e_{FL(B)} \in R{+}_{\pi_i/FL(B)}$ and $f(e_{FL(B)}) = [f(e)]$, then $e \in R{+}_{\pi_i}$.

(B) if $e_{FL(B)} \in R{+}_{\pi_1/FL(B)}$ iff $e_{FL(B)} \in R{+}_{\pi_2/FL(B)}$ and $f(e_{FL(B)}) = [f(e)]$, then $e \in R{+}_{\pi_1}$ iff $e \in R{+}_{\pi_2}$.

(C) if $e_{FL(B)} \in R{+}_{\pi_1/FL(B)} \cap R{+}_{\pi_2/FL(B)}$ and $f(e_{FL(B)}) = [f(e)]$, then $e \in R{+}_{\pi_1} \cap R{+}_{\pi_2}$.

(D) if $e_{FL(B)} \in R{+}_{\pi_1/FL(B)} - R{+}_{\pi_2/FL(B)}$ and $f(e_{FL(B)}) = [f(e)]$, then $e \in R{+}_{\pi_1} - R{+}_{\pi_2}$.

(iii) if $p_{FL(B)} \in P_{\pi_i/FL(B)}$ and $[s] = f(p_{FL(B)})$ and $[s'] = l(p_{FL(B)})$, then:

(A) if $p_{FL(B)} \in P_{\pi_i/FL(B)}$ and $f(p_{FL(B)}) = [f(p)]$, then $p \in P_{\pi_i}$.

(B) if $p_{FL(B)} \in P_{\pi_1/FL(B)}$ iff $p_{FL(B)} \in P_{\pi_2/FL(B)}$ and $f(p_{FL(B)}) = [f(p)]$, then $p \in P_{\pi_1}$ iff $p \in P_{\pi_2}$.

(C) if $p_{FL(B)} \in P_{\pi_1/FL(B)} \cap P_{\pi_2/FL(B)}$ and $f(p_{FL(B)}) = [f(p)]$, then $p \in P_{\pi_1} \cap P_{\pi_2}$.

(D) if $p_{FL(B)} \in P_{\pi_1/FL(B)} - P_{\pi_2/FL(B)}$ and $f(p_{FL(B)}) = [f(p)]$, then $p \in P_{\pi_1} - P_{\pi_2}$.

(4) For all $R_x(\pi_1, \pi_2, \pi_3, \pi_4) \in FL(B)$, $v_s(R_x(\pi_1, \pi_2, \pi_3, \pi_4)) = 1$ iff $v_{s/FL(B)}(R_x(\pi_1, \pi_2, \pi_3, \pi_4)) = 1$.

Proof (Sketch) Clauses 1 and 2 are textbook. Simultaneous induction over proper immediate subprotocols and subformulae seems to give us interesting cases. For the sake of convenience, adopt the following:

Definition 107 (Filtration Conventions)

(1) For any state, s, in an EDG, g, denote the corresponding state in $g_{FL(A)}$ with $s_{FL(A)}$.
(2) For any state, t, in an EDG, g, denote the corresponding transition in $g_{FL(A)}$ with $t_{FL(A)}$.
(3) For any edge, e, in an EDG, g, denote the corresponding edge in $g_{FL(A)}$ with $e_{FL(A)}$.
(4) For any path, p, in an EDG, g, denote the corresponding path in $g_{FL(A)}$ with $p_{FL(A)}$.

The preservation of graph-theoretic structure built into the definition of filtrations entails that the restriction rules are verified. Filtrations are thus EDGs, as the structures of states are unspecified—maximal consistent sets of formulae and equivalence classes of arbitrary indices or parameters are as admissible as states as anything else. More, the requirement that the paths in the filtration of g have isomorphic counterparts in g and vice versa extends the preservation of verification properties of filtrations to the graph-theoretic protocol equivalence formulae by Lemma L4 in the SUT and PGIIRC proofs. So, I illustrate some cases (not nearly all):

(1) Let $A \in FL(B)$.

(a) (Base)
  (i) LtoR. Let $A \in FL(B)$ and $v_s(A) = 1$. Then, immediately, for any $s' \in [s]$, $v_{s'/FL(B)}(A) = 1$.
  (ii) RtoL. Let $A \in FL(B)$. Let $v_{s'/FL(B)}(A) = 1$ for some $s' \in [s]$. Then, immediately, $v_s(A) = 1$.
(b) (Induction) Assume that clauses 1 and 2 both hold for all proper immediate subformulae of B and all proper immediate subprotocols of π. The reasoning is mechanical from the obvious extension of $v_{s/FL(B)}$ to complex formulae by FLL. Here are some illustrative cases:
  (i) Let $A = C_1 \rightarrow C_2 \in FL(B)$ and $v_s(C_1 \rightarrow C_2) = 1$. Then from FLL, $C_1 \in FL(B)$ and $C_2 \in FL(B)$. By the induction hypothesis, $v_{s/FL(B)}(C_1) = 1$ iff $v_s(C_1) = 1$ and $v_{s/FL(B)}(C_2) = 1$ iff $v_s(C_2) = 1$. Thus, $v_{s/FL(B)}(C_1 \rightarrow C_2) = 1$ iff $v_s(C_1 \rightarrow C_2) = 1$. Essentially the same reasoning applies to the other propositional connectives.
  (ii) Let $A = [\pi]_x C \in FL(B)$ and $v_s([\pi]_x C) = 1$. Assume that 1 and 2 hold for all proper immediate subprotocols of π and proper immediate subformulae of B. From the FLL, $A \in FL(B)$. By inductive hypothesis, if s verifies $[\pi]_x A$, then for any t, such that $([s], [t]) \in T_{\pi/FL(B)}$, t verifies A. Conversely, if, for all t, such that $([s], [t]) \in T_{\pi/FL(B)}$, t verifies A, then $(s, t) \in T_\pi$ and s verifies $[\pi]_x A$. Thus, s verifies $[\pi]_x A$ iff for all t, such that $([s], [t]) \in T_{\pi/FL(B)}$, t verifies A. But, by definition of filtrations (and induction hypothesis), t verifies A iff $[t]$ verifies A. Of course, $[s]$ verifies $[\pi]_x A$ iff for all t, such that $([s], [t]) \in T_{\pi/FL(B)}$, $[t]$ verifies A. Thus, s verifies $[\pi]_x A$ iff $[s]$ verifies $[\pi]_x A$ ($v_s([\pi]_x A) = v_{s/FL(B)}([\pi]_x A)$).

(2) Let $[\pi]_x A \in FL(B)$. The modifications introduced by the structure preserving clauses of the definition of filtrations should handle any complications introduced by the new path-theoretic and edge-theoretic expressions of PLEN.

(a) (Base) For atomic protocols, part (a) is immediate from the definition of filtrations. Part (b) requires using the reasoning for clause 1 and the definition of filtrations. Let $([s], [s']) \in T_{\pi/FL(B)}$ and $v_{s/FL(B)}([\pi]_x A) = 1$. Then $v_{s'/FL(B)}(A) = 1$. Thus, $v_{x'}(A) = 1$.
(b) (Induction) Assume that clauses 1 and 2 both hold for all proper immediate subformulae of B and all proper immediate subprotocols of π. The argument is mechanical from the definitions of $T_{\pi/FL(B)}$ for complex protocols. Here is an illustrative case:
  (i) Let $\pi = \pi_1; \pi_2$. By induction hypothesis, (a) and (b) hold for π_1 and π_2. For (a), let $(s, s') \in T_\pi$. Then there is some $x \in S$ such that $(s, x) \in T_{\pi_1}$ and $(x, s') \in T_{\pi_2}$. But then there must be $([s], [x]) \in T_{\pi_1/FL(B)}$ and $([x], [s']) \in T_{\pi_2/FL(B)}$. Thus, $([s], [s']) \in T_{\pi_1;\pi_2/FL(B)}$. $[\pi_1][\pi_2]_x A \in FL(B)$ and $[\pi_2]_x A \in FL(B)$. For (b), let $([s], [s']) \in T_{\pi/FL(B)}$ and $v_s([\pi_1; \pi_2]_x A$. Then there must be $([s], [x]) \in T_{\pi_1/FL(B)}$ and $([x], [s']) \in T_{\pi_2/FL(B)}$ and $v_{s/FL(B)}([\pi_1]_x[\pi_2]_x A) = 1$. Thus, $v_{x/FL(B)}([\pi_2]_x A) = 1$. Thus, $v_x([\pi_2]_x A) = 1$. Thus, $v_{s'}(A) = 1$.

(3) Let $R_x(\pi_1, \pi_2) \in FL(B)$. The modifications introduced by the structure preserving clauses of the definition of filtrations should handle any complications introduced by the new path-theoretic and edge-theoretic expressions of PLEN.

(a) (Base) For atomic, π_i, part (a) is immediate from the definition of filtrations. For part (b), the identification of $R{+}_\pi$ and R_π and the definition of filtrations give the result immediately, when coupled with the execution rules.

(b) (Induction) For complex π_i, suppose that the result holds for all proper immediate subprotocols. Part (a) is immediate from the definition of filtrations. Part (b) makes use of the obvious structure preserving nature of filtrations:

(i) if $e_{1FL(B)} \rightleftharpoons e_{2FL(B)}$ and $f(e_{1FL(B)}) = f[(e_1)]$, then $e_1 \rightleftharpoons e_2$.
(ii) if $e_{1FL(B)}e_{2FL(B)}$ and $f(e_{iFL(B)}) = [f(e_i)]$, then e_1e_2.
(iii) if $f(e_{1FL(B)}) = f(e_{2FL(B)})$ and $f(e_{FL(B)}) = [f(e)]$, then $f(e_1) = f(e_2)$.
(iv) if $p_{1FL(B)} \rightleftharpoons p_{2FL(B)}$ and $f(p_{FL(B)}) = [f(p)]$, then $p_1 \rightleftharpoons p_2$.
(v) if $f(p_{1FL(B)}) = f(p_{2FL(B)})$ and $f(p_{iFL(B)}) = [f(p_i)]$, then $f(p_1) = f(p_2)$.
(vi) if $p_{1FL(B)}p_{2FL(B)}$ and $f(p_{iFL(B)}) = [f(p_i)]$, then p_1p_2.
(vii) if $p_{1FL(B)} \sqsubseteq p_{2FL(B)}$ and $f(p_{FL(B)}) = [f(p)]$, then $p_1 \sqsubseteq p_2$.
(viii) if $e_{1FL(B)} \rightleftharpoons e_{2FL(B)}$ and $f(t_{1FL(B)}) = [f(t_1)]$, then $t_1 \rightleftharpoons t_2$.

(c) Mechanically, the definition of filtrations replicates the structural relations among transitions, edges, and paths. This mechanically implies the results. I reason through some illustrative clauses:

(i) if $t_{FL(B)} \in T_{\pi_i/FL(B)}$ and $f(t_{FL(B)}) = [f(t)]$, then $t \in T_{\pi_i}$.
(ii) if $t_{FL(B)} \in T_{\pi_i/FL(B)}$ iff $t_{FL(B)} \in T_{\pi_i/FL(B)}$ and $f(t_{FL(B)}) = [f(t)]$, then $t \in T_{\pi_i}$ iff $t \in T_{\pi_i}$.
(iii) if $t_{FL(B)} \in T_{\pi_1/FL(B)} \cap T_{\pi_2/FL(B)}$ and $f(t_{FL(B)}) = [f(t)]$, then $t \in T_{\pi_1} \cap T_{\pi_2}$.
(iv) if $t_{FL(B)} \in T_{\pi_1/FL(B)} - T_{\pi_2/FL(B)}$ and $f(t_{FL(B)}) = [f(t)]$, then $t \in T_{\pi_1} - T_{\pi_2}$.
(v) if $e_{FL(B)} \in R{+}_{\pi_i/FL(B)}$ and $f(e_{FL(B)}) = [f(e)]$, then $e \in R{+}_{\pi_i}$.
(vi) if $e_{FL(B)} \in R{+}_{\pi_1/FL(B)}$ iff $e_{FL(B)} \in R{+}_{\pi_2/FL(B)}$ and $f(e_{FL(B)}) = [f(e)]$, then $e \in R{+}_{\pi_1}$ iff $e \in R{+}_{\pi_2}$.

(A) Let $e_{FL(B)} \in R{+}_{\pi_1/FL(B)}$ iff $e_{FL(B)} \in R{+}_{\pi_2/FL(B)}$. LtoR: Let $e \in R{+}_{\pi_1}$. Then there is some $(s, t) \in T_{\pi_1}$ by the execution rules. But then $([s], [t]) \in T_{\pi_1/FL(B)}$. Thus, by definition of filtrations, there is an $e_{FL(B)} = ([s], \pi_1, [t]) \in R{+}_{\pi_1/FL(B)}$. By assumption, $e_{FL(B)} = ([s], \pi_1, [t]) \in R{+}_{\pi_2/FL(B)}$. Thus, by the definition of filtrations, there is an $e \in R{+}_{\pi_2}$. The right to left direction simply inverts the order of π_1 and π_2 in the prior reasoning.

(vii) if $e_{FL(B)} \in R{+}_{\pi_1/FL(B)} \cap R{+}_{\pi_2/FL(B)}$ and $f(e_{FL(B)}) = [f(e)]$, then $e \in R{+}_{\pi_1} \cap R{+}_{\pi_2}$.

(viii) if $e_{FL(B)} \in R+_{\pi 1/FL(B)} - R+_{\pi 2/FL(B)}$ and $f(e_{FL(B)}) = [f(e)]$, then $e \in R+_{\pi_1} - R+_{\pi_2}$.

(ix) if $p_{FL(B)} \in P_{\pi_i/FL(B)}$ and $f(p_{FL(B)}) = [f(p)]$, then $p \in P_{\pi_i}$.

(x) if $p_{FL(B)} \in P_{\pi 1/FL(B)}$ iff $p_{FL(B)} \in P_{\pi 2/FL(B)}$ and $f(p_{FL(B)}) = [f(p)]$, then $p \in P_{\pi_1}$ iff $p \in P_{\pi_2}$.

(A) Let $p_{FL(B)} \in P_{\pi 1/FL(B)}$ iff $p_{FL(B)} \in P_{\pi 2/FL(B)}$. LtoR: Let $p \in P_{\pi_1}$. Then there is some $(s, t) \in T_{\pi_1}$ by the execution rules. But then $([s], [t]) \in T_{\pi 1/FL(B)}$ where $[s] = f(p_{FL(B)})$ and $[t] = l(p_{FL(B)})$. Thus, by definition of filtrations (Clauses 1, 5, and 6 by induction over proper immediate subprotocols), there is a $p_{FL(B)} \in P_{\pi 1/FL(B)}$. By assumption, $p_{FL(B)} \in P_{\pi 2/FL(B)}$. Thus, by the definition of filtrations (Clauses 1, 5, and 6 by induction over proper immediate subprotocol) and the execution rules, there is a $p \in R+_{\pi_2}$. The right to left direction simply inverts the order of π_1 and π_2 in the prior reasoning.

(xi) if $p_{FL(B)} \in P_{\pi 1/FL(B)} \cap P_{\pi 2/FL(B)}$ and $f(p_{FL(B)}) = [f(p)]$, then $p \in P_{\pi_1} \cap P_{\pi_2}$.

(A) Let $p_{FL(B)} \in P_{\pi 1/FL(B)} \cap P_{\pi 2/FL(B)}$. Then $([s], [t]) \in T_{\pi 1/FL(B)} \cap T_{\pi 2/FL(B)}$ where $[s] = f(p_{FL(B)})$ and $[t] = l(p_{FL(B)})$. Then there is some $(s, t) \in T_{\pi_1} \cap T_{\pi_2}$. Thus, $p \in P_{\pi_1} \cap P_{\pi_2}$.

(xii) if $p_{FL(B)} \in P_{\pi 1/FL(B)} - P_{\pi 2/FL(B)}$ and $f(p_{FL(B)}) = [f(p)]$, then $p \in P_{\pi_1} - P_{\pi_2}$.

(4) Let $v_s(R_x(\pi_1, \pi_2, \pi_3, \pi_4)) = 1$. The only formula of this form is syntactic replacement, which, for any $\pi_1 ... \pi_4$ is true independent of the structure of any EDG. Thus, the definition (Clause 1) of filtrations gives $v_{s/FL(B)}(R_x(\pi_1, \pi_2, \pi_3, \pi_4)) = 1$. Let $v_{s/FL(B)}(R_x(\pi_1, \pi_2, \pi_3, \pi_4)) = 1$. Then, for the same reason, the definition (Clause 1) of filtrations gives $v_s(R_x(\pi_1, \pi_2, \pi_3, \pi_4)) = 1$.

□

The Small Model Property and Miscellaneous Steps on the Way to Completeness

Conjecture 108 (Small Model Property for PLEN) If A is a verifiable formula of L_{PLEN}, then A is verifiable in an EDG, g, such that

$$|S| \leq 2^{(len(\pi_1)+...+len(\pi_4)+|FL_\pi([\pi_1]_x \top)|+...+|FL_\pi([\pi_4]_x \top)|)}$$

Proof (Sketch) Let A be a verifiable protocol-theoretic formula. Then there is an EDG, g, such that A is verifiable at some s: $v_s(A) = 1$ for some $s \in S$. Let $FL(A)$ be the Fischer-Ladner Closure of A. By the filtration lemma, there is a filtration of g by $FL(A)$, $g_{FL(A)}$, such that there is a $[s]$ such that $v_{s/FL(A)}(A) = 1$ (i.e., $[s]$

verifies A). By construction, there are no more states in $g_{FL(A)}$ than there are distinct truth assignments to formulae in $FL(A)$ (the definition of filtrations collapses these states into $[s]$). Thus, from FLL3:

(1) If A is a propositional formula, then $|FL(A)| \leq len(A)$.
(2) If $A = [\pi]_x B$, then $|FL_\pi([\pi]_x B)| \leq len(\pi)$.
(3) If $A = R_x(\pi_1, \pi_2)$, then

$$|FL(A)| \leq (len(\pi_1) + len(\pi_2) + |FL_\pi([\pi_1]_x \top)| + |FL_\pi([\pi_2]_x \top)|)$$

(4) If $A = R_x(\pi_1, \pi_2, \pi_3, \pi_4)$,

$$|FL(A)| \leq (len(\pi_1) + \ldots + len(\pi_4) + |FL_\pi([\pi_1]_x \top)| + \ldots + |FL_\pi([\pi_4]_x \top)|)$$

Thus, the number of distinct truth assignments to formulae in $FL(A)$ can be no more than:

$$2^{(len(\pi_1)+\ldots+len(\pi_4)+|FL_\pi([\pi_1]_x \top)|+\ldots+|FL_\pi([\pi_4]_x \top)|)}$$

□

Definition 109 Let Σ be a set of protocol-theoretic formulae of PLEN. Σ is a *maximal consistent set* iff it is the largest consistent set such that the addition of any further formulae would be inconsistent.

Conjecture 110 Let Σ be a set of protocol-theoretic formulae of PLEN. Then:

(1) Set Σ is consistent iff either $\Sigma \cup \{A\}$ is consistent or $\Sigma \cup \{\neg A\}$ is consistent.
(2) If Σ is consistent, then Σ is contained in a maximal consistent set.

Proof Mechanical from the definition of consistency and the truth rules for protocol-theoretic formulae (especially negation). □

Conjecture 111 Let s and s' be maximal consistent sets. Then the following conditions are equivalent:

(1) For all formulae, A, if $A \in s'$, then $\langle a \rangle_x A \in s$.
(2) For all formulae, A, if $[a]_x A \in s$, then $A \in s'$.

Proof From 1 to 2 and from 2 to 1.

(1) (1to2) Assume 1. Let $[a]_x A \in s$. Then, by the prior lemma, $\neg \langle a \rangle_x A \notin s$ (given the complementary definition of the truth rules for the protocol modalities). From the assumption that 1, $\neg A \notin s'$. Thus, from the prior lemma, $A \in s'$.
(2) (2to1) Assume 2. Let $A \in s'$. Then, by the prior lemma, $\neg A \notin s'$. From the assumption that 1, $[a]_x \neg A \notin s$. Thus, from the prior lemma, $\langle a \rangle_x A \in s$ (given the complementary definition of the truth rules for the protocol modalities).

□

Definition 112 (Nonstandard Models) Let g be an EDG, and let s and t be states in S. Let g_N be a nonstandard EDG iff

$$g_N = \left\langle N, \Pi_g, R, R_0, \Pi_g^+, R+, R+_0, T_{\pi/N}, c, v_N \right\rangle$$

where:

(1) N is the set of maximal consistent sets of protocol-theoretic formulae in PLEN.
(2) Abusing notation a bit, where $s_{/N} \in N$, let $v_{s/N}(A) = 1$ ($s_{/N}$ verifies A) iff $A \in s_{/N}$.
(3) $T_{a/N} = \left\{(s_{/N}, s'_{/N}) \mid \text{for all } A, \text{ if } A \in s'_{/N}, \text{then} \langle a \rangle_x A \in s_{/N}\right\}$
(4) The structure of g_N (as determined by R and $R+$) satisfies all of the structural conditions for g_N to be an EDG, except that $T_{\pi*}$ need not be the reflexive, transitive closure of T_π but only be a reflexive, transitive relation containing T_π that satisfies the axioms:

(a) $([\pi]_t[\pi*]_t A) \wedge A \equiv [\pi*]_t A$
(b) $(A \wedge [\pi*]_t(A \rightarrow [\pi]_t A)) \rightarrow [\pi*]_t A$

Conjecture 113 (PLEN Filtration Lemma for Nonstandard Models) Let g_N be a nonstandard model and let $g_{FL(B)}$ be the filtration of g_N with respect to B. Then:

(1) For all $A \in FL(B)$, $v_{s/N}(A) = 1$ iff $v_{s/FL(B)}(A) = 1$.
(2) For all $[\pi]_x A \in FL(B)$:

(a) if $(s_{/N}, s'_{/N}) \in T_\pi$, then $([s_{/N}], [s'_{/N}]) \in T_{\pi/FL(B)}$
(b) if $([s_{/N}], [s'_{/N}]) \in T_{\pi/FL(B)}$ and $v_{s/N}([\pi]_x A) = 1$, then $v_{s'/N}(A) = 1$.

(3) for all $R_x(\pi_1, \pi_2) \in FL(B)$:

(a) for each π_i,
(i) if $t \in T_{\pi_i}$ with $s_{/N} = f(t)$ and $s'_{/N} = l(t)$, then $t_{FL(B)} \in T_{\pi_i/FL(B)}$ and $[s_{/N}] = f(t_{FL(B)})$ and $[s'_{/N}] = l(t_{FL(B)})$.
(ii) if $e \in R+_{\pi_i}$ with $s_{/N} = f(e)$ and $s'_{/N} = l(e)$, then $e_{FL(B)} \in R+_{\pi_i/FL(B)}$ and $[s_{/N}] = f(e_{FL(B)})$ and $[s'_{/N}] = l(e_{FL(B)})$.
(iii) if $p \in P_{\pi_i}$ with $s_{/N} = f(p)$ and $s'_{/N} = l(p)$, then $p_{FL(B)} \in P_{\pi_i/FL(B)}$ and $[s_{/N}] = f(p_{FL(B)})$ and $[s'_{/N}] = l(p_{FL(B)})$.

(b) for each π_i, the following conditionals hold as well as their converses:
(i) if $t_{FL(B)} \in T_{\pi_i/FL(B)}$ and $[s_{/N}] = f(t_{FL(B)})$ and $[s'_{/N}] = l(t_{FL(B)})$, then:
(A) if $t_{FL(B)} \in T_{\pi_i/FL(B)}$ and $f(t_{FL(B)}) = [f(t)]$, then $t \in T_{\pi_i}$.
(B) if $t_{FL(B)} \in T_{\pi_1/FL(B)}$ iff $t_{FL(B)} \in T_{\pi_2/FL(B)}$ and $f(t_{FL(B)}) = [f(t)]$, then $t \in T_{\pi_1}$ iff $t \in T_{\pi_2}$.
(C) if $t_{FL(B)} \in T_{\pi_1/FL(B)} \cap T_{\pi_2/FL(B)}$ and $f(t_{FL(B)}) = [f(t)]$, then $t \in T_{\pi_1} \cap T_{\pi_2}$.

(D) if $t_{FL(B)} \in T_{\pi_1/FL(B)} - T_{\pi_2/FL(B)}$ and $f(t_{FL(B)}) = [f(t)]$, then $t \in T_{\pi_1} - T_{\pi_2}$.

(ii) if $e_{FL(B)} \in R{+}_{\pi_i/FL(B)}$ and $[s] = f(e_{FL(B)})$ and $[s'] = l(e_{FL(B)})$, then:

(A) if $e_{FL(B)} \in R{+}_{\pi_i/FL(B)}$ and $f(e_{FL(B)}) = [f(e)]$, then $e \in R{+}_{\pi_i}$.

(B) if $e_{FL(B)} \in R{+}_{\pi_1/FL(B)}$ iff $e_{FL(B)} \in R{+}_{\pi_2/FL(B)}$ and $f(e_{FL(B)}) = [f(e)]$, then $e \in R{+}_{\pi_1}$ iff $e \in R{+}_{\pi_2}$.

(C) if $e_{FL(B)} \in R{+}_{\pi_1/FL(B)} \cap R{+}_{\pi_2/FL(B)}$ and $f(e_{FL(B)}) = [f(e)]$, then $e \in R{+}_{\pi_1} \cap R{+}_{\pi_2}$.

(D) if $e_{FL(B)} \in R{+}_{\pi_1/FL(B)} - R{+}_{\pi_2/FL(B)}$ and $f(e_{FL(B)}) = [f(e])$, then $e \in R{+}_{\pi_1} - R{+}_{\pi_2}$.

(iii) if $p_{FL(B)} \in P_{\pi_i/FL(B)}$ and $[s] = f(p_{FL(B)})$ and $[s'] = l(p_{FL(B)})$, then:

(A) if $p_{FL(B)} \in P_{\pi_i/FL(B)}$ and $f(p_{FL(B)}) = f(p)$, then $p \in P_{\pi_i}$.

(B) if $p_{FL(B)} \in P_{\pi_i/FL(B)}$ iff $p_{FL(B)} \in P_{\pi_i/FL(B)}$ and $f(p_{FL(B)}) = [f(p])$, then $p \in P_{\pi_i}$ iff $p \in P_{\pi_i}$.

(C) if $p_{FL(B)} \in P_{\pi_1/FL(B)} \cap P_{\pi_2/FL(B)}$ and $f(p_{FL(B)}) = [f(p)]$, then $p \in P_{\pi_1} \cap P_{\pi_2}$.

(D) if $p_{FL(B)} \in P_{\pi_1/FL(B)} - P_{\pi_2/FL(B)}$ and $f(p_{FL(B)}) = [f(p)]$, then $p \in P_{\pi_1} - P_{\pi_2}$.

(4) For all $R_x(\pi_1, \pi_2, \pi_3, \pi_4) \in FL(B)$, $v_{s/N}(R_x(\pi_1, \pi_2, \pi_3, \pi_4)) = 1$ iff $v_{s/FL(B)}(R_x(\pi_1, \pi_2, \pi_3, \pi_4)) = 1$.

Proof (Sketch) The result follows mechanically by the argument for the Filtration Lemma, but modifying the reasoning for cases involving iteration, $*$. I omit those steps, as they are familiar from textbook presentations. The modifications introduced by the structure preserving clauses of the definition of filtrations should handle any complications introduced by the new path-theoretic and edge-theoretic expressions of PLEN. □

Proof of Completeness

Conjecture 114 (Verifiability Lemma) If A is consistent, then A is verifiable in some EDG, g.

Proof If A is consistent, then it's part of some maximal consistent set, s_N. Thus, s_N is a state in the nonstandard model, g_N, and s_N verifies A. Thus, by the PLEN Filtration Lemma for Nonstandard Models, there is a filtration of g_N by $FL(A)$, $g_{N/FL(A)}$, such that $[s_N]$ verifies A. From the Small Model Theorem, $g_{N/FL(A)}$ is finite. Thus, from the definition of filtrations in PLEN, there is a finite EDG, $g_{N/FL(A)}$ itself, in which A is verifiable. □

Definition 115 For any protocol-theoretic formula, A, is refutable iff $\perp$ is derivable from A.

Fact 116 *For any protocol-theoretic formula, A, is refutable iff* $\neg A$ *is derivable from no premises.*

Proof Immediate application of Negation Introduction. □

Definition 117 For any protocol-theoretic formula, A, A is consistent iff it is not possible to derive $\bot$ from A.

Fact 118 *For any protocol-theoretic formula, A, A is consistent iff* $\neg A$ *is not derivable iff* $A \to \bot$ *is not derivable iff* $A \to \bot$ *is refutable iff* $\neg A$ *is refutable.*

Proof Immediate from definitions of refutability and consistency and the prior Fact. □

Conjecture 119 (Completeness Theorem) If $X \models_{PLEN} A$, then $X \vdash_{PLEN} A$.

Proof Completeness for formulae (Weak Completeness) is proven using the Verifiability Lemma, then it's extended to logical consequence (Strong Completeness) by the Deduction Theorem.

(1) $\models_{PLEN} A$ iff $\neg A$ is unverifiable
   (a) Immediate from definition of verification; valuations are classical.
(2) $\models_{PLEN} A$ iff $\neg A$ is not consistent.
   (i) Verifiability Lemma
(3) $\models_{PLEN} A$ iff $\neg A$ is refutable.
(4) $\models_{PLEN} A$ iff $\vdash_{PLEN} A$
   (a) If $\neg A$ is refutable, then $\neg A \vdash_{PLEN} \bot$.
   (b) Thus, $\models_{PLEN} A$ iff $\neg A \vdash_{PLEN} \bot$.
   (c) Thus, $\models_{PLEN} A$ iff $\vdash_{PLEN} \neg\neg A$
      (i) By Proof by Contradiction
   (d) Thus, $\models_{PLEN} A$ iff $\vdash_{PLEN} A$
      (i) By DNE
(5) Extend completeness for formulae to completeness for logical consequence by Deduction Theorem.
   (a) Let $X \models_{PLEN} A$ where $X = \{B_1, ..., B_n\}$. Two elementary proofs are available.
   (b) Proof 1:
      (i) $\emptyset \models_{PLEN} B_1 \to (...(B_n \to A))$ by Deduction Theorem
      (ii) $\vdash_{PLEN} B_1 \to (...(B_n \to A))$ by Weak Completeness
      (iii) $X \vdash_{PLEN} A$ by Deduction Theorem.
   (c) Proof 2:
      (i) $\models_{PLEN} X^\wedge \to A$ by the soundness of conditional proof
         (A) Logical consequence is preservation of verification at all states of all EDGs. So, for any g and any s in g, if X is all verified at s, then A is verified at s. Suppose that $X^\wedge$ is verified at s. Then,

immediately, all $B_i \in X$ are verified at s, and it follows that A is verified at s.

(ii) $\vdash_{PLEN} X^\wedge \rightarrow A$ by Weak Completeness

(iii) $X^\wedge \vdash_{PLEN} A$ by Deduction Theorem.

(iv) For any set of formulae, X, and A, $\{X^\wedge\} \vdash_{PLEN} A$ iff $X \vdash_{PLEN} A$, since PLEN does not employ a substructural logic. Thus, immediately, $X \vdash_{PLEN} A$.

□

Definition 120 (Segerberg Deductive System) The rules for Conditional Elimination and Generalization plus the following axiom schemata:

(Segerberg Deductive System)
all propositional tautologies (or enough of them),
$\langle\pi\rangle_t \perp \equiv \perp$
$\langle\pi_1;\pi_2\rangle_t A \equiv \langle\pi_1\rangle_t \langle\pi_2\rangle_t A$
$\langle\pi_1 \cup \pi_2\rangle_t A \equiv (\langle\pi_1\rangle_t A \vee \langle\pi_2\rangle_t A)$
$\langle\pi_1\rangle_t (A \vee B) \equiv (\langle\pi_1\rangle_t A \vee \langle\pi_1\rangle_t B)$
$\langle\pi*\rangle_t A \equiv (\langle\pi\rangle_t \langle\pi*\rangle_t A) \vee A$
$(A \wedge \langle\pi*\rangle_t (A \rightarrow \langle\pi\rangle_t A)) \rightarrow \langle\pi*\rangle_t A$

Definition 121 (Segerberg Deductive System PLEN Counterpart) The rules for Conditional Elimination and Generalization plus the following axiom schemata:

(Segerberg Deductive System)
all propositional tautologies (or enough of them),
$\langle\pi\rangle_t \perp \equiv \perp$
$\langle\pi_1;\pi_2\rangle_t A \equiv \langle\pi_1\rangle_t \langle\pi_2\rangle_t A$
$\langle\pi_1 \cup \pi_2\rangle_t A \rightarrow (\langle\pi_1\rangle_t A \vee \langle\pi_2\rangle_t A)$
$\langle\pi_1\rangle_t (A \vee B) \equiv (\langle\pi_1\rangle_t A \vee \langle\pi_1\rangle_t B)$
$\langle\pi*\rangle_t A \equiv (\langle\pi\rangle_t \langle\pi*\rangle_t A) \vee A$
$(A \wedge \langle\pi*\rangle_t (A \rightarrow \langle\pi\rangle_t A)) \rightarrow \langle\pi*\rangle_t A$

Conjecture 122 (Segerberg Completeness) The counterpart system of the Segerberg Deductive System in PLEN is complete with respect to the semantics of PLEN.

Proof The elementary (Kozen and Parikh) proof can be directly lifted for the shared control operators; the Small Model Theorem holds for the PLEN semantics, and the case by case reasoning applies unmodified except for the case of choice, which only works in one direction. □

Decidability, Complexity, and Compactness

Decidability Conjecture

Definition 123 The verifiability problem of PLEN is decidable iff, for any $A \in L_{PLEN}$, there is an effective procedure, M, for determining whether there is any EDG, g, in which A is verifiable.

Conjecture 124 (Decidability (PDL)) PDL is decidable by the Small Model Theorem for PDL, which gives an effective procedure for determining whether a given formula of PDL is satisfiable.

Definition 125 (Naive PLEN Verification Procedure) Let $A \in L_{PLEN}$. To determine whether A is verifiable:

(1) Construct all finite EDGs such that $|S| =$

$$2^{(len(\pi_1)+...+len(\pi_4)+|FL_\pi([\pi_1]_x\top)|+...+|FL_\pi([\pi_4]_x\top)|)}$$

(2) Check each state for verification of A.

Conjecture 126 The Naive PLEN Verification Procedure is effective.

Proof (Sketch) By the Small Model Theorem for PLEN, if A is verifiable, then it's verifiable in one of the EDGs constructed by Step 1. Step 2 executes an exhaustive search of all states in all such models. Thus, the procedure terminates with an affirmative verdict on A or it exhausts the set of EDGs constructed by Step 1, and by the Small Model Theorem, A is not verifiable. □

Conjecture 127 (Decidability (PLEN)) The verifiability problem of PLEN is decidable.

Proof (Sketch) The effectiveness of the Naive PLEN Verification Procedure suffices immediately. □

Complexity Conjecture

Conjecture 128 The verifiability problem for PLEN is at least EXPTIME-complete.

Proof (Sketch) The satisfiability problem for PDL is EXPTIME-complete, and PLEN is at least as complex. □

Compactness

Conjecture 129 PLEN is non-compact.

Proof (Sketch) PDL is non-compact, and PLEN contains the formulae that stand as witnesses for the non-compactness of PDL. □

References

1. Alchourron, C., P. Gardenfors, and D. Makinson. 1985. On the logic of theory change: Partial meet contraction and revision functions. *The Journal of Symbolic Logic* 50(2):510–530.
2. Baltag, A., and L.S. Moss. 2004. Logics for epistemic programs. *Synthese* 139:165–224.
3. Baltag, A., and S. Smets. 2008. Probabilistic dynamic belief revision. *Synthese* 165(2):179–202.
4. Beall, J. 2011. Multiple-conclusion LP and default classicality. *Review of Symbolic Logic* 4(2):326–336.
5. Beall, J. 2015. Free of detachment: Logic, rationality, and gluts. *Noûs* 49(2):410–423.
6. Beall, J.C., and G. Restall. 2005. *Logical pluralism*. Oxford University Press.
7. Evans, J., and K.E. Stanovich. 2013. Dual-process theories of higher cognition advancing the debate. *Perspectives on Psychological Science* 8(3):223–241.
8. Fagin, R., J. Halpern, Y. Moses, and M. Vardi. 2004. *Reasoning about knowledge*. MIT Press.
9. Field, H. 2009a. Epistemology without metaphysics. *Philosophical Studies* 143(2):249–290.
10. Field, H. 2009b. Pluralism in logic. *Review of Symbolic Logic* 2(2):342–359.
11. Gardenfors, P. 2008. *Knowledge in flux*. Reprint. Original: 1988. London: College Publications.
12. Geach, P. 1965. Assertion. *Philosophical Review* 74:449–465.
13. Girard, P., J. Seligman, and F. Liu. 2012. General dynamic dynamic logic. In *Advances in modal logic,* Vol 9, 239–260, ed. T. Bolander, T. Braüner, S. Ghilardi, and L. Moss. CSLI Publications.
14. Goldman, A. 1986. *Epistemology and cognition*. Harvard University Press.
15. Harel, D., D. Kozen, and J. Tiuryn. 2000. *Dynamic logic*. MIT Press.
16. Hakli, R., and S. Negri. 2012. Does the deduction theorem fail for modal logic? *Synthese* 187(3):849–867.
17. Hansson, S.O. 2017. Logic of belief revision. *The Stanford Encyclopedia of Philosophy* (Winter 2017 Edition), ed. Edward N. Zalta, forthcoming. https://plato.stanford.edu/archives/win2017/entries/logic-belief-revision/.
18. Hoshi, T. 2009. *Epistemic dynamics and protocol information*. ILLC Dissertation Series
19. Hoshi, T., and A. Yap. 2009. Dynamic epistemic logic with branching temporal structures. *Synthese* 169(2):259–281.
20. Jorgensen, J. 1937. Imperatives and logic. *Erkenntnis* 7:288–296.
21. Kahneman, D., P. Slovic, A. Tversky, Eds. 1982. *Judgment under uncertainty: Heuristics and biases*. Cambridge University Press.
22. Lam, B. Unpublished. The Dynamic Foundations of Epistemic Rationality. Ph.D Dissertation: Vassar College.
23. Leitgeb, H., and K. Segerberg. 2007. Dynamic doxastic logic: why, how, and where to? *Synthese* 155(2):167–190.
24. Lindström, S., and W. Rabinowicz. 1999. DDL unlimited: Dynamic doxastic logic for introspective agents. *Erkenntnis* 50(2-3):353–385.
25. Pacuit, E., and S. Simon. 2010. Reasoning with protocols under imperfect information. *Review of Symbolic Logic* 4(3):412–444.
26. Parikh, R. 2002. Social software. *Synthese* 132(3):187–211.

27. Pavese, C. 2015. Practical senses. *Philosophers' Imprint* 15:1–25.
28. Pollock. 2009. A recursive semantics for defeasible reasoning. In *Argumentation in artificial intelligence,* 173–197, ed. Simari G., and I. Rahwan. Boston, MA: Springer.
29. Schafer, K. 2014. Doxastic planning and epistemic internalism. *Synthese* 191(12):2571–2591.
30. Schmidt, R.A. & D. Tishkovsky. 2007. On combinations of propositional dynamic logic and doxastic modal logics. *Journal of Logic, Language and Information* 17(1):109–129.
31. Stanovich, K.E. 2005. *The Robot's rebellion: Finding meaning in the age of Darwin*. University of Chicago Press.
32. Stenning, K., and M. van Lambalgen. 2008. *Human reasoning and cognitive science*. Boston, USA: MIT Press.
33. van Benthem. J. 1996. *Exploring logical dynamics*. Stanford: CSLI Publications.
34. van Ditmarsch, H., Halpern, J., Van Der Hoek, W., and B. Kooi. 2007. *Dynamic epistemic logic*. Springer.
35. van Eijck, J., and Y. Wang. 2008. Propositional dynamic logic as a logic of belief revision. In: *Logic, language, information and computation,* WoLLIC 2008, eds. Hodges, W., de Queiroz, R. Lecture Notes in Computer Science, vol 5110. Berlin, Heidelberg: Springer.
36. van Lee, H.S., R.K. Rendsvig, and S. van Wijk. 2019. Intensional protocols for dynamic epistemic logic. *Journal of Philosophical Logic* 48(6):1077–1118.
37. Wang, Y. 2010. *Epistemic modeling and protocol dynamics*. Ph.D. Dissertation.

Chapter 5
The Revised PLEN Framework I: Protocol Equivalence (Useful Lemmata)

Abstract The prior chapter defined the formal system, PLEN. This and the next two chapters develop the protocol logic of PLEN by introducing specialized machinery for more closely analyzing the procedural structure of protocols and for analyzing protocol equivalence relations. The central constructions are protocol graphs, which are epistemic directed graphs that are mapped (in isomorphism classes) by novel "structural display" rules to each regular protocol. Protocol graphs are central to PLEN because they bind the syntactic form of protocols more closely than the execution rules to the graph-theoretic properties of epistemic directed graphs (EDGs), thereby providing a link between the syntax of protocols and well-known process equivalence relations (bisimulation, isomorphism) over dynamic process models (Kripke frames, graphs, LTSs, EDGs). The result is a novel and fine-grained protocol equivalence relation defined in terms of protocol graph isomorphism. The main work of this chapter is, first, presenting the core technical revisions that empower PLEN to analyze protocol equivalence, and then, second, deriving the main lemmata for the core results of the system which are derived in the next chapter.

5.1 Revising the PLEN Framework

Chapters 5, 6, and 7 comprise a dense trio of chapters that elaborate on PLEN in order to further develop its logic of protocols (hence, the logic of epistemic norms in PLEN). The core additions of these chapters are specialized constructions that deepen the analysis of the procedural properties of protocols and undergird a range of protocol equivalence relations not visible in the initial iteration of the system. These chapters discuss a suite of results that codify the properties of these structures, and then build a theory comprising valid replacement and implication rules codifying the logic of the resulting range of protocol equivalence relations. Ultimately, a sound derivation system is built atop the existing derivation system of PLEN that encompasses this deeper analysis, and provides a framework in which to analyze the logics of different protocol equivalence relations, each of which encodes an overall account or approach to modeling epistemic norms and dynamics. As a

R. Jenkins, *A Protocol-theoretic Framework for the Logic of Epistemic Norms*, Synthese Library 467, https://doi.org/10.1007/978-3-031-08597-0_5

nice bonus, this revised derivation system has a (plausibly) strongly complete and decidable fragment.

Of the added constructions, the most important are protocol graphs, which are specialized EDGs with graph-theoretic structures that are closely bound to the syntactic forms of protocols. These structures provide analyses of procedural properties that are invisible on the basis of the execution rules alone. In particular, they provide a finer-grained analysis of the procedural structures of protocols, i.e., protocol graphs can be used to distinguish (non-identical) protocols that dictate different sequences of choices among actions while procedural execution rules elide this distinction. Protocol graphs then provide a link between protocols, as syntactic objects, and the familiar range of process equivalence relations from the study of logical dynamics and modal transition systems. These process equivalence relations occupy different levels of granularity, and process equivalences between protocol graphs can then be used to define protocol equivalence relations of different levels of granularity. Accordingly, the plan of the next three chapters is straightforward.

First, I briefly motivate the focus on protocol equivalence relations. The main idea is simply reiterating some of the lessons of Part I—analyzing protocol equivalence relations is of paramount importance to the logic of protocols. Second, I revise the PLEN Framework by defining protocol graphs. This requires identifying a number of semantic and syntactic relations among protocols, substructures of EDGs, operations on EDGs, and process equivalence relations. On the basis of all of the foregoing things, protocol graphs are defined, and then the range of protocol equivalence relations in PLEN is defined. The theory of protocol equivalence is then articulated in the form of a system of "meta-level rules" and their object-level counterparts, which are incorporated into the derivation system of PLEN by revising the protocol-theoretic language from L_{PLEN} to L^R_{PLEN}. The chapter closes with an analysis of some important properties of protocols and protocol graphs, drawing attention to the properties of the proper immediate subprotocol relation, the structural uniqueness of protocol graphs, and some relations between protocol graphs and the process equivalence of isomorphism. This suffices for a partial proof of the soundness of the revised derivation system, and, more important, all of the results in this chapter are used, in turn, for understanding and deriving the results of Chap. 6.

5.2 Protocol Equivalence Relations 0

The logical analysis of equivalence relations is central to any framework for representing and reasoning about a domain of objects. The replacement of equivalents in a context (i.e., deriving the properties of x from premises about what properties are possessed by x's equivalents) is as fundamental and important an inferential move as detachment of implication (intersubstitution of identity being a special case of it), and the logical analysis of equivalence relations codifies this move. This was among the upshots of the arguments in Chap. 3. To summarize, a logic that doesn't empower us to tell, in a given context, whether and how thing X is different from thing Y or what follows if they are equivalent is a logic that lacks the inferential power to do much of interest.

The *syntactic replacement operations* on protocols and protocol-theoretic formulae accordingly frame the theory of protocol equivalence developed in PLEN; syntactic replacement in the context of protocols and protocol-theoretic formulae is the core syntactic criterion of all protocol equivalence relations. If I'm right in the prior paragraph, equivalence relations are governed by valid replacement or intersubstitution in various contexts; equivalent protocols must be equivalent with respect to some set of properties. Thus, the valid replacement rules that substitute equivalent protocols across contexts (that codify the properties of protocols) codify the power of equivalence relations to preserve protocol-theoretic properties and determine the strictness, fine-grainedness, and informativeness of equivalence relations. These properties determine the equivalence relation's quality as an individuation criterion. The analysis of systems of equivalence relations thus proceeds by enumerating the properties preserved by various classes of instances of syntactic replacement. The protocol-theoretic formulae defined in PLEN codify dynamic and procedural properties of protocols, the modal or dynamic properties of protocols (via modal operators indexed to each protocol), and pre and postconditions of protocols. Protocol equivalences can thus be formalized in PLEN by the valid syntactic replacement rules, i.e., the rules governing syntactic replacements that preserve: execution conditions, properties and relations of protocols (mappings to specialized constructions and properties of specialized constructions), and semantic values of classes of formulae.

Of course, distinct equivalence relations may have different *granularities*, in the sense that they support replacement across more or less inclusive classes of contexts in a language. Truth-functional equivalents are intersubstitutable in extensional contexts but not intensional contexts. Intensional equivalents are intersubstitutable in extensional and intensional contexts but not hyperintensional contexts (contexts for which replacement of logical equivalents does not preserve semantic value). Hyperintensional equivalents are intersubstitutable in extensional, intensional, and hyperintensional contexts but not in contexts of identity.[1] Thus, a systematic theory

[1] This may appear incorrect at first sight. After all, attitude reports seem to be hyperintensional precisely because identical objects seem not to be intersubstitutable salva veritate in attitude reporting contexts. For instance, Hal Jordan is the Green Lantern of Sector 2814, but while it's a priori or necessary (or...) that Hal Jordan is Hal Jordan, it's (under some comic book continuities) quite a discovery that Hal Jordan is the Green lantern of Sector 2814, and one can coherently believe the former without believing the latter. While it will obviously be inadequate to address this in a footnote, for the purposes of this book, let's say that it's the substitution of semantic items instead of syntactic items that generates this appearance. It's obvious that the name "Hal Jordan" and the descriptive expression "Green lantern of Sector 2814" are not identical syntactic items (mutatis mutandis the relevant sentences), and it's plausible that, were they identical, they would be intersubstitutable in attitude reports just as "Hal Jordan" and "Hal Jordan" are. Generally, it is syntactic, representational, or semantic information carrying constructions such as names, expressions, or sentences (etc.) that are intersubstituted in any context of a logically regimented

of equivalence for a class of objects requires examining a range of equivalence relations of varying granularity. This chapter builds a theory of protocol equivalence in PLEN that examines a range of protocol equivalence relations. This is the core groundwork for further developing the logic of protocols in PLEN, and so the core groundwork for developing the logic of epistemic norms in PLEN.

In PLEN, epistemic processes are analyzed by means of procedural paths, concatenated sequences of edges, each representing an epistemic action. Procedural epistemic norms are conceptualized as abstract procedural instructions that define restrictions on epistemic dynamics, and so represented with protocols. For each protocol, procedural execution rules define restrictions on epistemic processes, binary execution rules define restrictions on epistemic transitions, and ternary execution rules define unique restrictions on what can be conceptualized as the abstract action of executing a specific norm.

PLEN, as defined in Chap. 4, supports local execution-theoretic protocol equivalence relations or equivalences between protocols defined in terms of relations between their execution sets at a given state in a given EDG. Execution-theoretic protocol equivalence (hence, norm equivalence) relations possess varying levels of granularity. Protocol equivalence, $\leftrightarrow_r$, defined in terms of ternary relations (edge equivalence) in PLEN, amounts to identity. Protocol equivalence, $\leftrightarrow_t$, defined in terms of binary relations (transitional equivalence) is much less fine-grained, holding between protocols that are executable at the same states and that converge on the same terminal states. Protocol equivalence, $\leftrightarrow_p$, defined in terms of procedural paths (procedural equivalence) is more fine-grained than transitional equivalence, but falls short of implying protocol identity.

This is a good first step, but there is a wide gap between procedural equivalence and edge equivalence in terms of granularity. What can fill this gap? A first pass can be developed in terms of *non-local execution-theoretic protocol equivalence relations*.

5.3 Protocol Equivalence Relations I: Non-local Execution-Theoretic Protocol Equivalence

Protocol equivalence relations are defined at the meta-level, first. By this, I mean that I present protocol equivalence relations that are defined in such a way that they are not expressed in the object language of PLEN, L_{PLEN}, and they are not assigned truth values at individual states in individual EDGs. Rather, these equivalences are expressed in the meta-language notation of PLEN and defined in terms of meta-level properties of protocols, properties themselves defined in terms of the syntax of protocols, quantification over all EDGs, or quantification over all protocol graphs.

language, not the referents or truthmakers thereof. The sequence of claims in the main text can thus be interpreted harmlessly over syntactic items.

The meta-level theory of protocol equivalence, then, is defined in terms of rules governing these meta-level protocol equivalence relations. These meta-level rules are deployed for the same purpose as the string-algebraic form of PLEN's syntax—utility in stating and proving results, and each meta-level protocol equivalence relation and rule will be associated with object-level counterparts, defined over restricted subsets of the structures over which meta-level equivalences and rules are defined. This enables interpretation of protocol equivalence formulae at states of EDGs, which in turn provides a way of moving the meta-level protocol equivalences into the object language of PLEN, and thereby building protocol equivalence rules into the derivation system of PLEN.

More explicitly, the idea is that there are object-level counterparts to each meta-level equivalence relation in the following sense:

Definition 130 (Object-Level Counterparts) Let $\equiv_M$ be a meta-level protocol equivalence, and let $\equiv_O$ be an object-level protocol equivalence. Then $\equiv_O$ is the *object-level counterpart* of $\equiv_M$ iff for arbitrary protocols, π_1 and π_2, ($\models_{PLEN}$ $(\pi_1 \equiv_O \pi_2)$ iff $\pi_1 \equiv_M \pi_2$ (i.e., $\pi_1 \equiv_M \pi_2$ is true at the meta-level)).

This two-level format will be applied to the rules that codify the properties of protocol equivalences, as well; there will be sets of rules that characterize the meta-level protocol equivalence relations and sets of rules that characterize the object-level protocol equivalence relations. These sets of rules will be articulated in the meta language and object language, respectively.

5.3.1 Execution-Theoretic Protocol Equivalence: Meta-Language

Claim 131 The following protocol equivalence relations are all execution-theoretic protocol equivalence relations.

Definition 132 (Relation-Theoretic Protocol Equivalences)

(1) (Binary/Transitional Equivalence) Protocols π and π' are *transitionally equivalent* $(\pi \equiv_t \pi')$ iff for all EDGs, g, $T_\pi = T_{\pi'}$.

(2) (Ternary/Edge Equivalence) Protocols π and π' are *edge equivalent* $(\pi \equiv_r \pi')$ iff for all EDGs, g, $R+_\pi = R+_{\pi'}$.

Definition 133 (Path-Theoretic Protocol Equivalences)

(1) (Procedural/Path Equivalence) Protocols π and π' are *procedurally equivalent* or *path-theoretically equivalent* $(\pi \equiv_p \pi')$ iff for all EDGs, g, $P_\pi = P_{\pi'}$.

(2) (Path Correspondence Equivalence) Protocols π and π' are *correspondence equivalent* $(\pi \equiv_c \pi')$ iff for all EDGs, g:

(a) if $p \in P_\pi$, then there is a $p' \in P_{\pi'}$ such that $p \rightleftharpoons p'$, and
(b) if $p' \in P_\pi$, then there is a $p \in P_{\pi'}$ such that $p \rightleftharpoons p'$.

(3) (Full Path Correspondence Equivalence) Protocols π and π' are *correspondence equivalent* ($\pi \equiv_{fc} \pi'$) iff for all EDGs, g:

(a) if $p \in P_\pi$, then there is a $p' \in P_{\pi'}$ such that $p \rightleftarrows p'$, and
(b) if $p' \in P_\pi$, then there is a $p \in P_{\pi'}$ such that $p \rightleftarrows p'$.

5.3.2 Execution-Theoretic Protocol Equivalence: Object-Language Counterparts

Definition 134 (Relation-Theoretic Protocol Equivalences)

(1) (Binary/Transitional Equivalence) For an EDG, g, protocols π and π' are *transitionally equivalent at s* ($\pi \equiv_t^s \pi'$) iff for all $s' \in S$ such that there is any $p \in P_g$ with $f(p) = s$ and $l(p) = s'$, for all $t \in T_g$, if $s' = f(t)$, then $t \in T_\pi$ iff $t \in T_{\pi'}$.
(2) (Ternary/Edge Equivalence) For an EDG, g, protocols π and π' are *edge equivalent at s* ($\pi \equiv_r^s \pi'$) iff for all $s' \in S$ such that there is any $p \in P_g$ with $f(p) = s$ and $l(p) = s'$, for all $e \in R{+}_0$ such that $s = f(e)$ and $s' = l(e)$, $e \in R{+}_\pi$ iff $e \in R{+}_{\pi'}$.

Definition 135 (Path-Theoretic Protocol Equivalences)

(1) (Procedural/Path Equivalence) Protocols π and π' are *procedurally equivalent* or *path-theoretically equivalent at s* ($\pi \equiv_p^s \pi'$) iff for all $s' \in S$ such that there is any $p \in P_g$ with $f(p) = s$ and $l(p) = s'$, for all $p' \in P_g$, if $s' = f(p')$, then $p' \in P_\pi$ iff $p' \in P_{\pi'}$.
(2) (Path Correspondence Equivalence) For an EDG, g, protocols π and π' are *correspondence equivalent at s* ($\pi \equiv_c^s \pi'$) iff for all $s' \in S$ such that there is any $p \in P_g$ with $f(p) = s$ and $l(p) = s'$, for all p_1 and p_2 in g:

(a) if $p_1 \in P_\pi$ and $s' = f(p_1)$, then there is a $p_2 \in P_{\pi'}$ such that $s' = f(p_2)$ and $p_1 \rightleftharpoons p_2$,
(b) if $p_2 \in P_\pi$ and $s' = f(p_2)$, then there is a $p_1 \in P_{\pi'}$ such that $s' = f(p_1)$ and $p_1 \rightleftharpoons p_2$.

(3) (Full Path Correspondence Equivalence) For an EDG, g, protocols π and π' are *correspondence equivalent at s* ($\pi \equiv_{fc}^s \pi'$) iff for all $s' \in S$ such that there is any $p \in P_g$ with $f(p) = s$ and $l(p) = s'$, for all p_1 and p_2 in g:

(a) if $p_1 \in P_\pi$ and $s' = f(p_1)$, then there is a $p_2 \in P_{\pi'}$ such that $s' = f(p_2)$ and $p_1 \rightleftarrows p_2$,
(b) if $p_2 \in P_\pi$ and $s' = f(p_2)$, then there is a $p_1 \in P_{\pi'}$ such that $s' = f(p_1)$ and $p_1 \rightleftarrows p_2$.

Remark 136 Let $\Box_{S4}$, $\Box_{S5}$, and $\Box_{PLEN}$ refer, respectively, to the necessity operators in $S4$, $S5$, and $PLEN$. For systems $S4$ and $S5$, the behavior of the box corresponds to conditions on frames; the accessibility relation of $S4$ is reflexive and transitive, while the accessibility relation on $S5$ is an equivalence relation. In PLEN, the box is just a universal necessity. In virtue of the restriction rules defining EDGs, T_g always defines a reflexive, transitive relation, as does $[\pi *]_t$. Thus, where $\Pi_g = \{a_1, \ldots, a_n\}$, $[(a_1 \cup \ldots \cup a_n)*]_t(\pi \equiv_x^s \pi')$ is equivalent to $\Box_{S4}(\pi \leftrightarrow_x \pi')$ is equivalent to $[\pi *]_t(\pi \leftrightarrow_x \pi')$, for each object level counterpart equivalence relation. The stronger $\Box(\pi \leftrightarrow_x \pi')$ follows from the truth of the meta-level $\pi \equiv_x \pi$, as does $\Box_{S4}(\pi \leftrightarrow_x \pi')$, though the converses fail.

This is a stronger step toward an interesting analysis of protocol equivalence than the last. The meta-level protocol relations are equivalent to logically true, necessary object-level protocol equivalences. Nonetheless, there is quite a gap in granularity remaining between these protocol equivalence relations and protocol identity.[2] What can fill this second gap? *Graph-theoretic protocol equivalence relations*—protocol equivalence relations defined in terms of process equivalences between protocol graphs. However, as that kernel of an explication suggests, there is a lot of machinery underlying the definition and examination of these relations, including the signature constructions of PLEN, protocol graphs. Let me present this machinery incrementally, eventually building up to a revision of PLEN.

5.4 Preliminaries

Remark 137 The core novel construction of PLEN is the *protocol graph*, a specialized form of EDG that bears graph-theoretic properties that closely match the syntax of the regular protocols that a given protocol graph is mapped to by the rules that define it. But these are highly specialized constructions, and there are many features of EDGs and protocols that must be pointed out, as well as additional novel constructions and operations on the structures of PLEN that need to be defined before protocol graphs can be brought into focus. This is what occupies us in Sect. 5.4. In this section, I frontload the presentation of the revised system with definitions of the specialized constructions that the definitions and pivotal results refer to.

First, recall the important syntactic constructions: regular protocols, proper immediate subprotocols, and syntactic replacement operations.[3] Second, add the following definitions of specialized syntactic and semantic structures:

---

[2] This can be seen in the replacement rules and in hyperintensional contexts. See Sects. 6.4 and 6.5 of the next chapter.

[3] Let the syntactic replacement operations be suitably revised for the revisions of the protocol and protocol-theoretic languages introduced in this chapter.

Definition 138 (Syntactic Substructures)

(1) (Atomic Sequence) Let $At(\pi)*$ be the set of all sequences of atomics as they appear in π. If $\pi = x_1 \frown \ldots \frown x_n$, then let $\langle a_{1+k}, \ldots, a_m \rangle$ (for some $k \in \mathbb{N}$) be the *maximal atomic sequence* of π iff:

(a) $m \leq n$,
(b) $a_{1+k} = x_{1+k}$,
(c) for every a_i where $i \in \langle 1+k, \ldots, m \rangle$, there is some x_j where $j \in \langle 1, \ldots, n \rangle$ such that $a_i = x_j$,
(d) for all a_i and x_j, if $a_i = x_j$, then for all $a_{i'}$ and $x_{j'}$, $i \geq i'$ iff $j \geq j'$,
(e) each $a_i \in \Pi_0$, and for all x_j not in $At(\pi)*$, $x_j \in Prot \cup Prot+$.

(2) (Control Sequence) Denote the set of all control operators in π as $\circ(\pi)$. Let $\circ(\pi)*$ be the set of all sequences of control operators as they appear in π. If $\pi = x_1 \frown \ldots \frown x_n$, then let $\langle \circ_1, \ldots, \circ_m \rangle$ be the *maximal control sequence* of π iff:

(a) $m \leq n$,
(b) $\circ_1 = x_1$,
(c) for every $\circ_i$ where $i \in \langle 1, \ldots, m \rangle$, there is some x_j where $j \in \langle 1, \ldots, n \rangle$ such that $\circ_i = x_j$,
(d) for all $\circ_i$ and x_j, if $\circ_i = x_j$, then for all $\circ_{i'}$ and $x_{j'}$, $i \geq i'$ iff $j \geq j'$,
(e) each $\circ_i \in Prot \cup Prot+$, and for all x_j not in $\circ(\pi)*$, $x_j \in At(\pi)*$.

(3) (Embedding) Let X_π be an arbitrary protocol such that $\pi \in Sub(X_\pi)$. Let $X_{\pi'}$ be an arbitrary protocol such that $\pi' \in Sub(X_{\pi'})$. Let $\circ_i \in \circ(X_\pi) \cap \circ(X_{\pi'})$ such that $\circ_i$ operates on π or a further subprotocol of X_π that contains π, and $\circ_i$ operates on π' or a further subprotocol of $X_{\pi'}$ that contains π'. Then: *π and π' are embedded at equal depth in X_π and $X_{\pi'}$ (respectively) with respect to* $\circ_i$ iff

(a) $X_\pi \frown \circ_1 \frown \ldots \frown \circ_i \frown \ldots \frown \pi \ldots$
(b) $X_{\pi'} \frown \circ'_1 \frown \ldots \frown \circ_i \frown \ldots \frown \pi' \ldots$
(c) $\langle \circ_1, \ldots, \circ_i \rangle \in \circ(X_\pi)*$
(d) $\langle \circ'_1, \ldots, \circ_i \rangle \in \circ(X_{\pi'})*$
(e) $len(\langle \circ_1, \ldots, \circ_i \rangle) = len(\langle \circ'_1, \ldots, \circ_i \rangle)$

Example 139 So, for instance, $(a; (b; c)) \cap (d; (e \cup (f; g)))$ is a protocol in which $(b; c)$ and $(e \cup (f; g))$ are embedded at equal depth with respect to $\cap$, as are b and e, b and $(f; g)$, c and e, and c and $(f; g)$.

(4) (Main Control Operator Preserving Reductions) Let π_i be a *main control operator preserving reduction* of π iff $\pi_i \neq \pi$, $At(\pi_i) \subseteq At(\pi)$, $MC(\pi_i) = MC(\pi)$, $\circ(\pi_i) \subseteq \circ(\pi)$, $len(\pi_i) < len(\pi)$, and the order of control operators in $\circ(\pi)*$ is preserved in $\circ(\pi_i)*$ ($\circ(\pi_i)* \subseteq \circ(\pi)*$). Denote the set of all main control operator preserving reductions of π with $Reduct(\pi)$. Abbreviate "main control operator preserving reduction of π'" to "reduct of π'".

Definition 140 (Important Protocol Constructions) Assume the definitions of PLEN. The following definitions identify properties of protocols and relations among protocols on the basis of their syntactic and semantic properties. The following definitions are operative in the forthcoming proofs:

1. (Preconditions) For any protocol, π, let $Pre(\pi) = \{A \in L^R_{PLEN} \mid \langle\pi\rangle_p \top \rightarrow A\}$ where L^R_{PLEN} is defined as in Sect. 5.6.
   (a) This is the set of preconditions of π-the formulae that must be true if π is executable. This can be understood relative to an EDG or set thereof, which will be explicitly noted. By default, it should be understood with respect to all EDGs. That is, $Pre(\pi)$, without further elaboration, should always be understood as the set of universal preconditions of π, which, for most protocols, will be a rather small set (e.g., the preconditions of non-deterministic choices will contain the formulae encoding the executability of their proper immediate subprotocols and elaborations on these).
2. (Dictation) If π is executable in any EDG, g, then:
   (a) π *dictates* π', $\pi \Rightarrow \pi'$, iff for all g, every execution of π contains or corresponds to an execution of π'.
   (b) If π is executable in g, then π *dictates* π' in g, $\pi \Rightarrow_g \pi'$, iff every execution of π in g contains or corresponds to an execution of π'; else, π dictates nothing in g.
   (c) If π is executable in g at s, then π *dictates* π' at s in g, $\pi \Rightarrow_{g,s} \pi'$, iff every execution of π starting at s in g contains or corresponds to an execution of π'; else, π dictates nothing at s in g.
   (d) Else, π dictates nothing.
3. (Implementations) For any protocol, π, and regular protocol, π', π *implements* π', $\pi \mapsto \pi'$, iff:
   (a) (Semantic Condition) For all EDGs, g, for all $p \in P_g$, if $p \in P_\pi$, then $p \in P_{\pi'}$.
   (b) (Syntactic Conditions)
      (i) $\forall \pi_i (\pi_i \in Sub(\pi) \cap \Pi_0 \rightarrow Sub(\pi'))$,
      (ii) If $MC(\pi') \in \{;, \cap\}$, then $\pi \in \Pi_0$ or for every $\circ_i$ in $\pi = x_1 \frown \ldots \frown x_n$, $\circ_i \in \{;, \cap\}$,
      (iii) If $MC(\pi') \in \{\cup\}$, then, where $\pi_i \in PISub(\pi')$:
         (A) Either $\pi \mapsto \pi_i$ or $\pi \Rightarrow \pi_i$,
         (B) For all EDGs, g, if π is procedurally executable, then π_j is procedurally executable where $\pi_j \in PISub(\pi) - \{\pi_i\}$.
      (iv) If $MC(\pi') = *$, then $\pi' = \pi''*$ (for some π''), and $\pi \mapsto \pi''$ or (exclusively) there is some n such that $\pi \mapsto ;_1 \frown \ldots \frown ;_n \frown \pi''_1 \frown \ldots \frown \pi''_{n+1}$ where π''_i is the ith occurrence of π'' in a protocol formed from n occurrences of sequential composition (i.e., π implements some finite length of sequential compositions of π'').

(v) If $* \notin \circ(\pi')$, then:

$$\neg\exists\pi_i\left((\pi_i \in (PSub(\pi) \cup Reduct(\pi))) \wedge \pi_i \mapsto \pi'\right)$$

(vi) If $MC(\pi') \neq \cup$, then:

$$\forall\pi_1[\pi_1 \in PISub(\pi') \rightarrow \exists\pi_1 + (\pi_1 + \in (PSub(\pi) \cup Reduct(\pi)) \wedge \pi_1 + \mapsto \pi_1)]$$

(c) Let $Imp(\pi)$ be the set of *implementations of* π, let $Rimp(\pi) = Imp(\pi) \cap (String(\Pi_0 \cup Prot)$ be the set of all *regular implementations of* π, and let $Simp(\pi)$ be the set of all *sufficient implementations of* π-the set of all protocols that satisfy all conditions but syntactic condition v.

Remark 141 Dictation is a kind of execution-theoretic implication. Roughly, the idea is that one protocol dictates another if the execution conditions of the first imply that the other is or will be executable at some state in the execution of the first. The definitions above present a range of dictation relations, from strongest (quantifying over all executions in all EDGs) to weakest (quantifying over executions at a state in an EDG). The final clause rules out the possibility of vacuous dictation. The philosophical motivation for this condition is that executability is a litmus test for protocols that are to serve as models of procedural epistemic norms. Protocols (hence, procedural epistemic norms) that can never be executed—in any model of epistemic dynamics—are, in principle, impossible to apply or comply with. Thus, they can, as a conceptual matter, impose no legitimate requirements on action.

Remark 142 Implementation is a concept that is central to the deployment of protocols as models of the procedural information contained in or expressed by epistemic rationality norms. Many epistemic rationality norms can be complied with by carrying out different procedures which themselves can be carried out by executing different courses of action (at a finer grain than the procedure). For instance, a norm that requires that, if A implies B, then if one believes A, then one believes B, can be complied with by either revising belief in A or coming to believe B, and either of these may be carried out by different procedures for belief change (revisiting the evidence or analysis of evidence for A or B, acquiring new data, some Pascalian procedure for driving belief change into tacitness). Such a norm specifies very little in terms of procedure; you're on your own to figure out how to comply with the norm. Other norms may be more exacting. A learning theoretic norm for testing hypotheses might specify a very specific sequencing of tests and rule out others. Norms that require complex procedures for detecting validity can be carried out by multiple different derivation or model-checking algorithms. Examples can be multiplied at will. The essential point is that procedural epistemic rationality norms may be implemented by multiple, distinct procedures; they may be indeterministic.

The procedures that one is committed to carrying out when one is committed to complying with a norm are thus frequently multifarious. This feature of norms is encoded by protocols. Every protocol that contains a choice can be carried out by different protocols composed by subprotocols; sequences and combinations of more basic procedures. The distinct sequences of action that suffice to carry out a protocol that contains a non-deterministic choice can be thought of as ways of implementing it; specific, concrete realizations of the program-like instructions of the protocol.

The thought that procedural norms can be carried out by following "more basic" procedures is captured by implementations. The regular implementation set of π is intended to provide the set of procedural structures that suffice to execute π while also being minimal with respect to the atomic protocols and control operators out of which it is built; for regular implementations, these sets are always subsets or proper subsets of those of the original protocol, and regular implementations are always of equal or lower type. For illustration, the protocol, $a; (b \cap c)$, implements itself and $a; (c \cap b)$, as well as $(a; b) \cap (a; c)$ (which also implements $(a; c) \cap (a; b)$). Note that there are no new atomics among these implementations (no matter how structurally complex or higher type they get), nor introductions of new choices, though the implementations are syntactically distinct. Similarly, the protocol, $(((((a \cup b) \cap (a \cup c)) \cap b) \cap c) \cup a) \cup (b \cap c)$, is implemented by $b \cap c$, which is a lower type regular protocol built from a proper subset of atomics and a proper subset of control operators. Note that $((a \cap b) \cap c)$ does have $b \cap c$ as a main control operator preserving reduct, and so doesn't implement $(((((a \cup b) \cap (a \cup c)) \cap b) \cap c) \cup a) \cup (b \cap c)$ (i.e., implementation of $(((((a \cup b) \cap (a \cup c)) \cap b) \cap c) \cup a) \cup (b \cap c)$ doesn't require execution of a, though $((a \cap b) \cap c)$ is a sufficient implementation) but it does implement $((((a \cup b) \cap (a \cup c)) \cap b) \cap c)$. By contrast, the type 1 protocol, $a; b$, will be implemented only by $a; b$ itself; the protocol, $(a; b) \cap (a; b)$, satisfies most conditions of implementation, but contains a proper subprotocol $(a; b)$ that implements $a; b$, and the even more complicated parallel executions (e.g.), $(a; b) \cap ((a \cap a); b)$ and $((a \cap a); b) \cap (a; b)$, have $(a; b) \cap (a; b)$ as a main control operator preserving reduct. These are mechanical observations; I leave them as exercises.

Implementation for nondeterministic choice introduces complexities. Choice protocols, $\pi = \pi_1 \cup \pi_2$, are not immediately implementable by their proper immediate subprotocols, π_1 and π_2, in virtue of their execution rules and the requirements of implementation. If π' implements π_1 but the executability of π' doesn't imply the executability of π_2, π' will fail to suffice to execute π, and so fail to implement it. Moreover, for regular protocols formed from sequential composition or parallel execution, their implementations are regular protocols, as well; for any regular π, any non-regular protocol that satisfies the other conditions with respect to π will always have a proper subprotocol or reduct that suffices to implement π. Not so for choice. Protocols of the forms $\nabla(\langle \pi_2 \rangle_p \top)?; \pi_1$ and $\nabla(\langle \pi_1 \rangle_p \top)?; \pi_2$ are (non-regular) implementations of (arbitrary) $\pi_1 \cup \pi_2$ of higher type. These non-regular implementations may introduce new control operators - namely, sequential composition, test, and fixed point. In any case, implementation is a tool for finding a kind of minimal set of protocols; it eliminates choice branches.

For a sense of context, note that the idea of implementations of protocols is similar to that of computation sequences of programs in PDL [1]. Formally, computation sequences are the strings of atomic protocols that suffice to execute a complex protocol. Informally, they represent the basic steps that can be strung together to execute a program. Similarly, implementations are protocols that abstract away from choices and suffice for executing other protocols; that is, they are sequential compositions of more basic steps that can be strung together to execute a protocol. These implementations and computation sequences capture somewhat the same idea. In fact, though the implementations of π may not all be computation sequences (they are protocols, not merely sequences of atomics), each protocol formed exclusively from choices and sequential compositions defines a set of computation sequences that suffice to execute it. The subprotocol subsets and sets of implementations are thus related to the Fischer-Ladner Closure of programs, but neither is identical to it.[4]

5.4.1 Process Equivalence Relations

Remark 143 Process equivalence relations among transition systems, LTSs, or models of dynamic logics are well-known and well-studied. The main idea is that, taking transition systems, LTSs, or models of dynamic logics (multi-modal Kripke frames) to be models of processes, structure preserving mappings of various kinds are used to define process equivalences of varying degrees of granularity. The usual bestiary contains relations like trace equivalence, bisimulation, generated graph isomorphism, and isomorphism. It's straightforward to define these relations among EDGs.

Definition 144 (Process Equivalence Relations) For all definitions, consider two arbitrary epistemic directed graphs:

$$g = \left\langle S, \Pi_g, R, R_0, \Pi_g^+, R+, R+_0, c, v \right\rangle$$

and

$$g' = \left\langle S', \Pi_g', R', R_0', \Pi_g'^+, R'+, R'+_0, c', v' \right\rangle$$

(1) (Trace Equivalence) Graphs g and g' are *(finite) trace equivalent*, $g \sim_{fte} g'$ iff there is a binary relation, $fte \subseteq S \times S'$, such that, for every s in g there is a s' in g' such that: if $s = first(p)$ and $s(fte)s'$, then there is a q in $P_{g'}$ such that $s' = first(q)$, and for any π, $s(p) \in seq_g(\pi)$ iff $s(q) \in seq_{g'}(\pi)$, and converse.

[4] See Ch. 6 of [1] for discussion of the Fischer-Ladner Closure. It is trivial to prove nonidentity; the parenthetical comment above suffices.

(2) (Bisimulation) Graphs g and g' are *bisimilar*, $g \sim g'$, iff, for every s in g there is an s' in g' such that there is a (non-empty) binary relation, $\sim \subseteq S \times S'$, such that, for all atomic protocols:

  (a) if $s_1 \sim s_1'$, $first(e) = s_1$, and $s_2 = last(e)$ such that $e \in R_\pi$, then there is an $e' \in R_\pi'$ such that $s_1' = first(e)$, $s_2' = last(e')$, and $s_2 \sim s_2'$, and
  (b) if $s_1' \sim s_1$, $first(e') = s_1'$, and $s_2' = last(e')$ such that $e' \in R_\pi'$, then there is an $e \in R_\pi$ such that $s_1 = first(e)$, $s_2 = last(e)$, and $s_2 \sim s_2'$, and
  (c) the converse holds; for every s' in g', there is an s in g that satisfies (a) and (b).

(3) (Isomorphism) Graphs g and g' are *isomorphic*, $g \simeq g'$, iff there is a pair of bijections, $f_e : R \longrightarrow R'$ and $f_s : S \longrightarrow S'$ such that:

  (a) $f_e(e) = e'$ iff $\forall \pi \in \Pi_0 \left(e \in R_\pi \leftrightarrow e' \in R'_{\pi},\right)$
  (b) $f_s(s) = s'$ iff

$$\forall e \in R_0 \left((f_e(e) = e') \rightarrow \bigwedge \begin{bmatrix} (s = first(e)) \rightarrow (f_s(first(e)) = first(f_e(e')) = s') \\ (s = last(e)) \rightarrow (f_s(last(e)) = last(f_e(e')) = s') \end{bmatrix} \right)$$

(4) (Generated Graph Isomorphism) Graphs g and g' are *generated graph isomorphic*, $g \equiv_{GGI} g'$, iff, for every s in g there is a s' in g' such that there is a (non-empty) binary relation, $\equiv_{GGI} \subseteq S \times S'$, such that, if $s \equiv_{GGI} s'$, then for all pointed subgraphs $\langle g*, s \rangle$ and $\langle g'*, s' \rangle$ of g and g', respectively, $\langle g*, s \rangle \simeq \langle g'*, s' \rangle$, and converse.

(5) (Ternary Trace Equivalence) Graphs g and g' are *ternary trace equivalent*, $g \equiv_{TTE} g'$ iff there is a binary relation, $\equiv_{TTE} \subseteq S \times S'$, such that, if $s = first(e)$, $s' = first(e')$, and $s \equiv_{TTE} s'$, then $e \in R_\pi$ iff $e' \in R_\pi'$ for any π.

Remark 145 As is well-established, GGI implies bisimulation, which implies trace equivalence. Isomorphism trivially implies GGI, and for many structures the converse fails. Ternary trace equivalence is less well-studied, and will play some role in our analysis below.

Strictly speaking, the relations above that go by the well-known names of "bisimulation" and "generated graph isomorphism" are not identical to those well-known relations. Bisimulation and Generated Graph Isomorphism, as usually defined, are relations among states in LTSs rather than relations among graphs, and they require the preservation of interpretations for atomic formulae. See [1, 2] for standard definitions. This difference makes no difference to some results below (e.g., the failure of bisimulation invariance for formulae in PLEN), and is essential to others (e.g., that the protocol equivalence relations defined over these process equivalences over EDGs define equivalence classes for each protocol that can vary by valuation). The core idea for all of these process equivalence relations is that they identify a kind of procedural equivalence among (epistemic) processes (as represented by EDGs). They make no requirements on the outcomes internal or external to epistemic state.

The relations above are thus not identical to those relations for, among other things, they have distinct relata. However, the relations defined above hold between two graphs, g and g', wherever there are (e.g.) (standardly defined) bisimulations or generated graph isomorphisms relating all states in g with some state in g'. The converse follows wherever interpretations of atomics are preserved. There are, however, pairs of EDGs that are (e.g.) bisimilar or isomorphic as defined above but that are not bisimilar as standardly defined in virtue of varying with respect to valuations of atomics not involved in the precondition set of any protocol defined over their shared atomics. Further relations between these graph-theoretic relations and the standard bisimulation and GGI relations among states are interesting and somewhat mechanical to explore. It's trivial, for instance, that where g and g' contain states, s and s' (resp.), that are generated graph isomorphic, it doesn't follow that g and g' are GGI as defined in the text, as there may be states other than s in g that are not generated graph isomorphic to any s' in g'. Just the same for each of the other relations. None of the foregoing observations plays any important role in the results explored in this or future chapters.

The foregoing suite of process equivalences will be the key to graph-theoretic protocol equivalence relations. If there are distinctive sets of EDGs mapped to each protocol, then process equivalences between these graphs may be used to define equivalence relations between protocols. First, however, we must find a way to map protocols uniquely to EDGs. This function is served by protocol graphs, which will take just one more set of constructions—graph reduction operations—to define.

5.4.2 Graph Reduction Operations

Remark 146 Graph reduction operations are operations mapping EDGs to other EDGs that structurally resemble them in certain ways, while also being smaller or more minimal in some specified way. In Technical Appendix 2, I define a family of graph reduction operations with obvious relations to each other. Here, I present the operations of direct use in defining protocol graphs.

Definition 147 (Collapsing Paths) Let p_1 and p_2 be branching paths in arbitrary EDG, g. Let r and r' be arbitrary paths in g. Let p'' be an arbitrary path in arbitrary g'. Then *an operation on g that produces g'* ($g' \rightarrow g$ or, informally, g' is produced by the operation $\rightarrow$ on g) *collapses p_1 and p_2* iff there is a mapping, ϱ, of P_g into $P_{g'}$ such that:

(1) For any arbitrary regular π, there is a $p'' = \varrho(p_1) = \varrho(p_2)$ in g' such that $p_1 \in P_\pi$ iff $p_2 \in P_\pi$ iff $p'' \in P'_\pi$.
(2) For any converging paths, r and r':
    (a) if $l(r) = f(p_1)$ and $l(r') = f(p_2)$, then $f(p'') = l(\varrho(r)) = l(\varrho(r'))$.

(b) if $f(r) = l(p_1)$ and $f(r') = l(p_2)$, then $l(p'') = f(\varrho(r)) = f(\varrho(r'))$.
(c) for all $s, s' \in g$ and $s'' \in g'$, if $\varrho(s) = \varrho(s') = s''$, then for any π_i such that and π_i is executable at either s or s', for all formulae, $A \in Pre(\pi_i)$, $v_{s''}(A) = 1$ and $c'(s'') = c(s) \cup c(s')$ (i.e., $X \cup X'$, $Y \cup Y'$, $\Pi_M \cup \Pi_{M'}$).

(3) $\forall q \in P_g(\exists q' \in P_{g'}(q \simeq q'))$
(4) $\forall q' \in P_{g'}(\exists q \in P_g(q \simeq q'))$

Remark 148 The basic idea of collapsing paths is straightforward. An operation on graphs collapses two paths, p_1 and p_2, in a graph, g, when it replaces them with one path, while otherwise preserving the structure of the graph. This may induce new branches; where there were paths that continued on from the terminal states of p_1 and p_2, those paths now branch from the root state of p'. It may induce convergence; where there were paths that prefixed p_1 and p_2, those prefixes now share a terminal state (the root state of p'').

Operations that collapse paths can be used to reduce the sizes of graphs in various ways. Here is the basic form of graph reduction:[5]

Definition 149 (Partial π Graph Reduction at s) For arbitrary EDGs, g and g', and arbitrary atomic protocol, π, g' is a *partial π graph reduction of g at s* ($g' \xrightarrow[s,P]{\pi} g$) iff if there are any pairs of branching isomorphic executions of π at s in g, then for some such pair, p_1 and p_2, there is a mapping, ϱ, of P_g into $P_{g'}$ such that there is a single path, $p'' = \varrho(p_1) = \varrho(p_2)$, that executes π in g' and:

(1) For any converging paths, r and r':

(a) if $l(r) = f(p_1)$ and $l(r') = f(p_2)$, then $f(p'') = l(\varrho(r)) = l(\varrho(r'))$.
(b) if $f(r) = l(p_1)$ and $f(r') = l(p_2)$, then $l(p'') = f(\varrho(r)) = f(\varrho(r'))$.
(c) for all $s, s' \in g$ and $s'' \in g'$, if $\varrho(s) = \varrho(s') = s''$, then for any π_i such that and π_i is executable at either s or s', for all formulae, $A \in Pre(\pi_i)$, $v_{s''}(A) = 1$ and $c'(s'') = c(s) \cup c(s')$ (i.e., $X \cup X'$, $Y \cup Y'$, $\Pi_M \cup \Pi_{M'}$).

(2) $\forall q \in P_g(\exists q' \in P_{g'}(q \simeq q'))$
(3) $\forall q' \in P_{g'}(\exists q \in P_g(q \simeq q'))$

Remark 150 Here is the one that is used to define protocol graphs:

Definition 151 (Basic Graph Reduction) For arbitrary EDGs, g and g', g' *is a basic graph reduction of* g, ($g' \xrightarrow[\Pi_C]{\Pi_0} g$) iff there is some regular protocol, $\pi \in \Pi_0 \cup \Pi_C$, such that, for some pair of branching isomorphic paths, p and q, that execute π in g, there is a mapping, ϱ, of P_g into $P_{g'}$ such that there is a single path, $p'' = \varrho(p) = \varrho(q)$, that executes π in g' and:

(1) For any converging paths, r and r':

(a) if $l(r) = f(p)$ and $l(r') = f(q)$, then $f(p'') = l(\varrho(r)) = l(\varrho(r'))$.

[5] It is used later in this chapter to define an incremental graph reduction procedure.

(b) if $f(r) = l(p)$ and $f(r') = l(q)$, then $l(p'') = f(\varrho(r)) = f(\varrho(r'))$.
(c) for all $s, s' \in g$ and $s'' \in g'$, if $\varrho(s) = \varrho(s') = s''$, then for any π_i such that and π_i is executable at either s or s', for all formulae, $A \in Pre(\pi_i)$, $v_{s''}(A) = 1$ and $c'(s'') = c(s) \cup c(s')$ (i.e., $X \cup X'$, $Y \cup Y'$, $\Pi_M \cup \Pi_{M'}$).

(2) $\forall p \in P_g(\exists p' \in P_{g'}(p \simeq p'))$
(3) $\forall p' \in P_{g'}(\exists p \in P_g(p \simeq p'))$

Remark 152 Basic Graph Reduction takes an EDG, g, and produces a different EDG, g', in which some pair of branching, isomorphic executions of some protocol are collapsed into one, resulting in new correspondences and branches, as long as those new structures don't interfere with pathwise isomorphism of g and g'. The idea is that this reduction operation removes paths that are redundant for determining which protocols are executable in an EDG. As we see below, this operation can be used to define protocol graphs in a way that makes them minimal in the sense that they contain no such redundant paths.

5.4.3 *Protocol Graphs*

Remark 153 The protocol graphs of a given protocol are defined as those EDGs that satisfy a set of *structural display rules* with respect to that protocol defined in terms of the execution rules, implementations, and graph reduction. The structural display rules bind the graph-theoretic structure of protocol graphs closely but not uniquely to the syntactic form of protocols. Protocols are mapped to isomorphism classes of protocol graphs, and the syntactic form of each protocol determines a nearly unique, abstract, graph-theoretic structure.

Definition 154 (Terminological and Notational Conventions 3)

(1) For arbitrary EDG, g, let P_g be the set of all paths of any length in g.
(2) Let $\frown(X)$ denote the set of all concatenations of elements of arbitrary set or class, X.
(3) Let $(\ldots \circ \pi_1 \circ \ldots) \cap (\ldots \circ \pi_2 \circ \ldots)$ be a meta-variable for any protocol such that π_1 and π_2 are embedded at equal depth with respect to parallel execution.
(4) Let $seq_g(\pi)$ be the set of all state-sequences of all executions of π in arbitrary g.
(5) Call the procedural executions of a protocol, π, in the maximal path set of a graph, g, the *maximal executions* of π in g.
(6) Let tree-like epistemic directed graphs, (g, s) or $\langle g, s\rangle$, be any pointed, connected EDGs, g; i.e., g is a tree-like or pointed, connected EDG iff g is connected, and s is the root state of all maximal paths in g.

Definition 155 (Structural Display Rules) Define *structural display relations* as relations of the form, $F_\pi(g) \subseteq EDG \times P_{PLEN}$, defined by the following *structural display rules*:

(1) (SD1) $SD1_\pi(g)$ iff $SD1(1)_\pi(g)$, $SD1(2)_\pi(g)$, $SD1(3)_\pi(g)$, and $SD1(4)_\pi(g)$ where:

(a) (SD1(1)) $SD1(1)_\pi(g)$ iff $\forall \pi_i \in Sub(\pi)[P_{\pi i} \neq \emptyset]$,

(b) (SD1(2)) $SD1(2)_\pi(g)$ iff (i) if π contains no iterations, then $P_\pi = maxpath(g)$, and (ii) if π contains iterations $\pi_1*, \ldots, \pi_n*$, then there are also executions of each $\pi_1*, \ldots, \pi_n*$ that are detached loops.

(c) (SD1(3)) $SD1(3)(g)$ iff

$$\forall \pi_1, \pi_2 \left(\begin{array}{l} \left(\dfrac{\pi_1, \pi_2 \in Sub(\pi), \exists \pi_i \in Sub(\pi)\, (\pi_i = \pi_1 \cup \pi_2 \vee \pi_i = \pi_1 \cap \pi_2)}{\exists p, q, s(s = first(p) = first(q), p \in P_{\pi 1}, q \in P_{\pi 2})} \right), \\ \left(\dfrac{\left(\neg(\pi_1, \pi_2 \in Sub(\pi)), \exists \pi_i \in Sub(\pi)(\pi_1 \mapsto \pi_i, \pi_2 \mapsto \pi_i), \right)}{\exists p, q, s(s = first(p) = first(q), p \in P_{\pi 1}, q \in P_{\pi 2})} \right), \\ \left(\dfrac{\exists p, q, s(s = first(p) = first(q), p \in P_{\pi 1}, q \in P_{\pi 2})}{\bigvee \left(\begin{array}{c} \neg(\pi_1, \pi_2 \in Sub(\pi)), \exists \pi_i \in Sub(\pi) \left(\bigvee \begin{array}{c} \pi_1 \mapsto \pi_i, \pi_2 \mapsto \pi_i \\ \pi_i \Rightarrow_g \pi_1, \pi_i \Rightarrow_g \pi_2 \end{array} \right) \\ \pi_1, \pi_2 \in Sub(\pi), \exists \pi_i \in Sub(\pi)(\pi_i = ((\ldots \circ \pi_1 \circ \ldots) \cup (\ldots \circ \pi_2 \circ \ldots))) \\ \pi_1, \pi_2 \in Sub(\pi), \exists \pi_i \in Sub(\pi)(\pi_i = ((\ldots \circ \pi_1 \circ \ldots) \cap (\ldots \circ \pi_2 \circ \ldots))) \end{array} \right)} \right) \end{array} \right)$$

(d) (SD1(4)) $SD1(4)_\pi(g)$ iff for any π_1 and π_2, if there is no $\pi' \in Sub(\pi)$ such that $\pi_1 \cap \pi_2 \mapsto \pi''$ or $\pi'' = (\ldots \circ \pi_1 \circ \ldots) \cap (\ldots \circ \pi_2 \circ \ldots)$, then if $\nvDash_{PLEN} \Box(\pi_1 \leftrightarrow_c \pi_2)$, then $g \nvDash_{PLEN} (\neg\Box(\pi_1 \leftrightarrow_c \pi_2) \wedge \neg\Box(\pi_1 \leftrightarrow_t \pi_2))$.

(2) (SD2) $SD2_\pi(g)$ iff for all regular π', if $\pi' \mapsto \pi$, then $P_{\pi'} \neq \emptyset$.

(3) (SD3) $SD3_\pi(g)$ iff $\Pi_g = \Pi_0 \cap Sub(\pi)$

(4) (NST) $NST_\pi(g)$ iff there is no g' such that (a) $SD1(1,2,3,4)_\pi(g') \wedge SD2_\pi(g')$, and $RC_\pi(g')$, and (b) $g' \underset{\Pi_C}{\overset{\Pi_0}{\longrightarrow}} g$ or $g' \sqsubset g$.

(5) (RC) $RC_\pi(g)$ iff $\forall \pi_i \in PSub(\pi)[\exists g'(g' \sqsubset g \wedge g' \in G^{\pi_i})]$.

(6) (EXC) $EXC_\pi(g)$ iff if $SD1_\pi(g) \wedge SD2_\pi(g) \wedge NST_\pi(g) \wedge RC_\pi(g)$, then for all π', if $\nvDash_{PLEN} \Box(\pi' \leftrightarrow_c \pi)$, then $g \notin G^{\pi'}$.

Definition 156 (Protocol Graphs) For arbitrary, regular protocol, π, and arbitrary EDG, g, g is a *protocol graph of* π ($g \in G^\pi$), iff

$$[SD1_\pi(g) \wedge SD2_\pi(g) \wedge SD3_\pi(g) \wedge NST_\pi(g) \wedge RC_\pi(g) \wedge EXC_\pi(g)]$$

(Protocol Graph Set)

$$G^\pi = \left\{ g \in EDG \mid \begin{pmatrix} SD1_\pi(g), & SD2_\pi(g), & SD3_\pi(g), \\ NST_\pi(g), & RC_\pi(g), & EXC_\pi(g) \end{pmatrix} \right\}$$

Fact 157 *Protocol graphs are EDGs such that:*

(1) If g is a protocol graph of a regular protocol, π, then there is a proper subgraph of g, g', that contains no loops and is a multi-tree, where g' is a multi-tree iff g'

is a connected, directed, acyclic multi-graph with a selected state, s_0*, such that* $s_0 = first(p_i)$ *for all* p_i *in* $maxpath(g')$.

(2) $R+$ *and* R_0+ *can be safely ignored for all foregoing and forthcoming graph-theoretic results.*

Proof (Sketch)

(1) NST coupled with the rest of the structural display rules implies (1) directly; any detached paths or converging or mirrored trees can be omitted without violating the structural display rules.

(2) For any two g and g' in G^{π}, the isomorphism of the structures of g and g' necessary for satisfaction of the structural display rules (which is established by means of the structural uniqueness theorem of Sect. 5.7.4.1) implies the isomorphism of $R+$ and $R'+$ and $R+_0$ and $R'+_0$, by the restriction rules and so can safely be ignored.

□

Remark 158 The definition of protocol graphs makes no use of content or valuation functions. These are left entirely arbitrary across protocol graphs. This means, for instance, that two isomorphic protocol graphs may verify different atomic formulae at any given state. This is a matter of irrelevance for the protocol logic, as will be visible below.

Protocol graphs are, alongside protocols and EDGs, among the central objects of PLEN. The main idea of protocol graphs is that, for each protocol, there is a set of EDGs that most perspicuously displays the procedural structure of the protocol with elements of epistemic dynamics. In order to most perspicuously display such procedural structure, five intuitive requirements must be met.

The first requirement on such EDGs is that they execute every syntactic subprotocol of a complex protocol; else, there is procedural information missing. A graph in which not all syntactic subprotocols of π are executed cannot display all of the possible processes resulting from execution of or compliance with π nor can it display all of the possible processes (sequences of actions) that can execute or comply with π. This information is necessary to understand exactly what π requires.

The second requirement is that the graph bear graph-theoretic properties that somehow distinguish radically syntactically distinct protocols. For instance, where there is a non-deterministic choice of subprotocols, $\pi_i \cup \pi_j$, each constituent subprotocol must be executable by different paths so long as they are not equivalent or linked by a parallel execution to another protocol. Else, the parallel execution $\pi_i \cap \pi_j$ is executable wherever both choices are, and the graph doesn't show us how these clearly distinct protocols are distinct. This condition generally ensures that (e.g.) non-deterministic choices do not end up being confused with one of the choices or with parallel executions. This requirement also extends to implementations. A graph that fully displays the procedural content of a protocol should show us each and every possible implementation of the protocol. This encodes the possible ways in which the protocol can be carried out. Distinct implementations will typically be a way of distinguishing syntactically distinct protocols given that syntax of protocols codifies the procedural operations on action that the protocol requires. However, implementations may not always distinguish syntactically distinct protocols, and where they don't, it shows us that the syntax doesn't capture everything there is to know about the procedural structure of the protocol. Note that this requirement is not implied by the first requirement, though it does imply the first.

The third requirement is that the EDG must be maximally informative about procedural structure; all of the conditions met for π must be met for each syntactic subprotocol of π; else, there is procedural structure left unanalyzed. This requirement may be derivable from the first two. That is, if a graph displays executions for each subprotocol and implementation of π, it must already display executions for every subprotocol and every implementation of each subprotocol of π. The definition of subprotocols implies the claim for the first requirement; each subprotocol of each subprotocol of π is a subprotocol of π. Hence, every subprotocol of every subprotocol of π must be executable in g by the first requirement. To see the point for the second requirement, let $\pi = \pi_1; \pi_2$ be an arbitrary protocol such that π_1 has two distinct implementations, π_3 and π_4. Clearly, $\pi_3; \pi_2$ is a way of implementing π, as is $\pi_4; \pi_2$. There is certainly no way a graph can meet the second requirement with respect to π without meeting the second requirement for π_1. The argument will work similarly for any other syntactic form.

The fourth requirement is that these EDGs be structurally minimal while meeting the other requirements. Call this the "no-smaller-than" requirement; it is the requirement that there be no proper subgraph that achieves maximally parsimonious structural display. This must be close to an analytic truth. For instance, the longest paths in such an EDG must be those that execute π; else, the graph contains extraneous information. This is not simply a matter of computational costs; embedding protocol graphs inside larger graphs is useful for showing, inter alia, whether and to what degree the dictates of a protocol π can be carried out. If embedding fails due to π's protocol graphs containing extraneous information like useless states, additional edges, etc., this will mislead us concerning the properties of π.

The fifth and final condition is that protocol graphs must be unique per protocol. That is, they must not also display all of the structural information for other, clearly distinct protocols where a protocol is clearly distinct if it isn't itself at least procedurally or transitionally equivalent to the one under analysis. If this condition fails, then the basic goal of protocol graphs—defining structures more informative than procedural and transitional equivalence—also fails. That is, protocol graphs must provide some way of distinguishing syntactically distinct protocols that are procedurally or transitionally equivalent.

All five of the foregoing requirements are met, with respect to a given regular protocol, π, by the protocol graphs of π; the conditions that define protocol graphs correspond to the intuitive requirements of procedural display in rather self-explanatory ways. SD1-NST, for instance, directly require that a protocol graph of π execute all subprotocols and all implementations of π without any redundant structural features. RC requires that a protocol graph of π do the same for each subprotocol of π thereby providing maximal structural information about π. And

EXC explicitly requires that a protocol graph of π not be a protocol graph of any protocol that fails even to be merely "correspondence equivalent" to π.

The essential feature of protocol graphs is that their graph-theoretic structure is closely related to the syntax of the protocols that they graph. Consider the following protocol graphs of $a \cup (b \cap c)$, $(((((a \cup b) \cap (a \cup c)) \cap b) \cap c) \cup a) \cup (b \cap c)$, $(a \cup b); (c \cup d)$, and $(a; (c \cup d)) \cup (b; (c \cup d))$, respectively:

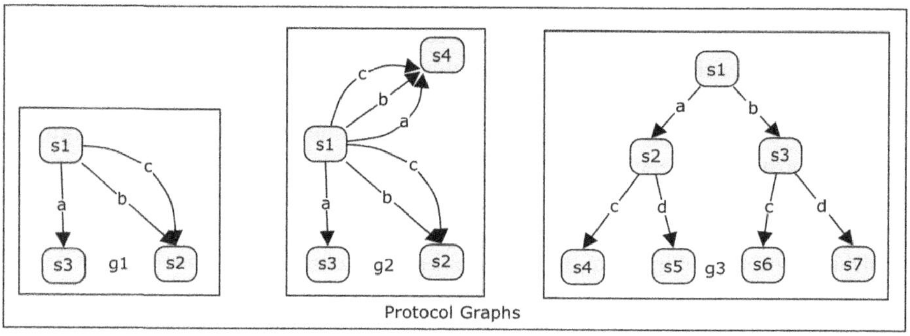

From the left, the first graph is a protocol graph of $a \cup (b \cap c)$, the second is a protocol graph of $(((((a \cup b) \cap (a \cup c)) \cap b) \cap c) \cup a) \cup (b \cap c)$, and the third is a protocol graph of both $(a \cup b); (c \cup d)$ and $(a; (c \cup d)) \cup (b; (c \cup d))$. It's not hard to mechanically check that these EDGs are protocol graphs of the respective protocols, given that they are formed from atomics.

Note that the set of executions of $a \cup (b \cap c)$ and $(((((a \cup b) \cap (a \cup c)) \cap b) \cap c) \cup a) \cup (b \cap c)$ are identical: if $(((((a \cup b) \cap (a \cup c)) \cap b) \cap c) \cup a) \cup (b \cap c)$ is executable in a graph, then so is $a \cup (b \cap c)$ and the converse also holds. However, that the middle graph is not a protocol graph of $a \cup (b \cap c)$ follows from NST—there are superfluous parts, namely state s_4 and the edges leading to it. The leftmost graph is not a graph of $(((((a \cup b) \cap (a \cup c)) \cap b) \cap c) \cup a) \cup (b \cap c)$ because there is no execution of the subprotocol and implementation, $((((a \cup b) \cap (a \cup c)) \cap b) \cap c)$. Thus, the structure of these graphs tracks the syntax of these protocols more closely than the set of executions. This, coupled with the foregoing observation suggest something interesting: the closer mapping of the structure of protocol graphs to the syntax of protocols has something to do with how well the graph displays procedural information like the sequence of decision problems that agents face if following them.

Protocol graphs are useful for displaying the plan or sequence of decision problems that agents executing them face, which is distinct from the set of executions of protocols. That is, protocol graphs visually or, technically, graph-theoretically represent the procedural operations encoded by a protocol's syntax in a way that is perspicuous for thinking intuitively about how the procedural operations will shape choices over time. At s_1, an agent executing $(a \cup b); (c \cup d)$ or $(a; (c \cup d)) \cup (b; (c \cup d))$ ultimately must execute two decisions: a choice of a and b followed by a choice of c and d. That is, both protocols require committing to a

or b prior to committing to c or d. The protocol graph of these equivalent[6] protocols visually displays this sequence of choices via the flow from s_1 along the arrows. However, the protocols $(a \cup b);\ (c \cup d)$ and $((a \cup b);\ c) \cup ((a \cup b);\ d)$ define distinct decision sequences. The first allows the choice between c and d after the choice between a and b, but the latter requires that one decide between c and d before choosing between a and b. In each of the foregoing cases, the pairs of protocols have identical execution sets. Protocol graphs distinguish these protocols just as they run together the prior ones. And this is as it should be given their mapping into decision sequences.

Note, however, that all of the foregoing was defined over regular protocols. There is a reason for that:

Proposition 159 *Protocols formed with irregular control operators cannot coherently be assigned protocol graphs.*

Proof (Proof-Outline) Protocol graphs can't coherently be assigned to the irregular protocols due to the unusual combination of syntax and execution rules; each of them contains either a subprotocol that the execution rules in conjunction with the structural display rules prohibit being part of their execution, or they break protocol graphs in some other way.

Let $g \in G^{-\pi}$. Then there is a proper subgraph of g, g', such that g' is a protocol graph of π by RC. SD1(3) rules out any graph of this form. SD1(1) requires that π be executable in g. However, $maxpath(g)$ cannot contain any executions of π by the execution rules, so either all such executions must be contained in maximal paths or branch from some states in maximal paths. The first is ruled out by the execution rules. The second implies that $\pi \cup \pi'$ for some π' executed by a subpath of a maximal path be executable at some state, s, in g. But this protocol does not have any of the syntactic relations to $-\pi$ required by SD1(3). Thus, SD1(3) fails, and g is not a protocol graph of $-\pi$, contradicting the assumption.

For test, reversal, and fixed point, the problem is that these protocols must, by the execution rules, be executed by loops. But $maxpath(g)$ cannot contain loops. Suppose that p is a maximal path and that p contains a loop. Then p is a proper subpath of a concatenated path of the form qp, pq, $p'qp''$, or pp, (where, respectively, the loop starts and ends at $first(p)$, $last(p)$, somewhere in the middle of p, or p itself loops) contradicting the definition of maximal paths. Thus, any putative protocol graph of a test or fixed point, g, violates SD1(2) with respect to that protocol since $P_\pi \not\subseteq maxpath(g)$. Test and fixed point obviously require loops, but so does reversal. The execution rules require that the executions of $\curvearrowleft \pi$ start at the end state of an execution of π and terminate at the root state of that execution of π; a loop. Thus, the execution rules for reversal break the requirements of maximal paths. □

Thus, all of the core results of PLEN quantify over only regular protocols, save for those laws that explicitly mention irregular protocol operators. Indeed, all results formulated in terms of graph-theoretic protocol equivalence relations—and graph-

[6] As we will see below.

theoretic protocol equivalence relations themselves—apply only to regular protocols in virtue of this limitation.

5.5 Protocol Equivalence Relations II: Graph-Theoretic Protocol Equivalence

Remark 160 The canonical form of *protocol equivalence* in PLEN is defined in terms of protocol graph isomorphism: protocols π and π' are graph-theoretically equivalent iff each protocol graph of one is isomorphic to some protocol graph of the other. This equivalence relation is the strictest, finest grained, and most informative protocol equivalence relation defined in PLEN that does not imply syntactic identity; syntactic replacement of equivalent protocols preserves truth for all protocol-theoretic formulae but for those for which truth-preserving replacement implies syntactic identity. We will see this in Sects. 6.3–6.5.

5.5.1 Graph-Theoretic Protocol Equivalence: Meta-Language

Definition 161 (Graph-theoretic Protocol Equivalence Relations)

(1) (Bisimulation) For any regular protocols, π and π', *π and π' are protocol graph bisimulation equivalent* ($\pi \equiv_B \pi'$) iff:

(a) $\forall g \in G^{\pi} \exists g' \in G^{\pi'}[g \sim g']$,
(b) $\forall g' \in G^{\pi'} \exists g \in G^{\pi}[g \sim g']$.

(2) (Generated Graph Isomorphism) For any regular protocols, π and π', *π and π' are protocol graph generated graph isomorphism equivalent* ($\pi \equiv_{GGI} \pi'$) iff:

(a) $\forall g \in G^{\pi} \exists g' \in G^{\pi'}[g \equiv_{GGI} g']$,
(b) $\forall g' \in G^{\pi'} \exists g \in G^{\pi}[g \equiv_{GGI} g']$.

(3) (Isomorphism) For any regular protocols, π and π', *π and π' are protocol graph isomorphism equivalent* ($\pi \equiv_G \pi'$) iff:

(a) $\forall g \in G^{\pi} \exists g' \in G^{\pi'}[g \simeq g']$,
(b) $\forall g' \in G^{\pi'} \exists g \in G^{\pi}[g \simeq g']$.

Definition 162 ((Canonical) Graph-theoretic Protocol Equivalence) For any regular protocols, π and π', *π and π' are (canonically) graph-theoretically protocol equivalent* iff *π and π' are protocol graph isomorphism equivalent* ($\pi \equiv_G \pi'$) iff:

(1) $\forall g \in G^{\pi} \exists g' \in G^{\pi'}[g \simeq g']$,
(2) $\forall g' \in G^{\pi'} \exists g \in G^{\pi}[g \simeq g']$.

This set of definitions formulates a repertoire of protocol equivalence relations and selects one as the canonical protocol equivalence of PLEN. Accordingly, the terms "protocol graph isomorphism equivalence" and "graph-theoretic protocol equivalence" will be used interchangeably. The object-level counterparts of these graph-theoretic equivalences are as follows.

5.5.2 Graph-Theoretic Protocol Equivalence: Object Language Counterparts

Definition 163 (Graph-Theoretic Protocol Equivalence Relations)

(1) (Bisimulation) For any regular protocols, π and π', and any EDG, g, *π and π' are protocol graph bisimulation equivalent at s in g* ($\pi \equiv^s_B \pi'$) iff for all (g_1, s) and (g_2, s) among the subgraphs of g:

(a) $\forall (g_1, s) \in G^{\pi} \exists (g_2, s) \in G^{\pi'}[(g_1, s) \sim (g_2, s)]$,
(b) $\forall (g_2, s) \in G^{\pi'} \exists (g_1, s) \in G^{\pi}[(g_1, s) \sim (g_2, s)]$.

(2) (Generated Graph Isomorphism) For any regular protocols, π and π', and any EDG, g, *π and π' are protocol graph generated graph isomorphism equivalent at s in g* ($\pi \equiv^s_{GGI} \pi'$) iff for all (g_1, s) and (g_2, s) among the subgraphs of g:

(a) $\forall (g_1, s) \in G^{\pi} \exists (g_2, s) \in G^{\pi'}[(g_1, s) \equiv_{GGI} (g_2, s)]$,
(b) $\forall (g_2, s) \in G^{\pi'} \exists (g_1, s) \in G^{\pi}[(g_1, s) \equiv_{GGI} (g_2, s)]$.

(3) (Isomorphism) For any regular protocols, π and π', and any EDG, g, *π and π' are protocol graph isomorphism equivalent at s in g* ($\pi \equiv^s_G \pi'$) iff for all (g_1, s) and (g_2, s) among the subgraphs of g:

(a) $\forall (g_1, s) \in G^{\pi} \exists (g_2, s) \in G^{\pi'}[(g_1, s) \simeq (g_2, s)]$,
(b) $\forall (g_2, s) \in G^{\pi'} \exists (g_1, s) \in G^{\pi}[(g_1, s) \simeq (g_2, s)]$.

Definition 164 ((Canonical) Graph-Theoretic Protocol Equivalence) For any regular protocols, π and π', and any EDG, g, *π and π' are (canonically) graph-theoretically equivalent at s in g* iff *π and π' are protocol graph isomorphism equivalent at s in g* ($\pi \equiv^s_G \pi'$) iff for all (g_1, s) and (g_2, s) among the subgraphs of g:

(1) $\forall (g_1, s) \in G^{\pi} \exists (g_2, s) \in G^{\pi'}[(g_1, s) \simeq (g_2, s)]$,
(2) $\forall (g_2, s) \in G^{\pi'} \exists (g_1, s) \in G^{\pi}[(g_1, s) \simeq (g_2, s)]$.

In the next subsection, I'll add the object-level counterparts of the execution and graph-theoretic protocol equivalences to L_{PLEN}, thereby defining a revised protocol-theoretic language, L^R_{PLEN}, define a semantics for them, and then present systems of rules governing both the meta-level and object-level protocol equivalence relations. The latter rules will be directly built into an extension of the derivation system of PLEN. The former rules comprise the main results of PLEN. The rest of this chapter and the next two are spent exploring these results, connecting them to the properties of the derivation system, and then using them to explore PLEN's properties as a logic of protocols.

5.6 Revised Derivation System: The R-Derivation System

The core of PLEN's logic of protocols is the *theory of protocol graph isomorphism equivalence*, the system of meta-level rules and axioms and their object-level counterparts governing the properties of the protocol equivalence relations in PLEN. The system distinguishes the protocol graph isomorphism equivalence relation from other protocol equivalence relations defined in PLEN. The derivation system is sound and plausibly complete, but it is intended to be a *basic* system in the sense that it is a starting point for the protocol-theoretic logic of epistemic norms. As such, it is somewhat modular; the propositional, modal, and epistemic logics can be replaced or modified ad hoc without disturbing the protocol logic. For instance, simple modification transforms the logic of the object-level rules into a the weak relevant logic, B (see Sect. 7.5). As this basic and revisable-with-the-logic-B derivation system is developed iteratively, the initial derivation system of PLEN goes unnamed or referred to simply as the PLEN derivation system, and the revised derivation system, complete with object-level counterparts of protocol equivalence rules and axioms, is accordingly called the *R-Derivation system.*

Claim 165 The R-Derivation System is the result of codifying the theory of protocol graph isomorphism equivalence and building it into the derivation system of PLEN. This is done by first developing the theory of protocol graph isomorphism equivalence at the meta-level as a system of rules and axioms governing the canonical protocol equivalence in PLEN and its relations to other meta-level protocol equivalences. The rules and axioms of protocol equivalence at the meta-level are then used to prove the soundness of more local protocol equivalence rules and axioms that are built into the object-language of PLEN and given truth rules that verify their constituent schematic formulae relative to states in EDGs.

5.6.1 *Revising the Syntax of PLEN:* (L^R_{PLEN})

Both sorts of syntactic objects in PLEN must be simultaneously revised in order to add the equivalence rules to the derivation system. Letting the revised protocol-theoretic formulae language of PLEN be L^R_{PLEN}, redefine P_{PLEN} over L^R_{PLEN}.

Definition 166 (String Algebraic Protocols of Type i (P^i_{PLEN}))
Let $\Pi_0 = \{a_1, \ldots, a_n, \ldots\}$ be a countable, non-empty set of atomic protocols.
Let $Prot = \{;, \cap, \cup, *\}$ and $Prot+ = \{(-)?, -, \frown, \nabla\}$ be the *regular* and *irregular control operators*, respectively.

The arity of each control operator is denoted $a(\circ)$ and explicitly specified for each control operator: if $\circ \in Prot \cup Prot + / \{*, (A)?, -, \frown, \nabla\}$, then $a(\circ) = 2$; else, $a(\circ) = 1$ and $\circ \frown y_1 \frown y_{a(\circ)}$ is elliptical for $\circ \frown y_1$.

(Protocols of Type i) The set of protocols of type i, P^i_{PLEN}, is the smallest subset of $String(\Pi_0 \cup Prot \cup Prot + \cup \bigcup_{j=1}^{i-1} L^{Rj}_{PLEN})$ closed under the following rules:

(1) If $x \in \Pi_0$, then $x \in P^0_{PLEN}$.
(2) If $x = \circ \frown y_1 \frown y_{a(\circ)}$, where each $y_j = \langle x_1 \frown \ldots \frown x_n \rangle$, each $y_j \in \bigcup_{j=1}^{i-1} P^j_{PLEN}$, some $y_j \in P^{i-1}_{PLEN}$, and $\circ \in Prot \cup Prot+$, then $x \in P^i_{PLEN}$.
   (a) For any protocol, x, $x = \circ \frown y_1 \frown y_{a(\circ)}$ or $x = \circ \frown y_{a(\circ)}$ iff $\circ$ is the *main control operator* of x.
      (i) iff $MC(x) = \circ$.
      (ii) iff $\circ$ is the first symbol in the string.
(3) If A is a metavariable for a formula of L^{Ri-1}_{PLEN}, then $(A)? \in P^i_{PLEN}$.

Definition 167 (String Algebraic Protocol Language (P_{PLEN}))

$$P_{PLEN} = \bigcup_{i=1}^{\omega} P^i_{PLEN}$$

Definition 168 (Infix Notation Protocol Language (P_{PLEN})) Where $a_i \in \Pi_0$, π_i is a metavariable for an arbitrary protocol with no occurrences of self-reference, and A is a metavariable for an arbitrary protocol-theoretic formula with no occurrences of self-reference, P_{PLEN} can be defined by the following grammar:

$$(a_i) \mid (\pi_1; \pi_2) \mid (\pi_1 \cap \pi_2) \mid (\pi_1 \cup \pi_2) \mid (\pi_1 *) \mid (A)? \mid (-\pi_1) \mid (\frown \pi_1) \mid (\nabla \pi_1)$$

Second, revise L_{PLEN} by adding a set, $Pop+$, of protocol equivalence operators.

Definition 169 (Protocol-Theoretic Formulae of Type i (L^{Ri}_{PLEN}))
Let $AForm = \{P_i, Q_i, \ldots, \top, \ldots\}$ be a countable, non-empty set of atomic protocol-theoretic formulae, and define the following sets of operators for building complex expressions and embedding protocol expressions where $x \in \{p, r, t\}$ and π_i is a meta-variable for an arbitrary protocol:

$$Syn = \{\in Sub, \in PSub, \in ISub, \rho(/)() =, M\}$$

$$Pop+ = \left\{\equiv^s_r, \equiv^s_G, \equiv^s_p, \equiv^s_t, \equiv^s_c, \mapsto_s\right\}$$

$$Op = \left\{[\pi_i]_x, \langle \pi_i \rangle_x, \Box, b, d\right\}$$

$$Conn = \{\wedge, \vee, \rightarrow, \equiv, \neg\}$$

$$Pop = \{\leftrightarrow_x, \curlywedge_x, \curlyvee_x\}$$

The arity of each control operator is denoted $a(\circ)$ and explicitly specified for each operator and connective:

(1) if $\circ \in Syn - \{\rho(/)() =, M\}$, then $a(\circ) = 2$; else,

(a) $\circ = M, a(\circ) = 1$, and $\circ \frown y_1 \frown y_{a(\circ)}$ is elliptical for $\circ \frown y_1$ OR
(b) $a(\circ) = 4$ and $\circ \frown y_1 \frown y_{a(\circ)}$ is elliptical for $\circ \frown y_1 \frown y_2 \frown y_3 \frown y_4$.

(2) if $\circ \in Op \cup Conn/\{\neg, \Box, b, d\}$, then $a(\circ) = 2$; else, $a(\circ) = 1$ and $\circ \frown y_1 \frown y_{a(\circ)}$ is elliptical for $\circ \frown y_1$,
(3) if $\circ \in Pop \cup Pop+$, then $a(\circ) = 2$.

(Protocol-theoretic Formulae of Type i) The set of protocol-theoretic formulae of type i, L^{Ri}_{PLEN}, is the smallest subset of

$$String(AForm \cup Conn \cup Op \cup Pop \cup Pop + \cup Syn \cup \bigcup_{j=1}^{i-1} P^{j}_{PLEN})$$

closed under the following rules:

(1) If $x \in AForm$, then $x \in L^{R0}_{PLEN}$.
(2) If $x = \circ \frown y_1 \frown y_{a(\circ)}$, where each $y_j = \langle x_1 \frown \ldots \frown x_n \rangle$, each $y_j \in \bigcup_{j=1}^{i-1} L^{Rj}_{PLEN}$, some $y_j \in L^{Ri-1}_{PLEN}$, and $\circ \in Conn$, then $x \in L^{Ri}_{PLEN}$.
(3) If $x = \circ \frown y_1 \frown y_{a(\circ)}$, where each $y_j = \langle x_1 \frown \ldots \frown x_n \rangle$, $y_1 \in P^{i-1}_{PLEN}$, $y_{a(\circ)} \in \bigcup_{j=1}^{i-1} L^{Rj}_{PLEN}$, and $\circ \in Op/\{\Box, b, d\}$, then $x \in L^{Ri}_{PLEN}$.
(4) If $x = \circ \frown y_1 \frown y_{a(\circ)}$, where each $y_j = \langle x_1 \frown \ldots \frown x_n \rangle$, each $y_j \in \bigcup_{j=1}^{i-1} L^{Rj}_{PLEN}$, some $y_j \in L^{Ri-1}_{PLEN}$, and $\circ \in \{\Box, b, d\}$, then $x \in L^{Ri}_{PLEN}$.
(5) If $x = \circ \frown y_1 \frown \ldots \frown y_{a(\circ)}$, where each $y_j = \langle x_1 \frown \ldots \frown x_n \rangle$, each $y_j \in \bigcup_{j=1}^{i-1} P^{j}_{PLEN}$, some $y_j \in P^{i-1}_{PLEN}$, and $\circ \in Pop \cup Pop + \cup Syn$, then $x \in L^{Ri}_{PLEN}$.

(a) If $a(\circ) = 1$, then $x = \circ \frown y_{a(\circ)}$.
(b) For $\langle x_1 \frown \ldots \frown x_n \rangle$, if $n = 1$, then $\langle x_1 \frown \ldots \frown x_n \rangle = \langle x_1 \rangle$.

(6) If $x = \circ \frown y_1 \frown y_{a(\circ)}$ or $x = \circ \frown y_{a(\circ)}$, then $\circ$ is the *main operator* of x.

Definition 170 (String Algebraic Protocol-Theoretic Language (L^{R}_{PLEN}))

$$L^{R}_{PLEN} = \bigcup_{i=1}^{\omega} L^{Ri}_{PLEN}$$

Remark 171 Formulae formed from metalevel operators aren't assigned truth conditions that vary by state or EDG. *Pop* is the set of object level operators; these are assigned truth conditions that vary by state and by EDG. For the protocol-theoretic dynamic operators in *Op*, the string-algebraic syntax works by treating them just like protocol control operators, but taking elements of both languages.

For example: $[x]_p \frown \pi \frown A$. The metalevel operators work the same. For the four-place $\rho(/)() =, \mapsto$, $\rho(\pi/\pi')(\pi'') = \pi'''$ will be $\rho \frown \pi \frown \pi' \frown \pi'' \frown \pi'''$.

Definition 172 (Infix Notation Protocol Language (L^R_{PLEN})) Where $P_i \in AForm$, π_i is a metavariable for an arbitrary protocol with no occurrences of self-reference, $A, B, \ldots$ are metavariables for arbitrary protocol-theoretic formulae with no occurrences of self-reference, and distinct subscripts $x \in \{G, p, r, t, c\}$ and $y \in \{p, r, t\}$ in any of the formulae that take them produce p, r, t, c, and G variants of those formulae, L^R_{PLEN} can be defined by the following grammar:

$$(\pi_1 \in Sub(\pi_2)) \mid (\pi_1 \in PSub(\pi_2)) \mid (\pi_1 \in ISub(\pi_2)) \mid (\rho(\pi_1/\pi_2)(X_{\pi_1}) = X_{\pi_2})$$

$$\mid (\pi_1 \mapsto_s \pi_2) \mid (M(\pi))$$

$$(\pi_1 \equiv^s_x \pi_2) \mid (\pi_1 \leftrightarrow_y \pi_2) \mid (\pi_1 \curlywedge_y \pi_2) \mid (\pi_1 \curlyvee_y \pi_2) \mid ([\pi_i]_y A) \mid (\langle \pi_i \rangle_y A)$$

$$(P_i) \mid (A \wedge B) \mid (A \vee B) \mid (A \rightarrow B) \mid (A \equiv B) \mid (\neg A)$$

$$(\Box A) \mid (b(A)) \mid (d(A))$$

5.6.2 *Revising the Semantics of PLEN: (L^R_{PLEN})*

Definition 173 (Truth Rules for Object-Level Counterparts) The truth rules for the object-level counterparts of meta-level formulae do not extend valuations; they are defined in terms of the graph-theoretic properties of substructures of EDGs that contain a specified state, s:

(1) $(\pi \equiv^s_G \pi')$ $s \Vdash \pi \equiv^s_G \pi$ iff for all pointed EDGs, (g', s), if $(g', s) \sqsubseteq g$, then if $(g', s) \in G^\pi$, then there is a $(g'', s) \in G^{\pi'} \cap Sub(g)$ such that $(g', s) \simeq (g'', s)$, and the converse is also true.
(2) $(\pi \equiv^s_p \pi')$ $s \Vdash \pi \equiv^s_p \pi$ iff for all $s' \in S$ such that there is any $p \in P_g$ with $f(p) = s$ and $l(p) = s'$, for all $p' \in P_g$, if $s' = f(p')$, then $p' \in P_\pi$ iff $p' \in P_{\pi'}$.
(3) $(\pi \equiv^s_t \pi')$ $s \Vdash \pi \equiv^s_t \pi'$ iff for all $s' \in S$ such that there is any $p \in P_g$ with $f(p) = s$ and $l(p) = s'$, for all $t \in T_g$, if $s' = f(t)$, then $t \in T_\pi$ iff $t \in T_{\pi'}$.
(4) $(\pi \equiv^s_c \pi')$ $s \Vdash \pi \equiv^s_c \pi'$ iff iff for all $s' \in S$ such that there is any $p \in P_g$ with $f(p) = s$ and $l(p) = s'$, for all p_1 and p_2 in g:

   (a) if $p_1 \in P_\pi$ and $s' = f(p_1)$, then there is a $p_2 \in P_{\pi'}$ such that $s' = f(p_2)$ and $p_1 \rightleftharpoons p_2$,
   (b) if $p_2 \in P_\pi$ and $s' = f(p_2)$, then there is a $p_1 \in P_{\pi'}$ such that $s' = f(p_1)$ and $p_1 \rightleftharpoons p_2$.

(5) ($\pi \equiv_r^s \pi'$) $s \Vdash \pi \equiv_r^s \pi''$ for all $s' \in S$ such that there is any $p \in P_g$ with $f(p) = s$ and $l(p) = s'$, for all $e \in R+_0$, if $s' = f(e)$, then $e \in R+_\pi$ iff $e \in R+_{\pi'}$. for all $e \in R+_0$ such that $s = f(e)$, $e \in R+_\pi$ iff $e \in R+_{\pi'}$.

(6) ($\pi \mapsto_s \pi'$) $s \Vdash \pi \mapsto_s \pi'$ iff:

(a) (Semantic Condition) For all $p \in P_g$, if $s = first(p)$ and $p \in P_\pi$, then $p \in P_{\pi'}$.

(b) (Syntactic Conditions)

(i) $\forall \pi_i (\pi_i \in Sub(\pi) \cap \Pi_0 \rightarrow Sub(\pi'))$,

(ii) If $MC(\pi') \in \{;\ , \cap\}$, then $\pi \in \Pi_0$ or for every $\circ_i$ in $\pi = x_1 \frown \ldots \frown x_n$, $\circ_i \in \{;\ , \cap\}$,

(iii) If $MC(\pi') \in \{\cup\}$, then, where $\pi_i \in PISub(\pi')$:

(A) Either $\pi \mapsto_s \pi_i$ or $\pi \Rightarrow_{g,s} \pi_i$,

(B) For all EDGs, g, if π is procedurally executable at s, then π_j is procedurally executable at s where $\pi_j \in PISub(\pi) - \{\pi_i\}$.

(iv) If $MC(\pi') = *$, then $\pi' = \pi''*$ (for some π''), and $\pi \mapsto_s \pi''$ or (exclusively) there is some n such that $\pi \mapsto_{s};_1 \frown \ldots \frown;_n \frown \pi''_1 \frown \ldots \frown \pi''_{n+1}$ where π''_i is the ith occurrence of π'' in a protocol formed from n occurrences of sequential composition (i.e., π implements some finite length of sequential compositions of π'').

(v) If $* \notin \circ(\pi')$, then:

$$\neg\exists \pi_i \left((\pi_i \in (PSub(\pi) \cup Reduct(\pi))) \wedge \pi_i \mapsto \pi'\right)$$

(vi) If $MC(\pi') \neq \cup$, then:

$$\forall \pi_1 [\pi_1 \in PISub(\pi') \rightarrow \exists \pi_1 + (\pi_1 + \in (PSub(\pi) \cup Reduct(\pi)) \wedge \pi_1 + \mapsto \pi_1)]$$

(c) Let $Imp_s(\pi)$ be the set of *local implementations of* π, let $Rimp_s(\pi) = Imp_s(\pi) \cap (String(\Pi_0 \cup Prot)$ be the set of all *regular local implementations of* π, and let $Simp_s(\pi)$ be the set of all *sufficient local implementations of* π—the set of all protocols that satisfy all conditions but syntactic condition v.

5.6.3 The R-Derivation System

The derivation system for Revised PLEN adds the theory of protocol graph isomorphism equivalence to the basic derivation system in the form of the object-level counterparts of the following meta-level rules and axioms:

Definition 174 (Meta-Level Protocol Graph Isomorphism Equivalence Rules) Let π and π' be arbitrary regular protocols. Let X_π be a variable for any regular protocol such that π is a (regular) subprotocol of it. That is, for any $x \in P_{PLEN}$, $X_\pi = x$ iff $\pi \in Sub(x)$. Let formula meta-variable with embedded protocol meta-variable, $A(X_\pi)$, stand in for any protocol-theoretic formula that does not embed X_π directly in the following schematic formulae: $M(X_\pi)$, $X_\pi \leftrightarrow_r \ldots$, $X_\pi \curlywedge_r \ldots$, $X_\pi \curlyvee_r \ldots$, or any of the purely syntactic expressions (Sub, $PSub$, $ISub$, $\rho()() =$), save where $\cap \notin o(X_\pi)$. In this latter case, $A(X_\pi)$ can embed X_π directly in $X_\pi \curlyvee_r \pi''$ in GRT and $GRTC$. Then the following sets of rules and axioms constitute the core rules of the theory of protocol equivalence in the Revised PLEN Framework:

(1) (Protocol Graph Isomorphism Equivalence Rules: Meta-level)

(a)

$$\left[\begin{array}{ll}
\dfrac{\pi \equiv_G \pi',\ \rho(\pi/\pi')(X_\pi) = X_{\pi'},\ \pi \in ISub(X_\pi)}{X_\pi \equiv_G X_{\pi'}}(RGSRT) & \dfrac{\pi \equiv_G \pi'}{\pi \equiv_p \pi'}(PRT1.1) \\
\dfrac{\pi \equiv_G \pi',\ \rho(\pi/\pi')(X_\pi) = X_{\pi'}}{X_\pi \equiv_G X_{\pi'}}(GSRT) & \dfrac{\pi \equiv_p \pi'}{\pi \equiv_{fc} \pi'}(PRT1.2) \\
\dfrac{\pi \equiv_G \pi',\ \pi' \equiv_G \pi''}{\pi \equiv_G \pi''}(PGET) & \dfrac{\pi \equiv_{fc} \pi'}{\pi \equiv_t \pi'}(PRT1.3) \\
\dfrac{}{\pi \equiv_G \pi}(PGER) & \dfrac{\pi \leftrightarrow_r \pi'}{\pi = \pi'}(PI) \\
\dfrac{\pi \equiv_G \pi'}{\pi' \equiv_G \pi}(PGES) &
\end{array}\right]$$

(b)

$$\left[\begin{array}{l}
\dfrac{\pi \equiv_G \pi',\ \rho(\pi/\pi')(X_\pi) = X_{\pi'},\ A_{(X\pi)}}{A_{(X\pi')}}(GRT) \\
\dfrac{\pi \equiv_G \pi',\ A_{(\pi)}}{A_{(\pi')}}(GRTC)
\end{array}\right]$$

(c)

$$\left[\begin{array}{l}
\dfrac{\pi \equiv_p \pi'}{\Box(\pi \leftrightarrow_p \pi')}(PNEC) \\
\dfrac{\pi \equiv_t \pi'}{\Box(\pi \leftrightarrow_t \pi')}(TNEC) \\
\dfrac{\pi \equiv_r \pi'}{\Box(\pi \leftrightarrow_r \pi')}R(NEC)
\end{array}\right]$$

(2) (Meta-level Graph-theoretic Protocol Equivalence Axioms)

(a) $(\pi_1; \pi_2); \pi_3 \equiv_G \pi_1; (\pi_2; \pi_3)$
(b) $\pi_1; (\pi_2; \ldots; (\pi_{n-1}; \pi_n)) \equiv_G ((\pi_1; \pi_2); \ldots;)\pi_{n-1}); \pi_n \ (n < \aleph_0)$
(c) $\pi_1 \cap \pi_2 \equiv_G \pi_2 \cap \pi_1$
(d) $\pi_1 \cup \pi_2 \equiv_G \pi_2 \cup \pi_1$
(e) $(\pi_1; \pi_2); \pi_3 \equiv_G \pi_1; (\pi_2; \pi_3)$
(f) $\pi_1; (\pi_2 \cap \pi_3) \equiv_G ((\pi_1; \pi_2) \cap (\pi_1; \pi_3))$

(g) $\pi_1; (\pi_2 \cup \pi_3) \equiv_G ((\pi_1; \pi_2) \cup (\pi_1; \pi_3))$
(h) $(\pi_1 \cap \pi_2); (\pi_3 \cap \pi_4) \equiv_G (\pi_2 \cap \pi_1); (\pi_4 \cap \pi_3)$
(i) $(\pi_1 \cup \pi_2); (\pi_3 \cup \pi_4) \equiv_G (\pi_2 \cup \pi_1); (\pi_4 \cup \pi_3)$
(j) $(\pi_1 \cap \pi_2) \cap \pi_3 \equiv_G \pi_1 \cap (\pi_2 \cap \pi_3)$
(k) $(\pi_1 \cup \pi_2) \cup \pi_3 \equiv_G \pi_1 \cup (\pi_2 \cup \pi_3)$
(l) $(\pi_1 \cup \pi_2) \cap (\pi_3 \cup \pi_4) \equiv_G ((\pi_1 \cap \pi_3) \cup (\pi_1 \cap \pi_4)) \cup ((\pi_2 \cap \pi_3) \cup (\pi_2 \cap \pi_4))$

(3) (Meta-level Execution-theoretic Equivalence Axioms)

(a) $\big((\pi_1 \cap \pi_2); (\pi_3 \cap \pi_4) \equiv_p ((\pi_1; \pi_3) \cap (\pi_2; \pi_4) \cap ((\pi_1; \pi_4) \cap (\pi_2; \pi_3))\big)$
(b) $(\pi_1 \cup \pi_2); (\pi_3 \cup \pi_4) \equiv_p ((\pi_1; \pi_3) \cup (\pi_1; \pi_4)) \cup ((\pi_2; \pi_3) \cup (\pi_2; \pi_4))$
(c) $\pi_1 \cup (\pi_2 \cap \pi_3) \equiv_p ((((\pi_1 \cup \pi_2) \cap (\pi_1 \cup \pi_3) \cap \pi_2) \cap \pi_3) \cup \pi_1) \cup (\pi_2 \cap \pi_3)$
(d) $(\pi_1 \cap \pi_2) \cup (\pi_3 \cap \pi_4) \equiv_p ((\pi_1 \cup \pi_3) \cap (\pi_1 \cup \pi_4)) \cap ((\pi_2 \cup \pi_3) \cap (\pi_2 \cup \pi_4))$

(4) (Meta-level Implementation Rules)

(a)

$$\left[\begin{array}{l}
\dfrac{\pi \equiv_G \pi',\ \rho(\pi/\pi')(X_\pi) = X_{\pi'},\ \pi \in ISub(X_\pi)}{\pi+ \mapsto X_\pi \equiv \pi+ \mapsto X_{\pi'}}(NSTLL) \\
\dfrac{\begin{array}{r}\pi+ \mapsto \pi_1, (\pi++ \in Sub(\pi+) \to \neg(\pi++ \mapsto \pi_2)), pre(\pi+) \\ \to \big\langle((\langle\pi_2\rangle_p \top)?\big\rangle_p \top))\end{array}}{\pi+ \mapsto (\pi_1 \cup \pi_2)}(I1.1) \\
\dfrac{\begin{array}{r}\pi+ \mapsto \pi_1 \wedge pre(\pi+) \to \big\langle((\langle\pi_2\rangle_p \top)?\big\rangle_p \top)), \pi++ \mapsto \pi_2 \wedge pre(\pi+) \\ \to \big\langle((\langle\pi_1\rangle_p \top)?\big\rangle_p \top))\end{array}}{\neg((\pi+ \cup \pi++) \mapsto (\pi_1 \cup \pi_2))}(I1.2) \\
\dfrac{\pi \mapsto (\pi_1 \cup \pi_2)}{\bigvee\left[\begin{array}{l}(\pi \mapsto \pi_1 \wedge pre(\pi+) \to \big\langle((\langle\pi_2\rangle_p \top)?\big\rangle_p \top) \\ (\pi \mapsto \pi_2 \wedge pre(\pi+) \to \big\langle((\langle\pi_1\rangle_p \top)?\big\rangle_p \top)\end{array}\right]}(I1.3) \\
\dfrac{\pi+ \mapsto \pi_1, \pi++ \mapsto \pi_2}{\pi+; \pi++ \mapsto \pi_1; \pi_2}(I2) \\
\dfrac{\pi+; \pi++ \mapsto \pi_1; \pi_2}{\pi+ \mapsto \pi_1 \wedge \pi++ \mapsto \pi_2}(I2C) \\
\dfrac{\pi+ \mapsto \pi_1, \pi++ \mapsto \pi_2}{\pi+ \cap \pi++ \mapsto \pi_1 \cap \pi_2}(I3) \\
\dfrac{\pi+ \cap \pi++ \mapsto \pi_1 \cap \pi_2}{\pi+ \mapsto \pi_1 \wedge \pi++ \mapsto \pi_2}(I3C)
\end{array}\right]$$

Definition 175 (Object-Level Counterparts of Meta-Level Rules) Let π and π' be arbitrary regular protocols. Let X_π be a variable for any regular protocol such that π is a (regular) subprotocol of it. That is, for any $x \in P_{PLEN}$, $X_\pi = x$ iff $\pi \in Sub(x)$. Let formula meta-variable with embedded protocol meta-variable, $A(X_\pi)$, stand in for any protocol-theoretic formula that does not embed X_π directly in the following schematic formulae: $M(X_\pi)$, $X_\pi \leftrightarrow_r \ldots$, $X_\pi \curlywedge_r \ldots$, $X_\pi \curlyvee_r \ldots$, or any of the purely syntactic expressions (Sub, $PSub$, $ISub$, $\rho()() =$), save where $\cap \notin o(X_\pi)$. In this latter case, $A(X_\pi)$ can embed X_π directly in $X_\pi \curlyvee_r \pi''$ in GRT and $GRTC$. Then the following sets of rules and axioms constitute the core rules of the theory of protocol equivalence in the Revised PLEN Framework:

(1) (Protocol Graph Isomorphism Equivalence Rules: Object-level)

(a)

$$\left[\begin{array}{ll}\dfrac{\pi \equiv^s_G \pi',\ \rho(\pi/\pi')(X_\pi) = X_{\pi'},\ \pi \in ISub(X_\pi)}{X_\pi \equiv^s_G X_{\pi'}}(RGSRT) & \dfrac{\pi \equiv^s_G \pi'}{\pi \equiv^s_p \pi'}(PRT1.1)\\ \dfrac{\pi \equiv^s_G \pi',\ \rho(\pi/\pi')(X_\pi) = X_{\pi'}}{X_\pi \equiv^s_G X_{\pi'}}(GSRT) & \dfrac{\pi \equiv^s_p \pi'}{\pi \equiv^s_{fc} \pi'}(PRT1.2)\\ \dfrac{\pi \equiv^s_G \pi',\ \pi' \equiv^s_G \pi''}{\pi \equiv^s_G \pi''}(PGET) & \dfrac{\pi \equiv^s_{fc} \pi'}{\pi \equiv^s_t \pi'}(PRT1.3)\\ \dfrac{}{\pi \equiv^s_G \pi}(PGER) & \dfrac{\pi \leftrightarrow_r \pi'}{\pi = \pi'}(PI)\\ \dfrac{\pi \equiv^s_G \pi'}{\pi' \equiv^s_G \pi}(PGES) & \end{array}\right]$$

(b)

$$\left[\begin{array}{c}\dfrac{\pi \equiv^s_G \pi',\ \rho(\pi/\pi')(X_\pi) = X_{\pi'},\ A_{(X\pi)}}{A_{(X\pi')}}(GRT)\\ \dfrac{\pi \equiv^s_G \pi',\ A_{(\pi)}}{A_{(\pi')}}(GRTC)\end{array}\right]$$

(c)

$$\left[\dfrac{\langle\pi\rangle_r A}{\langle\pi\rangle_p A}(RP)\quad \dfrac{\langle\pi\rangle_p A}{\langle\pi\rangle_r A}(PR)\right]$$

(d)

$$\left[\begin{array}{c}\dfrac{\pi \equiv^s_p \pi'}{\Box(\pi \leftrightarrow_p \pi')}(PNEC)\\ \dfrac{\pi \equiv^s_t \pi'}{\Box(\pi \leftrightarrow_t \pi')}(TNEC)\\ \dfrac{\pi \equiv^s_r \pi'}{\Box(\pi \leftrightarrow_r \pi')}R(NEC)\end{array}\right]$$

(2) (Object-level Graph-theoretic Protocol Equivalence Axioms)

(a) $(\pi_1; \pi_2); \pi_3 \equiv^s_G \pi_1; (\pi_2; \pi_3)$
(b) $\pi_1; (\pi_2; \ldots; (\pi_{n-1}; \pi_n)) \equiv^s_G ((\pi_1; \pi_2); \ldots;)\pi_{n-1}); \pi_n \ (n < \aleph_0)$
(c) $\pi_1 \cap \pi_2 \equiv^s_G \pi_2 \cap \pi_1$
(d) $\pi_1 \cup \pi_2 \equiv^s_G \pi_2 \cup \pi_1$
(e) $(\pi_1; \pi_2); \pi_3 \equiv^s_G \pi_1; (\pi_2; \pi_3)$
(f) $\pi_1; (\pi_2 \cap \pi_3) \equiv^s_G ((\pi_1; \pi_2) \cap (\pi_1; \pi_3))$
(g) $\pi_1; (\pi_2 \cup \pi_3) \equiv^s_G ((\pi_1; \pi_2) \cup (\pi_1; \pi_3))$
(h) $(\pi_1 \cap \pi_2); (\pi_3 \cap \pi_4) \equiv^s_G (\pi_2 \cap \pi_1); (\pi_4 \cap \pi_3)$
(i) $(\pi_1 \cup \pi_2); (\pi_3 \cup \pi_4) \equiv^s_G (\pi_2 \cup \pi_1); (\pi_4 \cup \pi_3)$
(j) $(\pi_1 \cap \pi_2) \cap \pi_3 \equiv^s_G \pi_1 \cap (\pi_2 \cap \pi_3)$

(k) $(\pi_1 \cup \pi_2) \cup \pi_3 \equiv^s_G \pi_1 \cup (\pi_2 \cup \pi_3)$
(l) $(\pi_1 \cup \pi_2) \cap (\pi_3 \cup \pi_4) \equiv^s_G ((\pi_1 \cap \pi_3) \cup (\pi_1 \cap \pi_4)) \cup ((\pi_2 \cap \pi_3) \cup (\pi_2 \cap \pi_4))$

(3) (Object-level Execution-theoretic Equivalence Axioms)

(a) $\Big((\pi_1 \cap \pi_2); (\pi_3 \cap \pi_4) \equiv^s_p ((\pi_1; \pi_3) \cap (\pi_2; \pi_4) \cap ((\pi_1; \pi_4) \cap (\pi_2; \pi_3))\Big)$
(b) $(\pi_1 \cup \pi_2); (\pi_3 \cup \pi_4) \equiv^s_p ((\pi_1; \pi_3) \cup (\pi_1; \pi_4)) \cup ((\pi_2; \pi_3) \cup (\pi_2; \pi_4))$
(c) $\pi_1 \cup (\pi_2 \cap \pi_3) \equiv^s_p ((((\pi_1 \cup \pi_2) \cap (\pi_1 \cup \pi_3) \cap \pi_2) \cap \pi_3) \cup \pi_1) \cup (\pi_2 \cap \pi_3)$
(d) $(\pi_1 \cap \pi_2) \cup (\pi_3 \cap \pi_4) \equiv^s_p ((\pi_1 \cup \pi_3) \cap (\pi_1 \cup \pi_4)) \cap ((\pi_2 \cup \pi_3) \cap (\pi_2 \cup \pi_4))$

(4) (Object-level Implementation Rules)

(a)

$$
\left[
\begin{array}{l}
\dfrac{\pi \equiv^s_G \pi',\ \rho(\pi/\pi')(X_\pi) = X_{\pi'},\ \pi \in ISub(X_\pi)}{\pi+ \mapsto_s X_\pi \equiv \pi+ \mapsto_s X_{\pi'}}\,(NSTLL) \\
\dfrac{\begin{array}{l}\pi+ \mapsto_s \pi_1, (\pi++ \in Sub(\pi+) \to \neg(\pi++ \mapsto_s \pi_2)),\ pre(\pi+) \\ \qquad \to \langle((\langle\pi_2\rangle_p \top)?\rangle_p \top))\end{array}}{\pi+ \mapsto (\pi_1 \cup \pi_2)}\,(I1.1) \\
\dfrac{\begin{array}{l}\pi+ \mapsto_s \pi_1 \wedge pre(\pi+) \to \langle((\langle\pi_2\rangle_p \top)?\rangle_p \top)),\ \pi++ \mapsto_s \pi_2 \wedge pre(\pi+) \\ \qquad \to \langle((\langle\pi_1\rangle_p \top)?\rangle_p \top))\end{array}}{\neg((\pi+ \cup \pi++) \mapsto_s (\pi_1 \cup \pi_2))}\,(I1.2) \\
\dfrac{\pi \mapsto_s (\pi_1 \cup \pi_2)}{\bigvee \begin{bmatrix} (\pi \mapsto_s \pi_1 \wedge pre(\pi+) \to \langle((\langle\pi_2\rangle_p \top)?\rangle_p \top) \\ (\pi \mapsto_s \pi_2 \wedge pre(\pi+) \to \langle((\langle\pi_1\rangle_p \top)?\rangle_p \top) \end{bmatrix}}\,(I1.3) \\
\dfrac{\pi+ \mapsto_s \pi_1, \pi++ \mapsto_s \pi_2}{\pi+; \pi++ \mapsto_s \pi_1; \pi_2}\,(I2) \\
\dfrac{\pi+; \pi++ \mapsto_s \pi_1; \pi_2}{\pi+ \mapsto_s \pi_1 \wedge \pi++ \mapsto_s \pi_2}\,(I2C) \\
\dfrac{\pi+ \mapsto_s \pi_1, \pi++ \mapsto_s \pi_2}{\pi+ \cap \pi++ \mapsto \pi_1 \cap \pi_2}\,(I3) \\
\dfrac{\pi+ \cap \pi++ \mapsto \pi_1 \cap \pi_2}{\pi+ \mapsto_s \pi_1 \wedge \pi++ \mapsto_s \pi_2}\,(I3C)
\end{array}
\right]
$$

Remark 176 The object level laws governing the remaining forms of protocol equivalence (generated graph isomorphism equivalence (local graph-theoretic equivalence), procedural equivalence, ...) complete the full theory of protocol equivalence. I omit their presentation, here, for lack of space and interest with respect to the main lines of the project.

With these revisions and elaborations of PLEN in hand, we can now start examining the main results of the Revised PLEN system, starting with some useful features of the syntax, properties of implementation, and some interactions between protocol graphs and isomorphism.

5.7 Useful Lemmata

Claim 177 The following lemmata provide an inferential path to the proof of the centerpiece of the theory of (canonical) protocol graph isomorphism equivalence by way of PLEN's distinctive invariance theorem. The results presented in this section are central features of the analysis of protocols in Revised PLEN as well as useful for the derivation of the main results of the theory of protocol graph isomorphism equivalence. Appendix 2 houses the derivations of some minor lemmata of these lemmata.

5.7.1 Proper Immediate Subprotocols and Well-Foundedness

Remark 178 As it's been remarked in several places, the proper immediate subprotocol relation over regular protocols is an important syntactic structure, but it's not immediately clear why. There are three related reasons. The first is that the proper immediate subprotocol relation is a well-founded relation, and so we find that it's useful in deriving almost all of the main results of PLEN. The second reason is that the prior utility makes the relation useful for stating some of the main results (or lemmata thereof) in ways that enable relatively elementary proofs. The third reason is that it's useful for defining procedures on protocols that facilitate proofs of the foregoing kinds.

5.7.1.1 Methodological Observations

The main method of proof for many of the following results is structural induction over proper immediate subprotocols, and the secondary method is induction over protocol types. As the first is a form of structural induction over formulae, and both are forms of well-founded induction, they need no defense as reliable proof methods other than showing that they are both forms of well-founded induction. Nonetheless, let me make these proof styles explicit and comment on them, as their centrality makes the proper immediate subprotocol relation central to the system.

Claim 179 The well-foundedness of the proper immediate subprotocol relation ensures the applicability of the following forms of well-founded induction to protocols:

Definition 180 (Structural Induction Over $PISub(\pi)$ for Regular π) For any proposition, P, that quantifies over regular protocols, an induction over the proper immediate subprotocols of π is a proof of the following form:

(1) (Base) Prove that P holds for any arbitrary, regular protocol, π, with atomic immediate subprotocols.

   (a) (Atomics) Prove that P holds for atomics.
   (b) (Atomic Proper Immediate Subprotocols) Prove that P holds for all π with atomic, proper, immediate subprotocols.

(2) (Induction) Prove that if P holds for all proper immediate subprotocols of arbitrary regular protocol, π, then P holds for π.

Definition 181 (Induction Over Regular Protocol Types) For any proposition, P, that quantifies over regular protocols, an induction over protocol types is a proof of the following form:

(1) (Base) Prove that P holds for any arbitrary, regular protocol,$\pi \in \bigcup_{j=1}^{1} P_{PLEN}^{j}$.

   (a) (P_{PLEN}^{0}) Prove that P holds for atomics
   (b) (P_{PLEN}^{1}) Prove that P holds for all π with atomic, proper, immediate subprotocols.

(2) (Induction (From P_{PLEN}^{i-1} to P_{PLEN}^{i})) Prove that if P holds for all protocols, $\pi_{i-1} \in \bigcup_{j=1}^{i-1} P_{PLEN}^{j}$, then it holds for all $\pi \in \bigcup_{j=1}^{i} P_{PLEN}^{j}$.

Claim 182 The soundness of these proof procedures follows from the well-foundedness of the proper immediate subprotocol relation.

Proposition 183 *$PISub(\pi)$ is a well-founded relation*[7] *over P_{PLEN}.*

Proof Let X be any set of protocols. Consider any arbitrary $\pi \in X$. Either $PISub(\pi) \cap X = \emptyset$ or not. If it's empty, then π is $PISub$-minimal. If there is a $\pi' \in PISub(\pi)$, then either $PISub(\pi') \cap X = \emptyset$ or not. Iteration of this reasoning terminates with $PISub$ minimal protocols, or it iterates forever. The latter is impossible, as every protocol is a finite concatenated string. □

Claim 184 In virtue of this property, structural induction over proper immediate subprotocols and induction over protocol types are both valid forms of reasoning about protocols:

Proposition 185 *Structural induction over $PISub(\pi)$ is sound, and induction over protocol types is sound.*

Proof Structural induction over $PISub(\pi)$ is a form of well-founded induction due to the well-foundedness of $PISub$. From the soundness of this method of structural induction, it follows immediately that induction over protocol types is sound. The reasoning is that, for any $\pi \in \bigcup_{j=1}^{i} P_{PLEN}^{j}$, the proper immediate subprotocols of π will be in $\bigcup_{j=1}^{i-1} P_{PLEN}^{j}$, so, by the structural induction method, P holds for π. □

[7] Strictly, $PISub(\pi)$ is a well-founded, irreflexive, transitive, and asymmetric relation over $Sub(\pi)$ where the metavariable, π, names an arbitrary protocol. The proper immediate subprotocol relation itself is a well-founded, irreflexive, transitive, and asymmetric relation over P_{PLEN}.

5.7.2 Immediate Subprotocol Hierarchies

Remark 186 The proper immediate subprotocols of regular protocols will be exploited relentlessly in deriving the core results of PLEN. For now, we note three things. First, the relation enables the definition of a simple procedure for building immediate subprotocol hierarchies (sequences of subprotocols of a given protocol that are ordered by the immediate subprotocol relation). This procedure makes some derivations mechanical, and we walk through an example with mild independent interest, noting that the procedure and the result will play a role in further results. Finally, this particular example can be used to make the well-foundedness of $PISub(\pi)$ immediately visible, bringing us, in some sense, full circle.

Start with immediate subprotocol hierarchies:

Definition 187 (Immediate Subprotocol Hierarchies) Let π be an arbitrary, regular protocol. Then, for any arbitrary sequence of protocols, $\langle \pi_1, \ldots, \pi_n \rangle$:

(1) Arbitrary sequence of protocols, $\langle \pi_1, \ldots, \pi_n \rangle$, is an *immediate subprotocol hierarchy* over π's subprotocols ($\langle \pi_1, \ldots, \pi_n = \pi \rangle \in ISH(\pi)$) iff

$$\forall i \leq n, \pi_i, \pi_{i+1}[\pi_i, \pi_{i+1} \in Sub(\pi), (\pi_i \in ISub(\pi_{i+1})]$$

(2) Arbitrary sequence of protocols, $\langle \pi_1, \ldots, \pi_n \rangle$, is a *proper immediate subprotocol hierarchy* over π's subprotocols ($\langle \pi_1, \ldots, \pi_n = \pi \rangle \in PISH(\pi)$) iff

$$\forall i \leq n, \pi_i, \pi_{i+1}[\pi_i, \pi_{i+1} \in Sub(\pi), (\pi_i \in PISub(\pi_{i+1})]$$

(3) If $\langle \pi_1, \ldots, \pi_n \rangle$ is a proper immediate subprotocol hierarchy, then $\langle \pi_1, \ldots, \pi_n \rangle$ is an immediate subprotocol hierarchy.

With the notion of immediate subprotocol hierarchies in hand, one can define a number of mechanical procedures for constructing them. Here is an outline of one:

Definition 188 (Proper Immediate Subprotocol Hierarchy Construction Procedure) For arbitrary X_{π_i} and arbitrary subprotocol thereof, π_i, the string-algebraic syntax implies that $X_{\pi_i} = x_1 \frown \ldots \frown x_n$ and $\pi_i = x_j \frown \ldots \frown x_k$ where $j \geq 1$ and $k \leq n$:

$$X_{\pi_i} = x_1 \frown \ldots \frown x_j \frown \ldots \frown x_k \frown \ldots \frown x_n$$

Thus, π_i is a substring of X_{π_i}, by definition of subprotocols. If $\pi_i \neq X_{\pi_i}$, then construct a proper immediate subprotocol hierarchy, $\langle \pi_i, \ldots, X_{\pi_i} \rangle$, by the following procedure:

(1) Find π_i in the string.
(2) Track left through the string until you come to the first control operator, $\circ_1$, such that the operator embeds π_i as a proper immediate subprotocol of a more complex (i.e., higher type) protocol, π_{i+1}.

(a) (Tracking Procedure)
  (i) Start at x_j;
  (ii) move one symbol to the left; rename X
  (iii) if X is a control operator that scopes over π_i, then produce (π_i, π_{i+1}) where:
    (A) $\pi_{i+1} = \circ_1 \frown \pi_i$ if there is no protocol to the right of x_k in the scope of $\circ_1$
    (B) $\pi_{i+1} = \circ_1 \frown \pi_i \frown \pi_{a(\circ_1)}$ if there is a protocol, $\pi_{(a(\circ_1))}$, to the right of x_k in the scope of $\circ_1$
    (C) $\pi_{i+1} = \circ_1 \frown \pi_j \frown \pi_i$ if there is a protocol to the left of x_k in the scope of $\circ_1$
  (iv) if X is not a control operator that scopes over π_i, then go to ii.
(b) The first symbol to the left of π_i is either:
  (i) a control operator and there is no protocol in the scope of $\circ_1$ to the right of π_i,
  (ii) a control operator and there is a protocol in the scope of $\circ_1$ to the right of π_i, or
  (iii) a protocol in the scope of $\circ_1$ which is the next symbol in the leftward sequence.
(c) Thus, π_{i+1} has one of the following syntactic forms: $\circ_1 \frown \pi_i$, $\circ_1 \frown \pi_i \frown \pi_{a(\circ 1)}$, or $\circ_1 \frown \pi_j \frown \pi_i$.
(d) Thus, $\pi_i \in PISub(\pi_{i+1})$.

(3) Iterate this procedure by replacing π_i with π_{i+1} (applying the procedure to the protocol that is identified in the prior step) until arriving at the main control operator of X_{π_i}.

  (a) The procedure always terminates because all protocols are finite length.
  (b) The output of the procedure is a sequence, $\langle \pi_i, \pi_{i+1}, \pi_{i+1+1}, \ldots, X_{\pi_i} \rangle$.
  (c) By the definition of proper immediate subprotocols, $\langle \pi_i, \pi_{i+1}, \pi_{i+1+1}, \ldots, X_{\pi_i} \rangle \in PISH(X_{\pi_i})$.
  (d) If $\pi_i = X_{\pi_i}$, then the procedure cannot be applied, but π_i is an immediate subprotocol of itself, and the immediate subprotocol hierarchy $\{\pi_i\}$ can be constructed trivially.

Remark 189 This procedure will come in handy in thinking about protocol equivalence. Section 6.4 employs the procedure to generalize one of the major rules of protocol equivalence in PLEN. For now, just note that, using this procedure (or others like it), it's mechanical to show that (e.g.) the following theorem is true:

Theorem 190 (Immediate Subprotocol Hierarchy Theorem (IHT)) *Let X_π be a regular protocol. Then: for any regular π, $\pi \in Sub(X_\pi)$ iff there is an $\langle \pi_1, \ldots, \pi_n \rangle \in ISH(X_\pi)$, such that $\pi \in ISub(\pi_1)$ and $\pi_n = X_\pi$.*

Proof The LtoR proof is given by construction and the finite generativity of P_{PLEN}. The RtoL proof is immediate.

(1) (RtoL) For any π and π', if $\pi \in ISub(\pi')$, then $\pi \in Sub(\pi')$. Subprotocol inclusion is transitive. Thus, if $\langle\pi_1, \ldots, \pi_n\rangle \in ISH(X_\pi)$, $\pi \in ISub(\pi_1)$, and $\pi_n = X_\pi$, it immediately follows that each $\pi_i \in Sub(\pi_{i+1})$, so $\pi \in Sub(\pi_1)$, $\pi \in Sub(\pi_n)$, and thus $\pi \in Sub(X_\pi)$.

(2) (LtoR) Assume the string algebraic definition of P_{PLEN}. Protocols, on this definition, are concatenated functions from sequences of natural numbers into the set of atomics and control operators. The outputs of these functions are concatenated sequences of symbols. Immediate subprotocol hierarchies satisfying the theorem can be constructed by the Proper Immediate Subprotocol Hierarchy Construction Procedure. Applying this procedure to arbitrary protocols, X_π, and π, there are two cases.

   (a) Either $PISub(X_\pi) = \emptyset$ or not.
   (b) (Base) If the proper immediate subprotocol set is empty, then X_π is atomic and $\pi = X_\pi$, and any sequence comprising only X_π itself is an immediate subprotocol hierarchy—e.g., $\langle a\rangle$ and $\langle a, a, a\rangle$.
   (c) (Induction) If $PISub(X_\pi)$ is not empty, then a trivial induction supplies the result. For arbitrary, regular X_π of type $i \geq 1$, assume that IHT holds for all proper immediate subprotocols of X_π. For protocols with atomic proper immediate subprotocols, a single iteration of the procedure suffices; applying the procedure to an arbitrary atomic proper immediate subprotocol, a_i $(=\pi)$, generates X_{a_i} on the first step, and $\langle a_i, X_{a_i}\rangle \in PISH(X_{a_i})$, and, trivially, $\langle a_i, X_{a_i}\rangle \in IISH(X_{a_i})$. By finite generativity, every protocol is a sequence of symbols of finite length, so the procedure terminates after some $n \in \mathbb{N}$ iterations. Suppose that the procedure has been iterated $n - 1$ times starting with a proper subprotocol, π_1, of protocol X_{π_1} and terminating with a proper immediate subprotocol, π_{n-1}, of X_{π_1}. From the induction assumption, if π_1 is a subprotocol of π_{n-1}, then there is an ISH, $\langle\pi_1, \ldots \pi_{n-1}\rangle$. The proper immediate subprotocol hierarchy procedure produces a sequence, $\langle\pi_1, \ldots, \pi_{n-1}\rangle$, such that for each pair, $\pi_i, \pi_{i+1}, \pi_i \in PISub(\pi_{i+1})$ by the definition of the procedure. By definition, this is also an ISH, as required by the induction assumption; the induction assumption requires that there is such a sequence and the procedure built one for us. Iterating the procedure the nth time by starting at π_{n-1} in the string, X_{π_1}, generates the sequence, $\langle\pi_1, \ldots, \pi_{n-1}, X_{\pi_1}(= \pi_n)\rangle$, such that for each pair, $\pi_i, \pi_{i+1}, \pi_i \in PISub(\pi_{i+1})$ by the definition of the procedure. Thus, if π_1 is a proper subprotocol of X_{π_1}, the resulting sequence, $\langle\pi_1, \ldots, X_{\pi_1}\rangle$, is a PISH. By definition, $\langle\pi_1, \ldots, X_{\pi_1}\rangle \in ISH(X_{\pi_1})$.

(3) Thus, application of the procedure to X_π starting with π generates an immediate subprotocol hierarchy satisfying the conditions of the theorem, as required.

□

Remark 191 This result is itself moderately interesting, as it supports the further construction of hierarchical trees of subprotocols:

Definition 192 (Proper Immediate Subprotocol Hierarchy Trees) For arbitrary regular protocol, π, any trees over the subprotocols of π, $\langle \pi, Sub(\pi), R_{PISH(\pi)} \rangle$, are *proper immediate subprotocol hierarchy trees* iff π is the root node of the tree and $R_{PISH(\pi)} \subseteq Sub(\pi) \times Sub(\pi)$ such that, for any π' and π'' in the subprotocols of π, $\langle \pi', \pi'' \rangle \in R_{PISH(\pi)}$ iff $\pi' \in PISub(\pi'')$, and every branch of the tree extends from π itself to an atomic subprotocol of π.

Remark 193 In other words, these trees are such that every branch is a proper immediate subprotocol hierarchy of π, $\langle \pi', \ldots, \pi \rangle$, with π as the root of the branch and some atomic subprotocol of π as its leaf.

Corollary 194 *Proper immediate subprotocol hierarchy trees can be constructed by applying the proper immediate subprotocol hierarchy construction procedure.*

Proof (Outline) The procedure is a proper immediate subprotocol hierarchy tree branch construction procedure. Run the procedure starting with every atomic subprotocol, a_i, of π to generate a set of ISHs, $\langle a_1, \ldots, \pi \rangle, \ldots, \langle a_n, \ldots, \pi \rangle$. Build a tree so that π is the root and each a_i is a leaf, and then, working down from the root, π, perform a graph reduction (combine every identical node while preserving branches from them). This produces a unique tree with π as the root and each atomic subprotocol of π as a leaf that maps out the syntax of π in terms of the proper immediate subprotocol relation; every branch is an ISH from a_i to π. □

Claim 195 Proper immediate subprotocol hierarchy trees make the well-foundedness of the $PISub(\pi)$ relation visually appreciable; constructing such a tree over the subprotocols of any protocol or even a set of trees over any arbitrary set of protocols constructs a set of leaf nodes, which are the $PISub$ minimal elements. See Fig. 5.1.

5.7.3 Implementation Rules

Remark 196 Implementations are also unique structures in PLEN. The implementations of a protocol, π, are protocols that identify a kind of minimal alternative procedure for executing π. As such, their properties are independently interesting, and they also have some important features of use to us in considering protocol equivalence.

5.7.3.1 Canonical Implementation Rules

Claim 197 The following rules hold for all regular protocols. The reasoning is mechanical or immediate from the definition of implementations for each of them.

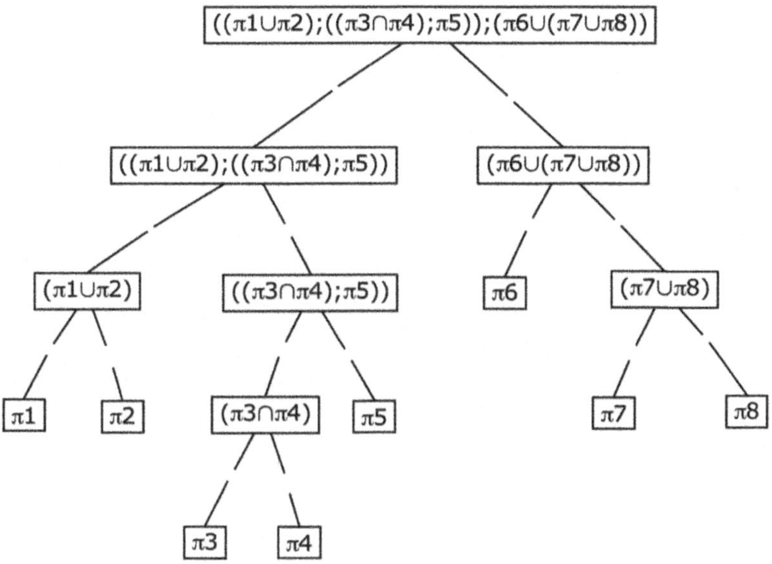

Fig. 5.1 Proper immediate subprotocol hierarchy trees: Proper immediate subprotocol hierarchy trees provide a useful way of seeing that the proper immediate subprotocol relation over any set of subprotocols is a well-founded relation, which is a fact that underlies the methods of proof used in this chapter—induction over the structure of protocols. For any protocol, if the procedure is carried out all the way to atomics, the leaves of the tree are the $PISub(\pi)$-minimal elements of $Sub(\pi)$

Theorem 198 (I1) *If* $\pi = \pi_1 \cup \pi_2$, *then:*

(1) For all implementations, $\pi+$, *of (arbitrarily chosen)* $\pi_i \in \{\pi_1, \pi_2\}$, *if* $\pi+$ *does not contain any proper subprotocol or main control operator preserving reduction that implements (the other proper immediate subprotocol)* π_j, *and it is axiomatic that* $\langle \pi+\rangle_p \top \rightarrow \left\langle ((\langle \pi_j\rangle_p \top)?\right\rangle_p \top$, *then* $\pi+ \mapsto \pi$.

(2) For all implementations, $\pi+$, *of* π, *if* $\pi+$ *implements* π_1 *and* $\langle \pi+\rangle_p \top \rightarrow \langle ((\langle \pi_2\rangle_p \top)?\rangle_p \top$ *while* $\pi++$ *implements* π_2 *and* $\langle \pi+\rangle_p \top \rightarrow \langle ((\langle \pi_1\rangle_p \top)?\rangle_p \top$, *then* $\pi + \cup \pi ++$ *fails to implement* π.

(3) For all implementations, $\pi+$, *of* π, $\pi+$ *implements either* π_1 *or* π_2 *and* $\langle \pi+\rangle_p \top \rightarrow \left\langle ((\langle \pi_j\rangle_p \top)?\right\rangle_p \top$ *where* $j = 1$ *if* $\pi+ \mapsto \pi_2$, *while* $j = 2$ *if* $\pi+ \mapsto \pi_1$.

(a) Mutatis mutandis for dictation.

Proof (Proof-plan) Direct arguments for each clause.

(1) Let $\pi+$ implement π_i, and satisfy the premises. Then: syntactic form is guaranteed, execution sets of $\pi+$ are subsets of π by the execution rules ($\pi+$ executions branch from some π_j execution if $\langle \pi+\rangle_p \top \rightarrow \left\langle ((\langle \pi_j\rangle_p \top)?\right\rangle_p \top$,

thereby making the executions of $\pi+$ that branch from π_j executions into executions of π), all atomics are shared, and no proper subprotocol or protocol formed from the atomics and operators of $\pi+$ with the same main control operator of $\pi+$ can suffice to execute π. Thus, $\pi+$ implements π.

(a) To clarify the final condition: Consider $\pi' = (((\pi_1 \cap \pi_2) \cap \pi_3) \cup \pi_1)$. It is immediately visible that $\langle(\pi_1 \cap \pi_2) \cap \pi_3\rangle_p \top \rightarrow \left\langle(\langle\pi_1\rangle_p \top)?\right\rangle_p \top$. Thus, $(\pi_1 \cap \pi_2) \cap \pi_3)$ implements $((\pi_1 \cap \pi_2) \cap \pi_3)$ itself and π'. It's tempting to think that $((\pi_1 \cap \pi_2) \cap \pi_3)$ doesn't implement π', given that the proper subprotocol of $(\pi_1 \cap \pi_2) \cap \pi_3)$, π_1, implements π' if $\langle\pi_1\rangle_p \top \rightarrow \left\langle(\langle(\pi_1 \cap \pi_2) \cap \pi_3)\rangle_p \top)?\right\rangle_p \top$. However, this condition need not hold.

(2) This result is trivial: the definition of implementations (condition (c)) eliminates choice operators under exactly the conditions in the antecedent.
(3) Immediate from the definition of implementations.

□

Theorem 199 (I2) *If $\pi = \pi_1; \pi_2$, then if $\pi+$ implements π_1 while $\pi++$ implements π_2, $\pi+; \pi++$ implements π.*

Proof (Proof-Plan) Mechanical check: Let $\pi+$ implement π_1 and $\pi++$ implement π_2. Consider $\pi+; \pi++$. Syntactic form is guaranteed, execution sets of $\pi+; \pi++$ are subsets of π by the execution rules, all atomics are shared, by the reasoning in the subproof, no proper subprotocol of $\pi+; \pi++$ or main control operator preserving reduction of $\pi+; \pi++$ can suffice to implement π, and, from the assumptions, every proper subprotocol is implemented by some proper subprotocol or reduct of $\pi+$. Thus, all conditions are met for $\pi+; \pi++$ to implement π.

(Subproof) Let $\pi-$ be a proper subprotocol of $\pi+; \pi++$ that implements π. By the definition of syntactic subprotocols, $\pi-$ is a subprotocol of $\pi+$ or a subprotocol of $\pi++$. But it's not possible for a subprotocol of $\pi+$ to implement π. Suppose that $\pi-$ implements π and is a subprotocol of $\pi+$. There are two cases:

(1) $(\pi- = \pi+)$ If $\pi- = \pi+$, then our conditions are contradicted; $\pi+$ implements π_1 and $\pi_1; \pi_2$:

(a) If $\pi+$ implements $\pi_1; \pi_2$ there must be a proper subprotocol of $\pi+$ that implements π_1, immediately contradicting the assumption that $\pi+$ implements π_1.

(2) $(\pi- \in PSub(\pi+))$ If $\pi-$ is a proper subprotocol of $\pi+$, then by the final condition of the definition of implementations, there must be a proper subprotocol or reduct of $\pi-$, $\pi--$, that implements π_1.

(a) If $\pi--$ is a proper subprotocol of $\pi-$ that suffices to implement π_1, then $\pi--$ is a proper subprotocol of $\pi+$ that implements π_1. Thus, $\pi+$ does not implement π_1, contradicting our assumption.
(b) If $\pi--$ is a reduct of $\pi-$ that suffices to implement π_1, then, again, our assumption that $\pi+$ implements π_1 is contradicted. There are two subcases:

(i) If $\pi --$ is a reduct of a proper immediate subprotocol of $\pi+$ that suffices to implement π_1, then there is a reduct of $\pi+$ that suffices to implement π_1 of the form $\pi -- \circ \pi_i$ or $\pi_i \circ \pi --$ where π_i is the other proper immediate subprotocol of $\pi+$ and $\circ = MC(\pi+)$. Thus, $\pi+$ does not implement π_1, contradicting our assumption.

(ii) If $\pi-$ is embedded at greater depth in $\pi+$, just construct the reduct of $\pi+$ inductively, starting at the first protocol that embeds $\pi-$ as a proper immediate subprotocol. For a schematic instance, where $\pi+$ has the form $\circ \frown \ldots \circ_1 \frown \pi- \frown \pi_{a(\circ_1)} \frown \ldots$, just construct the reduct of $\pi+$: $\circ \frown \ldots \circ_1 \frown \pi -- \frown \pi_{a(\circ_1)} \frown \ldots$. This reduct implements π_1, contradicting the assumption that $\pi+$ implements π_1.

Thus, in any case, a subprotocol of $\pi+$ could not be an implementation of π without contradicting the definition of implementation or our assumptions. Run the same argument for $\pi ++$. Thus, there can be no proper subprotocol of $\pi+; \pi ++$ that implements π.

The same reasoning applies (mutatis mutandis) to the main control operator preserving reduction case. By the definition of main control operator preserving reductions, if there is a reduct of $\pi+; \pi ++$ that implements $\pi_1; \pi_2$, then this reduct must be a protocol of the form: $\pi-; \pi --$, where $\pi-$ is a reduct of $\pi+$ and $\pi --$ is a reduct of $\pi ++$, $\pi-; \pi ++$ where $\pi-$ is a reduct of $\pi+$, or $\pi+; \pi --$ where $\pi --$ is a reduct of $\pi ++$. Thus, there must be a proper subprotocol or reduct of this reduct that suffices to implement π_1 (mutatis mutandis, π_2). Suppose there is a reduct of $\pi+; \pi++$ that implements $\pi_1; \pi_2$. If the reduct is $\pi-; \pi++$ or $\pi-; \pi --$, and $\pi-$ implements π_1, then $\pi+$ doesn't implement π_1, contradicting our assumptions. If the reduct is $\pi+; \pi --$ or $\pi-; \pi --$, and $\pi --$ implements π_2, then $\pi ++$ doesn't, contradicting our assumptions. If neither antecedent holds, then the final condition of the definition of implementations fails for this reduct of $\pi+; \pi ++$, contradicting our assumption. Thus, there can be no such reduct.

□

Remark 200 In step 2 of the subproof above, even if it were the case that $\pi- = \pi --$, which is ruled out, then there is a proper subprotocol of $\pi+$ that implements π_1, contradicting our assumption. Allowing that $\pi- = \pi+$, in step 2b, either $\pi --$ is itself a reduct of $\pi+$ because $\pi- = \pi+$ or it's a reduct of a proper subprotocol of $\pi+$ because $\pi-$ is a proper subprotocol of $\pi+$. The latter case is dealt with in the argument, but note that if $\pi --$ is a reduct of $\pi-$ (hence, $\pi+$) that implements π_1, then the contradiction is immediate; $\pi+$ can't be an implementation of π_1, after all.

Theorem 201 (I3) *If* $\pi = \pi_1 \cap \pi_2$, *then if* $\pi+$ *implements* π_1 *while* $\pi ++$ *implements* π_2, $\pi + \cap \pi ++$ *implements* π.

Proof (Sketch) The argument for I2, with mild modifications, suffices. See Appendix 2.

□

Theorem 202 (Converse of I2) *If $\pi+; \pi++$ implements $\pi = \pi_1; \pi_2$, then $\pi+$ implements π_1 and $\pi++$ implements π_2.*

Proof (Proof-Plan) Induction over proper immediate subprotocols.

(1) (Base) All implementations of sequential compositions of atomic protocols are identical to them. The result follows immediately.
(2) (Induction) Assume that the result holds for all proper immediate subprotocols of π. Let $\pi+; \pi++$ implement $\pi = \pi_1; \pi_2$. Suppose that $\pi+$ doesn't implement π_1. This could be due to the failure of any clause of the definition of implementations, but the failure of any clause suffices to show that $\pi+; \pi++$ doesn't implement π. If $\pi+$ doesn't suffice to execute π_1, doesn't have the right syntactic form, doesn't share atomics with π_1, or contains a proper subprotocol or reduct that does, then $\pi+; \pi++$ doesn't implement π. The same argument works for $\pi++$ with respect to π_2.

□

Theorem 203 (Converse of I3) *If $\pi+ \cap \pi++$ implements $\pi = \pi_1 \cap \pi_2$, then $\pi+$ implements π_1 and $\pi++$ implements π_2.*

Proof (Sketch) The argument for I2 Converse, with mild modifications, suffices. See Appendix 2. □

Claim 204 Together, these results bind the structures of implementations to the structures of protocols that they implement. This, in turn, proves useful in exploring the properties of protocol graphs, as is shown in the reasoning for the main results below. These implementation rules can, for instance, be used in reasoning to the invariance theorem or replacement rules discussed in Sects. 6.3 and 6.4.

5.7.3.2 Implementations and Protocol Graphs

A further rule relates regular protocols, syntactic replacement, immediate subprotocols, and implementations.

Lemma 205 (NSTL Lemma) *For all regular π and π', if $\pi \equiv_G \pi'$, $\rho(\pi/\pi')(X_\pi) = X_{\pi'}$, and $\pi \in ISub(X_\pi)$, then $\pi+ \mapsto X_\pi$ iff $\pi+ \mapsto X_{\pi'}$.*

Proof (Proof-Plan) Case by case. Assume the antecedent. Let $\pi+$ implement X_π. Prove: $\pi+$ implements $X_{\pi'}$. Let $X_\pi = \pi \circ \pi_{a(\circ)}$. By syntactic replacement, $X_{\pi'} = \pi' \circ \pi_{a(\circ)}$. There are two cases to consider:

Case 1. $(* \notin \circ(X_\pi))$ By induction over proper immediate subprotocols:

(1) (Base) For protocols of types 0 or 1, graph-theoretic equivalence implies identity, commutation, some redundant iteration of parallel executions and choice, or main control operator preserving expansions (i.e., many protocols are equivalent to their reducts). For instance, $a; b$ is equivalent to protocols of the form $(a; b) \cap (a; b)$, $(a; b) \cup (a; b)$, $a; (b \cap b)$, and $a; (b \cup b)$, as well as

further elaborations on this structure. All regular implementations, then, will be syntactically identical protocols, commutations, or reducts of equivalent protocols.

(a) The case that's almost interesting is type 1 choice protocols: $a \cup b$. These are implemented by fixed point tests for executability of their proper immediate subprotocols: $\nabla(\langle a\rangle_p \top)?; b$ and $\nabla(\langle b\rangle_p \top)?; a$. Of course, any X_π and $X_{\pi'}$ of thse forms that satisfy the premises will be such that $\pi = \pi'$ and $X_\pi = X_{\pi'}$, and so any implementation of one will immediately suffice to implement the other.

(Induction) If the result holds for all proper immediate subprotocols of X_π, then the result holds for X_π. Assume the induction premise. Let $\pi+$ implement X_π.

(a) Semantic Condition: By PRT1 (sec. 6.2), $\pi \equiv_G \pi'$ implies $P_\pi = P_{\pi'}$. Thus, by replacement of identical sets in the execution rules of $P_{X\pi}$ and $P_{X\pi'}$, $P_{X\pi} = P_{X\pi'}$. Thus, any protocol that suffices to execute X_π also suffices to execute $X_{\pi'}$.

(b) Syntactic Conditions:

(i) By definition, $\pi+$ has the right syntactic form on pain of contradicting the assumption that it's an implementation of X_π; by syntactic replacement, $MC(X_\pi) = MC(X_{\pi'})$, and so any syntactic features of $\pi+$ conditioned on the main control operator of $X_{\pi'}$ are already satisfied by $\pi+$ in virtue of its being an implementation of X_π.

(ii) By construction, $\pi+$ must be formed from atomic subprotocols that are among the atomic subprotocols of $X_{\pi'}$. Each path p that executes $\pi+$ is a sequence of edges. The atomics that label these edges must, by the execution rules and the uniqueness of edge labels, be among the atomic subprotocols of $\pi+$. Since p must also execute $X_{\pi'}$ (from PRT1 or (i)), those atomics in $\pi+$ must also be atomic subprotocols of $X_{\pi'}$. The case that's interesting is where $\pi+$ is a non-regular implementation of X_π and $MC(X_\pi) = \cup$. But then $MC(X_{\pi'}) = \cup$ as well, and, from the induction assumption, if $\pi+$ implements π and its executability implies that the other proper immediate subprotocol of X_π is executable, then it implements π' and (by syntactic replacement) its executability implies that the other immediate subprotocol of $X_{\pi'}$ is executable, as required. Thus, any non-regular implementation of X_π that implements π will be a non-regular implementation of $X_{\pi'}$ that implements π'. If $\pi+$ merely dictates π and its executability implies that the other proper immediate subprotocol of X_π is executable, then (by PRT1), it dictates π' and its executability implies the executability of the other proper immediate subprotocol of $X_{\pi'}$, as required. Thus, any non-regular implementation of X_π that dictates π will be a non-regular implementation of $X_{\pi'}$ that dictates π'.

(iii) Suppose that there is a proper subprotocol of $\pi+$, $\pi++$, that implements $X_{\pi'}$. Then $\pi++$'s executions are X_π executions (from the first step), its atomics are subprotocols of X_π (from the reasoning in the second step), it has the right syntactic form (from the third step) and suffices to implement $X_{\pi'}$'s proper immediate subprotocols with its proper subprotocols or reducts, and, if there were a proper subprotocol of $\pi++$ that implemented X_π, it would be a proper subprotocol of $\pi+$, contradicting the assumption that $\pi+$ implements X_π. Thus, there is no proper subprotocol of $\pi++$ that implements X_π. Thus, $\pi++$ implements X_π because all conditions are met. This contradicts the assumption that $\pi+$ implements X_π. Thus, there can be no proper subprotocol of $\pi++$ that implements $X_{\pi'}$. The exact same argument runs for main control operator preserving reductions of $\pi+$.

(iv) From the assumption that $\pi+$ implements X_π, it follows that $\pi+$ contains a reduct or proper subprotocol, $\pi-$, that implements π and another (not necessarily distinct) that implements each $\pi_{a(\circ)}$. From the induction premise, since $\pi \equiv_G \pi'$, $\rho(\pi/\pi')(\pi) = \pi'$, and $\pi \in ISub(\pi)$, $\pi-$ implements π'. By syntactic replacement, $\pi+$ contains a proper subprotocol or reduct that implements $\pi_{a(\circ)}$. Thus, $\pi+$ contains proper subprotocols or reducts that implement each proper immediate subprotocol of $X_{\pi'}$.

Case 2. ($* \in \circ(X_\pi)$) Direct argument for arbitrary protocols. By definition, if $X_\pi = \pi*$, then $\pi+$ is a finite length sequential composition of $\pi++$ where $\pi++$ implements π. There are two cases:

(1) If there is no iteration inside of π, then use the reasoning for Case 1 to show that $\pi++$ implements π'. In that case, it's trivial that $\pi++$ implements $X_{\pi'} = \pi'*$.

(2) If there are iterations inside of π, then the argument can just be iterated. Start at the level of the largest protocols containing no iterations. Apply the argument for Case 1. From there, the reasoning above for the case can be directly applied at every level at which iterations appear. Thus, by the definition of implementations, $\pi+$ implements $X_{\pi'}$. Transpose for the converse.

Thus, by the definition of implementations, $\pi+$ implements $X_{\pi'}$. Transpose for the converse. □

Remark 206 This rule plays a role in deriving the core syntactic replacement rules of PLEN in Sect. 6.4.

5.7.4 *Protocol Graphs and Isomorphism*

Prior to delving into the theory of the canonical graph-theoretic protocol equivalence in PLEN, I want to cash out the promise that protocol graphs are unique in the way they bind graph-theoretic structures to the syntax of protocols.

5.7.4.1 Structural Uniqueness

Claim 207 Protocol graphs for regular protocols are unique up to isomorphism.

That is, the protocol graphs of a given π are all isomorphic, thereby showing that there is a unique, abstract graph-theoretic structure that is mapped to the syntax of a given protocol. This is a result of central importance in PLEN as it plays a role in the derivation of many of PLEN's core results, and it also underlies the hyperintensionality of protocol-theoretic contexts in Revised PLEN (see Sects. 6.3–6.5).

More precisely:

Theorem 208 (Structural Uniqueness Theorem (SUT)) *For all regular protocols, π, and all EDGs, g and g', if $g \in G^{\pi}$ and $g' \in G^{\pi}$, then $g \simeq g'$.*

(Rule Form)

$$\forall \pi \in P_{PLEN} \cap String(\Pi_0 \cup Prot)$$
$$\forall g, g' \left[\frac{g \in G^{\pi},\ g' \in G^{\pi}}{g \simeq g'} \right]$$

The proof of SUT turns out to be interesting in the context of the rest of the system, so I present it below. In particular, the proof utilizes a number of lemmata that are of direct use in proving the invariance theorem in Sect. 6.3. More, the pattern of reasoning in that proof is very similar to that in the proof for SUT (beyond being structural inductions over proper immediate subprotocols).

Proof (Proof-Plan) The derivation of SUT from the definitions of Revised PLEN proceeds by structural induction over the proper immediate subprotocols of arbitrary regular protocols. First, however, we need to derive some more or less immediate lemmata concerning the relations among protocols and protocol graphs. □

Lemma 209 *The following six properties relate paths, protocol graphs, and isomorphism:*

(L1) For every regular protocol, $\pi = \pi_1 \circ \pi_2$, if $g \in G^{\pi}$, then:

(1) For every maximal execution, p, of π in g, there are subgraphs of g, g_1 and g_2, such that $g_1 \in G^{\pi_1}$, $g_2 \in G^{\pi_2}$, and p comprises maximal executions of π_1 and π_2 in g_1 and g_2, respectively, in distinctive ways with respect to the main control operator of π:

- *(a) ($\pi = \pi_1; \pi_2$) For any maximal execution of π in of g, p, there is a maximal execution of π_1 in g_1, q, and a maximal execution of π_2 in g_2, r, such that $p = qr$.*
- *(b) ($\pi = \pi_1 \cap \pi_2$) For any maximal execution of π in g, p, there is a maximal execution of π_1 in g_1, q, and a maximal execution of π_2 in g_2, r, such that $p = q$ and $q \rightleftharpoons r$ or $p = r$ and $q \rightleftharpoons r$.*

(c) ($\pi = \pi_1 \cup \pi_2$) *For any maximal execution of* π *in* g, p, *there is a maximal execution of* π_1 *in* g_1, q, *and a maximal execution of* π_2 *in* g_2, r, *such that* $p = q$ *or* $p = r$ *and* $first(p) = first(q) = first(r)$.

(d) ($\pi = \pi_1*$) *For any maximal execution of* π_1* *in* g, p, *for every* $n \geq 1$, p *is composed of* n *concatenation of executions of* π_1, *every maximal execution of* g *branches from every other maximal execution of* g, *save for a detached loop with* π_1* *as the middle term, and every implementation of* π_1 *is executed* n *concatenated times across all of* g*'s branches.*

(2) *For all protocol graphs of* π_1 *among the subgraphs of* g, *there is a protocol graph of* π_2 *among the subgraphs of* g, *such that the maximal paths of* g_1 *and* g_2 *all concatenate, iterate, correspond, or branch as in the operator rules above, composing maximal executions of* π *in* g, *and converse. More precisely:*

(a) *For every maximal execution,* p_1, *of* π_1 *in* g_1 *there is a maximal execution,* p_2, *of* π_2 *in* g_2 *such that* $p_1 \oplus p_2$ *relates to a* $p \in P_\pi$ *accordingly:* $p = p_1 \oplus p_2$, $p = p_1$ *and* $p_1 \oplus p_2$, *or* $p = p_2$ *and* $p_1 \oplus p_2$.

(b) *For every maximal execution,* p_2, *of* π_2 *in* g_2 *there is a maximal execution,* p_1, *of* π_1 *in* g_1 *such that* $p_1 \oplus p_2$ *relates to a* $p \in P_\pi$ *accordingly:* $p = p_1 \oplus p_2$, $p = p_1$ *and* $p_1 \oplus p_2$, *or* $p = p_2$ *and* $p_1 \oplus p_2$.

Proof (Proof-Plan) Direct arguments for both subclaims of the theorem.

(1) The execution rules coupled with the definitions of protocol graphs suffice for each result.

(a) First, the execution rules require that the maximal executions of π in g have the required graph-theoretic structures; sequential composition is bound to concatenation of paths, parallel execution is bound to corresponding paths, choice is bound to branching paths, and iteration is bound to branching concatenated paths and loops. Thus, where π has a regular syntactic form, $\pi = \circ \frown \pi_1 \frown \pi_2$ or $\pi = * \frown \pi_1$, the execution rules imply that there must be executions of π_1 and π_2 (or iterations of executions of π_1), call them q and r (or $q_1, \ldots, q_n$), that are related as in the result

(i) For iteration, note that if the maximal executions of π (where $\pi = \pi_1*$ for proper immediate subprotocol, π_1) in g don't have the form described, then either there are looping paths among the non-vacuous (looping, single edges with π_1* the middle term) executions of π or there are disconnected, non-looping executions. The latter case violates NST—there is a proper subgraph without them that satisfies the structural display rules. The former case results in the violation of $SD1(2)_{\pi_1*}(g)$, as there will be executions of π_1* that are not maximal paths (as any loop results in a path that is always a proper subpath of another path).

(b) Given the prior step, it suffices to show that there are protocol graphs of π_1 and π_2 (or just π_1) in which the q and r so concatenated, branched, iterated, or correspondent are maximal paths. This is straightforward from the definition of protocol graphs. The execution rules and NST require that, for arbitrary π_i and $g_i \in G^{\pi_i}$, every maximal path of g_i is an execution of π_i or one that results from the interleaving of paths due to parallel execution (this is the only construction that could result in there being maximal paths in any protocol graph of any regular π that are not executions of π_i); any maximal paths not thus formed can be excised while leaving the structural display relations intact, implying that there is a proper subgraph or a graph reduction that suffices. More precisely, because there must be maximal paths in g of the requisite form, and there must (by RC) be at least one protocol graph of each of π_1 and π_2 among the subgraphs of g, and such protocol graphs must (by SD1(2)) contain maximal executions of π_1 and π_2 (respectively), the only condition in which this lemma fails is where at least one of the proper immediate subpaths of at least one maximal execution of π in g somehow lies outside of all of the path sets of the protocol graphs of the relevant proper immediate subprotocol (i.e., some execution of (e.g.) π_1 lies outside of any protocol graph of π_1). If there is such a path, however, then either it branches from some maximal path in g or it doesn't. Suppose that the problematic path is an execution of π_1. Call it path q. If q doesn't branch from any maximal path in g, then it's a disconnected path, and so it violates NST as there must be a proper subgraph without q that satisfies the structural display rules. If q does branch from a maximal path in g, then it violates NST, as a graph reduction collapses it to a maximal execution of a protocol graph of π_1. Thus, there is no such path.

(2) The execution rules deliver the result that any such paths execute π. The result follows by a trivial reductio: if any other condition in the clause fails, the assumption that g is a protocol graph of π is contradicted. In particular, the graph reduction condition required by $NST_\pi(g)$ eliminates any redundant paths that might suffice to be counterexamples. If there are protocol graphs of π_1 and π_2 in which the maximal executions of π_1 (resp. π_2) are not structured as required by the lemma above, then they are redundant; there is no structural display rule that requires them to be in g, and so NST is violated. In particular:

   (a) If the redundant paths are detached, then there is a proper subgraph of g that satisfies the structural display rules.
   (b) If the redundant paths are connected by branching from the root state of g, then there is a reduction of g that satisfies the structural display rules.

□

(L2) For arbitrary procedural paths, p, q, and r, if $p \simeq q$ and $r \sqsubseteq p$, then there is an r' such that $r' \sqsubseteq q$ and $r \simeq r'$.

Proof (Proof-Plan) This is mechanical from the definition of isomorphism; the original pair of bijections suffices. □

(L3) For every g and p in P_g, if $g \in G^{\pi}$, then there is a maximal execution of π in g, q, such that $p \sqsubseteq q$ or there are $q_1, \ldots, q_n$ in $Sub(maxpath(g) \cap P_{\pi})$ out of which p is formed ($p = q_1 \ldots q_n$) or p is a subgraph of a detached iteration loop.

Proof (Proof-Plan) Suppose that p is not a detached iteration loop or a subpath thereof. The SD1(2) and NST conditions require that the set of executions of π in g exhausts P_g in exactly this sense; the graph reduction definition eliminates any possible counterpaths (paths that would falsify the conditions of the lemma (i.e., paths that are not composed of subpaths of maximal paths)). Were there any counterpaths in g, g would violate NST, as there would be some protocol such that branching isomorphic executions of it can be collapsed without violating SD1, SD2, or RC, producing a graph reduction or proper subgraph of g satisfying these structural display rules. Thus, the execution rules and NST require that every maximal path of g is an execution of π or one that results from the interleaving of paths due to parallel execution (this is the only construction that results in there being maximal paths in any protocol graph of π that are not executions of π). In the latter case, the resulting paths are as described in the latter disjunct of the lemma. (Note: this does not imply that the set of executions of implementations of π exhausts P_g; re: the counterexamples for PRT2, below.) □

(L4) For every g and g', that are protocol graphs of π, such that if (i) for every p in P_g there is a p' in $P_{g'}$ such that $p \simeq p'$, and (ii) for every p' in $P_{g'}$ there is a p in P_g such that $p \simeq p'$, then $g \simeq g'$.

Proof (Proof-Plan) This is mechanical from the set of possible graph-theoretic structural relations among paths in P_g. All paths either concatenate, correspond, or branch in a protocol graph of any non-iterative protocol. Trivially, isomorphism of paths replicates these relations (explicitly, L6). The only way for isomorphism to fail is for the graph to contain isomorphic subgraphs, but this is ruled out by NST. Graph reduction eliminates any redundant or mirrored (i.e., isomorphic) paths that could suffice as a counterexample. Thus, the isomorphism of all executions of π in g with some execution of π in g' and converse imply the result. □

(L5) For all g and g', if $g \simeq g'$, then (iii) every maximal path of g is isomorphic to some maximal path of g', and (iv) every maximal path of g' is isomorphic to some maximal path of g.

Proof (Proof-Plan) This is mechanical from the definition of isomorphism; the original pair of bijections suffices for arbitrary paths. □

Lemma 210 (L6) *Let $\oplus$ stand in for the graph-theoretic operations of concatenation, branching, iteration, or correspondence on procedural paths; i.e., for paths p and q, $p \oplus q$ is the graph that results formed from the concatenation, branching, iteration, or correspondence of p and q. Then: For all p, p', q, q', $p \simeq p' \wedge q \simeq q'$ iff $p \oplus q \simeq p' \oplus q'$ (mutatis mutandis with respect to $p_1, \ldots, p_m$ for path iteration).*

Proof (Proof-plan) *Each case is mechanical[8] for arbitrary paths:*

(1) LtoR:

(a) $(p \simeq p' \wedge q \simeq q') \rightarrow (pq \simeq p'q')$*: The original pair of isomorphisms suffices to define an isomorphism between the resulting graphs.*
(b) $(p \simeq p' \wedge q \simeq q') \rightarrow (p \rightleftharpoons q \simeq p' \rightleftharpoons q')$*: The original pair of isomorphisms suffices to define an isomorphism between the resulting graphs.*
(c) $(p \simeq p' \wedge q \simeq q') \rightarrow (p \cup q \simeq p' \cup q')$*: The original pair of isomorphisms suffices to define an isomorphism between the resulting graphs.*
(d) $(p \simeq p') \rightarrow (p^{m}_{i \leq m}* \simeq p'^{m}_{i \leq m}*)$*: Iteration protocol graphs reduce to branching and concatenation; thus, the original pair of isomorphisms suffices to define an isomorphism between the resulting graphs.*

(2) RtoL: The converses follow immediately from L2 and the definitions of the graph-theoretic operations.

□

With those lemmata in hand, we can proceed to the induction. First, say that paths, p and p', are *isomorphic in the sense of L6* iff either: (a) $p = qr$ and $p' = q'r'$ where $q \simeq q'$ and $r \simeq r'$, (b) $p = q$, $p' = q'$, and $p \oplus r \simeq q' \oplus r'$, where $\oplus$ is branching or concatenation (mutatis mutandis for $p = r$ and $p' = r'$), or (c) $p = q^{m}_{i \leq m}*$ and $p' = q'^{m}_{i \leq m}*$ and $q^{m}_{i \leq m}* \simeq q'^{m}_{i \leq m}*$. That is, either p and p' are directly isomorphic paths, or they are parts of basic graph-theoretic constructions (concatenated, branching, iterated, or corresponding paths) that are isomorphic. Second, say that $p = q \oplus r$ iff $p = qr$, $p = q$ and $q \oplus r$, or $p = r$ and $q \oplus r$.

(1) (Base) For all regular protocols with atomic immediate subprotocols, the result holds. There are five cases.

Case 1. If π is atomic, then the result holds. First, for any $\pi \in \Pi_0$, every protocol graph of π is a single edge with distinct root and terminal states. This can be established easily by two lines of reasoning. First, there can be only one edge in any graph of an atomic (from NST). Protocol graphs of atomics cannot be looping edges. Suppose that the looping edge, $\langle s, \pi, s \rangle$, was a protocol graph of π. Then it would also be a protocol graph of $\pi; \pi$, as every subprotocol and implementation is executed, the maximal paths are the $\pi; \pi$ paths, there are no choices and no corresponding subprotocols, there can be no smaller graph, and each proper subprotocol of $\pi; \pi$ has a protocol graph as a subgraph (by assumption). But π is not correspondence equivalent to $\pi; \pi$. Thus, EXC fails for π and for $\pi; \pi$. Contradiction. Thus, atomic π can not have a looping edge as a protocol graph. From another direction, if $g = \langle s, \pi, s \rangle$, then there are no maximal paths in g since each path is a proper subpath of another path. Thus, $P_\pi \not\subseteq maxpath(g) = \emptyset$, violating

[8] See Appendix II for details.

$SD1(2)_\pi(g)$. Thus, g and g' are both edges with the middle term π. Such edges are trivially isomorphic; construct a bijection on edges, f_e, such that $f_e(g) = g'$ and a bijection on states, f_s, such that $f_s(f(g)) = f(g')$ and $f_s(l(g)) = l(g')$.

Case 2. If $(ISub(\pi) \cap PSub(\pi)) \subseteq \Pi_0$ and $MC(\pi) =;$, then the result holds. Let $\circ =;$. Then any protocol graph of π is a concatenated path of edges executing π_1 and π_2: $g = \langle s, \pi_1, s', \pi_2, s'' \rangle$. Any graph of this form but differing in states is isomorphic to any other, by L6 or by an immediate construction of a pair of bijections satisfying the definition of isomorphism.

Case 3. If $(ISub(\pi) \cap PSub(\pi)) \subseteq \Pi_0$ and $MC(\pi) = *$, then the result holds. Let $\circ = *$. Any graph of $\pi*$ is a complex graph: for each $n \geq 1$, there is a path formed out of n length sequences of concatenated π executions branching from a root state, s. This structure is required by the condition that all implementations of $\pi*$ are executable in $\pi*$'s protocol graphs. There is thus a path formed out of concatenated π executions for each finite length sequential composition of π. But since both g and g' have such paths and each execution of π is an edge, there is a pair of bijections among the edges of g and g' that preserve concatenation and branching (i.e., they preserve iteration of edges), by L6 or by an immediate construction of a pair of bijections satisfying the definition of isomorphism.

Case 4. If $(ISub(\pi) \cap PSub(\pi)) \subseteq \Pi_0$ and $MC(\pi) = \cap$, then the result holds. Let $\circ = \cap$. Then any protocol graph of π is a corresponding pair of edges executing π_1 and π_2: $g = (\langle s, \pi_1, s' \rangle, \langle s, \pi_2, s' \rangle)$. Any graph of this form but differing in states is isomorphic to any other, by L6 or by an immediate construction of a pair of bijections satisfying the definition of isomorphism.

Case 5. If $(ISub(\pi) \cap PSub(\pi)) \subseteq \Pi_0$ and $MC(\pi) = \cup$, then the result holds. Let $\circ = \cup$. Then any protocol graph of π is a corresponding pair of edges executing π_1 and π_2: $g = (\langle s, \pi_1, s' \rangle, \langle s, \pi_2, s'' \rangle)$ where s' is not identical to s''. Any graph of this form but differing in states is isomorphic to any other, by L6 or by an immediate construction of a pair of bijections satisfying the definition of isomorphism.

(2) (Induction) If, for all proper immediate subprotocols, π_1 and π_2, of regular protocol, π, if $g_i \in G^{\pi_i}$ and $g_i' \in G^{\pi_i}$, then $g_i \simeq g_i'$, then for all protocol graphs of π, g and g', $g \simeq g'$. The foregoing lemmata (L1–L6) establish the induction.

(Induction Premise) Let it be the case that, *for all proper immediate subprotocols, π_1 and π_2, of regular protocol, π, if $g_i \in G^{\pi_i}$ and $g_i' \in G^{\pi_i}$, then $g_i \simeq g_i'$.*

(a) Let g and g' be protocol graphs of arbitrary, regular protocol, π. Thus, by RC, there are EDGs, g_1 and g_2, in the proper subgraphs of g such that $g_1 \in G^{\pi_1}$ and $g_2 \in G^{\pi_2}$, and there are EDGs, g_3 and g_4, among the proper subgraphs of g' such that $g_3 \in G^{\pi_1}$ and $g_4 \in G^{\pi_2}$. By the induction premise, $g_1 \simeq g_3$ and $g_2 \simeq g_4$.

(i) There may be multiple protocol graphs of any proper subprotocols of a given π_i in π's protocol graphs, and so the induction premise implies that there are isomorphism classes of g_1 and g_2 among the subgraphs of g, and there are isomorphism classes of g_3 and g_4 among the subgraphs of g'. The remaining steps of the argument are formulated in terms of pairs of protocol graphs of π's proper immediate subprotocols in the subgraphs of g and g'. Everything said about the protocol graphs in these pairs (especially g_2 and g_4) individually can be reformulated as statements about every graph in their isomorphism classes; the derivations of the reformulated claims from their individual cases is immediate.

(b) From (a) by L1, the maximal executions of π_1 and π_2 in g_1 and g_2 (resp.) relate to the maximal executions of π in g according to the control operator rule given in the relevant clause of the lemma, and the maximal executions of π_1 and π_2 in g_3 and g_4 (resp.) relate to the maximal executions of π in g' according to the same rule.

(c) From (a) by L5, each maximal execution of π_1 in g_1 is isomorphic to some maximal execution of π_1 in g_3, and each maximal execution of π_2 in g_2 is isomorphic to some maximal execution of π_2 in g_4, and the converses hold.
   (i) From the fact that every maximal path of g_1 is isomorphic to some maximal path of g_3, and every maximal path of g_2 is isomorphic to some maximal path of g_4 and the converses.

(d) From (b) and (c) by L1 and L6, it's mechanical to show that every maximal execution of π in g is isomorphic to some maximal execution of π in g', and converse.
   (i) Let p execute π in g. Then, from (b), $p = q \oplus r$ for some $q \in P^1_{\pi_1} \cap maxpath(g_1)$ and $r \in P^2_{\pi_2} \cap maxpath(g_2)$ by L1. From (b), (c), the execution rules, and L1, there must be some $p' \in P'_\pi$ such that $p' = q' \oplus r'$, $q' \in P^3_{\pi_1} \cap maxpath(g_3)$, $r' \in P^4_{\pi_2} \cap maxpath(g_4)$, $q \simeq q'$, and $r \simeq r'$.
      (A) Assume $\pi = \circ \frown \pi_1 \frown \pi_2$. Modify as necessary for the case where $\pi = * \frown \pi_1$. Given that $p = q \oplus r$ for some $q \in P^1_{\pi_1} \cap maxpath(g_1)$ and $r \in P^2_{\pi_2} \cap maxpath(g_2)$, there must, by L5 via the induction assumption, be a $q' \in P^3_{\pi_1} \cap maxpath(g_3)$ and an $r' \in P^4_{\pi_2} \cap maxpath(g_4)$, such that $q \simeq q'$ and $r \simeq r'$. From L6, for any p', if $p' = q' \oplus r'$, then $q \oplus r \simeq q' \oplus r'$ (p is isomorphic to p' in the sense of L6).
      (B) From the execution rules, any execution, p'', of π in g' must be formed from executions, q'' and r'', of π_1 and π_2 (respectively) by means of the same graph-theoretic structure as p; if $p'' \in P'_\pi$, then $p'' = q'' \oplus r''$ where $\oplus$ is concatenation, branching, iteration, or correspondence.
      (C) The definition of protocol graphs requires that q'' and r'' be maximal executions of π_1 and π_2 in g_3 and g_4, respectively; NST

and SD1(2) rule out the possibility that the executions be proper subpaths of other paths in their respective protocol graphs.

(D) From the induction assumption and L5, all protocol graphs of π_1 have pairwise isomorphic maximal execution sets, and the same holds for π_2.

(E) Thus, for any $q'' \in P^3_{\pi_1} \cap maxpath(g_3)$ and $r'' \in P^4_{\pi_2} \cap maxpath(g_4)$, either there is a $p'' \in P'_\pi$ such that $p'' = q'' \oplus r''$, or there is some $p''' = q''' \oplus r'''$ where $p''' \in P'_\pi$ while $q''' \simeq q''$ and/or $r''' \simeq r''$.

(F) Thus, if p' is an execution of π in g', then $p' = q' \oplus r'$ for some $q' \in P^3_{\pi_1} \cap maxpath(g_3)$ and an $r' \in P^4_{\pi_2} \cap maxpath(g_4)$, and $q' \simeq q$ and $r' \simeq r$. SD1(1) requires that the set of π' executions in g' be non-empty.

(ii) Thus, by L6, $q \oplus r \simeq q' \oplus r'$.

(iii) Since this worked for arbitrary executions of π, it works for maximal executions.

(iv) Transposition delivers the converse.

(e) From (a) by L3, every path in g is a subgraph of some maximal execution of π or concatenated from subpaths of maximal executions of π or a subpath of a detached iteration loop, and the same holds for g'.

(i) Detached iteration loops can be ignored from this point, as all detached iteration loops that execute the same iterated protocol are isomorphic.

(f) From (a) by L2, if a maximal execution of π in g, p, is isomorphic to a maximal execution of π in g', q, then for every subpath of p, there is a subpath of q that is isomorphic to it, and the converse holds

(i) From the generality of L2.

(g) From (d), (e), and (f), every path in g is isomorphic to some path in g' and converse.

(i) If p is a subpath of some maximal execution of π in g, p', then from steps (d) and (f), there is some maximal execution of π in g', q', such that $p' \simeq q'$. By L2, there must be some $q \sqsubseteq q'$ such that $p \simeq q$.

(ii) If p is formed by concatenation of subpaths, $p_1, \ldots, p_n$, of some maximal executions of π in g, $p'_1, \ldots, p'_m$, then, by L6, if there is a path, q, in g' formed by concatenation of subpaths, $q_1, \ldots, q_n$, with each q_i isomorphic to the correlated p_i, then $p \simeq q$. From (a), there must be such a q in g', as L1 implies that all maximal executions of π in g' are formed out of maximal executions of π_1 and π_2 in g_3 and g_4 by the same graph-theoretic operations (concatenation, branching, iteration, and correspondence) as their isomorphic counterparts in g_1 and g_2.

(iii) By (e), this exhausts the possible subpaths.

(h) Thus, from (g) by L4, $g \simeq g'$.

Remark 211 Note that content states and valuation functions are irrelevant to an EDG's status as a protocol graph of a given π. This is what ensures that the isomorphism class of protocol graphs of π is an isomorphism class; there are as many protocol graphs of a given π as there are combinations of valuations of atomics and content states. Note also that protocol graphs are not truly unique; syntactically non-identical protocols can have identical protocol graph sets. This is what allows protocol graphs to form the basis of protocol equivalences stronger than procedural equivalence and weaker than identity, as we shall see in Chap. 6.

5.7.4.2 ISOP and EQID

Claim 212 Isomorphism is a powerful process equivalence. It preserves many properties, and all structural properties relevant to the execution of protocols:

Lemma 213 (Isomorphism of Protocol Graphs Lemma (ISOPL)) *For all arbitrary EDGs, g, g', and arbitrary procedural paths, p, p', if $p \in P_g$, $p' \in P_{g'}$, $g \simeq g'$, and $p \simeq p'$, then for all regular protocols, π, $p \in P_\pi$ iff $p' \in P'_\pi$.*

The proof is mechanical (structural induction over proper immediate subprotocols) and is reserved for Appendix 2. The essential point is that, if we isolate two isomorphic paths that are embedded in isomorphic graphs, those paths must have the same execution-theoretic properties—they must execute exactly the same protocols. This serves as a lemma for the more interesting result that:

Claim 214 Isomorphism preserves the structural display relations and thus membership in the set of protocol graphs for arbitrary regular protocols.

Lemma 215 *Isomorphism of Protocol Graphs Proposition (ISOP)) For arbitrary EDGs, g, g', and regular protocol, π, if $g \simeq g'$ and $g \in G^\pi$, then $g' \in G^\pi$.*

(Rule Form)

$$\forall \pi \in P_{PLEN} \cap String(\Pi_0 \cup Prot)$$
$$\forall g, g' \left[\frac{g \simeq g',\ g \in G^\pi}{g' \in G^\pi} \right]$$

Isomorphism is a powerful enough process equivalence that it even preserves the elaborate structural relations between protocol graphs and protocol syntax. The proof is a rather mechanical structural induction over proper immediate subprotocols by means of the definition of protocol graphs, and so is saved for Appendix 2, as well.

Claim 216 Together, SUT and ISOP imply, by elementary derivation from the relevant definitions, that graph-theoretically equivalent protocols share protocol graph sets:

Corollary 217 (Equivalent Protocols, Identical Protocol Graph Sets (EQID)) *For all regular π and π', $\pi \equiv_G \pi'$ iff $G^\pi = G^{\pi'}$.*

In short, isomorphism is the measure of uniqueness for protocol graphs. Note, however, that protocols that are not syntactically identical may share protocol graphs (see the protocol graph isomorphism equivalence rules and axioms of the next chapter). Protocol graph isomorphism equivalence, then, turns out to be a very fine-grained protocol equivalence relation that falls between identity and weaker protocol equivalences. The next chapter explores the theory of protocol graph isomorphism equivalence.

Appendix 2

Remark 218 This appendix spells out the results discussed in Chap. 5.

Simultaneous Recursion and Countability

An early commenter on this book had worries about the size of PLEN's language. If the language is too large, then some of my definitions are imprecise, at best; among other things, I'm using functions where I should be speaking, more loosely, of operations or mappings among protocols. I've followed those recommendations, generally, but also note that, given the simultaneous recursion, this shouldn't be an issue:

Claim 219 (Simultaneous Recursion for PLEN's Revised Languages) As with the base version of the syntax, the revised syntactic system of PLEN is defined so that there is a pair of interweaved ascending hierarchies of pairs of protocol and protocol-theoretic languages:

Protocol type	Protocols are strings in:
0	Π_0
1	$String(\Pi_0 \cup Prot \cup Prot + \cup L^{R0}_{PLEN})$
2	$String(\Pi_0 \cup Prot \cup Prot + \cup L^{R0}_{PLEN} \cup L^{R1}_{PLEN})$
i	$String(\Pi_0 \cup Prot \cup Prot + \cup \bigcup_{j=1}^{i-1} L^{Rj}_{PLEN})$
$i+1$	$String(\Pi_0 \cup Prot \cup Prot + \cup \bigcup_{j=1}^{i} L^{Rj}_{PLEN})$

Formula type	Protocol-theoretic formulae are strings in:
0	$AForm$
1	$String(AForm \cup Conn \cup Op + \cup Pop \cup Pop + \cup Syn \cup P^0_{PLEN})$
2	$String(AForm \cup Conn \cup Op + \cup Pop \cup Pop + \cup Syn \cup P^1_{PLEN} \cup P^0_{PLEN})$
i	$String(AForm \cup Conn \cup Op + \cup Pop \cup Pop + \cup Syn \cup \bigcup_{j=1}^{i-1} P^j_{PLEN})$
$i+1$	$String(AForm \cup Conn \cup Op + \cup Pop \cup Pop + \cup Syn \cup \bigcup_{j=1}^{i} P^j_{PLEN})$

Proposition 220 *$P_{PLEN} \cup L^{R}_{PLEN}$ is countable.*

Proof Elementary set-theoretic facts: Finite products over countable sets are countable. Countable unions of countable sets are countable. Simultaneous induction suffices.

(1) (Base)

   (a) It's immediate by construction that $\Pi_0 = P^0_{PLEN}$ and $AForm = L^{R0}_{PLEN}$ are both countable.
   (b) Every set of strings of finite length, k, over a countable set S is a finite product $(S_1 \times \ldots \times S_k)$ over a countable set, and thus countable.
   (c) $String(\Pi_0 \cup Prot \cup Prot + \cup L^{R0}_{PLEN})$ is the countable union of all sets of finite strings over the finite union of countable sets, $(\Pi_0 \cup Prot \cup Prot + \cup L^{R0}_{PLEN})$.
   (d) $String(\Pi_0 \cup Prot \cup Prot + \cup L^{R0}_{PLEN})$ is countable.
   (e) $P^1_{PLEN} \subseteq String(\Pi_0 \cup Prot \cup Prot + \cup L^{R0}_{PLEN})$.
   (f) Thus, P^1_{PLEN} is countable.
   (g) By the same reasoning, L^{R1}_{PLEN} is countable.

(2) (Induction)

   (a) For every $i > 1$:

$$P^i_{PLEN} \subseteq String(\Pi_0 \cup Prot \cup Prot + \cup \bigcup_{j=1}^{i-1} L^{Rj}_{PLEN})$$

   where $\bigcup_{j=1}^{i-1} L^{Rj}_{PLEN}$ is a finite union of countable sets. Thus,

$$String(\Pi_0 \cup Prot \cup Prot + \cup \bigcup_{j=1}^{i-1} L^{Rj}_{PLEN})$$

   is a countable union of finite products over countable sets. Thus,

$$String(\Pi_0 \cup Prot \cup Prot + \cup \bigcup_{j=1}^{i-1} L^{Rj}_{PLEN})$$

   is countable. Thus, P^i_{PLEN} is countable.
   (b) P_{PLEN} is the countable union of all P^i_{PLEN}, each of which is countable.
   (c) Thus, P_{PLEN} is countable.
   (d) By the same reasoning, L^{R}_{PLEN} is countable.

(3) $P_{PLEN} \cup L^{R}_{PLEN}$ is a finite union of countable sets.

□

Additional Graph Reduction Operations

To cash out the remark made in Chap. 5, there are ranges of other graph reduction operations that can be defined on the basis of the basic notion of operations that collapse edges.

Definition 221 (Full π Graph Reduction at s) For arbitrary EDGs, g and g', and arbitrary atomic protocol, π, g' is a *full π graph reduction of g at s* ($g' \xrightarrow[s,F]{\pi} g$) iff for all branching isomorphic pairs executions of π at s in g, p_1 and p_2, there is a mapping, ϱ, of P_g into $P_{g'}$ such that there is a single path, $p'' = \varrho(p_1) = \varrho(p_2)$, that executes π in g' and:

(1) For any converging paths, r and r':

  (a) if $l(r) = f(p_1)$ and $l(r') = f(p_2)$, then $f(p'') = l(\varrho(r)) = l(\varrho(r'))$.
  (b) if $f(r) = l(p_1)$ and $f(r') = l(p_2)$, then $l(p'') = f(\varrho(r)) = f(\varrho(r'))$.

(2) $\forall q \in P_g(\exists q' \in P_{g'}(q \simeq q'))$
(3) $\forall q' \in P_{g'}(\exists q \in P_g(q \simeq q'))$

Definition 222 (Partial π Graph Reduction) For arbitrary EDGs, g and g', and arbitrary atomic protocol, π, g' is a *partial π graph reduction of g* ($g' \xrightarrow[S,P]{\pi} g$) iff for every $s \in S$, if there are any branching isomorphic pairs executions of π at any s in g, then for some such pair, p_1 and p_2, there is a mapping, ϱ, of P_g into $P_{g'}$ such that there is a single path, $p'' = \varrho(p_1) = \varrho(p_2)$, that executes π in g' and:

(1) For any converging paths, r and r':

  (a) if $l(r) = f(p_1)$ and $l(r') = f(p_2)$, then $f(p'') = l(\varrho(r)) = l(\varrho(r'))$.
  (b) if $f(r) = l(p_1)$ and $f(r') = l(p_2)$, then $l(p'') = f(\varrho(r)) = f(\varrho(r'))$.

(2) $\forall q \in P_g(\exists q' \in P_{g'}(q \simeq q'))$
(3) $\forall q' \in P_{g'}(\exists q \in P_g(q \simeq q'))$

Definition 223 (Full π Graph Reduction) For arbitrary EDGs, g and g', and arbitrary atomic protocol, π, g' is a *full π graph reduction of g* ($g' \xrightarrow[s,F]{\pi} g$) iff for any branching isomorphic pairs executions of π, p_1 and p_2, in g, there is a mapping, ϱ, of P_g into $P_{g'}$ such that there is a single path, $p'' = \varrho(p_1) = \varrho(p_2)$, that executes π in g' and:

(1) For any converging paths, r and r':

  (a) if $l(r) = f(p_1)$ and $l(r') = f(p_2)$, then $f(p'') = l(\varrho(r)) = l(\varrho(r'))$.
  (b) if $f(r) = l(p_1)$ and $f(r') = l(p_2)$, then $l(p'') = f(\varrho(r)) = f(\varrho(r'))$.

(2) $\forall q \in P_g(\exists q' \in P_{g'}(q \simeq q'))$
(3) $\forall q' \in P_{g'}(\exists q \in P_g(q \simeq q'))$

Definition 224 (Set Graph Reductions) For arbitrary EDGs, g and g', and arbitrary set of protocols $\{\pi_1, \ldots, \pi_n\}$:

(1) g' *is a partial* $\{\pi_1, \ldots, \pi_n\}$ *reduction of* g *at* s, ($g' \xrightarrow[s,P]{\{\pi_1,\ldots,\pi_n\}} g$) iff for each π_i, $g' \xrightarrow[s,P]{\pi_i} g$.

(2) g' *is a full* $\{\pi_1, \ldots, \pi_n\}$ *reduction of* g *at* s, ($g' \xrightarrow[s,F]{\{\pi_1,\ldots,\pi_n\}} g$) iff for each π_i, $g' \xrightarrow[s,F]{\pi_i} g$.

(3) g' *is a partial* $\{\pi_1, \ldots, \pi_n\}$ *reduction of* g, ($g' \xrightarrow[s,P]{\{\pi_1,\ldots,\pi_n\}} g$) iff for each π_i, $g' \xrightarrow[s,P]{\pi_i} g$.

(4) g' *is a full* $\{\pi_1, \ldots, \pi_n\}$ *reduction of* g, ($g' \xrightarrow[s,F]{\{\pi_1,\ldots,\pi_n\}} g$) iff for each π_i, $g' \xrightarrow[s,F]{\pi_i} g$.

Definition 225 (Atomic Graph Reductions) For arbitrary EDGs, g and g':

(1) g' *is a partial atomic graph reduction at* s ($g' \xrightarrow[s,P]{\Pi_0} g$) iff for all atomic protocols, a_i, $g' \xrightarrow[s,P]{a_i} g$.

(2) g' *is a full atomic graph reduction at* s ($g' \xrightarrow[s,F]{\Pi_0} g$) iff for all atomic protocols, a_i, $g' \xrightarrow[s,F]{a_i} g$.

(3) g' *is a partial atomic graph reduction* ($g' \xrightarrow[s,P]{\Pi_0} g$) iff for all atomic protocols, a_i, $g' \xrightarrow[s,P]{a_i} g$.

(4) g' *is a full graph reduction* ($g' \xrightarrow[s,F]{\Pi_0} g$) iff for all atomic protocols, a_i, $g' \xrightarrow[s,F]{a_i} g$.

Definition 226 (Basic Graph Reduction at s) For arbitrary EDGs, g and g', and state s in g, g' *is a basic graph reduction of* g at s, ($g' \xrightarrow[\Pi_C]{\Pi_0} g, s$) iff there is some regular protocol, $\pi \in \Pi_0 \cup \Pi_C$, such that, for some pair of branching isomorphic paths, p and q, that execute π at s in g, there is a mapping, ϱ, of P_g into $P_{g'}$ such that there is a single path, $p'' = \varrho(p) = \varrho(q)$, that executes π in g' and:

(1) For any converging paths, r and r':

(a) if $l(r) = f(p)$ and $l(r') = f(q)$, then $f(p'') = l(\varrho(r)) = l(\varrho(r'))$.
(b) if $f(r) = l(p)$ and $f(r') = l(q)$, then $l(p'') = f(\varrho(r)) = f(\varrho(r'))$.

(2) $\forall p \in P_g(\exists p' \in P_{g'}(p \simeq p'))$
(3) $\forall p' \in P_{g'}(\exists p \in P_g(p \simeq p'))$

Definition 227 (Total Graph Reduction) For arbitrary EDGs, g and g', g' *is a total graph reduction of* g, ($g' \xrightarrow[\forall \Pi_C]{\forall \Pi_0} g$) iff for all regular protocols, π, for all pairs of branching isomorphic paths, p and q, that execute $\pi \in \Pi_0 \cup \Pi_C$ in g, there is a single path, p'', that executes π in g' such that there is a mapping, ϱ, of P_g into $P_{g'}$ such that there is a single path, $p'' = \varrho(p) = \varrho(q)$, that executes π in g' and:

(1) For any converging paths, r and r':

 (a) if $l(r) = f(p)$ and $l(r') = f(q)$, then $f(p'') = l(\varrho(r)) = l(\varrho(r'))$.
 (b) if $f(r) = l(p)$ and $f(r') = l(q)$, then $l(p'') = f(\varrho(r)) = f(\varrho(r'))$.

(2) $\forall p \in P_g(\exists p' \in P_{g'}(p \simeq p'))$
(3) $\forall p' \in P_{g'}(\exists p \in P_g(p \simeq p'))$

Remark 228 The Incremental Graph Reduction Procedure eventually produces a total graph reduction, if iterated enough times.

Implementation Rules

Claim 229 The following rules hold for all regular protocols.

Theorem 230 (I3) *If* $\pi = \pi_1 \cap \pi_2$, *then if* $\pi+$ *implements* π_1 *while* $\pi++$ *implements* π_2, $\pi + \cap \pi ++$ *implements* π.

Proof (Proof-Plan) Mechanical check: Let $\pi+$ implement π_1 and $\pi++$ implement π_2 and let $\pi+ \neq \pi++$. Consider $\pi + \cap \pi ++$. Syntactic form is guaranteed, execution sets of $\pi + \cap \pi ++$ are subsets of π by the execution rules, all atomics are shared, by the reasoning in the subproof, no proper subprotocol or main control operator preserving reduction of $\pi + \cap \pi ++$ can suffice to execute π, and, from the assumptions, every proper subprotocol is implemented by some subprotocol or reduct of $\pi+$. Thus, all conditions are met for $\pi + \cap \pi ++$ to implement π.

(Subproof) Let $\pi-$ be a proper subprotocol of $\pi + \cap \pi ++$ that implements π. By the definition of syntactic subprotocols, $\pi-$ is a subprotocol of $\pi+$ or a subprotocol of $\pi ++$. But it's not possible for a subprotocol of $\pi+$ to implement π. Suppose that $\pi-$ implements π and is a subprotocol of $\pi+$. There are two cases:

(1) ($\pi- = \pi+$) If $\pi- = \pi+$, then our conditions are contradicted; $\pi+$ implements π_1 and $\pi_1 \cap \pi_2$, but this implies that there is a proper subprotocol of $\pi+$ that implements π_1.
(2) ($\pi- \in PSub(\pi+)$) If $\pi-$ is a proper subprotocol of $\pi+$, then by the final condition of the definition of implementations, there must be a proper subprotocol or reduct of $\pi-$, $\pi--$, that implements π_1.

 (a) If $\pi--$ is a proper subprotocol of $\pi-$ that suffices to implement π_1, then $\pi--$ is a proper subprotocol of $\pi+$ that implements π_1. Thus, $\pi+$ does not implement π_1, contradicting our assumption.

(b) If $\pi --$ is a reduct of $\pi-$ that suffices to implement π_1, then, again, our assumption that $\pi+$ implements π_1 is contradicted. There are two subcases:
   (i) If $\pi --$ is a reduct of a proper immediate subprotocol of $\pi+$ that suffices to implement π_1, then there is a reduct of $\pi+$ that suffices to implement π_1 of the form $\pi -- \circ \pi_i$ or $\pi_i \circ \pi --$ where π_i is the other proper immediate subprotocol of $\pi+$ and $\circ = MC(\pi+)$. Thus, $\pi+$ does not implement π_1, contradicting our assumption.
   (ii) If $\pi-$ is embedded at greater depth in $\pi+$, just construct the reduct of $\pi+$ inductively, starting at the first protocol that embeds $\pi-$ as a proper immediate subprotocol. For a schematic instance, where $\pi+$ has the form $\circ \frown \ldots \circ_1 \frown \pi- \frown \pi_{a(\circ_1)} \frown \ldots$, just construct the reduct of $\pi+$: $\circ \frown \ldots \circ_1 \frown \pi -- \frown \pi_{a(\circ_1)} \frown \ldots$. This reduct implements π_1, contradicting the assumption that $\pi+$ implements π_1.

Thus, in any case, a subprotocol $\pi+$ could not be an implementation of π_1 without contradicting the definition of implementation or our assumptions. Run the same argument for $\pi ++$. Thus, there can be no proper subprotocol of $\pi+; \pi ++$ that implements π.

The same reasoning applies (mutatis mutandis) to the main control operator preserving reduction case. By the definition of main control operator preserving reductions, if there is a reduct of $\pi+; \pi ++$ that implements $\pi_1 \cap \pi_2$, then this reduct must be a protocol of the form: $\pi - \cap \pi --$, where $\pi-$ is a reduct of $\pi+$ and $\pi --$ is a reduct of $\pi ++$, $\pi - \cap \pi ++$ where $\pi-$ is a reduct of $\pi+$, or $\pi + \cap \pi --$ where $\pi --$ is a reduct of $\pi ++$ (or commutations thereof). Thus, there must be a proper subprotocol or reduct of this reduct that suffices to implement π_1 (mutatis mutandis, π_2). Suppose there is a reduct of $\pi + \cap \pi ++$ that implements $\pi_1 \cap \pi_2$. If the reduct is $\pi - \cap \pi ++$ or $\pi - \cap \pi --$ (or a commutation thereof), and $\pi-$ implements π_1, then $\pi+$ doesn't implement π_1, contradicting our assumptions. If the reduct is $\pi + \cap \pi --$ or $\pi - \cap \pi --$ (or a commutation thereof), and $\pi --$ implements π_2, then $\pi ++$ doesn't, contradicting our assumptions. If neither antecedent holds, then the final condition of the definition of implementations fails for this reduct of $\pi + \cap \pi ++$, contradicting our assumption. Thus, there can be no such reduct. □

Theorem 231 (Converse of I3) *If $\pi + \cap \pi ++$ implements $\pi = \pi_1 \cap \pi_2$, then $\pi+$ implements π_1 and $\pi ++$ implements π_2.*

Proof (Proof-Plan) Induction over proper immediate subprotocols.

(1) (Base) All non-identical implementations of parallel executions of atomics are commutations of them. The result follows immediately.
(2) (Induction) Assume that the result holds for all proper immediate subprotocols of π. Let $\pi + \cap \pi ++$ implement $\pi = \pi_1 \cap \pi_2$. Suppose that $\pi+$ doesn't implement π_1. This could be due to the failure of any clause of the definition of implementations, but the failure of any clause suffices to show that $\pi + \cap \pi ++$ doesn't implement π. If $\pi+$ doesn't suffice to execute π_1, doesn't have the right syntactic form, doesn't share atomics with π_1, or contains a proper subprotocol

or reduct that does, then $\pi + \cap\pi ++$ doesn't implement π. The same argument works for $\pi ++$ with respect to π_2.

□

5.7.5 Details for SUT Lemma L6

The derivation of L6 claims that the original pair of bijections that sufficed for isomorphism between pairs of subgraphs, p and p' and q and q', of graph-theoretic structures, $p \oplus q$ and $p' \oplus q'$, also suffices to define a pair of bijections that satisfies the conditions of isomorphism between the graph-theoretic structures. For each graph-theoretic structure, I present below a pair of bijections that suffices for isomorphism based on the original pair of bijections that sufficed for the isomorphism of the separate paths:

1. $(p \simeq p' \wedge q \simeq q') \rightarrow (pq \simeq p'q')$: The original pair of isomorphisms suffices to define an isomorphism between the resulting graphs with one revision. The new isomorphism consists of a pair of bijections that preserve the mappings of the original pairs, but that $f_s(l(p)) = f_s(f(q))$ because, by construction, $l(p) = f(q)$ and, accordingly, $l(p') = f(q')$. Explicitly:

   a. Let f_{s_1} be a bijection among states defining the isomorphism between p and p', let f_{s_2} be the bijection defining the isomorphism between q and q'. Let f_{e_1} and f_{e_2} be defined accordingly as bijections between edges satisfying the conditions of isomorphism. Then the conditions for isomorphism are satisfied by the following pair of bijections, f_s and f_e, between pq and $p'q'$ where:

      i. for any $e_i \sqsubseteq pq$: $f_e(e_i) = \begin{cases} f_{e_1}(e_i) & \text{if } e_i \sqsubseteq p \\ f_{e_2}(e_i) & \text{if } e_i \sqsubseteq q \end{cases}$

      ii. for any $s_i \sqsubseteq pq$: $f_s(s_i) = \begin{cases} f_{s_1}(s_i) & \text{if } s_i \sqsubseteq p \wedge s_i \neq l(p) \\ f_{s_2}(s_i) & \text{if } s_i \sqsubseteq q \wedge s_i \neq f(q) \\ l(p) = f(q) & \text{if } s_i = l(p) \vee s_i = f(q) \end{cases}$

2. $(p \simeq p' \wedge q \simeq q') \rightarrow (p \rightleftharpoons q \simeq p' \rightleftharpoons q')$: The original pair of isomorphisms suffices to define an isomorphism between the resulting graphs with one revision. The new isomorphism consists of a pair of bijections that preserve the mappings of the original pairs, but that $f_s(f(p)) = f_s(f(q))$ and $f_s(l(p)) = f_s(l(q))$ because, by construction, $f(p) = f(q)$ and $l(p) = l(q)$, and, accordingly, $f(p') = f(q')$ and $l(p') = l(q')$. Explicitly:

   a. Let f_{s_1} be a bijection among states defining the isomorphism between p and p', let f_{s_2} be the bijection defining the isomorphism between q and q'. Let f_{e_1} and f_{e_2} be defined accordingly as bijections between edges satisfying the conditions of isomorphism. Then the conditions for isomorphism are satisfied by the following pair of bijections, f_s and f_e, between $p \rightleftharpoons q$ and $p' \rightleftharpoons q'$ where:

i. For any $e_i \sqsubseteq p \rightleftharpoons q$: $f_e(e_i) = \begin{cases} f_{e_1}(e_i) & \text{if } e_i \sqsubseteq p \\ f_{e_2}(e_i) & \text{if } e_i \sqsubseteq q \end{cases}$

ii. For any $s_i \sqsubseteq p \rightleftharpoons q$:

$$f_s(s_i) = \begin{cases} f_{s_1}(s_i) & \text{if } s_i \sqsubseteq p \wedge s_i \notin \{f(p), l(p)\} \\ f_{s_2}(s_i) & \text{if } s_i \sqsubseteq q \wedge s_i s_i \notin \{f(q), l(q)\} \\ f(p) = f(q) & \text{if } s_i \in \{f(p), f(q)\} \\ l(p) = l(q) & \text{if } s_i \in \{l(p), l(q)\} \end{cases}$$

3. $(p \simeq p' \wedge q \simeq q') \rightarrow (p \cup q \simeq p' \cup q')$: The original pair of isomorphisms suffices to define an isomorphism between the resulting graphs with one revision. The new isomorphism consists of a pair of bijections that preserve the mappings of the original pairs, but that $f_s(f(p)) = f_s(f(q))$ because, by construction, $f(p) = f(q)$ and, accordingly, $f(p') = f(q')$. Explicitly:

   a. Let f_{s_1} be a bijection among states defining the isomorphism between p and p', let f_{s_2} be the bijection defining the isomorphism between q and q'. Let f_{e_1} and f_{e_2} be defined accordingly as bijections between edges satisfying the conditions of isomorphism. Then the conditions for isomorphism are satisfied by the following pair of bijections, f_s and f_e, between $p \cup q$ and $p' \cup q'$ where

      i. For any $e_i \sqsubseteq p \cup q$: $f_e(e_i) = \begin{cases} f_{e_1}(e_i) & \text{if } e_i \sqsubseteq p \\ f_{e_2}(e_i) & \text{if } e_i \sqsubseteq q \end{cases}$

      ii. For any $s_i \sqsubseteq p \cup q$: $f_s(s_i) = \begin{cases} f_{s_1}(s_i) & \text{if } s_i \sqsubseteq p \wedge s_i \neq f(p) \\ f_{s_2}(s_i) & \text{if } s_i \sqsubseteq q \wedge s_i \neq f(q) \\ f(p) = f(q) & \text{if } s_i = f(p) \vee s_i = f(q) \end{cases}$

4. $(p \simeq p') \rightarrow (p^m_{i \leq m} * \simeq p'^m_{i \leq m} *)$: Iteration paths reduce to branching and concatenation; thus, the original pair of isomorphisms suffices to define an isomorphism between the resulting graphs by combining the maneuvers for concatenation and branching. In virtue of the structure of path iterations, there are now isomorphisms between p and every branching path in $p^m_{i \leq m} *$ of any length $m \geq 1$, and just so for p' and $p'^m_{i \leq m} *$. The new isomorphism thus consists of a pair of bijections that basically preserves the mappings of the original pair, but that $f_s(f(p^m_{i \leq m})) = f_s(f(p^{m+1}_{i \leq m+1}))$ for every $m \geq 1$, and $f_s(l(p)) = f_s(f(p_n))$ for every p_n in any branch of $p^m_{i \leq m} \cup p^{m-1}_{i \leq m-1} \cup \ldots \cup p^1_{i \leq 1}$. This, because, by construction, $f(p^m_{i \leq m}) = f(p^{m+1}_{i \leq m+1})$ and, accordingly, $f(p'^m_{i \leq m}) = f(p'^{m+1}_{i \leq m+1})$ for all $m \geq 1$, and $f_s(l(p)) = f_s(f(p_n))$ for every

p_n in any branch of $p^m_{i\leq m} \cup p^{m-1}_{i\leq m-1} \cup ... \cup p^1_{i\leq 1}$ and $f_s(l(p')) = f_s(f(p'_n))$ for every p'_n in any branch of $p'^m_{i\leq m} \cup p'^{m-1}_{i\leq m-1} \cup ... \cup p'^1_{i\leq 1}$. Explicitly:

a. Let f_{s_1} be a bijection among states defining the isomorphism between p and p'. Let f_{e_1} be defined accordingly as a bijection between edges satisfying the conditions of isomorphism between p and p'. Then the conditions for isomorphism are satisfied by the following pair of bijections, f_s and f_e, between $p^m_{i\leq m}*$ and $p'^m_{i\leq m}*$ where:
   i. For each, $p^n_{i\leq n}$ for each $n \leq m$, let f_{e_k} be a bijection between the edges in p and an isomorphic subpath of $p^n_{i\leq n}$. Since $p^n_{i\leq n}$ is n iterations of p long, there will be n many f_{e_k}. For each f_{e_k}, for each $e_i \sqsubseteq p^n_{i\leq n}$: $f_e(e_i) = f_{e_k}(e_i)$ for some f_{e_k}.
   ii. For each, $p^n_{i\leq n}$ for each $n \leq m$, let f_{s_l} be a bijection between the states in p and an isomorphic subpath of $p^n_{i\leq n}$. Since $p^n_{i\leq n}$ is n iterations of p long, there will be n many f_{s_l}. For any $s_i \sqsubseteq p_1 p_2$ where each p_i is isomorphic to p:

$$f_s(s_i) = \begin{cases} f_{s_l}(s_i) & \text{if } s_i \sqsubseteq p_1 \wedge s_i \neq l(p_1) \\ f_{s_l}(s_i) & \text{if } s_i \sqsubseteq p_2 \wedge s_i \neq f(p_2) \\ l(p_1) = f(p_2) & \text{if } s_i = l(p_1) \vee s_i = f(p_2) \end{cases}$$

   for some f_{s_l}.
   iii. For any $s_i \sqsubseteq p^m_{i\leq m} \cup p^{m-1}_{i\leq m-1} \cup ... \cup p^1_{i\leq 1}$, for any $n, n+1 \leq m$:

$$f_s(s_i) = \begin{cases} f_{s_l}(s_i) & \text{if } s_i \sqsubseteq p^n_{i\leq n} \wedge s_i \neq f(p^n_{i\leq n}) \\ f(p^n_{i\leq n}) = f(p^{n+1}_{i\leq n+1}) & \text{if } s_i = f(p^n_{i\leq n}) \end{cases}$$

5. These bijections immediately satisfy the conditions of isomorphism: Given that p and p' are isomorphic and q and q' are isomorphic, it immediately follows - for each case - that p and p' have the same numbers of edges (that execute the same atomic protocols and tests) and states (that satisfy the same preconditions), and thus that the mappings defined above are bijections that satisfy the conditions of isomorphism. Thus, the pairs of bijections defined above are isomorphisms in each case; $p \oplus q \simeq p' \oplus q'$ (mutatis mutandis with respect to p_1, ..., p_m for path iteration).

Lemmata Relating Isomorphism and Protocol Graph Sets: ISOPL, ISOP, and EQID

Lemma 232 (ISOPL) *For all arbitrary EDGs, g, g', and arbitrary procedural paths, p, p', if $p \in P_g$, $p' \in P_{g'}$, $g \simeq g'$, and $p \simeq p'$, then for all regular protocols, π, $p \in P_\pi$ iff $p' \in P'_\pi$.*

Proof By structural induction over the proper immediate subprotocols of arbitrary regular protocols. Assume $p \simeq p'$, $g \simeq g'$, $p \in P_\pi$.

(1) (Base) For all regular protocols, π, with atomic immediate subprotocols, the result holds.

(a) (Atomics) Let π be atomic. Then p is an edge with π the middle term and so is p'. Thus, $p' \in R_\pi$.

(b) (Atomic Immediate Proper Subprotocols) If $\pi = \pi_1 \circ \pi_2$, then all paths of π are composed of edges. For sequential composition, concatenated edges. For parallel execution, corresponding edges. For choice, branching edges. For iteration, branching sequences of edges. Consider $\pi_1; \pi_2$. Any execution of $\pi_1; \pi_2$ will be a path formed from concatenated edges, $p = e_1e_2$. Any isomorphic path formed from concatenated edges, $p' = e'_1e'_2$, will execute π by the execution rules for sequential composition. For π formed with any other regular control operator, minor changes to the reasoning suffice; simply replace the structure of p to reflect the execution conditions for the other operators. For the $\pi = \pi_1 \cap \pi_2$ case: let π_1, π_2 be atomics. Let $p \in P_{\pi_1 \cap \pi_2}$. Let $p \simeq q$. Let g and g' be such that $p \in P_g$, $q \in P_{g'}$, and $g \simeq g'$. Then, mechanically, $p \in P_{\pi_1 \cap \pi_2}$ iff $q \in P'_{\pi_1 \cap \pi_2}$ because $p \in P_{\pi_i}$ iff there is a p' that corresponds to p and executes π_j and $q \in P'_{\pi_i}$ iff there is a q' that corresponds to q and executes π_j. But all paths in this case are atomic, so q must have the same middle term as p, and there must be such corresponding edges in g and g' by isomorphism. Thus, $q \in P'_{\pi_1 \cap \pi_2}$.

(2) (Induction) For any $\pi = \pi_1 \circ \pi_2$, if, for all immediate subprotocols of π, π_i, if p_i, p'_i, g, g' are such that $g \simeq g'$, and $p_i \simeq p'_i$, then $p_i \in P_{\pi_i}$ iff $p'_i \in P'_{\pi_i}$, then if p, p', g, g' are such that $g \simeq g'$, and $p \simeq p'$, then $p \in P_\pi$ iff $p' \in P'_\pi$. Assume that $\pi = \pi_1 \circ \pi_2$ and that the result holds for each proper immediate subprotocol, π_i. Consider a path, p, that executes π. If π is a sequential composition, then $p = qr$ such that q executes π_1 and r executes π_2. Since p is isomorphic to p', it follows that:

(a) $p' = q'r'$, $q' \simeq q$, and $r' \simeq r$ (from L6 and L2),
(b) q' executes π_1, and r' executes π_2 from the induction premise.

But then p' executes $\pi_1; \pi_2$ by the execution rules of PLEN. For π formed with any other regular control operator, minor changes to the reasoning suffice; simply replace the graph-theoretic structure of p (from concatenation to branching, correspondence, or iteration) to reflect the execution conditions for the other operators, and L6 shows that isomorphism preserves the concatenation, correspondence, and branching properties, thereby preserving executions. Transposition completes the proof.

□

Claim 233 For p and p' that are proper subgraphs of non-isomorphic graphs, this result may fail. Suppose that $g = p$ and that g' is a graph composed of two corresponding paths, p' and q'. Let p and p' execute π_1 and q' execute π_2. Then isomorphism of p and p' does not preserve executed protocols, since p' executes $\pi_1 \cap \pi_2$ and p doesn't.

Lemma 234 (ISOP) *For arbitrary EDGs, g, g', and regular protocol, π, if $g \simeq g'$ and $g \in G^\pi$, then $g' \in G^\pi$.*
(Rule Form)

$$\forall \pi \in P_{PLEN} \cap String(\Pi_0 \cup Prot)$$
$$\forall g, g' \left[\frac{g \simeq g',\ g \in G^\pi}{g' \in G^\pi} \right]$$

Proof (Proof-Plan) By structural induction over the proper immediate subprotocols of arbitrary regular protocols. Let $g \in G^\pi$ and $g \simeq g'$.

(1) (Base) The result holds for all regular protocols with atomic immediate protocols.

   (a) (Atomics) If π is atomic, the result is trivial: for any atomic protocol, π, all protocol graphs of π are non-looping edges with π as the middle term (EXC and SD1(2) rule out looping edges, as any protocol graph that contains (or is identical to) a looping edge implies that the atomic protocol it executes corresponds to every n-length iteration of itself, but it's impossible for any atomic to correspond to any $n > 2$-length iteration of itself across all graphs, and there are no maximal paths at all in any looping edge). Let $e \in G^\pi$. Let $e' \simeq e$. That e' satisfies the structural display rules mechanically follows from the fact that π is atomic.

   (b) (Atomic Immediate Proper Subprotocols) If $\pi = \pi_1 \circ \pi_2$, then all graphs of π are composed of edges. For sequential composition, concatenated edges. For parallel execution, corresponding edges. For choice, branching edges. For iteration, a combination of those graph operations. Thus, the result is mechanical; simply consider a graph isomorphic to any such graph and step through each condition of protocol graphs. Consider $\pi_1; \pi_2$. The result is mechanical. Any protocol graph, g, of $\pi_1; \pi_2$ will be a concatenated pair of edges. Any isomorphic concatenated pair edges, g', will execute all subprotocols of π because the execution of the atomics, π_1 and π_2 is directly secured by the isomorphism, and the concatenation of those executions will compose an execution of $\pi_1; \pi_2$ itself (from the execution rules). The entire graph is its own maximal path set, which is the sole execution of $\pi_1; \pi_2$. There are no choices. There are no parallel executions and no corresponding paths. The sole implementation of $\pi_1; \pi_2$ is itself. The atomics in π are the atomics in $\Pi_{g'}$. There is no proper subgraph that even executes π. The only proper subprotocols are atomic, each of which is assigned a subgraph (the edge that executes it). And, finally, no non-identical protocol can have g' as a protocol graph, let alone a non-corresponding protocol. Thus, $g' \in G^{\pi_1;\pi_2}$. For π formed with any other regular control operator, minor changes to the foregoing reasoning suffice; simply let the structure of g reflect the execution conditions for the other operators.

(2) (Induction) If, for all proper immediate subprotocols, π_i, of regular protocol, π, if $g_i \simeq g_i'$ and $g_i \in G^{\pi_i}$, then $g_i' \in G^{\pi_i}$, then if g and g' are isomorphic and g is a protocol graph of π, then g' is a protocol graph of π.

   (a) (Proof-plan) Assume the antecedent: For all proper subprotocols, π_i, of π, if $g_i \simeq g_i'$ and $g_i \in G^{\pi_i}$, then $g_i' \in G^{\pi_i}$.
   (b) From ISOPL, g and g' are equivalent in all execution properties.
      (i) Every concatenation, correspondence, iteration, and branching in g is replicated in g' and vice versa by L6. Thus, every execution of any protocol in g is isomorphic to a path in g' that satisfies the same structural properties (e.g., concatenation, correspondence, iteration, or branching). Thus, every protocol, π, executed by p in g is executed by some isomorphic p' in g' and converse by ISOPL. From ISOPL and the restriction rules, it further follows that g and g' have identical execution properties with respect to procedural, correspondence, and transition executions. Thus, $SD1_\pi(g')$, $SD2_\pi(g')$, $SD3_\pi(g')$, and $EXC_\pi(g')$ hold as a mechanical consequence of isomorphism.
   (c) For the remaining structural display rules:
      (i) (NST) Reductio. Let g'' be such that $g'' \xrightarrow[\Pi_C]{\Pi_0} g'$ and $SD1(1,2,3,4)_\pi(g'')$, $SD2_\pi(g'')$, and $RC_\pi(g'')$. From the isomorphism of g' and g and the definition of $g' \xrightarrow[\Pi_C]{\Pi_0} g$, it is immediate that $g'' \simeq g'''$ for some $g''' \xrightarrow[\Pi_C]{\Pi_0} g$. From ISOPL, $SD1(1,2,3,4)_\pi(g''')$ and $SD2_\pi(g''')$. From the assumption of the induction step, if $g'' \in G^{\pi_i}$ for any $\pi_i \in PSub(\pi)$, then $g''' \in G^{\pi_i}$. More, for each subgraph of g'' that is a protocol graph of some proper subprotocol of π, there is a subgraph of g''' that is a protocol graph of that proper subprotocol—again, from the induction assumption. But, from isomorphism, every subgraph of g'' is isomorphic to some subgraph of g''' and vice versa. Thus, every protocol graph of any proper subprotocol of π contained in g'' is isomorphic to some subgraph of g'''. From the induction assumption,

then, each such subgraph is a protocol graph of the same proper subprotocols of π. Thus, from the fact that every proper subprotocol of π has a protocol graph among the subgraphs of g'', it follows from the assumption that every proper subprotocol of π has a protocol graph among the subgraphs of g'''. Thus, $RC_{\pi}(g''')$. Thus, a reduction of g with respect to atomic and complex protocols satisfies $SD1(1, 2, 3, 4)$, $SD2$, and RC, contradicting the assumption that $g \in G^{\pi}$. Thus, there can be no such $g'' \xrightarrow[\Pi_C]{\Pi_0} g'$. Thus, $NST_{\pi}(g')$. A simpler variant of the same argument applies for the subgraph case.

(ii) (RC) Let $\pi_i \in PSub(\pi)$. Then there is a $g'' \sqsubset g$ such that $g'' \in G^{\pi_i}$. From isomorphism, there is a $g''' \sqsubset g'$ such that $g''' \simeq g''$. From the assumption of the induction step, $g''' \in G^{\pi_i}$. π_i was arbitrary. Thus, for all $\pi_i \in PSub(\pi)$ there is a $g''' \sqsubset g'(g''' \in G^{\pi_i})$. Thus, $RC_{\pi}(g')$.

(d) By the definition of protocol graphs, $g' \in G^{\pi}$.

□

Corollary 235 (EQID) *For all regular π and π', $\pi \equiv_G \pi'$ iff $G^{\pi} = G^{\pi'}$.*

Proof Direct argument for each direction.

(1) (LtoR) Assume that π and π' are regular protocols. Then:

(a) $\pi \equiv_G \pi'$ (Premise)
 (i) $g \in G^{\pi}$ (Assumption for Conditional Introduction)
 (ii) $\therefore \exists y(y \in G^{\pi'} \wedge y \simeq g)$ (Def. $\equiv_G$: (a); UI (x/g): (a).i)
 (iii) $\therefore g' \in G^{\pi'} \wedge g' \simeq g$ (EI (y/g'): (a).ii)
 (iv) $\therefore g \in G^{\pi'}$ (ISOP: (a).iii)

(b) $\therefore g \in G^{\pi} \rightarrow g \in G^{\pi'}$ (Conditional Introduction: (a).i–(a).iv, Assumption Discharged)
 (i) $g \in G^{\pi'}$ (Assumption for Conditional Introduction; Universal Instantiation)
 (ii) $\therefore \exists y(y \in G^{\pi} \wedge y \simeq g)$ (Def. $\equiv_G$: (b); UI (x/g): (b).i)
 (iii) $\therefore g' \in G^{\pi} \wedge g' \simeq g$ (EI (y/g'): (b).ii)
 (iv) $\therefore g \in G^{\pi}$ (ISOP: (b).iii)

(c) $\therefore g \in G^{\pi'} \rightarrow g \in G^{\pi}$ (Conditional Introduction: (b).i–(b).iv, Assumption Discharged)

(d) $\therefore (g \in G^{\pi'} \equiv g \in G^{\pi})$ (Biconditional Introduction: (b), (c))

(e) $\therefore \forall z(z \in G^{\pi} \equiv z \in G^{\pi'})$ (UG (g/z): (d))

(f) $\therefore G^{\pi} = G^{\pi'}$ (Set Theory: (e))

(2) (RtoL) Let $G^{\pi} = G^{\pi'}$. Then let $g \in G^{\pi}$. Then g is identical to a $g' \in G^{\pi'}$, namely, g. Thus, there is a $g' \in G^{\pi'}$ that is isomorphic to g. As g was arbitrary, this result holds for each and every $g \in G^{\pi}$. Transpose for the converse. Thus, $\pi \equiv_G \pi'$.

□

References

1. Harel, D., D. Kozen, and J. Tiuryn. 2000. *Dynamic logic*. MIT Press.
2. van Benthem, J., and J. Bergstra. 1995. Logic of transition systems. *Journal of Logic, Language, and Information* 3(4):247–283

Chapter 6
The Revised PLEN Framework II: Protocol Equivalence (Core Results)

Abstract The main work of this chapter is presenting the core technical results of the PLEN Framework, the proofs of which are mainly elementary, but presented in detail. The centerpiece of these results is a novel invariance theorem for the regular control operators: there is a graph-construction operation that produces isomorphic protocol graphs of protocols formed with regular control operators out of protocol graphs of their proper immediate subprotocols. This result implies or enables the derivation of the main replacement rules governing protocol graph isomorphism equivalence, which comprise a novel theory of protocol equivalence. It is shown, additionally, that the protocol graph isomorphism equivalence relation is the finest grained protocol equivalence relation in PLEN short of identity; if two protocols are equivalent, then they can be replaced salva veritate in all contexts in PLEN save for those in which such salva veritate replacement would imply identity of the protocols. This property, perhaps surprisingly, connects protocol graph isomorphism equivalence to hyperintensional equivalence. The results of this chapter deploy the results of the prior chapter, and so these two chapters are necessary companions.

6.1 The Theory of Protocol Graph Isomorphism Equivalence

Chapters 5, 6, and 7 comprise a dense trio of chapters that elaborate on PLEN in order to further develop its logic of protocols (hence, the logic of epistemic norms in PLEN). Chapter 5 defined protocol graphs, protocol isomorphism equivalence, built the R-Derivation System, and partially established its soundness in the form of a range of results. All of the results proven in the last chapter function as lemmata for the core results of the entire Revised PLEN Framework, which are articulated and derived in this chapter. The main result is a distinctive invariance theorem for PLEN, which then undergirds the main syntactic replacement rules, which themselves undergird results concerning the relative granularity of the range of equivalence relations in PLEN.

Remark 235 This section develops the theory of protocol graph isomorphism equivalence in the Revised PLEN Framework. The theory is formulated by means of

R. Jenkins, *A Protocol-theoretic Framework for the Logic of Epistemic Norms*, Synthese Library 467, https://doi.org/10.1007/978-3-031-08597-0_6

the meta-level derivation system of PLEN; it consists of valid rules governing a set of graph-theoretic and execution-theoretic protocol equivalence relations.[1] These are rules that embed expressions of L^R_{PLEN} that are interpreted over the general graph-theoretic properties of EDGs and protocol graphs rather than the graphs containing the states at which object language expressions are interpreted.

The key thought is that protocol graph isomorphism equivalence relates protocols much more closely than other equivalences while also displaying important relations between the structure of processes that execute protocols and their syntactic forms. No other equivalence relation does both. This is seen in a number of theorems below.

This section proceeds in three steps. First, I present a set of results that are theoretically important themselves and useful for proving the core results of the theory. Second, I present the core results of the theory and some simple corollaries. Then I examine the protocol equivalence relations in PLEN in terms of their relative granularity, highlighting the interesting fact that graph-theoretic protocol equivalence is hyperintensional. The proofs, or at least proof-plans, of some of the minor lemmata for these results are presented in Technical Appendix 3.

6.1.1 Roadmap of the Core Results of PLEN

Claim 236 The theory of protocol graph isomorphism equivalence is defined by the rules PGER, PGES, PGET, the PRT variants, PI, PGIIRC, RGSRT, GSRT, and GRT:[2]

$$\left[\begin{array}{ll}
\dfrac{\pi \equiv_G \pi',\ \rho(\pi/\pi')(X_\pi) = X_{\pi'},\ \pi \in ISub(X_\pi)}{X_\pi \equiv_G X_{\pi'}}(RGSRT) & \dfrac{\pi \equiv_G \pi',\ \pi' \equiv_G \pi''}{\pi \equiv_G \pi''}(PGET) \\
\dfrac{\pi \equiv_G \pi',\ \rho(\pi/\pi')(X_\pi) = X_{\pi'}}{X_\pi \equiv_G X_{\pi'}}(GSRT) & \dfrac{\pi \equiv_G \pi'}{\pi \equiv_p \pi'}(PRT1.1) \\
\dfrac{\pi \equiv_G \pi',\ \rho(\pi/\pi')(X_\pi) = X_{\pi'},\ A_{(X\pi)}}{A_{(X\pi')}}(GRT) & \dfrac{\pi \equiv_p \pi'}{\pi \equiv_c \pi'}(PRT1.2) \\
\dfrac{}{\pi \equiv_G \pi}(PGER) & \dfrac{\pi \equiv_c \pi'}{\pi \equiv_t \pi'}(PRT1.3) \\
\dfrac{\pi \equiv_G \pi'}{\pi' \equiv_G \pi}(PGES) & \dfrac{\pi \leftrightarrow_r \pi'}{\pi = \pi'}(PI)
\end{array}\right]$$

$$\left[\forall\pi \left[\begin{array}{c}\forall\pi_1, \pi_2 \in P_{PLEN} \cap String(\Pi_0 \cup Prot), \forall g_1, g_3 \in G^{\pi_1} \forall g_2, g_4 \in G^{\pi_2} \\ \left[\dfrac{[[\pi = \circ \frown \pi_1 \frown \pi_2] \vee [\pi = \circ \frown \pi_1] \wedge (\circ \in \{;, \cap, \cup, *\})]}{[(\oplus^\pi_\circ(g_1, g_2) \simeq \oplus^\pi_\circ(g_3, g_4)) \wedge (\oplus^\pi_\circ(g_1, g_2), \oplus^\pi_\circ(g_3, g_4) \in G^\pi)]}\right]\end{array}\right]\right]$$
$$\times (PGIIRC)$$

---

[1] In particular, the theory amounts to the system comprising SUT, ISOP, EQID, PRT, RGSRT, PGIIRC, GSRT, GRT, and the trio of PGER, PGES, and PGET. The meaning of the acronyms used to denote these results will become clear as they are introduced.

[2] These abbreviations are explained in turn below.

Informally, each rule displays some property of protocol equivalence. PGER, PGES, and PGET show that $\equiv_G$ is, respectively, reflexive, symmetric, and transitive; a proper equivalence relation. The PRT variants display implication relations among protocol equivalences. PI shows that edge equivalence implies syntactic identity. PGIIRC is the distinctive invariance result of PLEN. RGSRT shows that protocol graph isomorphism equivalence is preserved by syntactic replacement of protocol graph isomorphism equivalent proper immediate subprotocols. The GSRT and the GRT are the core of the replacement properties of the canonical protocol equivalence. GSRT generalizes RGSRT to the preservation of graph-theoretic protocol equivalence by the syntactic replacement of arbitrary equivalent subprotocols. GRT shows that syntactic replacement of arbitrary equivalent subprotocols preserves truth for a large class of L^R_{PLEN} formulae. The object language counterparts (where applicable) of these rules are the core of the protocol logic of PLEN, the R-Derivation System. The soundness of the system, then, implies the soundness of the theory of graph-theoretic protocol equivalence.

Claim 237 PGIIRC has no object language counterpart, but is a core result of PLEN. Informally, it states that if a complex protocol, $\pi = \circ \frown \pi_1 \frown \pi_2$, is constructed using regular control operators, $\circ \in \{; , *, \cap, \cup\}$, then any epistemic directed graphs, $\oplus^{\pi}_{\circ}(g_1, g_2)$ and $\oplus^{\pi}_{\circ}(g_3, g_4)$, constructed out of arbitrary protocol graphs of π's proper immediate subprotocols (π_1 and π_2) by the operation, $\oplus^{\pi}_{\circ}$ (defined by the $\oplus^{\pi}_{\circ}$ Procedure built out of operations of concatenation, iteration, correspondence, branching, and graph reduction) are isomorphic protocol graphs of π. I show that the derivation of PGIIRC from the "useful lemmata" above utilizes all of the central metalevel results of PLEN, and can then be used to derive RGSRT (and thus GSRT). Interestingly, RGSRT and GRT can be directly derived using almost all of the foregoing lemmata, and, from there, RGSRT can be used to bypass critical steps of the derivation of PGIIRC.

Claim 238 As protocols are the analyzans of procedural epistemic norms in PLEN, the theory of protocol graph isomorphism equivalence is the core of the protocol-theoretic logic of epistemic norms. Thus, the proofs of the lemmata used in the derivation of these results, the interderivation of PGIIRC and RGSRT, and the derivation of GRT are the technical and philosophical centerpieces of the formalism.

Remark 239 The reasoning presented in this chapter deploys the core definitions of PLEN from Chap. 4 as well as the important syntactic constructions and graph-theoretic structures highlighted there and the specialized constructions and revised syntax, truth rules, and derivation rules introduced in Chap. 5. In particular, all of the following constructions will be operative in the derivation of the results below:

- regular protocols (Sect. 4.2.1.2),
- proper immediate subprotocols (Sect. 4.2.1.2),
- syntactic replacement operations (Sect. 4.2.1.2),
- the induced graph-theoretic structures associated with execution rules (path concatenation, path branching, path correspondence, path iteration) (Sect. 4.2.2.1),
- implementations (Sect. 5.4),

- dictation (Sect. 5.4),
- protocol graphs (Sect. 5.4.3)
- the various process equivalences (isomorphism, in particular) (Sect. 5.4)
- the various protocol equivalences (protocol graph isomorphism equivalence, in particular) (Sect. 5.2–5.5)

Additionally, the following bits of terminology will be useful:

Definition 240 (Graph-Theoretic Structures and Constructions) The following definitions are operative in the forthcoming proofs:

(1) Shorten "$maxpath(g)$" to "$mxpth(g)$" where space is prohibitive.
(2) Define $maxpath(g) \cap P_\pi$ to be the set of *maximal executions of* π *in* g.
(3) Let the terms $\pi_i maxpath(g_i)$ and $\pi_i mxpth(g_i)$ accordingly denote $maxpath(g_i) \cap P_{\pi_i}$ where space is prohibitive.
(4) For any tree, g, let $Max(len(s, p_1), ..., len(s, p_n))$ be the maximum of the lengths of all paths that terminate in s for any state in g.
(5) The level of any state, s, in g, $level(s) = Max(len(s, p_1), \ldots, len(s, p_n))$.
(6) For any g, where X is a set of paths in g, let $Sub(X)$ be the set of all subpaths of any path in X, and let $last(X)$ be the set of all terminal states of all paths in that set.
(7) For any pointed (tree) g, let $@g \in S$ be the root state or the point of g.
(8) Let g be a duplicate of g' iff $g \simeq g'$.
(9) The operation of *duplicating* a graph, g, is constructing an isomorphic graph, g'.
(10) Denote indexed duplicates of g with $d_i(g)$ where they aren't explicitly named.
(11) Generally, instructions for operations or procedures that modify EDG, g, can be understood as defining maps to EDGs, $g'_1, \ldots, g'_n$, with specified structures.

    (a) Strictly, the duplication operation maps g to an isomorphic graph, g'.
    (b) Duplicating g some n times is mapping g to n many isomorphic $g'_1, \ldots, g'_n$.
    (c) Affixing g to g' means mapping the pair, g, g', to g'' where g and g' are isomorphic to proper subgraphs of g'' that are attached in some specified way, e.g., sharing a path or state by means of concatenation, iteration, correspondence, or branching of paths.
    (d) Removing a path, p, from g is exactly what is sounds like: mapping g to g' where $P_{g'} = P_g - \{p\}$.
    (e) Further modifications may be specified (i.e., as in the graph reduction procedure, where any graphs that would have extended from the terminal states of a removed path are reattached to some other path).

(12) If $g \in EDG$ and $p \in P_g$, then p is redundant with respect to some property P iff P does not depend on the fact that $p \in P_g$ (i.e., removing p from g does not result in P failing for g).
(13) The term "protocol graph isomorphism" denotes the conjunction of isomorphism between graphs and the satisfaction of the structural display rules with respect to the same protocol; i.e., g and g' are protocol graph isomorphic iff $g \simeq g'$ and for some arbitrary π, $g \in G^\pi$ and $g' \in G^\pi$.

6.2 Properties of Protocol Equivalence

6.2.1 Formal Equivalence

Claim 241 The canonical protocol equivalence relation is a genuine equivalence relation, as typically understood; it is a reflexive, symmetric, and transitive binary relation over protocols:

Lemma 242 (Protocol Equivalence Reflexivity, Symmetry, and Transitivity (PGER, PGES, PGET)) *Protocol Graph Isomorphism Equivalence is reflexive, symmetric, and transitive. In Rule Form:*

$$\left[\frac{}{\pi \equiv_G \pi}\right](PGER),\ \left[\frac{\pi \equiv_G \pi'}{\pi' \equiv_G \pi}\right](PGES),\ \left[\frac{\pi \equiv_G \pi',\ \pi' \equiv_G \pi''}{\pi \equiv_G \pi''}\right](PGET)$$

Remark 243 The derivation of these properties is mechanical given the definition of protocol graph isomorphism equivalence and the reflexivity, symmetry, and transitivity of isomorphism itself. For rigor's sake, the reasoning is presented, as the theory of protocol graph isomorphism equivalence is central to PLEN's protocol logic.

Proof Direct arguments for each rule.

(1) (Reflexivity) Every $g \in G^\pi$ is isomorphic to a $g' \in G^\pi$—itself. Thus, for each graph of π, there is a graph of π isomorphic to it.
(2) (Symmetry) Let $\pi \equiv_G \pi'$. Then: (i) $\forall g \in G^\pi \exists g' \in G^{\pi'}[g \simeq g']$, (ii) $\forall g' \in G^{\pi'} \exists g \in G^\pi [g \simeq g']$. But then, by the commutativity of conjunction, (iii) $\forall g' \in G^{\pi'} \exists g \in G^\pi [g \simeq g']$ and (iv) $\forall g \in G^\pi \exists g' \in G^{\pi'}[g \simeq g']$. Thus, $\pi' \equiv_G \pi$.
(3) (Transitivity) Let $\pi \equiv_G \pi'$ and $\pi' \equiv_G \pi''$. Then: (i) $\forall g \in G^\pi \exists g' \in G^{\pi'}[g \simeq g']$, (ii) $\forall g' \in G^{\pi'} \exists g \in G^\pi [g \simeq g']$, (iii) $\forall g' \in G^{\pi'} \exists g'' \in G^{\pi''}[g \simeq g']$, and (iv) $\forall g'' \in G^{\pi''} \exists g' \in G^{\pi'}[g \simeq g']$. From (i) and (iii), it follows trivially that (v) $\forall g \in G^\pi \exists g'' \in G^{\pi''}[g \simeq g'']$. From (ii) and (iv), it follows trivially that (vi) $\forall g'' \in G^{\pi''} \exists g \in G^\pi [g \simeq g'']$. Thus, by definition, $\pi \equiv_G \pi''$.

□

Corollary 244 *Protocol Graph Isomorphism Equivalence is an equivalence relation.*

Proof Any relation, R, is an equivalence relation iff R is reflexive, symmetric, and transitive. From PGER, PGES, and PGET, $\equiv_G$ is reflexive, symmetric, and transitive. Thus, $\equiv_G$ is an equivalence relation. □

6.2.2 PRT

Claim 245 The canonical protocol equivalence relation is the strongest equivalence relation in PLEN that can hold between nonidentical protocols, in the sense that it implies every other protocol equivalence short of identity. Letting the unadorned turnstyle, $\models$, indicate meta-level logical implication, the following results state the implication relations among the meta-level protocol equivalence relations—and some of the object-level relations—defined above.

Lemma 246 (Protocol (Equivalence) Relation Transmission (PRT)) *For all regular protocols, π and π':*

PRT1	PRT2	PRT3
(i) $\pi \equiv_G \pi' \models \pi \equiv_p \pi'$ (ii) $\pi \equiv_p \pi' \models \pi \equiv_{fc} \pi'$ (iii) $\pi \equiv_{fc} \pi' \models \pi \equiv_t \pi'$	(i) $\pi \equiv_p \pi' \not\models \pi \equiv_G \pi'$ (ii) $\pi \equiv_{fc} \pi' \not\models \pi \equiv_p \pi'$ (iii) $\pi \equiv_t \pi' \not\models \pi \equiv_p \pi'$	(i) $\pi \equiv_r \pi' \models \pi \equiv_G \pi'$ (ii) $\pi \equiv_G \pi' \not\models \pi \equiv_r \pi'$

Proof None of these proofs is particularly interesting. PRT1 is most central to the coming results, so I present the proof here, which is a direct proof: □

(1) PRT1(i) First, a definition:

Definition 247 (Unfolding of p with Respect to π) For all regular protocols, π, and all procedural paths, p, in all EDGs, g, arbitrary EDG, g', is an *unfolding* of p with respect to π iff $p \in P_\pi$, g' is isomorphic to the smallest EDG containing no loops such that there is a maximal path, p', of g' that executes π, and p is bisimilar to p'.

(a) Second, two lemmata:

(i) (PRTL1) For all regular protocols, π, and all procedural paths, p, in all EDGs, g, $p \in P_\pi$ iff p has an *unfolding* with respect to π.

Proof (LtoR) Structural induction over proper immediate subprotocols gives the result mechanically by way of the execution rules. Assume p executes arbitrary, regular π. For atomics, p is an edge with π the middle term. If it does not loop, it is its own unfolding with respect to π. If it loops, then a non-looping edge with π the middle term is an unfolding with respect to π. For type 1 protocols, the details will vary with main control operator. For sequential compositions, p contains loops or it doesn't. If it doesn't, it is its own unfolding with respect to π. If it does, then the path concatenation of the unfoldings of whichever proper immediate subprotocols are executed by loops is an unfolding of p with respect to π. For all other main control operators, replace "path concatenation" with the requisite graph-theoretic structure. In particular, the execution rules require that if π is a parallel execution, then any looping executions of π are such that there is are two edges, each executing a proper immediate subprotocol of π, that loop back to the root state. The unfolding is simply the pair of corresponding edges with the proper immediate subprotocols of π as middle terms. For the induction step, assume that the result holds for all proper immediate subprotocols of π, then iterate the prior reasoning for each control operator, replacing all occurrences of the term "edges" with "paths". (RtoL) Assume p has an unfolding with respect to π, g'. The first condition implies that $p \in P_\pi$. □

(ii) (PRTL2) For all regular protocols π, all protocol graphs of π, all EDGs, g, and all procedural paths in g, p, $p \in P_\pi$ iff there is an unfolding of p with respect to π that is isomorphic (in the sense[3] of Lemma L6) to at least one maximal execution of π, p', for each protocol graph of π.

Proof ***(LtoR)*** Structural induction over proper immediate subprotocols gives the result mechanically by way of the definition of protocol graphs. Assume p executes arbitrary, regular π. From PRTL1, there is an unfolding, g_π, of p with respect to π. The maximal executions of π in g_π contain no loops. They are thus isomorphic (in the sense of L6) to maximal paths in arbitrary protocol graphs of π. For atomics and type 1 protocols, this is immediate. Assume the induction hypothesis, and let g' be a protocol graph of π. Then all maximal executions of π in g' and g_π, respectively, are formed out of isomorphic executions of π's proper immediate subprotocols by the graph-theoretic structures associated to the main control operator of π by L1, and neither contain any loops. Isomorphism immediately follows by L6. (RtoL). Assume there is an unfolding of p with respect to π that is isomorphic to some maximal execution of π in each protocol graph of π. From PRTL1, the result immediately follows (a fortiori), because there is an unfolding of p with respect to π. □

(b) Third, the derivation: Let π and π' be regular protocols. Assume $\pi \equiv_G \pi'$. Let p execute π in g. By PRTL2, it immediately follows that p has an unfolding, g_p, with respect to π that is isomorphic (in the sense of L6) to some maximal execution of π in every protocol graph of π. Let g_π be a protocol graph of π, and let p' be the maximal execution of π in g_π that is isomorphic to g_p (the existence of which we infer from PRTL2). From the assumption that $\pi \equiv_G \pi'$ and EQID, it follows that g_π is also a protocol graph of π', and thus, from SD1(2), that $p' \in P_{\pi'}$. Thus, g_p is isomorphic to a maximal execution of π' in g_π. Thus, by ISOPL, g_p executes π', as p' must have the requisite graph-theoretic structure to execute π'. Moreover, g_p contains no loops (being isomorphic to a subgraph containing maximal paths in a protocol graph), and there will be no smaller

[3] More precisely, the unfolding of π with respect to π is isomorphic to at least one subgraph of each protocol graph containing a maximal execution, p', of π. For instance, where $\pi = \pi_1 \cap \pi_2$, each maximal execution of π in any protocol graph of π is an execution of either π_1 or π_2, where that execution corresponds to an execution of the other proper immediate subprotocol, so the unfolding is isomorphic to the graph containing the corresponding executions of π_1 and π_2.

graph, as it must only have the structure necessary to execute π'. It follows that p is bisimilar to g_p. For protocols that contain no parallel executions, it immediately follows that p executes π', as required. In the case that π contains parallel executions, the bisimilarity of graphs does not transfer executability in general. It could be the case, granting that g_p and p are bisimilar, that g_p executes π' while p does not—p could execute one proper immediate subprotocol of π' while branch from an execution of the other proper immediate subprotocol rather than corresponding. However, granting that π and π' are protocol graph isomorphism equivalent, if π' is a parallel execution, then π is either a parallel execution as well, embeds parallel executed protocols at equal depth with respect to another control operator, or π is identical to both proper immediate subprotocols of π'. In all cases, the bisimilarity of g_p suffices to show that $p \in P_{\pi'}$ precisely because it sufficed to execute π; in the case of parallel execution, the structural properties of the graph surrounding p must suffice to execute π (i.e., p must execute a proper immediate subprotocol of π there must be a corresponding q that executes the other proper immediate subprotocol), and this structure will work for similarly structured π'. Thus, if p executes π in g for arbitrary g, p executes π' in g. The converse is given by transposition of π with π' in the reasoning above.

(2) PRT1(ii) Let $seq(\pi) = \{\langle s_1, ..., s_n\rangle \mid \exists p \in P_\pi(\langle s_1, ..., s_n\rangle \in s(p))\}$. Assume $P_\pi = P_{\pi'}$ for all g. Let $s(p) \in seq(\pi)$. Then $p \in P_\pi$. But, from the assumption, $p \in P_{\pi'}$. But then, $s(p) \in seq(\pi')$. That's LtoR. Transpose for RtoL. All π, p, $s(p)$, and g were arbitrary. Thus, $\pi \equiv_C \pi'$ for all g.

(3) PRT1(iii) Let $seq(\pi) = \{\langle s_1, ..., s_n\rangle \mid \exists p \in P_\pi(\langle s_1, ..., s_n\rangle \in s(p))\}$. Suppose that $seq(\pi) = seq(\pi')$. By definition, $T_\pi = \{t \mid \exists p \in P_\pi(t = \langle first(p), last(p)\rangle)\}$. (This is explicit for parallel executions, given the revised transition execution rules, and it's mechanical for every other main control operator.) Let $t \in T_\pi$. Then there is a p that executes π, and $t = \langle first(p), last(p)\rangle$. Thus, $t = \langle first(s(p)), last(s(p))\rangle$ for some $s(p)$. But $s(p) \in P_{seq(\pi')}$ by assumption. Thus, there is a $p' \in P_{\pi'}$ such that $s(p') = s(p)$. Thus,

$$t = \langle first(s(p)), last(s(p))\rangle = \langle first(s(p')), last(s(p'))\rangle = \langle first(p'), last(p')\rangle$$

Thus, $t \in T_{\pi'}$. Transpose for LtoR. Of course, π, t, and g were all arbitrary. Thus, $\pi \equiv_t \pi'$.

Remark 248 The converses of PRT1, as noted above, all fail. None of these proofs are particularly interesting. Counterexamples are presented in Appendix 2.

The properties of the various protocol and process equivalence relations coupled with PRT1, PRT2, and PRT3, imply the following corollaries (again letting the unadorned turnstyle, $\models$, indicate meta-level logical implication):

Corollary 249 (Protocol (Equivalence) Relation Transmission Corollary (PRTC)) *For all regular π and π':*

(1) $\pi \equiv_p \pi' \nvDash \pi \equiv_r \pi'$, *and* $\pi \equiv_t \pi' \nvDash \pi \equiv_r \pi'$
(2) $\pi \equiv_r \pi' \models \pi = \pi'$; *thus,* $\pi \equiv_r \pi' \models \pi \equiv_G \pi'$, $\pi \equiv_r \pi' \models \pi \equiv_p \pi'$, *and* $\pi \equiv_r \pi' \models \pi \equiv_t \pi'$
(3) $\pi \equiv_G \pi'$ *iff* $\pi \equiv_{GGI} \pi'$
(4) $\pi \equiv_G \pi'$ *iff* $\pi \equiv^s_{GGI} \pi'$
(5) $\pi \equiv^s_G \pi'$ *iff* $\pi \equiv^s_{GGI} \pi'$

Corollary 250 *Parallel implication relations hold for the graph-relative variants of the protocol equivalences above. More precisely, for all regular protocols,* π *and* π'*:*

PRT4	PRT5
(i) $\pi \equiv^s_G \pi' \models \pi \equiv^s_p \pi'$ (ii) $\pi \equiv^s_p \pi' \models \pi \equiv^s_{fc} \pi'$ (iii) $\pi \equiv^s_{fc} \pi' \models \pi \equiv^s_t \pi'$	(i) $\pi \equiv^s_p \pi' \nvDash \pi \equiv^s_G \pi'$ (ii) $\pi \equiv^s_{fc} \pi' \nvDash \pi \equiv^s_p \pi'$ (iii) $\pi \equiv^s_t \pi' \nvDash \pi \equiv^s_p \pi'$
PRT6	PRT7
(i) $\pi \equiv^s_r \pi' \models \pi \equiv^s_G \pi'$ (ii) $\pi \equiv^s_G \pi' \nvDash \pi \equiv^s_r \pi'$	(i) $\pi \equiv_p \pi' \models \Box\pi \leftrightarrow_p \pi'$ (ii) $\Box\pi \leftrightarrow_p \pi' \nvDash \Box\pi \leftrightarrow_r \pi'$ (iii) $\pi \equiv_p \pi' \nvDash \Box\pi \leftrightarrow_r \pi'$ (iv) $\Box\pi \leftrightarrow_p \pi' \nvDash \pi \equiv_p \pi'$ (v) $\Box\pi \leftrightarrow_r \pi' \models \Box\pi \leftrightarrow_p \pi'$

Remark 251 These implication relations among protocols are interesting as parts of the theory of protocol equivalence. After all, they provide some of the inferential mechanisms by which later results are derived, and they form part of the framework for comparative analysis of different accounts of protocol equivalence, as they relate closely to the relative granularities of different protocol equivalence relations. Sections 6.3 through 6.5 bear this out. Nonetheless, the proofs are all quite mechanical or immediate from prior results. As such, they're all housed in Appendix 2.

6.3 Protocol Graph Isomorphism Invariance for Regular Constructions

Remark 252 Now, let me present the distinctive invariance theorem of PLEN. Bisimulation invariance is a familiar notion from the modal logic of transition systems or from logical dynamics more generally. The basic thought is that a formula, A, is invariant for bisimulation iff wherever A is true at a state s in a graph-theoretic structure, g, and the state s' in g' is bisimilar to s, it follows that A is true at s, and vice versa. This is a kind of semantic invariance over bisimilar structures.

Protocols, on the other hand, can be shown to be invariant in a more complicated way. SUT already showed that regular protocols are invariant over protocol graph isomorphic EDGs. Similarly, ISOP shows that the structural display rules defining protocol graphs for arbitrary, regular π are invariant over isomorphic EDGs. It turns out that the regular protocol control operators of PLEN correspond to graph-theoretic operations that preserve protocol graph isomorphism.

Claim 253 Regular constructions are those expressions in P_{PLEN} built out of regular control operators. Protocol graph isomorphism invariance for regular constructions, then, is a property of preservation; as we build more complex (i.e., higher type) protocols out of lower type protocols by means of regular control operators, we can also build isomorphic protocol graphs of the newly constructed protocols out of the protocol graphs of the lower type protocols. In other words, if two protocols, π_1 and π_2, are fused with a regular control operator $\circ$ into a higher type protocol, $\pi_1 \circ \pi_2$, then any pairs of isomorphic protocol graphs of π_1 and π_2 can be fused into isomorphic protocol graphs of $\pi_1 \circ \pi_2$ by a corresponding operation, $\oplus_\circ^\pi$. In more exact terms:

Definition 254 (Protocol Graph Isomorphism Invariance) Let $\pi, \pi_1, \pi_2 \in P_{PLEN} \cap String(\Pi_0 \cup Prot)$, $MC(\pi) = \circ, \pi_1, \pi_2 \in PISub(\pi)$, and $g_1, g_3 \in G^{\pi_1}$ and $g_2, g_4 \in G^{\pi_2}$. Then:

(1) *Protocol graph isomorphism is invariant for protocol graph operation,* $\oplus_\circ^\pi$, iff $\oplus_\circ^\pi(g_1, g_2) \simeq \oplus_\circ^\pi(g_3, g_4)$ and $\oplus_\circ^\pi(g_1, g_2), \oplus_\circ^\pi(g_3, g_4) \in G^\pi$.
(2) Protocol graph operator $\oplus_\circ^\pi$ *preserves/is safe for protocol graph isomorphism* iff $\oplus_\circ^\pi(g_1, g_2) \simeq \oplus_\circ^\pi(g_3, g_4)$ and $\oplus_\circ^\pi(g_1, g_2), \oplus_\circ^\pi(g_3, g_4) \in G^\pi$.
(3) *Protocol graph isomorphism is invariant for regular control operator,* $\circ$, iff $\oplus_\circ^\pi(g_1, g_2) \simeq \oplus_\circ^\pi(g_3, g_4)$ and $\oplus_\circ^\pi(g_1, g_2), \oplus_\circ^\pi(g_3, g_4) \in G^\pi$.
(4) Regular control operator $\circ$ *preserves/is safe for protocol graph isomorphism* iff $\oplus_\circ^\pi(g_1, g_2) \simeq \oplus_\circ^\pi(g_3, g_4)$ and $\oplus_\circ^\pi(g_1, g_2), \oplus_\circ^\pi(g_3, g_4) \in G^\pi$.

Theorem 255 (Protocol Graph Isomorphism Invariance for Regular Constructions) *The regular control operators preserve protocol graph isomorphism; protocol graph isomorphism is invariant for regular control operators. More precisely:*

$$\left[\forall \pi \left[\begin{array}{c}\forall \pi_1, \pi_2 \in P_{PLEN} \cap String(\Pi_0 \cup Prot), \forall g_1, g_3 \in G^{\pi_1} \forall g_2, g_4 \in G^{\pi_2} \\ \left[\dfrac{[[\pi = \circ \frown \pi_1 \frown \pi_2] \vee [\pi = \circ \frown \pi_1] \wedge (\circ \in \{;, \cap, \cup, *\})]}{[(\oplus_\circ^\pi(g_1, g_2) \simeq \oplus_\circ^\pi(g_3, g_4)) \wedge (\oplus_\circ^\pi(g_1, g_2), \oplus_\circ^\pi(g_3, g_4) \in G^\pi)]}\right]\end{array}\right]\right]$$

Remark 256 In order to derive this result, we need an excursion into graph reduction and graph building procedures. The major challenge in understanding PGIIRC is finding an operation, $\oplus_\circ^\pi$, by which the protocol graphs of immediate subprotocols of $\pi_1 \circ \pi_2$ can themselves be combined into a new EDG. The idea of the result being that it would suffice to establish the theorem to show that there is an operation on the protocol graphs of π_1 and π_2 that always produces isomorphic graphs that are protocol graphs of $\pi_1 \circ \pi_2$. I define $\oplus_\circ^\pi$ in two steps presently.

First, I define an incremental graph reduction operation that simply identifies, at each step, a graph that is reduced with respect to some pair of branching, isomorphic executions of some protocol. Second, I define a recursive procedure for constructing EDGs out of protocol graphs of arbitrary proper immediate subprotocols of a given protocol conditional on its main control operator. This procedure deploys the incremental graph reduction operation as a step. The operation in question, then, is just a mapping from the inputs of this procedure with respect to a given protocol and an arbitrary pair of protocol graphs of its proper immediate subprotocols to the procedure's outputs. With that machinery in hand, we'll be able to derive the invariance theorem.

6.3.1 Incremental Graph Reduction Procedure

Definition 257 (Incremental Graph Reduction Procedure) The basic graph reduction operation defines a natural base from which to recursively construct incremental graph reductions. Define the incremental graph reduction procedure by the following rules:

(1) Let g be an arbitrary EDG.

(2) For any atomic protocol, π, if there is some finite number, n, of edges that execute π at $@g$, then, for some $0 \leq i \leq n$ executions of π, construct an EDG, $g_{@}$, such that $g_{@} \xrightarrow[@g,P]{\pi} g$; else, move down one level or halt.

   (a) :: Collapse some $0 \leq i \leq n(n-1)/2$ pairs of executions of π at $@g$; else, move down one level or halt.

(3) For $g_{@}$, for any atomic protocol, π, if there is some finite number, n, of edges that execute π at arbitrary level 1 state, s_1, then, for some $0 \leq i \leq n$ executions of π, construct an EDG, g_1 such that $g_1 \xrightarrow[s_1,P]{\pi} g_{@}$; else, move down one level or halt.

   (a) :: Collapse some $0 \leq i \leq n(n-1)/2$ pair of executions of π at s_1; else, move down one level or halt.

(4) For $g_{i\geq 1}$, for any atomic protocol, π, if there is some finite number, n, of edges that execute π at arbitrary level $i+1$ state, s_{i+1}, then, for some $0 \leq i \leq n$ executions of π, construct an EDG, g_{i+1} such that $g_{i+1} \xrightarrow[s_{i+1},P]{\pi} g_i$; else, move down one level.

   (a) :: Collapse some $0 \leq i \leq n(n-1)/2$ pair of executions of π at s_{i+1}; else, move down one level or halt.

(5) Denote the result of executing the incremental graph reduction procedure once up to level m: $g^1_m \xrightarrow[\Pi_C]{\Pi_0} g$.

(6) Denote the result of iterating the incremental graph reduction procedure k times, each time up to level m_k: $g^{1,\ldots,k}_{m_1,\ldots,m_k} \xrightarrow[\Pi_C]{\Pi_0} g$.

Remark 258 This procedure, by itself, makes a very small structural change to any arbitrary EDG. The iteration of it multiple times, however, can radically alter the structure of an EDG. See the visualizations of these procedures below.

6.3.2 The $\oplus^{\pi}_{\circ}$ Procedure and Protocol Graph Operation $\oplus^{\pi}_{\circ}$

Definition 259 (Protocol Graph Operation $\oplus^{\pi}_{\circ}$) Let $\oplus^{\pi}_{\circ}$ be an operation on epistemic directed graphs, $\oplus^{\pi}_{\circ} : G^{\pi_1} \times G^{\pi_2} \longrightarrow EDG$ such that: for any regular protocol, π, with main control operator, $\circ$, $\oplus^{\pi}_{\circ}$ maps protocol graphs, g_1 and g_2, of π's proper immediate subprotocols, π_1 and π_2 (where, if $\circ = *$, $\pi_1 = \pi_2$ and $g_1 = g_2$) to epistemic directed graphs constructed out of g_1 and g_2 by the following $\oplus^{\pi}_{\circ}$ Procedure.

Definition 260 ($\oplus^{\pi}_{\circ}$ Procedure) The following inductive, recursive procedure defines protocol graph operations, $\oplus^{\pi}_{\circ}$, for each regular control operator and regular protocol.

(1) (Base Rule) If $\pi \in \Pi_0$ and $g \in G^{\pi}$, then duplicate g one time; $\oplus^{\pi}_{\emptyset}(g) = d(g)$.
(2) (Core Operator Rules) The following rules define protocol graph sub-operations, $\oplus^{\pi}_{;}$, $\oplus^{\pi}_{\cap}$, $\oplus^{\pi}_{\cup}$, and $\oplus^{\pi}_{*}$ corresponding to each regular control operator.
  (a) ($\oplus^{\pi}_{;}$) If $\pi =; \frown \pi_1 \frown \pi_2$, $g_1 \in G^{\pi_1}$, and $g_2 \in G^{\pi_2}$, then:
    (i) Let $n = |last(maxpath(g_1))|$.
    (ii) Duplicate an arbitrary protocol graph of π_2, g_2, n times: $d_1(g_2)$, ..., $d_n(g_2)$.
    (iii) Affix each duplicate of g_2, $d_i(g_2)$, to exactly one $p \in maxpath(g_1)$ so that $@d_i(g_2) = last(p)$.
    (iv) The resulting graph is $\oplus^{\pi}_{;}(g_1, g_2)$.
  (b) ($\oplus^{\pi}_{*}$) If $\pi = * \frown \pi_1$ and $g_1 \in G^{\pi_1}$, then:
    (i) Define an i-depth tree with respect to a protocol π as a tree such that the maximal paths are concatenations of i many executions of π. For π_1*, duplicate an arbitrary protocol graph of π_1, g_1, countably many times so that, for every $i \in \mathbb{N}$, there are enough g_1 duplicates, $d_1(g_1)$, ..., $d_k(g_1)$, to construct an i-depth tree of g_1 duplicates, $m_i = \langle @m_i, d_1(g_1), \ldots, d_k(g_1)\rangle$, by iterating the procedure for ;. More precisely:
      (A) Start with the $i = 1$ case: an arbitrary protocol graph of π_1, g_1, such that $m = |last(maxpath(g_1))|$; g_1 is a 1-depth tree.
      (B) For every step of the procedure for each $i > 1$, denote the graph that results from step $i - 1$ as g_{i-1}; $m^{i-1} = |last(maxpath(g_{i-1}))|$.

(C) For each $i > 1$, duplicate g_{i-1} once, producing $d(g_{i-1})$; produce m^{i-1} duplicates of g_1 ($d_1(g_1), ..., d_{m^{i-1}}(g_1)$); affix exactly one g_1 duplicate, $d_i(g_1)$, to each of the m^{i-1} terminal states of the maximal paths of $d(g_{i-1})$ so that, for each g_1 duplicate, $d_i(g_1)$, $@d_i(g_1) = last(p)$ for some maximal p in $d(g_{i-1})$.

(D) The result is an i-depth tree, m_i, as the maximal paths of m_i are composed of i many executions of π_1.

(ii) For each $i > 1$, affix the resulting i-depth tree of g_1 duplicates, m_i, to the result of step $i-1$ (call it m_{i-1}) so that $@m_i = @m_{i-1}$, thereby building a massive (infinite) tree, $(@m_1, m_1, ..., m_i, m_{i+1}, ...)$, composed out of each i-depth tree, m_i, for each $i \in \mathbb{N}$.

(iii) Build a disconnected, single loop, e, such that $e \in P_{\pi_1 *}$ for the $n = 0$ case execution of $\pi_1 *$ that is not attached to any $@m_i$ for any such tree.

(iv) The resulting graph is $\oplus^{\pi}_{*}(g_1) = (@m_1, m_1, ..., m_n, e)$.

(c) ($\oplus^{\pi}_{\cap}$, $\oplus^{\pi}_{\cup}$) If $\pi = \cap \frown \pi_1 \frown \pi_2$ or $\pi = \cup \frown \pi_1 \frown \pi_2$, $g_1 \in G^{\pi_1}$, and $g_2 \in G^{\pi_2}$, then:

(i) If $\circ = \cap$ or $\circ = \cup$, and $MC(\pi_i) = \cup$ for exactly one of π_1 and π_2 (denote it with π_i), then duplicate each p in $maxpath(g_j) \cap P_{\pi_j}$ (for the other proper immediate subprotocol of π, π_j) a total of n times where $n = \left|last(maxpath(g_i)) \cap P_{\pi_i}\right|$, and affix each p in $maxpath(g_j) \cap P_{\pi_j}$ to each q in $maxpath(g_i) \cap P_{\pi_i}$ so that:

(A) If $\circ = \cap$, then $p \rightleftharpoons q$.

(B) If $\circ = \cup$, then $first(p) = first(q)$.

(ii) Else, select either of g_i and g_j arbitrarily, and duplicate and affix its maximal executions as in the prior step.

(iii) Call the result g.

(iv) (Guarded Incremental Graph Reduction) Iterate the Incremental Graph Reduction Procedure on g until the structural display relations hold with respect to π_1 and π_2 for some pair of proper subgraphs, g_1 and g_2, of the result of the procedure; iterate the Incremental Graph Reduction Procedure on the result of the prior step until no edges can be collapsed without violating the structural display rules for either π_1 or π_2.

(v) The resulting graph is $\oplus^{\pi}_{\cap}(g_1, g_2)$ (resp. $\oplus^{\pi}_{\cup}(g_1, g_2)$).

(d) (Restriction Rules) Ensure that $\oplus^{\pi}_{\circ}(g_1, g_2)$ satisfies the Restriction Rules defining EDGs by adding edges for any $p \in P_{\pi_1}$ and $q \in P_{\pi_2}$:

(i) For every pq in the resulting graph that wasn't in either of g_1 or g_2, add an edge of the form $\langle f(p), \pi_1; \pi_2, f(q)\rangle$.

(ii) For every $p \cup q$ in the resulting graph that wasn't in either of g_1 or g_2, add edges of the form $\langle f(p), \pi_1 \cup \pi_2, f(p)\rangle$, $\langle f(q), \pi_1 \cup \pi_2, f(q)\rangle$.

(iii) For every $p^n_{i \leq n}$ in the resulting graph that wasn't in either of g_1 or g_2, add an edge of the form $\left\langle f(p^n_{i \leq n}), \pi_i *, l(p^n_{i \leq n})\right\rangle$.

(iv) For every $p \cap q$ in the resulting graph that wasn't in either of g_1 or g_2, add an edge of the form $\langle f(p), \pi_1 \cap \pi_2, f(p)\rangle$.

(e) (Valuations) Define the valuation function of $\oplus_{\circ}^{\pi}(g_1, g_2)$, $v_{\oplus_{\circ}^{\pi}(g_1,g_2)}$:
 (i) For any $s_1 \in S_{g_1}$ and $s_2 \in S_{g_2}$, and arbitrary formula, A:
  (A) If $s \in S_{d(g_1)\sqsubseteq\oplus_{\circ}^{\pi}(g_1,g_2)}$, $s_1 \equiv_{GGI} s$, $\neg(s_2 \equiv_{GGI} s)$, then $v_{\oplus_{\circ}^{\pi}(g_1,g_2)}(A, s) = v_{g_1}(A, s_1)$.
  (B) If $s \in S_{d(g_2)\sqsubseteq\oplus_{\circ}^{\pi}(g_1,g_2)}$, $s_2 \equiv_{GGI} s$, $\neg(s_1 \equiv_{GGI} s)$, then $v_{\oplus_{\circ}^{\pi}(g_1,g_2)}(A, s) = v_{g_2}(A, s_2)$.
  (C) If $s \in S_{d(g_2)\sqsubseteq\oplus_{\circ}^{\pi}(g_1,g_2)}$, $s_1 \equiv_{GGI} s$, $s_2 \equiv_{GGI} s$, then $v_{\oplus_{\circ}^{\pi}(g_1,g_2)}(A, s) = v_{g_1}(A, s_1)$ or $v_{\oplus_{\circ}^{\pi}(g_1,g_2)}(A, s) = v_{g_2}(A, s_2)$ arbitrarily.
  (D) If $\neg(s_1 \equiv_{GGI} s)$, $\neg(s_2 \equiv_{GGI} s)$, $MC(\pi) \in \{\cup, \cap\}$, then, if $A = \langle\pi_i\rangle_p \top$ for any $\pi_i \in PISub(\pi)$, then $v_{\oplus_{\circ}^{\pi}(g_1,g_2)}(A, s) = 1$; else, $v_{\oplus_{\circ}^{\pi}(g_1,g_2)}(A, s) = v_{g_1}(A, s_1)$ or $v_{\oplus_{\circ}^{\pi}(g_1,g_2)}(A, s) = v_{g_2}(A, s_2)$ arbitrarily.

(f) (Content Functions) Define the content function of $\oplus_{\circ}^{\pi}(g_1, g_2)$, $c_{\oplus_{\circ}^{\pi}(g_1,g_2)}$:
 (i) For any $s_1 \in S_{g_1}$ and $s_2 \in S_{g_2}$:
  (A) If $s_1 \in S_{g_1}$ and $s \in S_{d(g_1)\sqsubseteq\oplus_{\circ}^{\pi}(g_1,g_2)}$ where $s_1 \equiv_{GGI} s$ and $\neg(s_2 \equiv_{GGI} s)$, then $c_{\oplus_{\circ}^{\pi}(g_1,g_2)}(s) = c_{g_1}(s_1)$.
  (B) If $s_2 \in S_{g_2}$ and $s \in S_{d(g_2)\sqsubseteq\oplus_{\circ}^{\pi}(g_1,g_2)}$ where $s_2 \equiv_{GGI} s$ and $\neg(s_1 \equiv_{GGI} s)$, then $c_{\oplus_{\circ}^{\pi}(g_1,g_2)}(s) = c_{g_2}(s_2)$.
  (C) If $s_1 \in S_{g_1}$ and $s \in S_{d(g_1)\sqsubseteq\oplus_{\circ}^{\pi}(g_1,g_2)}$ where $s_1 \equiv_{GGI} s$ and $s_2 \equiv_{GGI} s$, then $c_{\oplus_{\circ}^{\pi}(g_1,g_2)}(s) = c_{g_1}(s_1)$ or $c_{\oplus_{\circ}^{\pi}(g_1,g_2)}(s) = c_{g_1}(s_2)$ arbitrarily.
  (D) If $\neg(s_1 \equiv_{GGI} s)$ and $\neg(s_2 \equiv_{GGI} s)$, then $c_{\oplus_{\circ}^{\pi}(g_1,g_2)}(s) = c_{g_1}(s_1)$ or $c_{\oplus_{\circ}^{\pi}(g_1,g_2)}(s) = c_{g_1}(s_2)$ arbitrarily.

(g) (Iterative Rule) For arbitrary regular protocol, π, where $\pi = \circ \frown \pi_1 \frown \pi_{a(\circ)}$, $\pi \in P^i_{PLEN}$, and $i \geq 1$, build $\oplus_{\circ}^{\pi}(g_1, g_{a(\circ)})$ (if $\pi = * \frown \pi_1$, then $a(\circ) = 1$, and $\oplus_{\circ}^{\pi}(g_1, g_1)$ is abbreviated with $\oplus_{\circ}^{\pi}(g_1)$) by iterating applications of the Control Operator Rules[4] upward over ascending types of

---

[4] The presentation in the main text elliptically presents an explicit iteration rule over types of proper subprotocols. The following equivalently presents an inductive rule over proper immediate subprotocols:

(Iterative Rule): For arbitrary regular protocol, π, where $\pi = \circ \frown \pi_1 \frown \pi_{a(\circ)}$, $\pi \in P^i_{PLEN}$, and $i \geq 1$, build $\oplus_{\circ}^{\pi}(g_1, g_{a(\circ)})$ (if $\pi = * \frown \pi_1$, then $a(\circ) = 1$, and $\oplus_{\circ}^{\pi}(g_1, g_1)$ is abbreviated with $\oplus_{\circ}^{\pi}(g_1)$) by iterating applications of the Control Operator Rules upward over ascending types of proper subprotocols up to the proper immediate subprotocols, π_1 and $\pi_{a(\circ)}$, and then finally to π itself:

(a) For all $\pi' \in PSub(\pi)$:
 (i) If $\pi' \in PSub(\pi) \cap \Pi_0$, then apply the Base Rule to π' using an arbitrary protocol graph of π'.
  (A) The result: $\oplus_{\circ}^{\pi'}(e) = d_1(e)$ for arbitrary $e \in G^{\pi'}$.
 (ii) If $\pi' \in PSub(\pi) \cap P^1_{PLEN}$, then apply the Base Rule to the proper immediate subprotocols of π', and then apply the COR Rules to π' using the results of the Base Rule for the proper immediate subprotocols of π'.
  (A) The result: $\oplus_{\circ}^{\pi'}(\oplus_{\circ}^{\pi'_1}(e_1), \oplus_{\circ}^{\pi'_{a(\circ)}}(e_{a(\circ)}))$ for arbitrary $e_i \in G^{\pi'_i}$ for each $\pi'_i \in PISub(\pi')$.

proper subprotocols up to the proper immediate subprotocols, π_1 and $\pi_{a(\circ)}$, and then finally to π itself:

(a) For all $\pi' \in PSub(\pi)$:

(i) If $\pi' \in PSub(\pi) \cap \Pi_0$, then apply the Base Rule to π'.

(ii) If $PISub(\pi') \subseteq P^0_{PLEN}$, then apply the Control Operator Rules to the graphs, $g_1, g_{a(MC(\pi'))}$, of each $\pi_i \in PISub(\pi')$ that result from applying the Base Rule to π''s proper immediate subprotocols.

(iii) If $PISub(\pi') \subseteq \bigcup_{j=1}^{1} P^j_{PLEN}$, then apply the Control Operator Rules to the graphs, $g_1, g_{a(MC(\pi'))}$, of each $\pi_i \in PISub(\pi')$ that result from applying the Base Rule or prior iterations of the Control Operator Rules to π''s proper immediate subprotocols.

(iv) If $PISub(\pi') \subseteq \bigcup_{j=1}^{2} P^j_{PLEN}$, then apply the Control Operator Rules to the graphs, $g_1, g_{a(MC(\pi'))}$, of each $\pi_i \in PISub(\pi')$ that result from applying the Base Rule or prior iterations of the Control Operator Rules to π''s proper immediate subprotocols.

(v) If $PISub(\pi') \subseteq \bigcup_{j=1}^{3} P^j_{PLEN}$, then apply the Control Operator Rules to the graphs, $g_1, g_{a(MC(\pi'))}$, of each $\pi_i \in PISub(\pi')$ that result from applying the Base Rule or prior iterations of the Control Operator Rules to π''s proper immediate subprotocols.

...

(vi) If $PISub(\pi') \subseteq \bigcup_{k=1}^{(i-1)-1} P^k_{PLEN}$, then apply the Control Operator Rules to the graphs, $g_1, g_{a(MC(\pi'))}$, of each $\pi_i \in PISub(\pi')$ that result from applying the Base Rule or prior iterations of the Control Operator Rules to π''s proper immediate subprotocols.

(b) For π: The final step of the inductive iteration of the rule applies the Control Operator Rules to the results of applying the procedure to the proper immediate subprotocols of π, π_1 and $\pi_{a(\circ)}$, and the protocol graphs of them that result from prior iteration of the Control Operator Rules to the proper immediate subprotocols of π_1 and $\pi_{a(\circ)}$ and protocol graphs thereof.

Remark 261 Call this, interchangeably, the $\oplus^{\pi}_{\circ}$ Procedure or the PGIIRC procedure. The basic idea of the procedure and the accompanying operation is that, for

---

(iii) If $\pi' \in PSub(\pi) \cap P^k_{PLEN}$ where $k \leq i-1$, then apply the Base Rule and COR Rules to π' using the results of the application of the Base and COR Rules for the proper immediate subprotocols of π'.

(A) The result: $\oplus^{\pi'}_{\circ}(\oplus^{\pi'_1}_{\circ}(g_1), \oplus^{\pi'_{a(\circ)}}_{\circ}(g_{a(\circ)}))$ for arbitrary $g_i \in G^{\pi'_i}$ for each $\pi'_i \in PISub(\pi')$.

(b) For π: The final step of the (structural) inductive iteration of the rule applies the Control Operator Rules to the results of applying the procedure to the proper immediate subprotocols of π, π_1 and $\pi_{a(\circ)}$, and the protocol graphs of them that result from prior iteration of the Control Operator Rules to the proper immediate subprotocols of π_1 and $\pi_{a(\circ)}$ and protocol graphs thereof.

any complex, regular protocol, $\pi = \circ \frown \pi_1 \frown \pi_2$, the main control operator, $\circ$, corresponds to an operation, $\oplus_\circ^\pi$, on the protocol graphs of π's proper immediate subprotocols that produces EDGs. The operation works according to the procedure, so if we take protocol graphs, g_1 and g_2, of π's proper immediate subprotocols, π_1 and π_2, the operation produces $\oplus_\circ^\pi(g_1, g_2)$, such that $\oplus_\circ^\pi(g_1, g_2) \in EDG$, and g_1 and g_2 are isomorphic to duplicates of proper subgraphs of $\oplus_\circ^\pi(g_1, g_2)$ linked by concatenation, iteration, correspondence, branching, and graph reduction. Perhaps more precisely, g_1 and g_2 are connected by the operation in such a way that their maximal paths are connected by operations of concatenation, iteration, correspondence, branching, and graph reduction. The procedure is somewhat complex, so see the following diagrams for help in visualizing the steps.

Remark 262 (Procedure Visualization)

(COR Rules): The procedure correlates the regular (main) control operators of arbitrary protocols, π, to operations on the protocol graphs of the proper immediate subprotocols of π. State labels are arbitrary, as they are irrelevant to graph-theoretic structure.

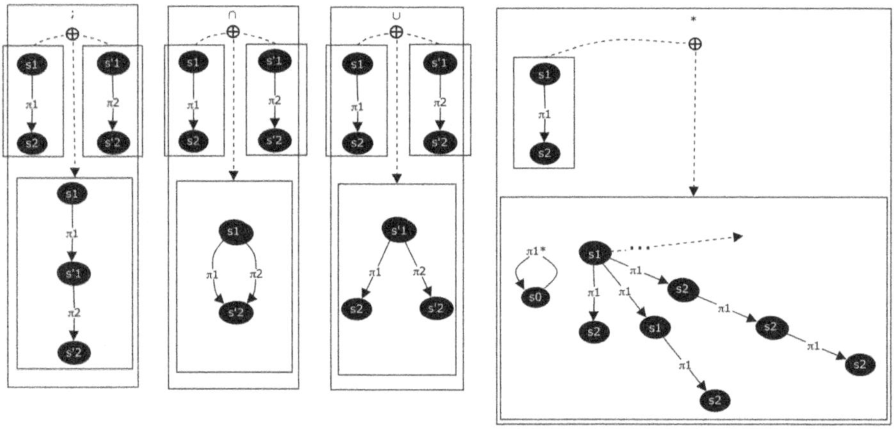

(Iteration COR Rule): Iteration requires large (infinite) structures that quickly become visually unworkable. The aim of PGIIRC is to define a construction procedure for protocol graphs of arbitrary regular π out of protocol graphs of its proper, immediate subprotocols. The requirement that all implementations of iterations be executable generates elementary but very large structures for even atomic proper subprotocols. Again, state labels are arbitrary, but note that the root state of each tree is identical to every other root connected by an edge labeled with " =". The sequence of steps indicates an infinite sequence of increasingly large trees.

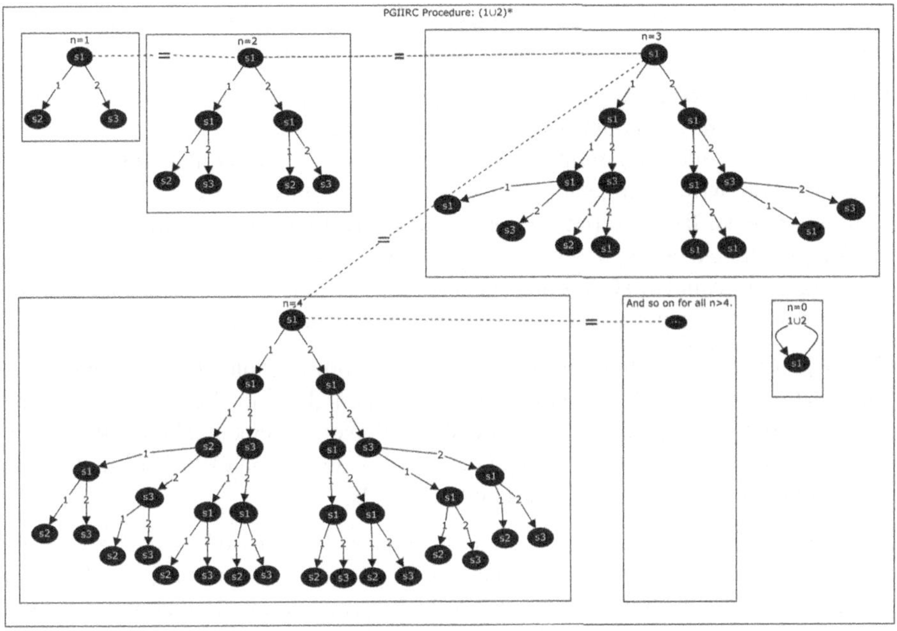

(Guarded Incremental Graph Reduction): Application of the $\oplus^{\pi}_{\circ}$ Procedure to $((\pi_1;\, (\pi_3 \cup \pi_4)) \cup (\pi_2;\, (\pi_5 \cup \pi_6))) \cap (\pi_7;\, \pi_8)$ proceeds in clockwise illustrations. Step 1 introduces arbitrary protocol graphs of $(\pi_1;\, (\pi_3 \cup \pi_4)) \cup (\pi_2;\, (\pi_5 \cup \pi_6))$ and $\pi_7;\, \pi_8$. Step 2 carries out the relevant duplication and affixing steps. Step 3 carries out the graph reduction step, which eliminates redundant structures. Note that collapsing any of the remaining π_8 edges would result in violation of SD1(4).

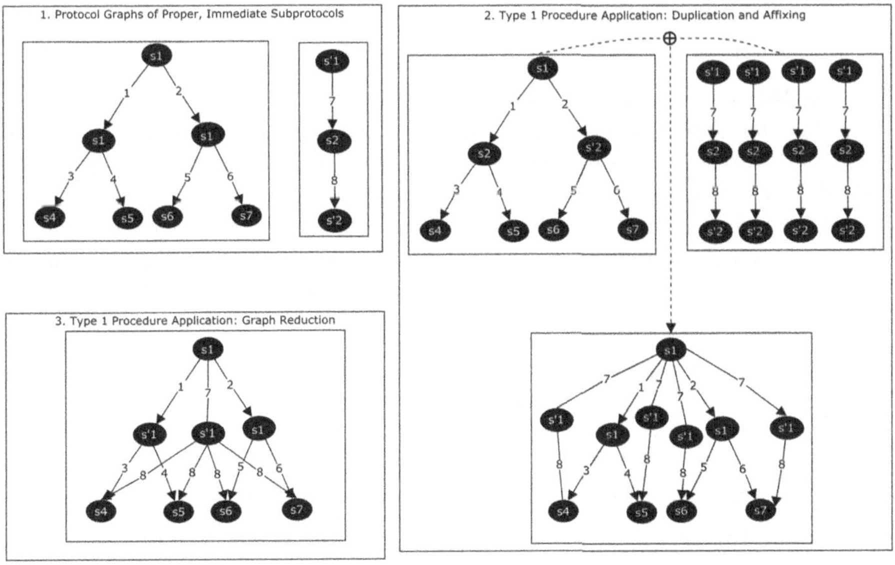

Remark 263 Interestingly, the Core Operator Rules specify procedures that can be codified with protocols, given a generous notion of (relatively) basic protocols.

- The sequential composition rule is a sequential composition of graph duplication and affixing subprocedures on the protocol graphs of π_1 and π_2.
- The iteration rule is an iteration, n times for all $n \in \mathbb{N}$, of sequentially composed duplication and affixing subprocedures on the outputs of prior steps, with each iteration and affixing step itself composed with a loop construction step.
- The parallel execution and choice rules are sequential compositions of choices of duplication and affixing subprocedures sequentially composed with a guarded iteration of the Incremental Graph Reduction procedure on the result.

The reduction step of the rule is a sequential composition of: (1) a sequential composition of (1a) guarded iteration of Incremental Graph Reduction on g and (1b) a sequential composition of a guarded iteration of Incremental Graph Reduction and a reversal on the result of 1a, and (2) a guarded iteration of Incremental Graph Reduction on the result of 1b. A little more exactly:

(1) While it is not the case that there are protocol graphs of π_1 and π_2 among the proper subgraphs of g_i, where g_i results from 0 or more iterations of Incremental Graph Reduction, iterate Incremental Graph Reduction on g_i until a graph, g_{i+1}, is constructed that contains subgraphs that satisfy the structural display rules for both π_1 and π_2;
(2) While there are proper subgraphs of g_{i+1} that are protocol graphs of π_1 and π_2; iterate Incremental Graph Reduction until a graph, g_{i+2}, is constructed that contains no proper subgraphs for one of π_1 or π_2;
(3) Reverse the last step to reinstate the condition;
(4) While there are proper subgraphs of the result of step 3 that are protocol graphs of π_1 and π_2, iterate 2-3 on the result of step 3 (or the result of prior iterations of 1-3) until no edges can be collapsed without violating the structural display rules for π_1 or π_2.

Lemma 264 *Two immediate properties of the PGIIRC Procedure:*

(L0) By the duplication of edges, valuations, and content functions by the COR Rules and Restriction Rules step of the $\oplus^{\pi}_{\circ}$ Procedure, every $\oplus^{\pi}_{\circ}(g_1, g_2)$ is an epistemic directed graph.

(L0.5) L0 is immediate from the construction. For L0.5: If applied to protocol graphs, g_1 and g_2, of regular $\pi = \circ \frown \pi_1 \frown \pi_2$ where $\circ \in \{\cap, \cup\}$, the guarded incremental graph reduction step of the COR rules produces g'_1 and g'_2 such that $g'_1 \simeq g'_2$.

Proof Suppose that g_1 and g_2 are isomorphic protocol graphs of π, and the procedure is applied to them (to completion) to produce g'_1 and g'_2. Suppose that g'_1 and g'_2 are not isomorphic. Then there must at least be some redundant or detached edge or loop that breaks the bijection among edges or a detached state that breaks the bijection among states. Suppose that g'_i contains some bijection breaking redundant or detached structure. But whatever structure it is must either be contained in (i.e., must be a subgraph of) a proper subgraph of g_i' that is a protocol graph of either π_1 or π_2 or it must be outside of any protocol graphs of π_1 and π_2. If it's outside, then the procedure should have removed it, contradicting the assumption that the procedure was applied. If the structure is inside a graph of π_1 (resp. π_2), then there are non-isomorphic graphs of π_1 (resp. π_2), contradicting SUT. □

Claim 265 The basic idea of PGIIRC, then, is that, for any complex, regular protocol, $\pi = \circ \frown \pi_1 \frown \pi_2$, if the $\oplus_\circ^\pi$ procedure is applied to π and protocol graphs of π_1 and π_2, g_1 and g_2, it outputs $\oplus_\circ^\pi(g_1, g_2)$, and if it is applied to π and protocol graphs of π_1 and π_2, g_3 and g_4, it outputs $\oplus_\circ^\pi(g_3, g_4)$, and $\oplus_\circ^\pi(g_1, g_2)$ and $\oplus_\circ^\pi(g_3, g_4)$ are isomorphic protocol graphs of π. This result can equivalently be read as a result about the $\oplus_\circ^\pi$ Procedure to the effect that it produces protocol graphs of π from protocol graphs of π's proper immediate subprotocols.

6.3.3 Derivation of the Theorem

Remark 266 The proof (naturally) requires deriving (I) ($\oplus_\circ^\pi(g_1, g_2) \simeq \oplus_\circ^\pi(g_3, g_4)$) and (II) ($\oplus_\circ^\pi(g_1, g_2) \in G^\pi$) and $\oplus_\circ^\pi(g_3, g_4) \in G^\pi$) for arbitrary, regular π (given pairs, g_1 and g_3, g_2 and g_4, of protocol graphs of π's proper immediate subprotocols) from the assumption that the premises hold for π. Structural induction over the proper, immediate subprotocols of arbitrary, regular π suffices.

There is an interesting parallel between the induction for SUT and the inductive step of PGIIRC. The structural induction step for PGIIRC relies on Lemmata L2, L5, and L6 that were used to prove SUT, and it also uses modified variants of L1, L3, and L4. Given these lemmata, the reasoning for the induction rule in the following proof proceeds in nearly the same inferential steps as the inductive step for SUT, but replacing L1, L2, and L4 with L1', L3', and L4'.

Proof (Sketch[5])

(1) Derive Lemmata L1', L3', and L4', and recall L2, L5, and L6.
(2) For arbitrary regular $\pi \in P_{PLEN}^i$:

---

[5] ***Proof*** **(*Protocol*)** There is a second path for this derivation that intertwines the induction with simultaneous executions of the $\oplus_\circ^\pi$ Procedure:

(1) Derive Lemmata L1', L3', and L4', and recall Lemmata L2, L5, and L6.
(2) For arbitrary, regular $\pi \in P_{PLEN}^i$ where $\pi = \circ \frown \pi_1 \frown \pi_2$, execute the $\oplus_\circ^\pi$ Procedure on π twice in parallel, while deriving (I) and (II) for each type of proper subprotocol at each stage of iteration.

   (a) (Base) For arbitrary $\pi' \in PSub(\pi)$ of types 0 or 1:
      (i) If π' is atomic ($i = 0$), iterate the $\oplus_\circ^\pi$ Procedure (Base Rule) on π' twice using arbitrary edges, g_1 and g_2, with π' the middle term to construct duplicate edges with π' the middle term: $\oplus_\emptyset^{\pi'}(g_1)$ and $\oplus_\emptyset^{\pi'}(g_2)$.

(a) (Base)

(i) If π is atomic ($i = 0$), iterate the $\oplus^{\pi}_{\circ}$ Procedure (Base Rule) twice to construct an arbitrary pair of edges with π as the middle term.

(A) Derive (I) for the results of the Base Rule by direct construction of bijections or immediately from the definition of edges.

(B) Derive (II) for the results of the Base Rule by showing, stepwise, that the edges satisfy the structural display rules w.r.t. π.

(A) Derive (I) for the outputs of the Base Rule ($\oplus^{\pi'}_{\emptyset}(g_1) \simeq \oplus^{\pi'}_{\emptyset}(g_2)$) by direct construction of bijections; Derive (II) for the outputs of the Base Rule ($\oplus^{\pi'}_{\emptyset}(g_1) \in G^{\pi'}$, $\oplus^{\pi'}_{\emptyset}(g_2) \in G^{\pi'}$) by showing, stepwise, that $\oplus^{\pi'}_{\emptyset}(g_1)$ and $\oplus^{\pi'}_{\emptyset}(g_2)$ satisfy the structural display rules w.r.t. π'.

(ii) If π' has atomic proper immediate subprotocols ($i = 1$), π'_1 and π'_2, iterate the $\oplus^{\pi}_{\circ}$ Procedure (COR Rules) on π' twice in parallel; the first iteration produces the pairs of protocol graphs, $\oplus^{\pi'_1}_{\emptyset}(g_1)$ and $\oplus^{\pi'_2}_{\emptyset}(g_2)$ and $\oplus^{\pi'_1}_{\emptyset}(g_3)$ and $\oplus^{\pi'_2}_{\emptyset}(g_4)$, of π''s proper immediate subprotocols by the Base Rule. The results of the iteration that uses those protocol graphs: $\oplus^{\pi'}_{\circ'}(\oplus^{\pi'_1}_{\emptyset}(g_1), \oplus^{\pi'_2}_{\emptyset}(g_2))$ and $\oplus^{\pi'}_{\circ'}(\oplus^{\pi'_1}_{\emptyset}(g_3), \oplus^{\pi'_2}_{\emptyset}(g_4))$.

(A) Derive (I) for the outputs of the COR Rules ($\oplus^{\pi'}_{\circ'}(\oplus^{\pi'_1}_{\emptyset}(g_1), \oplus^{\pi'_2}_{\emptyset}(g_2)) \simeq \oplus^{\pi'}_{\circ'}(\oplus^{\pi'_1}_{\emptyset}(g_3), \oplus^{\pi'_2}_{\emptyset}(g_4))$) by direct construction of bijections or immediately by L6; Derive (II) for the outputs of the COR Rules ($\oplus^{\pi'}_{\circ'}(\oplus^{\pi'_1}_{\emptyset}(g_1), \oplus^{\pi'_2}_{\emptyset}(g_2)) \in G^{\pi'}$ and $\oplus^{\pi'}_{\circ'}(\oplus^{\pi'_1}_{\emptyset}(g_3), \oplus^{\pi'_2}_{\emptyset}(g_4)) \in G^{\pi'}$) by showing, stepwise, that the structural display rules w.r.t. π' are satisfied.

(b) (Induction) A more exacting rule induces the results for arbitrary proper subprotocols of π', including π_1 and π_2:

(Restricted Induction Rule): *For arbitrary $\pi' \in PSub(\pi)$ of arbitrary type $1 < k \leq i - 1$, including π_1 and π_2, if the $\oplus^{\pi}_{\circ}$ Procedure has been iterated $k - 1$ times twice in parallel, and (I) and (II) have been shown to hold for the results of each of the $k - 1$ iterations (the procedure thus producing pairs of isomorphic protocol graphs for the proper immediate subprotocols of π' ($\pi' = \circ \frown \pi'_1 \frown \pi'_2$) from the results of prior iterations (themselves produced from the results of prior iterations)), then (I) and (II) hold for π'.*

(i) If $\pi' \in P^k_{PLEN}$ where $k > 1$, then iterate the $\oplus^{\pi}_{\circ}$ Procedure (COR Rules) on π' twice in parallel using the protocol graphs of π''s proper immediate subprotocols, π'_1 and π'_2, ($\oplus^{\pi'_1}_{\circ'_1}(g_1)$ and $\oplus^{\pi'_2}_{\circ'_2}(g_2)$, $\oplus^{\pi'_1}_{\circ'_1}(g_3)$ and $\oplus^{\pi'2}_{\circ'_2}(g_4)$) resulting from step $k-1$ (or steps preceding it).

(A) The results: $\oplus^{\pi'}_{\circ'}(\oplus^{\pi'_1}_{\circ'_1}(g_1), \oplus^{\pi'2}_{\circ'_2}(g_2))$ and $\oplus^{\pi'}_{\circ'}(\oplus^{\pi'_1}_{\circ'_1}(g_3), \oplus^{\pi'_2}_{\circ'_2}(g_4))$ where each $\oplus^{\pi'_n}_{\circ'_n}(g_m)$ is elliptical for

$$\oplus^{\pi'_n}_{\circ'_n}(\oplus^{\pi_{x_1}}_{\circ_{x_1}}(\oplus^{\pi_y}_{\circ_y}(...(\oplus^{\pi_{z-1}}_{\circ_{z-1}}(\oplus^{\pi_z}_{\circ_z}(\oplus^{a_1}_{\emptyset}(e_1), \oplus^{a_2}_{\emptyset}(e_2)), ...), ...), ...), \oplus^{\pi_{x_2}}_{\circ_{x_2}}(...)))$$

(ii) If π has atomic proper immediate subprotocols ($i = 1$), iterate the $\oplus_{\circ}^{\pi}$ Procedure (COR Rules) twice to arbitrary (edge) protocol graphs of π's proper immediate subprotocols.
   (a) Derive (I) for the results of the COR Rules by direct construction of bijections or immediately by L6.
   (b) Derive (II) for the results of the COR Rules by showing, stepwise, that the structural display rules w.r.t. π are satisfied.

(b) (Induction) The $\oplus_{\circ}^{\pi}$ Procedure and the induction principle are intertwined.
(Induction Rule): *For all regular protocols, $\pi \in P^{i}_{PLEN}$, if both (I) and (II) hold for results of applying the $\oplus_{\circ}^{\pi}$* Procedure *to the proper immediate subprotocols of π, then the conjunction of (I) and (II) holds for the results of applying the $\oplus_{\circ}^{\pi}$* Procedure *to π.*
   (i) If $\pi \in P^{i}_{PLEN}$ where $i > 1$ and $\pi = \circ \frown \pi_1 \frown \pi_2$ (mutatis mutandis $* \frown \pi_1$), then iterate the $\oplus_{\circ}^{\pi}$ Procedure (Base Rule; COR Rules) twice over the proper subprotocols of π (from π's atomic

---

which results from applying the $\oplus_{\circ}^{\pi}$ Procedure iteratively over the proper immediate subprotocols of π_1' and π_2', where: $\oplus_{\emptyset}^{a_o}(e_o)$ are the results of base rule application to atomic subprotocols of π_n' using arbitrary edges with the relevant atomics the middle term, and $\circ_{x_1}$ is the second control operator in the control sequence of π_n', i.e., $\circ_n'$ is the first and main control operator, $\circ_{x_1}$ is the main control operator of the first of π_n''s proper immediate subprotocols, and $\circ_{x_2}$ is the main control operator of the second (if applicable).
   (ii) From the induction premise: derive (I) for the outputs of the COR Rules $(\oplus_{\circ'}^{\pi'}(\oplus_{\circ_1'}^{\pi_1'}(g_1), \oplus_{\circ_2'}^{\pi_2'}(g_2)) \simeq \oplus_{\circ'}^{\pi'}(\oplus_{\circ_1'}^{\pi_1'}(g_3), \oplus_{\circ_2'}^{\pi_2'}(g_4)))$ by means of lemmata L1', L2, L3', L4', L5, and L6.
   (iii) From the induction premise: derive (II) for the outputs of the COR Rules $(\oplus_{\circ'}^{\pi'}(\oplus_{\circ_1'}^{\pi_1'}(g_1), \oplus_{\circ_2'}^{\pi_2'}(g_2)) \in G^{\pi'}$ and $\oplus_{\circ'}^{\pi'}(\oplus_{\circ_1'}^{\pi_1'}(g_3), \oplus_{\circ_2'}^{\pi_2'}(g_4)) \in G^{\pi'})$ by showing, stepwise, that the structural display rules w.r.t. π' are satisfied.

(3) The penultimate iteration of the $\oplus_{\circ}^{\pi}$ Procedure (COR Rules) to π produces pairs of protocol graphs of π's proper immediate subprotocols, $\oplus_{\circ_1}^{\pi_1}(g_1)$ and $\oplus_{\circ_2}^{\pi_2}(g_2)$, $\oplus_{\circ_1}^{\pi_1}(g_3)$ and $\oplus_{\circ_2}^{\pi_2}(g_4)$. The prior stages of proof showed that $\oplus_{\circ_1}^{\pi_1}(g_1) \simeq \oplus_{\circ_1}^{\pi_1}(g_3)$, $\oplus_{\circ_2}^{\pi_2}(g_2) \simeq \oplus_{\circ_1}^{\pi_1}(g_4)$, $\oplus_{\circ_1}^{\pi_1}(g_1), \oplus_{\circ_1}^{\pi_1}(g_3) \in G^{\pi_1}$ and $\oplus_{\circ_2}^{\pi_2}(g_2), \oplus_{\circ_2}^{\pi_2}(g_r) \in G^{\pi_2}$. The final iteration produces $\oplus_{\circ}^{\pi}(\oplus_{\circ_1}^{\pi_1}(g_1), \oplus_{\circ_2}^{\pi_2}(g_2))$ and $\oplus_{\circ}^{\pi}(\oplus_{\circ_1}^{\pi_1}(g_3), \oplus_{\circ_2}^{\pi_2}(g_4))$.

   (a) Derive (I) for the final outputs of the COR Rules $(\oplus_{\circ}^{\pi}(\oplus_{\circ_1}^{\pi_1}(g_1), \oplus_{\circ_2}^{\pi_2}(g_2)) \simeq \oplus_{\circ}^{\pi}(\oplus_{\circ_1}^{\pi_1}(g_3), \oplus_{\circ_2}^{\pi_2}(g_4)))$ by means of lemmata L1', L2, L3', L4', L5, and L6.
   (b) Derive (II) for the final outputs of the COR Rules $(\oplus_{\circ}^{\pi}(\oplus_{\circ_1}^{\pi_1}(g_1), \oplus_{\circ_2}^{\pi_2}(g_2)) \in G^{\pi}$ and $\oplus_{\circ}^{\pi}(\oplus_{\circ_1}^{\pi_1}(g_3), \oplus_{\circ_2}^{\pi_2}(g_4)) \in G^{\pi})$ by showing, stepwise, that the structural display rules w.r.t. π are satisfied.

□

subprotocols to the proper immediate subprotocols of π's proper immediate subprotocols) all the way up to the proper immediate subprotocols, π_1 and π_2.

(A) The results are pairs of EDGs: g_1 and g_3 from application of $\oplus_\circ^\pi$ Procedure to π_1, and g_3 and g_4 from application of $\oplus_\circ^\pi$ Procedure to π_2.

(B) From the induction assumption: $g_1 \simeq g_3$, $g_2 \simeq g_4$, $g_1, g_3 \in G^{\pi_1}$ and $g_2, g_4 \in G^{\pi_2}$.

(ii) Iterate the $\oplus_\circ^\pi$ Procedure on π twice, using the protocol graphs of π's proper immediate subprotocols constructed by prior iterations of the COR Rules.

(A) The results: $\oplus_\circ^\pi(g_1, g_2)$ and $\oplus_\circ^\pi(g_3, g_4)$.

(iii) Derive the consequent of the induction rule on the basis of the results of the iterated $\oplus_\circ^\pi$ Procedure and Lemmata L1', L2, L3', L4', L5, and L6:

(A) From the induction premise: show that (I) $\oplus_\circ^\pi(g_1, g_2) \simeq \oplus_\circ^\pi(g_3, g_4)$ by means of L1', L2, L3', L4', L5, and L6;

(B) From the induction premise: show that (II) $\oplus_\circ^\pi(g_1, g_2) \in G^\pi$ and $\oplus_\circ^\pi(g_3, g_4) \in G^\pi$ by showing, stepwise, that the structural display rules w.r.t. π are satisfied.

□

Remark 267 The presentation of the details of the derivation thus proceeds in two phases. First, I derive modified Lemmata L1', L3', and L4' for later use in the argument. Then, I execute the simultaneous structural induction over proper immediate subprotocols for arbitrary regular π satisfying the premises.

6.3.3.1 Modified Lemmata

Lemma 268 *(L1') If $\pi = \circ \frown \pi_1 \frown \pi_2$, $g_1 \in G^{\pi_1}$, and $g_2 \in G^{\pi_2}$, then the Regular Control Operations Table (immediately below) displays the properties of the maximal executions of π in $\oplus_\circ^\pi(g_1, g_2)$.*

(Regular Control Operations Table)

If $\oplus^{\pi}_{;} : G^{\pi_1} \times G^{\pi_2} \longrightarrow EDG$ *such that* $\oplus^{\pi}_{\circ}(g_1, g_2) = g$, *then* $P_g = Sub(\pi maxpath(g))$ *and, among the subgraphs of* g *:* • $i \leq \lvert last(maxpath(g_1))\rvert$ • there are duplicates of g_1 and g_2 s.t.: $\forall p \in \pi mxpth(g) \exists p' \in \pi_1 mxpth(d_i(g_1)) \exists p'' \in mxpth(d_i(g_2))(p = p'p'')$ • for all duplicates of g_1 there is a duplicate of g_2 s.t.: $\forall p' \in \pi_1 mxpth(g_1) \exists p'' \in \pi_2 mxpth(d_i(g_2)) \exists p \in \pi mxpth(g)(p = p'p'')$ • for all duplicates of g_2 there is a duplicate of g_1 s.t.: $\forall p' \in \pi_2 mxpth(d_i(g_2)) \exists p'' \in \pi_1 mxpth(d_i(g_1)) \exists p \in \pi mxpth(g)(p = p'p'')$ • *For every maximal path,* p, *in* g_1, *let* $first(p) = @d_i(g_2)$ *for some duplicate of* g_2. • *There are no redundant paths in* P_g.	*If* $\oplus^{\pi}_{*} : G^{\pi_1} \times G^{\pi_1} \longrightarrow EDG$ *such that* $\oplus^{\pi}_{\circ}(g_1) = g$, *then* $P_g = Sub(\pi maxpath(g))$ *and:* • *For every* $i \in \mathbb{N}$, *there is an* i*-depth sequence (tree),* $m_i = \langle @m_i, d_1(g_1), ..., d_k(g_1)\rangle$, *of duplicates of* g_1, *where the depth,* i, *is the length of concatenations of maximal paths in the* m_i. • *For every* $i \in \mathbb{N}$, *there is an* i*-length sequence (tree/path),* $p_i = d_1(p_j)...d_k(p_j) = d_{i \leq n}(p_j)^n$, *of duplicates of some maximal execution of* π_1, p_j, *where the depth,* i, *is the length of concatenations of maximal paths* p_i. • *Every* m_i *is a subgraph of* g; $g = (m_1, ..., m_i)$ *for each* $i \in \mathbb{N}$. • *For every* $m_i, m_j \in \langle m_1, ..., m_n\rangle$, $@m_i = @m_j = @g$.
If $\oplus^{\pi}_{\cup} : G^{\pi_1} \times G^{\pi_2} \longrightarrow EDG$ *such that* $\oplus^{\pi}_{\circ}(g_1, g_2) = g$, *then* $P_g = Sub(\pi maxpath(g))$ *and for all pairs of duplicates of* g_1 *and* g_2 *among the subgraphs of* g *:* • *if duplicating* g_j, *then* $k \leq \lvert maxpath(g_i)\rvert$ • $\pi maxpath(g) = \pi_1 maxpath(g_i) \cup \pi_2 maxpath(d_k(g_j))$ • $\forall p' \in \pi_1 mxpth(d_k(g_i)) \exists p'' \in \pi_2 mxpth(d_k(g_j)) \exists p \in \pi mxpth(g)(p' \cup p'')$ • $\forall p' \in \pi_2 mxpth(d_k(g_j)) \exists p'' \in \pi_1 mxpth(d_k(g_i)) \exists p \in \pi mxpth(g)(p' \cup p'')$ • *There are no redundant paths in* P_g.	*If* $\oplus^{\pi}_{\cap} : G^{\pi_1} \times G^{\pi_2} \longrightarrow EDG$ *such that* $\oplus^{\pi}_{\circ}(g_1, g_2) = g$, *then* $P_g = Sub(\pi maxpath(g))$ *and for all pairs of duplicates of* g_1 *and* g_2 *among the subgraphs of* g *:* • *if duplicating* g_j, *then* $k \leq \lvert maxpath(g_i)\rvert$ • $\forall p \in \pi mxpth(g) \exists p' \in \pi_1 mxpth(d_k(g_i)) \exists p'' \in \pi_2 mxpth(d_k(g_j))(p' \rightleftharpoons p'')$ • $\forall p' \in \pi_1 mxpth(d_k(g_i)) \exists p'' \in \pi_2 mxpth(d_k(g_j)) \exists p \in \pi mxpth(g)(p' \rightleftharpoons p'')$ • $\forall p' \in \pi_2 mxpth(d_k(g_j)) \exists p'' \in \pi_1 mxpth(d_k(g_i)) \exists p \in \pi mxpth(g)(p' \rightleftharpoons p'')$ • *There are no redundant paths in* P_g.

Proof (Proof-Plan) Inductive iteration of the PGIIRC procedure and then induction over (regular) protocol structure by mechanical or even trivial derivation steps.

(1) (Base) The Base Rule produces edges for atomic protocols. The iteration of the Control Operator Rules for protocols of type 1 generate graphs for protocols, π, with atomic proper immediate subprotocols on the basis of the main control operator of π:

Case 1. (;) Let $\pi = \pi_1; \pi_2$. Let $g_1 \in G^{\pi_1}$ and $g_2 \in G^{\pi_2}$. Since $\pi_1, \pi_2 \in \Pi_0$, there are edges, e_1 and e_2, such that $g_1 = e_1$ and $g_2 = e_2$. The Control Operator Rules output $e'e''$ such that $e' \simeq e_1$ and $e'' \simeq e_2$, as required.

Case 2. ($*$) Let $\pi = \pi_1*$. Let $g_1 \in G^{\pi_1}$. Since $\pi_1 \in \Pi_0$, there is an edge, e_1, such that $g_1 = e_1$. The Control Operator Rules output a graph in which all paths are either the single, detached looping edge, $e*$, that executes π_1* or a finite (i—length for some $i \in \mathbb{N}$) sequence, $e_1...e_1$, such that, for all such sequences, any pair, p_i and p_j are such that $f(p_i) = f(p_j)$, and each e_i executes π_1. Each such path is a tree plus a detached loop, as is each edge, thus satisfying the conditions in the Regular Control Operations Table.

Case 3. ($\cap$) Let $\pi = \pi_1 \cap \pi_2$. Let $g_1 \in G^{\pi_1}$ and $g_2 \in G^{\pi_2}$. Since $\pi_1, \pi_2 \in \Pi_0$, there are edges, e_1 and e_2, such that $g_1 = e_1$ and $g_2 = e_2$. The Control Operator Rules output $e' \rightleftharpoons e''$ such that $e' \simeq e_1$ and $e'' \simeq e_2$, as required.

Case 4. ($\cup$) Let $\pi = \pi_1 \cup \pi_2$. Let $g_1 \in G^{\pi_1}$ and $g_2 \in G^{\pi_2}$. Since $\pi_1, \pi_2 \in \Pi_0$, there are edges, e_1 and e_2, such that $g_1 = e_1$ and $g_2 = e_2$. The Control Operator Rules output $f(e') = f(e'')$ such that $e' \simeq e_1$ and $e'' \simeq e_2$, as required.

(2) (Induction) If L1' holds for the immediate proper subprotocols of π, then it holds for π. The base step shows that L1' holds for all regular type 0 and type 1 protocols. Assume that, for any protocol, π, of type i, L1' holds for all proper immediate subprotocols of π.

Case 1. (;) Let $\pi = \pi_1; \pi_2$. Let $g_1 \in G^{\pi_1}$ and $g_2 \in G^{\pi_2}$. Let p_1 be a maximal execution of π_1 in g_1. Let p_2 be a maximal execution of π_2 in g_2. The Control Operator Rules output $p'p''$ such that $p' \simeq p_1$ and $p'' \simeq p_2$, for every pair of arbitrary maximal executions of π_1 and π_2, as required.

Case 2. ($*$) Let $\pi = \pi_1*$. Let $g_1 \in G^{\pi_1}$. The Control Operator Rules output a graph in which all paths are either the single, detached looping edge, $e*$, that executes π_1* or a finite sequence, $p_1...p_1$, such that, each p_1 duplicate executes π_1 and for all such sequences, any pair, $d_i(p_1)$ and $d_i(p_1)$ are such that $first(d_i(p_1)) = first(d_j(p_1))$. Each such path is an i-length tree for some $i \in \mathbb{N}$, thus satisfying the conditions in the Regular Control Operations Table.

Case 3. ($\cap$) Let $\pi = \pi_1 \cap \pi_2$. Let $g_1 \in G^{\pi_1}$ and $g_2 \in G^{\pi_2}$. Let p_1 be a maximal execution of π_1 in g_1. Let p_2 be a maximal execution of π_2 in g_2. The Control Operator Rules output $p' \simeq p_1$ and $p'' \simeq p_2$ such that $p' \rightleftharpoons p''$, for every pair of arbitrary maximal executions of π_1 and π_2, as required. Graph reduction eliminates any possible counterpaths (paths that could suffice as counterexamples to the result), and can never remove any correlation among maximal paths.

Case 4. (∪) Let $\pi = \pi_1 \cup \pi_2$. Let $g_1 \in G^{\pi_1}$ and $g_2 \in G^{\pi_2}$. Let p_1 be a maximal execution of π_1 in g_1. Let p_2 be a maximal execution of π_2 in g_2. The Control Operator Rules output $p' \simeq p_1$ and $p'' \simeq p_2$ such that $f(p') = f(p'')$, for every pair of arbitrary maximal executions of π_1 and π_2, as required. Graph reduction eliminates any possible counterpaths (paths that could suffice as counterexamples to the result), and can never remove any branching among maximal paths.

□

Lemma 269 *(L3') For arbitrary EDG, g, and arbitrary p in* P_g, *if* $g = \oplus_\circ^\pi(g_1, g_2)$, $g_1 \in G^{\pi_1}$, *and* $g_2 \in G^{\pi_2}$, *then there is a maximal execution of* π *in g, q, such that* $p \sqsubseteq q$ *or there are* $q_1, \ldots, q_n$ *in* $Sub(maxpath(g) \cap P_\pi)$ *such that* $p = q_1 \ldots q_n$ *or p is a subgraph of a detached iteration loop.*

Proof (Proof-Plan) Suppose that p is not a detached iteration loop or subpath thereof. The SD1(2), NST conditions, and construction procedure require that the set of executions of π in g exhausts P_g in exactly this sense. By assumption, NST held for g_1 and g_2, which implied that the lemma held for them, and the construction procedure cannot result in any paths that are extraneous to the structural display rules in a way that violates this lemma; the incremental graph reduction procedure rules out or eliminates any possible counterpaths. □

Lemma 270 *(L4') For arbitrary EDGs, g and g', such that:* $g = \oplus_\circ^\pi(g_1, g_2)$, $g' = \oplus_\circ^\pi(g_3, g_4)$, $g_1, g_3 \in G^{\pi_1}$, *and* $g_2, g_4 \in G^{\pi_2}$, *if (i) for every p in* P_g *there is a p' in* $P_{g'}$ *such that* $p \simeq p'$, *and (ii) for every p' in* $P_{g'}$ *there is a p in* P_g *such that* $p \simeq p'$, *then* $g \simeq g'$.

Proof (Proof-Plan) The guarded incremental graph reduction procedure immediately rules out any paths that would violate the structural display rules with respect to π_1 and π_2. Thus, given the premises, the only way for isomorphism of g and g' to fail is for one or the other of g and g' to contain at least two duplicates of some proper subgraph of the other graph, but this is ruled out by NST; a proper subgraph satisfying SD1, SD2, and RC can be constructed by removing one of the duplicates or a graph reduction is possible without disturbing the structural display relations of π_1 and π_2. Thus, the isomorphism of all executions of π in g with some execution of π in g' and converse imply the result. □

6.3.3.2 Structural Induction

Proof (Proof-Plan[6]) Assume: $\circ \in Prot, \pi, \pi_2, \pi_3 \in P_{PLEN} \cap String(\Pi_0 \cup Prot)$, $\pi = \circ \frown \pi_1 \frown \pi_2$, $\pi = \circ \frown \pi_2 \frown \pi_1$, or $\pi = \circ \frown \pi_1$. Let $g_1, g_3 \in G^{\pi_1}$, $g_2, g_4 \in$

[6] (π/π') For the second proof-plan/derivation protocol, iterate the Base, Isomorphism Induction, and Protocol Graph Induction steps of the following derivation first for arbitrary proper subprotocol, π', inductively over ascending protocol types all the way up to the proper immediate subprotocols of π, π_1 and π_2, then reiterate the Isomorphism Induction and Protocol Graph Induction steps for π itself on the basis of the results for π_1 and π_2.

G^{π_2}. Finally, let $g = \oplus_\circ^\pi(g_1, g_2)$ and $g' = \oplus_\circ^\pi(g_3, g_4)$. Simultaneous structural induction over the proper immediate subprotocols of arbitrary, regular protocols establishes that $(\oplus_\circ^\pi(g_1, g_2) \simeq \oplus_\circ^\pi(g_3, g_4))$ and $(\oplus_\circ^\pi(g_1, g_2), \oplus_\circ^\pi(g_3, g_4) \in G^\pi)$ for all regular protocols.

(1) (Base) For all regular protocols with atomic immediate subprotocols (regular type 0 or type 1 protocols), both conjuncts of the result hold. There are five cases.

(a) (Atomics)

(i) (Isomorphism) For any $\pi \in \Pi_0$, every protocol graph of π is a single edge with distinct root and terminal states. Let g and g' be arbitrary protocol graphs of π; since π is atomic g and g' are non-looping edges with π as the middle term, $g \simeq g'$. (Again, protocol graphs of atomics cannot be looping edges. Suppose that the looping edge, $\langle s, \pi, s\rangle$, was a protocol graph of π. Then it would also be a protocol graph of $\pi; \pi$, as every subprotocol and implementation is executed, the maximal paths are the $\pi; \pi$ paths, there are no choices and no corresponding subprotocols, there can be no smaller graph, and each proper subprotocol of $\pi; \pi$ has a protocol graph as a subgraph (by assumption). But π is not correspondence equivalent to $\pi; \pi$. Thus, EXC fails for π and for $\pi; \pi$. Contradiction. From another direction, if $g = \langle s, \pi, s\rangle$, then there are no maximal paths in g since each path is a proper subpath of another path. Thus, $P_\pi \neq maxpath(g) = \emptyset$, violating $SD1(2)_\pi(g)$. Thus, atomic π can not have a looping edges as protocol graphs.) Then, by the Base Rule defining $\oplus_\circ^\pi$, $\oplus_\circ^\pi(g) = g$ and $\oplus_\circ^\pi(g') = g'$. Thus, $g \simeq g'$.

(ii) (Protocol Graphs) Edges with π the middle term are trivially protocol graphs of π; the structural display rules cannot fail, and there are no smaller graphs (reductions or subgraphs) that suffice.

(b) (;)

(i) (Isomorphism) If $PISub(\pi) \subseteq \Pi_0$ and $MC(\pi) =;$, then the result holds. Let $\circ =;$. By the Control Operator Rules, $g = \oplus_\circ^\pi(g_1, g_2)$ is a concatenated path of edges executing π_1 and π_2: $g = \langle s, \pi_1, s', \pi_2, s''\rangle$. By the Control Operator Rules, $g' = \oplus_\circ^\pi(g_3, g_4) = \langle s''', \pi_3, s'''', \pi_4, s'''''\rangle$. Given that the immediate subprotocols of π are atomic, $\pi_1 = \pi_3$ and $\pi_2 = \pi_4$ by the uniqueness of edge labels. These graphs are thus isomorphic, by an immediate construction of bijections. Here is a pair of bijections that does the trick: $f_e(g_1) = g_3$, $f_e(g_2) = g_4$, $f_s(s) = first(g_3)$, $f_s(s') = last(g_3) = first(g_4)$, $f_s(s'') = last(g_4)$. Lemma L6 also implies the result.

(ii) (Protocol Graphs) The result is mechanical, graphs of this form are shown to be protocol graphs of type 1 sequential compositions in the ISOP proof.

(c) $(*)$

(i) (Isomorphism) If $PISub(\pi) \subseteq \Pi_0$ and $MC(\pi) = *$, then the result holds. Let $\circ = *$. Let g_1 and g_2 be arbitrary protocol graphs of π_1. Each is an edge with π_1 the middle term. By the Control Operator Rules, $g = \oplus_\circ^\pi(g_1)$ and $g' = \oplus_\circ^\pi(g_2)$. Thus, each of g and g' is a complex graph: for each $i \geq 1$, there is a path formed out of n length sequences of concatenated π_1 graphs branching from a root state, s, and there is a detached, looping edge with π_1* as its middle term. Each execution of π_1 in either of the graphs constructed is an edge. Each is thus isomorphic, and, for each finite $i \in \mathbb{N}$, there is a pair of bijections among the edges of g and g' preserving their concatenation and branching (i.e., preserving their iteration), as required. The bijections can be mechanically constructed. Lemma L6 also implies the result.

(ii) (Protocol Graphs) The result is mechanical, graphs of this form are shown to be protocol graphs of type 1 iterations in the ISOP proof.

(d) $(\cup)$

(i) (Isomorphism) If $PISub(\pi) \subseteq \Pi_0$ and $MC(\pi) = \cup$, then the result holds. Let $\circ = \cup$. By the Control Operator Rules, $g = \oplus_\circ^\pi(g_1, g_2)$ is a branching pair of edges executing π_1 and π_2: $g = \oplus_\circ^\pi(g_1, g_2) = (\langle s, \pi_1, s' \rangle, \langle s, \pi_2, s'' \rangle)$ where s' is not identical to s''. Just so for $g' = \oplus_\circ^\pi(g_3, g_4) = (\langle s''', \pi_3, s'''' \rangle, \langle s''', \pi_4, s''''' \rangle)$. Given that the immediate subprotocols of π are atomic, $\pi_1 = \pi_3$ and $\pi_2 = \pi_4$ by the uniqueness of edge labels. These graphs are thus isomorphic, by immediate construction of bijections. Here is a pair of bijections that does the trick: $f_e(g_1) = g_3$, $f_e(g_2) = g_4$, $f_s(s) = first(g_3) = first(g_4)$, $f_s(s') = last(g_3)$, $f_s(s'') = last(g_4)$. Lemma L6 also implies the result.

(ii) (Protocol Graphs) The result is mechanical, graphs of this form are shown to be protocol graphs of type 1 nondeterministic choices in the ISOP proof.

(e) $(\cap)$

(i) (Isomorphism) If $PISub(\pi) \subseteq \Pi_0$ and $MC(\pi) = \cap$, then the result holds. Let $\circ = \cap$. By the Control Operator Rules, $g = \oplus_\circ^\pi(g_1, g_2)$, is a corresponding pair of edges executing π_1 and π_2: $g = \oplus_\circ^\pi(g_1, g_2) = (\langle s, \pi_1, s' \rangle, \langle s, \pi_2, s' \rangle)$ where s' is not identical to s''. Just so for $g' = \oplus_\circ^\pi(g_3, g_4) = (\langle s'', \pi_3, s''' \rangle, \langle s'', \pi_4, s''' \rangle)$. Given that the immediate subprotocols of π are atomic, $\pi_1 = \pi_3$ and $\pi_2 = \pi_4$ by the uniqueness of edge labels. These graphs are thus isomorphic, by an immediate construction of bijections. Here is a pair of bijections that does the trick: $f_e(g_1) = g_3$, $f_e(g_2) = g_4$, $f_s(s) = first(g_3) = first(g_4)$, $f_s(s') = last(g_3) = last(g_4)$. Lemma L6 also implies the result.

(ii) (Protocol Graphs) The result is mechanical, graphs of this form are shown to be protocol graphs of type 1 parallel executions in the ISOP proof.

(2) (Induction 1: Isomorphism) *If* $\circ \in Prot$, $\pi, \pi_2, \pi_3 \in P_{PLEN} \cap String(\Pi_0 \cup Prot)$, $\pi = \circ \frown \pi_1 \frown \pi_2$, $\pi = \circ \frown \pi_2 \frown \pi_1$, *or* $\pi = \circ \frown \pi_1$, *and for all proper immediate subprotocols of* π, π_1 *and* π_2, *the result holds, then if* $g_1, g_3 \in G^{\pi_1}$, $g_2, g_4 \in G^{\pi_2}$, $g = \oplus^{\pi}_{\circ}(g_1, g_2)$, *and* $g' = \oplus^{\pi}_{\circ}(g_3, g_4)$, *then* $g \simeq g'$. Proof-plan: Derivation from L1', L2, L3', L4', L5, and L6.

(a) (Induction Premise) Let $\pi = \circ \frown \pi_1 \frown \pi_2$ (modulo the other syntactic forms), let PGIIRC hold for π_1 and π_2, and let g and g' be graphs such that $g = \oplus^{\pi}_{\circ}(g_1, g_2)$ and $g' = \oplus^{\pi}_{\circ}(g_3, g_4)$, $g_1 \in G^{\pi_1}$ and $g_2 \in G^{\pi_2}$, and $g_3 \in G^{\pi_1}$ and $g_4 \in G^{\pi_2}$.

(b) Let graphs g_1 and g_3 result from prior iterations of the $\oplus^{\pi}_{\circ}$ Procedure on $\pi_1 = \circ' \frown \pi'_1 \frown \pi'_2$ (iteration of the Base and COR Rules on the proper immediate subprotocols of π_1), and g_2 and g_4 similarly result from prior iterations of the $\oplus^{\pi}_{\circ}$ Procedure on $\pi_2 = \circ'' \frown \pi''_1 \frown \pi''_2$.

(i) For some $g'_1, g'_3 \in G^{\pi'_1}$ and $g'_2, g'_4 \in G^{\pi'_2}$ for proper immediate subprotocols of π_1, π'_1 and π'_2, $\oplus^{\pi_1}_{\circ}(g'_1, g'_2) = g_1$, and $\oplus^{\pi_1}_{\circ}(g'_3, g'_4) = g_3$.

(ii) For some $g'_5, g'_7 \in G^{\pi''_1}$ and $g'_6, g'_8 \in G^{\pi''_2}$ for proper immediate subprotocols of π_2, π''_1 and π''_2, $\oplus^{\pi_2}_{\circ}(g'_5, g'_6) = g_2$, and $\oplus^{\pi_2}_{\circ}(g'_7, g'_8) = g_4$.

(c) From (b) by the induction premise (a), it follows that $g_1 \simeq g_3$ and $g_2 \simeq g_4$.

(i) For $\pi_i \in PISub(\pi)$, if graphs, $\oplus^{\pi}_{\circ}(g_j, g_k)$ and $\oplus^{\pi}_{\circ}(g_l, g_m)$, are constructed out of arbitrary protocol graphs of π_i's proper immediate subprotocols (assign them arbitrarily), then $\oplus^{\pi}_{\circ}(g_j, g_k) \simeq \oplus^{\pi}_{\circ}(g_l, g_m)$ and $\oplus^{\pi}_{\circ}(g_j, g_k), \oplus^{\pi}_{\circ}(g_l, g_m) \in G^{\pi_i}$.

(A) Thus: $\oplus^{\pi_1}_{\circ}(g'_1, g'_2) = g_1 \simeq \oplus^{\pi_1}_{\circ}(g'_3, g'_4) = g_3$, and $\oplus^{\pi_1}_{\circ}(g'_1, g'_2) = g_1, \oplus^{\pi_1}_{\circ}(g'_3, g'_4) = g_3 \in G^{\pi_1}$

(B) Thus: $\oplus^{\pi_2}_{\circ}(g'_5, g'_6) = g_2 \simeq \oplus^{\pi_2}_{\circ}(g'_7, g'_8) = g_4$, and $\oplus^{\pi_2}_{\circ}(g'_5, g'_6) = g_2, \oplus^{\pi_2}_{\circ}(g'_7, g'_8) = g_4 \in G^{\pi_2}$.

(ii) By SUT and ISOP, all protocol graphs of π_1 and π_2 are isomorphic to the graphs constructed by application of $\oplus^{\pi}_{\circ}$.

(iii) As with the SUT derivation, the remaining steps of the argument are formulated in terms of pairs of protocol graphs of π's proper immediate subprotocols that, by the induction premise, were constructed by the $\oplus^{\pi}_{\circ}$ procedure. In any protocol graph of any proper subprotocol of π, there may be multiple protocol graphs of any proper subprotocols of that proper subprotocol. Thus, everything said about the protocol graphs in these pairs (especially g_2 and g_4) individually can be reformulated as statements about every graph in their isomorphism classes; the derivations of the reformulated claims from their individual cases is immediate.

(d) From (a) by L1', the maximal executions of π in g are composed of maximal executions of π_1 in g_1 (or a duplicate thereof) and maximal executions of π_2 in g_2 (or a duplicate thereof) according to the relevant control operator rule, and the maximal executions of π in g' are composed of maximal executions of π_1 in g_3 (or a duplicate thereof) and the maximal executions of π_2 in g_4 (or a duplicate thereof) according to the same rule.

(e) From (c) by L5, each maximal path of g_1 is isomorphic to some maximal path of g_3, and each maximal path of g_2 is isomorphic to some maximal path of g_4, and the converses hold. Isomorphism implies that this holds for all duplicates thereof.

(f) From (d) and (e) by L1' and L6, it's mechanical to show that each of the maximal executions of π in g is isomorphic to some maximal execution of π in g' and converse.

(i) Let p execute π in g. Then, from (b), $p = q \oplus r$ for some $q \in P^1_{\pi_1} \cap maxpath(g_1)$ (or a duplicate thereof) and $r \in P^2_{\pi_2} \cap maxpath(g_2)$ (or a duplicate thereof) by L1'. From (d), (e), the COR rules, and L1', there must be some $p' \in P'_\pi$ such that $p' = q' \oplus r'$, $q' \in P^3_{\pi_1} \cap maxpath(g_3)$ (or a duplicate thereof), $r' \in P^4_{\pi_2} \cap maxpath(g_4)$ (or a duplicate thereof), $q \simeq q'$, and $r \simeq r'$.

A. Assume $\pi = \circ \frown \pi_1 \frown \pi_2$. Modify as necessary for the case where $\pi = * \frown \pi_1$. Given that $p = q \oplus r$ for some $q \in P^1_{\pi_1} \cap maxpath(g_1)$ (or a duplicate thereof) and $r \in P^2_{\pi_2} \cap maxpath(g_2)$ (or a duplicate thereof), there must, by L5 by the induction assumption, be a $q' \in P^3_{\pi_1} \cap maxpath(g_3)$ (or a duplicate thereof) and an $r' \in P^4_{\pi_2} \cap maxpath(g_4)$ (or a duplicate thereof), such that $q \simeq q'$ and $r \simeq r'$. From L6, for any p', if $p' = q' \oplus r'$, then $q \oplus r \simeq q' \oplus r'$ (p is isomorphic to p' in the sense of L6).

B. From the COR rules, any maximal path, p'', in g' must be formed from executions, q'' and r'', of π_1 and π_2 (respectively) by means of the same graph-theoretic structure as p; if $p'' \in P'_\pi$, then $p'' = q'' \oplus r''$ where $\oplus$ is concatenation, branching, iteration, or correspondence.

C. The definition of protocol graphs requires that q'' and r'' be maximal executions of π_1 and π_2 in g_3 and g_4, respectively; NST and SD1(2) rule out the possibility that the executions be proper subpaths of other paths in their respective protocol graphs.

D. From the induction assumption and L5, all protocol graphs of π_1 have pairwise isomorphic maximal execution sets, and the same holds for π_2.

E. Thus, for any $q'' \in P^3_{\pi_1} \cap maxpath(g_3)$ (or a duplicate thereof) and $r'' \in P^4_{\pi_2} \cap maxpath(g_4)$ (or a duplicate thereof), either there is a $p'' \in P'_\pi$ such that $p'' = q'' \oplus r''$, or there is some $p''' = q''' \oplus r'''$ where $p''' \in P'_\pi$ while $q''' \simeq q''$ and/or $r''' \simeq r''$.

F. Thus, if p' is an execution of π in g', then $p' = q' \oplus r'$ for some $q' \in P^3_{\pi_1} \cap maxpath(g_3)$ (or a duplicate thereof) and an $r' \in P^4_{\pi_2} \cap maxpath(g_4)$ (or a duplicate thereof), and $q' \simeq q$ and

$r' \simeq r$. The construction of g' requires that the set of maximal executions in g' be non-empty.

(ii) Thus, by L6, $q \oplus r \simeq q' \oplus r'$.

(iii) Since this worked for arbitrary executions of π, it works for maximal executions.

(iv) Transposition delivers the converse.

(g) From (a) by L3', every path in g is a subgraph of some maximal execution of π or concatenated from subpaths of maximal executions of π or a subpath of a detached iteration loop, and the same holds for g'.

(i) Detached iteration loops can be ignored from this point, as all detached iteration loops that execute the same iterated protocol are isomorphic.

(h) From (a) by L2, if a maximal execution of π in g, p, is isomorphic to a maximal execution of π in g', q, then for every subpath of p, there is a subpath of q that is isomorphic to it, and the converse holds.

(i) From the generality of L2.

(i) From (f), (g), and (h) by L6, every path in g is isomorphic to some path in g' and converse.

(i) If p is a subpath of some maximal execution of π in g, p', then from (f), there is some maximal execution of π in g', q', such that $p' \simeq q'$. By L2 and (h), there must be some $q \sqsubseteq q'$ such that $p \simeq q$.

(ii) If the definition of protocol graphs were revised with SD1(2*): $P_\pi \subseteq maxpath(g)$ and $maxpath(g) \subseteq^\frown Sub(P_\pi)$, step (i) would need to be amended with the following:

(A) If p is formed by concatenation of subpaths, $p_1, ..., p_n$, of some maximal executions of π in g, $p'_1, ..., p'_m$, then, by L6, if there is a path, q, in g' formed by concatenation of subpaths, $q_1, ..., q_n$, with each q_i isomorphic to the correlated p_i, then $p \simeq q$. From (a), there must be such a q in g', as L1' implies that all maximal executions of π in g' are formed out of maximal executions of π_1 and π_2 in g_3 and g_4 by the same graph-theoretic operations (concatenation, branching, iteration, and correspondence) as their isomorphic counterparts in g_1 and g_2.

(B) In short, there is no operation in the $\oplus^\pi_\circ$ Procedure that changes the order of concatenations.

(iii) By (g), this exhausts the possible cases.

(j) Thus, from (i) by L4', $g \simeq g'$.

(3) (Induction 2: Protocol Graphs) *If* $\circ \in Prot, \pi, \pi_2, \pi_3 \in P_{PLEN} \cap String(\Pi_0 \cup Prot)$, $\pi = \circ \frown \pi_1 \frown \pi_2$, $\pi = \circ \frown \pi_2 \frown \pi_1$, *or* $\pi = \circ \frown \pi_1$, *and for all proper immediate subprotocols of* π, π_1 *and* π_2, *the result holds, then if* $g_1, g_3 \in G^{\pi_1}$, $g_2, g_4 \in G^{\pi_2}$, $g = \oplus^\pi_\circ(g_1, g_2)$, *and* $g' = \oplus^\pi_\circ(g_3, g_4)$, *then* g *and* g' *can be mechanically shown to be protocol graphs of* π. Assume the induction premise, then start with g and transpose for g'.

(a) (SD1) Each clause is mechanical or immediate:

(i) (SD1(1)) $Sub(\pi) = Sub(\pi_1) \cup Sub(\pi_2) \cup \{\pi\}$. From the fact that g was built from protocol graphs of π_1 and π_2, every protocol in $Sub(\pi_1) \cup Sub(\pi_2)$ is executable in g. The execution rules suffice to show that P_π is non-empty; $P_\pi = \{p \oplus^\pi_\circ q \mid p \in P_{\pi_1}, q \in P_{\pi_2}\}$. L1' and ISOPL suffice.

(ii) (SD1(2)) This comes for free from L1' or mechanically from the rules for $\oplus^\pi_\circ$; the maximal paths of g are all built out of the maximal paths of g_1 and g_2, and the maximal paths of g' are all built out of the maximal paths of g_2 and g_4. By L6, these are exclusively what compose the maximal executions of π. Thus, the maximal paths in g are exhausted by the executions of π in the way required by SD1(2).

(iii) (SD1(3)) For the proof of the first two rules of SD1(3), let $\pi_j \in Sub(\pi)$. Then $\pi_j \in Sub(\pi_1) \cup Sub(\pi_2) \cup \{\pi\}$. In the first two cases, the assumption that g was built from protocol graphs of π_1 and π_2 coupled with the rules for $\oplus^\pi_\circ$ suffice for the result because the construction of g by $\oplus^\pi_\circ$ doesn't remove any branching paths that are required by the structural display rules. In the third case, the subprotocols of π with the requisite syntactic form are the immediate subprotocols of π, and the main control operator of π will be $\cup$ or $\cap$. These subprotocols will then be executed by branching paths by the execution rules (corresponding paths are, of course, branching paths). For the third rule, let g contain branching p and q that execute π_p and π_q, respectively. The construction of g by $\oplus^\pi_\circ$ preserves the structural display properties of g_1 and g_2, and any branching introduced by the construction will always execute protocols with the requisite properties (the only cases where new branches appear in g that didn't appear in g_1 or g_2 is when the main control operator of π or some immediate subprotocol thereof is $\cap$, $\cup$, or $*$; but in each case, there is some subprotocol of π–π itself—that satisfies the syntactic and semantic conditions required by SD1(3): $\pi = (...\circ\pi_p\circ...) \cap (...\circ\pi_q\circ...)$, $\pi = (...\circ\pi_p\circ...) \cup (...\circ\pi_q\circ...)$, $\pi = \pi_1*$ where $\pi_p = \pi_1$ and $\pi_q = ((\pi_1; ...); \pi_1)$ (or vice versa), or $\pi = \pi_1*$ where $\pi_p = \pi_1+$ and $\pi_q = ((\pi_1 ++; ...); \pi_1 +++)$ (or vice versa) where $\pi_1 ++$ and $\pi_1 +++$ are arbitrary implementations of π_1, or, finally, π dictates π_p and π_q. Thus, branching paths occur in g only where there is a subprotocol of π of the requisite syntactic or semantic properties.

(iv) (SD1(4)) The only corresponding paths in g will be those that were corresponding in g_1 or g_2 (and thereby satisfy the rule) or those that are maximal paths linked by the $\oplus^\pi_\circ$ operation. In the latter case, the rule holds trivially, as any protocols not correspondence equivalent to π_1 or π_2 will not be linked by corresponding paths by the operation.

(b) (SD2) Let $\pi+ \mapsto \pi$.

(i) If π is a choice of π_1 and π_2, then, by I1(3), $\pi+$ implements either π_1 or π_2, the preconditions of $\pi+$ imply the executability of whichever protocol $\pi+$ doesn't implement, and the result holds, as every implementation of π_1 and π_2 must be executed in their respective protocol graphs, and the procedure branches them.

(ii) If π is a sequential composition of π_1 and π_2, then, if $MC(\pi+) = ;$, it follows from I2 and the Converse of I2 that $\pi+$ will have the form of a sequential composition of implementations of π_1 and π_2, and the result holds; the procedure affixes all executions of the relevant implementations of π_1 and π_2 in their respective protocol graphs in a way that ensures that the sequential composition of those implementations is executed by maximal paths in g.

(iii) If π is an iteration of π_1, then, from the definition of implementations, $\pi+$ will have the form of a sequential composition of some finite number of instances of implementations of π_1. Thus, since all implementations are, by assumption, executed in g_1, and the $\oplus_\circ^\pi$ Procedure produces a graph such that the maximal paths in it are either finite concatenations of executions of implementations of π_1 or else edges with π_1* as the middle term. Thus, the procedure ensures that the result holds.

(iv) If π is a parallel execution of π_1 and π_2, then, from the Converse of I3, $\pi+$ will be a parallel execution of implementations of π_1 and π_2 and the result holds, as any executions of implementations of π_1 and π_2 will be affixed in a way that ensures correspondence in g.

(c) (SD3) No atomics are added or deleted by $\oplus_\circ^\pi$.

(d) (NST) The removal of redundancies by the graph reduction step while preserving the structural display properties of g_1 and g_2 preserves NST for g; if it didn't, then it would produce redundant, isomorphic paths, contrary to the definition of the procedure.

(e) (RC) By assumption, $g_1 \in G^{\pi_1}$ and $g_2 \in G^{\pi_2}$, and $\oplus_\circ^\pi$ preserves structural display properties, so every subprotocol of π_1 or π_2 is graphed by some subprotocol of g and so are π_1 and π_2 themselves. Thus, all proper subprotocols of π are graphed by subgraphs of g; else, the definition of the procedure is contradicted.

(f) (EXC) For arbitrary π', suppose that π' is not correspondence equivalent to π. It follows that $P_\pi \neq P_{\pi'}$. Thus, $maxpath(g) \neq P_{\pi'}$. Thus, $g \notin G^{\pi'}$.

□

Remark 271 The significance of this theorem can be seen by comparison with bisimulation invariance. Bisimulation invariance can be taken to be a criterion of process individuation, and the set of bisimulation invariant formulae comprise a theory of the process. The Revised PLEN Framework provides a language in which processes can be individuated at a finer grain than bisimulation invariance; the regular protocol constructions in P_{PLEN} can be embedded in protocol-theoretic constructions of L^R_{PLEN} that are not bisimulation invariant. This is due, in part, to the distinctive execution conditions of parallel execution and choice in PLEN, and, in part, to the demanding conditions of the semantics of some protocol-theoretic formulae. The protocol-theoretic syntactic contexts that are not bisimulation invariant include (at least) local execution-theoretic procedural equivalence, $\leftrightarrow_p$, the non-local process equivalences, $\equiv_p$ and $\equiv^s_p$, the graph-theoretic equivalences, $\equiv_G$ and $\equiv^s_G$, any formulae that entail syntactic identity of protocols (e.g., $\curlywedge_r$), and, of course, any protocol-theoretic contexts that entail the foregoing. Witnesses are mechanical to construct, but see the following example and the results linking PRT and GRT, below for verification.

Example 272 Let g and g' be arbitrary bisimilar but non-isomorphic EDGs. Then g can be a protocol graph of π while g' isn't.

Example 273 Consider arbitrary EDGs, g and g' defined over the following sets of states and atomic edge sets:

$$g = \left\{ \begin{array}{c} S = (s_1, s_2, s_3) \\ R = ((s_1, a, s_3), (s_1, c, s_1), (s_1, b, s_2)) \end{array}, \right\}$$

and

$$g' = \left\{ \begin{array}{c} S' = (s'_1, s'_2) \\ R' = ((s'_1, a, s'_2), (s'_1, c, s'_2), (s'_1, b, s'_1)) \end{array}, \right\}$$

It's mechanical to show that $g \sim g'$ with $s_1 \sim s'_1$, while $g \simeq g'$ and $s_1 \equiv_{GGI} s'_1$ fail. Consider the pair of formulae: $b \cup c* \leftrightarrow_p b$ and $b * \cup c \leftrightarrow_p c$. If valuations for protocol-theoretic formulae are added to g and g', it's mechanical to show that:

(1) $s_1 \Vdash b \cup c* \leftrightarrow_p b$
(2) $s_1 \nVdash b * \cup c \leftrightarrow_p c$
(3) $s'_1 \Vdash b * \cup c \leftrightarrow_p c$
(4) $s'_1 \nVdash b \cup c* \leftrightarrow_p b$

Remark 274 Of course, the failure of local procedural equivalence implies the failure of the stronger equivalences. Thus, the foregoing classes of formulae in PLEN are not bisimulation invariant. Additionally, mere procedural executability is not bisimulation invariant in PLEN. These formulae are, however, isomorphism invariant; isomorphic graphs satisfy the same formulae at generated graph isomorphic states. This can be extrapolated from the core results of the system in this chapter, and it can also be mechanically verified. Thus, the syntactic replacement of protocols that are protocol graph bisimulation equivalent fails to preserve at least some of the protocol-theoretic properties that replacement of protocol graph isomorphism equivalent protocols preserves.

These observations show that the languages of PLEN and the Revised PLEN Framework are finer grained than bisimulation invariant modal languages. The resulting theory of epistemic processes, protocols, and protocol equivalence is thus capable of drawing distinctions and validating inferences (at least in the fragment containing only regular protocols) that cannot be drawn or validated in intensional modal languages while avoiding the erroneous validation of inference that occurs in them (re: the problems pointed out in Chaps. 1–4). This claim, which is one of the core virtues of the Revised PLEN Framework, is borne out in the rest of the chapter, which draws a throughline from the protocol graph isomorphism invariance of regular constructions through protocol graph isomorphism equivalence to preservation results for all but the protocol-theoretic contexts that entail syntactic identity of protocols, and, finally, to the hyperintensionality of protocol graph isomorphism equivalence.

6.4 Protocol Equivalence and Syntactic Replacement

Claim 275 It turns out that the foregoing invariance result supports the derivation of the central syntactic replacement rules of the theory of protocol graph isomorphism equivalence in PLEN. The first replacement rule states that the syntactic replacement of graph-theoretically equivalent proper immediate subprotocols preserves graph-theoretic equivalence. This is interderivable with PGIIRC.

6.4.1 RGSRT

Claim 276 The restricted graph-theoretic syntactic replacement theorem (RGSRT) establishes the basic rule of syntactic replacement in PLEN. The rule essentially states that replacement of protocol graph isomorphism equivalent proper immediate subprotocols preserves protocol graph isomorphism equivalence. Equivalently, the idea is that all regular protocols are protocol graph isomorphism equivalent to the protocols produced by syntactic replacement of protocol graph isomorphism equivalent proper immediate subprotocols.

Theorem 277 (Restricted Graph-Theoretic Syntactic Replacement Theorem (RGSRT)) *For all regular protocols, π, π', X_π, and $X_{\pi'}$: If $\rho(\pi/\pi')(X_\pi) = X_{\pi'}$, $\pi \equiv_G \pi'$, and $\pi \in ISub(X_\pi)$, then $X_\pi \equiv_G X_{\pi'}$. In rule form (with implicit quantification over regular protocols):*

$$\left[\forall \pi, \pi', X_\pi, X_{\pi'} \left[\frac{\left[\begin{array}{l} \pi \in Sub(X_\pi), \pi' \in Sub(X\pi'), \\ \pi, \pi', X_\pi, X_{\pi'} \in P_{PLEN} \cap String(\Pi_0 \cup Prot), \\ \pi \equiv_G \pi', \rho(\pi/\pi')(X_\pi) = X_{\pi'}, \pi \in ISub(X_\pi) \end{array} \right]}{X_\pi \equiv_G X_{\pi'}} \right] \right] \quad (RGSRT)$$

Claim 278 PGIIRC and RGSRT are interderivable.

Corollary 279 (From PGIIRC to RGSRT) *Given the definitions of the PLEN Framework, IHT, I1-I3 and their Converses, SUT, ISOPL, and ISOP, if the rule form of PGIIRC is sound, then the rule form of RGSRT is sound.*

Proof Premises: $\pi \equiv_G \pi'$, $\rho(\pi/\pi')(X_\pi) = X_{\pi'}$, $\pi \in ISub(X_\pi)$, and $\circ \in Prot$. Let $g_1 \in G^\pi$, $g_3 \in G^{\pi'}$, and $g_2, g_4 \in G^{\pi_i}$.

(1) By syntactic replacement, the premises directly imply one of the following conjunctions:

(a) $X_\pi = \pi \circ \pi_i$ (String algebraic form: $X_\pi = \circ \frown \pi \frown \pi_i$) and $X_{\pi'} = \pi' \circ \pi_i$ (String algebraic form: $X_{\pi'} = \circ \frown \pi' \frown \pi_i$)
(b) $X_\pi = \pi_i \circ \pi$ (String algebraic form: $X_\pi = \circ \frown \pi_i \frown \pi$) and $X_{\pi'} = \pi_i \circ \pi'$ (String algebraic form: $X_{\pi'} = \circ \frown \pi_i \frown \pi'$)
(c) $X_\pi = \circ\pi$ (String algebraic form: $X_\pi = \circ \frown \pi$) and $X_{\pi'} = \circ\pi'$ (String algebraic form: $X_{\pi'} = \circ \frown \pi'$)
(d) $X_\pi = \pi$ and $X_{\pi'} = \pi'$

(2) From the premises, $g_1 \simeq g_3$ and $g_2 \simeq g_4$. (SUT, ISOP (EQID))
(3) From 2 and the premises, all protocol graphs of π are protocol graphs of π' and converse. (ISOP (by way of EQID))

(a) Thus, $g_1, g_3 \in G^\pi$, and $g_1, g_3 \in G^{\pi'}$.

(4) By SUT, all protocol graphs of π_i are isomorphic for each π_i.
(5) From 1 and 2: $\oplus_\circ^{X\pi}(g_1, g_2) \simeq \oplus_\circ^{X\pi'}(g_3, g_4)$, $\oplus_\circ^{X\pi}(g_1, g_2) \in G^{X\pi}$, $\oplus_\circ^{X\pi'}(g_3, g_4) \in G^{X\pi}$, $\oplus_\circ^{X\pi}(g_1, g_2) \in G^{X\pi'}$, and $\oplus_\circ^{X\pi'}(g_3, g_4) \in G^{X\pi'}$. (By Iteration of PGIIRC)

(a) Iterate PGIIRC twice for each syntactic case:

$$\left[(C1) \left[\frac{[X_\pi = \circ \frown \pi \frown \pi_i]}{\wedge \left[\begin{array}{c} (\oplus_\circ^{X\pi}(g_1, g_2) \simeq \oplus_\circ^{X\pi}(g_3, g_4)) \\ (\oplus_\circ^{X\pi}(g_1, g_2), \oplus_\circ^{X\pi}(g_3, g_4) \in G^{X_\pi}) \end{array} \right]} \right] \atop \left[\frac{[X_{\pi'} = \circ \frown \pi' \frown \pi_i]}{\wedge \left[\begin{array}{c} (\oplus_\circ^{X\pi'}(g_1, g_2) \simeq \oplus_\circ^{X\pi'}(g_3, g_4)) \\ (\oplus_\circ^{X\pi'}(g_1, g_2), \oplus_\circ^{X\pi'}(g_3, g_4) \in G^{X_{\pi'}}) \end{array} \right]} \right] \right]$$

$$\left[(C2) \left[\frac{[X_\pi = \circ \frown \pi_i \frown \pi]}{\wedge \left[\begin{array}{c} (\oplus_\circ^{X\pi}(g_1, g_2) \simeq \oplus_\circ^{X\pi}(g_3, g_4)) \\ (\oplus_\circ^{X\pi}(g_1, g_2), \oplus_\circ^{X\pi}(g_3, g_4) \in G^{X_\pi}) \end{array} \right]} \right] \atop \left[\frac{[X_{\pi'} = \circ \frown \pi_i \frown \pi']}{\wedge \left[\begin{array}{c} (\oplus_\circ^{X\pi'}(g_1, g_2) \simeq \oplus_\circ^{X\pi'}(g_3, g_4)) \\ (\oplus_\circ^{X\pi'}(g_1, g_2), \oplus_\circ^{X\pi'}(g_3, g_4) \in G^{X_{\pi'}}) \end{array} \right]} \right] \right]$$

$$\left[\begin{array}{l} (C3) \left[\dfrac{[X_\pi = \circ \frown \pi]}{\bigwedge\left[\begin{array}{c}(\oplus_\circ^{X_\pi}(g_1,g_2) \simeq \oplus_\circ^{X_\pi}(g_3,g_4))\\ (\oplus_\circ^{X_\pi}(g_1,g_2), \oplus_\circ^{X_\pi}(g_3,g_4) \in G^{X_\pi})\end{array}\right]}\right] \\ \left[\dfrac{[X_{\pi'} = \circ \frown \pi']}{\bigwedge\left[\begin{array}{c}(\oplus_\circ^{X_{\pi'}}(g_1,g_2) \simeq \oplus_\circ^{X_{\pi'}}(g_3,g_4))\\ (\oplus_\circ^{X_{\pi'}}(g_1,g_2), \oplus_\circ^{X_{\pi'}}(g_3,g_4) \in G^{X_{\pi'}})\end{array}\right]}\right]\end{array}\right]$$

(b) From syntactic replacement, $MC(X_\pi) = MC(X_{\pi'})$. Thus, $\oplus_\circ^{X_\pi}(g_1, g_2) = \oplus_\circ^{X_{\pi'}}(g_1, g_2)$, and $\oplus_\circ^{X_\pi}(g_3, g_4) = \oplus_\circ^{X_{\pi'}}(g_3, g_4)$.

(c) From each pair of iterations of the $\oplus_\circ^\pi$ Procedure, $\oplus_\circ^{X_\pi}(g_1, g_2) \simeq \oplus_\circ^{X'_{\pi'}}(g_3, g_4)$, $\oplus_\circ^{X_\pi}(g_1, g_2) \in G^{X_\pi}$, $\oplus_\circ^{X_{\pi'}}(g_3, g_4) \in G^{X_\pi}$, $\oplus_\circ^{X_\pi}(g_1, g_2) \in G^{X_{\pi'}}$, and $\oplus_\circ^{X_{\pi'}}(g_3, g_4) \in G^{X_{\pi'}}$ are derivable by conjunction decomposition and composition.

(6) From 5: $X_\pi \equiv_G X_{\pi'}$ (SUT, ISOPL, ISOP (EQID))

(a) From SUT, all protocol graphs of X_π are isomorphic to $\oplus_\circ^{X_\pi}(g_1, g_2)$, and all protocol graphs of $X_{\pi'}$ are isomorphic to $\oplus_\circ^{X'_{\pi'}}(g_3, g_4)$. Thus, ISOP implies that all protocol graphs of X_π are protocols of $X_{\pi'}$ and converse. Thus, from EQID, the equivalence holds.

□

Corollary 280 (From RGSRT to PGIIRC) *Given the definitions of the PLEN Framework, IHT, I1-I3 and their Converses, SUT, ISOPL, and ISOP, if the rule form of RGSRT is sound, then the rule form of PGIIRC is sound.*

Claim 281 Granted the further lemma, NSTL, RGSRT can be derived directly from the Useful Lemmata of prior sections, without PGIIRC.

Lemma 282 (NSTL) *For all π, X_π formed from regular control operators: If $\rho(\pi/\pi')(X_\pi) = X_{\pi'}$, $\pi \equiv_G \pi'$, and $\pi \in ISub(X_\pi)$, then an EDG, g, satisfies $SD1(1,2,3,4)$, $SD2$, and RC for X_π iff g satisfies $SD1(1,2,3,4)$, $SD2$, and RC for $X_{\pi'}$.*

Remark 283 NSTL has no uses outside of the following derivation, and the proof of NSTL is mechanical by way of NSTLL. So, the proof is saved for Appendix 2.

Remark 284 The direct proof of RGSRT is mechanical, then, but central to the theory of protocol graph isomorphism equivalence, and so presented in detail.

Proof Assume that $\pi \equiv_G \pi'$, $\rho(\pi/\pi')(X_\pi) = X_{\pi'}$, and $\pi \in ISub(X_\pi)$ for arbitrary, regular π, π', X_π, and $X_{\pi'}$. For an arbitrary g, let $g \in G^{X_\pi}$. Let $X_\pi =$

$\pi \circ \pi_{a(\circ)}$, and adjust the reasoning to other syntactic forms as necessary. The plan of the proof is to establish that $g \in G^{X\pi'}$ by showing that each of $SD1(1-4)_{X\pi'}(g)$, $SD2_{X\pi'}(g)$, $SD3_{X\pi'}(g)$, $NST_{X\pi'}(g)$, $RC_{X\pi'}(g)$, and $EXC_{X\pi'}(g)$ follows from the assumptions.

(1) (SD1(1)) For arbitrary $\pi, \pi', X_\pi, X_{\pi'}$, prove that, given the premises, $SD1(1)_{X\pi'}(g)$ holds for any g such that $g \in G^{X\pi}$. The proof is schematic for all of the possible syntactic forms of X_π and $X_{\pi'}$. The converse follows by transposition. Assume $\pi \equiv_G \pi'$, $\rho(\pi/\pi')(X_\pi) = X_{\pi'}$, and $\pi \in ISub(X_\pi)$. Let $g \in G^{X\pi}$. By syntactic replacement, $X_{\pi'} = \pi' \circ \pi_{a(\circ)}$. $P_{X\pi} = \{p \mid p = q \oplus_\circ^\pi r, q \in P_\pi, r \in P_{\pi_i}\}$ where $\oplus_\circ^\pi$ corresponds to the relations among executions of the immediate subprotocols of $\pi' \circ \pi_{a(\circ)}$. For instance, sequential composition corresponds to path concatenation, parallel execution corresponds to path correspondence, and choice corresponds to branching from a shared root state. By PRT, $P_\pi = P_{\pi'}$. Thus, by substitution of identical sets, $P_{X\pi} = P_{X\pi'}$. Thus, $P_{X\pi} \neq \emptyset$ iff $P_{X\pi'} \neq \emptyset$, $P_\pi \neq \emptyset$ iff $P_{\pi'} \neq \emptyset$, and $P_{\pi i} \neq \emptyset$ iff $P_{\pi i} \neq \emptyset$. By the syntax of X_π, $Sub(X_\pi) = \{Sub(\pi) \cup Sub(\pi_{a(\circ)}) \cup \{\pi \circ \pi_{a(\circ)}\}\}$. Thus, by syntactic replacement, $Sub(X_{\pi'}) = \{Sub(\pi') \cup Sub(\pi_{a(\circ)}) \cup \{\pi' \circ \pi_{a(\circ)}\}\}$. By $SD1(1)_{X\pi}(g)$, $\forall \pi_j[\pi_j \in Sub(\pi_{a(\circ)}) \rightarrow P_{\pi j} \neq \emptyset]$. By $RC_{X\pi}(g)$, $\exists g'[g' \sqsubset g \wedge g' \in G^\pi]$. By EQID, $g' \in G^{\pi'}$. Thus, by $SD1_{\pi'}(g')$, $\forall \pi_k[\pi_k \in Sub(\pi') \rightarrow P'_{\pi k} \neq \emptyset]$. Thus, since $g' \sqsubset g$, $\forall \pi_k[\pi_k \in Sub(\pi') \rightarrow P_{\pi k} \neq \emptyset]$. Thus, $\forall \pi_l[\pi_l \in Sub(X_{\pi'}) \rightarrow P_{\pi l} \neq \emptyset]$. Thus, $SD1(1)_{X\pi'}(g)$.

(2) (SD1(2)) Assume $\pi \equiv_G \pi'$, $\rho(\pi/\pi')(X_\pi) = X_{\pi'}$, and $\pi \in ISub(X_\pi)$. Let $g \in G^{X\pi}$. By definition, $P_{X_\pi} = maxpath(g)$ and all maximal paths in g are formed out of subpaths of maximal executions of X_π. By PRT, $P_{X\pi} = P_{X\pi'}$. Thus, $P_{X\pi'} = maxpath(g)$ and all maximal paths in g are formed out of subpaths of maximal executions of $X_{\pi'}$ in g. Thus, $SD1(2)_{X\pi'}(g)$.

(3) (SD1(3)) Assume $\pi \equiv_G \pi'$, $\rho(\pi/\pi')(X_\pi) = X_{\pi'}$, and $\pi \in ISub(X_\pi)$. Let $g \in G^{X\pi}$. There are three rules and, accordingly, three subproofs that follow the same plan:

(a) (Rule 1) Let g satisfy the conditions of the premise of the first rule: there is a parallel execution or a choice formed from π_1 and π_2 among the subprotocols of $X_{\pi'}$. Call this protocol π_i. First, note that, given that $\pi' \in ISub(X_{\pi'})$, the possible syntactic forms of $X_{\pi'}$ are: $X_{\pi'} = \pi' \circ \pi_{a(\circ)}$, $X_{\pi'} = \pi_{a(\circ)} \circ \pi'$, $X_{\pi'} = \circ\pi'$, or $X_{\pi'} = \pi'$. Thus, for some $\pi_i \in Sub(X_{\pi'})$, $\pi_1, \pi_2 \in Sub(\pi_i)$. Given $\pi_i \in PSub(X_{\pi'})$, $\pi_i \in Sub(\pi') \cup Sub(\pi_{a(\circ)})$. Thus, $\{\pi_1, \pi_2\} \subseteq Sub(\pi') \cup Sub(\pi_{a(\circ)})$. So, there are three cases: $\pi_i \in Sub(\pi_{a(\circ)})$, $\pi_i \in Sub(\pi')$, or $\pi_i = X_{\pi'}$.

Case 1. Suppose $\pi_i \in Sub(\pi_{a(\circ)})$. But since $\pi_{a(\circ)} \in ISub(X_\pi)$ by syntactic replacement, $SD1(3)_{X\pi(g)}$ directly implies that $SD1(3)_{X\pi'}(g)$ holds.

Case 1. Suppose $\pi_i \in Sub(\pi')$. Thus, since $\pi \equiv_G \pi'$, EQID and $RC_{X\pi}(g)$ imply that $SD1(3)_{X\pi'}(g)$. From RC, there is a $g' \sqsubset g$ such that $SD1(3)_{\pi}(g')$. From EQID, g' is a protocol graph of π'. Thus, $SD1(3)_{\pi'}(g')$. Thus, $SD1(3)_{X\pi'}(g')$.

Case 1. Suppose that $\pi_i = X_{\pi'}$. Then from the SD1(1) arguments and PRT1, the result follows mechanically.

(b) (Rule 2) Let g satisfy the conditions of the premise of the second rule: π_1 and π_2 implement some subprotocol, π_i, of $X_{\pi'}$ or they are dictated by π_i. First, note that, given that $\pi' \in ISub(X_{\pi'})$, the possible syntactic forms of $X_{\pi'}$ are: $X_{\pi'} = \pi' \circ \pi_{a(\circ)}$, $X_{\pi'} = \pi_{a(\circ)} \circ \pi'$, $X_{\pi'} = \circ\pi'$, or $X_{\pi'} = \pi'$. Given $\pi_i \in PSub(X_{\pi'})$, $\pi_i \in Sub(\pi') \cup Sub(\pi_{a(\circ)})$. So, there are three cases: $\pi_i \in Sub(\pi_{a(\circ)})$, $\pi_i \in Sub(\pi')$, or $\pi_i = X_{\pi'}$.

Case 1. Suppose $\pi_i \in Sub(\pi_{a(\circ)})$. But since $\pi_{a(\circ)} \in ISub(X_{\pi})$ by syntactic replacement, $SD1(3)_{X\pi(g)}$ directly implies that $SD1(3)_{X\pi'}(g)$ holds.

Case 2. Suppose $\pi_i \in Sub(\pi')$. Thus, since $\pi \equiv_G \pi'$, EQID and $RC_{X\pi}(g)$ imply that $SD1(3)_{X\pi'}(g)$. From RC, there is a $g' \sqsubset g$ such that $SD1(3)_{\pi}(g')$. From EQID, g' is a protocol graph of π'. Thus, $SD1(3)_{\pi'}(g')$. Thus, $SD1(3)_{X\pi'}(g')$.

Case 3. Suppose that $\pi_i = X_{\pi'}$. Then from the SD1(1) arguments, the NSTL Lemma, and PRT, the result follows mechanically.

(c) (Rule 3) Let g satisfy the conditions of the premise of the third rule: there are p and q that execute some π_1 and π_2, respectively, and $first(p) = first(q)$. It follows that π_1 and π_2 satisfy the syntactic conditions with respect to $X_{\pi'}$. First, since $g \in G^{X\pi}$, the conclusion of the rule holds for X_{π}: for any such π_1 and π_2 executed by p and q with $s = first(p) = first(q)$, π_1 and π_2 are subprotocols of X_{π}, and there is a π_i in the subprotocols of X_{π} such that $\pi_i = \pi_1 \cup \pi_2$, $\pi_i = \pi_1 \cap \pi_2$, π_i is some parallel execution of $(... \circ \pi_1 \circ ...)$ and $(... \circ \pi_2 \circ ...)$, π_i dictates both π_1 and π_2 in g, or π_1 and π_2 implement π_i. Second, given that $\pi \in ISub(X_{\pi})$, $X_{\pi} = \pi \circ \pi_{a(\circ)}$, $X_{\pi} = \pi_{a(\circ)} \circ \pi$, $X_{\pi} = \circ\pi$, or $X_{\pi} = \pi$. Thus, $\pi_i \in Sub(\pi_{a(\circ)})$, $\pi_i \in Sub(\pi)$, or $\pi_i = X_{\pi}$. So, there are three cases:

Case 1. Suppose that $\pi_i \in Sub(\pi_{a(\circ)})$. By syntactic replacement, $\rho(\pi/\pi')(\pi \circ \pi_{a(\circ)}) = \pi' \circ \pi_{a(\circ)}$, $\rho(\pi/\pi')(\pi_{a(\circ)} \circ \pi) = \pi_{a(\circ)} \circ \pi'$, $\rho(\pi/\pi')(\circ\pi) = \circ\pi'$, and $\rho(\pi/\pi')(\pi) = \pi'$. The last two cases are ruled out by assumption. The first two cases are such that there are protocols—π_1 and π_2—among the subprotocols of $X_{\pi'}$ satisfying the syntactic or semantic (implementation and dictation) conditions of the rule. So, if π_i is a subprotocol of π_j, then $SD1(3)_{X\pi'}(g)$ follows directly, since $\pi_{a(\circ)} \in ISub(X_{\pi'})$.

Case 2. Suppose that $\pi_i \in Sub(\pi)$. Recall that g is a protocol graph of X_{π}. If $\pi_i \in Sub(\pi)$, then there is some choice or parallel execution of π_1 or π_2 in π, some parallel execution of $(... \circ \pi_1 \circ ...)$ and $(... \circ \pi_2 \circ ...)$ in π, some subprotocol of π implemented by π_1 and π_2, or some subprotocol of π that dictates π_1 and π_2. By RC, there is a protocol graph of π, g', among the proper subgraphs of g, containing the branching paths, p and q. Thus, by definition, $SD1(3)_{\pi}(g')$. By

EQID, every protocol graph of π is a protocol graph of π'. Thus, g' is a protocol graph of π'. So, $SD1(3)_{\pi'}(g')$. Thus, the syntactic conditions with respect to $X_{\pi'}$ must hold; there must be some choice or parallel execution of π_1 and π_2 in π', some parallel execution of $(... \circ \pi_1 \circ ...)$ and $(... \circ \pi_2 \circ ...)$ in π', some subprotocol that dictates π_1 and π_2, or some π_i in $Sub(\pi')$ implemented by π_1 and π_2. Thus, $SD1(3)_{X\pi'}(g)$, since $g' \sqsubset g$.

Case 3. Suppose that $\pi_i = X_\pi$. Then $SD1(3)_{X\pi}(g)$. But by NSTL Lemma, it follows that $Imp(X_\pi) = Imp(X_{\pi'})$. Thus, by the implementation rules, if π_1 and π_2 are implementations of X_π, then they are implementations of $X_{\pi'}$; thus, there are $\pi+$ and $\pi++$ (π_1 and π_2 themselves) that implement an immediate subprotocol of $X_{\pi'}$ ($X_{\pi'}$ itself) and are executed by the branching paths that execute π_1 and π_2. Else, X_π and $X_{\pi'}$ have the same syntactic form, sharing main control operators and diverging only over the structure of π and π', respectively. Thus, if X_π has the requisite syntactic form with respect to π_1 and π_2, so does $X_{\pi'}$. More, from PRT1, it follows that X_π and $X_{\pi'}$ dictate the same protocols generally and thus, a fortiori, in g.

Thus, $SD1(3)_{X\pi'}(g)$.

(4) (SD1(4)) Assume $\pi \equiv_G \pi'$, $\rho(\pi/\pi')(X_\pi) = X_{\pi'}$, and $\pi \in ISub(X_\pi)$. Let $g \in G^{X\pi}$. Let some π_1 and π_2 correspond in g but not generally. Then there is a subprotocol, $\pi_!$, of X_π of the form $(... \circ \pi_1 \circ ...) \cap (... \circ \pi_2 \circ ...)$ or $\pi_1 \cap \pi_2$ implements $\pi_!$. Given the assumptions, $X_\pi = \pi \circ \pi_{a(\circ)}$, $X_\pi = \pi_{a(\circ)} \circ \pi$, $X_\pi = \pi*$, or $X_\pi = \pi$. Thus, $Sub(X_\pi) = Sub(\pi) \cup Sub(\pi_{a(\circ)}) \cup \{X_\pi\}$. By syntactic replacement, $X_{\pi'} = \pi' \circ \pi_{a(\circ)}$, $X_{\pi'} = \pi_{a(\circ)} \circ \pi'$, $X_{\pi'} = \pi'*$, or $X_{\pi'} = \pi'$. Thus, $Sub(X_{\pi'}) = Sub(\pi') \cup Sub(\pi_{a(\circ)}) \cup \{X_{\pi'}\}$. Thus, if $\pi_!$ is in $Sub(\pi_{a(\circ)})$, then $\pi_! \in Sub(X_{\pi'})$, as required by SD1(4). If $\pi_!$ is in $Sub(\pi)$, then by RC and EQID, there is a protocol, $\pi_{!!}$, of the requisite form in $Sub(\pi')$, as there is a $g' \in G^\pi$ among the subgraphs of g that contains the corresponding executions of π_1 and π_2, and it must also be a protocol graph of π'. Thus, $SD1(4)_{\pi'}(g')$, and this implies that there must be a protocol of the requisite form, $\pi_{!!}$, in $Sub(X_{\pi'})$. Finally, if $\pi_! = X_\pi$, then, where $\pi_!$ implements X_π, NSTLL implies that $\pi_!$ implements $X_{\pi'}$, as required. If $\pi_! = X_\pi = (... \circ \pi_1 \circ ...) \cap (... \circ \pi_2 \circ ...)$, then $X_{\pi'}$ has the same main control operator, and PRT and the execution rules suffice for the result. Let $\pi_i = (... \circ \pi_2 \circ ...)$ and $\pi = (... \circ \pi_1 \circ ...)$. Given the syntactic form and the execution rules, π' must have the form $(... \circ \pi_1 \circ ...)$ and correspond with $(... \circ \pi_2 \circ ...)$. If π' didn't have the form $(... \circ \pi_1 \circ ...)$, then the execution rules would be violated. The requisite correspondence is immediate.

(5) (SD2) Assume $\pi \equiv_G \pi'$, $\rho(\pi/\pi')(X_\pi) = X_{\pi'}$, and $\pi \in ISub(X_\pi)$. Let $g \in G^{X\pi}$. By NSTL Lemma, any protocol, $\pi+$, is an implementation of X_π iff it is an implementation of $X_{\pi'}$. Thus, from $SD2_{X\pi}(g)$, $SD2_{X\pi'}(g)$ follows. For the restricted NSTL Lemma given by using syntactic implementations, the result still holds. Either π is a proper subprotocol of X_π or not. If it is, then, by the restricted NSTL Lemma, $\pi+$ syntactically implements X_π iff it implements

$X_{\pi'}$, and so the result holds. If π is not a proper subprotocol of X_π, then $X_\pi = \pi$, so $X_{\pi'} = \pi'$ and $X_\pi \equiv_G X_{\pi'}$ follows by syntactic replacement. Then by EQID, $SD2_{X\pi'}(g)$ follows.

(6) (SD3) Assume $\pi \equiv_G \pi'$, $\rho(\pi/\pi')(X_\pi) = X_{\pi'}$, and $\pi \in ISub(X_\pi)$. Let $g \in G^{X\pi}$. By the proof of the NSTL Lemma or by deploying PRT and the uniqueness of edge labels, $At(\pi) = At(\pi')$. Thus, $At(X_\pi) = At(X_{\pi'})$ by syntactic replacement. Thus, by $SD3_{X\pi}(g)$, $\Pi_g = At(X_\pi)$. Thus, $SD3_{X\pi'}(g)$ follows.

(7) (NST) Assume $\pi \equiv_G \pi'$, $\rho(\pi/\pi')(X_\pi) = X_{\pi'}$, and $\pi \in ISub(X_\pi)$. Let $g \in G^{X\pi}$. Suppose $NST_{X\pi'}(g)$ fails. Then there is some $g' \sqsubset g$ or $g' \xrightarrow[\Pi_C]{\Pi_0} g$ such that $SD1(1,2,3,4)_{X\pi'}(g')$, $SD2_{X\pi'}(g')$, and $RC_{X\pi'}(g')$. By NSTL, $SD1(1,2,3,4)_{X\pi}(g')$, $SD2_{X\pi}(g')$, and $RC_{X\pi}(g')$. Thus, g fails NST. Thus, $g \notin G^{X\pi}$. $\perp$ Thus, by reductio, $NST_{X\pi'}(g)$.

(8) (RC) Assume $\pi \equiv_G \pi'$, $\rho(\pi/\pi')(X_\pi) = X_{\pi'}$, and $\pi \in ISub(X_\pi)$. Let $g \in G^{X\pi}$. $PSub(X_{\pi'}) = \{Sub(\pi') \cup Sub(\pi_{a(\circ)})\}$ (syntactic replacement). $RC_{X\pi}(g)$ implies that all protocols π_j are such that if π_j is a subprotocol of π_i, then there is a subgraph of g, g', such that g' is a protocol graph of π_j. EQID implies that $G^\pi = G^{\pi'}$. Thus, $RC_{X\pi}(g)$ implies that all protocols π_k are such that if π_j is a subprotocol of π', then there is a subgraph of g, g', such that g' is a protocol graph of π_k. Thus, $RC_{X\pi'}(g)$.

(9) (EXC) Assume $\pi \equiv_G \pi'$, $\rho(\pi/\pi')(X_\pi) = X_{\pi'}$, and $\pi \in ISub(X_\pi)$. Let $g \in G^{X\pi}$. From the foregoing proofs, $P_{X\pi} = P_{X\pi'}$ (or by PRT1). Thus, $\pi'' \equiv_c X_\pi$ fails iff $\pi'' \equiv_c X_{\pi'}$ fails. Thus, for all graphs, g, $g \in G^{X\pi} \rightarrow g \notin G^{\pi''}$ iff $g \in G^{X\pi'} \rightarrow g \notin G^{\pi''}$. Thus, $EXC_{X\pi'}(g)$.

Thus, from the definition of protocol graphs, $g \in G^{X\pi'}$.

□

Claim 285 From there, we can bypass step 2 or step 3 of the inductive step of the proof of PGIIRC.

Proof Suppose that RGSRT is sound. Let X_π be an arbitrary regular protocol such that $\pi \in PISub(X_\pi)$: $X_\pi = \circ \frown \pi \frown \pi_{a(\circ)}$ (modulo syntactic form). Let $\rho(\pi/\pi')(X_\pi) = X_{\pi'} = \circ \frown \pi' \frown \pi_{a(\circ)}$. Assume $\pi = \pi'$. Let it be the case that: $\circ \in Prot$, $\pi, \pi', X_\pi, X_{\pi'} \in P_{PLEN} \cap String(\Pi_0 \cup Prot)$. Let $g_1 \in G^\pi$, $g_3 \in G^{\pi'}$, $g_2, g_4 \in G^{\pi_{a(\circ)}}$, and, finally, $g = \oplus_\circ^{X_\pi}(g_1, g_2)$ and $g' = \oplus_\circ^{X_\pi}(g_3, g_4)$. The derivation is nearly immediate:

(1) It follows that $\pi \equiv_G \pi'$ from their identity.

  (a) From SUT and ISOP, $g_1, g_3 \in G^\pi$ and $g_1, g_3 \in G^{\pi'}$.
  (b) From RGSRT, $X_\pi \equiv_G X_{\pi'}$.
  (c) Thus, the following rules are immediate:
    (i) If $g = \oplus_\circ^\pi(g_1, g_2) \simeq g' = \oplus_\circ^\pi(g_3, g_4)$, then (by ISOP) $g = \oplus_\circ^\pi(g_1, g_2) \in G^{X_\pi}$ iff $g' = \oplus_\circ^\pi(g_3, g_4) \in G^{X_{\pi'}}$

(ii) If $g = \oplus_{\circ}^{\pi}(g_1, g_2) \in G^{X_\pi}$ and $g' = \oplus_{\circ}^{\pi}(g_3, g_4) \in G^{X_{\pi'}}$, then (by SUT) $g = \oplus_{\circ}^{\pi}(g_1, g_2) \simeq g' = \oplus_{\circ}^{\pi}(g_3, g_4)$

(d) Thus, all that's required to arrive at the conclusion of PGIIRC is to derive the antecedents of either of these rules. The reasoning in either the Isomorphism Step of the PGIIRC proof or the reasoning in the Protocol Graph Step of the PGIIRC proof suffices, obviating the remainder of the PGIIRC proof.

(e) Thus:

(i) $g = \oplus_{\circ}^{X_\pi}(g_1, g_2) \simeq g' = \oplus_{\circ}^{X_\pi}(g_3, g_4)$, $g = \oplus_{\circ}^{X_\pi}(g_1, g_2) \in G^{X_\pi}$, and $g' = \oplus_{\circ}^{X_\pi}(g_3, g_4) \in G^{X_\pi}$.

(ii) $g = \oplus_{\circ}^{X_{\pi'}}(g_1, g_2) \simeq g' = \oplus_{\circ}^{X_{\pi'}}(g_3, g_4)$, $g = \oplus_{\circ}^{X_{\pi'}}(g_1, g_2) \in G^{X_{\pi'}}$, and $g' = \oplus_{\circ}^{X_{\pi'}}(g_3, g_4) \in G^{X_{\pi'}}$.

□

6.4.2 Graph Syntactic Replacement Theorems

With the two threads of PLEN represented by PGIIRC and RGSRT properly bound, we can then derive the most powerful results of the theory of protocol graph isomorphism equivalence by either of two paths:

(Path 1): From Revised PLEN: derive the Useful Lemmata IHT, Implementation Rules, SUT, ISOP, and EQID; derive PGIIRC; derive RGSRT; generalize RGSRT to GSRT; derive GRT.

(Path 2) From Revised PLEN: derive the Useful Lemmata IHT, Implementation Rules, SUT, ISOP, EQID, PRT, NSTLL, and NSTL; derive RGSRT; generalize RGSRT to GSRT; derive GRT.

The main replacement rule of Revised PLEN is GRT:

Theorem 286 (GRT) *If $A(\pi)$ is any formula of L^R_{PLEN} that does not embed arbitrary, regular π in $M(\pi)$ or any other strictly syntactic expression (subprotocol expressions, syntactic replacement itself) nor in a subset of ternary expression ($\pi \leftrightarrow_r$ ..., $\pi \curlywedge_r$ ..., or, if X_π contains parallel executions, $\pi \curlyvee_r$ ...), then the following rule is valid: For all regular protocols, for all $s \in S$, for all g, if $\pi \equiv_G \pi'$ and $s \Vdash A(\pi)$, then $s \Vdash A(\pi')$.*

Remark 287 Protocol graph isomorphism equivalence is a novel protocol equivalence relation stronger than bisimulation equivalence or any execution-theoretic equivalence short of identity, and it is weaker than syntactic identity itself. As the granularity results below show, this has something to do with protocols being hyperintensional objects, and protocol graph isomorphism equivalence being a hyperintensional context. Ultimately, this is a virtue of PLEN's analysis of procedural epistemic norms, as it's increasingly apparent that normative contexts are hyperintensional. GRT bears this claim out.

Claim 288 The basic thought of GRT allows the syntactic replacement of protocol graph isomorphism equivalent protocols in all protocol-theoretic contexts in PLEN short of those in which truth-preserving replacement implies protocol identity. GRT itself shows the positive half of that claim. The granularity results of the next section show the negative half. If correct, this shows that protocol graph isomorphism equivalence is an individuation criterion among protocols that is weaker than identity and stronger than any other protocol equivalence in PLEN. Section 6.5.2 connects protocol graph isomorphism equivalence to hyperintensional equivalence by means of the former's satisfaction of recently proposed criteria on the latter.

To prove GRT, we need to generalize RGSRT. The proper immediate subprotocol hierarchy construction procedure (from all the way back in Sect. 5.7.2) can be used to mechanically derive the following rule:

Lemma 289 (GSRTL) *For all regular protocols, if* $\rho(\pi/\pi')(X_\pi) = X_{\pi'}$*, then:*

$$\forall \langle \pi = \pi_1, ..., X_\pi \rangle \in ISH(X_\pi) \exists \left\langle \pi' = \pi'_1, ..., X_{\pi'} \right\rangle \in ISH(X_{\pi'})$$
$$\left[\forall \pi_i, \pi_{i+1} \forall \pi'_i, \pi'_{i+1} (\rho(\pi_i/\pi'_i)(\pi_{i+1}) = \pi'_{i+1})\right]$$

Proof Direct argument from application of the Proper Immediate Subprotocol Hierarchy Construction Procedure.

(1) Let $\pi \in Sub(X_\pi)$. Then $X_\pi = o_1 \frown x_1 \frown ... \frown x_n$, where $\pi = x_i \frown ... \frown x_j$ for some $i \geq 1$ and $j \leq n$.

 (a) By the string-algebraic definition of subprotocols.

(2) Let $\rho(\pi/\pi')(X_\pi) = X_{\pi'}$. Then $X_{\pi'} = o'_1 \frown x'_1 \frown ... \frown x'_m$, where $\pi' = x'_k \frown ... \frown x'_l$ for some $k \geq 1, l \leq m$, and $d(i-1) = d(k-1)$ and $d(j+1) = d(l+1)$.

 (a) By the string-algebraic definition of subprotocols and the definition of syntactic replacement of protocols.

(3) From the assumption that $\rho(\pi/\pi')(X_\pi) = X_{\pi'}$, it follows by the definition of syntactic replacement of π with π' that, for every pair of elements of these concatenated strings, x_z and $x'_{z'}$, if $z < i$ and $z' < k$ or $z > j$ and $z > l$, and $d(z) = d(z')$, then $x_z = x'_{z'}$.

(4) Applying the Proper Immediate Subprotocol Hierarchy Construction Procedure to arbitrary X_π by starting at π produces a proper immediate subprotocol hierarchy, $\langle \pi, ..., \pi_x, X_\pi \rangle$ where $\pi_x \in PISub(X_\pi)$, and the proper immediate subprotocol hierarchy is of some arbitrary length.

(5) Applying the procedure to $X_{\pi'}$ by starting at π' produces a proper immediate subprotocol hierarchy, $\left\langle \pi', ..., \pi'_y, X_{\pi'} \right\rangle$ where $\pi'_y \in PISub(X_{\pi'})$, and from 3, it follows that $\langle \pi, ..., \pi_x, X_\pi \rangle$ and $\left\langle \pi', ..., \pi'_y, X_{\pi'} \right\rangle$ have the same length.

 (a) The sequence of control operators between the main control operator of π and the main control operator of X_π is identical to the sequence of control operators between the main control operator of π' and that of $X_{\pi'}$ by syntactic replacement. The length of this sequence of control operators determines the length of the proper immediate subprotocol hierarchy.

(6) It suffices for the result to show that, given $\rho(\pi/\pi')(X_\pi) = X_{\pi'}$, for any protocols, π_z and π'_z that follow π and π' in their respective proper immediate subprotocol hierarchies at equal distance, if $\rho(\pi_{z-1}/\pi'_{z-1})(\pi_z) = \pi'_z$, then $\rho(\pi_z/\pi'_z)(\pi_{z+1}) = \pi'_{z+1}$.

(a) Let $\pi_z \in \langle \pi, ..., \pi_x, X_\pi \rangle$ where π_z comes immediately after π in $\langle \pi, ..., \pi_x, X_\pi \rangle$, and let $\pi'_z \in \left\langle \pi', ..., \pi'_y, X_{\pi'} \right\rangle$ where π'_z comes immediately after π' in $\left\langle \pi', ..., \pi'_y, X_{\pi'} \right\rangle$.

(b) First, it trivially holds that $\rho(\pi/\pi')(\pi) = \pi'$.

(c) From 3, it immediately follows that $\rho(\pi/\pi')(\pi_z) = \pi'_z$.

(i) Since $\pi \in PISub(\pi_z)$, it follows by definition that $\pi_z = \circ \frown \pi \frown \pi_{a(\circ)}$, $\pi_z = \circ \frown \pi$, or $\pi_z = \circ \frown \pi_{a(\circ)} \frown \pi$, and $\pi'_z = \circ' \frown \pi' \frown \pi'_{a(\circ)}$, $\pi'_z = \circ' \frown \pi'$, or $\pi'_z = \circ' \frown \pi'_{a(\circ)} \frown \pi'$. Thus, from 3, $\circ = \circ'$ and $\pi_{a(\circ)} = \pi'_{a(\circ)}$. Thus, $\rho(\pi/\pi')(\pi) = \pi'$ and $\rho(\pi/\pi')(\pi_z) = \pi'_z$, and the result holds.

(ii) In this case, $\pi = \pi_{z-1}$ and $\pi' = \pi'_{z-1}$; thus, $\rho(\pi_{z-1}/\pi'_{z-1})(\pi_z) = \pi'_z$.

(iii) From 3, it immediately follows for any π_{z+1} and π'_{z+1} that follow this particular pair of starting protocols (that followed π and π' immediately in their respective ISHs), that $\rho(\pi_z/\pi'_z)(\pi_{z+1}) = \pi'_{z+1}$.

(A) Since $\pi_z \in PISub(\pi_{z+1})$, it follows by definition that $\pi_{z+1} = \circ \frown \pi_z \frown \pi_{a(\circ)}$, $\pi_{z+1} = \circ \frown \pi_z$, or $\pi_{z+1} = \circ \frown \pi_{a(\circ)} \frown \pi_z$, and $\pi'_{z+1} = \circ' \frown \pi'_z \frown \pi'_{a(\circ)}$, $\pi'_{z+1} = \circ' \frown \pi'_z$, or $\pi'_{z+1} = \circ' \frown \pi'_{a(\circ)} \frown \pi'_z$. Thus, from 3, $\circ = \circ'$ and $\pi_{a(\circ)} = \pi'_{a(\circ)}$. Thus, $\rho(\pi_z/\pi'_z)(\pi_{z+1}) = \pi'_{z+1}$, and the result holds.

(d) Suppose that $\rho(\pi_{z-1}/\pi'_{z-1})(\pi_z) = \pi'_z$ for any arbitrary π_z and π'_z that follow π and π' at equal distance in $\langle \pi, ..., \pi_x, X_\pi \rangle$ and $\left\langle \pi', ..., \pi'_y, X_{\pi'} \right\rangle$, respectively. Consider π_{z+1} and π'_{z+1}, which immediately follow π_z and π'_z, respectively.

(e) From 3, it immediately follows that $\rho(\pi_z/\pi'_z)(\pi_{z+1}) = \pi'_{z+1}$.

(i) Since $\pi_z \in PISub(\pi_{z+1})$ (and re-using notation for control operators and arbitrary, proper immediate subprotocols distinct from those we are considering), it follows by definition that $\pi_{z+1} = \circ \frown \pi_z \frown \pi_{a(\circ)}$, $\pi_{z+1} = \circ \frown \pi_z$, or $\pi_{z+1} = \circ \frown \pi_{a(\circ)} \frown \pi_z$, and $\pi'_{z+1} = \circ' \frown \pi'_z \frown \pi'_{a(\circ)}$, $\pi'_{z+1} = \circ' \frown \pi'_z$, or $\pi'_{z+1} = \circ' \frown \pi'_{a(\circ)} \frown \pi'_z$. Thus, from 3, $\circ = \circ'$ and $\pi_{a(\circ)} = \pi'_{a(\circ)}$. Thus, $\rho(\pi_z/\pi'_z)(\pi_{z+1}) = \pi'_{z+1}$, and the result holds.

□

Remark 290 The idea is that, wherever π is a subprotocol of X_π, and syntactic replacement of π with π' produces $X_{\pi'}$, the protocol π' occupies a node on $X_{\pi'}$'s proper immediate subprotocol hierarchy tree that is the same number of edges away from $X_{\pi'}$ as π is from X_π, and every pair of corresponding nodes on these branches are products of syntactic replacement of the prior nodes. That is, if x is on the branch between π and X_π and y is on the branch between π' and $X_{\pi'}$, and x and y are the same distance from the root of their respective proper immediate subprotocol hierarchy trees, then y is the product of syntactic replacement of x's immediate predecessor on the branch $(x-1)$ with y's immediate predecessor on the branch $(y-1)$, and vice versa.

Claim 291 In virtue of the properties of proper immediate subprotocols, this rule underlies a rule for iterating applications of RGSRT:

Lemma 292 (Iteration Rule) *For all regular protocols:*

$$\left[\frac{(\pi \equiv_G \pi',\ \rho(\pi/\pi')(X_\pi) = X_{\pi'})}{\forall i \leq n, [(\pi_i \in \langle \pi_1, ..., \pi_n(= X_\pi)\rangle \wedge \pi'_i \in \langle \pi'_1, ..., \pi'_n(= X_{\pi'})\rangle) \rightarrow (\pi_i \equiv_G \pi'_i)]}\right]$$

Proof (Proof-Plan) Iteration of RGSRT over the immediate subprotocol hierarchies linked by syntactic replacement by GSRTL. Assume $\pi \equiv_G \pi'$ and $\rho(\pi/\pi')(X_\pi) = X_{\pi'}$.

(1) From GSRTL, there are two immediate subprotocol hierarchies, $\langle \pi_1, ..., X_\pi\rangle$ and $\langle \pi'_1, ..., X_{\pi'}\rangle$, such that: $\pi \in ISub(\pi_1)$, and for each i, $\rho(\pi_{i-1}/\pi'_{i-1})(\pi_i) = \pi'_i$, and, in particular $\rho(\pi/\pi')(\pi_1) = \pi'_1$. Thus, given the assumption, the premises of RGSRT hold. Apply RGSRT to derive $\pi_1 \equiv_G \pi'_1$. Formally, the argument deploys the following instance of RGSRT:

$$\left[\frac{\begin{bmatrix} \pi \equiv_G \pi', \\ \rho(\pi/\pi')(\pi_1) = \pi'_1, \\ \pi \in ISub(\pi_1) \end{bmatrix}}{\pi_1 \equiv_G \pi'_1}\right] (RGSRT)$$

(2) For $\langle \pi_1, ..., X_\pi\rangle$ and $\langle \pi'_1, ..., X_{\pi'}\rangle$, for each $i > 1$, replace π with π_{i-1}, π' with π'_{i-1}, π_{i-1} with π_i, and π'_{i-1} with π'_i in the RGSRT schema:

$$\left[\frac{\begin{bmatrix} \pi_{i-1} \equiv_G \pi'_{i-1}, \\ \rho(\pi_{i-1}/\pi'_{i-1})(\pi_i) = \pi'_i, \\ \pi_{i-1} \in ISub(\pi_i) \end{bmatrix}}{\pi_i \equiv_G \pi'_i}\right] (RGSRT)$$

(3) Derive $\pi_i \equiv_G \pi'_i$ by RGSRT for each $1 \leq i \leq n$ by successive iterations of steps 1 and 2.

For π_1, from the assumption of the antecedent of RGSRT, $\pi_1 \equiv_G \pi_1'$ follows. Deploy Step 2 for each $i \leq n$, to derive $\pi_i \equiv_G \pi_i'$. This step is undergirded by GSRTL; the application of RGSRT only works if each π_i' is the result of syntactic replacement of π_{i-1} with π_{i-1}', but this is the result of GSRTL. For induction, let $\pi_i \equiv_G \pi_i'$. Then for π_{i+1} in the relevant ISH, $\pi_i \in ISub(\pi_{i+1})$. By GSRTL, $\rho(\pi_i/\pi_i')(\pi_{i+1}) = \pi_{i+1}'$. Thus, plugging these premises into the relevant schema generated by Step 2, we apply RGSRT to derive $\pi_{i+1} \equiv_G \pi_{i+1}'$. Thus, by iterative application of RGSRT, $\pi_i \equiv_G \pi_i'$ follows for every $i \leq n$. □

Claim 293 This iteration rule then immediately implies the full generalization of RGSRT:

Theorem 294 (GSRT) *For all regular protocols: If $\rho(\pi/\pi')(X_\pi) = X_{\pi'}$ and $\pi \equiv_G \pi'$, then $X_\pi \equiv_G X_{\pi'}$. In rule form:*

$$\left[\frac{\begin{bmatrix}\pi \equiv_G \pi', \\ \rho(\pi/\pi')(X_\pi) = X_{\pi'}\end{bmatrix}}{X_\pi \equiv_G X_{\pi'}}\right]$$

Remark 295 The idea is that, wherever a protocol, π, falls in the subprotocol hierarchy of X_π (i.e., however long the branch is between the node π and the root, X_π, in the proper immediate subprotocol hierarchy tree of X_π), every protocol between π and X_π is equivalent to a protocol between π' and $X_{\pi'}$ in virtue of an immediate application of RGSRT. Thus, working one's way up from π, we eventually arrive at the final step (root of the tree), and apply RGSRT once more to establish the conclusion.

By means of the foregoing results, we can see the full range of protocol-theoretic contexts in which graph-theoretic equivalents can be validly replaced. It's simple enough to prove a more general proposition, of which GRT is an instance:

Proposition 296 (GRT+) *Let $A(X_\pi)$ be any formula of L^R_{PLEN} that embeds arbitrary, regular X_π in either the dynamic operators or the connectives of L^R_{PLEN}, except that the formula variable $A(X_\pi)$ cannot embed X_π in $M(X_\pi)$ or any other strictly syntactic expression (subprotocol expressions, syntactic replacement itself) nor in a subset of ternary expression ($\pi \leftrightarrow_r$ ..., $\pi \curlywedge_r$ ..., or, if X_π contains parallel executions or choices, $\pi \curlyvee_r$ ...). Then: For all regular protocols, for all $s \in S$, for all g, if $\rho(\pi/\pi')(X_\pi) = X_{\pi'}$, $\pi \equiv_G \pi'$, and $s \Vdash A(X_\pi)$, then $s \Vdash A(X_{\pi'})$. In rule form:*

$$\left[\frac{\pi \equiv_G \pi',\ \rho(\pi/\pi')(X_\pi) = X_{\pi'},\ A_{(X\pi)}}{A_{(X\pi')}}\right]$$

Proof (Proof-Plan) The proof proceeds by enumeration. I show that syntactic replacement of π with π' preserves verification at arbitrary state, s, in arbitrary EDG, g, for all protocol-theoretic formulae formed by embedding X_π in each of the novel operators of Revised PLEN. The extension of this result to all other protocol-theoretic contexts in which protocols are embedded up to arbitrary depth is mechanical.

(Schematic Argument) Assume $\pi \equiv_G \pi'$, $\rho(\pi/\pi')(X_\pi) = X_{\pi'}$, $A_{(X\pi)}$. Then by GSRT, $X_\pi \equiv_G X_{\pi'}$. Thus, by PRT, $P_{X\pi} = P_{X\pi'}$ for all graphs. Thus, trivially from PRT, $\models_{PLEN} \Box(X_\pi \leftrightarrow_p X_{\pi'})$ and, more directly, for all p for all s, if $first(p) = s$, then $p \in P_{X\pi}$ iff $p \in P_{X\pi'}$. Assume $s \Vdash A(X_\pi)$. Prove $s \Vdash A(X_{\pi'})$ from the definition of the execution and truth conditions. The truth rules for $A(X_{\pi'})$ imply $s \Vdash A(X_{\pi'})$, as required.

The last inferential step is mechanical to check. Thus:

(1) $s \Vdash [X_\pi]_p A$ iff $s \Vdash [X_{\pi'}]_p A$ because all paths of X_π from s are identical to paths of $X_{\pi'}$ from s and vice versa.
(2) $s \Vdash \langle X_\pi \rangle_p A$ iff $s \Vdash \langle X_{\pi'} \rangle_p A$ because the path executing X_π must be an execution of $X_{\pi'}$.
(3) $s \Vdash X_\pi \leftrightarrow_p \pi''$ iff $s \Vdash X_{\pi'} \leftrightarrow_p \pi''$ because $P_{X\pi} = P_{\pi''}$, and $P_{X\pi} = P_{X\pi'}$, so $P_{X\pi'} = P_{\pi''}$.
(4) $s \Vdash X_\pi \curlywedge_p \pi''$ iff $s \Vdash X_{\pi'} \curlywedge_p \pi''$ because $P_{X\pi} \cap P_{\pi''} \neq \emptyset$, and $P_{X\pi} = P_{X\pi'}$, so $P_{X\pi'} \cap P_{\pi''} \neq \emptyset$.
(5) $s \Vdash X_\pi \curlyvee_p \pi''$ iff $s \Vdash X_{\pi'} \curlyvee_p \pi''$ because $P_{X\pi}/P_{\pi''} \neq \emptyset$ or $P_{\pi''}/P_{X\pi} \neq \emptyset$, and $P_{X\pi} = P_{X\pi'}$, so $P_{X\pi'}/P_{\pi''} \neq \emptyset$ or $P_{\pi''}/P_{X\pi'} \neq \emptyset$.
(6) $s \Vdash [X_\pi]_r A$ iff $s \Vdash [X_{\pi'}]_r A$ because the restriction rules imply that every p that executes any protocol π is paired with an edge, $\langle s, \pi, s' \rangle$, such that $first(p) = s$ and $last(p) = s'$. The execution rules imply the converse. Let $e \in R_{X\pi}$. Then there is a $p \in P_{X\pi}$ such that $first(e) = first(p)$ and $last(e) = last(p)$. Thus, $p \in P_{X\pi'}$. But then there must be an $e' \in R_{X\pi'}$ such that $first(e') = first(p)$ and $last(e') = last(p)$ by the restriction rules. This holds for any arbitrary execution of X_π. Transpose for the converse result.
(7) $s \Vdash \langle X_\pi \rangle_r A$ iff $s \Vdash \langle X_{\pi'} \rangle_r A$ by parallel reasoning.
(8) For the case where X_π contains no parallel executions or choices, $s \Vdash X_\pi \curlyvee_r \pi''$ iff $s \Vdash X_{\pi'} \curlyvee_r \pi''$ by similar reasoning; any edge-theoretic divergence comes accompanied by a path divergence (this isn't the case for (e.g.) $a \cap b$ and $b \cap a$—these have edge divergences without path divergences). Thus, if X_π is path-divergent with π'' and path equivalent with $X_{\pi'}$, $X_{\pi'}$ will diverge in paths from π''.
(9) $s \Vdash X_\pi \equiv^s_G \pi''$ iff $s \Vdash X_{\pi'} \equiv^s_G \pi''$
    (a) First, note that, for any regular π and π', $\pi \equiv_G \pi'$ iff $\pi \equiv_{GGI} \pi'$.
        (i) RtoL: All protocol graphs for any regular protocol are trees or trees with detached loops with middle term $\pi_i *$ where π_i is any proper immediate subprotocol for any iteration subprotocol of π, in the case of protocols containing iterations. All loops with identical middle terms are isomorphic, trivially. For arbitrary, g, let $g \in G^\pi$. It follows immediately that:

(A) if π contains no iterations, then the tree $(g, @g)$ is itself a pointed subgraph of g and $(g, @g) \in G^{\pi}$, and
(B) since each loop is a pointed subgraph of g, if π contains n subprotocols with main control operator, $*$, then $\langle (g, @g), \{e_1, ..., e_n\} \rangle \in G^{\pi}$, where $\{e_1, ..., e_n\}$ is a possibly empty set of loops of the aforementioned forms.

(ii) Let $\pi \equiv_{GGI} \pi'$. Then, letting $\langle (g, @g), \{e_1, ..., e_n\} \rangle \in G^{\pi}$, there must be some

$$\left\langle (g', @g'), \left\{e_1', ..., e_m'\right\} \right\rangle \in G^{\pi'}$$

such that

$$\langle (g, @g), \{e_1, ..., e_n\} \rangle \simeq \left\langle (g', @g'), \left\{e_1', ..., e_m'\right\} \right\rangle$$

From SUT, $\left\langle (g', @g'), \left\{e_1', ..., e_m'\right\} \right\rangle$ is isomorphic to every graph of π'. Thus, from ISOP (by means of the EQID Corollary) $\pi \equiv_G \pi'$.

(iii) LtoR: Let $\pi \equiv_G \pi'$. It follows that every protocol graph of π is isomorphic to some protocol graph of π' and converse. Thus, from isomorphism, each pointed subgraph of every protocol graph of π is isomorphic to some pointed subgraph of some protocol graph of π' and converse.

(b) Thus, there are two simple proofs.

(i) The relation, $\equiv_G$, is transitive and symmetric from PGET and PGES. Thus, from LtoR: $X_{\pi} \equiv_G^s \pi''$ implies $X_{\pi} \equiv_G \pi''$. $X_{\pi} \equiv_G X_{\pi'}$ follows from RGSRT (or iteration of PGIIRC), and $X_{\pi} \equiv_G X_{\pi'}$ implies $X_{\pi'} \equiv_G^s X_{\pi}$, which implies $X_{\pi'} \equiv_G X_{\pi}$. Thus, $X_{\pi'} \equiv_G \pi''$ by PGET. Thus, $X_{\pi'} \equiv_G^s \pi''$. Transposition gives the RtoL.

(ii) Assume $\pi \equiv_G \pi'$, $\rho(\pi/\pi')(X_{\pi}) = X_{\pi'}$, $A_{(X\pi)}$. Then by RGSRT (or iteration of PGIIRC), $X_{\pi} \equiv_G X_{\pi'}$. First, it follows that $X_{\pi} \equiv_G^s X_{\pi'}$ for any s and any g. Let $(s, g') \in G^{X\pi}$. Then by EQID, $(s, g') \in G^{X\pi'}$. Now, let $s \Vdash X_{\pi} \equiv_G^s \pi''$ for some g. Then it follows that $s \Vdash X_{\pi'} \equiv_G^s \pi''$. Let $(s, g') \sqsubseteq g$ and $(s, g') \in G^{X\pi'}$. Then, by EQID, $(s, g') \in G^{X\pi}$. From the assumption, any g' that is a protocol graph of X_{π} is a protocol graph of π'' and vice versa. Thus, $(s, g') \in G^{\pi''}$. Let $(s, g') \in G^{\pi''}$. Then, from the assumption, $(s, g') \in G^{X\pi}$. Thus, by EQID, $(s, g') \in G^{X\pi'}$. Thus, $(s, g') \in G^{X\pi'}$ iff $(s, g') \in G^{\pi''}$, and, by definition: $X_{\pi'} \equiv_G^s \pi''$.

(10) $s \Vdash X_{\pi} \equiv_p \pi''$ iff $s \Vdash X_{\pi'} \equiv_p \pi''$ from GSRT, PGET, and PRT1.

(a) $s \Vdash X_{\pi} \equiv_p^s \pi''$ iff $s \Vdash X_{\pi'} \equiv_p^s \pi''$ from 10.

(11) $s \Vdash X_{\pi} \equiv_t \pi''$ iff $s \Vdash X_{\pi'} \equiv_t \pi''$ from GSRT, PGET, and PRT1.

(a) $s \Vdash X_{\pi} \equiv_t^s \pi''$ iff $s \Vdash X_{\pi'} \equiv_t^s \pi''$ from 11.

(12) $s \Vdash X_\pi \equiv_c \pi''$ iff $s \Vdash X_{\pi'} \equiv_c \pi''$ from GSRT, PGET, and PRT1.

   (a) $s \Vdash X_\pi \equiv_c^s \pi''$ iff $s \Vdash X_{\pi'} \equiv_c^s \pi''$ from 12.

(13) $s \Vdash \pi + \mapsto X'_\pi$ iff $s \Vdash \pi + \mapsto X_{\pi'}$ from GSRT and the NSTL Lemma.

   (a) $s \Vdash \pi + \mapsto_s X'_\pi$ iff $s \Vdash \pi + \mapsto_s X_{\pi'}$ from 13.

From PRT1, any preservation results for procedural, correspondence, and transitional equivalence holds for graph-theoretic equivalence. From there, normal extensional and intensional equivalence of semantic conditions supplies the rest of the consequent of the rule. This completes the proof for GRT+. □

Claim 297 This is a central result as far as the development of the theory of protocol equivalence goes. It can be shown that protocol graph isomorphism equivalence is the strongest and finest grained protocol equivalence relation definable in PLEN that doesn't imply identity of protocols; GRT and PRT are essential parts of this result.

6.5 Granularity

Claim 298 Graph-theoretic protocol equivalence is the finest grained protocol equivalence relation in PLEN short of syntactic identity. I explicitly explore this property in terms of strictness (the inclusiveness of the extension of the protocol equivalence), informativeness (the inclusiveness of the set of protocol-theoretic contexts for which replacement of graph-theoretic equivalents is valid), and adequacy as a criterion of hyperintensional equivalence for protocols. The strictness of protocol graph isomorphism equivalence is codified by PRT; its informativeness is codified by GRT.

6.5.1 *Strictness and Informativeness*

Claim 299 Graph-theoretic protocol equivalence is the strictest and most informative protocol equivalence relation defined in PLEN that doesn't imply syntactic identity. It is coarse-grained enough that non-identical protocols can be equivalent, but fine-grained enough that equivalent protocols are intersubstitutable in all contexts in PLEN where intersubstitution doesn't imply syntactic identity. As such, given PLEN as a logical framework for protocols (or procedural epistemic norms), graph-theoretic protocol equivalence is a theoretically interesting and useful criterion of protocol individuation. The following results spell this out.

Corollary 300 (Identity Preserving Formulae) *For any instance of schematic formula, $A(\pi)$, suppose that $A(\pi)$ embeds π via any operator in the following set* *of operators of* L^R_{PLEN}:

$$\{Sub, ISub, PISub, \rho()() =, \curlywedge_r, \curlyvee_r, \leftrightarrow_r, M()\}$$

Then: If $A(\pi) \equiv A(\pi')$, *then* $\pi = \pi'$.

Proof Direct argument schemata for each operator.

(1) For each of the subprotocol variants, $M()$, and syntactic replacement, the result is trivial; each is explicitly a syntactic formula.
(2) The definition of edges mechanically provides for $\curlywedge_r$ and $\leftrightarrow_r$; shared edges means identical middle terms.
(3) For $\curlyvee_r$: Suppose that $(\pi \curlyvee_r \pi'') \equiv (\pi' \curlyvee_r \pi'')$ for some π''. This is equivalent to: $\neg(\pi \curlyvee_r \pi'') \equiv \neg(\pi' \curlyvee_r \pi'')$. But, for any protocols, π_1 and π_2, $\neg(\pi_1 \curlyvee_r \pi_2)$ implies $\pi_1 = \pi_2$; if there are no edges that are not shared between π_1 and π_2, then they share all edges, and so these edges have identical middle terms, and it follows that $\pi_1 = \pi_2$. Thus, $\pi = \pi'' \equiv \pi' = \pi''$. Thus, $\pi = \pi'$.

□

Claim 301 Coupling the foregoing with the explicit restrictions on GRT, it's trivial that:

Corollary 302 (GRT Strength) *For any schematic formula,* $A(\pi)$, *and protocols,* π *and* π', *if* $\pi \equiv_G \pi'$ *and* $A(\pi)$ *don't imply* $A(\pi')$, *then* $A(\pi) \equiv A(\pi')$ *implies* $\pi = \pi'$.

Proof The schematic formulae for which GRT fails are exactly those for which the Identity Preserving Formulae corollary holds.

(1) $(\rho()() =)$ Counterexample: Let $\pi \equiv_G \pi'$ and $\rho(a/\pi) = X_\pi$. It trivially does not follow that $\rho(a/\pi') = X_\pi$.
(2) (Sub): Counterexample: Let $a \cap b \equiv_G b \cap a$ and $s \Vdash a \cap b \in Sub((a \cap b); c)$. It trivially does not follow that $s \Vdash b \cap a \in Sub((a \cap b); c)$.
(3) $(ISub)$: Counterexample: Let $a \cap b \equiv_G b \cap a$ and $s \Vdash a \cap b \in ISub((a \cap b); c)$. It trivially does not follow that $s \Vdash b \cap a \in ISub((a \cap b); c)$.
(4) $(PISub)$: Counterexample: Let $a \cap b \equiv_G b \cap a$ and $s \Vdash a \cap b \in PISub((a \cap b); c)$. It trivially does not follow that $s \Vdash b \cap a \in PISub((a \cap b); c)$.
(5) $(\rho()() =)$: Counterexample: Let $a \cap b \equiv_G b \cap a$ and $s \Vdash \rho(c/a \cap b)(c; d) = ((a \cap b); d)$. It trivially does not follow that $s \Vdash \rho(c/b \cap a)(c; d) = ((a \cap b); d)$.
(6) $(M())$: Counterexample: Let $a \cap b \equiv_G b \cap a$ and $s \Vdash M(a \cap b)$. It trivially does not follow that $s \Vdash M(b \cap a)$.
(7) $(\curlywedge_r)$: Counterexample: Let $a \cap b \equiv_G b \cap a$ and $s \Vdash a \cap b \curlywedge_r \pi$ for some arbitrary π. By definition, it follows that there is some $e = \langle s, a \cap b, s' \rangle$ and $\langle s, a \cap b, s' \rangle \in R_\pi$. Thus, by definition, $e = \langle s, \pi, s' \rangle$. Thus, $\pi = a \cap b$. Thus, if $s \Vdash b \cap a \curlywedge_r \pi$, then $a \cap b = b \cap a$. Thus, $s \nVdash b \cap a \curlywedge_r \pi$.
(8) $(\curlyvee_r)$: Counterexample: Let $a \cap b \equiv_G b \cap a$. Let $\pi = b \cap a$. Let $s \Vdash a \cap b \curlyvee_r b \cap a$. By definition, it follows that there is some $e = \langle s, a \cap b, s' \rangle$ and $\langle s, a \cap b, s' \rangle \notin R_{b \cap a}$. Thus, $s \nVdash b \cap a \curlyvee_r b \cap a$.

(9) ($\leftrightarrow_r$): Counterexample: Let $a \cap b \equiv_G b \cap a$ and $s \Vdash a \cap b \leftrightarrow_r \pi$ for some arbitrary π. By definition, it follows that every execution of $a \cap b$, $e = \langle s, a \cap b, s' \rangle$, is also in R_π. Thus, by definition, $e = \langle s, \pi, s' \rangle$. Thus, $\pi = a \cap b$. Thus, if $s \Vdash b \cap a \leftrightarrow_r \pi$, then $a \cap b = b \cap a$. Thus, $s \nVdash b \cap a \leftrightarrow_r \pi$.

□

Given these results, we can start to at least qualitatively compare the strictness and informativeness of different protocol equivalence relations.

Definition 303 (Strictness and Informativeness)
Let R_1 and R_2 be protocol equivalence relations.

Let L_{R_i} be the set of expressions of L^R_{PLEN} of the form $A(\pi)$ over which syntactic replacement of protocols related by R_i preserves truth. That is, let L_{R_i} be the set of distinct schematic formulae, $A(\pi)$, such that the following rule is sound:

$$\left[\frac{R_i(\pi, \pi'),\ \rho(\pi/\pi')(X_\pi) = X_{\pi'},\ A_{(X\pi)}}{A_{(X\pi')}}\right]$$

Then:

(1) (Strictness) R_1 *is stricter than* R_2 ($R_1 \gg_S R_2$) iff the extension of R_1 is a subset of the extension of R_2.
(2) (Informativeness) R_1 *is more informative than* R_2 ($R_1 \ggg_I R_2$) iff $L_{R_2} \subset L_{R_1}$.

Proposition 304 (Strictness Corollary)

(1) $\equiv_r \gg_S \equiv_G \gg_S \equiv_p \gg_S \equiv_{fc} \gg_S \equiv_t$,
(2) $\equiv_r \gg_S \equiv^S_G \gg_S \equiv^S_p \gg_S \equiv^S_{fc} \gg_S \equiv^S_t$,
(3) $\leftrightarrow_r \gg_S \leftrightarrow_p \gg_S \leftrightarrow_t$.

Proof This is mechanically extractable from PRT. □

Proposition 305 (Informativeness Corollary)

(1) $\equiv_R \ggg_I \equiv_G \ggg_I \equiv_p \ggg_I \equiv_{fc} \ggg_I \equiv_t$,
(2) $\equiv_r \ggg_I \equiv^S_G \ggg_I \equiv^S_p \ggg_I \equiv^S_{fc} \ggg_I \equiv^S_t$,
(3) $\leftrightarrow_r \ggg_I \leftrightarrow_p \ggg_I \leftrightarrow_t$.

Proof (Sketch) Each clause is either immediate or mechanically extracted from the GRT Strength results and the reasoning in the Soundness proof. I outline the extraction of the third clause:

(1) Consider the SEP_x rules in the proof of R-Derivation System Soundness. In particular, consider the following schematic rule or its corresponding axiom schema:

(a) (SEP_x Rule Schema) $\dfrac{(\pi \leftrightarrow_x \pi'),\ A(\pi)}{A(\pi')}$
(b) (SEP_x Axiom Schema) $\pi \leftrightarrow_x \pi' \rightarrow (A(\pi) \equiv A(\pi'))$

(2) For each protocol equivalence, the rule/law is valid for distinct classes of L^R_{PLEN} of the form $A(\pi)$. An array of such classes of formulae is presented in the following table:

(a)

$\leftrightarrow_r$	$\equiv^s_G$	$\leftrightarrow_p$	$\leftrightarrow_t$
$M(\pi)$	$\ldots \mapsto \pi$	$\pi \curlywedge_p \ldots$	
$\ldots \mapsto \pi$	$\ldots \equiv^s_g \pi$	$\pi \curlyvee_p \ldots$	
$\ldots \equiv^s_g \pi$	$\pi \curlywedge_p \ldots$	$\pi \leftrightarrow_p \ldots$	
$\pi \curlywedge_p \ldots$	$\pi \curlyvee_p \ldots$		
$\pi \curlyvee_p \ldots$	$\pi \leftrightarrow_p \ldots$		
$\pi \leftrightarrow_p \ldots$			
$\pi = \ldots$			
$\ldots \in Sub(\pi)$			

(3) From left to right, the state-relative equivalence relations flow from strictest to laxest. The columns present a (non-exhaustive) list of classes of schematic formulae of L^R_{PLEN} that can be plugged into SEP_x to generate a valid instance. The information in each column is mechanical to derive.

(a) (Examples) Ternary equivalence implies syntactical identity. Thus, syntactic replacement preserves truth for all contexts in L^R_{PLEN}, including methodological expressions and ternary variants of expressions. This is not so for graph-theoretic (and thus procedural) equivalence; GRT fails in the context of ternary variant expressions. For example, let $\pi \equiv_G \pi'$. Let $\pi \curlywedge_r \pi''$. Thus, from the ternary truth rules, π and π'' share an edge. But by the execution rules, this requires that $\pi = \pi''$. Now, if $\pi \neq \pi'$, then it is not the case that $\pi' \curlywedge_r \pi''$ because it would follow that $\pi' \neq \pi''$. Thus, the premises $\pi \equiv_G \pi'$ and $\pi \curlywedge_r \pi''$ don't imply $\pi' \curlywedge_r \pi''$. Similarly, the premises $\pi \equiv_G \pi'$ and $M(\pi)$ don't imply $M(\pi')$. Just so for procedural equivalents. This can all be restated in terms of tautologies. Thus, for instance, $(\pi \equiv_G \pi' \wedge \pi \curlywedge_r \pi'') \to \pi' \curlywedge_r \pi''$ is not a tautology, though $(\pi \equiv_G \pi' \wedge \pi \curlywedge_p \pi'') \to \pi' \curlywedge_p \pi''$ and $(\pi \equiv_p \pi' \wedge \pi \curlywedge_p \pi'') \to \pi' \curlywedge_p \pi''$ are. On the other hand, $(\pi \equiv_r \pi' \wedge \pi \curlywedge_r \pi'') \to \pi' \curlywedge_r \pi''$ is tautologous. For another example, and $(\pi \equiv_r \pi' \wedge M(\pi)) \to M(\pi')$ is a tautology but $(\pi \equiv_G \pi' \wedge M(\pi)) \to M(\pi')$ is assuredly not—use the example $\pi_1; (\pi_2 \cap \pi_3) \equiv_G ((\pi_1; \pi_2) \cap (\pi_1; \pi_3))$. Given that graph-theoretic equivalence doesn't preserve truth for these classes of formulae, PRT implies that state-relative procedural or binary equivalence cannot preserve truth for these classes of formulae; if they could, then, contrary to PRT, graph-theoretic equivalence would, as well.

□

Claim 306 In sum, the protocol-theoretic contexts (contexts in L^R_{PLEN} that take protocols) for which replacement of (canonical) graph-theoretic protocol equivalence doesn't preserve truth are exactly those for which syntactic replacement of protocols salva veritate implies protocol identity. Moreover, (canonical) graph-theoretic equivalence doesn't imply protocol identity. Canonical graph-theoretic protocol equivalence is thus the strictest and most informative protocol equivalence relation defined in PLEN short of protocol identity. It's not clear whether there could be a stricter form of protocol equivalence that doesn't imply protocol identity. If these results are correct, then, in this iteration of PLEN, there can't be.

Remark 307 These results may take on a clearer importance in Sects. 7.4 and 7.5. For now, the final part of this excursion into the study of protocol equivalence relations in PLEN connects the protocol logic of PLEN to the analysis of normative contexts by way of hyperintensionality.

6.5.2 *Hyperintensional Equivalence*

Recent work on normativity[7] (indeed, epistemic normativity) has discovered normative contexts that are, apparently, hyperintensional.[8]

Definition 308 (Hyperintensional Contexts) A sentential context, $C(x)$, is *hyperintensional* if, where $\vdash$ denotes logical truth, $\vdash \Box(A \equiv B)$ does not entail $\vdash \Box(C(A) \equiv C(B))$.

Which is to say that there are normative contexts for which, much like opaque belief contexts, replacement of logically equivalent sentences (even necessarily logically equivalent sentences) does not preserve truth.

This observation has prompted work on notions of hyperintensional equivalence, a form of equivalence between sentential and non-sentential objects (e.g., properties) that is governed by rules of truth-preserving replacement over hyperintensional contexts.

Definition 309 (Hyperintensional Equivalence) Expressions, x and y, are *hyperintensionally equivalent* iff x and y can be substituted salva veritate in a hyperintensional context.

The logic of norms must, after all, be able to account for the hyperintensionality of some normative contexts—at least modeling the failure of replacement of logical equivalents in hyperintensional equivalents and the success of replacement of hyperintensional equivalents in them.

---

[7] I follow Faroldi's [1] definitions of hyperintensionality, hyperintensional equivalence, and criteria of hyperintensionality in this section, and I provide protocol-theoretic adaptations of them. See chapters 1–3 of [1], especially chapter 3.

[8] See [1] and the sources referenced therein.

While PLEN is not designed to account for the hyperintensionality of all normative contexts, it turns out that hyperintensionality is a feature of some protocol-theoretic contexts in L^R_{PLEN}. In particular, protocol graph isomorphism equivalence is a hyperintensional context.

Let me make that explicit by adapting the relevant definitions. Logical and necessary equivalence are not transparently applicable notions to protocols, given Jorgensen's dilemma [2]; protocols are not themselves truth-evaluable. They are syntactic objects whose "semantics" consist of execution conditions, which, in turn, can be used to assign truth values to protocol-theoretic formulae. Nonetheless, taking EDGs as normal modal frames, we can identify varieties of execution-theoretic, necessary execution-theoretic, logically execution-theoretic, and logically necessary execution-theoretic protocol equivalence: , $\leftrightarrow_x$, $\Box(... \leftrightarrow_p ...)$, and their logically true variants (where $\equiv_x$ is elliptical for $\models_{PLEN} \Box(... \leftrightarrow_x ...)$). Explicitly:

Definition 310 (Extensional Protocol Equivalences) Let π_1 and π_2 be *extensionally protocol equivalent* iff $\pi_1 \leftrightarrow_x \pi_2$ for some execution-theoretic protocol equivalences that do not imply syntactic identity.

Definition 311 (Logical Protocol Equivalences) Let π_1 and π_2 be *logically protocol equivalent* iff $\vdash_{PLEN} \pi_1 \leftrightarrow_x \pi_2$ for some execution-theoretic protocol equivalences that do not imply syntactic identity.

Definition 312 (Intensional Protocol Equivalences) Let π_1 and π_2 be *necessarily protocol equivalent* iff $\Box(\pi_1 \leftrightarrow_x \pi_2)$ for execution-theoretic protocol equivalences that do not imply syntactic identity.

Definition 313 (Logical Necessary Protocol Equivalences) Let π_1 and π_2 be *logically necessarily protocol equivalent* iff $\vdash_{PLEN} \Box(\pi_1 \leftrightarrow_x \pi_2)$ for some execution-theoretic protocol equivalences that do not imply syntactic identity.

Definition 314 (Hyperintensional Protocol Contexts) A protocol-theoretic context, $C(x)$, is *hyperintensional* if $\vdash_{PLEN} \Box(\pi_1 \leftrightarrow_x \pi_2)$ does not entail $\vdash_{PLEN} \Box(C(\pi_1) \equiv C(\pi_2))$.

Definition 315 (Hyperintensional Protocol Equivalence) Protocols, π_1 and π_2, are *hyperintensionally protocol equivalent* iff π_1 and π_2 can be substituted salva veritate in a hyperintensional context in L^R_{PLEN}.

Claim 316 Given these definitions, it's almost immediate that, where $\vdash_{PLEN}$ is the syntactic validity relation of the R-Derivation System:

Proposition 317 *Graph-theoretic protocol equivalence (protocol graph isomorphism equivalence) is a hyperintensional protocol context.*

Proof (Proof-Plan) Direct enumeration of cases suffices; there are π_1 and π_2 such that:

(1) $\vdash_{PLEN} \Box(\pi_1 \leftrightarrow_p \pi_2)$ and the following equivalence fails: $(\pi_3 \equiv_G \pi_1)$ iff $(\pi_3 \equiv_G \pi_2)$

(2) $\vdash_{PLEN} \Box(\pi_1 \equiv^s_p \pi_2)$ and the following equivalence fails: $(\pi_3 \equiv_G \pi_1)$ iff $(\pi_3 \equiv_G \pi_2)$
(3) $\vdash_{PLEN} \Box(\pi_1 \leftrightarrow_t \pi_2)$ and the following equivalence fails: $(\pi_3 \equiv_G \pi_1)$ iff $(\pi_3 \equiv_G \pi_2)$
(4) $\vdash_{PLEN} \Box(\pi_1 \leftrightarrow_p \pi_2)$ and $\nvdash_{PLEN} \Box((\pi_3 \equiv^s_G \pi_1) \equiv (\pi_3 \equiv^s_G \pi_2))$
(5) $\vdash_{PLEN} \Box(\pi_1 \equiv^s_p \pi_2)$ and $\nvdash_{PLEN} \Box((\pi_3 \equiv^s_G \pi_1) \equiv (\pi_3 \equiv^s_G \pi_2))$
(6) $\vdash_{PLEN} \Box(\pi_1 \leftrightarrow_t \pi_2)$ and $\nvdash_{PLEN} \Box((\pi_3 \equiv^s_G \pi_1) \equiv (\pi_3 \equiv^s_G \pi_2))$
(7) $\vdash_{PLEN} \Box(\pi_1 \equiv^s_t \pi_2)$ and $\nvdash_{PLEN} \Box((\pi_3 \equiv^s_G \pi_1) \equiv (\pi_3 \equiv^s_G \pi_2))$
(8) ...

□

Now, given that there are normative contexts that are hyperintensional, it is therefore a virtue, in devising a logical framework for normative language, to be able to provide some criterion of hyperintensional equivalence of the objects of hyperintensional contexts. Moreover, it's a virtue to give such a criterion that is, as far as possible, not ad hoc because, of course, it's always desirable that analyses of the logical properties of our domain of study be non-ad-hoc. The criteria for hyperintensional equivalence should, accordingly, account for its granularity and generality:[9]

Definition 318 (General Conditions on Criteria of Hyperintensional Equivalence) Let $Crit$ be a set of conditions expressible in our logically regimented language. A criterion, $Crit$, (or set thereof) for hyperintensional equivalence of expressions x and y must satisfy the following conditions:

- G1: $Crit$ is finer-grained than logical or necessary equivalence; there are logically equivalent expressions that fail to satisfy $Crit$.
- G2: $Crit$ is coarse-grained enough in order to not collapse on to identity; the satisfaction of $Crit$ by expressions x and y does not imply $x = y$.
- G3: $Crit$ is general enough to apply to all hyperintensional contexts; if x and y satisfy $Crit$, then they are intersubstitutable salva veritate in a hyperintensional context.

Adapting the foregoing criteria for protocols and protocol-theoretic contexts, then:

Definition 319 (Conditions on Criteria of Hyperintensional Equivalence for Protocols) Let $PCrit$ be a set of conditions expressible in the meta or object languages of L^R_{PLEN}. A criterion, $PCrit$, (or set thereof) for hyperintensional equivalence of protocols π_1 and π_2 must satisfy the following conditions:

- G1: $PCrit$ is finer-grained than logical or necessary execution-theoretic equivalence; there are logically and necessarily execution-theoretically equivalent protocols that fail to satisfy $PCrit$.
- G2: $PCrit$ is coarse-grained enough in order to not collapse on to identity; the satisfaction of $PCrit$ by expressions π_1 and π_2 does not imply $\pi_1 = \pi_2$.

[9] I am closely following Faroldi's [1] (Chapter 3), here.

- G3: $PCrit$ is general enough to apply to all hyperintensional protocol contexts; if π_1 and π_2 satisfy $PCrit$, then they are intersubstitutable salva veritate in all hyperintensional contexts in PLEN.

Obviously, this set of criteria is not entirely true to the intent of the original set. The set of criteria on hyperintensional equivalence above is intended to be truly general, identifying criteria of hyperintensional equivalence for all hyperintensional objects for all hyperintensional contexts, including attitude reports and non-causal grounding relations. It's not clear whether there can be a criterion of hyperintensional equivalence of that level of generality.[10] Thus, I restrict focus to the hyperintensional equivalence of protocols and hyperintensional protocol-theoretic contexts. If the intended interpretation of PLEN is right, then protocol graph isomorphism equivalence captures a notion of hyperintensional equivalence for procedural epistemic norms for the chunk of normative language captured by the protocol-theoretic language of PLEN (however comprehensive it ultimately is).

From the main results of the theory of protocol graph isomorphism equivalence and the foregoing results of strictness and informativeness, it follows that the canonical graph-theoretic protocol equivalence relation of PLEN satisfies these criteria:

Proposition 320 *Graph-theoretic protocol equivalence satisfies the conditions, G1, G2, and G3, on criteria of hyperintensional equivalence for protocols.*

Proof Let protocols, π_1 and π_2, be hyperintensionally equivalent iff $\pi_1 \equiv_G \pi_2$. PRT and PRTC together suffice for the satisfaction of G1 and G2. RGSRT, GRT, GRT Strength, and Identity Preserving Formulae suffice for G3; graph-theoretically equivalent protocols, π_1 and π_2 are intersubstitutable in the context of graph-theoretically equivalent superprotocols, X_{π_1} and X_{π_2}, and this results in their truth-preserving intersubstitutability in all contexts in PLEN for which truth-preserving intersubstitution implies identity. □

In other words, protocols are objects that are capable of being equivalent or inequivalent in ways that outstrip the analysis provided by intensional properties. The protocol-theoretic language of PLEN, L^R_{PLEN}, has the expressive power to encode some of these intensional properties and to record their logical interactions. Protocol graph isomorphism equivalence preserves truth for all hyperintensional contexts in PLEN that don't imply protocol identity. Protocol graph isomorphism equivalence among protocols (hence, norms) is thus a criterion for hyperintensional equivalence of procedural epistemic norms that explains the failure of truth preserving intersubstitution of even necessarily transition equivalent actions and processes.

The hyperintensionality of epistemic normative contexts in PLEN is accounted for by the hyperintensionality of the structures of protocol graphs; the procedural structure or content of protocols analyzed by protocol graphs carries more information than can be accounted for intensionally. Thus, given the intended

[10] See the upshot of Chapter 3 of [1] for a persuasive argument, the litigation of which goes beyond the scope of this section and, indeed, book.

interpretation of PLEN, procedural epistemic norms are hyperintensional objects, dictating finer (procedural) restrictions than can be analyzed in terms of extensional or intensional contexts. Generally, if protocols capture the procedural structure of epistemic norms, and protocols are hyperintensional objects, this accounts for the hyperintensionality of procedural epistemic normative contexts by means of the claim that procedural epistemic norms are themselves hyperintensional objects. Thus, procedural requirements may be too fine-grained to be accounted for by means of epistemic transitions. The normative contexts pointed out in the Honest Twin case, for example, are thus hyperintensional precisely because they are procedural.

6.6 The Logic of Protocols and Protocol Graph Isomorphism Equivalence

The development of PLEN and then Revised PLEN built to the theory of protocol graph isomorphism equivalence. The uses to which this system is put in the next few chapters hopefully justify the suspense of incrementally building up to this point. The theory of protocol graph isomorphism equivalence is the core of the protocol logic of the Revised PLEN Framework. It is on the basis of this theory of protocol equivalence that Revised PLEN can be used as a framework for formalizing the kinds of reasoning that are central to epistemic proceduralism. It is, further, on the basis of this theory that Revised PLEN can be used for studying different sublogics of protocols. This is borne out in Chap. 7.

Appendix 3

Remark 321 This appendix spells out the results discussed in Chap. 6.

Protocol (Equivalence) Relation Transmission: PRT

Protocol Relation Transmission

Lemma 322 (PRT) *For all regular protocols, π and π':*

Proof (Proof-Plan) Direct arguments for each group of rules. PRT1 is derived in Chap. 6. The remaining clauses are presented here:

(1) PRT2(i) Let $\pi = \pi_1 \cup (\pi_2 \cap \pi_3)$ and $\pi' = (((\pi_1 \cap \pi_2) \cap \pi_3) \cup \pi_1) \cup (\pi_2 \cap \pi_3)$ such that each $\pi_i \in \Pi_0$.

PRT1	PRT2	PRT3
(i) $\pi \equiv_G \pi' \models \pi \equiv_p \pi'$ (ii) $\pi \equiv_p \pi' \models \pi \equiv_{fc} \pi'$ (iii) $\pi \equiv_{fc} \pi' \models \pi \equiv_t \pi'$	(i) $\pi \equiv_p \pi' \not\models \pi \equiv_G \pi'$ (ii) $\pi \equiv_{fc} \pi' \not\models \pi \equiv_p \pi'$ (iii) $\pi \equiv_t \pi' \not\models \pi \equiv_p \pi'$	(i) $\pi \equiv_r \pi' \models \pi \equiv_G \pi'$ (ii) $\pi \equiv_G \pi' \not\models \pi \equiv_r \pi'$

(a) The following schema is valid:

$$\pi_1 \cup (\pi_2 \cap \pi_3) \equiv_P (((\pi_1 \cap \pi_2) \cap \pi_3) \cup \pi_1) \cup (\pi_2 \cap \pi_3)$$

(i) LtoR Let $p \in P_\pi$. Then $p \in P_{\pi_1} \cup P_{\pi_2 \cap \pi_3}$. If $p \in P_{\pi_2 \cap \pi_3}$, then either $p \in P_{\pi 2}$ and $\exists q \in P_{\pi_3}(p \rightleftharpoons q)$ or $p \in P_{\pi 3}$ and $\exists q \in P_{\pi_2} \wedge p \rightleftharpoons q)$. If $p \in P_{\pi_1}$, then $p \in P_{\pi'}$ by the execution rules for choice; if p executes π_1, then it executes π' by executing π_1. If $p \in P_{\pi_2 \cap \pi_3}$, then $p \in P_{\pi'}$ by the execution rules for choice; if p executes $\pi_2 \cap \pi_3$, then it executes π' by executing $\pi_2 \cap \pi_3$. Thus, if $p \in P_\pi$, then $p \in P_{\pi'}$.

(ii) RtoL Let $p \in \pi'$. Then $p \in P_{\pi_1} \cup P_{\pi_2 \cap \pi_3} \cup P_{\pi_1 \cap (\pi_2 \cap \pi_3)}$. If $p \in P_{\pi_2 \cap \pi_3}$, then either $p \in P_{\pi_2}$ and $\exists q \in P_{\pi_3}(p \rightleftharpoons q)$ or $p \in P_{\pi_3}$ and $\exists q \in P_{\pi_2}(p \rightleftharpoons q)$. If $p \in P_{\pi_1}$, then $p \in P_\pi$ by the execution rules for choice; if p executes π_1, then it executes π by executing π_1. If $p \in P_{\pi_2 \cap \pi_3}$, then $p \in P_\pi$ by the execution rules for choice; if p executes $\pi_2 \cap \pi_3$, then it executes π by executing $\pi_2 \cap \pi_3$. If $p \in P_{\pi_1 \cap (\pi_2 \cap \pi_3)}$, then $p \in P_{\pi_1}$ and $\exists q \in P_{\pi_2 \cap \pi_3}(p \rightleftharpoons q)$ (thus, $\exists r(p \rightleftharpoons q \rightleftharpoons r)$ and $q \in P_{\pi_2}$ and $r \in P_{\pi_3}$ or vice versa) or $p \in P_{\pi_2 \cap \pi_3}$ and $\exists q \in P_{\pi_1}(p \rightleftharpoons q)$ (thus, $\exists r(p \rightleftharpoons q \rightleftharpoons r)$ and $p \in P_{\pi_2}$ and $r \in P_{\pi_3}$ or vice versa). In the first case, p executes π_1, and so executes π. In the second case, it executes $\pi_2 \cap \pi_3$, and so executes π.

(iii) Another counterexample. Let $\pi = (\pi_1 \cap \pi_2) \cup (\pi_3 \cap \pi_4)$ and $\pi' = ((\pi_1 \cup \pi_3) \cap (\pi_1 \cup \pi_4)) \cap ((\pi_2 \cup \pi_3) \cap (\pi_2 \cup \pi_4))$. It's easy to prove that $(\pi_1 \cap \pi_2) \cup (\pi_3 \cap \pi_4) \equiv_P ((\pi_1 \cup \pi_3) \cap (\pi_1 \cup \pi_4)) \cap ((\pi_2 \cup \pi_3) \cap (\pi_2 \cup \pi_4))$. It's easy to see that these protocols are not graph-theoretically equivalent; a graph of π' must have corresponding paths of π_1, π_2, and π_4 and a graph of π need not.

(b) The following schema is not valid:

$$\pi_1 \cup (\pi_2 \cap \pi_3) \equiv_G (((\pi_1 \cap \pi_2) \cap \pi_3) \cup \pi_1) \cup (\pi_2 \cap \pi_3)$$

(i) Let $g = \langle S, C, \Pi_g, R, R_0, \Pi{+}_g, R{+}, R{+}_0, c, v \rangle$ such that $S = \{s_1, s_2, s_3\}$ and

$$R_\pi = \{\langle s_1, \pi_1, s_2 \rangle, \langle s_1, \pi_2, s_3 \rangle, \langle s_1, \pi_3, s_3 \rangle\}$$

Let $g' = \langle S', C', \Pi_{g'}, R', R'_0, \Pi{+}_{g'}, R'{+}, R'{+}_0, c', v' \rangle$ such that $S' = \{s_1, s_2, s_3, s_4\}$ and

$$R_{\pi'} = \left\{ \begin{array}{l} \langle s_1, \pi_1, s_2 \rangle \,,\ \langle s_1, \pi_1, s_3 \rangle \,,\ \langle s_1, \pi_2, s_3 \rangle \,, \\ \langle s_1, \pi_3, s_3 \rangle \,,\ \langle s_1, \pi_2, s_4 \rangle \,,\ \ \langle s_1, \pi_3, s_4 \rangle \end{array} \right\}$$

(ii) First, prove that g is a protocol graph of π, and g' is a protocol graph of π'. The set of maximal paths in g is the set of executions of π and π'. The set of maximal paths in g' is the set of executions of π and π'. Every subprotocol of π is executed in g. Every subprotocol of π' is executed in g'. Every implementation of π is executed in g. Every implementation of π' is executed in g'. Every pair of constituents of each choice in π is executable at some state in g. Every pair of constituents of each choice in π' is executable at some state in g'. No non parallel execution linked subprotocols of π are correspondent in g. No non parallel execution linked subprotocols of π' are correspondent in g'. Every atomic subprotocol of π has a protocol graph in g (the relevant edges), and every complex subprotocol does, as well, which is easily seen. Every atomic subprotocol of π' has a protocol graph in g' (the relevant edges), and every complex subprotocol does, as well, which is easily seen. There is no smaller graph that executes all implementations of π. There is no smaller graph that executes all implementations of π'. There is no non-corresponding protocol of π that is graphed by g. There is no non-corresponding protocol of π' that is graphed by g'.

(iii) Second, prove that g is not a protocol graph of π': there is no execution of $(\pi_1 \cap \pi_2) \cap \pi_3)$, which implements itself. Thus, RC fails with respect to π' (there is a subprotocol of π', $(\pi_1 \cap \pi_2) \cap \pi_3)$, such that not all implementations of it are executed in g precisely because $(\pi_1 \cap \pi_2) \cap \pi_3)$ is not executable in g). Also, g' is not a protocol graph of π: π_1 and π_3 correspond in g' where there is no implementation or subprotocol of π that executes them in parallel. Thus, SD1(4) fails with respect to π. Thus, by EQID, π and π' are not graph-theoretically equivalent. Additionally, g and g' are not isomorphic, and so, by the proof of EQID, π and π' are not equivalent.

(2) PRT2(ii) Convert the proof for PRT2(iii) to state sequences from transitions.

(3) PRT2(iii) Let $\pi = (a; b) \cap (c; d)$ and $\pi' = (a; d) \cap (c; b)$.

(a) It follows that $(a; b) \cap (c; d) \equiv_T (a; d) \cap (c; b)$. Let $t \in T_{(a;b)\cap(c;d)}$. Then there is a $p \in P_{(a;b)\cap(c;d)}$. By the definition of the execution rules, either $p \in P_{a;b}$ and there is a $q \in P_{c;d}$ such that $p \rightleftharpoons q$ or vice versa. Then, any such g has the graph-theoretic structure described below. Thus, there is a $p' \in P_{a;d}$ and a $q' \in P_{c;b}$ and $p' \rightleftharpoons q'$. Thus, $t = \langle first(p), last(p) \rangle = \langle first(q), last(q) \rangle = \langle first(p'), last(p') \rangle = \langle first(q'), last(q') \rangle$. Thus, there is a $p \in P_{(a;d)\cap(c;b)}$ such that $t = \langle first(p), last(p) \rangle$ namely p' (or q'). Thus, $t \in T_{(a;d)\cap(c;b)}$.

   (b) It also follows that $\neg((a; b) \cap (c; d) \equiv_P (a; d) \cap (c; b))$. Let $p = qr$ where q executes a and r executes b such that there is a $p' = q'r'$ where q' executes c and r' executes d where p and p' correspond. Thus, any such g has the foregoing structure. Thus, by the execution rules, qr' executes $a; d$ and $q'r$ executes $c; b$ and $qr' \rightleftharpoons q'r$. Thus, qr' executes $(a; d) \cap (c; b)$ and so does $q'r$. But qr' does not execute $(a; b) \cap (c; d)$, nor does $q'r$.

(4) PRT3(i) The requirement of edge-based execution conditions implies that the middle term of each edge that executes π is π itself. This implies that any edge equivalent protocols are identical.
(5) PRT3(ii) Consider the sequential compositions $(a \cup b); c$ and $(b \cup a); c$. These are graph-theoretically equivalent but syntactically nonidentical. Thus, they are not edge equivalent (i.e., syntactically identical).

□

Corollary 323 (PRTC)

(1) $\pi \equiv_p \pi' \nvDash \pi \equiv_r \pi'$, *and* $\pi \equiv_t \pi' \nvDash \pi \equiv_r \pi'$
(2) $\pi \equiv_r \pi' \models \pi = \pi'$; *thus,* $\pi \equiv_r \pi' \models \pi \equiv_G \pi'$, $\pi \equiv_r \pi' \models \pi \equiv_p \pi'$, *and* $\pi \equiv_r \pi' \models \pi \equiv_t \pi'$
(3) $\pi \equiv_G \pi'$ *iff* $\pi \equiv_{GGI} \pi'$
(4) $\pi \equiv_G \pi'$ *iff* $\pi \equiv^s_{GGI} \pi'$
(5) $\pi \equiv^s_G \pi'$ *iff* $\pi \equiv^s_{GGI} \pi'$

Proof Clause by clause direct arguments.

(1) First, procedurally equivalent and transitionally equivalent but syntactically nonidentical protocol pairs exist—see the protocol equivalence axioms. Thus, they are not edge equivalent (i.e., syntactically identical). This suffices to show the negative claims, but PRT3 suffices to imply this result, as well.
(2) Second, the rest of the clauses follow mechanically—even trivially—from the properties of identity. Let $\pi \equiv_r \pi'$. Then $\pi = \pi'$, as any edge executing π must have π as it's middle term, and just so for π'. Thus, if the edges executing π also execute π', then $\pi = \pi'$. Thus, every other equivalence follows by substitution of identicals.
(3) Third, note that all protocol graphs but those for protocols containing iterations are tree-like EDGs (i.e., connected EDGs with a distinguished state, s, that is the root state of all maximal paths in g). For these, the result is immediate.

   (a) RtoL: Assume $\pi \equiv_{GGI} \pi'$. Let g be an arbitrary protocol graph of π, and let g' an arbitrary protocol graph of π'. By assumption, $g \equiv_{GGI} g'$. Given the definition of protocol graphs, g is either a tree-like EDG (i.e., connected EDGs with a distinguished state, s, that is the root state of all maximal paths in g), or a tree-like EDG with detached loops for each iteration contained in π. Thus, g is itself a pointed subgraph of a protocol graph of π or there is a core tree-like EDG in g that is a pointed subgraph of itself and there are some finite number of iteration loops that are pointed subgraphs as well. The same conditions hold for g'. Thus, if g and g' are generated graph isomorphic, they are themselves isomorphic, and the result follows from SUT and ISOP.

(b) LtoR: Assume $\pi \equiv_G \pi'$. By SUT and ISOP, all of π's protocol graphs are isomorphic to all of π''s protocol graphs. Thus, each of the pointed subgraphs of any of π's protocol graphs is isomorphic to some pointed subgraph of one of π''s protocol graphs and vice versa.

(4) Fourth:

(a) RtoL: Assume $\pi \equiv^s_{GGI} \pi'$. Let (g_1, s) be a pointed subgraph of some arbitrary g. By definition, there must be some (g_2, s) among the pointed subgraphs of g such that $(g_1, s) \simeq (g_2, s)$. But then: all protocol graphs of π are isomorphic to (g_2, s) by SUT and the transitivity of isomorphism, and all protocol graphs of π' are isomorphic to (g_2, s) by SUT. Thus, by EQID, $\pi \equiv_G \pi'$.

(b) LtoR: Assume $\pi \equiv_G \pi'$. Let (g_1, s) and (g_2, s) be arbitrary pointed subgraphs of g such that $(g_1, s) \in G^{\pi}$ and $(g_2, s) \in G^{\pi'}$. Immediately, the result follows because the pointed subgraphs of g must also obey the definition of protocol equivalence. Thus, $(g_1, s) \simeq (g_2, s)$, as these were arbitrary pointed subgraphs.

(5) Finally, it is immediate from prior steps that $\equiv^s_{GGI}$ and $\equiv^s_G$ are equivalent.

□

Corollary 324 *Parallel implication relations hold for the graph-relative variants of the protocol equivalences above. More precisely, for all regular protocols, π and π':*

PRT4	PRT5
(i) $\pi \equiv^s_G \pi' \models \pi \equiv^s_p \pi'$, (ii) $\pi \equiv^s_p \pi' \models \pi \equiv^s_{fc} \pi'$, (iii) $\pi \equiv^s_{fc} \pi' \models \pi \equiv^s_t \pi'$	(i) $\pi \equiv^s_p \pi' \not\models \pi \equiv^s_G \pi'$, (ii) $\pi \equiv^s_{fc} \pi' \not\models \pi \equiv^s_p \pi'$, (iii) $\pi \equiv^s_t \pi' \not\models \pi \equiv^s_p \pi'$
PRT6	**PRT7**
(i) $\pi \equiv^s_r \pi' \models \pi \equiv^s_G \pi'$ (ii) $\pi \equiv^s_G \pi' \not\models \pi \equiv^s_r \pi'$	(i) $\pi \equiv_p \pi' \models \Box\pi \leftrightarrow_p \pi'$, (ii) $\Box\pi \leftrightarrow_p \pi' \not\models \Box\pi \leftrightarrow_r \pi'$, (iii) $\pi \equiv_p \pi' \not\models \Box\pi \leftrightarrow_r \pi'$, (iv) $\Box\pi \leftrightarrow_p \pi' \not\models \pi \equiv_p \pi'$, (v) $\Box\pi \leftrightarrow_r \pi' \models \Box\pi \leftrightarrow_p \pi'$.

Proof (Proof-Plan) Direct arguments for each group of rules.

(1) (PRT4) For (i) and (ii), simply convert the proofs of the PRT1 clauses. For (iii), assume that $\pi \equiv^s_{fc} \pi'$, and, for reductio, $T_\pi \neq T_{\pi'}$. Thus, there is a $t \in T_\pi$ and $t \notin T_{\pi'}$. Thus, there is a p that executes π such that $t =$

$\langle first(p), last(p)\rangle$. But then, from the assumption, there must be some q that full path corresponds to p such that q executes π'. But then, by the definition of full path correspondence, $t = \langle first(q), last(q)\rangle$. Thus, $t \in T_{\pi'}$, contradicting our assumption. This holds for arbitrary t.

(2) (PRT5) The same pairs of protocols works for PRT5 as worked for PRT2. Just convert the proofs.
(3) (PRT6) Convert the proofs for PRT3.
(4) (PRT7)

(a) (i) Let $\pi \equiv_p \pi'$. Then for all EDGs, g, $P_\pi = P_{\pi'}$. Let s be a state in any EDG. Then, if $s = first(p)$, for any p, $p \in P_\pi$ iff $p \in P_{\pi'}$. Thus, $s \Vdash \pi \leftrightarrow_p \pi'$. If $s = first(p)$ for no p, then it's vacuously the case that $p \in P_\pi$ iff $p \in P_{\pi'}$. Thus, $s \Vdash \pi \leftrightarrow_p \pi'$. Thus, for any state in any EDG, $s \Vdash \pi \leftrightarrow_p \pi'$. Thus, for any state at any EDG, $s \Vdash \Box\pi \leftrightarrow_p \pi'$.
(b) (ii) Given the Restriction Rules, everywhere there is a path that executes π, there is an edge that executes π such that the middle term of the edge is π. However, $\pi \leftrightarrow_r \pi'$ only if $\pi = \pi'$. Thus, there are countermodels: Consider the sequential compositions $a \cup b; c$ and $b \cup a; c$. The relational execution rules for sequential composition require that, in any g in which $a; c$ can be executed, there be at least three edges: $\langle s, a, s'\rangle$, $\langle s', c, s''\rangle$, and $\langle s, a; c, s''\rangle$. However, the path $\langle s, a, s', c, s''\rangle$ suffices to procedurally execute $a; c$. Consider a graph consisting only of $\langle s, a, s', c, s''\rangle$, $\langle s, b, s''', c, s''\rangle$, $\langle s, a; c, s''\rangle$, $\langle s, b \cup a, s'\rangle$, $\langle s, a \cup b, s'\rangle$, $\langle s, b \cup a, s'''\rangle$, $\langle s, a \cup b, s'''\rangle$, $\langle s, b \cup a; c, s''\rangle$, $\langle s, a \cup b; c, s''\rangle$, and $\langle s, b; c, s''\rangle$. Strictly speaking, this graph violates the restriction rules—or at least incompletely realizes them. At s, $a \cup b; c$ and $b \cup a; c$ are path-theoretically equivalent, but not ternary equivalent because $a \cup b; c \neq b \cup a; c$, and so the edges labeled with them cannot execute both protocols.
(c) (iii) The same countermodel that invalidated (ii) invalidates (iii).
(d) (iv) This result is trivial. The set of paths in any graph, g, is a subset of all paths in all graphs. Two protocols, π and π' can be equivalent for the paths in g, but inequivalent for the paths in other graphs.
(e) (v) Ternary equivalence of π and π', even at an individual state, s, implies that $\pi = \pi'$. So $\pi \leftrightarrow_p \pi'$, immediately.

□

NSTL

Lemma 325 (NSTL) *For all π, X_π formed from regular control operators: If $\rho(\pi/\pi')(X_\pi) = X_{\pi'}$, $\pi \equiv_G \pi'$, and $\pi \in ISub(X_\pi)$, then an EDG, g, satisfies $SD1(1, 2, 3, 4)$, $SD2$, and RC for X_π iff g satisfies $SD1(1, 2, 3, 4)$, $SD2$, and RC for $X_{\pi'}$.*

Proof (Proof-Plan) This proof anticipates the method of direct proof of RGSRT. See that section for further development. LtoR. Assume $\rho(\pi/\pi')(X_\pi) = X_{\pi'}$, $\pi \equiv_G \pi'$, $\pi \in ISub(X_\pi)$, and that g satisfies $SD1(1, 2, 3, 4)$, $SD2$, RC for X_π. Prove $\therefore$ g satisfies $SD1(1, 2, 3, 4)$, $SD2$, RC for $X_{\pi'}$.

(1) (SD1(1)) See the subproof of SD1(1) in the RGSRT proof. This shows by PRT1 that $P_{X\pi} = P_{X\pi'}$. Thus, $X_{\pi'}$ has executions in g. From the fact that $\pi' \in ISub(X_{\pi'})$, it follows that $Sub(X_{\pi'}) = \{Sub(\pi') \cup Sub(\pi_{a(\circ)}) \cup \{X_{\pi'}\}\}$ (where π_i may be nothing) from the definition of subprotocols. Trivially, all subprotocols of π_i are executed in g. By EQID, all subgraphs of g that are graphs of π are graphs of π', so all subprotocols of π' are executed. Done.
(2) (SD1(2)) As shown in the subproofs of SD1(1), $P_{X\pi} = P_{X\pi'}$ in g follows from the premises for any syntactic form of X_π. Thus, since SD1(2) holds with respect to X_π for g, it follows trivially that SD1(2) holds with respect to $X_{\pi'}$ for g.
(3) (SD1(3)) For the first two rules, suppose that the subprotocol of X_π of the requisite form is in π_i. Then by syntactic replacement, it's also in $X_{\pi'}$, as required. Suppose that it's a subprotocol of π. Then by EQID, there is a subgraph of g that satisfies SD1(3) for π and it's a protocol graph of π', too, so it satisfies SD1(3) for π'. But then, there must be a subprotocol of the requisite form in $X_{\pi'}$ after all—any subprotocol of the subprotocol π' is a subprotocol of $X_{\pi'}$, by definition. Suppose that it's X_π itself. Then PRT1 suffices. For the third rule, suppose that there are branching paths, p and q, that execute subprotocols of X_π, π_p and π_q. Then, from SD1(3), it follows that there must be some subprotocol, π_{pq}, of X_π of the requisite form or implementation and dictation properties. But this protocol is either a subprotocol of π_i, π, or is X_π itself. In the first case, syntactic replacement gives the intended results. In the second case, it's EQID. In the third case, it's PRT1 and NSTLL. A more belabored version of this reasoning takes place in the proof of RGSRT.
(4) (SD1(4)) Suppose that some π_1 and π_2 correspond in g but not generally. Then there is a protocol of the requisite form in X_π by assumption. But that protocol is either in the subprotocols of π_i or those of π or it is X_π itself. In the first case, syntactic replacement implies that $X_{\pi'}$ has a subprotocol of the requisite form. In the second case, EQID implies that $X_{\pi'}$ has a subprotocol of the requisite form. In the third case, NSTLL and the execution rules suffice to give the result. A more belabored version of this reasoning takes place in the proof of RGSRT.
(5) (SD2) Let $\pi+ \mapsto X_{\pi'}$. The NSTL Lemma implies that $\pi+ \mapsto X_\pi$. Thus, $\pi+$ is executed in g. $\pi+$ was arbitrary, thus, by UG, all implementations of $X_{\pi'}$ are executed in g. Thus, $SD2_{X\pi'}(g)$.
(6) (RC) This is mechanical and nearly trivial: π_i must have a graph in g on pain of contradiction, and, from EQID, any graph of π is also a graph of π'; thus, the protocol graph of π in g suffices as a graph of π'. Thus, both proper immediate subprotocols of $X_{\pi'}$ have protocol graphs in the subgraphs of g. Iterate downward for protocol graphs of π and π_i. Thus, all proper subprotocols of $X_{\pi'}$ have graphs in g.

Thus, $SD1(1, 2, 3, 4)$, $SD2$, and RC hold for g with respect to $X_{\pi'}$. Transpose for RtoL. □

References

1. Faroldi, F. 2019. *Hyperintensionality and normativity*. Springer.
2. Jorgensen, J. 1937. Imperatives and logic. *Erkenntnis* 7:288–296.

Chapter 7
The Revised PLEN Framework III: An Adaptable Protocol Logic

Abstract The prior chapters developed a theory of protocol equivalence that bound the syntax of protocols to well-known and novel process equivalence relations (bisimulation, isomorphism, protocol graph isomorphism) over dynamic process models (Kripke frames, graphs, LTSs, EDGs). The result was a novel and fine-grained protocol equivalence relation defined in terms of protocol graph isomorphism. The resulting theory of protocol equivalence was built into a revised derivation system for PLEN. This chapter shows that, conditional on the correctness of protocols as representations of procedural epistemic norms, the revised derivation system provides a formally regimented system for reasoning about epistemic norms and a framework for comparing different formal approaches for representing and reasoning about epistemic norms. On the basis of this system, it is shown that different philosophical orientations to modeling epistemic dynamics and epistemic norms can be formalized and qualitatively compared. This is a result of the relative independence of the protocol logic, the propositional and modal logics, and the properties of the contents of epistemic states in PLEN. As a result, PLEN is a sort of modular, formal laboratory in which sublogics encoding different theories of procedural epistemic norms can be built and studied.

7.1 The Revised Protocol Logic

The core results of the prior chapter serve two purposes. The first is the exploration of the theory of graph-theoretic protocol equivalence in PLEN. That's done. The second is partially extending the soundness theorem of PLEN to the R-Derivation System by showing that the meta-level protocol equivalence rules are valid at the meta-level. This, in turn, implies the soundness of the object-level counterparts, and provides Revised PLEN with a sound protocol logic. This system houses sublogics that encode different models of protocols (i.e., procedural epistemic norms): the graph-theoretic, procedural, ternary, and binary models. These models of protocols and dynamics are defined, ultimately, by their protocol equivalence relations. The results thus far derived thus give us a framework for investigating them. The protocol logic of PLEN comes apart from its other subsystems, and so the modularity of

R. Jenkins, *A Protocol-theoretic Framework for the Logic of Epistemic Norms*, Synthese Library 467, https://doi.org/10.1007/978-3-031-08597-0_7

PLEN allows us to adapt it by slotting in different sets of structures, focusing on restricted subsets of structures or sublanguages, or varying the underlying modal and propositional logics of PLEN.

This chapter presents the extension of the standard metatheorems to this elaborated version of PLEN and embarks on the formalization of epistemic proceduralism. The excursion into the revised metatheorems takes up Sect. 7.2. Section 7.3 is, for the purpose of the main argument of the book, the central section; it begins the formalization of epistemic proceduralism by identifying the valid argument forms by which the R-Derivation System codifies the distinctive forms of reasoning identified in Chap. 3. In virtue of the extension of the soundness (et al) results, these forms of reasoning are valid in PLEN. Section 7.4 briefly examines some sublogics in the Revised PLEN Framework, primarily concerning the granularity of their respective protocol equivalence relations and their resources for handling some of the distinctions introduced in Chaps. 1–3. Section 7.5 briefly comments on some of the results of slotting in different subsystems of structures or logics. The upshot is that the Revised PLEN Framework is exactly that: a protocol-theoretic framework for building logics of procedural epistemic norms. The protocol logic, the propositional logic, and the logic of epistemic states in PLEN can be largely disentangled, allowing revisions to one or the other with some independence. This makes the PLEN framework adaptable for multiple philosophical purposes both within and beyond the scope of this book's project.

7.2 Standard Metalogic Revised

The soundness and deduction theorems for the basic derivation system of PLEN can be extended to the R-Derivation System directly, with only minor modifications to the relevant definitions to account for the new object-level counterparts of the metalevel rules. For instance, the definition of derivations must be defined over L^R_{PLEN} and revised to countenance the new three premise rules introduced as counterparts for (e.g.) RGSRT. That's all straightforward, so, the usual, stepwise procedure for establishing soundness works for the R-Derivation System—all that's required is showing that the derivation rules preserve truth in PLEN's models and the axioms are true at all states in all of PLEN's models, and this is mechanical. The only special elaboration on the way to the soundness of the R-Derivation System is the proof of the meta-level rules introduced in Sect. 5.6.2, the (metalogical) soundness of which then directly implies the soundness of their object-level counterparts. Most of those rules are shown sound in Chaps. 5 and 6.

Due to the simultaneous development of the theory of protocol graph isomorphism equivalence at two levels (i.e., the meta-level and the object-level), I take a mildly unorthodox road to this result. Rather than the straightforward approach above, focusing only on the object-level counterparts of the rules for protocol equivalence, I want to establish the soundness of a broader notion of derivability defined by both the meta-level rules and axioms governing protocol graph isomorphism equivalence and their object-level counterparts. This will imply the soundness of the R-Derivation system, as defined in Chap. 5, as well as demonstrate the use of Revised PLEN as a formal framework.

What I mean, concerning the latter point, has to do with an elementary observation about the usefulness of formal logics. One important use of formal logics is the regimentation of reasoning by means of the theory of validity (in a given system) and the use of the attendant technical procedures for modeling arguments and determining their validity (in a given system). That is, logics provide algorithmic deduction systems over their object languages that we can use to reason validly (relative to the logical system) in those object languages. The tools for this kind of regimentation are what we teach our students in introductory logic classes. Call this the use of logics for reasoning in accordance with principles of validity.

Another—more philosophically central—use of formal logics is to provide frameworks in which we can reason in a regimented way, by sound patterns and methods of mathematical construction and inference, *about* (putative) principles of valid reasoning (encoded by means of the object language and derivation system), as well as about modeling assumptions concerning the domain of a logic.

The theory of protocol graph isomorphism equivalence coupled with the R-Derivation system provides a framework in which to reason about the principles governing protocol equivalence as well as in accordance with the principles governing protocol equivalence. The usefulness of the framework, then, is best established by showing that both kinds of reasoning are sound. This can be done in one—admittedly sort of odd—maneuver: expanding the notion of derivability to encompass meta-level and object-level rules; showing that meta-level and object-level notions of soundness converge (at least in this domain); and finally showing that each of both sorts of rules and axioms are sound in one sense or the other.

Definition 325 Let L^{R+}_{PLEN} be the language that results from adding the operators:

$$Pop++ = \{\equiv_r, \equiv_G, \equiv_p, \equiv_t, \equiv_c, \mapsto\}$$

The resulting formulae are interpreted as in the meta-language, i.e., in terms of syntactic conditions, and quantification over specialized constructions and EDGs.

Definition 326 (Revised Derivation) For any $X \subseteq L^{R+}_{PLEN}$ and $B \in L^{R+}_{PLEN}$ a *derivation* (in the R-Derivation System) of protocol-theoretic formula B from a possibly empty set of premises (protocol-theoretic formulae), $X = \{\emptyset, A_1, ..., A_n\}$, is any tree, $\langle L, X, R, B\rangle$, such that:

(1) $(L \cup X \cup \{B\}) \subseteq L^{R+}_{PLEN}$ is the set of nodes.
(2) The elements of X are in arbitrary, linear order.
(3) R is a Rule-based Ordering over $L \cup \{B\}$ with respect to the derivation rules of PLEN including the meta-level rules of 6.1.8, and for all $x \in L \cup \{B\}$, XRx.
(4) If $x \in L \cup \{B\} - X$, then for all $A_i \in X$ it holds that $A_i Rx$.
(5) The sequence of nodes in X is the *trunk* of the tree, and each sequence of R-ordered nodes is a *branch*. Each branch extends from the trunk, but

distinguish a *core branch* composed only of the trunk, axiom nodes, and nodes added by derivation rules (including conditional and negation introduction; no assumptions and no formulae derived from assumptions can be on the core branch).

(6) B is the terminal node of the core branch (for all x on the core branch such that $x \neq B, xRB$.

Remark 327 The following definitions state the broader notions of derivability and theoremhood deployed in the following soundness proof. These correspond to broader notions of soundness and validity, presented immediately afterward.

Definition 328 (Revised Derivability) Let $A \in L^{R+}_{PLEN}$, $X \subseteq L^{R+}_{PLEN}$, and $\{\emptyset, A_1, ..., A_n\} = X$. Then:

(1) A is *object-derivable* from X in PLEN ($X \vdash^{O}_{PLEN} A$) iff there is a derivation of A from X:

$$\langle L, X = \{\emptyset, A_1, ..., A_n\}, R, A\rangle$$

where $A \in L^{R}_{PLEN}$, $X \subseteq L^{R}_{PLEN}$, and R is characterized only by object-level rules and axioms.

(2) A is *meta-derivable* from X in PLEN ($X \vdash^{M}_{PLEN} A$) iff there is a derivation of A from X:

$$\langle L, X = \{\emptyset, A_1, ..., A_n\}, R, A\rangle$$

where R is characterized only by meta-level (counterpart) rules and axioms.

(3) A is *derivable* from X in PLEN ($X \vdash_{PLEN} A$) iff there is a derivation of A from X:

$$\langle L, X = \{\emptyset, A_1, ..., A_n\}, R, A\rangle$$

where R is characterized by either or both meta-level or object-level rules and axioms.

Definition 329 (Revised Theorem) For any $A \in L^{R+}_{PLEN}$:

(1) A is an *object-level theorem* in PLEN ($\vdash^{O}_{PLEN} A$) iff $A \in L^{R}_{PLEN}$ and A is an instance of an object-level axiom or A can be derived from the object-level axioms of PLEN.

(2) A is a *meta-level theorem* in PLEN ($\vdash^{M}_{PLEN} A$) iff A is a meta-level (counterpart) axiom or A can be derived from the meta-level axioms of PLEN.

(3) A is a *theorem* in PLEN ($\vdash_{PLEN} A$) iff A is an axiom of either kind or A can be derived from axioms of either kind in PLEN.

Definition 330 (Logical Truth/Tautology) For any $A \in L^{R+}_{PLEN}$, A is a *logical truth/tautology* in PLEN ($\models_{PLEN} A$) iff A is verified by all states in all EDGs, i.e., every state, s, in every EDG, g, is such that $s \Vdash A$.

Definition 331 (Logical Consequence) For any $X \subseteq L^{R+}_{PLEN}$ and $A \in L^{R+}_{PLEN}$, A is a *logical consequence in PLEN* of $X = \{\emptyset, A_1, ..., A_n\}$ ($X \models_{PLEN} A$) iff for every g and every s in S, if $\forall B_i \subseteq X(s \Vdash B_i)$, then $s \Vdash A$.

Definition 332 (Rule Soundness) A meta-level derivation rule of PLEN, $\dfrac{B_1, ..., B_n}{A}$, is *rule sound* ($X \models^M_{PLEN} A$) iff there is a meta-level proof (a proof via mathematical (i.e., set-theoretical or type theoretical) reasoning) of A from a possibly empty set of the premises, B_1,...,B_n.

(1) Both the rule and the proof can be partially formalized or regimented with the syntax of PLEN, formalized in Type Theory, formalized in first-order logic (where applicable), or formulated entirely in the meta-level language of PLEN.
(2) A derivation of A from X is rule sound in PLEN ($X \models^M_{PLEN} A$) iff every derivation rule applied to construct it is sound.
(3) Meta-level axiom, A, is rule sound in PLEN ($\models^M_{PLEN} A$) iff there is a meta-level proof of A from no premises.

Remark 333 Given a formalization of PLEN in the logic of Type Theory, this just means that the rule is valid in all general models of the logic of Type Theory. All of the forthcoming formalized meta-theoretical reasoning omits mention of types, as they are generally irrelevant to the results being derived.

Remark 334 The core results of the PLEN Framework are proofs of the soundness of meta-level rules; the proofs of PGIIRC, RGSRT, GRT, etc., are proofs of the rule soundness of those results (in their rule form). The soundness proof below relies on the observation that the rule soundness of a meta-level rule implies the rule validity, defined as follows, of their object-level counterparts.

Definition 335 (Rule Validity) An object-level derivation rule or object-level counterpart of a meta-level derivation rule of PLEN, $\dfrac{B_1, ..., B_n}{A}$, is *valid* iff A is a logical consequence in PLEN of $\{B_1, ..., B_n\}$.

(1) If a derivation of A from possibly empty set of premises, X, is formed by means of valid rules only, then A is a logical consequence of X.
    (a) Logical consequence is, as defined, transitive.

Definition 336 (Revised Soundness) If X is a set of protocol-theoretic formulae formed using connectives in $Pop++$, then let X^O be the set of protocol-theoretic formulae in which all occurrences of connectives in $Pop++$ are replaced with connectives in $Pop+$. If X is a set of protocol-theoretic formulae formed using connectives in $Pop+$, then let X^M be the set of protocol-theoretic formulae in which all occurrences of connectives in $Pop+$ are replaced with connectives in $Pop++$. The R-Derivation System is *sound* with respect to the revised PLEN semantics iff:

(1) If $X \vdash^O_{PLEN} A$, then $X \models_{PLEN} A$.
(2) If $X \vdash^M_{PLEN} A$, then $X \models^M_{PLEN} A$.
(3) If $X \vdash_{PLEN} A$, then either $X^M \models_{PLEN} A^M$ or $X^O \models_{PLEN} A^O$.

7.2.1 Revised Soundness Theorem

Lemma 337 *The following rules hold for derivability:*

(1) If $X \vdash^{O}_{PLEN} A$, then $X \vdash_{PLEN} A$.
(2) If $X \vdash^{M}_{PLEN} A$, then $X \vdash_{PLEN} A$.
(3) If $X_1 \vdash^{O}_{PLEN} A_1, ..., X_k \vdash^{O}_{PLEN} A_k$, and $X_l \vdash^{M}_{PLEN} A_l, ..., X_m \vdash^{M}_{PLEN} A_m$, then $X_1, ..., X_m \vdash_{PLEN} A_1, ..., X_1, ..., X_m \vdash_{PLEN} A_m$, and, for any $A_i, ..., A_j \in \{A_1, ..., A_m\}$, if $A_i, ..., A_j \vdash^{O}_{PLEN} B$ or $A_i, ..., A_j \vdash^{M}_{PLEN} B$, then $X_1, ..., X_m \vdash_{PLEN} B$.

Proof Clauses 1 and 2 of the lemma are immediate from the inclusivity of clause 3 of the definition of derivability. Assume the antecedent of clause 3. The first conjunct of the consequent of clause 3 is immediate from 1 and 2 by monotonicity. For the second conjunct of the consequent of clause 3, the result is immediate from the construction of a derivation of B from $X_1, ..., X_m$ by either of two procedures:

(1) Sequential composition of the proofs $X_1, ..., X_m \vdash_{PLEN} A_1, ..., X_1, ..., X_m \vdash_{PLEN} A_m$, and $A_i, ..., A_j \vdash^{O}_{PLEN} B$ or $A_i, ..., A_j \vdash^{M}_{PLEN} B$
(2) Explicitly treating $A_1, ..., A_n$ as lemmata in a derivation of B from $X_1, ..., X_m$

□

Theorem 338 (Revised Soundness Theorem) *If X is a set of protocol-theoretic formulae formed using connectives in $Pop++$, then let X^O be the set of protocol-theoretic formulae in which all occurrences of connectives in $Pop++$ are replaced with connectives in $Pop+$. If X is a set of protocol-theoretic formulae formed using connectives in $Pop+$, then let X^M be the set of protocol-theoretic formulae in which all occurrences of connectives in $Pop+$ are replaced with connectives in $Pop++$. Then:*

(1) If $X \vdash^{O}_{PLEN} A$, then $X \models_{PLEN} A$.
(2) If $X \vdash^{M}_{PLEN} A$, then $X \models^{M}_{PLEN} A$.
(3) If $X \vdash_{PLEN} A$, then either $X^M \models_{PLEN} A^M$ or $X^O \models_{PLEN} A^O$.

Proof (Proof-Plan) Direct proof that each meta-level rule is sound, that each meta-level axiom is true, that each object-level rule is valid, and that each object-level axiom is logically true.

From the definition of R-Derivations, it follows that, if there is a derivation of A from X and every rule and axiom schema is sound/valid, then the truth of each element of X necessarily preserves the truth of A in both the object-level semantics of PLEN and in the meta-theory (i.e., Type Theory). □

(1) If $X \models^{M}_{PLEN} A$, then $X^O \models_{PLEN} A^O$.

   (a) First, the rule soundness of the meta-level derivation rules governing protocol graph isomorphism equivalence and implementation mechanically entails the rule validity of the object-level counterparts of the meta-level derivation rules; the derivations are often just a matter of instantiation. Overall, the rule validity of each object-level rule is derivable from the soundness of the corresponding meta-level rule;[1] in other words, if each step of the meta-level derivation is rule-sound, then the translation of the derivation into object-level counterparts will produce a derivation in which each step is rule valid.

(b) Here is the outline of one path: PRT allows the sound derivation of $\pi_1 \equiv_p \pi_2$ from $\pi_1 \equiv_G \pi_2$, while $\pi_1 \equiv_p \pi_2$ can be used to soundly derive the truth of $\pi_1 \equiv^s_p \pi_2$ at every state of every EDG. Further, PRT shows that the truth of $\pi_1 \equiv^s_G \pi_2$ at any state of any EDG can, in turn, can be used to soundly derive $\pi_1 \equiv_G \pi_2$ (and converse). Suppose that $\pi_1 \equiv^s_G \pi_2$ is true at some s of some g. The foregoing path of sound meta-level derivation shows that $\pi_1 \equiv^s_p \pi_2$ is true at every state of every EDG, and thus also at s in g. Thus, the derivation of $\pi_1 \equiv^s_p \pi_2$ from $\pi_1 \equiv^s_G \pi_2$ is valid (at the object-level). Generally, where premise sets formed from main operator meta-level operators can be used to soundly derive a meta-level formula, A_M, the meta-level premise set is soundly derivable from its object-level counterpart (by PRT), and A_M can be used to soundly derive its object-level counterpart, A_O, it follows immediately that the object-level counterpart of the premise set can be used to soundly derive A_O. This provides a plan for deriving the validity of object-level rules from their meta-level counterparts that works for many of the protocol graph isomorphism rules; it does not work for all of the rules (e.g., the implementation rules), which can be proven valid independently.

(c) Here is the outline of another path: The object-level rules can be proven valid independently. Meta-level formulae are true or false independent of valuations at states; each instance is true for all states in all EDGs. This fact, together with the definition of object-level counterparts renders the result almost immediate. The meta-level truth of X preserves meta-level truth for A (i.e., they're related by derivability or consequence in a higher order logic). From the definition of object-level counterparts, if each $B_i \in X$ is true, then each $B_i^O \in X^O$ is such that $\models_{PLEN} B_i^O$. The same follows for A and A^O. Thus, suppose $X \models^M_{PLEN} A$. Let it be the case that X^O is true at some state of some EDG. If each $B_i \in X$ is true, it immediately follows that: $\models_{PLEN} X^O$, A (from the truth of the antecedent), and $\models_{PLEN} A^O$ (from the truth of A), as required. If some $B_i \in X$ is false, then A^O will follow from X^O by means of rule valid object-level rules governing the operators in $Pop+$.

(2) If $X \models_{PLEN} A$, then $X^M \models^M_{PLEN} A^M$.

(a) The meta-level counterparts of object-level formulae recast the semantics of the latter in terms of the graph-theoretic structures of EDGs, the properties of their substructures (e.g., paths, edges, and transitions), and the properties of their valuation functions by means of the formulae (optionally abstracting out the protocols in the formulae). For example, for a set of

[1] The plan of derivation is indicated below.

protocol-theoretic formulae, X, and formula, A, the semantics of X and A are, at the meta-level, really statements to the (abstract) effect that:

(i) For any EDG, g, and state s, if $X(g, s)$, then $A(g, s)$ where these constructions denote some relation holding between X (resp. A) and the pointed EDG (g, s), e.g., verification at s.

(ii) For any class of EDGs, $\{g_1, ..., g_n\}$, and states $s_1^i, ..., s_m^i \in S_i$, if $X(g_1, ..., g_n, s_1, ..., s_m)$, then $A(g_1, ..., g_n, s_1, ..., s_m)$.

(iii) For any protocols, π_1,...,π_n, class of EDGs, $\{g_1, ..., g_n\}$, and states $s_1^i, ..., s_m^i \in S_i$, if

$$X(\pi_1, ..., \pi_k, g_1, ..., g_n, s_1, ..., s_m)$$

then $A(\pi_1, ..., \pi_k, g_1, ..., g_n, s_1, ..., s_m)$.

(b) Thus, every valid object level rule can be recast as a sound meta-level rule characterizing the graph-theoretic structures of EDGs, the properties of their substructures (e.g., paths, edges, and transitions), and the properties of their valuation functions by means of the formulae (optionally abstracting out the protocols in the formulae).

(3) If a derivation of A from possibly empty set of premises, X, is formed by means of sound rules and axioms only, then A is a logical consequence or theorem of X.

(a) From the prior Lemmata, if there are derivations of $A_1, ..., A_n$ from X and B from $A_1, ..., A_n$, a derivation of B from X can be built—either by sequential composition of the proofs or otherwise explicitly treating $A_1, ..., A_n$ as lemmata in a derivation of B from X. For meta-level derivation steps, the proofs that make them sound can themselves be chained by composition or lemmatizing. Rule Soundness is thus transitive by construction, and the claim follows from 1 and the transitivity of logical consequence or rule validity.

Thus, given the revised definition of derivations, it suffices for a derivation of soundness to show that:

(1) If $X \vdash^M_{PLEN} A$, then $X \models^M_{PLEN} A$.
(2) If $X \vdash^O_{PLEN} A$, then $X \models_{PLEN} A$.

These clauses of the soundness theorem are immediate from the core results of the prior chapters:

(Meta-level Graph-theoretic Protocol Equivalence Rules): The core protocol equivalence rules are shown sound in Chap. 6 or are immediate from prior results.

RGSRT: Derived from PGIIRC, which is derived from the definitions of PLEN in prior sections, and proven directly by means of the lemmata derived from PLEN in Sects. 6.3 and 6.4.

GSRT: Derived from RGSRT and IHT, each derived from prior results or directly from the definitions of PLEN in Sect. 6.4.

GRT: Derived from RGSRT, PRT, and the protocol equivalence properties proven in Sect. 6.4.
PRT: Derived directly from the definitions of PLEN in Sect. 6.2.
PGER: Derived directly from the definitions of PLEN in Sect. 6.2.
PGES: Derived directly from the definitions of PLEN in Sect. 6.2.
PGET: Derived directly from the definitions of PLEN in Sect. 6.2.
PR & RP: Trivial from Restriction Rules.
PI: Trivial from Restriction Rules.
PNEC, TNEC, RNEC: Trivial quantificational implications.

(Object-level Graph-theoretic Protocol Equivalence Rules): (Proof-plan) The core graph-theoretic protocol equivalence results of PLEN directly imply their object-language counterparts by means of the useful lemmata and other core results of PLEN. Recall that $\pi \equiv_G \pi'$ iff $\pi \equiv^s_G \pi'$ iff $\pi \equiv_{GGI} \pi'$ by PRTC. Thus, each meta-level rule (save the implementation rules) implies its object-level counterpart by immediate quantificational reasoning, PRTC, and PRT4:

$$\left[\begin{array}{ll}
\dfrac{\pi \equiv_G \pi',\ \rho(\pi/\pi')(X_\pi) = X_{\pi'},\ \pi \in ISub(X_\pi)}{X_\pi \equiv_G X_{\pi'}} & \dfrac{\pi \equiv_G \pi',\ \pi' \equiv_G \pi''}{\pi \equiv_G \pi''} \\
\times(RGSRT) & \times(PGET) \\[1em]
\dfrac{\pi \equiv_G \pi',\ \rho(\pi/\pi')(X_\pi) = X_{\pi'}}{X_\pi \equiv_G X_{\pi'}}(GSRT) & \dfrac{\pi \equiv_G \pi'}{\pi \equiv_p \pi'}(PRT1.1) \\[1em]
\dfrac{\pi \equiv_G \pi',\ \rho(\pi/\pi')(X_\pi) = X_{\pi'},\ A_{(X_\pi)}}{A_{(X_{\pi'})}}(GRT) & \dfrac{\pi \equiv_p \pi'}{\pi \equiv_c \pi'}(PRT1.2) \\[1em]
\dfrac{}{\pi \equiv_G \pi}(PGER) & \dfrac{\pi \equiv_c \pi'}{\pi \equiv_t \pi'}(PRT1.3) \\[1em]
\dfrac{\pi \equiv_G \pi'}{\pi' \equiv_G \pi}(PGES) & \dfrac{\pi \leftrightarrow_r \pi'}{\pi = \pi'}(PI)
\end{array}\right]$$

$$\Downarrow$$

$$\left[\begin{array}{ll}
\dfrac{\pi \equiv^s_G \pi',\ \rho(\pi/\pi')(X_\pi) = X_{\pi'},\ \pi \in ISub(X_\pi)}{X_\pi \equiv^s_G X_{\pi'}} & \dfrac{\pi \equiv^s_G \pi',\ \pi' \equiv^s_G \pi''}{\pi \equiv^s_G \pi''} \\
\times(RGSRT) & \times(PGET) \\[1em]
\dfrac{\pi \equiv^s_G \pi',\ \rho(\pi/\pi')(X_\pi) = X_{\pi'}}{X_\pi \equiv^s_G X_{\pi'}}(GSRT) & \dfrac{\pi \equiv^s_G \pi'}{\pi \equiv^s_p \pi'}(PRT1.1) \\[1em]
\dfrac{\pi \equiv^s_G \pi',\ \rho(\pi/\pi')(X_\pi) = X_{\pi'},\ A_{(X_\pi)}}{A_{(X_{\pi'})}}(GRT) & \dfrac{\pi \equiv^s_p \pi'}{\pi \equiv^s_c \pi'}(PRT1.2) \\[1em]
\dfrac{}{\pi \equiv^s_G \pi}(PGER) & \dfrac{\pi \equiv^s_c \pi'}{\pi \equiv^s_t \pi'}(PRT1.3) \\[1em]
\dfrac{\pi \equiv^s_G \pi'}{\pi' \equiv^s_G \pi}(PGES) & \dfrac{\pi \leftrightarrow_r \pi'}{\pi = \pi'}(PI)
\end{array}\right]$$

(Meta-level Implementation Rules): The proofs of the Useful Lemmata contain derivations of each meta-level implementation rule, thereby establishing their rule soundness:

$$\left[\begin{array}{l}
\dfrac{\pi \equiv_G \pi',\ \rho(\pi/\pi')(X_\pi) = X_{\pi'},\ \pi \in ISub(X_\pi)}{\pi{+} \mapsto X_\pi \equiv \pi{+} \mapsto X_{\pi'}}(NSTLL) \\[2ex]
\dfrac{\pi{+} \mapsto \pi_1,\ (\pi{+}{+} \in Sub(\pi{+}) \rightarrow \neg(\pi{+}{+} \mapsto \pi_2)),\ pre(\pi{+}) \rightarrow \langle(\langle\pi_2\rangle_p \top)?\rangle_p \top)}{\pi{+} \mapsto (\pi_1 \cup \pi_2)} \\
\times(I1.1) \\[2ex]
\dfrac{\begin{array}{r}\pi{+} \mapsto \pi_1 \wedge pre(\pi{+}) \rightarrow \langle(\langle\pi_2\rangle_p \top)?\rangle_p \top)),\ \pi{+}{+} \mapsto \pi_2 \wedge pre(\pi{+}) \\ \rightarrow \langle(\langle\pi_1\rangle_p \top)?\rangle_p \top))\end{array}}{\neg((\pi{+} \cup \pi{+}{+}) \mapsto (\pi_1 \cup \pi_2))} \\
\times(I1.2) \\[2ex]
\dfrac{\pi \mapsto (\pi_1 \cup \pi_2)}{\bigvee\left[\begin{array}{l}(\pi \mapsto \pi_1 \wedge pre(\pi{+}) \rightarrow \langle(\langle\pi_2\rangle_p \top)?\rangle_p \top) \\ (\pi \mapsto \pi_2 \wedge pre(\pi{+}) \rightarrow \langle(\langle\pi_1\rangle_p \top)?\rangle_p \top)\end{array}\right]}(I1.3) \\[2ex]
\dfrac{\pi{+} \mapsto \pi_1,\ \pi{+}{+} \mapsto \pi_2}{\pi{+};\pi{+}{+} \mapsto \pi_1;\pi_2}(I2) \\[2ex]
\dfrac{\pi{+};\pi{+}{+} \mapsto \pi_1;\pi_2}{\pi{+} \mapsto \pi_1 \wedge \pi{+}{+} \mapsto \pi_2}(I2C) \\[2ex]
\dfrac{\pi{+} \mapsto \pi_1,\ \pi{+}{+} \mapsto \pi_2}{\pi{+} \cap \pi{+}{+} \mapsto \pi_1 \cap \pi_2}(I3) \\[2ex]
\dfrac{\pi{+} \cap \pi{+}{+} \mapsto \pi_1 \cap \pi_2}{\pi{+} \mapsto \pi_1 \wedge \pi{+}{+} \mapsto \pi_2}(I3C)
\end{array}\right]$$

(Object-level Implementation Rules): The rule validity of all of the object-level implementation rules follows from the rule soundness of their meta-level counterparts. The proofs are mechanical applications of the same reasoning as in the derivations of their intended meta-level counterparts:[2]

[2] This is true in spite of $\mapsto_s$ holding for pairs of choice protocols for which $\mapsto$ fails in virtue of the premise sets. The thought is that, for some protocols, π and π', the precondition condition holds at a state s that does not hold generally. However, the premises explicitly state that the precondition condition holds, and so the premises of the object level rule may hold for pairs of protocols for which the premises of the meta-level rule don't. Nonetheless, the reasoning that shows each instance of the meta-level rule sound also shows each instance of the object-level rule valid.

$$\left[\begin{array}{l}
\dfrac{\pi \equiv^{s}_{G} \pi',\ \rho(\pi/\pi')(X_{\pi}) = X_{\pi'},\ \pi \in ISub(X_{\pi})}{\pi + \mapsto_{s} X_{\pi} \equiv \pi + \mapsto_{s} X_{\pi'}}(NSTLL) \\[2ex]
\dfrac{\begin{array}{l}\pi + \mapsto_{s} \pi_1, (\pi + + \in Sub(\pi+) \rightarrow \neg(\pi + + \mapsto_{s} \pi_2)), pre(\pi+) \\ \qquad \rightarrow \left\langle (\langle \pi_2 \rangle_p \top)?\right\rangle_p \top))\end{array}}{\pi + \mapsto (\pi_1 \cup \pi_2)}(I1.1) \\[2ex]
\dfrac{\begin{array}{l}\pi + \mapsto_{s} \pi_1 \wedge pre(\pi+) \rightarrow \left\langle (\langle \pi_2 \rangle_p \top)?\right\rangle_p \top)), \pi + + \mapsto_{s} \pi_2 \wedge pre(\pi+) \\ \qquad \rightarrow \left\langle (\langle \pi_1 \rangle_p \top)?\right\rangle_p \top))\end{array}}{\neg((\pi + \cup \pi + +) \mapsto_{s} (\pi_1 \cup \pi_2))}(I1.2) \\[2ex]
\dfrac{\pi \mapsto_{s} (\pi_1 \cup \pi_2)}{\bigvee \left[\begin{array}{l}(\pi \mapsto_{s} \pi_1 \wedge pre(\pi+) \rightarrow \left\langle (\langle \pi_2 \rangle_p \top)?\right\rangle_p \top) \\ (\pi \mapsto_{s} \pi_2 \wedge pre(\pi+) \rightarrow \left\langle (\langle \pi_1 \rangle_p \top)?\right\rangle_p \top)\end{array}\right]}(I1.3) \\[2ex]
\dfrac{\pi + \mapsto_{s} \pi_1, \pi + + \mapsto_{s} \pi_2}{\pi +; \pi + + \mapsto_{s} \pi_1; \pi_2}(I2) \\[2ex]
\dfrac{\pi +; \pi + + \mapsto_{s} \pi_1; \pi_2}{\pi + \mapsto_{s} \pi_1 \wedge \pi + + \mapsto_{s} \pi_2}(I2C) \\[2ex]
\dfrac{\pi + \mapsto_{s} \pi_1, \pi + + \mapsto_{s} \pi_2}{\pi + \cap \pi + + \mapsto \pi_1 \cap \pi_2}(I3) \\[2ex]
\dfrac{\pi + \cap \pi + + \mapsto \pi_1 \cap \pi_2}{\pi + \mapsto_{s} \pi_1 \wedge \pi + + \mapsto_{s} \pi_2}(I3C)
\end{array}\right]$$

(Graph-theoretic Protocol Equivalence Axioms):

(1) $(\pi_1; \pi_2); \pi_3 \equiv_G \pi_1; (\pi_2; \pi_3)$
(2) $\pi_1; (\pi_2; ...; (\pi_{n-1}; \pi_n)) \equiv_G ((\pi_1; \pi_2); ...;)\pi_{n-1}); \pi_n \ (n < \aleph_0)$
(3) $\pi_1 \cap \pi_2 \equiv_G \pi_2 \cap \pi_1$
(4) $\pi_1 \cup \pi_2 \equiv_G \pi_2 \cup \pi_1$
(5) $(\pi_1; \pi_2); \pi_3 \equiv_G \pi_1; (\pi_2; \pi_3)$
(6) $\pi_1; (\pi_2 \cap \pi_3) \equiv_G ((\pi_1; \pi_2) \cap (\pi_1; \pi_3))$
(7) $\pi_1; (\pi_2 \cup \pi_3) \equiv_G ((\pi_1; \pi_2) \cup (\pi_1; \pi_3))$
(8) $(\pi_1 \cap \pi_2); (\pi_3 \cap \pi_4) \equiv_G (\pi_2 \cap \pi_1); (\pi_4 \cap \pi_3)$
(9) $(\pi_1 \cup \pi_2); (\pi_3 \cup \pi_4) \equiv_G (\pi_2 \cup \pi_1); (\pi_4 \cup \pi_3)$
(10) $(\pi_1 \cap \pi_2) \cap \pi_3 \equiv_G \pi_1 \cap (\pi_2 \cap \pi_3)$
(11) $(\pi_1 \cup \pi_2) \cup \pi_3 \equiv_G \pi_1 \cup (\pi_2 \cup \pi_3)$
(12) $(\pi_1 \cup \pi_2) \cap (\pi_3 \cup \pi_4) \equiv_G ((\pi_1 \cap \pi_3) \cup (\pi_1 \cap \pi_4)) \cup ((\pi_2 \cap \pi_3) \cup (\pi_2 \cap \pi_4))$

(Proof-plan): The $\oplus^{\pi}_{\circ}$ procedure and the PGIIRC, direct RGSRT, and ISOP proofs provide a template for establishing that protocols are graph-theoretically equivalent, which gives the meta-level axioms. The plan involves mechanical demonstration of the structural display rules. The meta-level axioms will then imply their object-level counterparts by means of the same clauses of PRTC used above: $\pi \equiv_G \pi'$ iff $\pi \equiv^{s}_{G} \pi'$ iff $\pi \equiv_{GGI} \pi'$.

Example 339 Prove that $\pi_1; (\pi_2 \cup \pi_3) \equiv_G ((\pi_1; \pi_2) \cup (\pi_1; \pi_3))$ by direct model-theoretic reasoning. For arguments establishing graph-theoretic equivalence by model-theoretic reasoning, see the proofs of ISOP and RGSRT.

Proof Plan: First, LtoR, then RtoL.

(1) LtoR: Let $g \in G^{\pi_1;(\pi_2 \cup \pi_3)}$. Show that the structural display rules hold for g with respect to $(\pi_1; \pi_2) \cup (\pi_1; \pi_3)$.

   (a) (SD1(1) For reductio, suppose that SD1(1) fails for g with respect to $((\pi_1; \pi_2) \cup (\pi_1; \pi_3))$. Then some $\pi_i \in Sub((\pi_1; \pi_2) \cup (\pi_1; \pi_3))$ isn't executable in g. But either π_i is a proper immediate subprotocol of $((\pi_1; \pi_2) \cup (\pi_1; \pi_3))$ or not. If it's a proper immediate subprotocol of $((\pi_1; \pi_2) \cup (\pi_1; \pi_3))$, then it's either $\pi_1; \pi_2$ or $\pi_1; \pi_3$. Trivially, if neither is executable in g, then neither is $\pi_1; (\pi_2 \cup \pi_3)$. $\bot$. If π_i is not an immediate subprotocol of $((\pi_1; \pi_2) \cup (\pi_1; \pi_3))$, then it's a subprotocol of π_1, π_2, or π_3. But then, by the definition of subprotocols, π_i is a subprotocol of $\pi_1; (\pi_2 \cup \pi_3)$. Thus, if π_i is not executable, SD1(1) fails for g with respect to $\pi_1; (\pi_2 \cup \pi_3)$. $\bot$ The remaining subprotocol of $(\pi_1; \pi_2) \cup (\pi_1; \pi_3)$ is itself, which is immediately executable from the execution rules. That gives SD1(1).

   (b) (SD1(2)) Let p execute $\pi_1; (\pi_2 \cup \pi_3)$. Then it is a concatenation of executions of π_1 and $\pi_2 \cup \pi_3$. Any execution of $\pi_1 \cup \pi_3$ is a branching of executions of π_1 and π_3. Thus, $p = qr$ where q executes π_1, r executes π_2 and there is an execution, r', that executes π_3 that branches from the terminal state of q, or $p = qr$ where q executes π_1, r executes π_3 and there is an execution, r', that executes π_2 that branches from the terminal state of q. Thus, $p = qr$ executes $\pi_1; \pi_2$ and $p' = qr'$ executes $\pi_1; \pi_3$ (or vice versa). Thus, p suffices to execute $((\pi_1; \pi_2) \cup (\pi_1; \pi_3))$. Thus, any maximal execution of $\pi_1; (\pi_2 \cup \pi_3)$ in g is an execution of $((\pi_1; \pi_2) \cup (\pi_1; \pi_3))$ in g. The converse is easy to see. Let $p = qr$ where q executes π_1, r executes π_2, and p branches from a distinct $p' = q'r'$ where q' executes π_1 and r' executes π_3. Then, by NST, $q = q'$, so at the root state of q, $\pi_2 \cup \pi_3$ is executable. Thus, any maximal execution of $((\pi_1; \pi_2) \cup (\pi_1; \pi_3))$ in g is an execution of $\pi_1; (\pi_2 \cup \pi_3)$ in g. SD1(2) follows immediately.

   (c) (SD1(3)) From the prior arguments, $\pi_1; \pi_2$ and $\pi_1; \pi_3$ are executable at the root state of any protocol graph of $\pi_1; (\pi_2 \cup \pi_3)$. However, g also contains a state, call it s, at which π_2 and π_3 are both executed by means of distinct paths that branch from a shared root state. Thus, $SD1(3)_{((\pi_1;\pi_2) \cup (\pi_1;\pi_3))}(g)$ holds; though there is no subprotocol such that π_2 and π_3 are joined by parallel execution or choice, nor any subprotocol implemented by both, and $(\pi_1; \pi_2) \cup (\pi_1; \pi_3)$ dictates neither π_2 nor π_3 (nor $\pi_2 \cup \pi_3$), π_2 and π_3 are embedded in $(\pi_1; \pi_2) \cup (\pi_1; \pi_3)$ at equal depth with respect to $\cup$, and so the branching executions of π_2 and π_3 are not superfluous to a protocol graph of $(\pi_1; \pi_2) \cup (\pi_1; \pi_3)$.

(d) (SD1(4)) There are no corresponding executions of any subprotocols of either $\pi_1; (\pi_2 \cup \pi_3)$ or $((\pi_1; \pi_2) \cup (\pi_1; \pi_3))$ that are not shared subprotocols.
(e) (SD2) The implementations of $((\pi_1; \pi_2) \cup (\pi_1; \pi_3))$ are itself and $\pi_1; (\pi_2 \cup \pi_3)$. The converse holds for $\pi_1; (\pi_2 \cup \pi_3)$.
(f) (SD3) The set of atomics in both protocols is identical, so SD3 holds.
(g) (RC) holds trivially; it holds for π_1, π_2, and π_3, and protocol graphs among the subgraphs of g for these subprotocols are linked in such a way that every maximal path of any graph of π_1 is concatenated with a maximal path of a graph of π_2 (same or π_3). The resulting graphs—containing graphs of π_1 and π_2 (resp. π_3)—are such that all of the structural display relations hold for them. Thus, the immediate subprotocols of $((\pi_1; \pi_2) \cup (\pi_1; \pi_3))$ have protocol graphs among the proper subgraphs of g.
(h) (NST) To establish NST, there can be no smaller graph of $((\pi_1; \pi_2) \cup (\pi_1; \pi_3))$ than g; a smaller graph would violate NST with respect to $\pi_1; (\pi_2 \cup \pi_3)$. Suppose that NST fails for g with respect to $((\pi_1; \pi_2) \cup (\pi_1; \pi_3))$. Then there is some proper subgraph of g that satisfies SD1(1,2), SD2, and RC with respect to $((\pi_1; \pi_2) \cup (\pi_1; \pi_3))$. Call that proper subgraph g'. From the prior arguments, SD1(1,2), SD2, and RC all hold for g'. Thus, g fails NST with respect to $\pi_1; (\pi_1 \cup \pi_3)$. $\perp$
(i) (EXC) Any protocol that is correspondence equivalent to $\pi_1; (\pi_2 \cup \pi_3)$ is correspondence equivalent to $((\pi_1; \pi_2) \cup (\pi_1; \pi_3))$, trivially. So, that, finally, gives EXC.

(2) RtoL: Graph reduction will collapse any branching π_1 executions in any graph of $((\pi_1; \pi_2) \cup (\pi_1; \pi_3))$, and no structural display rules will be violated. The resulting graph will immediately satisfy the structural display rules for $\pi_1; (\pi_2 \cup \pi_3)$.

□

The proofs of the implementation rules in 5.7 and the soundness of the basic PLEN derivation system completes[3] the rest of the proof of the R-Derivation System's soundness.

So much for this elaborate, unusual approach to extending the soundness result of PLEN to Revised PLEN. I return to the use of PLEN as a formal framework for examining protocol-theoretic reasoning and modeling assumptions in Sects. 7.3, 7.4, and 7.5, after very briefly speculating about the completeness of the R-Derivation System.[4]

---

[3] I omit the proofs of the modal and propositional rules, as these are standard parts of soundness proofs for modal and propositional logics. I omit proofs for most graph-theoretic protocol equivalence axiom schemata, leaving them to the reader for mechanical tests. Generally, I focus on proofs for rules or axioms are interesting (e.g., in cases where the logics for control operators diverge from familiar dynamic logics). I omit almost all proofs for the binary axioms, as they are easily extractable from the proofs of the other axioms or familiar from PDL.

[4] The system just shown sound is surely not complete. An unusual and undesirable feature of the blending of meta and object-level rules is that there are instances of valid object-level rules that

7.2.2 *Revised Completeness?*

The extension of the completeness conjecture from PLEN to Revised PLEN is more limited. Where the regular fragment of the basic derivation system of PLEN is plausibly complete, the completeness theorem for the R-Derivation System PLEN certainly fails. The small model theorem fails, as no EDGs containing states that verify graph-theoretic equivalences for protocols containing iterations can be finite. This is because protocol graphs for iterations are not finite graphs. Thus, the naive effective procedure devised for the basic system never terminates for graphs verifying formulae as simple as $a* \equiv_G a*$. Thus, decidability fails and the argument for the verifiability lemma fails. Thus, the argument for completeness in Technical Appendix 1 breaks down. It's unclear whether other arguments would work.

However, we can make completeness plausible for a subsystem of the regular fragment of the revised PLEN framework. In particular, if we exchange the finite iteration operator, $*$, with a bounded iteration operator, $*_n$, that specifies a finite number of iterations instead of all finite iterations, we might be able to reinstate the relevant lemmata.

First, redefine P_{PLEN} into P^R_{PLEN} (in infix notation for the sake of brevity):

$$(a_i) \mid (\pi_1; \pi_2) \mid (\pi_1 \cap \pi_2) \mid (\pi_1 \cup \pi_2) \mid (\pi_1 *_n) \mid (A)? \mid (-\pi_1) \mid (\curvearrowleft \pi_1) \mid (\nabla \pi_1)$$

Claim 340 The equivalent string-algebraic version is straightforward; just replace $*$ in $Prot$ with $*_n$.

---

are seemingly non sequiturs. For instance, every true syntactic replacement formula entails every other. Consider the rule: $\dfrac{\rho(\pi_1/\pi_2)(\pi_1; \pi_3) = \pi_2; \pi_3}{\rho(\pi_4/\pi_5)(\pi_4 \cup \pi_6) = \pi_5 \cup \pi_6}$. This rule is object-level sound:

$$\rho(\pi_1/\pi_2)(\pi_1; \pi_3) = \pi_2; \pi_3 \models_{PLEN} \rho(\pi_4/\pi_5)(\pi_4 \cup \pi_6) = \pi_5 \cup \pi_6$$

All instances of the premise and all instances of the conclusion are true by the definition of syntactic replacement. The truth rules are defined in terms of syntactic replacement, and so there is no state in any EDG in which (any instance of) the premise is true and (any instance of) the conclusion is false. The rule is also rule sound, as it's a trivial consequence of the definition of syntactic replacement:

$$\rho(\pi_1/\pi_2)(\pi_1; \pi_3) = \pi_2; \pi_3 \models^M_{PLEN} \rho(\pi_4/\pi_5)(\pi_4 \cup \pi_6) = \pi_5 \cup \pi_6$$

The rule can be characterized as a tautological property of valuations at states; assuming that a valuation, v_s, at an arbitrary s in an arbitrary g gives (an arbitrary instance of) the premise the value 1, it it can be shown that v_s also gives (an arbitrary instance of) the conclusion the value 1. But there is no such rule in the derivation system defined over L^{R+}_{PLEN} by the meta-level and object-level derivation rules. Indeed, there are no rules governing inference to syntactic replacement in the system, and so there can't be a derivation in the system of this conclusion from this premise.

Now, a cascade of revisions of other definitions relevant to the completeness arguments presented in Chap. 4:

Definition 341 Let L^R_{PLEN} be defined over P^R_{PLEN}, and let the rest of the revised PLEN framework be defined as in prior sections of this appendix.

Definition 342 Revise the execution conditions for finite iterations for bounded iterations:

$$P_{\pi *_n} = \left\{ p \in P_g \mid \exists \langle q_1, ..., q_n \rangle \, (p = q_1 ... q_n, \forall i \leq n(q_i \in P_\pi)) \right\}$$

Definition 343 For bounded iterations, $\pi *_n$, revise the definition of formula length and the relevant clause of the definition of Fischer-Ladner Closure:

(1) $len(\pi *_n) = len(\pi) + n$
(2) $FL_\pi([\pi_1 *_n]_x A) = \{[\pi_1 *_1]_x A\} \cup ... \cup \{[\pi_1 *_n]_x A\} \cup FL_\pi([\pi_1]_x [\pi_1 *_{n-1}]_x A)$

Definition 344 (Revised Fischer-Ladner Closure) Redefine FL_π, so that:

$$FL_\pi([\pi_1 *_n]_x A) = \{[\pi_1 *_1]_x A\} \cup ... \cup \{[\pi_1 *_n]_x A\} \cup FL_\pi([\pi_1]_x [\pi_1 *_{n-1}]_x A)$$

Definition 345 (Revised Fischer-Ladner Lemma (RFLL)) RFLL is identical to FLL of Chap. 4 save for the iteration clause , which is replaced with:

(1) If $[\pi_1 *_n]_x A \in FL(B)$, then $[\pi_1 *_1]_x A, ..., [\pi_1 *_n]_x A, [\pi_1]_x [\pi_1 *_{n-1}]_x A \in FL(B)$.

Conjecture 346 (Revised Fischer-Ladner Lemma (RPFLL)) If the argument for FLL is sound, then the argument for RFLL, which is identical but for replacement of the iteration clause, is also sound.

Definition 347 (Revised EDG Filtration) Revise the notion of filtrations on EDGs with the clauses:

(1) $(*_n) \; T_{\pi *_n / FL(A)} =$

$$\left\{ ([s],[t]) \in T_{g/FL(A)} \mid \left[\begin{array}{l} \exists [x_1]...[x_n](\forall i \leq n([x_i] \in S_{FL(A)})) \\ \bigwedge \left[\begin{array}{l} [s] = [x_1], [s'] = [x_n] \\ \forall i \leq n-1 \, \langle [x_i], [x_{i+1}] \rangle \in T_{\pi/FL(A)} \end{array} \right] \end{array} \right] \right\}$$

(2) $(*_n) \; R{+}_{\pi * / FL(A)} =$

$$\left\{ \langle [s], \pi *_n, [s'] \rangle \in R{+}_{0/FL(A)} \mid \left[\begin{array}{l} \exists [x_1]...[x_n](\forall i \leq n(x_i \in S_{FL(A)})) \\ \bigwedge \left[\begin{array}{l} [s] = [x_1], [s' = x_n], \\ \forall i \leq n \, \langle [x_i], \pi, [x_{i+1}] \rangle \in R{+}_{\pi/FL(A)} \end{array} \right] \end{array} \right] \right\}$$

(3) $(*_n)$ $P_{\pi*/FL(A)} =$

$$\{p \in P_{g/FL(A)} \mid \exists \langle q_1, \ldots, q_n \rangle (p = q_1 \ldots q_n, \forall i \leq n(q_i \in P_{\pi/FL(A)}))\}$$

Lemma 348 (Revised PLEN Filtration Lemma (RPFL)) *If the argument for PFL is sound, then the argument for RPFL, which is identical but for replacement of the iteration clause, is also sound.*

Proof (Sketch) The preservation of graph-theoretic structure in the definition of filtrations for PLEN works unchanged; in particular, the requirement that the paths in the filtration of g have isomorphic counterparts in g and vice versa extends the preservation of verification properties of filtrations to the graph-theoretic protocol equivalence formulae by Lemma L4 in the SUT and PGIIRC proofs. □

In sum, the idea is that slotting these revised definitions into the arguments in Appendix 1 should preserve their validity. If the argument in Appendix 1 sufficed to establish (e.g.) the Filtration Lemma for PLEN, then the revised argument should suffice as well for the resulting (e.g.) Revised Filtration Lemma. If this all works, then the Small Model Property can be extended to the revised system, and, from there, the completeness and decidability arguments should, if they worked for the basic form of PLEN, work for the further revision of Revised PLEN with bounded iterations.

7.3 Protocol Logic and Distinctive Argument Forms

The formalization of D1-D3 is one of the main projects of this book, as it displays the usefulness of PLEN for representing and reasoning about epistemic norms. Given that protocols represent procedural epistemic rationality norms, and given that the R-Derivation System is sound, we are now in a place where we can start investigating PLEN's prospects for formalizing the distinctive forms of reasoning Chap. 3 argued are central to thinking about and with epistemic norms. In other words, this section explores ways of solving the problem of formalizing D3 and D3C1:

Problem 349 Formalize the following cluster of principles of epistemic proceduralism.

(1) (D3) Epistemic rationality norms can be reasoned about explicitly; terms denoting epistemic rationality norms and statements about the properties of epistemic norms can be embedded in valid arguments.
    (a) Epistemic rationality norms can be equivalent in multiple, distinct ways; reasoning about epistemic rationality must at least sometimes involve reasoning about these equivalence relations.

(b) The essential forms of reasoning about and with epistemic rationality norms include:
   (i) Arguments establishing norm equivalence and deriving facts about norms from premises about norm equivalence.
      (A) Arguments from the features of norms, norm-kernels, or restrictions to norm equivalence.
      (B) Arguments from protocol equivalence to propositions about protocols, protocol functions, and restrictions on procedural construction, processes, transitions, and states.
      (C) Arguments replacing embedded norms with equivalents in normative, procedural, deontic, modal, and propositional contexts.
   (ii) Arguments deriving normative verdicts (valences, deontic statuses, procedural requirements) from conditions of application, procedural structure, sentential or propositional contexts (including modal and deontic propositions).
   (iii) Arguments establishing dependence, sufficiency, and independence relations among norms.

(2) (D3C1) Where N_i is a term denoting an arbitrary norm, A_i is a formula embedding a norm term (letting $A(N_i, x)$ be a special case encoding N_i's verdict on an arbitrary epistemic factor, x), C_i is an arbitrary formula stating some condition, and $\equiv_x$ is an arbitrary form of normative equivalence, the D3C1 Table (re: Chap. 3) summarizes and explicates schematic inference rules for the foregoing kinds of reasoning.

Remark 350 The foregoing table displays some of the forms of reasoning about epistemic rationality norms the formalization of which is criterial for a logic of epistemic norms. Or so it's argued in Chaps. 2 and 3. The rules introduced in this section formalize many instances of these argument forms. I don't explicitly present illustrations for each form, but it's not difficult to apply the R-Derivation System to the task.

7.3.1 *Regimentation/Formalization of D3 and D3C1: The Protocol-Theoretic Logic of Epistemic Norms I*

The thought is that the formal assertion that corresponds to D3 and D3C1, as per the specification of adequacy, is this: for each specified form or class of arguments indicated in D3 and D3C1, the logic of PLEN provides a demonstrably valid argument form in which formal statements encoding assertions about protocols and their properties (syntactic, graph-theoretic, execution-theoretic, preconditions, postconditions, etc.) are embedded in the premises and conclusions. These argument forms come in both meta-level and object-level forms. Let me outline the way that reasoning about epistemic rationality norms is formalized in PLEN and then illustrate on a cluster of argument forms for doing so. Explicitly:

Solution 351 The theory of protocol equivalence in PLEN and the R-Derivation system immediately formalize the forms of reasoning articulated by D3 and D3C1:

(1) (D3F) Protocols can be reasoned about explicitly in PLEN; terms denoting protocols and statements about the properties of protocols can be embedded in valid arguments.

   (a) (D3.1F) Protocols can be equivalent in multiple, distinct ways; reasoning about protocols must at least sometimes involve reasoning about these equivalence relations.
   (b) (D3.2F) The essential forms of reasoning about and with protocols include:

      (i) Arguments establishing protocol equivalence of protocols and deriving facts about protocols from premises about protocol equivalence.

         (A) Arguments from the features of protocols, protocol functions, or restrictions on procedural construction and epistemic dynamics defined by protocols to protocol equivalence.
         (B) Arguments from protocol equivalence to propositions about protocol functions.
         (C) Arguments replacing embedded protocols with equivalents in meta-level, procedural, deontic, modal, and propositional contexts.

      (ii) Arguments deriving protocol function verdicts (valences, deontic statuses, procedural requirements) from conditions of application, procedural structure, sentential or propositional contexts (including modal and deontic propositions) concerning protocols.
      (iii) Arguments establishing dependence, sufficiency, and independence relations among protocols.

(2) (D3C1F) The R-Derivation System formalizes most of the schematic forms of reasoning articulated in the D3C1 Table, and provides a framework for building complex arguments therefrom. Some of the schematic argument forms are immediate, and the remaining forms can be immediately derived. See the D3C1F Table 7.1.

Interpreting the D3C1F Table: Each rule is a schematic argument form encoding the logical properties of sentential contexts embedding norm/protocol terms that encode the properties of norms/protocols.

(1) Column 1: General Logical Properties

   (a) Row 1: Argument form encoding how arbitrary sentential contexts for norms/protocols combine with arbitrary conditions to imply sentential contexts embedding norms/protocols

Table 7.1 D3C1F table

Basic forms, verdicts, and properties	Equivalence	(In) Dependence, sufficiency
$\frac{A_1(\pi_1) \quad C}{A_2(\pi_2)}$	$\left[\frac{\pi \equiv_G \pi',\ \rho(\pi/\pi')(X_\pi) = X_{\pi'},\ \pi \in ISub(X_\pi)}{X_\pi \equiv_G X_{\pi'}}\right.$ $\frac{\pi \equiv_G \pi',\ \rho(\pi/\pi')(X_\pi) = X_{\pi'}}{X_\pi \equiv_G X_{\pi'}}$ $\frac{\pi \equiv_G \pi',\ \rho(\pi/\pi')(X_\pi) = X_{\pi'},\ A_{(X_\pi)}}{A_{(X_{\pi'})}}$ $\left.\frac{\pi_1 \equiv_G \pi_2}{\pi_1 \supset_G \pi_2}, \frac{\pi_1 \equiv_G \pi_2}{\pi_2 \supset_G \pi_1}\right]$	$\frac{\pi_1 \supset_x \pi_2 \quad A_1(\pi_1)}{A_2(\pi_2)}$
$\frac{A_1(\pi_1) \quad C_1}{C_2}$	$\left[\frac{\pi \equiv_G \pi',\ \rho(\pi/\pi')(X_\pi) = X_{\pi'},\ \pi \in ISub(X_\pi)}{X_\pi \equiv_G X_{\pi'}}\right.$ $\frac{\pi \equiv_G \pi',\ \rho(\pi/\pi')(X_\pi) = X_{\pi'}}{X_\pi \equiv_G X_{\pi'}}$ $\frac{\pi \equiv_G \pi',\ \pi' \equiv_G \pi''}{\pi \equiv_G \pi''}$ $\frac{}{\pi \equiv_G \pi}$ $\left.\frac{\pi \equiv_G \pi'}{\pi' \equiv_G \pi}\right]$	$\frac{\neg(\pi_1 \supset_x \pi_2) \wedge \neg(\pi_2 \supset_x \pi_1) \quad A_1(\pi_1)}{A_2(\pi_2)}$
$\frac{C_1 \quad C_2 \quad \ldots}{A(\pi)}, \frac{C_1 \quad C_2 \quad \ldots}{A(\pi, x)}$	$\left[\frac{\pi \equiv_G \pi'}{\pi \equiv_p \pi'} \quad \frac{\pi_1 \supset_G \pi_2,\ \pi_2 \supset_G \pi_1}{\pi_1 \equiv_G \pi_2}\right.$ $\frac{\pi \equiv_p \pi'}{\pi \equiv_c \pi'}$ $\frac{\pi \equiv_c \pi'}{\pi \equiv_t \pi'}$ $\left.\frac{\pi \leftrightarrow_r \pi'}{\pi = \pi'}\right]$	$\left[\frac{\pi_1 \equiv_G \pi_2}{\pi_1 \supset_G \pi_2}, \frac{\pi_1 \equiv_G \pi_2}{\pi_2 \supset_G \pi_1}\right]$
$\frac{A(\pi_1, x) \quad C}{A(\pi_2, x)}$		$\frac{A_1(\pi_1) \quad A_2(\pi_2) \quad \ldots}{\neg(\pi_1 \supset_x \pi_2) \wedge \neg(\pi_2 \supset_x \pi_1)}$

(b) Row 2: Argument form encoding how arbitrary sentential contexts for norms/protocols combine with arbitrary conditions to imply other arbitrary conditions (not embedding norms/protocols)

   (c) Row 3: Argument form encoding how arbitrary conditions may be sufficient conditions for a given sentential context for norms/protocols, with the special case of the sufficient conditions for sentential contexts encoding normative verdicts
   (d) Row 4: Argument form encoding how the sentential contexts for norms/protocols encoding normative verdicts combine with arbitrary conditions to imply other sentential contexts for norms/protocols encoding normative verdicts
(2) Column 2: Logic of Norm/Protocol Equivalence
   (a) Row 1: Argument form encoding how sentential contexts for norms/protocols that encode different types of norm/protocol equivalence combine with arbitrary sentential contexts for norms/protocols to imply sentential contexts embedding norm/protocol
   (b) Row 2: Argument form encoding how arbitrary sentential contexts for norms/protocols encode sufficient conditions for norm/protocol equivalences of different types
   (c) Row 3: Argument form encoding logical relations among sentential contexts for norms/protocols that encode different types of norm/protocol equivalence
(3) Column 3: Logic of Norm/Protocol Dependence, Sufficiency, and Independence
   (a) Row 1: Argument form encoding how sentential contexts for norms/protocols that encode different types of norm implication/sufficiency/dependence combine with arbitrary sentential contexts for norms/protocols to imply sentential contexts embedding norms/protocols
   (b) Row 2: Argument form encoding how sentential contexts for norms/protocols that encode different types of norm implication failure/independence/insufficiency combine with arbitrary sentential contexts for norms/protocols to imply sentential contexts embedding norms/protocols
   (c) Row 3: Argument form encoding how arbitrary sentential contexts for norms/protocols may be sufficient for different types of norm/protocol implication/sufficiency/dependence
   (d) Row 4: Argument form encoding how arbitrary sentential contexts for norms/protocols may be sufficient for different types of norm/protocol implication failure/insufficiency/independence

Claim 352 Formalizing arguments of the forms in the D3C1 table for multiple kinds of protocol equivalence suffices to formalize the D3 Corollary; there are sets of valid rules and laws corresponding to the forms of argument involving norm equivalence. The R-Derivation System of PLEN provides the codification.

First, there are items (protocols) in the object language that represent norms, and these are embedded in assertions and sets of assertions in the object language that encode the logical properties of epistemic rationality norms (protocol-theoretic formulae). For now, it's taken for granted that protocols are apt constructions for representing procedural epistemic norms, but note that Chap. 9 explicitly defends this assumption. Protocols represent epistemic rationality norms, and, because protocols are directly built into expressions in L^R_{PLEN}, protocol-theoretic formulae are naturally read as statements about the properties of protocols, including local forms of equivalence, convergence, divergence, preconditions, and postconditions. Thus, protocol-theoretic formulae, under the intended interpretation of PLEN, are naturally read as statements about the corresponding properties of epistemic rationality norms. From this, it follows that the logic of PLEN—codified in part by the R-Derivation System—is a logic of epistemic rationality norms.

These meta-level rules and the R-Derivation System supported by them provide the logical apparatus in PLEN to formalize arguments establishing equivalence and deriving propositions from statements about equivalence. Note that the table above presents schematic rules; they represent the structure of classes of valid arguments constructed with propositions about protocol equivalence as premises. For instance, RGSRT, GSRT, and GRT are schematic rules that codify derivations of equivalence from premises that include statements about the syntactic properties of protocols. Taking the syntax of protocols to encode the internal structure or procedural content of an epistemic norm, these rules codify valid inference from propositions about the procedural content of epistemic norms to equivalence. From there, the rest of the deductive machinery of PLEN can supply derivative inferences from restrictions, preconditions and postconditions, etc. to equivalence and vice versa. For example, they can be coupled with the deductive framework of PLEN to construct complex derivations of propositions about protocols from premises about protocol equivalence. Additionally, these rules can be coupled with axiom schemata provable by metalogical reasoning to extend the range of derivable results. I'll demonstrate this mingling of meta-level and object-level reasoning below.

7.3.1.1 Derivations of Protocol Equivalence

To establish the equivalence of two protocols, one has numerous strategies available in PLEN. First, one can mechanically perform model-theoretic reasoning to see whether two protocols are graph-theoretically, procedurally, ternary, or binary equivalent. Multiple examples of this kind of reasoning, as structured by PLEN, are provided in the proofs of the central equivalence rules, RGSRT, GSRT, and the versions of PRT.

A second strategy, more central to the development of PLEN, is to develop a repertoire of rules and axioms. Then, one might perform a syntactic check to see whether protocols fit any of the rules or axioms. The following set of protocol graph isomorphism equivalence axiom schemata are mechanical to derive, and can be used to supply premises in the meta-level rules:

Definition 353 (Meta-Level Graph-Theoretic Protocol Equivalence Axioms)

(1) $(\pi_1; \pi_2); \pi_3 \equiv_G \pi_1; (\pi_2; \pi_3)$

(2) $\pi_1; (\pi_2; ...; (\pi_{n-1}; \pi_n)) \equiv_G ((\pi_1; \pi_2); ...;)\pi_{n-1}); \pi_n \ (n < \aleph_0)$
(3) $\pi_1 \cap \pi_2 \equiv_G \pi_2 \cap \pi_1$
(4) $\pi_1 \cup \pi_2 \equiv_G \pi_2 \cup \pi_1$
(5) $(\pi_1; \pi_2); \pi_3 \equiv_G \pi_1; (\pi_2; \pi_3)$
(6) $\pi_1; (\pi_2 \cap \pi_3) \equiv_G ((\pi_1; \pi_2) \cap (\pi_1; \pi_3))$
(7) $\pi_1; (\pi_2 \cup \pi_3) \equiv_G ((\pi_1; \pi_2) \cup (\pi_1; \pi_3))$
(8) $(\pi_1 \cap \pi_2); (\pi_3 \cap \pi_4) \equiv_G (\pi_2 \cap \pi_1); (\pi_4 \cap \pi_3)$
(9) $(\pi_1 \cup \pi_2); (\pi_3 \cup \pi_4) \equiv_G (\pi_2 \cup \pi_1); (\pi_4 \cup \pi_3)$
(10) $(\pi_1 \cap \pi_2) \cap \pi_3 \equiv_G \pi_1 \cap (\pi_2 \cap \pi_3)$
(11) $(\pi_1 \cup \pi_2) \cup \pi_3 \equiv_G \pi_1 \cup (\pi_2 \cup \pi_3)$
(12) $(\pi_1 \cup \pi_2) \cap (\pi_3 \cup \pi_4) \equiv_G ((\pi_1 \cap \pi_3) \cup (\pi_1 \cap \pi_4)) \cup ((\pi_2 \cap \pi_3) \cup (\pi_2 \cap \pi_4))$

From there, the equivalence rules of the R-Derivation System and the various rule-like technical results of PLEN (from the Useful Lemmata to the meta-level rules governing protocol graph isomorphism equivalence) can structure complex reasoning. One can build proofs of protocol equivalence by stringing together applications of meta-level and object-level rules and axioms.

A third strategy is to apply a mixed form of straightforward mathematical reasoning about the structures in PLEN, applying the central equivalence rules above and technical results wherever they provide shortcuts.

Here are three sample arguments that illustrate each strategy (perhaps more than once per sample):

Example 354 (Argument 1) Prove: $\pi_1; (\pi_2 \cap \pi_3) \equiv_G ((\pi_1; \pi_2) \cap (\pi_1; \pi_3))$

Proof Direct model-theoretic reasoning mirroring the argument used in 7.2.1 but with branching replaced with correspondence. For arguments establishing graph-theoretic equivalence by model-theoretic reasoning, see the proofs of ISOP and RGSRT.

(1) LtoR: Let $g \in G^{\pi_1;(\pi_2 \cap \pi_3)}$. Show that the structural display rules hold for g with respect to $(\pi_1; \pi_2) \cap (\pi_1; \pi_3)$.

    (a) (SD1(1) For reductio, suppose that SD1(1) fails for g with respect to $((\pi_1; \pi_2) \cap (\pi_1; \pi_3))$. Then some $\pi_i \in Sub((\pi_1; \pi_2) \cap (\pi_1; \pi_3))$ isn't executable in g. But either π_i is a proper immediate subprotocol of $((\pi_1; \pi_2) \cap (\pi_1; \pi_3))$ or not. If it's a proper immediate subprotocol of $((\pi_1; \pi_2) \cap (\pi_1; \pi_3))$, then it's either $\pi_1; \pi_2$ or $\pi_1; \pi_3$. Trivially, if either of those are not executable at the root state of g, then neither is $\pi_1; (\pi_2 \cap \pi_3)$. $\perp$. If π_i is not a proper immediate subprotocol of $((\pi_1; \pi_2) \cap (\pi_1; \pi_3))$, then it's a subprotocol of π_1, π_2, or π_3. But then, by the definition of subprotocols, π_i is a subprotocol of $\pi_1; (\pi_2 \cap \pi_3)$. Thus, if π_i is not executable, SD1(1) fails for g with respect to $\pi_1; (\pi_2 \cap \pi_3)$. $\perp$The remaining subprotocols of $((\pi_1; \pi_2) \cap (\pi_1; \pi_3))$ is itself, which is, by the execution rules, executable wherever $\pi_1; (\pi_2 \cap \pi_3)$ is. That gives SD1(1).
    (b) (SD1(2)) Let p execute $\pi_1; (\pi_2 \cap \pi_3)$. Then it is a concatenation of executions of π_1 and $\pi_2 \cap \pi_3$. Any execution of $\pi_1 \cap \pi_3$ is a corresponding

pair of executions of π_1 and π_3. Thus, $p = qr$ where q executes π_1, r executes π_2 and there is an execution, r', that executes π_3 that corresponds to r, or $p = qr$ where q executes π_1, r executes π_3 and there is an execution, r', that that corresponds to r. Thus, $p = qr$ executes $\pi_1; \pi_2$ and $p' = qr'$ executes $\pi_1; \pi_3$ (or vice versa), and $p \rightleftharpoons p'$. Thus, p suffices to execute $((\pi_1; \pi_2) \cap (\pi_1; \pi_3))$. Thus, any maximal execution of $\pi_1; (\pi_2 \cup \pi_3)$ in g is an execution of $((\pi_1; \pi_2) \cap (\pi_1; \pi_3))$ in g. The converse is easy to see. Let $p = qr$ where q executes π_1, r executes π_2, and p corresponds to a distinct $p' = q'r'$ where q' executes π_1 and r' executes π_3. Then, by NST, $q = q'$, or else there's a basic graph reduction of g with the same execution theoretic properties, so at the root state of q, $\pi_2 \cap \pi_3$ is executable. Thus, any maximal execution of $((\pi_1; \pi_2) \cap (\pi_1; \pi_3))$ in g is an execution of $\pi_1; (\pi_2 \cap \pi_3)$ in g. SD1(2) follows immediately.

(c) (SD1(3)) From the prior arguments, $\pi_1; \pi_2$ and $\pi_1; \pi_3$ are executable at the root state of any protocol graph of $\pi_1; (\pi_2 \cap \pi_3)$. However, g also contains a state, call it s, at which π_2 and π_3 are both executed by means of distinct paths that branch from a shared root state. Thus, $SD1(3)_{((\pi_1;\pi_2)\cap(\pi_1;\pi_3))}(g)$ holds; though there is no subprotocol such that π_2 and π_3 are joined by parallel execution or choice, nor any subprotocol implemented by both, and $(\pi_1; \pi_2) \cap (\pi_1; \pi_3)$ dictates neither π_2 nor π_3 (nor $\pi_2 \cap \pi_3$), π_2 and π_3 are embedded in $(\pi_1; \pi_2) \cap (\pi_1; \pi_3)$ at equal depth with respect to $\cap$, and so the branching executions of π_2 and π_3 are not superfluous to a protocol graph of $(\pi_1; \pi_2) \cap (\pi_1; \pi_3)$.

(d) (SD1(4)) There are no corresponding executions of any subprotocols of either $\pi_1; (\pi_2 \cap \pi_3)$ or $((\pi_1; \pi_2) \cap (\pi_1; \pi_3))$ that are not shared subprotocols.

(e) (SD2) The implementations of $((\pi_1; \pi_2) \cap (\pi_1; \pi_3))$ are itself and $\pi_1; (\pi_2 \cap \pi_3)$. The converse holds for $\pi_1; (\pi_2 \cap \pi_3)$.

(f) (SD3) The set of atomics in both protocols is identical, so SD3 holds.

(g) (RC) holds trivially; it holds for π_1, π_2, and π_3, and protocol graphs among the subgraphs of g for these subprotocols are linked in such a way that every maximal path of any graph of π_1 is concatenated with a maximal path of a graph of π_2 (same or π_3). The resulting graphs—containing graphs of π_1 and π_2 (resp. π_3)—are such that all of the structural display relations hold for them. Thus, the immediate subprotocols of $((\pi_1; \pi_2) \cap (\pi_1; \pi_3))$ have protocol graphs among the proper subgraphs of g.

(h) (NST) To establish NST, there can be no smaller graph of $((\pi_1; \pi_2) \cap (\pi_1; \pi_3))$ than g; a smaller graph would violate NST with respect to $\pi_1; (\pi_2 \cap \pi_3)$. Suppose that NST fails for g with respect to $((\pi_1; \pi_2) \cap (\pi_1; \pi_3))$. Then there is some proper subgraph of g that satisfies SD1(1,2), SD2, and RC with respect to $((\pi_1; \pi_2) \cap (\pi_1; \pi_3))$. Call that proper subgraph g'. From the prior arguments, SD1(1,2), SD2, and RC all hold for g'. Thus, g fails NST with respect to $\pi_1; (\pi_1 \cap \pi_3)$. $\perp$

(i) (EXC) Any protocol that is correspondence equivalent to $\pi_1; (\pi_2 \cap \pi_3)$ is correspondence equivalent to $((\pi_1; \pi_2) \cap (\pi_1; \pi_3))$, trivially. So, that, finally, gives EXC.

(2) RtoL: Graph reduction will collapse any branching π_1 executions in any graph of $((\pi_1; \pi_2) \cap (\pi_1; \pi_3))$, and no structural display rules will be violated. The resulting graph will immediately satisfy the structural display rules for $\pi_1; (\pi_2 \cap \pi_3)$.

□

Remark 355 Of course, arguments showing that such equivalences fail can be used to show establish the independence of protocols, and, in each case, reductio and negation introduction enable the R-Derivation System to support arguments establishing the negations of the protocol equivalence relations or other protocol-theoretic properties.

Example 356 (Argument 2) Prove:

$$(\pi_1; \pi_4) \cup (\pi_1; (\pi_2 \cap \pi_3)) \equiv_t ((\pi_1; \pi_4) \cup ((\pi_1; \pi_2) \cap (\pi_1; \pi_3)))$$

Proof Direct proof, using the meta-level rules and axioms.

(1) $\pi_1; (\pi_2 \cap \pi_3) \equiv_G ((\pi_1; \pi_2) \cap (\pi_1; \pi_3))$

(a) Equivalence Axiom

(2) ∴

$$\rho(\pi_1; (\pi_2 \cap \pi_3)/((\pi_1; \pi_2) \cap (\pi_1; \pi_3)))[(\pi_1; \pi_4) \cup \pi_1; (\pi_2 \cap \pi_3)] = ((\pi_1; \pi_4) \cup ((\pi_1; \pi_2) \cap (\pi_1; \pi_3)))$$

(a) From def. Syntactic Replacement

(3) ∴

$$\pi_1; (\pi_2 \cap \pi_3) \in ISub((\pi_1; \pi_4) \cup \pi_1; (\pi_2 \cap \pi_3))$$

and

$$((\pi_1; \pi_2) \cap (\pi_1; \pi_3)) \in ISub((\pi_1; \pi_4) \cup ((\pi_1; \pi_2) \cap (\pi_1; \pi_3)))$$

(a) From def. of immediate subprotocols

(4) ∴

$$(\pi_1; \pi_4) \cup (\pi_1; (\pi_2 \cap \pi_3)) \equiv_G ((\pi_1; \pi_4) \cup ((\pi_1; \pi_2) \cap (\pi_1; \pi_3)))$$

(a) From RGSRT applied to 1-3.

(b) Or: From iteration of PGIIRC. For arbitrary pairs of protocol graphs, (g_1, g_2) and (g_3, g_4) where g_1 and g_3 are protocol graphs of $\pi_1; \pi_4$ and g_2 and g_4 are protocol graphs of $\pi_1; (\pi_2 \cap \pi_3)$, construct $\oplus_{\circ}^{(\pi_1;\pi_4)\cup(\pi_1;(\pi_2\cap\pi_3))}(g_1, g_2)$ and $\oplus_{\circ}^{(\pi_1;\pi_4)\cup(\pi_1;(\pi_2\cap\pi_3))}(g_3, g_4)$. From PGIIRC, $\oplus_{\circ}^{(\pi_1;\pi_4)\cup(\pi_1;(\pi_2\cap\pi_3))}(g_1, g_2) \simeq \oplus_{\circ}^{(\pi_1;\pi_4)\cup(\pi_1;(\pi_2\cap\pi_3))}(g_3, g_4)$ and both are protocol graphs of $(\pi_1; \pi_4) \cup (\pi_1; (\pi_2 \cap \pi_3))$. By 1 and EQID, (g_1, g_3) and (g_2, g_4) are pairs of protocol graphs of $\pi_1; \pi_4$ and $(\pi_1; \pi_2) \cap (\pi_1; \pi_3)$, respectively. Thus, $\oplus_{\circ}^{((\pi_1;\pi_4)\cup((\pi_1;\pi_2)\cap(\pi_1;\pi_3)))}(g_1, g_2)$ and $\oplus_{\circ}^{((\pi_1;\pi_4)\cup((\pi_1;\pi_2)\cap(\pi_1;\pi_3)))}(g_3, g_4)$ are isomorphic and both are protocol graphs of $((\pi_1; \pi_4) \cup ((\pi_1; \pi_2) \cap (\pi_1; \pi_3))$. Then, from SUT and ISOP, the result follows.

(5) $\therefore$

$$(\pi_1; \pi_4) \cup (\pi_1; (\pi_2 \cap \pi_3)) \equiv_p ((\pi_1; \pi_4) \cup ((\pi_1; \pi_2) \cap (\pi_1; \pi_3)))$$

(a) From PRT1(i) applied to 4

(6) $\therefore$

$$(\pi_1; \pi_4) \cup (\pi_1; (\pi_2 \cap \pi_3)) \equiv_t ((\pi_1; \pi_4) \cup ((\pi_1; \pi_2) \cap (\pi_1; \pi_3)))$$

(a) From PRT1(i,ii) applied to 5

□

Example 357 (Argument 3) Define graph-theoretic implication and derive introduction and elimination rules for graph-theoretic equivalence by meta-level reasoning.

Proof Third strategy. Let $\pi_1 \supset_G \pi_2$ iff $\forall g[g \in G^{\pi_1} \rightarrow \exists g', g''((g' \in G^{\pi_2}) \wedge ((g \simeq g'') \wedge (g'' \sqsubseteq g'))]$

Introduction and Elimination rules for graph-theoretic equivalence formulated in terms of graph-theoretic implication can be shown to be valid.

First, note that $\supset_G$ is properly an implication or dependence relation because the subgraph relation is a preorder. It's reflexive, transitive, and anti-symmetric. If every graph of π_1 is a subgraph of some π_2 graph, and every graph of π_2 is a subgraph of some π_3 graph, then every graph of π_1 is a subgraph of some π_3 graph. It is reflexive (every graph is a subgraph of itself). It is anti-symmetric: if $g \sqsubseteq g'$ and $g' \sqsubseteq g$, then $g = g'$. From these facts, it follows that $\pi_1 \supset_G \pi_2$ and $\pi_2 \supset_G \pi_1$ imply $\pi_1 \equiv_G \pi_2$ and that $\pi_1 \equiv_G \pi_2$ implies $\pi_1 \supset_G \pi_2$ and $\pi_2 \supset_G \pi_1$:

Intro $\dfrac{\pi_1 \supset_G \pi_2,\ \pi_2 \supset_G \pi_1}{\pi_1 \equiv_G \pi_2}$

(Subproof) Let $\pi_1 \supset_G \pi_2$ and $\pi_2 \supset_G \pi_1$. Thus,

$$\forall g[g \in G^{\pi_1} \rightarrow \exists g', g''((g' \in G^{\pi_2}) \wedge ((g \simeq g'') \wedge (g'' \sqsubseteq g'))]$$

and

$$\forall g[g \in G^{\pi_2} \rightarrow \exists g', g''((g' \in G^{\pi_1}) \wedge ((g \simeq g'') \wedge (g'' \sqsubseteq g'))]$$

It suffices to show that $\forall g[(g \in G^{\pi_1}) \equiv (g \in G^{\pi_2})]$. Let $g \in G^{\pi_1}$. Then $\exists g', g''(g' \in G^{\pi_2} \wedge ((g \simeq g'') \wedge (g'' \sqsubseteq g')))$. But then $\exists g''', g''''(g''' \in G^{\pi_1} \wedge ((g' \simeq g'''') \wedge (g'''' \sqsubseteq g''')))$. And so on. Now, it can't be the case that g is isomorphic to a proper subgraph of g': if so, then g', a protocol graph of π_2, would contain a proper subgraph that was isomorphic to a protocol graph of π_2. This violates NST, as any graph of π_2 that contains a proper subgraph that's a protocol graph of π_2 can be reduced to that proper subgraph without violating the structural display rules. Contradiction. The same reasoning runs the other direction. Thus, either $g = g'$ (g is a subgraph of g' and vice versa) or $g \simeq g'$ (g is isomorphic to a subgraph of g' and vice versa). Thus, from EQID or ISOP, , g is a protocol graph of π_1 and π_2. But then, g is isomorphic to all protocol graphs of π_1 and π_2 by SUT. Thus, from ISOP, for arbitrary g, $(g \in G^{\pi_1}) \equiv (g \in G^{\pi_2})$. Basic set theory implies that $G^{\pi_1} = G^{\pi_2}$. Thus, by EQID, $\pi_1 \equiv_G \pi_2$.

Elim $\dfrac{\pi_1 \equiv_G \pi_2}{\pi_1 \supset_G \pi_2}, \dfrac{\pi_1 \equiv_G \pi_2}{\pi_2 \supset_G \pi_1}$

(Subproof) Let $\pi_1 \equiv_G \pi_2$. Then $\forall g[g \in G^{\pi_1} \rightarrow \exists g'(g' \in G^{\pi_2} \wedge g \simeq g')]$ and $\forall g[g \in G^{\pi_2} \rightarrow \exists g'(g' \in G^{\pi_1} \wedge g \simeq g')]$. Of course, graphs are all subgraphs of themselves, so g is isomorphic to a subgraph of a protocol graph of π_2. Thus, $\pi_1 \supset_G \pi_2$. Transposition on the same argument gives $\pi_2 \supset_G \pi_1$. □

7.3.1.2 Derivations from Protocol Equivalence

Derivations of various properties of protocols (hence, epistemic rationality norms) from protocol equivalence are also formalized in PLEN. In general, each rule is a schematic valid argument form for deriving conclusions about various properties of protocols from premises that contain assertions about equivalence. Reasoning from equivalence to executability, process structure, verdicts or, really almost any proposition about protocols expressible in PLEN can be structured by application of GRT. Explicitly, these derivations incorporate both the meta-level equivalence rules and object level expressions. Here are four sample derivations constructed with both the equivalence rules and the axioms of PLEN:

Example 358 (Argument 4) Prove:

$$(\pi_1; \pi_4) \cap (\pi_1; (\pi_2 \cap \pi_3)) \curlywedge_p \pi_4 \rightarrow ((\pi_1; \pi_4) \cap ((\pi_1; \pi_2) \cap (\pi_1; \pi_3))) \curlywedge_p \pi_4$$

Proof Direct proof using axioms, object-level logic, and meta-level rules.

(1)

$$(\pi_1; \pi_4) \cap (\pi_1; (\pi_2 \cap \pi_3)) \curlywedgeuparrow_p \pi_4$$

(a) Assumption 1

(2)

$$\pi_1; (\pi_2 \cap \pi_3) \equiv_G ((\pi_1; \pi_2) \cap (\pi_1; \pi_3))$$

(a) Equivalence Axiom

(3) ∴

$$(\pi_1; \pi_4) \cap (\pi_1; (\pi_2 \cap \pi_3)) \equiv_G ((\pi_1; \pi_4) \cap ((\pi_1; \pi_2) \cap (\pi_1; \pi_3)))$$

(a) From RGSRT on the Equivalence Axiom (or by iteration of PGIIRC)

(4) ∴

$$(\pi_1; \pi_4) \cap ((\pi_1; \pi_2) \cap (\pi_1; \pi_3))) \curlywedgeuparrow_p \pi_4$$

(a) From GRT applied to Assumption 1 and 3

(5) ∴

$$(\pi_1; \pi_4) \cap (\pi_1; (\pi_2 \cap \pi_3)) \curlywedgeuparrow_p \pi_4 \rightarrow ((\pi_1; \pi_4) \cap ((\pi_1; \pi_2) \cap (\pi_1; \pi_3))) \curlywedgeuparrow_p \pi_4$$

(a) Conditional Introduction: 1-4

□

Example 359 (Argument 5) Prove:

$$(\pi_5 \mapsto \pi_1; (\pi_2 \cap \pi_3) \wedge \langle \pi_5 \rangle_p \top) \rightarrow (\langle (\pi_1; \pi_2) \cap (\pi_1; \pi_3) \rangle_p \top \wedge \langle \pi_1 \rangle_p (\langle \pi_2 \rangle_p \top \wedge \langle \pi_3 \rangle_p \top))$$

Proof Direct Proof using axioms, object-level logic, and meta-level rules.

(1)

$$\pi_5 \mapsto \pi_1; (\pi_2 \cap \pi_3) \wedge \langle \pi_5 \rangle_p \top$$

(a) Assumption 1

(2) ∴

$$\pi_5 \mapsto \pi_1; (\pi_2 \cap \pi_3)$$

(a) Simplifying 1

(3) ∴

$$\langle \pi_5 \rangle_p \top$$

(a) Simplifying 1

(4) ∴

$$\pi_1; (\pi_2 \cap \pi_3) \equiv_G ((\pi_1; \pi_2) \cap (\pi_1; \pi_3))$$

(a) Equivalence Axiom

(5) ∴

$$\langle \pi_1; (\pi_2 \cap \pi_3) \rangle_p A \leftrightarrow \langle \pi_1 \rangle_p \langle \pi_2 \cap \pi_3 \rangle_p A$$

(a) Axiom

(6) ∴

$$\langle \pi_1; (\pi_2 \cap \pi_3) \rangle_p A \leftrightarrow \langle \pi_1 \rangle_p (\langle \pi_2 \rangle_p A \wedge \langle \pi_3 \rangle_p A)$$

(a) Axiom, Transitivity on 5

(7) ∴

$$\langle \pi_1; (\pi_2 \cap \pi_3) \rangle_p \top$$

(a) From 2 and 3 by definition of implementations

(8) ∴

$$\langle \pi_1 \rangle_p (\langle \pi_2 \rangle_p \top \wedge \langle \pi_3 \rangle_p \top)$$

(a) From 6 and 7.

(9) $\therefore$

$$\pi_5 \mapsto ((\pi_1; \pi_2) \cap (\pi_1; \pi_3))$$

(a) From GRT applied to Assumption 1 and 4

(10) $\therefore$

$$\langle(\pi_1; \pi_2) \cap (\pi_1; \pi_3)\rangle_p \top$$

(a) From 3 and 9 by definition of implementations

(11) $\therefore$

$$\langle(\pi_1; \pi_2) \cap (\pi_1; \pi_3)\rangle_p \top \wedge \langle\pi_1\rangle_p(\langle\pi_2\rangle_p \top \wedge \langle\pi_3\rangle_p \top)$$

(a) Conjunction Introduction on 8 and 10

(12) $\therefore$

$$(\pi_5 \mapsto \pi_1; (\pi_2 \cap \pi_3) \wedge \langle\pi_5\rangle_p \top) \rightarrow (\langle(\pi_1; \pi_2) \cap (\pi_1; \pi_3)\rangle_p \top \wedge \langle\pi_1\rangle_p(\langle\pi_2\rangle_p \top \wedge \langle\pi_3\rangle_p \top))$$

(a) Conditional Introduction 1-11, Assumption 1 Discharged

□

Example 360 (Derived Rules)

(1) From $\langle a; ((b \cap c); d)\rangle_p \top$, it follows that:

(a) $\therefore \langle a\rangle_p \langle((b \cap c); d)\rangle_p \top$
(b) $\therefore \langle a\rangle_p \langle b \cap c\rangle_p \langle d\rangle_p \top$
(c) $\therefore \langle a\rangle_p \langle b\rangle_p \langle d\rangle_p \top \wedge \langle a\rangle_p \langle c\rangle_p \langle d\rangle_p \top$
(d) $\therefore \langle a; (b; d)\rangle_p \top \wedge \langle a; (c; d)\rangle_p \top$

Example 361 (Argument 6) Prove:

$$(b(A) \rightarrow \langle a; ((b \cap c); d)\rangle_p \top) \rightarrow (b(A) \rightarrow \langle a\rangle_p \langle c\rangle_p \langle d\rangle_p \top)$$

Proof Direct proof by axioms, object-level logic, and meta-level rules.

(1)

$$b(A) \rightarrow \langle a; ((b \cap c); d)\rangle_p \top$$

(a) Assumption

(2) ∴

$$(b(A) \rightarrow (\langle a; (b; d)\rangle_p \top \wedge \langle a; (c; d)\rangle_p \top)$$

(a) Derived Rule d

(3) ∴

$$(b(A) \rightarrow (\langle a\rangle_p \langle b\rangle_p \langle d\rangle_p \top \wedge \langle a\rangle_p \langle c\rangle_p \langle d\rangle_p \top)$$

(a) Derived Rule c

(4) ∴

$$(b(A) \rightarrow \langle a\rangle_p \langle c\rangle_p \langle d\rangle_p \top)$$

(a) Logic

(5) ∴

$$(b(A) \rightarrow \langle a; ((b \cap c); d)\rangle_p \top) \rightarrow (b(A) \rightarrow \langle a\rangle_p \langle c\rangle_p \langle d\rangle_p \top))$$

(a) Conditional Introduction: 1-4, discharging the assumption.

□

Claim 362 From there, facts about process structure, executability, etc. can be derived, and from there, by the rules above and the machinery from Chaps. 5 and 6, normative verdicts including epistemic deontic verdicts. As this requires machinery that has not yet been presented, I'll return to this point in Chaps. 8 and 10.

7.3.1.3 Dependence, Independence, and Sufficiency

I omit explicit illustration of the argument forms establishing these relations among protocols, as this is straightforward. Where dependence and sufficiency relations among protocols can be thought of in terms of protocol-theoretic relations (equivalence, dictation, etc.), reasoning that establishes protocol equivalence or protocol implication suffices to establish dependence and sufficiency. Moreover, protocol equivalences can be shown to fail by the foregoing methods, and this suffices to establish independence or insufficiency. For examples, one can turn to Appendix 2 and the proofs of the remaining clauses of PRT and PRTC.

7.4 Investigating Sublogics of Epistemic Norms and Dynamics

7.4.1 Sublogics of PLEN and Theories of Protocol Equivalence

PLEN is deployed both (i) as a formal theory of epistemic rationality norms and (ii) as a formal framework for constructing and reasoning about formal theories of epistemic rationality norms. That is, the framework, PLEN, contains multiple, distinct formalisms for representing and reasoning about epistemic rationality norms as fragments, and the overarching protocol logic of PLEN provides a system for establishing and analyzing their properties. The core differences among these different systems can be encapsulated by the differences among the logical theories of protocol equivalence in the fragments.

PLEN is a unified framework in which each of these fragments (resp. strategies) can be defined and reasoned about rigorously. We can use the framework to generate distinct logics and compare them. L^R_{PLEN} contains indexed variants of a number of its connectives and operators; this is the device by which L^R_{PLEN} can be broken into different sub-languages. There are p, t, c, and r variants of protocol connectives and dynamic epistemic operators. A sub-language, L^x_{PLEN}, can be defined by taking P_{PLEN} and the set of atomic propositions, and defining a language as usual with the propositional connectives, epistemic operators, and an indexed set of protocol connectives and dynamic epistemic operators. Taking only the p variants of the connectives and operators, we get L^p_{PLEN}. Taking the ternary variants only, we get L^R_{PLEN}. And so on.

Expressions formed from distinct variants of the same operators are interpreted naturally over EDGs using the distinct sets of execution rules. The p, t, and r are interpreted using the procedural, transitional, and ternary or edge execution sets, respectively. The truth rules can be paired with these to interpret the respective L^x_{PLEN} variants. Additionally, one can generate some variation between sublogics of protocols by pairing sublanguages with distinct sets of execution rules in restricted subsets of the set of EDGs. Each pairing of a sublanguage, execution rules, and domain of EDGs defines a distinct protocol-theoretic logic of epistemic norms, and the resulting logics can be compared for (e.g.) strength of their deductive systems, properties of the native object-level protocol equivalence relations definable in them, relative expressive power, and so on.

For ease of reference, the following terms, separated by a slash, will be used interchangeably for some prominent fragments of PLEN.

Definition 363 (Binary/Transitional Fragment) The system resulting from interpretation of L^t_{PLEN} and P_{PLEN} over the set of all EDGs via the binary execution rules, the corresponding truth rules, and the corresponding object-level derivation rules (those formed using only the t variants of operators and connectives).

Definition 364 (Ternary/Edge Fragment) The system resulting from interpretation of L^R_{PLEN} and P_{PLEN} over the set of all EDGs via the ternary/edge execution rules, the corresponding truth rules, and the corresponding object-level derivation rules (those formed using only the r variants of operators and connectives).

Definition 365 (Procedural/Path-Theoretic Fragment) The system resulting from interpretation of L^p_{PLEN} and P_{PLEN} over the set of all EDGs via the procedural execution rules, the corresponding truth rules, and the corresponding object-level derivation rules (those formed using only the p variants of operators and connectives).

Definition 366 (Correspondence Fragment) The system resulting from interpretation of L^c_{PLEN} and P_{PLEN} over the set of all EDGs via the procedural execution rules, the corresponding truth rules, and the corresponding object-level derivation rules (those formed using only the c variants of operators and connectives[5]).

These fragments represent different strategies for formally representing epistemic rationality norms and epistemic dynamics. The binary or transitional fragment encodes the strategy of representing epistemic action by means of binary transition relations, and the logic of protocols (hence, epistemic norms) defined over this model will largely behave like the program logic of PDL with respect to (e.g.) protocol equivalence relations. This fragment and the resulting logic of protocols will suffer the processes problem articulated in Chap. 4, and will formally validate the erroneous inferences identified as errors that cascade from this problem. In contrast, the procedural fragment encodes a strategy that sharply distinguishes epistemic transitions from actions and processes, and thereby is free of the processes problem and the erroneous inferences licensed by the logic built over the transitional fragment.

The Revised PLEN framework is interpreted simultaneously over all sets of EDGs at once (by deploying a distinct set of execution rules for each sub-language), so that distinct fragments can be analyzed at once within the same logical system. It can very well be the case that (for an example of bedrock simplicity) $\pi \leftrightarrow_t \pi'$ while $\pi \leftrightarrow_p \pi'$ fails, or that $\Box(A_t(\pi) \rightarrow A_t(\pi'))$ is a schematic logical truth while $\Box(A_p(\pi) \rightarrow A_p(\pi'))$ fails. Additionally, we can mechanically identify, for each fragment, which laws hold over which restricted sets of EDG, X, Y, and Z, by defining restricted logical consequence relations, $\models^X_{PLEN}$, $\models^Y_{PLEN}$, and $\models^Z_{PLEN}$, over these sets of EDGs. From there, direct comparisons of the lists of validities between these consequence relations and with $\models_{PLEN}$ are mechanical. One can multiply examples as needed by recalling any clauses of PRT, GRT, GRT Strength, or, really, any of the consequences of the results thus far amassed. Thus, the protocol logic of PLEN (i.e., the full set of theorems of the R-Derivation system) provides a system in which to comparatively analyze the logics produced by these formalization strategies. Let me point out a set of quite abstract observations about this sort of comparative analysis.

---

[5] So-called for the fact that the protocol equivalence native to this fragment is defined over path correspondence.

First, a sublanguage of L^R_{PLEN} may obey different laws in different semantic fragments. That is, given a set of formulae, one subclass of EDGs may validate different theorems and even different inferences than another despite sharing identical execution and truth rules.[6]

Different sets of structures represent different general strategies for modeling epistemic dynamics as well as different arrays of philosophical commitments. For instance, one might impose a determinism requirement (exactly one reachable state by each atomic protocol per root state), a non-convergence requirement, or even a requirement that blurs the line between processes and procedures (consider the proper subclass of EDGs in which each path has a unique pair of root and terminal states). These different classes of structures formalize different conceptual claims about epistemic dynamics, and they clearly define different logics. These logics, then, correspond to different pre-theoretic modeling assumptions about epistemic dynamics.

Given that the operators and connectives encode properties of (protocols, hence) epistemic noms, distinct underlying models of epistemic dynamics generate distinct accounts of the properties of epistemic norms. These distinct underlying models are contained in the set of EDGs, and so can be compared rigorously in PLEN by determining their effects on a given sublanguage.

Moreover, these differences disappear when looking at only the logical fragments (those without dynamic epistemic operators or protocol connectives). So, the propositional and modal logics of these fragments are the same. Thus, the logical differences across different semantic fragments is made by the logic of protocols (hence, epistemic norms).

Second, different sublanguages may obey different laws in the same semantic fragment. That is, counterpart formulae (syntactically identical formulae but that some operator or connective is replaced uniformly throughout with a p, t, r, or c variant) and arguments formed from them may differ in their truth conditions and validity conditions within the same set of EDGs. As above, these differences disappear when removing the dynamic epistemic operators and protocol connectives. The counterpart sublanguages—coupled with the relevant execution and truth rules—represent different ways of formalizing the same set of properties of protocols. For instance, consider the differences between transitional vs procedural analyses of: executability conditions of protocols, postconditions of protocols, equivalence of protocols, convergence or divergence of protocols with respect to their executions, the normative verdicts of protocols, etc.. Thus, distinct analyses of the properties of protocols (built with protocol operators and connectives, execution rules, and truth rules) generate different accounts of the properties of epistemic norms given a fixed model of epistemic dynamics. These distinct analyses are contained in the syntax and semantics of PLEN, and so can be compared rigorously in PLEN by

[6] This should be immediate, but one could mechanically step through the soundness proofs to see which axioms are logical truths in the set of all DMSs without restriction rules as compared to those that hold in the proper subset of DMSs that are EDGs.

determining whether and which rules and laws are validated in the fragments defined by sublanguage, execution rule, and truth rule triples.

Together, entirely different fragments encode syntactically and semantically diverse ways of modeling epistemic dynamics and protocol-theoretic ways of formalizing epistemic norms and their properties. The different fragments articulated above (and others still) can be compared within the PLEN framework for (inter alia) the properties that they attribute to protocols and the analyses they provide of logical relations like protocol equivalence by means of the derivation system. This claim is spelled out rigorously in the protocol graph isomorphism equivalence results of prior sections and chapters. Consider again, the properties of the native protocol equivalence relations across fragments.

Proposition 367 *Consider the SEP_x rules in the proof of R-Derivation System Soundness. In particular, consider the following schematic rule or its corresponding axiom schema:*

(1) (SEP_x Rule Schema) $\dfrac{(\pi \leftrightarrow_x \pi'),\ A(\pi)}{A(\pi')}$

(2) (SEP_x Axiom Schema) $\pi \leftrightarrow_x \pi' \rightarrow (A(\pi) \equiv A(\pi'))$

For each protocol equivalence, the rule/law is valid for distinct classes of L^R_{PLEN} formulae of the form $A(\pi)$. An array of such classes of formulae is presented in the following table:

$\leftrightarrow_r$	$\equiv^s_G$	$\leftrightarrow_p$	$\leftrightarrow_t$
$M(\pi)$	$\ldots \mapsto \pi$	$\pi \curlywedge_p \ldots$	
$\ldots \mapsto \pi$	$\ldots \equiv^s_g \pi$	$\pi \curlyvee_p \ldots$	
$\ldots \equiv^s_g \pi$	$\pi \curlywedge_p \ldots$	$\pi \leftrightarrow_p \ldots$	
$\pi \curlywedge_p \ldots$	$\pi \curlyvee_p \ldots$		
$\pi \curlyvee_p \ldots$	$\pi \leftrightarrow_p \ldots$		
$\pi \leftrightarrow_p \ldots$			
$\pi = \ldots$			
$\ldots \in Sub(\pi)$			

The protocol equivalence relations of each fragment encode some of the core features of their respective approaches to representing and reasoning about procedural epistemic norms: the theory of protocol equivalence in a fragment encodes their individuation conditions for protocols (hence, epistemic rationality norms) with respect to everything the logic can say about them. The core results of PLEN thus provide a framework for comparing the procedural, transitional, ternary, and correspondence accounts of protocols, while also providing the novel theory protocol graph isomorphism equivalence that interacts with each of them. For one main lesson, note that transitional approaches to modeling the content of epistemic norms (i.e., transitional protocol equivalence) represent procedural epistemic norms as extensional objects, the equivalence of protocols preserving truth over intersubstitution only in extensional contexts. This can upgraded by the theory of necessary transition equivalence. Nonetheless, the resulting theory still has hyperintensional contexts (subprotocol, syntactic replacement, and implementation) for protocols, and even logically necessarily transition equivalent protocols will not be hyperintensionally equivalent. The procedural fragment is more fine-grained, but still doesn't get us to a criterion of hyperintensional protocol equivalence. And so on. The observations that can be drawn in this framework about the relative expressive and analytical powers of the fragments of PLEN far exceed the scope of this section, which is merely aimed at pointing out that the Revised PLEN Framework bears this usefulness.

7.5 Modularity and Further Expansions or Revisions

Further emphasizing PLEN's use as a framework for building logics of protocols (and thus, logics of procedural epistemic norms), I want to point out that PLEN is adaptable or, if you will, modular. The protocol logic and the propositional and modal logics of PLEN are largely autonomous. One can modify the protocol logic or the propositional and modal logics ad hoc with minimal consequences for the other side. Given the recursive definition of PLEN's syntax and the bifurcated nature of its semantics, this is a relatively straightforward process.

If one wants to add more protocol constructions (or delete some), one needs only to plug in (or remove) the novel protocol constructions and their execution rules to (from) the truth rules. If one wants to revise the propositional logic, modal logic, or epistemic logic, one need only add the suite of operators one wants, and then elaborate on the valuation functions, replacing them with relations and adding whatever accessibility relations or novel functions one needs. In the former case, the logic of propositions doesn't change. In the latter case, some of the dynamics rules will fail (e.g., $\dfrac{[\pi]_P(A \rightarrow B),\ [\pi]_P A}{[\pi]_P B}(MP6)$ is no longer valid if one adds a non-detaching conditional), but the protocol logic will be otherwise unaffected. In either case, the core results of PLEN (the theory of protocol graph isomorphism, the granularity results) will remain sound, as none of the core results save GRT depends on the propositional (et al) logic.

I take brief looks at two illustrations of this: making the propositional logic of PLEN relevant, and imposing conditions on the contents of epistemic states.

7.5.1 Modularity: Relevant PLEN

Remark 368 PLEN is, in principle, compatible with non-classical logics. All that needs to be done is to add some extra constructions to EDGs and define alternative interpretation rules for the relevant operators. For instance, here is a simple way to make PLEN relevant (to a degree) and paraconsistent, with an intensional conditional that preserves important rules such as conditional detachment:

Definition 369 Basic[7] relevant logic B can replace classical logic soundly in the R-Derivation System.

(1) Partition S into S_N and S_W; S_N are the normal worlds, S_W are the non-normal (weird) worlds.
(2) Add a ternary relation, $R_t \subseteq S \times S \times S$ such that if $s \in S_N$, then $R_t(s, s', s'')$ iff $s' = s''$ (Normality Condition).
(3) Add a star function, $* : S \longrightarrow S$ such that $s = **s$.
(4) Replace the valuation function, v, with a valuation relation, r such that $r \subseteq S \times L^R_{PLEN} \times \{0, 1\}$
(5) One gets the basic relevant logic B, as presented in Chap. 10 of [2], by re-defining truth for negation and conditional (or adding a relevant implication), letting $s \Vdash A$ iff $\langle s, A, 1\rangle \in r$:

    (a) $s \Vdash \neg A$ iff $*s \Vdash A$
    (b) $s \Vdash A \rightarrow B$ iff for all s, s', s'' such that $R_t(s, s', s'')$, if $s' \Vdash A$, then $s'' \Vdash B$.

(6) Replace the conditional and negation rules in the R-Derivation system with their relevant counterparts,[8] and call the resulting system the B-Derivation System (as opposed to the R-Derivation System).

Conjecture 370 The B-Derivation System is (almost) sound with respect to the revised, relevant semantics (i.e., over the normal worlds).

Roughly speaking, the protocol logic is unchanged by the revision, and the relevant logic added to PLEN is known to be sound over this Routley-Meyer-like semantics. The rule $\dfrac{[\pi]_p(A \rightarrow B),\ [\pi]_p A}{[\pi]_p B}(MP6)$ will fail over non-normal worlds, as it should—as modus ponens does.

7.5.2 *Modularity: Conditions on Epistemic States*

The content states of the PLEN system are bare, syntactic constructions: triples of sets of formulae and protocols $\langle X : Y : \Pi_M\rangle$. These are taken, in Chap. 4, to represent the temporary contents of epistemic judgments: X represents the set of accepted claims, Y the set of rejected claims, and Π_M the endorsed methodology of the single agent whose dynamics are modeled by a given EDG, g, and whose current state is s where $c(s) = \langle X : Y : \Pi_M\rangle$. This is a point of modularity in PLEN. PLEN can be used as the basis for building distinct theories of epistemic dynamics in two

[7] I'm following Priest's naming conventions, here [2].

[8] See Chapters, 8, 9, and 10 of [2].

ways: (1) by focusing on proper subsets of EDGs in which content states abide by some or another set of constraints, or by (2) swapping out content states for some other structure.

As for the first option, there are no constraints on acceptance or rejection built into PLEN. That is, X and Y need be neither exclusive nor exhaustive; they can have non-empty intersections and they need neither contain A or $\neg A$ for arbitrary A. Moreover, the acceptance set need not contain logical truths, nor need the rejection set contain logical falsehoods; the converse requirements that the negations of logical falsehoods be accepted and the negations of logical truths be rejected also fail. This is is, in fact, a virtue.

Aside from freeing PLEN from any commitment to logical omniscience, this feature of PLEN enables us to model all possible outcomes (on all or nothing acceptance and rejection[9]) of epistemic processes, and so to identify classes of processes that result in epistemic judgments that violate would-be epistemic norms. This is integral to modeling the restrictions on epistemic processes—we must be able to say that some processes result in absurd judgments, and that's why they're wrong. For instance, $[\pi]_p(b(A \wedge \neg A)$ or $[\pi]_p(b(A) \wedge d(A)$ might be used to identify actions and processes that result in unacceptable forms of inconsistency.

This feature enables PLEN to model numerous alternative versions of deductive closure norms without baking any of them into the system. Consider $[\pi_1]_p(b(A \wedge (A \rightarrow B)) \rightarrow b(B))$ as opposed to $[\pi_2]_p(b(A \wedge (A \rightarrow B)) \rightarrow \neg d(B))$ as opposed to $[\pi_3]_p((A \rightarrow B) \rightarrow (b(A) \rightarrow b(B))$ as opposed to $[\pi_4]_p((A \rightarrow B) \rightarrow (b(A) \rightarrow \neg d(B))$. None of these are valid in PLEN, but may be valid in fragments of PLEN under some set of restrictions on EDGs or under the assumption of some set of axioms characterizing content states. Thus, PLEN can provide a neutral framework in which to reason about distinct sets of philosophical assumptions about epistemic content and epistemic dynamics.

For instance, PLEN can house theories that build these sorts of results into their theory of epistemic states as subsystems. One can, for instance, easily devise the sorts of constraints on epistemic states that would generate AGM belief revision theory as a subsystem of PLEN. The basic procedure is simply to map the belief sets of AGM into the content states of the subset of EDGs, so that each belief set K gets mapped uniquely to an epistemic state, s, such that $c(s)$ satisfies the following rules:

(1) **Mapping**: $A \in K$ iff $A \in X$ and $A \notin Y$ and $\neg A \in Y$
(2) **Exhaustion**: For all $A \in L^B_{PLEN}$, $(b(A) \vee d(A))$
(3) **Exclusion**: For all $A \in L^B_{PLEN}$, $\neg b(A) \equiv d(A)$
(4) **Closure**: K is closed under entailment

[9] PLEN also models the revision of methodologies by epistemic processes, and enables us to analyze reasoning about protocols that dictate change in one's endorsed protocols. A robust theory of normative conflict and revision is in the offing in future work on PLEN.

Given these kinds of modifications, the properties of epistemic commitment functions can be embedded in PLEN as properties of protocols or their executions. For instance, let $\pi{+}_A$ and $\pi{-}_A$ encode expanding with A and contracting A, respectively, and let $b(A)$ mean that A is in X at the given state ($s \Vdash b(A)$ iff $a \in X$). The Recovery postulate of AGM could then be encoded with the schema: $b(A) \rightarrow [\pi{-}_A; \pi{+}_A]b(A)$.

From there, it's straightforward to map epistemic commitment functions, K_{+A}, K_{-A}, $K_{\div A}$, into protocols, π_{+A}, π_{-A}, $\pi_{\div A}$, in such a way that axioms mirroring the rationality postulates for each commitment function holds for the corresponding protocols. The following table gives one set of rules for doing so, allowing some abuse of notation:[10]

(-1): $K{-}_A = Cn(K{-}_A)$ **(-2):** $K{-}_A \subseteq K$ **(-3):** If $A \notin K$, then $K{-}_A = K$. **(-4):** If $A \in K{-}_A$, then $\vdash A$. **(-5):** $K \subseteq (K{-}_A){+}_A$ **(-6):** If A is equivalent to B, then $K{-}_A = K{-}_B$. **(-7):** $(K\, {-}_A \cap K{-}_B) \subseteq K{-}_{A \wedge B}$ **(-8):** If $A \notin K{-}_{A \wedge B}$, then $K{-}_{A \wedge B} \subseteq K{-}_A$.	**(-1):** For all s, if $s = last(t)$ and $t \in T_{\pi-A}$, then $c(s) = Cn(first(t){-}A)$. **(-2):** For all s, if $s = last(t)$ and $t \in T_{\pi-A}$, then $c(s) \subseteq c(first(t))$. **(-3):** For all s, $A \notin c(s)$, then for all s', if $s' = last(t)$ and $t \in T_{\pi-A}$, then $c(s') = c(s)$. **(-4):** For all s, if $s = last(t)$, $t \in T_{\pi-A}$, and $A \in c(s)$, then $\vdash A$. **(-5):** For all s, if $s = first(t)$ and $t \in T_{\pi-A;\pi+A}$, then $c(s) \subseteq c(last(t))$. **(-6):** If A is equivalent to B, then for all s, $s = last(t)$ and $t \in T_{\pi-A}$ iff $s = last(t)$ and $t \in T_{\pi-B}$. **(-7):** For all s, if $c(s) = c(last(t)) \cap c(last(t'))$ for $t \in T_{\pi-A}$ and $t' \in T_{-B}$, then $s = last(t)$ and $t \in T_{\pi-A \wedge B}$. **(-8):** For all s, if $s = c(last(t))$, $A \notin c(last(t))$, and $t \in T_{\pi-A\wedge B}$, then $c(last(t)) \subseteq c(last(t'))$ for some $t' \in T_{\pi-A}$.

An alternative way of doing this might be to find object-level laws that encode these features of the protocols that are mapped to epistemic commitment functions:

(-1) $B_1, \ldots, B_n \models C$ and $[\pi_{-A}]_t b(B_1, \ldots, B_n) \rightarrow [\pi_{-A}]_t b(C)$.
(-6) If A is equivalent to B, then $[\pi_{-A}]_t \neg b(A)$ iff $[\pi_{-B}]_t \neg b(B)$.
(-7) $[\pi_{-A} \cap \pi_{-B}]_t \neg b(A \wedge B)$.

[10] Where $c(s) = [X : Y : \Pi_M]$ and $c(s') = [X' : Y' : \Pi'_M]$, take $c(s) \subseteq c(s')$ to mean that $X \subseteq X'$, $Y \subseteq Y'$, and $\Pi_M \subseteq \Pi'_M$. Additionally, take $Cn(s)$ to be the set of classical logical closure of X.

The reduction of AGM to PLEN is a corollary of the already well-known reduction of AGM to PDL [3–5]. Thus, the fact that PDL can be embedded in a fragment of PLEN in a way that preserves its logic[11] suffices to show that AGM can be embedded in a fragment of PLEN in a way that preserves its properties. So, I'll only briefly state the main result of the conjecture:

Proposition 371 *Given (a complete version) of either translation scheme above, for every formula, A, of L^R_{PLEN}, there are protocols, π_{+A}, π_{-A}, $\pi_{\div A}$, such that there is a class of EDGs, EDG_{AGM}, in which π_{+A}, π_{-A}, and $\pi_{\div A}$ satisfy all of the rules that correspond to the rationality postulates for the corresponding epistemic commitment function.*

Corollary 372 *These protocols, in virtue of the mimicking of the properties of belief states by epistemic content states, satisfy the rationality postulates for each epistemic commitment function.*

Corollary 373 *These protocols, in virtue of the mimicking of the properties of belief states by epistemic content states, satisfy the entrenchment conditions for each epistemic commitment function.*

As for the second option, it's easy to see how simply replacing content states with belief sets would do much the same. More radical modules can be inserted. One can swap out content states for probability distributions over finite belief corpora, and use EDGs as the basis for a Bayesian network in which protocols define the causal pathways of epistemic processes.[12] This, of course, is already part of the work on probabilistic dynamic doxastic logics [1].

7.6 Toward the Formalization of Epistemic Proceduralism

These brief excursions into modularity concludes the core explorations and refinements of the Revised PLEN framework's protocol logic. The core developments of this and the prior chapter—the addition of protocol graphs, the derivation of the theory of protocol graph isomorphism equivalence, the elaboration of the derivation system, and finally, its modularity—provide the foundation for applying PLEN as the protocol-theoretic logical framework for representing and reasoning about epistemic rationality norms, as well as the foundation for arguing that it is theoretically interesting useful in this role. The next and final part of the book carries out this application and spells out this argument. In particular, it particular, the next three chapters build ways of translating D1–D3 and D1C1–D3C3 into PLEN, along with additional constructions that enable the derivation of those translations in PLEN's protocol logic.

[11] Replace the choice execution rule in the transitional fragment with the rule for choice in PDL, and the regular protocol fragment now contains the logic of PDL.

[12] This, of course, is a path left for future work.

Appendix 4

Remark 374 This appendix spells out the results discussed in Chap. 7.

Revised Deduction Theorem

Corollary 375 (Deduction Theorem for Revised PLEN) $X \cup \{B\} \vdash_{PLEN} A$ *iff* $X \vdash_{PLEN} B \rightarrow A$.

Proof The proof in Chap. 4 only requires modification for three-premise rules:

(1) If A is derivable from $X \cup \{B\}$ by any three-premise rule, (Dr), then there are some formulae, C_1, C_2, and C_3 of formulae such that:

(a) $X \cup \{B\} \vdash_{PLEN} C_1$, $X \cup \{B\} \vdash_{PLEN} C_2$, and $X \cup \{B\} \vdash_{PLEN} C_3$ by steps of the proof that precede or are equal to step n.
(b) C_1, C_2, and C_3 are instances of the premises of (Dr).
(c) A is an instance of the consequent of (Dr).
(d) Thus, $X \vdash_{PLEN} B \rightarrow C_1$ by the induction hypothesis and a.
(e) Thus, $X \vdash_{PLEN} B \rightarrow C_2$ by the induction hypothesis and a.
(f) Thus, $X \vdash_{PLEN} B \rightarrow C_3$ by the induction hypothesis and a.
(g) Trivially, if A follows by means of rule (Dr) from C_1, C_2, and C_3, then for any set of premises Y, $Y \cup \{C_1\} \cup \{C_2\} \cup \{C_3\} \vdash_{PLEN} A$ by monotonicity.
(h) Thus, $\emptyset \cup \{C_1\} \cup \{C_2\} \cup \{C_3\} \vdash_{PLEN} A$ from d, e, and f.
(i) Thus, $\emptyset \vdash_{PLEN} ((C_1 \wedge C_2) \wedge C_3) \rightarrow A$ from h by Conditional Proof.
(j) Thus, $X \vdash_{PLEN} ((C_1 \wedge C_2) \wedge C_3) \rightarrow A$ from i by monotonicity, since $X = X \cup \emptyset$.
(k) Thus, $X \vdash_{PLEN} B \rightarrow ((C_1 \wedge C_2) \wedge C_3)$ from d, e, and f, by propositional logic.
(l) Hypothetical Syllogism (HS) is a derived law in the R-Derivation System (Conditional Introduction establishes it):
 (i) $\emptyset \vdash_{PLEN} (A \rightarrow B) \rightarrow ((B \rightarrow C) \rightarrow (A \rightarrow C))$
 (ii) The relevant instance: $\emptyset \vdash_{PLEN} (B \rightarrow (C_1 \wedge C_2)) \rightarrow (((C_1 \wedge C_2) \rightarrow A) \rightarrow (B \rightarrow A))$.
(m) Thus, $X \vdash_{PLEN} B \rightarrow A$ from k and l by monotonicity and Conditional Elimination.

□

Revised Completeness: Bits and Pieces

Proposition 376 *The R-Derivation System is sound, but the completeness and decidability arguments for the PLEN derivation system fail.*

Proof The small model theorem fails, as no EDGs containing states that verify graph-theoretic equivalences for protocols containing iterations can be finite; protocol graphs for iterations are not finite graphs. Thus, the naive effective procedure devised for the basic system never terminates for graphs verifying formulae as simple as $a* \equiv_G a*$. Thus, decidability fails and the argument for the verifiability lemma fails. Thus, the argument for completeness breaks down. □

Revised Small Model Property

Conjecture 377 (Revised Small Model Property for PLEN) If A is a verifiable formula of L^R_{PLEN}, then A is verifiable in an EDG, g, such that

$$|S| \leq 2^{(len(\pi_1)+...+len(\pi_4)+|FL_\pi([\pi_1]_x \top)|+...+|FL_\pi([\pi_4]_x \top)|)}$$

Proof (Sketch) Let A be a verifiable protocol-theoretic formula defined over P^R_{PLEN}. Then there is an EDG, g, such that A is verifiable at some s: $v_s(A) = 1$ for some $s \in S$. Let $FL(A)$ be the Fischer-Ladner Closure of A. By the filtration lemma, there is a filtration of g by $FL(A)$, $g_{FL(A)}$, such that there is a $[s]$ such that $v_{s/FL(A)}(A) = 1$ (i.e., $[s]$ verifies A). By construction, there are no more states in $g_{FL(A)}$ than there are distinct truth assignments to formulae in $FL(A)$ (the definition of filtrations collapses these states into $[s]$). Thus, from FLL3:

(1) If A is a propositional formula, then $|FL(A)| \leq len(A)$.
(2) If $A = [\pi]_x B$, then $|FL_\pi([\pi]_x B)| \leq len(\pi)$.
(3) If $A = R_x(\pi_1, \pi_2)$, then

$$|FL(A)| \leq (len(\pi_1) + len(\pi_2) + |FL_\pi([\pi_1]_x \top)| + |FL_\pi([\pi_2]_x \top)|)$$

(4) If $A = R_x(\pi_1, \pi_2, \pi_3, \pi_4)$,

$$|FL(A)| \leq (len(\pi_1) + ... + len(\pi_4) + |FL_\pi([\pi_1]_x \top)| + ... + |FL_\pi([\pi_4]_x \top)|)$$

Thus, the number of distinct truth assignments to formulae in $FL(A)$ can be no more than:

$$2^{(len(\pi_1)+...+len(\pi_4)+|FL_\pi([\pi_1]_x \top)|+...+|FL_\pi([\pi_4]_x \top)|)}$$

□

Definition 378 (Revised Naive PLEN Verification Procedure) Let $A \in L^R_{PLEN}$ defined over bounded iterations. To determine whether A is verifiable:

(1) Construct all finite EDGs such that $|S| =$

$$2^{(len(\pi_1)+...+len(\pi_4)+|FL_\pi([\pi_1]_x \top)|+...+|FL_\pi([\pi_4]_x \top)|)}$$

(2) Check each state for verification of A.

Conjecture 379 The Revised Naive PLEN Verification Procedure is effective.

Proof (Sketch) By the Revised Small Model Theorem for PLEN, if A is verifiable, then it's verifiable in one of the EDGs constructed by Step 1. Step 2 executes an exhaustive search of all states in all such models. Thus, the procedure terminates with an affirmative verdict on A or it exhausts the set of EDGs constructed by Step 1, and by the Small Model Theorem, A is not verifiable. □

Conjecture 380 (Revised Decidability (PLEN)) The verifiability problem of PLEN is decidable.

Proof (Sketch) The effectiveness of the Naive PLEN Verification Procedure suffices immediately. □

Completeness for the R-Derivation System with Bounded Iterations

Conjecture 381 The R-Derivation System defined over bounded iterations is sound and plausibly complete with respect to the revised semantics.

Proof (Sketch) If the filtration and small model theorems work for the basic system, then they should extend to this revised system immediately, by means of the preservation of graph-theoretic structure in the definition of filtrations for PLEN. In particular, the requirement that the paths in the filtration of g have isomorphic counterparts in g and vice versa extends the preservation of verification properties of filtrations to the graph-theoretic protocol equivalence formulae by Lemma L4 in the SUT and PGIIRC proofs. □

References

1. Baltag, A., and S. Smets. 2008. Probabilistic dynamic belief revision. *Synthese* 165(2):179–202.
2. Priest, K. 2008. *If to is: An introduction to non-classical logics*. Cambridge: Cambridge University.
3. van Benthem. J. 1996. *Exploring logical dynamics*. Stanford: CSLI Publications.
4. van Benthem. J. 2007. Dynamic logic for belief revision. *Journal of Applied Non-Classical Logics* 17(2):129-155.
5. van Benthem. J. 2014. *Logical dynamics of information and interaction*. Cambridge: Cambridge University Press.

Part III
Applying PLEN

Chapter 8
Philosophical Results I: A Protocol-Theoretic Logic of Epistemic Deontics, Procedural Knowledge, and Norm Application

Abstract This chapter presents a penultimate layer of elaborations of PLEN. First, a protocol-theoretic logic of epistemic deontics is built into the Revised PLEN framework of Chaps. 4, 5, 6, and 7. The deontic logic is defended from several different directions. First, a number of immediate philosophical arguments, drawing on ideas from the inception of deontic logic, show the greater plausibility of protocol-theoretic deontics as compared to the intensional operators of standard deontic logic or even the merely dynamic operators of dynamic deontic logic. Second, the chapter observes, drawing on Chap. 6, that protocols are hyperintensional objects; indeed, protocol graph isomorphism equivalence is a form of hyperintensional equivalence. This draws a broad and formally important consonance between the protocol-theoretic analysis of epistemic norms and arguments for the hyperintensionality of normative concepts in general. Third, the chapter shows that the protocol-theoretic logic solves a wide array of paradoxes of deontic logic. The second system added by this chapter is a logic of procedural knowledge and norm application. Procedural knowledge is knowledge of the procedural requirements of a system of norms or rules, and this notion is built into the logic of norm application, per the analysis of the regulative role of norms in prior chapters. The latter additions are given a much less extensive defense, as their primary notions are derived from prior considerations raised about the regulative role of epistemic norms. Both systems are only minimally explored from a technical perspective, as they serve primarily to extend the expressive power of PLEN in order to capture some of the principles of epistemic proceduralism.

R. Jenkins, *A Protocol-theoretic Framework for the Logic of Epistemic Norms*, Synthese Library 467, https://doi.org/10.1007/978-3-031-08597-0_8

8.1 Introduction

This chapter builds the protocol-theoretic logic of epistemic deontics, procedural knowledge, and norm application into the Revised PLEN Framework.[1] The distinctive feature of these new systems is that they provide protocol-theoretic analyses of epistemic deontic operators, knowledge operators, and novel operators concerning norm application. These projects are, respectively, taken on in Sects. 8.2 and 8.3. I spend a significant portion of Sect. 8.2 defending and motivating the deontic logic. The logic of procedural knowledge and norm application is not as extensively defended, because it is based primarily on observations about the regulative role of epistemic norms that were defended in Chaps. 2 and 3.

8.2 A Protocol-Theoretic Logic of Epistemic Deontics (PLED)

8.2.1 PLED: The Idea

Deontic logic is the central approach to the logic of norms. The standard approach of deontic logic, in turn, is to directly adapt a modal logic, interpreting the accessibility relation over possible worlds as a kind of deontic possibility—possibility with respect to some set of norms, rules, or standards. More precisely, the idea is that modal frames, $\langle W, R\rangle$, where W is a set of worlds, and R is an arbitrary accessibility relation over W, analyze a kind of normative necessity and possibility. R tells us what worlds are reachable by courses of action that follow an unspecified normative protocol, and so the accessible worlds from w are those worlds that can be reached without violating whatever norms, rules, or standards are hidden behind R. Let's call this a deontic accessibility relation, and say that w' is rightly or correctly accessible from w when wRw'.

Standard deontic logics (SDLs), then, enable us to reason about deontic accessibility relations in modal logics built atop modal frames. Setting aside the various paradoxes or other objections to this approach, there is a fundamental limitation to it: it is norm-agnostic, and this renders it inappropriate as a logic of norms. The logic of necessity in a deontic logic is a norm-agnostic logic of obligation; it attempts to codify valid reasoning about the notion of obligation simpliciter, independent of the properties of norms, rules, or standards themselves. Each frame has a distinct accessibility relation, save for their similarities with respect to sets

[1] A terminological note: I'll sometimes use the phrase "epistemic deontic logic", and this is intended to refer to a logic of *epistemic deontics*, not a combination of epistemic and deontic logic. Of course, PLEN does contain doxastic operators, b and d, and this chapter even adds a knowledge operator. Still, the logic of these notions is minimally developed at this stage, and the deontic logic is independent of the knowledge operators while still being an epistemic deontic logic.

of formal constraints (e.g., transitivity, symmetry, reflexiveness). As such, each frame is a model of deontic accessibility with respect to some (any?) system of norms, rules, or standards. The valid inference forms and theorems of SDLs are those that characterize the rules that are valid with respect to all models satisfying whatever constraints are imposed on the accessibility relations of the frames of a given system of SDLs. The logic of SDLs thus characterizes obligation in a way that is independent of any account of the norms, rules, or standards that define deontic accessibility (other than that given by the constraints on the accessibility relation).

Now, analyzing the general notions of obligation and permission is itself an interesting project, but it makes SDLs fundamentally unfit as a logic of norms. There is no reasoning about norms in an SDL, and this is a fatal expressive limitation. There is no way, for instance, to say that norm system N_1 obligates doing a while N_2 obligates doing b, let alone to reason about what conditions imply this sort of thing or what follows if it's true. SDL just isn't a logic of norms because it doesn't provide a framework for reasoning about norms or reasoning that takes norms as objects of our reasoning. Additionally, it provides no real philosophical insight into the nature of the deontic accessibility relation. The accessible worlds from w are just the worlds that are "OK" in some unspecified normative sense.

It's natural, given the interest of a modal approach to the logic of norms and the recognition that sometimes normative systems can themselves be reasoned about, to then generalize this approach by means of multi-modal logic, in which we replace R with a set of deontic accessibility relations, $R_1, \ldots, R_n$, each representing what is rightly accessible with respect, again, to an unspecified set of norms, rules, or standards. Here, we actually get different notions of obligation, one corresponding to each modal sublogic, $L_{R_1}, \ldots, L_{R_n}$, for each deontic accessibility relation. This is a step forward, as it at least allows us to reason about different notions of obligation and permission with respect to distinct, unspecified normative systems.

Nonetheless, a multi-modal approach to deontic logic has three fatal failings.[2] First, the preceding expressive limitation persists. Even multi-modal logics of obligation are not really logics of norms because they do not formalize any forms of direct reasoning about norms. This is a result of their failure to be norm exogenous. By "direct reasoning about norms", I mean reasoning of the forms articulated by D3C1. Roughly, a logic of norms should provide a logical framework for analyzing and validating (or invalidating) reasoning of forms like "Norms N_1, N_2, ..., N_n

[2] Here, I'm treating Meyer's dynamic deontic logic, PDeL, as an instance of this multi-modal approach [15, 16]. It has the pleasing feature that modals are fundamentally action operators. That system identifies a violation constant, V, (understood, ultimately, as encoding the claim that the unspecified system of norms has been violated), and defines the deontic operators in terms of quantification over the deontically accessible worlds (with respect to the various accessibility relations in the logic, each mapped to PDL programs) at which the violation constant holds. Forbiddenness of a, for instance, means $[a]V$. I'm not treating Segerberg's dynamic deontic logic [21] as an instance. A comprehensive comparison of the virtues of PLED against those of rival systems is reserved for future work.

bear relation, F; therefore, ...". This directly entails norm exogeny, which SDLs do not have.

Second, the logic of norms in a multi-modal system is defined by binary accessibility relations, and this makes it impossible to recognize the litany of normative distinctions introduced in Chaps. 1–3 with respect to procedural norms. Indeed, the core of this problem is excised and expanded upon in Chap. 4 in the guise of the so-called processes problem.

Third, there is, again, no principled analysis of the norms that define deontic accessibility. While the R_i may provide different deontic accessibility relations with different "shapes" concerning what is obligatory or permissible, they still don't shed light on how norms define deontic accessibility or what deontic accessibility comes to.

The point of PLED, then, is to solve these problems.[3] The solution lies in explicitly formalizing the relativity that the multi-modal approach recognized (namely, that the logic of obligation and permission must be defined relative to systems of norms), while also providing a principled analysis of epistemic norms themselves that analyzes deontic accessibility relations in terms of norms. PLEN already provided a norm exogenous system. The task is thus to analyze deontic operators in terms of protocols. The resulting focus on action coheres better with the original intentions of deontic logic than SDL (or a multi-modal generalization), while providing means of solving some of the bestiary of paradoxes of SDL and even multi-modal or dynamic generalizations of SDL.

8.2.2 *Defining the PLED System*

Remark 382 This section defines a class of epistemic deontic operators in terms of protocols, which lays the foundation for a more full-throated protocol-theoretic epistemic deontic logic. While a full development of an epistemic deontic logic in PLEN is reserved for future work, a basic system is defined, and the deontic fragment of it is shown to have a number of useful features.[4] The basic thrust of the deontic logic in PLEN is that (epistemic) deontic statuses should be defined relative to procedural epistemic norms, which are themselves analyzed with protocols. Thus, the deontic operators in PLEN are defined in terms of the properties of protocols. In particular, the properties of the protocol modalities, $[\pi]_x$ and $\langle\pi\rangle_x$, as well as the graph-theoretic properties of the execution sets of protocols, are built into the definitions of suites of deontic operators. As such, PLEN can encode deontic operators that apply to states, transitions, processes, procedures, and propositions

[3] Any deontic logic should grapple with the foregoing problems, so, while PLED is a protocol-theoretic logic of *epistemic* deontics, it is nonetheless built to solve problems for deontic logic generally, and the defense of PLED approaches the matter accordingly.

[4] For the impatient, the useful features are articulated in Sect. 8.3 and Chaps. 9 and 10.

because protocols have structural relations to all of these objects that correspond neatly to intuitive notions of dictation and permission. I present the system entirely and directly before commenting on it.

Claim 383 (Deontic Properties of Protocols) The following dictation relations are the groundwork for adding deontic operators to PLEN. The essential thought, here, is that protocols dictate other protocols in the sense that the execution of one protocol requires the execution of other protocols.

Definition 384 (Meta-Level Protocol Dictation) For arbitrary π and π':

(1) If π is executable in any EDG, then π *dictates* π', $\pi \Rightarrow \pi'$, iff for every g in which π is executable, every execution of π contains or corresponds to an execution of π'; else, π dictates nothing.

(2) If π is executable in g, then π *dictates* π' in g, $\pi \Rightarrow_g \pi'$, iff every execution of π in g contains or corresponds to an execution of π'; else, π dictates nothing in g.

(3) If π is executable in g, then π *dictates* π' at s in g, $\pi \Rightarrow_{g,s} \pi'$, iff every execution of π starting at s in g contains or corresponds to an execution of π'; else, π dictates nothing at s in g

(4) If π is executable in any EDG, then π *dictates* $\pi_1, \ldots, \pi_n$ *exclusively* at s in g, $\pi \Rightarrow_{g,s,x} \pi_1, \ldots, \pi_n$, iff every execution of π starting at s in g contains or corresponds to executions of $\pi_1, \ldots, \pi_n$ and no other protocols; else, π exclusively dictates nothing at s in g.

(5) If π is executable in any EDG, then π *dictates* $\pi_1, \ldots, \pi_n$ *exclusively* in g, $\pi \Rightarrow_{gx} \pi_1, \ldots, \pi_n$, iff every execution of π in g contains or corresponds to executions of $\pi_1, \ldots, \pi_n$ and no other protocols; else, π exclusively dictates nothing in g.

(6) If π is executable in any EDG, then π *dictates* $\pi_1, \ldots, \pi_n$ *exclusively*, $\pi \Rightarrow_x \pi_1, \ldots, \pi_n$, iff every execution of π in any g contains or corresponds to executions of $\pi_1, \ldots, \pi_n$ and no other protocols; else, π exclusively dictates nothing.

(7) If π is executable in any EDG, then π *dictates* p, $\pi \Rightarrow p$, iff for all g, every execution of π contains or corresponds to a subpath isomorphic to p; else, π dictates nothing.

    (a) Exclusive, graph-relative, and state relative variants are straightforward.

(8) If π is executable in any EDG, then π *dictates* $c(s)$, $\pi \Rightarrow c(s)$, iff for all g, every execution of π terminates with a state, s', such that $c(s) = c(s')$; else, π dictates nothing.

    (a) Exclusive, graph-relative, and state relative variants are straightforward.

(9) If π is executable in any EDG, then π *dictates* s, $\pi \Rightarrow s$, iff for all g, every execution of π terminates with a state, s; else, π dictates nothing.

    (a) Exclusive, graph-relative, and state relative variants are straightforward.

(10) If π is executable in any EDG, then π *dictates* A, $\pi \Rightarrow A$, iff for all g, every execution of π terminates with a state, s, such that $s \Vdash A$; else, π dictates nothing.

(a) Exclusive, graph-relative, and state relative variants are straightforward.

Remark 385 The dictation relation codifies the idea that procedural epistemic norms define requirements on epistemic action, and these requirements can (as argued in Chaps. 1–3) be translated into requirements on other epistemic factors. Note, as in Chap. 5, that there are no vacuous dictations; a protocol that is not executable anywhere can not dictate any actions.

Claim 386 On the basis of this notion of dictation, we can derive a meta-level protocol-theoretic analysis of deontic operators defined in terms of the dictation properties of protocols. The idea, informally, is that protocols obligate other protocols (actions, processes, procedures) when they dictate them. The dictated protocols become non-optional with respect to the dictating protocol. Permission, then, is compatibility with a protocol, and prohibition or forbiddance is incompatibility. More formally:

Definition 387 (Meta-Level Protocol-Theoretic Deontic Status) For arbitrary π, π', and g in which π is executable:

(1) (Requirement/Obligation) π requires/obligates π', $\pi R \pi'$, iff $\forall p \exists q (p \in P_\pi \rightarrow (q \in P_{\pi'} \wedge (q \sqsubseteq p \vee q \leftrightharpoons p)))$ (iff π dictates π' in g).
(2) (Permission) π permits π', $\pi P \pi'$, iff $\exists p \exists q (p \in P_\pi \wedge (q \in P_{\pi'} \wedge (q \sqsubseteq p \vee q \leftrightharpoons p)))$ (iff π doesn't dictate $-\pi'$ in g).
(3) (Prohibition/Forbiddance) π prohibits/forbids π', $\pi F \pi'$, iff $\forall p \forall q ((p \in P_\pi \wedge q \in P_{\pi'}) \rightarrow \neg(q \sqsubseteq p \vee q \leftrightharpoons p))$ (iff π dictates $-\pi'$ in g).
(4) (State Requirement) π requires π', $\pi R \pi'$, at s iff $\forall p \exists q (s = first(p) \wedge p \in P_\pi \rightarrow (q \in P_{\pi'} \wedge (q \sqsubseteq p \vee q \leftrightharpoons p)))$.
(5) (State Permission) π permits π', $\pi P \pi'$, at s iff $\exists p \exists q (s = first(p) \wedge p \in P_\pi \wedge q \in P_{\pi'} \wedge (q \sqsubseteq p \vee q \leftrightharpoons p))$.
(6) (State Prohibition) π prohibits π', $\pi F \pi'$, at s iff $\forall p \forall q ((s = first(p) \wedge p \in P_\pi) \rightarrow \neg(q \sqsubseteq p \vee q \leftrightharpoons p))$.

Claim 388 With that foundation in place, add the following deontic operators and their semantics to L^R_{PLEN} to construct L^D_{PLEN}, the protocol-theoretic deontic language of PLED.

Definition 389 (Protocol-Theoretic Deontics: Syntax (L^D_{PLEN}) and Semantics)

(1) (Protocol-theoretic Deontics: Syntax) Defining P_{PLEN} and L^R_{PLEN} as normal, L^D_{PLEN} is the language resulting from the (inductive) addition of all formulae of the form:

(a) (Rational Obligation w.r.t. π) If $\pi, \pi' \in P_{PLEN}$, then $RO_\pi(\pi') \in L^D_{PLEN}$.
(b) (Rational Permission w.r.t. π) If $\pi, \pi' \in P_{PLEN}$, then $RP_\pi(\pi') \in L^D_{PLEN}$.

(c) (Rational Forbiddance w.r.t. π) If $\pi, \pi' \in P_{PLEN}$, then $RF_{\pi}(\pi') \in L^{D}_{PLEN}$.

(2) (Methodological Deontics: Syntax) Defining P_{PLEN} and L^{R}_{PLEN} as normal, L^{D}_{PLEN} is the language resulting from the (inductive) addition of all formulae of the form:

(a) (Methodological Obligation w.r.t. M) If $M \subseteq P_{PLEN}$ and $\pi \in P_{PLEN}$, then $RO_M(\pi) \in L^{D}_{PLEN}$.
(b) (Methodological Permission w.r.t. M) If $M \subseteq P_{PLEN}$ and $\pi \in P_{PLEN}$, then $RP_M(\pi) \in L^{D}_{PLEN}$.
(c) (Methodological Forbiddance w.r.t. M) If $M \subseteq P_{PLEN}$ and $\pi \in P_{PLEN}$, then $RF_M(\pi) \in L^{D}_{PLEN}$.

(3) (Protocol-theoretic Deontics: Semantics) Letting EDGs be defined as normal, and letting g be an arbitrary EDG:

(a) (Rational Obligation w.r.t. π) $s \Vdash RO_{\pi}(\pi')$ iff $\pi R \pi'$ where the definition of R quantifies over exactly the paths with s as their roots.
(i) (iff π dictates π' at s in g).
(b) (Rational Permission w.r.t. π) $s \Vdash RP_{\pi}(\pi')$ iff $\pi P \pi'$ where the definition of P quantifies over exactly the paths with s as their roots.
(i) (iff π doesn't dictate $-\pi'$ at s in g).
(c) (Rational Forbiddance w.r.t. π) $s \Vdash RF_{\pi}(\pi')$ iff $\pi F \pi'$ where the definition of F quantifies over exactly the paths with s as their roots.
(i) (iff π doesn't dictate $-\pi'$ at s in g).

(4) (Methodological Deontics: Semantics) Letting EDGs be defined as normal, and letting g be an arbitrary EDG:

(a) (Methodological Obligation w.r.t. M) $s \Vdash RO_M(\pi')$ iff for every p in the non-empty intersection of P_{π_i} for all π_i such that $s \Vdash M(\pi_i)$ is such that if $s = first(p)$, then there is a $q \in P_{\pi'}$ such that $q \sqsubseteq p$ or $q \leftrightharpoons p$; else, the intersection is empty and the methodology dictates/requires/obligates nothing.
(b) (Methodological Permission w.r.t. M) $s \Vdash RP_M(\pi')$ iff for some p in the non-empty intersection of P_{π_i} for all π_i such that $s \Vdash M(\pi_i)$ is such that if $s = first(p)$, then there is a $q \in P_{\pi'}$ such that $q \sqsubseteq p$ or $q \leftrightharpoons p$; else, the intersection is empty and the methodology permits nothing.
(c) (Methodological Forbiddance w.r.t. M) $s \Vdash RF_M(\pi')$ iff for no p in the non-empty intersection of P_{π_i} for all π_i such that $s \Vdash M(\pi_i)$ is such that if $s = first(p)$, then there is a $q \in P_{\pi'}$ such that $q \sqsubseteq p$ or $q \leftrightharpoons p$; else, the intersection is empty and the methodology forbids nothing.

Claim 390 The language of L^D_{PLEN} also sustains the application of protocol-theoretic deontic operators to objects other than protocols.[5] First, we must add a language that allows us to denote a fuller range of epistemic factors. This is not done with a full, first-order language, but by adding constants for objects.

Definition 391 (Dynamic and Static Objects: Syntax (L^D_{PLEN}) and Semantics)

(1) (Protocol-theoretic Deontics) Syntax

(a) (Objects) Add finite sets of names for paths, transitions, edges, and states to the language of L^D_{PLEN}:
 (i) $P_c = \{p^1_c, \ldots, p^n_c\}$, $T_c = \{t^1_c, \ldots, t^n_c\}$, $E_c = \{e^1_c, \ldots, e^n_c\}$, $S_c = \{s^1_c, \ldots, s^n_c\}$

(b) (Object: Conditions) If $\pi \in P_{PLEN}$, and if $A \in L^D_{PLEN}$, then $RO_\pi(A), RP_\pi(A), RF_\pi(A) \in L^D_{PLEN}$.

(c) (Object: Paths) If $\pi \in P_{PLEN}$, and $p^i_c \in P_c$, then $RO_\pi(p^i_c)$, $RP_\pi(p^i_c)$, $RF_\pi(p^i_c) \in L^D_{PLEN}$.

(d) (Object: Transitions) If $\pi \in P_{PLEN}$, and $t^i_c \in T_c$, then $RO_\pi(t^i_c)$, $RP_\pi(t^i_c)$, $RF_\pi(t^i_c) \in L^D_{PLEN}$.

(e) (Object: Edge) If $\pi \in P_{PLEN}$, and $e^i_c \in e_c$, then $RO_\pi(e^i_c)$, $RP_\pi(e^i_c)$, $RF_\pi(e^i_c) \in L^D_{PLEN}$.

(f) (Object: State) If $\pi \in P_{PLEN}$, and $s^i_c \in S_c$, then $RO_\pi(s^i_c)$, $RP_\pi(s^i_c)$, $RF_\pi(s^i_c) \in L^D_{PLEN}$.

(2) The extension of the syntax to methodologies is straightforward:

(a) (Object: Conditions) If $M \subseteq P_{PLEN}$, and if $A \in L^R_{PLEN}$, then $RO_M(A), RP_M(A), RF_M(A) \in L^D_{PLEN}$.

(b) (Object: Paths) If $M \subseteq P_{PLEN}$, and if $p \in P_g$ for some g, then $RO_M(p^i_c), RP_M(p^i_c), RF_M(p^i_c) \in L^D_{PLEN}$.

(c) (Object: Transitions) If $M \subseteq P_{PLEN}$, and if $t \in T_g$ for some g, then $RO_M(t^i_c), RP_M(t^i_c), RF_M(t^i_c) \in L^D_{PLEN}$.

(d) (Object: Edge) If $M \subseteq P_{PLEN}$, and if $e \in R{+}_0$ for some g, then $RO_M(e^i_c), RP_M(e^i_c), RF_M(e^i_c) \in L^D_{PLEN}$.

(e) (Object: State) If $M \subseteq P_{PLEN}$, and if $s \in S$ for some g, then $RO_M(s^i_c), RP_M(s^i_c), RF_M(s^i_c) \in L^D_{PLEN}$.

Definition 392 (Dynamic and Static Objects: Syntax (L^D_{PLEN}) and Semantics)

(1) (Protocol-theoretic Deontics) Semantics

(a) For every s in g, let $P_s = \{p_i \mid s = f(p_i)\}$, $T_s = \{t_i \mid s = f(t_i)\}$, and $E_s = \{e_i \mid s = f(e_i)\}$, $S_s = \{s_i \mid \exists p \in P_g(s = f(p), s_i = l(p))\}$.

[5] Indeed, we must add formulae to L^D_{PLEN} that do so in order to enable the expression of norm-kernels that take epistemic factors other than protocols as objects.

(b) Add interpretation functions that assign names to dynamic and static objects in g:

(i) $\twoheadrightarrow_p: P_c \longrightarrow P_g$
(ii) $\twoheadrightarrow_t: T_c \longrightarrow T_g$
(iii) $\twoheadrightarrow_e: E_c \longrightarrow E_g$
(iv) $\twoheadrightarrow_s: S_c \longrightarrow S_g$

(c) (Object: Conditions) Given that π is executable in g:

(i) $s \Vdash RO_\pi(A)$ iff $P_\pi \neq \emptyset$ and $\forall p \in P_s(p \in P_\pi \rightarrow last(p) \Vdash A)$.
(ii) $s \Vdash RP_\pi(A)$ iff $\exists p \in P_s(p \in P_\pi \wedge last(p) \Vdash A)$.
(iii) $s \Vdash RF_\pi(A)$ iff $P_\pi \neq \emptyset$ and $\forall p \in P_s(p \in P_\pi \rightarrow last(p) \nVdash A)$.

(d) (Object: Paths) Given that π is executable in g:

(i) $s \Vdash RO_\pi(p_c^i)$ iff $\forall p' \in P_s(p' \in P_\pi \rightarrow (\twoheadrightarrow_p (p_c^i) \sqsubseteq p' \vee \twoheadrightarrow_p (p_c^i) \rightleftharpoons p'))$.
(ii) $s \Vdash RP_\pi(p_c^i)$ iff $\exists p' \in P_s(p' \in P_\pi \wedge (\twoheadrightarrow_p (p_c^i) \sqsubseteq p' \vee \twoheadrightarrow_p (p_c^i) \rightleftharpoons p'))$.
(iii) $s \Vdash RF_\pi(p_c^i)$ iff $\forall p' \in P_s(p' \in P_\pi \rightarrow \neg(\twoheadrightarrow_p (p_c^i) \sqsubseteq p' \vee \twoheadrightarrow_p (p_c^i) \rightleftharpoons p'))$.

(e) (Object: Transitions) Given that π is executable in g:

(i) $s \Vdash RO_\pi(t_c^i)$ iff $\forall p \in P_s(p \in P_\pi \rightarrow \twoheadrightarrow_t (t_c^i) = \langle f(p), l(p)\rangle)$.
(ii) $s \Vdash RP_\pi(t_c^i)$ iff $\exists p \in P_s(p \in P_\pi \wedge \twoheadrightarrow_t (t_c^i) = \langle f(p), l(p)\rangle)$.
(iii) $s \Vdash RF_\pi(t_c^i)$ iff $\forall p \in P_s(p \in P_\pi \rightarrow \twoheadrightarrow_t (t_c^i) \neq \langle f(p), l(p)\rangle)$.

(f) (Object: Edges) Given that π is executable in g:

(i) $s \Vdash RO_\pi(e_c^i)$ iff $\forall p \in P_s(p \in P_\pi \rightarrow (f(\twoheadrightarrow_e (e_c^i)) = f(p) \wedge l(\twoheadrightarrow_e (e_c^i)) = l(p)))$.
(ii) $s \Vdash RP_\pi(e_c^i)$ iff $\exists p \in P_s(p \in P_\pi \wedge (f(\twoheadrightarrow_e (e_c^i)) = f(p) \wedge l(\twoheadrightarrow_e (e_c^i)) = l(p)))$.
(iii) $s \Vdash RF_\pi(e_c^i)$ iff $\forall p \in P_s(p \in P_\pi \rightarrow \neg(f(\twoheadrightarrow_e (e_c^i)) = f(p) \wedge l(\twoheadrightarrow_e (e_c^i)) = l(p)))$.

(g) (Object: States) Given that π is executable in g:

(i) $s \Vdash RO_\pi(s_c^i)$ iff $\forall p \in P_s(p \in P_\pi \rightarrow \twoheadrightarrow_s (s_c^j) = l(p))$.
(ii) $s \Vdash RP_\pi(s_c^i)$ iff $\exists p \in P_s(p \in P_\pi \wedge \twoheadrightarrow_s (s_c^j) = l(p))$.
(iii) $s \Vdash RF_\pi(s_c^i)$ iff $\forall p \in P_s(p \in P_\pi \rightarrow \twoheadrightarrow_s (s_c^j) \neq l(p))$.

(h) The extension of the semantics to methodologies is straightforward; I give only some illustrative clauses:

(i) (Object: Conditions) Given that π is executable in g:

(A) $s \Vdash RO_M(A)$ iff for every p in the intersection of P_{π_i} for all π_i such that $s \Vdash M(\pi_i)$ is such that if $s = first(p)$, then $last(p) \Vdash A$.
(B) $s \Vdash RP_M(A)$ iff for some p in the intersection of P_{π_i} for all π_i such that $s \Vdash M(\pi_i)$ is such that if $s = first(p)$, then $last(p) \Vdash A$.
(C) $s \Vdash RF_M(A)$ iff for no p in the intersection of P_{π_i} for all π_i such that $s \Vdash M(\pi_i)$ is such that if $s = first(p)$, then $last(p) \Vdash A$.

Remark 393 While the direct application of deontic operators to formulae will be objected to below, PLED can nonetheless recapture this sort of deontic operator while shaving off many of its vicious features. We will see this more fully below, but, for now, one can think of PLED as providing an analysis of the deontic properties of conditions (formulae) that is derivative from the deontic properties of actions and processes, rather than the other way around. Perhaps it would be best to drop deontic operators that take formulae as operands entirely, but this chapter proceeds with the recapture, if only to show that PLED can house the idea.

Claim 394 The prior deontic operators were defined in terms of procedural executions. Deontic operators with slightly different properties can be defined by means of the other execution rules.

Definition 395 (Binary and Ternary Deontics: Syntax (L^D_{PLEN}) and Semantics)

(1) (Transitional Deontics) Syntax

(a) (Rational Transitional Obligation w.r.t. π) If $\pi, \pi' \in P_{PLEN}$, then $RO^t_\pi(\pi') \in L^D_{PLEN}$.
(b) (Rational Transitional Permission w.r.t. π) If $\pi, \pi' \in P_{PLEN}$, then $RP^t_\pi(\pi') \in L^D_{PLEN}$.
(c) (Rational Transitional Forbiddance w.r.t. π) If $\pi, \pi' \in P_{PLEN}$, then $RF^t_\pi(\pi') \in L^D_{PLEN}$.

(2) (Transitional Deontics) Semantics

(a) (Rational Transitional Obligation w.r.t. π) $s \Vdash RO^t_\pi(\pi')$ iff for all t, if $first(t) = s$ and $t \in T_\pi$, then $t \in T_{\pi'}$.
(b) (Rational Transitional Permission w.r.t. π) $s \Vdash RP^t_\pi(\pi')$ iff for some t, $first(t) = s$, $t \in T_\pi$, and $t \in T_{\pi'}$.
(c) (Rational Transitional Forbiddance w.r.t. π) $s \Vdash RF^t_\pi(\pi')$ iff for all t, if $first(t) = s$ and $t \in T_\pi$, then $t \notin T_{\pi'}$.

(3) (Ternary Deontics) Syntax

(a) (Rational Ternary Obligation w.r.t. π) If $\pi, \pi' \in P_{PLEN}$, then $RO^r_\pi(\pi') \in L^D_{PLEN}$.
(b) (Rational Ternary Permission w.r.t. π) If $\pi, \pi' \in P_{PLEN}$, then $RP^r_\pi(\pi') \in L^D_{PLEN}$.
(c) (Rational Ternary Forbiddance w.r.t. π) If $\pi, \pi' \in P_{PLEN}$, then $RF^r_\pi(\pi') \in L^D_{PLEN}$.

(4) (Ternary Deontics) Semantics

(a) (Rational Ternary Obligation w.r.t. π) $s \Vdash RO^r_\pi(\pi')$ iff for all e, if $first(e) = s$ and $e \in R+_\pi$, then $e \in R+_{\pi'}$.
(b) (Rational Ternary Permission w.r.t. π) $s \Vdash RP^r_\pi(\pi')$ iff for some e, $first(e) = s$, $e \in R+_\pi$, and $e \in R+_{\pi'}$.

(c) (Rational Ternary Forbiddance w.r.t. π) $s \Vdash RF^r_\pi(\pi')$ iff for all e, if $first(e) = s$ and $e \in R{+}_\pi$, then $e \notin R{+}_{\pi'}$.

Claim 396 While obligation, permission, and prohibition (forbiddance) are the core deontic operators, other deontics have been recognized, and PLEN can accommodate them, as well.

Definition 397 (Protocol-Theoretic Deontics: Optionality, Omissibility, and Defeasible Reasons) Where P_{PLEN} is defined as normal add (inductively) the following formulae and their semantics:

(1) (Defeasible Reasons: Syntax) If $\pi, \pi' \in P_{PLEN}$ and $R_1, R_2, R_3, A \in L^D_{PLEN}$, then

$$RD_\pi(A, R_1), RD_\pi(\pi', R_1) \in L^D_{PLEN}$$

(2) (Defeasible Reasons: Semantics) Let R_1 be a putative reason for A—a propositional variable with a suggestive name, not to be confused with other uses of capital "R" in the notation. Let R_2 be some conjunction excluding R_1 but compatible with R_1 ($\neg((R_1 \wedge R_2) \rightarrow \bot)$). Let R_3 be some conjunction including R_1 (where R_3 may be a consistent superset of R_2 but is not necessarily a superset or even consistent with R_2). Then:

(a) $(RD_\pi(A, R_1))$ $s \Vdash RD_\pi(A, R_1)$ iff one of the following conjunctions of conditions holds:
  (i) (Rejection)
    (A) $s \Vdash RP_\pi(b(R_2) \wedge d(A))$,
    (B) $s \Vdash RO_\pi(b(R_1 \wedge R_2) \rightarrow \neg d(A))$,
    (C) $s \Vdash RP_\pi(b(R_3) \wedge d(A))$.
  (ii) (Acceptance)
    (A) $s \Vdash RP_\pi(b(R_2) \wedge \neg b(A))$,
    (B) $s \Vdash RO_\pi(b(R_1 \wedge R_2) \rightarrow b(A))$,
    (C) $s \Vdash RP_\pi(b(R_3) \wedge \neg b(A))$.
  (iii) Say that "the single abstract agent modeled by PLEN has a defeasible reason, R_1, for accepting A at s in g".

(b) $(RD_\pi(\pi', R_1))$ $s \Vdash RD_\pi(\pi', R_1)$ iff the following conjunction of conditions holds:
  (i) $s \Vdash RP_\pi(b(R_2) \wedge \neg RO_\pi(\pi'))$,
  (ii) $s \Vdash RO_\pi(b(R_1 \wedge R_2) \rightarrow RO_\pi(\pi'))$,
  (iii) $s \Vdash RP_\pi(b(R_3) \wedge \neg RO_\pi(\pi'))$.
  (iv) Say that "the single abstract agent modeled by PLEN has a defeasible reason, R_1, for doing π at s in g."

(3) (Optionality)

(a) Syntax: If $\pi, \pi' \in P_{PLEN}$, then $ROP_\pi(\pi') \in L^D_{PLEN}$.

(b) Semantics: π' is optional but not a matter of indifference under π, $s \Vdash ROP_{\pi}(\pi')$ iff $s \Vdash (RP_{\pi}(\pi') \wedge (\neg RO_{\pi}(\pi') \wedge RD_{\pi}(\pi', R_1)))$ for some R_1.

(4) (Omissibility)

(a) Syntax: If $\pi, \pi' \in P_{PLEN}$, then $ROM_{\pi}(\pi') \in L^{D}_{PLEN}$.
(b) Semantics: π' is omissible (a matter of indifference) under π, $s \Vdash ROM_{\pi}(\pi')$ iff $s \Vdash (RP_{\pi}(\pi') \wedge \neg RO_{\pi}(\pi'))$.

Remark 398 The reduction of defeasible reasons to obligation and permission is meant to formalize several intuitive ideas. The disjunctive sets of conditions captures the different flavors of defeasible reasons, reasons against rejection and reasons for acceptance.

The Rejection conditions formalize the idea that R_1 is a defeasible reason for A only if there are some configurations of beliefs such that R_1 is the difference-maker between the permissibility and forbiddance of rejecting A. This decomposes. First, in order for R_1 to count as reason for A at all, R_1 must be capable of carrying the day; there must be some configuration of beliefs that entails that, on the basis of R_1, one ought (rationally) not reject A. The definition captures the "on the basis of R_1" notion by setting R_2 to be a conjunction of claims that doesn't by itself imply the obligation not to reject A (w.r.t π) while the conjunction of R_1 and R_2 does. Second, in order for R_1 to count as defeasible, there must be some configuration of beliefs that removes R_1's ability to carry the day; there must be some configuration of beliefs that entail that, despite R_1, one is (rationally) permitted to reject A.

Similarly, the Acceptance conditions formalize the similar idea that R_1 is a defeasible reason for A only if there are some configurations of beliefs such that R_1 is the difference-maker between the permissibility and forbiddance of not accepting A.

The defeasible reasons for protocols operate similarly. In PLED, one has a defeasible reason, R_1, to carry out protocol π' (i.e., to follow or comply with π') only with respect to a procedural norm, π, that, given some configuration of reasons including R_1, obligates doing π'. Of course the defeasibility enters by means of the condition that R_1 doesn't suffice to obligate doing π' under all configurations of reasons, even those including R_1.

Remark 399 PLED treats indifference[6] as omissibility, and distinguishes it from optionality. The idea being that a norm or normative system may provide us with reasons to do something permissible (i.e., the normative system may do more than merely permit a course of action), while not obligating it. The definitions seem to capture at least the thrust of the distinction, as the optional is not a matter of indifference with respect to normative upshots, but it's still optional; one needn't

[6] See the discussion of supererogation and related notions in [14] for a brief introduction.

do it on pain of norm violation. Omissibility is very much a matter of indifference (w.r.t. a given π).

Definition 400 (PLED) The system that results from incorporating the foregoing revisions of L^R_{PLEN} and the corresponding semantic rules into PLEN and operating on it according to the R-Derivation System is the protocol-theoretic logic of epistemic deontics (PLED).

8.2.2.1 The Logic of PLED

I omit any serious exploration of the derivation system or validity relation of PLED, as this lies beyond the scope of this book. Rather, I focus on the expressive extension that PLED provides to PLEN. It is not an obscure matter how to incorporate PLED, as defined, into the R-Derivation system, and I explore some specific valid inferences in Chap. 10. The metatheory of PLED is untouched, here.

8.2.3 Immediate Philosophical Arguments

Even without a thorough technical investigation, PLED has a number of noteworthy virtues. First, there are a number of philosophical arguments available to us for supporting the thought that PLED is the proper approach to the logic of epistemic deontics.

The main arguments are straightforward: PLED solves the problems raised in Sect. 8.2.1. First, PLED is norm exogenous, and thereby formalizes numerous forms of direct reasoning about epistemic norms. Second, PLED solves the processes problem, readily distinguishing deontic verdicts on actions, processes, and procedures from those on states and transitions. Third, PLED provides a principled analysis of deontic accessibility. Where SDLs (and multi-modal generalizations) leave the underlying protocols that define deontic accessibility unspecified entirely, PLED provides a robust syntax for analyzing such protocols, and execution rule systems for deriving deontic accessibility relations (binary executions) from the content of the norms that protocols represent. Moreover, protocols are hyperintensional objects, and so PLED provides a system in which, in principle at least, deontic accessibility relations can be defined that systematically distinguish intensional from hyperintensional contexts.[7]

A secondary argument in favor of PLED is that it is action focused. SDLs are modal logics, in which deontic operators are modal operators. As such, deontics are applied to sentences (sentential expressions and contexts). This is a terrible fit. For whom are the things denoted by sentential expressions (facts, truths, propositions?)

[7] This is not one of the projects taken on in this chapter, and is left to future work.

permitted or obligated? Who but Descartes' God is in a position to consider whether the inverse square law is permitted? What can it even mean for it to be permissible that (e.g.) "polar bears are mammals"? Actions are obligatory or permitted, not statements, propositions, states of affairs, etc. Facts, states of affairs, truthmakers, propositions, or whatever else sentences are supposed to express, denote, or represent are not the sorts of things that deontic statuses are assigned to in normative thinking. States of affairs are never obligatory, though it may be obligatory to bring about a state of affairs.[8] As such, deontic logic should be a logic of operators that are appended to some kind of expression that denotes actions (even if only actions defined in terms of the states of affairs they bring about).

Now, of course, it's standard to understand obligation and permission in SDL in terms of truth in deontically accessible worlds. So, for instance, $Ob(P)$, says that (e.g.) it is obligatory for you that polar bears are mammals, and this really means that polar bears are mammals in all deontically accessible worlds, which are worlds that can be brought about by your acting in accordance with some unspecified normative protocol. Thus, the logical truths, being true in all possible worlds, are true in all deontically accessible worlds, and so they are obligatory in the sense that (a fortiori) if your actions accord with the aforementioned unspecified normative protocol, you'll end up in a world where the logical truths are true. So, one can rest assured that there is a world that is correct according to some unspecified system of norms in which polar bears are mammals, and that whatever we do, if we act properly, we won't un-mammal the mammals or square the circle, and so on. So, SDL countenances oblique reference to action, and can be interpreted in a way that somewhat diminishes the immediate absurdity of its application of deontic operators to sentences.

This doesn't save the approach. Even setting aside the aforementioned failure to shed light on the underlying protocols that define deontic accessibility, the deontic operators simply fit better as guides to action because deontic language applies to actions in actual deliberation. We think about which courses of action are permitted, not which propositions are. In issuing norms, normative authorities don't permit or command propositions, but actions and courses of action. In giving normative judgments or regulative guidance, we recommend or command actions, not propositions. We don't deliberate about propositions, we deliberate about actions (including epistemic actions) that might bring them about. SDL just faces the wrong direction on the matter. From another angle, there are clearly two different senses of obligation and permission—that applying to actions and that, derivatively, applying to states of the world, conditions, or propositions. SDL wrongly collapses them.

---

[8] Which, of course, can be reasoned about in PLED very naturally by means of postcondition contexts or, more directly, by means of the protocol-theoretic deontics applied to conditions in PLED. For another approach that makes much the same point, see the STIT framework [7, 22, 24] in the logic of action.

The unsatisfactory nature of the SDL analysis of deontic notions is explicitly recognized by von Wright in the foundational work on deontic logic, though it somehow fell out of the picture in subsequent decades before being rediscovered [6]. PLED better fits von Wright's original, action-oriented approach to deontic logic. The fundamental deontic operators in PLED apply directly to protocols, which specify actions, processes, and procedures. Nonetheless, there are derivative deontic operators thrown in for good measure, just in case epistemic factors other than epistemic actions are properly the objects of deontics.

8.2.4 Protocol-Theoretic Logic and Hyperintensionality

I want to briefly consider the relations between deontics and hyperintensionality. It has been argued, persuasively, that many of the familiar paradoxes of deontic logic are products of the failure to recognize that deontic contexts are hyperintensional [4]. Now, recall that protocol graph isomorphism equivalence is a hyperintensional context and that protocol graph isomorphism equivalence is a criterion for hyperintensionality with respect to protocols that satisfies, for protocols at least, the central conditions on such criteria. This connects the analysis of deontic accessibility in PLED to hyperintensionality, but not in a way that accounts for the hyperintensionality of deontic contexts.

Deontic contexts in PLEN are not hyperintensional—necessarily logically (procedurally) equivalent protocols are intersubstitutable in deontic contexts in PLEN. Nonetheless, there is room for the definition of hyperintensional protocol-theoretic deontic contexts. One need only define deontic operators in terms of protocol graph isomorphism. For instance, replacing the procedural dictation relations in the definition of $RO_{\pi}(\pi')$ with $\supset_G$ would generate a context in which π_1 and π_2 such that $\Box(\pi_1 \leftrightarrow_p \pi_2)$ are not interchangeable.

More important, the failure of hyperintensionality of deontic contexts in PLED is not a fatal problem: the paradoxes that motivate the move to hyperintensionality fail in PLED even without recourse to explicit treatment of the deontic contexts as hyperintensional. The next section tackles this in a brief and, for many readers, likely unsatisfactory manner that is intended to inaugurate further work rather than settle matters.

8.2.5 Paradox Resolutions

Claim 401 Here, I outline, in a very sketchy form, the advantages of PLED as a deontic logic. What I want to indicate here is that PLED solves many of the core problems of standard deontic logics (and some of the paradoxes of dynamic deontic logics). Many of these solutions are grounded in the philosophical merits of the PLEN framework. These results are to be spelled out more in future work, but the main lines of these thoughts should be visible from general properties of PLEN. The thrust of the argument, here, is that the usual suite of paradoxes of deontic logic can be either resolved or dissolved in a principled, unified manner by the protocol-theoretic analysis of deontic operators. Strictly speaking, the paradoxes here are not specifically paradoxes of epistemic deontic logic, and so the solutions treat PLED as a deontic logic, rather than thinking of it strictly as a logic of epistemic deontics. The objective is thus to show that the protocol-theoretic logic of deontics in PLEN avoids these problems as a pure deontic logic. This provides a theoretically virtuous first step toward a more full-throated epistemic deontic logic.

8.2.5.1 Jorgensen's Dilemma

Jorgensen's Dilemma [8, 14] is resolved by PLED. Jorgensen's dilemma is typically received as a challenge to the very idea of deontic logic. The challenge is that, first, the following argument appears, prima facie, to be sound:

(1) Norms don't seem to be objects with truth conditions because they are (at least partly captured by) commands or imperatives.
(2) Logics are theories of what follows from what, and the operative notion of *following from* seems to be a relation that can only hold between objects that have truth conditions.
(3) ∴ Thus, there can be no logic of norms.

Second, however, it seems like norms can bear logical relations—including relations of following from and mutual following from (equivalence)—with one another. So, are we to explain away 1 or 2? Explaining away 1 seems to require rejecting the imperative content of norms, which would amount to rejecting their use in ante hoc regulation. Explaining away 2 requires a radical reconception of logical consequence that applies to imperatives.

The solution in PLED is to explain away 2, and to show how the logical relations among protocols (hence, norms) can be systematically incorporated into an old-fashioned truth preservation account of logical consequence. Protocols—like imperatives—have execution conditions rather than truth conditions, and these can be used to (i) define and investigate logical properties among norms, and (ii) to fill out the truth conditions of normative propositions. The execution conditions of protocols are compositional, and the execution rules in conjunction with the truth rules of protocol-theoretic expressions provides a systematic recipe for semantic interpretation of expressions in which protocol expressions are embedded. This solves at least a technical version of the embedding problem for normative expressions. Norms can thus bear logical relations defined in terms of relations between their execution conditions[9] and in terms of more straightforward logical

[9] This is all more fully fleshed out in Chaps. 4, 5, 6, and 7.

relations between embedding propositions, as in the protocol-theoretic logic of PLEN and the deontic logic of PLED.

8.2.5.2 The Necessity of Obligations

The Necessity of Obligations Problem [14, 23] is resolved by PLED. The Necessity of Obligations Problem is that it seems possible that there may be no obligations at all, but the definition of obligation in standard deontic logics renders this seemingly true thought incoherent; every necessary truth must be obligatory because the obligation operator is a necessity operator. Given that there are necessary truths, the thought that it's possible for there to be no obligations implies that each necessary truth, A, is both obligatory and not obligatory.

There are multiple routes to paradox, here. But none of them work in PLED. The protocol-theoretic analysis of deontics resolves this problem by analyzing deontics in a way that is relative to protocols and conditional on executability.

First, for every set of atomic or regular protocols without iteration, $\{\pi_1, \ldots, \pi_n\}$, there are epistemic directed graphs, g, and states in g, s, such that there is no protocol, process, state, or formula that is obligatory with respect to the individual π_i nor the methodology, $M = \{\pi_1, \ldots, \pi_n\}$, at any s in g. For the first case, just let $P_{\pi_i} = \emptyset$ for some π_i. Then $RO_{\pi_i}(x)$ may fail, even for necessary truths, as, explicitly, rational obligation with respect to π is defined in terms of dictation, and non-executable protocols dictate nothing (either generally or in specific EDGs). The paradox simply fizzles out for protocol-theoretic deontics. Methodological deontics similarly solve the problem. Let $P_{\pi_1} \cap \ldots \cap P_{\pi_n} = \emptyset$. Thus, $RO_M(x)$ may always fail, even for necessary truths, as they did for individual protocols. Explicitly, for necessary truth, A, the formulae $RO_\pi(A)$ and $RO_M(A)$ may fail because all deontics are conditioned on the executability of protocols and just so for methodologies.

This is not a universal feature of protocols, however. Iterations, $\pi*$, are executable at every s (that is, there are "zero length" iterations of $\pi*$ at every state or, in other words, executions of $\pi*$ that involve iterating π zero times), so these protocols do obligate necessary truths. Tests and fixed point protocols have this property as well. This doesn't undermine the solution of the necessity of obligations paradox. This brings us to the second solution that PLED provides.

Second, protocols encode the content of procedural norms and methodologies encode the content of systems of procedural norms. Some of these norms define requirements on action, and others don't. Some of these obligate necessities, and others don't. However, no particular norm or set of norms is stipulated in PLED to encode the "true requirements of epistemic normativity"; it is left open that there are actually no "true" obligations of epistemic normativity, whatever.

In other words, the puzzle is that there may be no obligations with respect to the correct system of norms (or perhaps that there is no correct system of norms), not that there are no obligations with respect to any norms. The latter would be demonstrably false; take any legal system or proposed norm (legitimate, coherent, or otherwise) as a decisive counterexample to the claim.

There are, of course, deontic operators relativized to any executable set of protocols, some of which do obligate every necessary truth, but these may be defined by protocols that are merely codifications of *possible* norms. PLED straightforwardly makes room for the possibility that there are no requirements of epistemic normativity, insofar as epistemic normativity is properly encoded by some particular set of protocols.

In sum, there are systems of norms in PLED for which necessary truths are not obligatory, and it's consistent with PLED that there may be no "real" epistemic obligations at all, even though there are some procedural norms that do obligate necessary truths.

8.2.5.3 Free Choice Permissions and Violability

The Free Choice Permission Paradox and Violability Puzzle [14, 19] are similarly resolved by the protocol-theoretic analysis of deontics in PLED. First, the Free Choice Permission Paradoxes take multiple forms. First, there is the problem that it seems true that if I am permitted to do x or y, then I am permitted to do both. More exactly, in the terms of SDL, if $P(A \vee B)$, then $P(A) \wedge P(B)$. But, letting P be our permission operator, $P(A \vee B)$ does not imply $P(A) \wedge P(B)$ in SDL. And if we added a rule or elaboration of semantics that validated that inference, the standard rules of disjunction introduction would trivialize the logic: for arbitrary B, $P(A)$ would imply $P(A \vee B)$ by disjunction introduction, which would imply $P(B)$ by the rule in question and conjunction simplification.

Second, there is the Violability Puzzle, which is also related to the interaction between disjunction and deontic operators. Letting O be our standard obligation operator, $O(A)$ implies $O(A \vee B)$, but this is absurd. Suppose that it's obligatory that I take care of my sister's cat. From this it follows that it's obligatory to take care of her cat or murder her cat. But then, if I don't take care of her cat, it's obligatory that I murder her cat. It's absurd that the obligation to take care of her cat somehow implies that I have an obligation that is violated by not murdering her cat in the case where I fail to take care of it; rather, it seems that such a murder is an obvious violation of the original obligation, not an obligation entailed by it. Worse, it is absurd that I can discharge the contracted obligation by neglecting and then murdering the cat.

The solutions to both of these problems are built into the protocol logic: the essential inferences for the paradoxes are ruled out on both formal and philosophical grounds. First, the deontic operators of PLED, RP_{π} and RO_{π}, do not obey rules parallel to disjunction introduction. Protocols are not formulae, and so $RO_{\pi}(\pi' \vee \pi'')$ and $RP_{\pi}(\pi' \vee \pi'')$ are ill-formed. The only construction in PLED that captures the intended free choice obligation or permissions are $RO_{\pi}(\pi' \cup \pi'')$ and $RP_{\pi}(\pi' \cup \pi'')$, respectively. The execution conditions of non-deterministic choice require nondeterminism; a choice is only executable if both choices are executable. Thus, the fact that an execution, p, of π contains an execution of π' does not entail that it contains an execution of $\pi' \cup \pi''$, as there may be no branching π'' path. Thus, $RO_\pi(\pi')$ does not imply $RO_\pi(\pi' \cup \pi'')$; that all executions of π contain executions of π' does not entail that any contain an execution of π'' by the same reasoning that shows the failure of this condition in the particular case of p. Thus, a fortiori, from the fact that $RP_\pi(\pi')$ holds at s, it does not follow that $RP_\pi(\pi' \cup \pi'')$ for any π'' let alone for arbitrary π''. Thus, the addition of a rule or law enabling inference of $RP_\pi(\pi') \wedge RP_\pi(\pi'')$ from $RP_\pi(\pi' \cup \pi'')$ would not trivialize the logic (by means of the argument that does so in standard deontic logic, at least).

Moreover, this is exactly as it should be, philosophically speaking, once we move away from the mistake of applying deontic operators to sentences; RO_π and RP_π function as protocol operators rather than merely sentential operators. That is, the protocol-theoretic deontics operate on the procedural recipes for procedures and actions, and this is a fundamentally different notion from deontics that operate on sentences. Deontic operators on procedures and actions fit naturally into the control or regulation of action where operators on sentences do not. The action controlling function of obligation is to restrict or determine admissible paths of action. To say that one is obligated to do a or b is to say that one is bound to choosing between a or b. Choosing between a or b or c is not an option; given the former obligation, choosing from this wider set is explicitly ruled out since we were to choose between a or b.

Consider, for instance, an epistemic or methodological norm that obligates the use of some specific formal procedure from a list of n procedures in the analysis of some data set or argument. The norm obligates free choice among the n procedures on the list; consider a logic handbook that provides numerous equipollent procedures for deduction, and a methodological norm that requires that one follow the handbook. Such a norm does not imply that one is obligated (or even permitted) to choose freely among all arbitrarily longer lists of $n + m$ procedures that contain the original list; the norm restricts one's choices to the original list! The original obligation implicitly requires that one not make a choice from the longer list. For another instance, note that obligation implies permission (in PLED and intuitively). Thus, if a norm, N, obligates some action, a, then, given a rule like disjunction introduction on actions, N also permits choosing from actions other than and incompatible with doing a (indeed, it permits choosing among all actions incompatible with doing a) despite the fact that it obligates a. This is just getting the concept of obligation wrong. A final instance: My students are obligated to do their final essays by a specified deadline. It does not follow from this obligation that they are (also) obligated to choose freely among an expanded list of options including any other freely chosen deadlines or skipping the essays entirely and submitting interpretive drawings instead. The original policy restricts the range of permissible choices, it should not entail the permission of broader ranges of choice.

The inference in question is no less absurd when applied to permissions. Imagine an epistemic (or other) agent saying, "You said I could come in late tomorrow, so I decided to burn down the church because it logically follows that I was permitted to choose between those options. Disjunction introduction and all that." For an epistemic example, consider the absurd: "This inference ticket permits me to conclude that Plato was mortal, since I fully accept that Plato was a biologically unexceptional human being, and all such humans are mortal. Thus, I've concluded that the moon is the fetus of an ancient space god, as the permission to infer either that Plato was mortal or the moon is the fetus of an ancient space god logically follows from the prior permission to infer the former disjunct." Note that these ludicrous maneuvers do not require the inference from $P(A \vee B)$ to $P(A) \wedge P(B)$ and then to $P(B)$. Given that $P(A \vee B)$ is an action-controlling construction, as normative verdicts should be (re: Chaps. 1–3), it tells agents that they are permitted to choose course of action (bringing it about that) A or course of action (bringing it about that) B. Thus, the foregoing agents are simply acting in compliance with the permission to choose from among a much larger range of options, and this permission is entailed by the original permission by disjunction introduction. In other words, the trivialization of permissibility is a direct consequence of disjunction introduction, and not the allegedly controversial inference from $P(A \vee B)$ to $P(A) \wedge P(B)$.

Generally, some procedural norms obligate or permit one very specific course of action, where the procedural structure of that course of action rules out other courses of action, some procedural norms obligate specific courses of action while explicitly permitting no others, and still other norms are merely silent on other courses of action. The obligation or permission of a procedure or action, π_1, by some procedural norm, π, thus does not, generally, imply that one is obligated or permitted to freely choose between π_1 and arbitrary other π_2. Additionally, procedural recipes for action don't expand the way that propositions do. Carrying out a doesn't imply that one carries out a nondeterministic choice among a or b; deliberation is a process and this process need not occur in doing a, and there may have been no choice to do b at all. All of these considerations undermine the plausibility of inferences parallel to disjunction introduction.

Given these observations, it is absurd to build it into our logic of norms that, upon discovering that some course of action is obligatory or permissible, one can freely infer the obligation or permission of arbitrarily wide ranges of free choices among courses of action by means of a rule parallel to disjunction addition. From another angle, the conceptual role of obligation is to restrict or constrain the range of options open to an agent. A disjunction introduction rule would undermine the basic function of the concept by exploding the range of options rather than constraining it.[10]

---

[10] Of course, disjunction introduction does still work in PLED. From $RP_\pi(\pi')$, it does follow that $RP_\pi(\pi') \vee RP_\pi(\pi'')$. However, this is quite distinct from $RP_\pi(\pi' \cup \pi'')$ in PLED. The former does not say that there is a permissible choice according to π; it says only that π permits π' or it permits π'' (quite possibly to the exclusion of the other). This conflation may be important, as the error that $P(A \vee B)$ expresses the permissibility of a choice may be what lies behind the intuition that it ought to imply the permissibility of both of the things we are permitted to choose among. More, this disjunction does not imply $RP_\pi(\pi' \cup \pi'')$, it does follow from $RP_\pi(\pi' \cup \pi'')$ and from $RP_\pi(\pi'')$, but it rather obviously does not and should not imply $RP_\pi(\pi') \wedge RP_\pi(\pi'')$.

Second, the protocol-theoretic logic of deontics rules out the inference of $RP_\pi(\pi') \wedge RP_\pi(\pi'')$ from $RP_\pi(\pi' \cup \pi'')$ (as well as the corresponding inference for RO_π) on formal and philosophical grounds. Recall that $RP_\pi(\pi' \cup \pi'')$ is true at s iff there is some execution, p, of π that contains or corresponds to an execution, q, of $\pi' \cup \pi''$. From this, it does not follow that there are executions of both π' and π'' that are contained in nor correspond to p, so it does not follow that $RP_\pi(\pi')$ and $RP_\pi(\pi'')$, though one of these conjuncts does follow and both may be true. For a countermodel, simply let some execution of π' be a subgraph of p where no execution of π'' is a subgraph or corresponds to any subgraph of p, and let q be an execution of π that contains a subgraph that contains or corresponds to an execution of π''. At no state in either p or q need $\pi' \cup \pi''$ be executable. Thus, the problematic law is invalidated in PLED straightforwardly by the execution rules.

Again, this is exactly as it should be. In general, norms may permit one or the other of actions π_1 and π_2 where doing π_1 renders it impermissible or impossible to do π_2 or converse. Consider, for instance, a classroom work submission policy of the form: "You may either take an Incomplete in the course or hand in your unfinished term paper on time as it is; if you do the former, you are not permitted to do the latter, and if you do the latter, you are not permitted to do the former." Such a policy permits choosing one or the other path while prohibiting the selection of both. Indeed, the permissibility of each is conditional on the other not being chosen, and the choice between them is permissible; $RP_\pi(\pi' \cup \pi'')$ is true but $RP_\pi(\pi' \cap \pi'')$ and $RP_\pi(\pi') \wedge RP_\pi(\pi'')$ are both false, where π encodes the policy in question. The idea is that $\pi' \cup \pi''$ is a procedural construction, and not a sentential construction; the choice procedure is permitted, and this permission doesn't and shouldn't push through to each option. Thus, that a protocol (i.e., procedural norm) permits choosing between π_1 and π_2 can't, as a matter of logic, imply that it permits both π_1 and π_2.

Third, the codification of norm compliance by means of execution conditions in PLEN removes some of what makes the violability puzzle so objectionable. Let us suppose, contrary to the logic of PLED, that RO_π and RP_π obeyed choice introduction: $RO_\pi(\pi_1)$ entails $RO_\pi(\pi_1 \cup \pi_2)$ for arbitrary π_2 and $RP_\pi(\pi_1)$ entails $RP_\pi(\pi_1 \cup \pi_2)$ for arbitrary π_2. One problem with the violability puzzle is the assumption that I can discharge my original normative obligation by executing an option that is supplied by choice introduction. But this is an error. Let $RO_\pi(\pi_1)$. From our contra-logical assumption, it follows that π dictates $\pi_1 \cup \pi_2$, as well as numerous other derivative obligations. A system that executes π_2 executes $\pi_1 \cup \pi_2$, but fails to execute π_1 and also fails to execute π in any situation in which π_1 is dictated by π and π_2 isn't. Thus, while one can derive arbitrarily many absurd derivative obligations from any obligation, it doesn't follow that one can even comply with the original norm by discharging those derivative obligations, let alone that one can discharge the original obligation. This is not so if we treat deontics both as sentential operators and as action-controlling constructions.

Again, this formal feature of PLED mirrors its domain. In general, norms may impose many obligations at once. Failure to discharge all of the obligations is failure to comply with the norm; π doesn't cease to obligate π_1, and so, after the failure to do π_1, I cannot comply with π by doing π_2. I can only comply with the derivative obligations. All of the obligations dictated by π are still in place, and so, if I fail to execute π_1 and then go on to do π_2, all I do is successfully discharge one among infinitely many derivative obligations. Thus, even given the absurd choice introduction rule, in the cat-sitting case, my failure to take care of the cat does (absurdly) imply the derivative obligation to choose whether to care for the cat or murder it, but it also implies the obligation to choose whether to train to be an astronaut or finish my reading backlog, and it certainly doesn't follow that I've discharged my original obligation or even complied with the norm dictating the original obligation by choosing among these other things.

8.2.5.4 The Good Samaritan

The Good Samaritan Paradox [14, 18] is resolved by PLED. Following [14], the Good Samaritan Paradox is as follows.

(1) It ought to be the case that Jones helps Smith who has been robbed.
(2) It ought to be the case that Smith has been robbed.

Now it seems that the following must be true: Jones helps Smith who has been robbed if and only if Jones helps Smith and Smith has been robbed.

Thus, from the intersubstitutivity of equivalents in the scope of deontic operators, the problem is this: $O(H \wedge R)$ implies $O(R)$. That is, it's obligatory that Jones helps smith and Smith has been robbed; thus, it's obligatory that Smith has been robbed.

This problem is solved in PLED by means of the way that PLED decomposes the notion of helping someone who has been robbed. There are numerous ways to do this that are non-paradoxical. Here are two: First, we may simply encode the action of Helping(Smith(who has been robbed)) by means of a primitive protocol, π_{SWHBR}. Then $RO_{\pi}(\pi_{SWHBR})$ simply doesn't decompose in the problematic way. Second, we may encode Helping(Smith(who has been robbed)) by means of a protocol or set of protocols with a specific set of preconditions. Let Smith's being robbed be R. Then $R \equiv \langle \pi_{SWHBR} \rangle_p \top$. While this implies that R holds at some state in any state sequence of any path of π extending from any state at which $RO_{\pi}(\pi_{SWHBR})$ holds, the execution of π_{SWHBR} merely presupposes R and does not necessarily result in a state in which R holds. Thus, π itself does not obligate R. Indeed, it could be the case that $[\pi_{SWHBR}]_p \neg R$. It may seem an abuse of language, but helping Smith might entail bringing it about that he is repaid—that he is "un-robbed". Explicitly, the following rule fails: $\langle \pi_{SWHBR} \rangle_p \top \equiv R, RO_{\pi}(\pi_{SWHBR}) \vDash_{PLEN} RO_{\pi}(R)$. So, no such absurdity as "It is obligatory that Smith is robbed" is implied. Again, the real problem appears to be that deontic operators are applied to sentences, the logic of which is abused to formalize the logic of actions.

8.2.5.5 Conflicting Obligations

The Conflicting Obligations Paradox [11, 14] is resolved by PLED. The Conflicting Obligations Problem is just what it sounds like: $O(A)$ and $O(B)$ where B implies $\neg A$ and A implies $\neg B$ are contradictories, but it seems that agents frequently have conflicting obligations of exactly this form. I may, as Sartre pointed out [20], have an obligation to take care of my ailing mother and an obligation to fight against the occupation, but these are mutually exclusive courses of action. The logic of norms that reduces to triviality in trying to codify such situations is no logic of norms.

PLED solves this problem outright by analyzing normative conflict without logical inconsistency; different protocols—and the deontics they define—can coherently conflict by defining exclusive restrictions on further protocols, processes, transitions, and states. On the basis of this fact, methodologies can be inconsistent, and machinery for resolving conflicts can be added easily, and defeasibility conditions can be defined in terms of preconditions. The pair of expressions, $RO_{\pi}(\pi')$ and $RO_{\pi''}(-\pi')$, is consistent under a large variety of cases as long as π and π'' are not procedurally equivalent, as is the triple of $RO_{\pi}(\pi')$, $RP_{\pi''}(-\pi')$, and $\neg RP_{\pi''}(\pi')$. The same holds for methodologies trivially. For formulae, as long as π and π' are not transition convergent, there is no problem of trivialization: i.e., $RO_{\pi}(A)$ and $RO_{\pi'}(\neg A)$ are consistent as long as $\pi \curlywedge_t \pi'$ fails. The latter would imply that A and $\neg A$ are true at some shared terminal state of executions of π and π'.

The machinery of PLED even allows the interpretation of expressions that encode the idea that a protocol (resp. methodology), π (resp. M), obligates the change of a methodology M'. The simplest version of such an expression: $(M(\pi) \wedge RO_{\pi}(\pi')) \wedge [\pi']_p \neg M(\pi)$. This says, roughly, that the current methodology (of the single agent modeled in PLEN) contains a norm that obligates revising to a methodology that does not contain that very norm. These can be taken to be meta-norms for resolving normative conflicts, thereby further analyzing the general phenomenon of norm conflict and conflict resolution.[11] Finally, L^D_{PLEN} has formulae of the form $C \equiv RO_{\pi}(\pi')$. These state necessary and sufficient conditions for the obligation of π' by π. These are essential to the process of specification [1] in diagnosing and resolving normative conflict.

[11] While a more detailed analysis of normative conflict is forthcoming, recall the remarks from Chap. 4 on the matter of protocol-theoretic formulae that encode statements about protocols determining, among other things, what can be learned about other protocols and even themselves. The fact that methodologies can change from state to state in PLEN immediately provides a framework for articulating and reasoning about (formal counterparts of) statements concerning protocols that result in methodology change. Thus, PLEN provides a framework in which to represent and reason about the properties of procedural epistemic norms with respect to their restrictions on what can be learned about procedural (and other) epistemic norms as well as their restrictions on what norms can be adopted into one's methodology. This is part of the machinery that can be leveraged to analyze norm conflict.

8.2.5.6 Chisolm's Paradox

Chisolm's Paradox [3, 14] is resolved by PLED. Letting g encode a statement to the effect that I am going to a party that I have promised to attend while t encodes the statement that I tell the party's organizer that I will be attending, Chisolm's paradox is the result of the following inconsistent tetrad:

(1) $OB(g)$ (it's obligatory to go since I made the promise to),
(2) $OB(g \rightarrow t)$ (it's obligatory to tell the party organizer that I am going if I am going),
(3´) $\neg g \rightarrow OB(\neg t)$ (if I am not going, then it's obligatory not to tell the party organizer that I am going),
(4) $\neg g$ (I'm not going).

These imply the contradiction that $OB(t)$ and $OB(\neg t)$ (by means of $OB(g) \rightarrow OB(t)$, which follows from 2). And this, of course, cannot be made sense of in standard deontic logic, as the contradiction trivializes. And yet, it seems a perfectly normal and sensible configuration of obligations.

In PLED, plausible translations of these formulae are all consistent and independent, where π_{go} is the protocol for attending the party, π_{tell} is the protocol for telling the party organizer that one plans to attend the party, and G and T are formulae for "I am going to attend" and "I am going to tell the party organizer that G", respectively:

(1´) $OB(g) :: ROM_M(\pi_{go})$,
(2´) $OB(g \rightarrow t) :: G \rightarrow ROM_M(\pi_{tell})$,
(3´) $\neg g \rightarrow OB(\neg t) :: \neg G \rightarrow ROM_M(-\pi_{tell})$,
(4´) $\neg g :: \neg G$.

Let $\langle\pi_{go}\rangle\top \equiv G$ and $\langle-\pi_{go}\rangle\top \equiv \neg G$, as this binds the relevant future-directed intentions to go to the party or not with the actions of going to the party or not, respectively. The translations, $1'$–$4'$, are consistent and satisfiable. There is no contradiction or trivialization under this translation.

It may, however, seem odd that one's norm system, M, dictates going to the party and also not telling that one is going to the party. After all, semantically, this means that some $p \in P_M$ contains a path, q, that is or corresponds to an execution of π_{go} and also an r that is or corresponds to an execution of $-\pi_{tell}$. This makes M very odd—the proper path, according to M, may well consist of not telling that you'll go to the party and then going, in that order.

How to solve this? Simple:

(2'+3') $OB(g \rightarrow t) \wedge \neg g \rightarrow OB(\neg t) :: ROM_M((((G)? \cap -(\neg G)?); \pi_{tell}) * \cup(((\neg G)? \cap -(G)?); -\pi_{tell}))*$
(2"+3") $OB(g \rightarrow t) \wedge \neg g \rightarrow OB(\neg t) :: ROM_M((((G)? \cap -(\neg G)?); \pi_{tell}) *_1 \cup(((\neg G)? \cap -(G)?); -\pi_{tell}))*_1$

What is really obligatory is the conditional choice of determining whether one is going or not, and then choosing the next action in sequence accordingly. Let $\neg G$ be true at s. Then the parallel test for G and non-test for $\neg G$ must fail, which means the iteration of it ends after zero steps, and so the only execution of the conditional choice available at s—and thus the only paths forward according to M given G—must execute $-\pi_{tell}$. I present both the unbounded iteration and bounded iteration formulations, as the former is sufficient but not as intuitive as the latter, but the latter is not built into the canonical version of PLEN. In any case, there are formalizations of the inconsistent tetrad defining Chisolm's paradox in PLED that are non-paradoxical.

8.2.5.7 The Gentle Murder

The Gentle Murder Paradox [5, 14] is resolved by PLED. The Gentle Murder paradox results from the decomposition of actions. The thought is that the following pentad implies an absurdity:

(1) one is obligated not to murder some person, X,
(2) one is conditionally obligated to murder X gently if one does murder X,
(3) gently murdering X implies murdering X (it's a logical truth),
(4) logical truths are always obligatory, and
(5) one in fact murders X.

From 1–5, it follows that it's obligatory to murder X. If this weren't bad enough, standard deontic logic then implies a contradiction and trivializes.

PLED solves this problem in multiple ways. First, let π_{GM} encode the action of gently murdering X and let π_M encode the action of murdering X. Suppose that $P_{\pi GM} \subset P_M$ for all graphs. (Processes of gentle murder are still processes of murder.) Then, from the fact that $RO_\pi(\pi_{GM})$ it does follow that $RO_\pi(\pi_M)$. And this is a problem. However, this does not properly decompose the relevant instructions for action. π_M should be thought of as the nondeterministic choice of $\pi_{M1}, \ldots, \pi_{Mn}$, for the n many ways to kill X. Thus, π_{GM} should be thought of as a choice formed from a proper subset of those protocols. Thus, from $RO_\pi(\pi_{GM})$ it does not follow that $RO_\pi(\pi_M)$ due to the execution rules for nondeterministic choice; a state at which a proper subset of $\pi_{M1}, \ldots, \pi_{Mn}$ are all executable is not a subset at which all of them are, which is what the execution rule for $\pi_M = \pi_{M1} \cup \ldots \cup \pi_{Mn}$ requires. Thus, even if one is committed to murdering X, and such a commitment entails the duty to murder X gently, it does not follow that there is a detachable general duty to murder X.

Second, claim (2) makes no sense from a protocol-theoretic perspective; it states no protocol that one could act on. One can't be obligated to retroactively gentle-fy the murder of X after one has done it. The conditional obligation can only be forward facing with respect to temporal indices (and they should, quite plausibly, not be conditioned on facts directly but on awareness of facts). Thus, (2) should be rejected outright as ill-formed. The proper translation of a conditional obligation in the neighborhood: If you must murder X, then you ought to murder X gently. There is no immediate way to formalize the "must" modality. For a fully satisfactory account, it's plausible that more elaborate temporal machinery is necessary. For now, here are two proposals, where π is an arbitrary protocol and π_M is a protocol for murdering X:

(Must 1) $(((\langle\pi\rangle_p \top \vee \langle-\pi\rangle_p \top) \rightarrow \langle\pi_M\rangle_p \top)$
(Must 2) $(\pi_M \curlywedge_p (\pi \cup -\pi))$

Must 1 tells us, informally, that the executability of either π or the complement of π (i.e., the executability of any action) implies the executability of π_M; if anything is executable, then the murder of X is. Must 2 tells us, informally, that the murder of X coincides with every process. Thus, consider:

(Must-Ob 1) $(((\langle\pi\rangle_p \top \vee \langle-\pi\rangle_p \top) \rightarrow \langle\pi_M\rangle_p \top) \rightarrow RO_\pi(\pi_{GM})$
(Must-Ob 2) $(\pi_M \curlywedge_p (\pi \cup -\pi)) \rightarrow RO_\pi(\pi_{GM})$

Do either of these imply $RO_\pi(\pi_M)$? No. If we decompose π_M as a non-deterministic choice of which π_{GM} is a sub-choice, neither has this consequence.

Third, PLED does not validate (4), as was shown concerning the Necessity of Obligations paradox. There are systems of norms in PLED that obligate nothing, including necessary truths. Moreover, PLED affords the distinction between the procedural obligation of action with respect to a norm and the "obligation" of truths at all deontically accessible worlds that are reachable in conformance with a norm. If we distinguish these, the derivation of the paradox is an equivocation.

Fourth, the paradox is a result of the failure of SDL to handle normative conflicts. If it's forbidden to murder X, then of course it's forbidden to murder X gently, as this implies murdering X. This is as it should be. Thus, the problem is simply that the forbiddenness of murdering X (in any way, by any means) directly conflicts with the detached obligation to murder X gently. That is, it's just a normative conflict. One can't consistently be obligated not to murder (simpliciter) while also being obligated to murder (however gently) under some conditions. The latter obligation is an exception to the former, since the former rules out the latter. The trivialization is the only real problem, here, and PLED does not trivialize under norm conflict, as shown above.

8.2.5.8 Paradox of the Miners

The Miners Paradox[12] has several solutions in PLED, and further solutions in PLEKNA, below. Let A stand for the claim that there are miners trapped in shaft A. Let B stand for the claim that there are miners trapped in shaft B. Both shafts are gradually filling with a determinate volume of water; just enough that if neither are blocked, one miner in each shaft dies, and if one is blocked, the other fills and

[12] See [9] (p. 115–116) for a presentation and analysis. The case is credited to earlier works by Derek Parfit and Donald Regan. Closer grappling with the case presented for the solution preferred by Kolodny and MacFarlane is left to future work. I simply present a path of solution derivable from the philosophical underpinnings and formal features of PLED and, in the next section, of PLEKNA.

all miners in the filled shaft die. We don't know which shaft the miners are trapped in: assume $A \vee B$. Letting π_A and π_B be the actions of blocking shafts A or B respectively, it seems that:

(1) if A, then $O(\pi_B)$,
(2) if B, then $O(\pi_A)$.

But then:

(3) $O(\pi_B) \vee O(\pi_A)$ follows by propositional logic.

However, it is supposedly the case that one ought to block neither shaft:

(4) $F(\pi_A)$ and $F(\pi_B)$.

Clearly, 3 and 4 are contradictories, and 4 is putatively, an obvious truth.

PLED provides two solutions, immediately.[13] The most immediate solution, building on the application of PLED to other paradoxes in Sect. 8.2, is to formalize 4 as the prohibition of a choice (with respect to a normative system M):

(4') $RF_M(\pi_A \cup \pi_B)$

This says that it is forbidden to choose among π_A and π_B. This does not conflict with 3, and so, granting the relevant reformulations of 1–3 with respect to M, the paradox disappears. This technical solution has some philosophical support, in that it seems clear that, not knowing whether A or B hold, one ought not make such a potentially deadly choice. This is a more plausible reading than 4, which says that I am forbidden to do both. Indeed, it's implausible that I'm forbidden to do both; one of these options is perfectly harmless given the facts. I just can't know which, and so can't responsibly decide. In other words, the idea is that 4 is false; it's not forbidden to block shaft A and also forbidden to block shaft B (one or the other is actually obligatory), it is forbidden to make this grave decision given our lack of knowledge concerning which shaft is filling with water.

A second solution is to formulate each of 1–3 and 4 with respect to different normative systems corresponding, respectively, to "objective" and "subjective" norms of assessment:[14]

(1') if A, then $RO_{M_O}(\pi_B)$,
(2') if B, then $RO_{M_O}(\pi_A)$.

But then:

(3') $RO_{M_O}(\pi_B) \vee RO_{M_O}(\pi_A)$ follows by propositional logic.

However, it is supposedly the case that one ought to block neither shaft:

(4") $RF_{M_S}(\pi_A)$ and $RF_{M_S}(\pi_B)$.

[13] Of course, both solutions can be combined for a third.

[14] Or, in MacFarlane's terms, observer and assessor relative assessments [9, 12, 13].

It seems, for instance, that 1 and 2 (and thus 3) hold from an objective or external perspective, while 4 seems to hold from a subjective or internal perspective. That is, from the God's eye view, 3 holds because either A or B hold, and there is clearly a right answer given the truth of one or the other disjunct. However, the agent in this situation cannot know either, and so, from a perspective relative to the agent's knowledge, there is an obligation to choose neither. Relativizing these deontics to different normative systems respectively encoding these differing evaluative perspectives, then, generates a conflict of norms. That such norm systems might conflict is unproblematic, given the prior analysis of normative conflict.

Remark 402 So, PLED is at least a prima facie plausible basis for the logic of epistemic deontics, and this supports the overall argument of the book for PLEN's adequacy as a logic of epistemic norms both directly and indirectly. Directly, deontic concepts are plausibly central to the logic of epistemic norms, so the better the deontic logic, the better the logic of such norms. Indirectly, the plausibility of the analysis of deontics makes the protocol function theorem of the next chapter more theoretically interesting and plausible; it's not a technical contrivance but a result that shows the relation between the properties of protocols that encode deontic concepts of requirement and the properties of protocols that encode normative verdicts.

8.3 A Protocol-Theoretic Logic of Procedural Knowledge and Norm Application (PLEKNA)

8.3.1 PLEKNA: The Idea

Claim 403 This section outlines a preliminary attempt at analyzing a species of epistemic operator in PLEN as well as some of the core notions of norm application hinted at in the discussion of the regulative role of epistemic norms (re: Chap. 2). This preliminary analysis is enough to ground the kernel of formalization of D3C3 in Chap. 10, and that chapter houses more elaborate discussion of the system than is presented in this section, though truly thorough justification of this procedural knowledge and norm application logic remains for future work.

Claim 404 Knowledge, in PLEN, is not a product of the modeling of epistemic states with uncertainty relations as in standard epistemic logic. Rather, knowledge is multi-modal; there is a knowledge operator associated with every protocol, π. The basic concept of knowledge in PLEN, which remains to be thoroughly defended, is that of factive beliefs that result from procedurally correct processes. In other words, given a protocol-theoretic analysis of rational justification (accordance with a protocol that encodes an epistemic norm), knowledge is procedurally justified true belief. The system allows for multiple variants of knowledge operators defined in terms of both individual protocols and methodologies. This provides a general framework in which to reason about different specifications of the basic knowledge operator in terms of different norms with respect to which justification can be defined. This basic proceduralization of knowledge operators works as the foundation for encoding the epistemic concepts operative in the application of epistemic norms. These include knowing and deliberate applications of a norms and a version of the kind of normative re-direction that the default interventionist model of rational cognition requires. I present the system with little commentary, as it directly responds to arguments in Chaps. 2 and 3 and it is applied in Chap. 10, where we see at least the outlines of some of its logical features.

Remark 405 The basic informal idea of the following system is that knowledge is procedurally justified true belief, that norms can be explicitly reasoned about, and that processes of deliberation (about both how to apply norms and which norms to apply) can be carried out, resulting in re-direction of one's natural or heuristic epistemic processes. The system simply adds operators for building and interpreting formulae in PLEN that express explicit reasoning,[15] deliberation, and application of norms. These are taken to be operators that express the relations between the single agent (whose dynamics are analyzed by EDGs) and the protocols in question.

8.3.2 Defining PLEKNA

Claim 406 On the basis of this procedural justification conception of knowledge, we can define pairs of modally distinct conceptions of norm compliance, norm implementation, restriction of epistemic factors according to one's endorsed norms, and adherence to norms (as distinct from mere compliance). The modal distinctions between these notions are those between feasible and actual compliance, adherence, (et al), and knowing compliance, adherence (et al). Thus, the epistemic norm application operators come in both knowing and unknowing forms, and feasible and actual forms.

Definition 407 (Protocol-Theoretic Epistemic Logic: Syntax (L^{DK}_{PLEN})) Add the following epistemic operators and their semantics to L^{D}_{PLEN} to construct L^{DK}_{PLEN}.

(1) (Knowledge) If $\pi \in P_{PLEN}$, and if $A \in L^{DK}_{PLEN}$, then $K(A) \in L^{DK}_{PLEN}$.

   (a) Example: (Knowing Dictation) If $\pi, \pi' \in P_{PLEN}$, then $K(RO_{\pi}(\pi')) \in L^{DK}_{PLEN}$.

[15] Technically, this was already possible in PLEN, given the *b* and *d* operators and that theory of protocol equivalence in the object-language, but more elaborate epistemic attitudes toward these kinds of propositions are added, and operators expressing agentic relations to protocols are added.

(2) (Feasible Norm Compliance) If $\pi \in P_{PLEN}$, then $Com(\pi) \in L^{DK}_{PLEN}$.

(a) :: π can be complied with

(3) (Feasible Knowing Compliance) If $\pi \in P_{PLEN}$, then $KCom(\pi) \in L^{DK}_{PLEN}$.

(a) :: a protocol, π', dictated by π can be knowingly complied with: $K(RO_{\pi}(\pi') \wedge KCom(\pi))$

(4) (Feasible Knowing Implementation) If $\pi \in P_{PLEN}$, then $KImp(\pi, \pi') \in L^{DK}_{PLEN}$.

(a) :: π can be knowingly implemented by doing π'

(5) (Feasible Knowing Restriction) If $\pi \in P_{PLEN}$, $(\pi_1, \ldots, \pi_n) \subseteq P_{PLEN}$, $(p^1_c, \ldots, p^n_c) \subseteq P_c$, $(t^1_c, \ldots, t^n_c) \subseteq T_c$, $(s^1_c, \ldots, s^n_c) \subseteq S_c$, and $x_1, \ldots, x_n$ is one of these sets, then $KRes_{\pi}(x_1, \ldots, x_n) \in L^{DK}_{PLEN}$.

(a) :: the set of epistemic factors $(x_1, \ldots, x_n)$ can be knowingly restricted in accordance with π

(6) (Feasible Adherence) If $\pi \in P_{PLEN}$, then $Adh(\pi) \in L^{DK}_{PLEN}$.

(a) :: π can be adhered to
(b) :: an agent whose EDG is g *can adhere to* π at s

(7) (Feasible Knowing Adherence) If $\pi \in P_{PLEN}$, then $KAdh(\pi) \in L^{DK}_{PLEN}$.

(b) :: π can be knowingly adhered to
(b) :: an agent whose EDG is g *can knowingly adhere* to π at s

(8) (Actual Norm Compliance) If $\pi \in P_{PLEN}$, then $ACom(\pi) \in L^{DK}_{PLEN}$.

(a) :: π has been complied with

(9) (Actual Knowing Compliance) If $\pi \in P_{PLEN}$, then $KACom(\pi) \in L^{DK}_{PLEN}$.

(a) :: a protocol, π', dictated by π has been knowingly complied with: $K(RO_{\pi}(\pi') \wedge KCom(\pi))$

(10) (Actual Knowing Implementation) If $\pi \in P_{PLEN}$, then $KAImp(\pi, \pi') \in L^{DK}_{PLEN}$.

(a) :: π has been knowingly implemented by doing π'

(11) (Actual Knowing Restriction) If $\pi \in P_{PLEN}$, $(\pi_1, \ldots, \pi_n) \subseteq P_{PLEN}$, $(p^1_c, \ldots, p^n_c) \subseteq P_c$, $(t^1_c, \ldots, t^n_c) \subseteq T_c$, $(s^1_c, \ldots, s^n_c) \subseteq S_c$, and $x_1, \ldots, x_n$ is one of these sets, then $KARes_{\pi}(x_1, \ldots, x_n) \in L^{DK}_{PLEN}$.

(a) :: the set of epistemic factors $(x_1, \ldots, x_n)$ has been knowingly restricted in accordance with π

(12) (Actual Adherence) If $\pi \in P_{PLEN}$, then $AAdh(\pi) \in L^{DK}_{PLEN}$.

(b) :: π has been adhered to

(b) :: an agent whose EDG is g *has adhered to* π at s

(13) (Actual Knowing Adherence) If $\pi \in P_{PLEN}$, then $KAAdh(\pi) \in L^{DK}_{PLEN}$.

(b) :: π has been knowingly adhered to
(b) :: an agent whose EDG is g *has knowingly adhere*d to π at s

Note two things about PLEKNA. First, all knowledge is procedural in PLEKNA; it's procedurally justified true belief. Second, one can have propositional knowledge of the procedural structures of protocols, their procedural requirements, and their outcomes. On the basis of this, one can knowingly comply, implement, or adhere to a protocol (hence, procedural epistemic norm).

The core thought of PLEKNA, then, is that procedural knowledge has two flavors: one forward looking (feasibility) and one backward looking (actuality). One can know what procedures one can carry out as well as what procedures were used to get to the current state. The language of PLEKNA gives us two operators for talking about these ways of enacting norms. On the one hand, we can say that it is feasible to knowingly comply, implement, or adhere to a norm (a forward-facing procedural operator). On the other, we can say that we have actually done so (a backward-facing procedural operator).

The semantics of these novel operators are defined over EDGs with the additional elaboration of epistemic histories. Epistemic histories are designated prefixes of paths with respect to states. That is, there is a set of paths, for each $s \in S$ for each g, that lead up to s. Among these epistemic histories, we single out a designated actual history per EDG. The idea, then, is that feasible epistemic norm application operators are evaluated at s in terms of what paths unfold from s, and the actual epistemic norm application operators are evaluated at s in terms of the actual history up to s.

Definition 408 (Protocol-Theoretic Epistemic Logic: (L^{DK}_{PLEN}) Semantics) Let g be an arbitrary EDG. For every s in g, let $H_s = \{p_i \in P_g \mid l(p_i) = s\}$. Call H_s the set of *possible epistemic histories up to s in g*. Let $f_h : S \longrightarrow H_s$ arbitrarily map each state to a unique, arbitrary $p_i \in H_s$. If $h \in H_s$ and $f_h(s) = h$, then call h the *actual epistemic history up to s in g*. In the remainder of the chapter, let $h, h' \in H_s$ and $f_h(s) = h$ for states, s, determined by context. Extend the semantics of PLEDNA to L^{DK}_{PLEN} by the following rules:

(1) (Knowledge) $s \Vdash K(A)$ iff there is some $h' \sqsubseteq h$ such that $h' \in P_M$, $s = l(h) = l(h')$, $c(f(h')) = \langle X : Y : \Pi_M \rangle$, $c(s) = \langle X' : Y' : \Pi_M \rangle$, $s \Vdash b(A) \wedge A$, $s \Vdash RO_M(b(A))$, and $first(h') \Vdash RO_M(b(A))$.
(2) (Feasible Norm Compliance) $s \Vdash Com(\pi))$ iff $c(s) = \langle X : Y : \Pi_M \rangle$, and $s \Vdash \langle \pi \rangle_p \top$.
(3) (Feasible Knowing Compliance) $s \Vdash KCom(\pi))$ iff $c(s) = \langle X : Y : \Pi_M \rangle$, and $s \Vdash K(\langle \pi \rangle_p \top)$.
(4) (Feasible Knowing Implementation) $s \Vdash Kimp(\pi, \pi')$ iff $c(s) = \langle X : Y : \Pi_M \rangle$, and $s \Vdash K((\pi' \mapsto \pi) \wedge \langle \pi' \rangle_p \top)$.

(5) (Feasible Knowing Restriction) $s \Vdash KRes_{\pi}(x_1, \ldots, x_n)$ iff $c(s) = \langle X : Y : \Pi_M \rangle$, and there is some subset, $\{x_i, \ldots, x_j\}$ of $\{x_1, \ldots, x_n\}$ such that $s \Vdash RP_{\pi}(x_i), \ldots, s \Vdash RP_{\pi}(x_j)$.
(6) (Feasible Adherence) $s \Vdash Adh(\pi)$ iff $c(s) = \langle X : Y : \Pi_M \rangle$, $s \Vdash \langle \pi \rangle_p \top \wedge M(\pi)$.
(7) (Feasible Knowing Adherence) $s \Vdash KAdh(\pi)$ iff $c(s) = \langle X : Y : \Pi_M \rangle$, $s \Vdash K(\langle \pi \rangle_p \top \wedge M(\pi))$.
(8) (Actual Norm Compliance) $s \Vdash ACom(\pi))$ iff there is some $h' \sqsubseteq h$ such that $h' \in P_{\pi}$, $s = l(h) = l(h')$, and $c(s) = \langle X : Y : \Pi_M \rangle$.
(9) (Actual Knowing Compliance) $s \Vdash KACom(\pi))$ iff there is some $h' \sqsubseteq h$ such that $h' \in P_{\pi}$, $s = l(h) = l(h')$, $c(s) = \langle X : Y : \Pi_M \rangle$, and $f(h) \Vdash K(\langle \pi \rangle_p \top)$.
(10) (Actual Knowing Implementation) $s \Vdash KAImp(\pi, \pi')$ iff there is some $h' \sqsubseteq h$ such that $h' \in P_{\pi'}$, $s = l(h) = l(h')$, $c(s) = \langle X : Y : \Pi_M \rangle$, and $f(h') \Vdash K((\pi' \mapsto \pi) \wedge \langle \pi' \rangle_p \top)$.
(11) (Actual Knowing Restriction) $s \Vdash KARes_{\pi}(x_1, \ldots, x_n)$ iff there is some $h' \sqsubseteq h$ such that $h' \in P_{\pi}$, $s = l(h) = l(h')$, $c(s) = \langle X : Y : \Pi_M \rangle$, there is some subset, $\{x_i, \ldots, x_j\}$ of $\{x_1, \ldots, x_n\}$ such that $f(h') \Vdash RP_{\pi}(x_i), \ldots, f(h') \Vdash RP_{\pi}(x_j)$, and:

   (a) for some x_i such that $RP_{\pi}(x_i)$, if $x_i = p^i_c$, then $\twoheadrightarrow_p (p^i_c) \sqsubseteq h'$
   (b) for some x_i such that if $RP_{\pi}(x_i)$, if $x_i = t^i_c$, then $f(h') = \twoheadrightarrow_t (t^i_c)$ and $l(h') = l(\twoheadrightarrow_t (t^i_c))$
   (c) for some x_i such that $RP_{\pi}(x_i)$, if $x_i = s^i_c$, then $l(h') = \twoheadrightarrow_s (s^i_c)$

(12) (Actual Adherence) $s \Vdash AAdh(\pi)$ iff there is some $h' \sqsubseteq h$ such that $h' \in P_{\pi}$, $s = l(h) = l(h')$, $c(s) = \langle X : Y : \Pi_M \rangle$, and $f(h') \Vdash \langle \pi \rangle_p \top \wedge M(\pi)$.
(13) (Actual Knowing Adherence) $s \Vdash KAAdh(\pi)$ iff there is some $h' \sqsubseteq h$ such that $h' \in P_{\pi}$, $s = l(h) = l(h')$, $c(s) = \langle X : Y : \Pi_M \rangle$, $f(h') \Vdash K(\langle \pi \rangle_p \top \wedge M(\pi))$.

Definition 409 The extension of the syntax to methodologies is straightforward, as is the semantics, which I omit:

(1) (Knowledge) If $M \subseteq P_{PLEN}$, and if $A \in L^{DK}_{PLEN}$, then $K(A) \in L^{DK}_{PLEN}$.

   (a) (Knowing Dictation) If $M \subseteq P_{PLEN}$ and $\pi' \in P_{PLEN}$, then $K(RO_M(\pi')) \in L^{DK}_{PLEN}$.

(2) (Knowing Compliance) If $M \subseteq P_{PLEN}$, then $KCom(M), KACom(M) \in L^{DK}_{PLEN}$.

   (a) :: protocol, π', dictated by M can be knowingly complied with: $K(RO_M(\pi') \wedge KCom(M))$
   (b) :: protocol, π', dictated by M has been knowingly complied with: $K(RO_M(\pi') \wedge KACom(M))$

(3) (Knowing Implementation) If $M \subseteq P_{PLEN}$, then $KImp(M), KAImp(M, \pi) \in L^{DK}_{PLEN}$.

(4) (Knowing Restriction) If $\pi \in P_{PLEN}$, $(\pi_1, \ldots, \pi_n) \subseteq P_{PLEN}$, $(p_c^1, \ldots, p_c^n) \subseteq P_c$, $(t_c^1, \ldots, t_c^n) \subseteq T_c$, $(s_c^1, \ldots, s_c^n) \subseteq S_c$, and $x_1, \ldots, x_n$ is one of these sets, then $KRes_M(x_1, \ldots, x_n)$, $KARes_M(x_1, \ldots, x_n) \in L^{DK}_{PLEN}$.

   (a) :: the set of epistemic factors $(x_1, \ldots, x_n)$ can be knowingly restricted in accordance with M
   (b) :: the set of epistemic factors $(x_1, \ldots, x_n)$ has been knowingly restricted in accordance with M

(5) (Knowing Adherence) If $M \subseteq P_{PLEN}$, then $KAdh(M), KAAdh(M) \in L^{DK}_{PLEN}$.

   (a) M can be knowingly adhered to
   (b) M has been knowingly adhered to

Claim 410 In addition to the analysis of defeasible reason as a deontic operator, PLEKNA attempts to analyze the abstract process of reasoning. For this, we first add structures in the protocol-theoretic language that encode the conditions under which protocol-theoretic formulae are counted as reasons for other formulae. The informal thought being that some formulae in the protocol-theoretic language count as reasons for another when the former bear some normative relation to the other, making it permissible, obligatory, forbidden, or defeasibly reasonable to accepting the latter.

Definition 411 (Reasons) Letting L^D_{PLEN} be defined as normal:

(1) (Syntax) If $A_1, .., A_n, B \in L^{DK}_{PLEN}$, then $R(A_1, \ldots, A_n \mid B) \in L^{DK}_{PLEN}$.
(2) (Semantics) $A_1, \ldots, A_n$ are reasons for the single agent modeled by a given EDG, g, (at the current state, state s), $v_s(R(A_1, \ldots, A_n \mid B)) = 1$, iff disjunctively:

   (a) $v_s((b(A_1) \wedge \ldots \wedge b(A_n)) \wedge (RD_M(B, A_1 \wedge \ldots \wedge A_n))) = 1$,
   (b) $v_s((b(A_1) \wedge \ldots \wedge b(A_n)) \rightarrow (RO_M(b(B)))) = 1$,
   (c) $v_s((b(A_1) \wedge \ldots \wedge b(A_n)) \rightarrow (RP_M(b(B)))) = 1$,
   (d) $v_s((b(A_1) \wedge \ldots \wedge b(A_n)) \rightarrow (RF_M(d((B)))) = 1$.

Claim 412 Reasoning processes are straightforwardly encoded in PLEKNA: reasoning processes are epistemic processes that add or delete reasons from the acceptance and rejection sets of epistemic states, as well as breaking apart the objective normative forces of them.

Definition 413 (Reasoning Processes) Let g be an arbitrary EDG. Arbitrary $p \in P_g$ is a *process of reasoning* iff:

(1) $(f(p) \nVdash R(A_1, \ldots, A_n \mid B)$ and $l(p) \Vdash R(A_1, \ldots, A_n \mid B)$ for some $A, \ldots, A_n, B \in L^{DK}_{PLEN}$, or vice versa,
(2) $(f(p) \nVdash b(R(A_1, \ldots, A_n \mid B))$ and $l(p) \Vdash b(R(A_1, \ldots, A_n \mid B))$ for some $A, \ldots, A_n, B \in L^{DK}_{PLEN}$, or $(f(p) \nVdash d(R(A_1, \ldots, A_n \mid B))$ and $l(p) \Vdash d(R(A_1, \ldots, A_n \mid B))$ for some $A, \ldots, A_n, B \in L^{DK}_{PLEN}$,

(3) for every subpath of p, q, the addition or subtraction of reasons at $l(q)$ (verifying or falsifying $R(A_1, \ldots, A_n \mid B)$ or $R(A_1, \ldots, A_n \mid B)$ at $l(p)$) or adding $R(A_1, \ldots, A_n \mid B)$ to X or Y at $l(p)$) is believed to be obligated or permitted by the methodology at $f(q)$.

Claim 414 On the basis of the foregoing machinery, we can reach our formalization destination: defining an operator for the application of epistemic norms. The informal idea is that, in accordance with the arguments of Chaps. 2 and 3, procedural epistemic norms can be deliberately applied. Again, we draw distinctions between actual and feasible (deliberate application requiring knowing adherence) application.

Definition 415 (Feasible Deliberate Application: Syntax) If $\pi \in P_{PLEN}$, then $Del(\pi) \in L^{DK}_{PLEN}$.

(1) :: π has been deliberately applied

Definition 416 (Feasible Deliberate Application: Semantics) π can be *feasibly deliberately applied* at s, $s \Vdash Del(\pi)$, iff $s \Vdash KAdh(\pi)$, and for some π', $s \Vdash K(\langle\pi'\rangle_p \top \wedge \pi' \curlyvee_p \pi)$, $s \Vdash RD_M((RP_M(\pi) \vee RO_M(\pi)) \wedge RF_M(\pi')), R_1)$, $s \Vdash b(R_1)$, and there is a process of reasoning, p, such that $l(p) = s$.

Definition 417 (Actual Deliberate Application: Syntax) If $\pi \in P_{PLEN}$, then $ADel(\pi) \in L^{DK}_{PLEN}$.

(1) :: π has been deliberately applied

Definition 418 (Actual Deliberate Application: Semantics) π has been *actually deliberately applied* to reach s, $s \Vdash ADel(\pi)$, iff there is some $h' \sqsubseteq h$ such that $h' \in P_\pi$, $s = \langle X : Y : \Pi_M \rangle = l(h)$, $s \Vdash KAAdh(\pi)$, and for some π', $f(h') \Vdash K(\langle\pi'\rangle_p \top \wedge \pi' \curlyvee_p \pi)$, $f(h') \Vdash RD_M((RP_M(\pi) \vee RO_M(\pi)) \wedge RF_M(\pi')), R_1)$, $f(h') \Vdash b(R_1)$, and there is a process of reasoning, $p \sqsubseteq h$, such that $l(p) = f(h')$.

The deliberate application of protocols (procedural epistemic norms) involves explicit processes of reasoning about the requirements (the deontic verdicts) of multiple procedural epistemic norms followed by knowing adherence to the one that is supported by the reasons adduced by the reasoning process.

In accordance with the default interventionist model of rationality appealed to in Chap. 2, as well as the litany of observations of normal epistemic practice, PLEKNA finally provides analysis of heuristic processes and their normative regulation. These are formalized as meta-level semantic notions rather than shunted into the object language. The purpose of this machinery is primarily to investigate the relation between deliberate application and normative regulation in Chap. 10, as well as to demonstrate that PLEN can be put to such uses. The gist of the machinery is that EDGs are elaborated with a subtree of heuristic processes, which prefix processes that execute protocols in one's methodology and, in particular, that prefix processes of knowing adherence or deliberate application.

Claim 419 The core notion of normative regulation, then, is *normative re-direction*, below.

Definition 420 (Heuristic Processes) For arbitrary EDG, g, identify an arbitrary subtree, g_n, of g such that no path in g constitutes a knowing adherence with any protocol in any methodology in g_n.

(1) This sub-tree may or may not converge with the actual epistemic histories of various states and the same holds with respect to the execution sets of any protocols.
(2) If $p \in P_{g_n}$, then say that p is a *heuristic process.*
(3) Let $g - g_N$ be the tree comprising all *effortful processes.*

Definition 421 (Effortful Halting) A heuristic process, q, is *halted by effortful cognition* at s iff $q \in P_{g_n}$, and $last(q) = first(r)$ for some $r \in P_g$ that is not in P_{g_n} that executes all π in Π_M at $last(q)$, and $first(r) \Vdash KAdh(\pi)$.

Definition 422 (Normative Re-direction) , h, for a state s, is a model of *normative re-direction* with respect to protocol π iff $h = qr$ where there is a $qq' \in P_g$ such that $last(q) = first(q') = first(r)$, and either:

(1) $first(r) \Vdash (KAdh(\pi)) \wedge (RF_{\pi}(q_c q_c') \vee RF_{\pi}(q_c'))$,
(2) $first(r) \Vdash (KAdh(\pi)) \wedge (RF_{\pi}((f(q_c), l(q_c'))) \vee RF_{\pi}((f(q_c'), l(q_c'))))$
(3) $first(r) \Vdash (KAdh(\pi)) \wedge (RF_{\pi}(l(q_c')))$,
(4) $last(r) \Vdash KAAdh(\pi)$

Remark 423 Normative re-direction is what it sounds like. One ends the heuristic processes one was previously executing and from that point carries out normative processes. Deliberate application straightforwardly results in normative re-direction, as well.

Definition 424 (PLEKNA) The system that results from incorporating the foregoing revisions of L^D_{PLEN} and the corresponding semantic rules into PLEN and operating on it according to the R-Derivation System is the protocol-theoretic logic of procedural knowledge and norm application (PLEKNA).

Remark 425 The semantics for individual protocols can be extended to methodological formulae by replacing all occurrences of the clauses naming individual protocols with clauses naming M and replacing the execution sets of protocols with the intersections of the execution sets of protocols in M.

Remark 426 The foregoing definitions add a knowledge operator and a suite of new expressions to PLEN. For each case in which the K operator is applied to a formula that is not in the object language of L^R_{PLEN}, there is a straightforward way to shunt these expressions into the object language. I omit the details, as there really ought not be any serious obstacles to doing this; see how it was done for PLED.

Remark 427 The main thought underlying this protocol-theoretic procedural knowledge and norm application logic is that the knowledge operator can be translated into a notion of factive, procedurally justified belief with respect to the methodology at the state the knowledge expression is interpreted at. Some of the expressions that K operates on in the above rules have already been added in

L^R_{PLEN}, and others haven't. These are worth only brief remarks. For meta-level expressions like "$p \in P_\pi$", these can be interpreted as elliptical for $RP_\pi(p)$, and will be necessary truths or falsehoods given the graph in which the expression is being interpreted. For the expressions $s \in R_{S\pi}$ and $t \in R_{T\pi}$, interpret these simply as s (resp. t) being rational according to the restriction on states derived from π by RRT2. Of course, these expressions are placeholders: if the non-local variants of dictation and implementation are added to the object language and interpreted (as necessary truths in EDGs or axioms), these expressions straightforwardly apply to them.

8.3.3 The Logic of PLEKNA

I omit any serious exploration of the derivation system or validity relation of PLEKNA, as this lies beyond the scope of this book. Rather, I focus on the expressive extension that PLEKNA provides to PLEN. It is not an obscure matter how to incorporate PLEKNA, as defined, into the R-Derivation system, and I explore some specific valid inferences in Chap. 10. The metatheory of PLEKNA, like that of PLED, is untouched, here.

8.3.4 Some Immediate Virtues

I reserve detailed investigation and defense of PLEKNA for future work, but some immediate comments are warranted. First, the epistemic logic of PLEN is a unique variant in which knowledge is not investigated directly as a modal operator but as a protocol-theoretic operator. This carries some of the virtues of PLED to PLEKNA. For instance, the deontic accessibility relation in PLED has a principled analysis in terms of the protocol logic of epistemic norms; the states reachable from any s are those that can be brought about by the execution of some protocols, i.e., the execution of some processes and courses of epistemic action that comply with some procedural norms. Of the basis of this feature of PLEN, PLEKNA can provide a principled, protocol-theoretic analysis of the logic of epistemic operators in terms of the protocol logic of epistemic norms. The behavior of the knowledge operator in PLEKNA can be defined by multiple, competing systems of protocol-theoretic axioms and then investigated within the logical framework of PLEKNA. For instance, the consequences of axioms like $(K(A) \rightarrow (K(A \rightarrow B) \rightarrow K(B)))$ (which are not tautologies in PLEKNA) can be drawn and pairs of (systems of) protocols satisfying axioms like $[\pi_1]_p(K(A) \rightarrow (K(A \rightarrow B) \rightarrow K(B)))$ and $[\pi_2]_p\neg(K(A) \rightarrow (K(A \rightarrow B) \rightarrow K(B)))$ can be compared.

The epistemic logic of PLEKNA also has a direct connection to normative justification (especially the procedural aspects of justification) and conditions of epistemic rationality. This sheds light on the normative aspects of the notion of knowledge, itself. In PLEKNA, it's clear why knowledge is normative; it's bound closely to one's epistemic norms. One knows what one truly believes when that belief is produced by proper application of one's epistemic methods. It might be good to add some kind of safety or invariance condition, and that's left open in PLEKNA.

Second, the knowledge operator in PLEKNA is, partially, a syntactic operator in the sense that its semantics are defined in part in terms of sets of formulae rather than in terms of deontic accessibility relations over worlds in frames.[16] The underlying doxastic operator, b, is interpreted over arbitrary sets of formulae mapped to epistemic states by the content function. Because $K(A)$ requires $b(A)$ and $b(A)$ obeys no extensional or intensional rules with respect to all EDGs (clearly, there are regularities among subsets of EDGs concerning how protocols impact the content of acceptance and rejection sets), the knowledge operator in PLEKNA handily evades the logical omniscience problem. It is immediate that $\vdash_{PLEN} A$ does not imply $\vdash_{PLEN} b(A)$ and that $(K(A) \rightarrow (K(A \rightarrow B) \rightarrow K(B)))$ is not a tautology. What is true is determined by logic, but what is believed, in PLEKNA, is determined by epistemic dynamics. What one knows is then determined by what is true, what is believed, and how it is that the belief is produced.

Third, the addition of epistemic operators to the epistemic deontic logic of PLED provides means of formalizing further resolutions to some of the earlier paradoxes of deontic logic. For an instance, recall the Miners Paradox above at the end of Sect. 8.2. PLED provided two solutions, and PLEKNA aids in formulating a third.

In addition to the solutions offered by PLED, a third—and more complicated—solution in PLEKNA identifies the problem as the assumption, underlying 1 and 2, that the operative system of norms—call it M—conditions its normative verdicts directly on external circumstances rather than on internally accessible states related to external circumstances, i.e., belief or knowledge. This property is incompatible with the use of M in ante hoc regulation, and so, the assumption that 1 and 2 characterize a system of norms, M, means that M is incompatible with the action-oriented approach to deontics. Thus, 1 and 2 ought to be rejected outright, for any norms in the domain of analysis by PLED (and thus PLEKNA), undermining the inference to 3. In other words, the formalization of the puzzling situation in PLEKNA would rule out anything directly corresponding to 1 and 2.

Taking an approach that is more amenable to the action-oriented perspective, PLEKNA supplies an alternative formulation of the puzzling tetrad:

(1") if $K(A)$, then $RO_M(\pi_B)$,
(2") if $K(B)$, then $RO_M(\pi_A)$,
(3") $(K(A) \vee K(B)) \rightarrow (RO_M(\pi_A \cup \pi_B)$,
(4"') $(\neg K(A) \wedge \neg K(B)) \rightarrow (RF_M(\pi_B) \wedge RF_M(\pi_A))$

[16] See, for instance, [10], for more on syntactic accounts of knowledge operators.

In PLEKNA, the following facts hold:

- 3" follows from 1" and 2", by propositional logic,.
- The more plausible 3" and 4"' are consistent with $(RF_M(\pi_B) \wedge RF_M(\pi_A))$.
- The consequent of 3" does not follow from the facts of the case.
- $RO_M(\pi_A \cup \pi_B)$ implies $(RO_M(\pi_B) \vee RO_M(\pi_A)$, and the converse fails.
- $RF_M(\pi_A) \wedge RF_M(\pi_B)$ implies $RF_M(\pi_A \cup \pi_B)$, and the converse fails.
- The case verifies $(\neg K(A) \wedge \neg K(B))$, so $RF_M(\pi_A) \wedge RF_M(\pi_B)$ and $RF_M(\pi_A \cup \pi_B)$ hold, which are, intuitively, the right judgments.

That is, given our state of ignorance, it's obligatory not to block either passage, but it would be obligatory to block one or the other (and thereby to choose among π_A and π_B) given the relevant knowledge. From another perspective, what PLEKNA says about this case is that, if there is no way for us to come to know whether A or B by rational means (with respect to methodology M), then we cannot detach either obligation nor an obligation to choose from the available options.

What motivates this reformulation? Recall that PLEN is a system for directly stating and characterizing procedural norms by means of its protocol logic, and that PLED and PLEKNA are systems for characterizing the properties of procedural norms in terms of deontic and epistemic operators and simultaneously characterizing these operators in terms of procedural norms. Thus, conditionals of the form 1 and 2, in the regimented language of PLEKNA, are not to be read as direct formalizations of colloquial formulations of normative advice, e.g., "If C, do π!". Rather, they should capture the restrictions on dynamics defined by the relevant procedural norms, e.g., "If C, then π' dictates doing π" or "all executions of π' require doing π when C". Such conditionals should characterize the conditions under which a norm assigns its verdicts, and thereby characterize the conditions of violation for the norm. That is, norms define restrictions on epistemic dynamics, which specify the courses of action that are to be undertaken and the conditions under which they are to be undertaken. Given this, three arguments for rejecting the assumption that norms condition verdicts on external circumstances rather than on features of epistemic state are immediate.

First, external circumstances cannot be what the mechanisms for cognitive or behavioral control are triggered by; in other words, agents can only act on their internal representations of external circumstances. The external circumstances aren't in our heads. External circumstances only impinge on deliberation mechanisms by means of layers of intervening representational states. Procedural norms are the programs for controlling one's deliberation mechanisms in ante hoc regulation. Thus, in order for procedural norms to be usable in ante hoc regulation (as per Chap. 2), they would have to condition their verdicts on the representational states that trigger deliberation mechanisms.

Pollock's account of normative regulation motivates the argument. The thought is that, in order to break out of a number of familiar regress arguments (the "intellectualist regress" re: [17, p. 125] or, perhaps, the familiar Carroll Regress [2]), the correct model of procedural normative regulation must involve the acquisition of plan-like norms that harness (at least somewhat) cognitive mechanisms to channel behavior along the lines of the restrictions they define. Such cognitive mechanisms can operate on or be triggered by only what is internally accessible. Nonetheless, procedural norms must direct these mechanisms in order to function in ante hoc regulation. Thus, procedural norms for directing these mechanisms must condition their directives on what such mechanisms can be triggered by, which is restricted to internally accessible mental states. As Pollock points out, concerning norms governing mundane activities like bike-riding:

> [W]e have been describing one of our bike-riding norms as telling us that if the bicycle leans to the right then we should turn the handlebars to the right, but that is not really what we learn when we learn to ride a bicycle. The automatic processing systems implemented in our neurology do not have access to whether the bicycle is leaning to the right. What they do have access to are things like our thinking that the bicycle is leaning to the right, and certain balance sensations emanating from our inner ear. What we learn is (roughly) to turn the handlebars to the right if we either experience those balance sensations or think on some other basis that the bicycle is leaning to the right. The circumstance-types to which our norms appeal in telling us to do something in circumstances of those types must be directly accessible to our system of cognitive processing. [17, p. 132]

Thus, a rule like "all executions of π' require doing π when C" should only take C that are sensitive to internal states, of which epistemic judgments of various kinds will often suffice. The basic thought, then, is that the fact of the miners' location is not the sort of thing that the actual cognitive mechanisms in play can use; a deliberation mechanism can only operate on an agent's thinking about where the miners are located. Thus, the norms in M, if they are to be usable in ante hoc regulation, must condition their verdicts on what we can know or believe about the circumstances, which undermines 1 and 2 as originally formulated.

Second, this position follows from the usual Kantian commitment to "ought implies can" style principles plus the account of deliberate action and norm application in PLEKNA. One cannot do what one cannot acquire the requisite knowledge to do. In this case, one cannot knowingly or deliberately save any miners without rationally discerning whether A or B, and one can't, given the protocol-theoretic or procedural notion of knowledge in PLEKNA, do that without some methodology one can use to do so. There is, in this case, no methodology to use to discern whether A or B. Thus, there is no way to knowingly or deliberately comply with the duties specified by 1 and 2, and, stronger, where there is no method for knowing the relevant external circumstances, there is no way to knowingly or deliberately comply with or apply any norms that dictate paths of action conditional only on external circumstances.

Third, given that agents can only ever act on their internal models of a set of circumstances, the relevant norms in situations concerned with managing or preventing loss are often sensitive to epistemic position. This is because, in addition to considerations about the feasibility of knowingly complying with norms, inaccuracy has costs. Consider any consequentialist norm that dictates minimizing losses. If one can't know whether A or B holds, then one can't know whether one could save more lives by blocking one of the shafts than one can by blocking neither. Thus, one cannot knowingly or deliberately follow this norm, and one cannot even reliably do so across cases in which this norm applies. Taking "a shot" may ultimately result in an even more egregious violation of the motivating norm, and one is certainly not permitted to carry out that violation. Thus, norms that do not condition their verdicts on strong epistemic states are likely to go sideways, and so be both less feasible to adhere to and more costly (i.e., morally worse overall) to adhere to. In other words, it seems a plausible meta-principle that, if I cannot, by any rational means, find out whether one of my available options will *kill everyone I intend to save*, then I cannot be obligated to carry out one of those options.

In sum, what I think is what I must act on, and the strength of my epistemic position with respect to the external circumstances—what I know or rationally think about them—is a determinant of whether I can knowingly or deliberately apply a norm and whether my attempting to do so will actually result in even more egregious violation of it.

Finally, a formalization of the puzzle that is grounded in all of the foregoing is available in PLEKNA:

(1''') if $M(\pi)$, $\langle\pi\rangle_p K(A)$, and $Del(\pi)$, then $RO_M(\pi_B)$,
(2''') if $M(\pi)$, $\langle\pi\rangle_p K(B)$, and $Del(\pi)$, then $RO_M(\pi_A)$,
(3''') $(RO_M(\pi_A \cup \pi_B)) \equiv ((M(\pi) \wedge (\langle\pi\rangle_p K(A) \wedge Del(\pi)) \vee (\langle\pi\rangle_p K(B) \wedge Del(\pi))))$,
(4'''') $((M(\pi) \wedge \neg(\langle\pi\rangle_p K(A) \wedge Del(\pi)) \wedge \neg(\langle\pi\rangle_p K(B) \wedge Del(\pi)))) \rightarrow (RF_M(\pi_A) \wedge RF_M(\pi_B))$.

This formulation tells us that the obligations or permissions, with respect to our normative system M, to block either shaft are conditional on there being some procedural norm in our methodology that we can deliberately apply that can produce knowledge whether A or B. Surely, the failure of any of these conditions entails the absence of such deontic verdicts. Again, it seems immediately plausible that if I cannot, by any rational means, find out whether each of my available options will *kill everyone I intend to save*, then I cannot be obligated to carry out one of those options.

8.4 A Protocol-Theoretic Logic of Epistemic Deontics, Procedural Knowledge, and Norm Application (PLEDKNA)

Definition 428 (PLEDKNA) The system that results from incorporating the revisions of L^R_{PLEN} and the corresponding semantic rules of PLED and PLEKNA into PLEN and operating on it according to the R-Derivation System is the protocol-theoretic logic of epistemic deontics, procedural knowledge, and norm application (PLEDKNA). Strictly speaking, PLEKNA is PLEDKNA; the "D" was just silent in prior discussion of PLEKNA so as not to distract from the already quite elaborate additions that define the procedural application logic.

Claim 429 It is beyond the scope of this book to fully evaluate the philosophical prospects of PLEDKNA. Indeed, I won't even comment on the logic of this complicated system beyond noting that it plugs directly into the protocol logic of PLEN, and its philosophical upshots are more easily seen when paired with the representation theorems of the next chapter. This entire chapter is, in some sense, an additional exercise in testing the modularity and adaptability of the Revised PLEN framework.

Of course, exploring PLEDKNA more thoroughly is clearly imperative for future work on PLEN, as the plausibility of the analyses offered by it as well as its formal coherence (i.e., non-paradoxicality) all impact the ultimate quality of the solution that PLEN offers to the problem of formalizing epistemic proceduralism. Nonetheless, it's important to note, again, that this project is about developing and defending the Revised PLEN framework as the core of a program for representing and reasoning about epistemic norms. There is room in future work for fuller defense of PLEDKNA (and thus, a more complete defense of PLEN); the objective in focus is showing that the Revised PLEN framework can support adaptations that allow us to derive formal translations of the principles in D3C2 and D3C3. This is the point of PLEDKNA, at this stage. We will see some of what PLEDKNA allows us to do in Chap. 10.

References

1. Beauchamp, T.L., and J. Childress. 1994. *Principles of biomedical ethics*. Oxford University Press.
2. Carroll, L. 1895. What the tortoise said to achilles. *Mind* 4(14):278–280.
3. Chisholm, R.M. 1963. Contrary-to-duty imperatives and deontic logic. *Analysis* 24:33–36.
4. Faroldi, F. 2019. *Hyperintensionality and normativity*. Springer.
5. Forrester, J.W. 1984. Gentle murder, or the adverbial samaritan. *Journal of Philosophy* 81:193–196.
6. Gabby, D., J. Horty, and X. Parent, Eds. 2013. *Handbook of deontic logic and normative systems*. College Publications.
7. Horty, J. 2001. *Agency and deontic logic*. Oxford: Oxford University Press.
8. Jorgensen, J. 1937. Imperatives and logic. *Erkenntnis* 7:288–296.
9. Kolodny, N. and J. MacFarlane. 2010. Ifs and oughts. *Journal of Philosophy* 107(3):115–143.
10. Konolige, K. 1986. *A deduction model of belief*. Morgan Kaufmann Publishers.
11. Lemmon, E.J. 1962. Moral dilemmas. *Philosophical Review* 71:139–158.
12. MacFarlane, J. 2004. In What Sense (If Any) Is Logic Normative for Thought? Unpublished. (Presented at the 2004 Central Division APA symposium on the normativity of logic.) Accessed online: http://johnmacfarlane.net/normativity_of_logic.pdf.
13. MacFarlane, J. 2017. Is Logic a Normative Discipline? Unpublished. (Conference on the Normativity of Logic University of Bergen) Accessed online: https://johnmacfarlane.net/normative.pdf.
14. McNamara, P. 2019. Deontic logic. *The Stanford Encyclopedia of Philosophy* (Summer 2019 Edition), ed. Edward N. Zalta, forthcoming. https://plato.stanford.edu/archives/sum2019/entries/logic-deontic/.
15. Meyer, J.-J. 1987. A different approach to deontic logic: Deontic logic viewed as a variant of dynamic logic. *Notre Dame Journal of Formal Logic* 29(1):109–136.

16. Meyer, J.-J. 2000. Dynamic logic for reasoning about actions and agents. In *Logic-based artificial intelligence,* ed. Jack Minker. Kluwer International Series In Engineering And Computer Science, vol. 597, 281–311. Norwell, MA, USA: Kluwer Academic Publishers.
17. Pollock, J., and J. Cruz. 1999. *Contemporary theories of lnowledge*, 2nd edn. Rowman & Littlefield.
18. Prior, A.N. 1958. Escapism: The logical basis of ethics. In *Essays in moral philosophy*, ed. A.I. Melden, 135–146. Seattle: University of Washington Press.
19. Ross, A. 1941. Imperatives and logic. *Theoria* 7:53–71.
20. Sartre, J.-P. 2007. *Existentialism is a humanism.* Yale University Press.
21. Segerberg, K. 2012. $D\triangle L$: Dynamic deontic logic. *Synthese* 185(S1):1–17.
22. Segerberg, K., J.-J. Meyer, and M. Kracht. 2016. The logic of action. *The Stanford Encyclopedia of Philosophy* (Winter 2016 Edition), ed. Edward N. Zalta. https://plato.stanford.edu/archives/win2016/entries/logic-action/.
23. von Wright, G.H. 1951. Deontic logic. *Mind* 60:1–15.
24. Xu, M. 2010. Combinations of stit and actions. *Journal of Logic, Language and Information* 19(4):485–503.

Chapter 9
Philosophical Results II: Representation Theorems and Rule-Following

Abstract Prior chapters (4, 5, 6, and 7) developed a formal logic of protocols under the presumption that it is useful and interesting as a logic of epistemic norms. Where the development of that system indirectly justified that presumption, this chapter attempts to directly justify it by arguing that protocols in PLEN represent the most important conceptual and logical features of procedural epistemic norms. This is done by means of two representation theorems. The first representation theorem shows that the syntactic form of protocols is logically bound to the verdicts of restrictions on epistemic dynamics. Thus, protocols are constructions for deriving the verdicts of epistemic norms that can be codified with restrictions on epistemic dynamics. The second representation theorem does the parallel work with norm-kernels by means of a protocol-theoretic analysis of epistemic deontic operators, showing that the deontic verdicts of norm-kernels that take elements of epistemic dynamics as objects can all be derived from the syntax and semantics of protocols. These representation theorems together bind protocols, via their syntax and semantics, to restrictions on epistemic dynamics and norm-kernels, which are, as argued in Chaps. 1–3, the main conceptual and natural language constructions for thinking with and about epistemic norms. Here, a problem needs to be solved: the argument up to this point has relied extensively on the idea that restrictions on epistemic dynamics and norm-kernels have intelligible relations that can be figured out in order to apply or knowingly follow norms. The rule-following paradox might thus be a direct refutation of that idea, and so an account of the intelligibility of rule-following pace the paradox is developed by appeal to the constructions of the prior chapter.

9.1 Representation Theorems

The discussion thus far has focused on two constructions that are central to accounting for the content of procedural epistemic norms and epistemic norms in general. I show in this chapter that protocols are closely bound by the semantics of PLEN to the most important features of both of these constructions.

R. Jenkins, *A Protocol-theoretic Framework for the Logic of Epistemic Norms*, Synthese Library 467, https://doi.org/10.1007/978-3-031-08597-0_9

Restrictions on epistemic dynamics are the first constructions. Chapter 1 argued that restrictions on epistemic dynamics are useful set-theoretic constructions for representing the content of procedural epistemic norms. In particular, they are useful for regimenting the sorts of reasoning about procedural epistemic norms that is central to epistemic regulation. Moreover, though the explicit construction of restrictions on epistemic dynamics is introduced in this book, the lines of argument in Chaps. 2 and 3 draw on multiple, independent strands of logical, epistemological, and psychological research. Restrictions on epistemic dynamics thus appear to be a robust conceptual tool for representing and reasoning about epistemic norms.

Norm-kernels are the second constructions. Their usefulness as natural language (or quasi-natural, logico-english) constructions for articulating the content of epistemic norms is supported by robust strands of epistemological and logical research. It's a central assumption of deontic logic going back to its inception, as well as a central assumption of the extant study of epistemic norms, that norm-kernels are the primary natural language constructions for representing the content of epistemic norms in general. It's difficult even to conceive of other constructions that could articulate the verdicts of epistemic norms other than restrictions on epistemic dynamics and norm-kernels.

This brief chapter articulates two representation theorems in order to show that, if the models of PLEN are useful for representing epistemic dynamics (and thus for representing the content of restrictions on epistemic dynamics and the conditions and objects of norm-kernels), then protocols bear close logical relations to both of these constructions; the verdicts of arbitrary restrictions on epistemic dynamics and norm-kernels (defined over the elements of EDGs) can be derived from the syntactic and semantic properties of protocols. Protocols are thus the tools by which PLEN represents (procedural and other) epistemic norms. This is not merely by interpretive fiat or stipulation for the sake of argument. Protocols are exceptionally natural and perspicuous formal objects with which to encode procedural epistemic norms and, indeed, epistemic norms in general because they can be used in PLEN to derive the conditional verdicts of both restrictions on epistemic dynamics and norm-kernels for any epistemic factors over which either can be defined that can also be represented by the parts of EDGs.

To summarize the chapter, then:

Claim 430 A pair of representation theorems maps the syntactic and semantic properties of protocols to the distinctive, regulative content of procedural epistemic norms, as regimented using the syntactic and semantic resources of PLEN.

(1) For clarity, some naive definitions:

   (a) Let the verdicts of an epistemic norm, N, be the assignments of a normative status (including deontic statuses) to N's objects that are implied by N.

(b) Let restrictions on epistemic dynamics be partitions of epistemic factors (states, transitions, processes, actions, procedures) by their properties along with a ranking on the cells and a designation of some cells in the partition as the rational cell and another as the irrational cell.
(c) Let norm-kernel functions be functions that capture the content of meta-language norm-kernels.
(d) Let protocol functions be functions that map the conditions and objects of norm-kernels to the protocol-theoretic analyses of deontic operators according to the syntactic and semantic relations of protocols to the objects of norm-kernels.

(2) The Restriction Representation Theorem maps each restriction on epistemic dynamics (defined over P_{PLEN} or the semantic structures of PLEN that satisfies a specified set of plausibility conditions) to some set of protocols in a way that enables the derivation of the verdicts or content of those restrictions from the syntactic and semantic properties of the protocols to which they're mapped.
(3) The Protocol Function Theorem maps norm-kernel functions defined over L^R_{PLEN}, P_{PLEN}, and the semantic structures of PLEN to a set of protocol functions that agree on all verdicts for all conditions and objects.

9.2 Representation Theorem 1

The first representation theorem states that the syntactic forms of protocols are, due to the binding of syntactic form to graph-theoretic structure in PLEN, closely bound to the contents of restrictions on epistemic dynamics. Every protocol defines a restriction on epistemic dynamics in virtue of its execution sets, and, more important for our purposes, the verdicts of every arbitrary restriction on epistemic dynamics definable over epistemic factors represented by the parts of EDGs can be derived from the syntax of some set of protocols. In other words, every restriction on dynamics (defined over the elements of epistemic dynamics represented by EDGs) is represented by some set of protocols in such a fine-grained way that all of its verdicts can be read off from the syntax of those protocols.

It's important to note that this provides a fine-grained enough way to read off the content of epistemic norms that we can readily distinguish between verdicts on epistemic processes and epistemic transitions, as required by the arguments at the end of Chap. 4. As in prior chapters, let me frontload the statement and discussion of the main results with definitions or reminders of definitions.

9.2.1 Restrictions on Epistemic Dynamics

Epistemic directed graphs are taken, in PLEN, to be abstract models of single-agent epistemic dynamics, with each (informal) element of epistemic dynamics mapped to a feature of epistemic directed graphs with the intent that the mapping produces a kind of structural resemblance. Recall the following table from Sect. 4.3.1.2, which makes that explicit:

Epistemic States: Judgment and Content	$Represented\ By \Longrightarrow$	States: $S = \{s_1, \ldots, s_n\}$, Content States: $\langle X : Y : \Pi_M \rangle$, Content functions: $c(s)$
Epistemic State Transitions	$Represented\ By \Longrightarrow$	Transitions: $t = \langle s, s' \rangle$
Epistemic Action Tokens	$Represented\ By \Longrightarrow$	Edges: $e = \langle s, a, s' \rangle$
Epistemic Action Types	$Represented\ By \Longrightarrow$	Execution Sets for Atomic Protocols: $R_a = \{e \in S \times \Pi_0 \times S \mid e \in S \times a \times S\}$ Edge Sets for Complex Protocols: $R{+}_\pi = \{e \in S \times \Pi_C \times S \mid e \in S \times \pi \times S\}$
Epistemic Process Tokens	$Represented\ By \Longrightarrow$	Procedural Paths: $p = \langle e_1, \ldots, e_n \rangle$ where for each e_i and e_{i+1}, $last(e_i) = first(e_{i+1})$ and there are π_i and π_{i+1} such that $e_i \in R_{\pi i}$ and $e_{i+1} \in R_{\pi i+1}$
Epistemic Process Types (Epistemic Procedures)	$Represented\ By \Longrightarrow$	Execution Sets for Complex Protocols: P_π such that $p \in P_\pi$ iff p satisfies the relevant execution rules with respect to π
Feasibility of a Transition under C	$Represented\ By \Longrightarrow$	Let t be the transition. Then t is feasible under C iff $first(t) \Vdash C$.
Feasibility of an Action under C	$Represented\ By \Longrightarrow$	Let e be the action token. Then e is feasible under C iff $first(e) \Vdash C$.
Feasibility of a Process under C	$Represented\ By \Longrightarrow$	Let p be the process token. Then p is feasible under C iff $first(p) \Vdash C$.

Capability of a Transition	*Represented By* $\Longrightarrow$	Let t be the transition. Then t is within the capacities of agent i iff g represents i's epistemic dynamics and there is some s such that $s = first(t)$.
Capability of an Action	*Represented By* $\Longrightarrow$	Let e be the action token. Then e is within the capacities of agent i iff g represents i's epistemic dynamics and there is some s such that $s = first(e)$ (iff $e \in R$).
Capability of a Process	*Represented By* $\Longrightarrow$	Let p be the process token. Then p is within the capacities of agent i iff g represents i's epistemic dynamics and there is some s such that $s = first(p)$ (iff $e \in P_g$).

The first representation theorem binds the syntax of protocols to the verdicts of restrictions on epistemic dynamics defined over EDGs. That is, for this result to get off the ground, we must take EDGs to represent the epistemic dynamics of single agents (as in Sect. 4.3.1.2), and consider the restrictions defined over the epistemic factors represented by the parts of EDGs.

Definition 431 (Elements of Epistemic Directed Graphs) Let g be an EDG. Then any subset of S, T_g, R, $R+$, R_0, $R+_0$, C, and P_g, is a set of elements of epistemic directed graphs.

Definition 432 (Restrictions on Epistemic Dynamics (over Elements of EDGs)) Let X be a finite, possibly empty set of elements of epistemic directed graphs, i.e., some set of states, transitions, procedural paths, or some specified subset thereof. Let $P(X)$ be the powerset of X. Call some designated element of $P(X)$, T_X, the standard of irrationality. Let $\preceq$ be a partial order on the elements, P_i, of $P(X)$. Now, let R_X and I_X be subsets of X. Then:

(Restrictions on Epistemic Dynamics) Every structure defined over X of the form

$$R_{\preceq}(X) = (P(X), \preceq, T_X, R_X, I_X)$$

is a *restriction on epistemic dynamics* (defined over EDGs) iff:

(1)

$$\forall P_i \in P(X) \forall x \in X \cap P_i (P_i \preceq T_X \equiv x \in I_X)$$

(2)

$$\forall P_i \in P(X) \forall x \in X \cap P_i (P_i \succ T_X \equiv x \in R_X)$$

(Restriction Conventions 2)

(1) Let R_X and I_X be the *verdict sets* of $R_{\preceq}(X)$.

(a) R_X is the rational set.
(b) I_X is the irrational set.

(2) Let T_X be the *threshold set* of $R_{\preceq}(X)$.
(3) Let $x \in R_X$, $x \in I_X$, and statements of the form $x \in P_i \succeq P_j$ be *verdicts* of $R_{\preceq}(X)$. Respectively, they say:

(a) $x \in R_X$ :: x is rational (rationally permissible).
(b) $x \in I_X$ :: x is not rational (rationally impermissible).
(c) $x \in P_i \succeq P_j$ :: x is more rational than any y in P_j.
(d) $x \succeq y$ :: x is more rational than y (i.e., $x \in P_i$, $y \in P_j$, and $P_i \succeq P_j$)

(4) Denote arbitrary restrictions on X with $R_{\preceq}(X)$ or, where the set of epistemic factors is irrelevant or determined by context, $R_{\preceq}$.

(a) Say that $R_{\preceq}(X)$ is defined over X, and, that any restriction defined over any set of elements of epistemic directed graphs is a restriction defined over the semantic structures of PLEN.
(b) Introduce indexing of restrictions on the same set, $R^i_{\preceq}(X)$, where necessary.

9.2.2 Restrictions on Procedural Construction

As D2 argues, procedural epistemic norms restrict not only epistemic processes and transitions, but also the ways of constructing epistemic procedures out of more basic epistemic actions. We formalize this by means of restrictions defined over ways of building epistemic procedures. First, we offer a construction that captures ways of building epistemic procedures:

Definition 433 (Syntactic Schemata)
Let a *syntactic schema* be any structure of the following form:

$$\langle \pi_v \in String(\Pi_v \cup Prot \cup Prot+), \Pi_0^{\pi v} \subseteq \Pi_0*, f_{\pi v} \rangle$$

where:

(1) π_v is a concatenated string of symbols satisfying the syntactic rules of P_{PLEN} but defined over a set of variables (Π_V) for elements of Π_0,

(a) By the syntactic rules of PLEN: $\pi_v = \langle o_1 \frown \ldots \frown \ldots x_1 \frown \ldots \frown o_m \ldots \frown x_n \rangle$
(b) Each $x_i \in \Pi_v$

(2) $\Pi_0^{\pi_v}$ is a set of sequences of atomic protocols, and
(3) $f_{\pi v} : \Pi_0^{\pi_v} \to P_{PLEN}$ is a function that acts as a filling instruction that maps sequences of atomics to protocols of the syntactic form modeled by π_v:

(a) $f_{\pi v}(\langle a_1, \ldots, a_n\rangle) = \pi$ iff
  (i) $\langle a_1, \ldots, a_n\rangle \in \Pi_0^{\pi_v}$,
  (ii) $\pi = \langle \circ_1 \frown \ldots \frown \ldots a_1 \frown \ldots \frown \circ_m \ldots \frown a_n\rangle$,
    (A) Thus, $At(\pi)* = \langle a_1, \ldots, a_n\rangle$
  (iii) $\circ(\pi)* = \circ(\pi_v)* = \langle \circ_1, \ldots, \circ_m\rangle$.

(4) Let $\Pi_v^\forall$ be the set of all syntactic schemata.

With a way of distinguishing and quantifying over ways of building epistemic procedures in hand, restrictions on procedural construction are straightforwardly defined over classes of such ways of building procedures: Every restriction on epistemic dynamics defined over $\Pi_v^\forall$ is a *restriction on procedural constructions* or a *syntactic restriction*. Explicitly:

Definition 434 (Restrictions on Procedural Construction) Every structure defined over $\Pi_v^\forall$ of the form

$$R_\preceq(\Pi_v^\forall) = (P(\Pi_v^\forall), \preceq, T_{\Pi_v^\forall}, R_{\Pi_v^\forall}, I_{\Pi_v^\forall})$$

is a *restriction on procedural constructions* or a *syntactic restriction* iff:

(1)

$$\forall P_i \in P(\Pi_v^\forall)\forall \pi_v \in \Pi_v^\forall \cap P_i(P_i \preceq T_{\Pi_v^\forall} \equiv \pi_v \in I_{\Pi_v^\forall})$$

(2)

$$\forall P_i \in P(\Pi_v^\forall)\forall \pi_v \in \Pi_v^\forall \cap P_i(P_i \succ T_{\Pi_v^\forall} \equiv \pi_v \in R_{\Pi_v^\forall})$$

(Restriction Conventions 3)

(1) Say that any restriction on procedural constructions is defined over the set of syntactic schemata.
(2) Restrictions defined over syntactic schemata also determine restrictions on protocols:

(a) Let

$$R_\preceq(\Pi_v^\forall) = (P(\Pi_v^\forall), \preceq, T_{\Pi_v^\forall}, R_{\Pi_v^\forall}, I_{\Pi_v^\forall})$$

be a restriction on procedural constructions.
(b) Define

$$R_\preceq(P_{PLEN}) = (P(P_{PLEN}), \preceq, T_{P_{PLEN}}, R_{P_{PLEN}}, I_{P_{PLEN}})$$

where $R_{P_{PLEN}}$ is the set of protocols that match any element in $R_{\Pi_\upsilon^\forall}$, $I_{P_{PLEN}}$ is the set of protocols that match any element in $I_{\Pi_\upsilon^\forall}$, etc..

(3) Given that restrictions defined over syntactic schemata also determine restrictions on protocols, say that any restriction on procedural constructions is defined over P_{PLEN}, as well.

Remark 435 The main thought underlying the definition of restrictions on procedural construction is that, if we identify a set or space of possible ways of building procedures out of a repertoire of basic actions, a restriction thereon is simply a partition on that set/space and a ranking of the cells of that partition that identifies a set of cells that are the correct or legitimate ways of building procedures. The space of possible ways of building procedures out of arbitrary repertoires is defined in PLEN by the control operators: $\Pi_\upsilon^\forall$ abstractly identifies the possible configurations of procedural constructions by specifying the possible sequences of control operators into which a repertoire of basic actions can be slotted.

Example 436 By replacing the subprotocols in a protocol with variables that themselves can be replaced with atomic protocols, we identify ways of building procedures. Take $x_1; x_2$. This schema identifies the way of building procedures by sequential composition of atomic actions. The more complex schema $(x_1; x_2) \cap (x_3; (x_1 \cup x_4))$ identifies the way of building procedures by parallel execution on a sequential composition of atomics and a sequential composition of an atomic and the first component of the first sequential composition and a new component as the second component of the second composition, e.g., $(a; b) \cap (c; (a \cup d))$ and $(a; a) \cap (a; (a \cup a))$.

Example 437 A restriction on procedural constructions is, intuitively, just a way of partitioning the possible ways to build protocols into rational and irrational sets. Thus, given that syntactic schemata analyze possible procedural constructions, a restriction on procedural constructions might, for instance, single out all sequential compositions as the rational ways of building procedures and the rest as irrational. This would mean that $R_{\Pi_\upsilon^\forall}$ contains $x_1; x_2$, $x_1; (x_2; x_3)$, $(x_1; x_2); x_3$, $(x_1 \cup x_2); x_3$, and so on for all of the possible ways of building a protocol such that sequential composition is the main control operator. Of course, more restrictive or subtle restrictions (e.g., all sequential compositions over some subset of atomics) are possible and definable using these conventions.

Remark 438 Now, a relation can be stated, relating protocols to syntactic schemata, and thus relating procedural epistemic norms to ways of building epistemic procedures:

Definition 439 (Protocol to Schema Matching) For any protocol, π, and syntactic schema, π_υ, π *matches* π_υ iff $f_{\pi\upsilon}(At(\pi)*) = \pi$.

Remark 440 The relation is symmetrical.

This relation extends the verdicts of restrictions on procedural construction to protocols, and can then be used as a bridge from procedural constructions to other epistemic factors. Given a restriction on procedural constructions, we might, for instance, be able to define a derivative restriction on processes that execute matching protocols in such a way that it corresponds to the original restriction on procedural constructions on the basis of what protocols match the rational schemata. This is a key idea in the first representation theorem.

9.2.3 Representation Theorem 1: Restriction Representation Theorem (RRT)

Definition 441 (Restriction Correspondence)
Let $R_{\Pi_v^\forall}$ be the rational set of syntactic schemata in a restriction where $X = \Pi_v^\forall$.
Let R_{Pg} be the rational set of processes in a restriction where $X = P_g$ for some g.
Let R_{Tg} be the rational set of transitions in a restriction where $X = T_g$.
Let R_{Sg} be the rational set of states in a restriction where $X = S_g$.
Define $I_{\Pi_v^\forall}$, I_{Pg}, I_{Tg}, and I_{S_g} by parallel clauses.
Define $T_{\Pi_v^\forall}$, T_{Pg}, T_{Tg}, and T_{S_g} by parallel clauses.
(Restriction Correspondence) Restrictions on different sets of elements, $R_{\preceq}(X_1)$ and $R_{\preceq}(X_2)$, *correspond* iff:

(1) $X_1, X_2 \in \{S_g, T_g, P_g, \Pi_v^\forall\}$
(2) For any π_v, π that matches π_v, and p such that $p \in P_\pi$, π_v is in $R_{\Pi_v^\forall}$ iff p is in R_{Pg}.
(3) For any p and any t such that $t = \langle first(p), last(p)\rangle$, p is in R_{Pg} iff t is in R_{Tg}.
(4) For any t and any s such that $s = last(t)$, t is in R_{Tg} iff s is in R_{Sg}.
(5) Counterparts of 2–4 define correspondence for $I_{\Pi_v^\forall}$, I_{Pg}, I_{Tg}, and I_{S_g}.
(6) Counterparts of 2–4 define correspondence for $T_{\Pi_v^\forall}$, T_{Pg}, T_{Tg}, and T_{S_g}.

Definition 442 (Reduction of Restrictions on Epistemic Dynamics) A class of restrictions on elements of epistemic dynamics, X_1, reduces to a class of restrictions on elements of epistemic dynamics, X_2, iff every restriction on X_1 corresponds to some restriction on X_2.

Example 443 The class of restrictions on epistemic states in connected EDGs reduces to the class of restrictions on transitions in connected EDGs iff every restriction defined over states in connected EDGs corresponds to some restriction defined over transitions in connected EDGs. Note that this need not be symmetrical; it could very well be the case that not all restrictions defined over transitions in connected EDGs correspond to some restriction defined over states in connected EDGs. Indeed, this and other asymmetries are proven below and are theoretically important to the logic of epistemic norms in PLEN.

Claim 444 The restrictions on procedural construction reduce all other restrictions defined over all other epistemic factors represented by the elements of EDGs, and protocols are syntactic structures that formally represent restrictions on procedural construction. More precisely:

Theorem 445 (Restriction Representation Theorem (RRT)) *Protocols define restrictions on every aspect of epistemic dynamics:*

(1) (RRT1) For every protocol, π, there is a restriction on procedural constructions, $R^{\pi}_{\preceq}(\Pi^{\forall}_{v})$, such that:

(a) $\pi_v \in R_{\Pi^{\forall}_{v}}$ iff π_v matches π
(b) $\pi_v \in I_{\Pi^{\forall}_{v}}$ iff π_v does not match π

(2) (RRT2) For every restriction on procedural constructions, $R_{\preceq}(\Pi^{\forall}_{v})$, there are sets of protocols, Π_R, Π_T, and Π_I, and a set of restrictions on procedural construction containing $R_{\preceq}(\Pi^{\forall}_{v})$, $R!$, satisfying the following rules:

(a) Every restriction that $R!$ contains is identical to $R_{\preceq}(\Pi^{\forall}_{v})$ with respect to its rational, irrational, and threshold sets (but may differ in its rank ordering).
(b) For all $R^{i}_{\preceq}(\Pi^{\forall}_{v})$ in $R!$ where $X^{i}_{\Pi^{\forall}_{v}} \in \left\{R^{i}_{\Pi^{\forall}_{v}}, I^{i}_{\Pi^{\forall}_{v}}, T^{i}_{\Pi^{\forall}_{v}}\right\}$, for any syntactic schema, π_v, $\pi_v \in X^{i}_{\Pi^{\forall}_{v}}$ iff π_v matches any protocol in Π_X.

(3) (RRT3) For any connected EDG, g:

(a) For every protocol, π, there is a set of restrictions defined over P_g, $R^{\pi}_{\preceq}(P_g)$, such that:
(i) $p \in R_{P_g}$ iff $p \in P_\pi$
(ii) $p \in I_{P_g}$ iff $p \notin P_\pi$
(b) For every protocol, π, there is a set of restrictions defined over T_g, $R^{\pi}_{\preceq}(T_g)$, such that:
(i) $t \in R_{T_g}$ iff $t \in T_\pi$
(ii) $t \in I_{T_g}$ iff $t \notin T_\pi$
(c) For every protocol, π, there is a set of restrictions defined over S, $R^{\pi}_{\preceq}(S_g)$, such that:
(i) $s \in R_{S_g}$ iff there is a $p \in P_\pi$ such that $s = first(p)$
(ii) $s \in I_{S_g}$ iff there is a $p \in P_g - P_\pi$ such that $s = first(p)$

(4) (RRT4) For every restriction, $R_{\preceq}(X)$, defined over a set, X, of elements of epistemic directed graph, g, there are sets of protocols, Π_R, Π_T, and Π_I, and a set of restrictions, $R_X!$, on elements of g containing $R_{\preceq}(X)$, that conjointly satisfy the following conditions:

(a) Every restriction that $R!$ contains is identical to $R_{\preceq}(\Pi^{\forall}_{v})$ with respect to its rational, irrational, and threshold sets (but may differ in its rank ordering).
(b) For all $R^{i}_{\preceq}(X)$ in $R!$, for any element, x, $x \in R_X$ iff:
(i) if $x \in P_g$, then there is a $\pi \in \Pi_R$ such that $x \in P_\pi$
(ii) if $x \in T_g$, then there is a $\pi \in \Pi_R$ such that $x \in T_\pi$

(iii) if $x \in S_g$, then there is a $\pi \in \Pi_R$ such that there is a $p \in P_\pi$ such that $s = first(p)$

(c) Parallel clauses define the irrational and threshold sets of the restrictions in $R!$.

Proof (Proof-Plan) Direct argument for the first three clauses by means of direct construction or lemmata proved below, and then derivation of the fourth clause from the first three clauses. I present the outline of the argument for each of the first three clauses and the argument for the fourth from the first three presently. The argument relies, at an essential point, on a Reduction Lemma the proof of which is reserved for Technical Appendix 5. □

(1) For RRT1, it suffices to map arbitrary protocols to restrictions on procedural construction such that the extension of the restriction's rational, irrational, and threshold sets can be determined by the definition of protocol to schema matching.

Proof Define an arbitrary restriction on procedural constructions

$$R^{\pi}_{\preceq}(\Pi^{\forall}_{\upsilon}) = (P(\Pi^{\forall}_{\upsilon}), \preceq, T_{\Pi^{\forall}_{\upsilon}}, R_{\Pi^{\forall}_{\upsilon}}, I_{\Pi^{\forall}_{\upsilon}})$$

(with arbitrary ordering relation) such that:
(Verdict Rule) For any π_υ, if $\pi = f_{\pi\upsilon}(At(\pi)*)$, then $\pi_\upsilon \in R_{\Pi^{\forall}_{\upsilon}}$; else, $\pi_\upsilon \in I_{\Pi^{\forall}_{\upsilon}} = T_{\Pi^{\forall}_{\upsilon}}$.
Thus, the elements of $R_{\Pi^{\forall}_{\upsilon}}$ (resp. $I_{\Pi^{\forall}_{\upsilon}}$) can be enumerated by checking to see whether the filling instructions for π_υ generate π when fed the sequence of atomics in π. Since the ranking relation was arbitrary, there is a set of restrictions to be read off of π by the verdict rule that have identical assignments into $R_{\Pi^{\forall}_{\upsilon}}$ and $I_{\Pi^{\forall}_{\upsilon}}$ but differing only in the ordering relation. Thus, every well-formed protocol in PLEN can be mapped to a restriction on procedural constructions (or a set thereof differing only ordering) that satisfies the Verdict Rule, as required. □

(2) For RRT2, it suffices to define a procedure that exploits protocol to schema matching for deriving the extension of the rational, irrational, and threshold sets of arbitrary restrictions on procedural construction from the syntax of some set of protocols.

Proof (Proof-Plan) First, a simple procedure for identifying a set of protocols is defined, then it's shown that the procedure produces sets of protocols with the requisite properties.

(a) (Reading Off Procedure 1) For every syntactic schema, π_υ, where $\Pi_0^{\pi_\upsilon} \neq \emptyset$, there is a non-empty set of protocols that matches it by the definition of protocol to schema matching (in the case with no control operators, all atomic protocols match the atomic schemata). Thus, there is a trivial procedure for building sets of protocols with the requisite properties:

(i) (Protocol Set Selection)

(A) Define a finite set of protocols:

$$\Pi_R = \left\{\pi \in P_{PLEN} \mid \exists \pi_v \in R_{\Pi_v^\forall}(\pi \text{ matches } \pi_v)\right\}$$

(B) Define a finite set of protocols:

$$\Pi_T = \left\{\pi \in P_{PLEN} \mid \exists \pi_v \in T_{\Pi_v^\forall}(\pi \text{ matches } \pi_v)\right\}$$

(C) Define a finite set of protocols:

$$\Pi_I = \left\{\pi \in P_{PLEN} \mid \exists \pi_v \in I_{\Pi_v^\forall}(\pi \text{ matches } \pi_v)\right\}$$

(ii) (Reading Off Rules) From these sets of protocols, derive verdict sets, $X_{\Pi_v^\forall}^{\Pi_X}$, over procedural construction from the syntax of the elements of each Π_X by the rules:

(A) For all π_v, $\pi_v \in R_{\Pi_v^\forall}^{\Pi_R}$ iff $\exists \pi_i \in \Pi_R$ such that π_v matches π_i.

(B) For all π_v, $\pi_v \in T_{\Pi_v^\forall}^{\Pi_T}$ iff $\exists \pi_i \in \Pi_T$ such that π_v matches π_i.

(C) For all π_v, $\pi_v \in I_{\Pi_v^\forall}^{\Pi_I}$ iff $\exists \pi_i \in \Pi_I$ such that π_v matches π_i.

(b) Let $R_\preceq(\Pi_v^\forall)$ be an arbitrary restriction on procedural constructions.

(i) Execute Reading Off Procedure 1 to define Π_R, Π_T, and Π_I.

(ii) For any restriction on schemata with non-empty $\Pi_0^{\pi_v}$, each element of the rational set is matched by a set of protocols that differ only in their respective sequences of atomics. Thus, each set of protocols, Π_R, Π_T, and Π_I, only needs to contain one matching protocol for each schema in $R_{\Pi_v^\forall}$, $I_{\Pi_v^\forall}$, and $T_{\Pi_v^\forall}$ (respectively) in order to read off exactly the verdict sets of $R_\preceq(\Pi_v^\forall)$ by the Reading Off Rules.

(c) From Π_R, Π_T, and Π_I, we can derive a set of restrictions, $R!$, containing exactly the restriction with which we started.

(i) Let $R!$ be a set of restrictions on procedural constructions derived by means of the Reading Off Rules; for each restriction on procedural construction, $R_\preceq^i(\Pi_v^\forall) \in R!$, let it be the case that:

(A) $R_{\Pi_v^\forall}^i = R_{\Pi_v^\forall}^{\Pi_R}$

(B) $T_{\Pi_v^\forall}^i = T_{\Pi_v^\forall}^{\Pi_T}$

(C) $I_{\Pi_v^\forall}^i = R_{\Pi_v^\forall}^{\Pi_I}$

(ii) Of course, $R_{\Pi_v^\forall}^i = R_{\Pi_v^\forall}^j$, $T_{\Pi_v^\forall}^i = T_{\Pi_v^\forall}^j$, and $I_{\Pi_v^\forall}^i = I_{\Pi_v^\forall}^j$ for all $R_\preceq^i(\Pi_v^\forall), R_\preceq^j(\Pi_v^\forall) \in R!$.

(A) Strictly, all that can be read off of the syntax of these protocols (resp. sets of protocols) is the triple of rational, threshold, and irrational sets of schemata, and, from there, arbitrarily many restrictions on schemata can be defined by adding distinct ordering relations.
(B) This gives clause (b).

(iii) For each $R^i_{\preceq}(\Pi^{\forall}_v) \in R!$, $R^i_{\Pi^{\forall}_v} = R_{\Pi^{\forall}_v}$, $T^i_{\Pi^{\forall}_v} = T_{\Pi^{\forall}_v}$, and $I^i_{\Pi^{\forall}_v} = I_{\Pi^{\forall}_v}$.
(A) This is immediate from Protocol Set Selection and the Reading Off Rules: Let π_v be in $X_{\Pi^{\forall}_v}$. By the Protocol Set Selection procedure, there is a matching π in Π_X. From the Reading Off Rules, $\pi_v \in X^{\Pi_X}_{\Pi^{\forall}_v}$. Let $\pi_v \in X^{\Pi_X}_{\Pi^{\forall}_v}$. By the Reading Off Rules, there is a matching π in Π_X. From the Protocol Set Selection Procedure, $\pi_v \in X_{\Pi^{\forall}_v}$. Thus, $X_{\Pi^{\forall}_v} = X^{\Pi_X}_{\Pi^{\forall}_v}$. Thus, $X_{\Pi^{\forall}_v} = X^i_{\Pi^{\forall}_v}$ for each verdict set of each $R^i_{\preceq}(\Pi^{\forall}_v)$ in $R!$ and each verdict set in $R_{\preceq}(\Pi^{\forall}_v)$.
(B) This gives clause (a).
(C) The resulting restrictions on procedural construction are identical with $R_{\preceq}(\Pi^{\forall}_v)$ with respect to its categorical verdicts (and, by the machinery in later sections, its deontic verdicts).

(iv) Thus, everything to be derived about the categorical verdicts of the target restriction on procedural construction, $R_{\preceq}(\Pi^{\forall}_v)$, can be derived from the set of restrictions on procedural construction, $R!$, and can thus be derived from the protocol sets, Π_R, Π_I, and Π_T by sequences of the Protocol Set Selection and Reading Off Rules.

□

(3) For RRT3, it suffices to show that arbitrary restrictions on epistemic dynamics definable over elements of connected EDGs can be reduced to restrictions on processes defined over connected EDGs, and all but a specified, proper subset of restrictions on processes defined over connected EDGs can be reduced to restrictions on procedural construction.

(a) This is immediate from the following lemma, which has a mechanical but lengthy proof in Appendix 5:

Lemma 446 (Reduction Lemma)

(a) For every restriction on the construction of procedures, there is some connected EDG, g, such that there is a corresponding restriction on processes defined over P_g.
(b) For arbitrary connected EDG, g:
(i) Let $R_{\preceq}(P_g)$ be a restriction on processes defined over P_g such that there are no isomorphic paths, p and q, such that p is in the rational set, q is in the irrational set, and, for all pairs of states, s and s' in p and q, respectively, that are located at the same indices with respect to

$s(p)$ and $s(q)$, $v_s = v_{s'}$ and $c_s = c_{s'}$. Then: there is a corresponding restriction on procedural constructions, $R_{\preceq}(\Pi_g)$, defined over Π_g.

(ii) For every restriction on transitions defined over T_g, there is a corresponding restriction on processes defined over P_g.

(iii) For every restriction on states defined over S_g, there is a corresponding restriction on transitions defined over T_g.

(iv) It is not the case that, for every restriction on processes defined over P_g, there is a corresponding restriction on transitions defined over T_g nor a corresponding restriction on states defined over S_g.

(v) It is not the case that, for every restriction on transitions defined over T_g, there is a corresponding restriction on states defined over S_g.

(4) Thus, from the reduction of restrictions on dynamics defined over connected EDGs to restrictions over procedural construction, the extensions of the rational, irrational, and threshold sets of any arbitrary restriction on epistemic dynamics definable over connected EDGs can be derived from the syntax of some set of protocols by means of procedures exploiting protocol to schema matching.

(5) From RRT2, each clause of RRT4 follows by means of the Reduction Lemma. The verdicts of any arbitrary restriction on processes, transitions, and states that corresponds to a restriction on procedural constructions can all be derived from the syntax of some set of protocols by deriving the restriction on procedural construction by means of the Reading Off Procedure articulated below and then deriving the other restrictions from it, by clauses 1–4 of the Reduction lemma.[1]

(6) This suffices for RRT4 and completes the proof of each clause of RRT.

Remark 447 In other words, this result says that every protocol can be used to define a corresponding restriction on epistemic dynamics defined over any class of epistemic factors (represented by elements of EDGs), and, more important, the verdicts of every restriction on epistemic dynamics defined over any class of epistemic factors can be derived from the syntax of some set of protocols. That is, every restriction on epistemic dynamics is represented by some set of protocols and every set of protocols is represented by some set of restrictions on epistemic dynamics. *The result shows that we can recover the syntax of protocols from restrictions and we can recover the verdicts of restrictions from the syntax of protocols.*

Claim 448 This Restriction Representation Theorem provides several ways in which we can use the protocol-theoretic analysis of procedural epistemic norms to distinguish the normative verdicts on processes from the normative verdicts on transitions. Define a restriction, $R_{\preceq}(P_g)$, and a separate restriction, $R_{\preceq}(T_g)$, both over some EDG, g. For some p and q, let $p \in R_{P_g}$ and $q \in I_{P_g}$, and let there

[1] It's also possible to run nearly the same argument while bypassing the restrictions on procedural construction. Simply note that each protocol is directly associated with a restriction on processes by the execution rules. From the relevant clauses of the reduction lemma, the rest of the restrictions can be read off of protocols anyway.

be a transition, $t \in T_g$, such that $t = \langle f(p), l(p)\rangle = \langle f(q), l(q)\rangle$. First, note that defining R_{P_g} and R_{T_g} over the parts of EDGs provides a direct means by which restrictions on dynamics can distinguish processes from transitions; there is nothing about the structure of restrictions on dynamics, especially the union of two independent restrictions, that requires that t be in either the rational or irrational set of $R_{\preceq}(T_g)$ given its relations to p and q. The restrictions on processes and transitions are autonomous. This entails that transition equivalent processes may be normatively distinct, as required. Second, to cash out the promise in the introduction of this section, note that there are sets of protocols from which $R_{\preceq}(P_g)$ and $R_{\preceq}(T_g)$ can be derived, and these sets of protocols are distinct. RRT4 entails that there is a triple of sets of protocols, Π^1_R, Π^1_I, Π^1_T such that, given p and q, there must be some $\pi_1 \in \Pi^1_R$ where $p \in P_{\pi_1}$, some $\pi_2 \in \prod^1_I$ where $q \in P_{\pi_2}$. RRT4 entails that there is a triple of sets of protocols, Π^2_R, Π^2_I, Π^2_T, where $t \in R_{T_g}$ implies that $\pi_1, \pi_2 \in \Pi^2_R$ (because $t \in T_{\pi_1} \cap T_{\pi_2}$) and $t \in I_{T_g}$ implies that $\pi_1, \pi_2 \in \Pi^2_I$ (because $t \in T_{\pi_1} \cap T_{\pi_2}$). These triples of sets of protocols may but need not be identical, they may but need not intersect, and they may but need not be disjoint while containing some pairs of protocols that are transition convergent. In any case, they can be used to derive the distinct verdicts with respect to paths and transitions, even where some transitions bookend the same paths. As long as $\{\Pi^1_R, \Pi^1_I, \Pi^1_T\}$ and $\{\Pi^2_R, \Pi^2_I, \Pi^2_T\}$ are not identical, the resulting sets of protocols can be used to derive verdicts in a way that distinguishes paths from transitions.

9.3 Representation Theorem 2

The second representation theorem connects protocols to epistemic norm-kernels in two ways. First, it defines an explicit translation of epistemic norm-kernels as conditionals in the language of PLED. Second, it shows that every protocol defines a function from conditions of application expressible in the language of PLED to verdicts on objects of norm-kernels that take elements of EDGs (or elements of epistemic dynamics represented by parts of EDGs) as objects. Moreover, every norm-kernel with conditions of application in the language of PLED that takes elements of EDGs as objects has its exact mapping of conditions of application to verdicts mirrored by some set of protocols.

In other words, every epistemic norm-kernel (defined over the language of PLED and the elements of epistemic dynamics represented by EDGs) is represented by some set of protocols in such a fine-grained way that all of its verdicts can be read off from the syntax of those protocols. As with the first representation theorem, the thought is that the verdicts of each norm-kernel can be recovered from the syntactic and semantic properties of some set of protocols and vice versa.

Again, it's important to note that this provides a fine-grained enough way to read off the content of epistemic norms that we can readily distinguish between verdicts on epistemic processes and epistemic transitions, as required by the arguments at the end of Chap. 4. As in prior chapters (and the prior section), let me frontload the statement and discussion of the main results with definitions or reminders of definitions.

9.3.1 *Protocol-Theoretic Norm-Kernels*

Given the centrality of deontic logics to the enterprise of building the logic of norms and, more relevant, the import of deontic operators in the formulation of norm-kernels, the binding of protocols to norm-kernels in the protocol-theoretic logic of procedural epistemic norms naturally requires a protocol-theoretic account of epistemic deontic operators. This is provided by PLED, in the prior chapter. I make free use of the language of PLED in defining norm-kernels and normative (deontic) verdicts.

Remark 449 Epistemic rationality norms are often encoded via norm-kernels, conditionals that assign normative verdicts to the objects of norms under their conditions of application. Norm-kernels have the following form:

(Norm Kernel) Sentences of the form, "If C, then $Ver(x)$" are norm-kernels where C is the condition of application, x is the object of the norm kernel, and $Ver(x)$ is a statement predicating a normative status to x (the normative verdict of the norm-kernel).

$Ver(x)$ is typically a deontic statement, e.g., a statement of the form "x is rationally obligatory (permissive, prohibited)". For the purposes of PLEN, normative statuses are exhausted by deontic statuses with respect to norms of rationality.

Remark 450 Norm-kernels of additional forms are readily available:

Norm-kernels	Sufficient condition	Necessary condition
Monadic	If C, then $Ver(x)$.	If $Ver(x)$, then C.
n-adic	If C, then $Ver(x_1, \ldots, x_n)$.	If $Ver(x_1, \ldots, x_n)$, then C.
Conjunctive monadic	If C, then $Ver(x)$ and $Ver'(y)$.	If $Ver(x)$ and $Ver'(y)$, then C
Conjunctive n-adic	If C, then $Ver(x_1, \ldots, x_n)$ and $Ver(y_1, \ldots, y_n)$.	If $Ver(x_1, \ldots, x_n)$ and $Ver(y, \ldots, y_n)$, then C.

Definition 451 (PLEN-Kernels) Norm-kernels are directly expressed in L^D_{PLEN} by means of conditionals and protocol-theoretic deontics:

(1) $C \to RO_\pi(\pi')$
(2) $C \to RP_\pi(\pi')$
(3) $C \to RF_\pi(\pi')$
(4) $C \to RO_M(\pi')$
(5) $C \to RP_M(\pi')$

(6) $C \rightarrow RF_M(\pi')$
(7) ...

Remark 452 I explicitly present only norm-kernels taking protocols as objects and identifying sufficient conditions for a verdict, below, but the extension of these formulae to other epistemic factors and the construction of other conditional constructions is straightforward. The incorporation of comparative normative verdicts requires more machinery than is immediately present in PLED and is reserved for future work.

Claim 453 Taking EDGs to model epistemic factors, and the objects of norm-kernels to be epistemic factors (including belief) as well as propositions, each norm-kernel corresponds to a norm function, mapping conditions of application and objects to deontic verdicts. This is a trivial feature of their conditional form.

Definition 454 Let g be an arbitrary EDG. Let N = (If C, then $Ver(x)$) be an arbitrary norm-kernel where $C, Ver(x) \in L^D_{PLEN}$ and

$$x \in (P_{PLEN} \cup S_g \cup T_g \cup P_g \cup L^D_{PLEN})$$

Then:
(Norm Functions) The *norm function* of N over g is any function of the form:

$$f_N : (Con(N), Dom(N, g), S) \longrightarrow Val(N)$$

where:

(1) the set of conditions of application of N is $Con(N) \subseteq L^D_{PLEN}$ (subscripts omitted where inconvenient),
(2) the domain of N's objects with respect to arbitrary g is $Dom(N, g) \subseteq P_{PLEN} \cup S_g \cup T_g \cup P_g \cup R{+}_0$ (subscripts omitted where inconvenient),
(3) $Val(N) \in L^D_{PLEN}$ is the set of possible verdicts that N assigns to its objects,
(4) $s \Vdash C \rightarrow Ver(x)$ iff $f_N(C, x, s) = Ver(x)$.

Claim 455 The final clause of the definition of f_N implies that norm-kernels in PLEN are exactly mirrored by their norm functions.

9.3.2 Representation Theorem 2: Protocol Function Theorem

Claim 456 Protocol functions map permutations of pairs of conditions of application expressible in L^R_{PLEN} and objects (elements of epistemic dynamics) to deontic valences, which are analyzed in terms of execution-theoretic properties of protocols. This provides a step toward codification of norm-kernels in PLEN, as well as further codification of the deontic concepts of dictation or obligation, permission, and forbiddance or prohibition in PLEN.

First, some elements necessary to defining protocol functions need introducing.

Definition 457 (Protocol-Theoretic Norm-Kernel Elements) Let π be an arbitrary protocol. Let g be an EDG. Then:

(1) Let $\pi Val = \{RO_\pi(x), RP_\pi(x), RF_\pi(x)\}$ and define the duals as usual. This is the set of possible[2] deontic verdicts for the protocol π.
(2) If x is a state, let $-x$ be the complement of x in g $(S - x)$.
(3) If x is a protocol π', let $-x = -\pi'$.
(4) If x is a formula of L^R_{PLEN}, then let $-x$ be the disjunction of every dual of x.

With those elements defined, define the following sets of elements for building and assigning norm-function-like functions to π:

Definition 458 (Elements of Protocol-Theoretic Norm-Kernels)

(1) (Conditions of Application for π)

$$Con(\pi, g) = \left\{A \mid \exists s \in S(s \Vdash (A \equiv \langle\pi\rangle_p \top))\right\}$$

(2) (Objects of π: Actions)

$$Dom(\pi)\Pi(g) = \left\{\pi' \mid \exists\pi'' \in Sub(\pi)(\pi'' \Rightarrow {}_g\pi')\right\}$$

(3) (Objects of π: Conditions)

$$Dom(\pi)Con(g) = \left\{B \mid \exists s \in S(s \Vdash ([\pi]_p B \vee \langle\pi\rangle_p B))\right\}$$

(4) (Objects of π: States)

$$Dom(\pi)Stat(g) = \{s \in S \mid \exists p \in P_\pi(s = last(p))\}$$

(5) (Objects of π: Transitions)

$$Dom(\pi)Tran(g) = \{t \mid \exists p \in P_\pi(t = \langle first(p), last(p)\rangle)\}$$

(6) (Objects of π: Processes)

$$Dom(\pi)Proc(g) = \{p \mid \exists q \in P_\pi(p \sqsubseteq q)\}$$

[2] For purposes of this representation theorem, we ignore defeasible reasons, optionality, and omissibility.

(7) (Domain of π)

$$Dom(\pi, g) = \bigcup \begin{bmatrix} Dom(\pi)\Pi(g), & Dom(\pi)Stat(g), & Dom(\pi)Proc(g), \\ Dom(\pi)Con(g), & Dom(\pi)Tran(g) & \end{bmatrix}$$

Claim 459 Protocol functions are then functions that map conditions of application and objects to deontic verdicts in some way that matches the dictation or other deontic properties of protocols. There are two kinds of protocol functions differentiated by their relative scope. Individual protocol functions map conditions of application and objects to deontic verdicts in a way that matches a protocol's deontic properties in a given EDG at a given state. Universal protocol functions map conditions of application and objects to deontic verdicts in a way that matches a protocol's deontic properties in all EDGs in which it can be executed.

Definition 460 (Individual Protocol Functions) For arbitrary EDG, g, and protocol, π, any function of the form:

$$f_{\pi}^{g} : (\pi Con(g), \pi Dom(R)(g), S) \longrightarrow \pi Val$$

is an *individual protocol function iff*

(1) If $A \in \pi Con(g)$ and $x \in Dom(\pi)\Pi(g)$, then f_{π}^{g} maps (A, x, s) to $RO_{\pi}(x)$ iff for all s, if $s \Vdash A$, then $\pi \Rightarrow_{gs} x$.
(2) If $A \in \pi Con(g)$ and $x \in Dom(\pi)\Pi(g)$, then f_{π}^{g} maps (A, x, s) to $RP_{\pi}(x)$ iff for some s, $s \Vdash A$, and $\pi \Rightarrow_{gs} x$.
(3) If $A \in \pi Con(g)$ and $x \in Dom(\pi)\Pi(g)$, then f_{π}^{g} maps (A, x, s) to $RF_{\pi}(x)$ iff for all s, if $s \Vdash A$, then $\pi \Rightarrow_{gsx} -x$.
(4) If $A \in \pi Con(g)$ and $x \in Dom(\pi)Con(g)$, then f_{π}^{g} maps (A, x, s) to $RO_{\pi}(x)$ iff for all s, if $s \Vdash A$, then $s \Vdash [\pi]_{p}x$.
(5) If $A \in \pi Con(g)$ and$x \in Dom(\pi)Con(g)$, then f_{π}^{g} maps (A, x, s) to $RP_{\pi}(x)$ iff for some s, $s \Vdash A$, and $s \Vdash \langle\pi\rangle_{p} x$.
(6) If $A \in \pi Con(g)$ and $x \in Dom(\pi)Con(g)$, then f_{π}^{g} maps (A, x, s) to $RF_{\pi}(x)$ iff for all s, if $s \Vdash A$, then $s \Vdash ([\pi]_{p} - x$.
(7) If $A \in \pi Con(g)$ and $x \in Dom(\pi)Stat(g)$, then f_{π}^{g} maps (A, x, s) to $RO_{\pi}(x)$ iff for all $p \in P_{\pi}$ such that $s = first(p)$, if $s \Vdash A$, then $x = last(p)$.
(8) If $A \in \pi Con(g)$ and $x \in Dom(\pi)Stat(g)$, then f_{π}^{g} maps (A, x, s) to $RP_{\pi}(x)$ iff for some $p \in P_{\pi}$ such that $s = first(p)$, $s \Vdash A$, and $x = last(p)$.
(9) If $A \in \pi Con(g)$ and $x \in Dom(\pi)Stat(g)$, then f_{π}^{g} maps (A, x, s) to $RF_{\pi}(x)$ iff for all $p \in P_{\pi}$ such that $s = first(p)$, if $s \Vdash A$, then $-x = last(p)$.
(10) If $A \in \pi Con(g)$ and $x \in Dom(\pi)Tran(g)$, then f_{π}^{g} maps (A, x, s) to $RO_{\pi}(x)$ iff for all $p \in P_{\pi}$ such that $s = first(p)$, if $s \Vdash A$, then $x = \langle s, last(p)\rangle$.
(11) If $A \in \pi Con(g)$ and $x \in Dom(\pi)Tran(g)$, then f_{π}^{g} maps (A, x, s) to $RP_{\pi}(x)$ iff for some $p \in P_{\pi}$ such that $s = first(p)$, $s \Vdash A$, and $x = \langle s, last(p)\rangle$.

(12) If $A \in \pi Con(g)$ and $x \in Dom(\pi)Tran(g)$, then f_π^g maps (A, x, s) to $RF_\pi(x)$ iff for no $p \in P_\pi$ such that $s = first(p)$, if $s \Vdash A$, then $x = \langle s, last(p) \rangle$.
(13) If $A \in \pi Con(g)$ and $x \in Dom(\pi)Proc(g)$, then f_π^g maps (A, x, s) to $RO_\pi(x)$ iff for all $p \in P_\pi$ such that $s = first(p)$, if $s \Vdash A$, then x is a subpath of p or corresponds to it.
(14) If $A \in \pi Con(g)$ and $x \in Dom(\pi)Proc(g)$, then f_π^g maps (A, x, s) to $RP_\pi(x)$ iff for some $p \in P_\pi$ such that $s = first(p)$, $s \Vdash A$, and x is a subpath of p or corresponds to it.
(15) If $A \in \pi Con(g)$ and $x \in Dom(\pi)Proc(g)$, then f_π^g maps (A, x, s) to $RF_\pi(x)$ iff for no $p \in P_\pi$ such that $s = first(p)$, if $s \Vdash A$, then x is a subpath of p or corresponds to it.

Remark 461 Informally, individual protocol functions are functions that conditionally map conditions of application for a protocol (preconditions) and epistemic factors (related to the protocol by the way that protocol syntax and graph-theoretic structure are bound in the semantics of PLEN) to deontic operators.

Remark 462 Note that, in defining protocol functions, it was assumed that classes of states can be defined by conditions verified at them, e.g., $\{s \mid s \Vdash A\}$. Thus, the definitions extend to classes of states; classes of states can be mapped to objects in πVal or be declared, by a norm, to be obligatory, permissible, or forbidden.

Definition 463 (Universal Protocol Function)
Let $f_\pi^{g_1}, \ldots, f_\pi^{g_n}$ be a set of n protocol functions for π and graphs $g_1, \ldots, g_n$ in which π is executable.
Let $\pi Con = \bigcup_{i=1}^{n} Con(\pi, g_i)$ and $Dom(\pi) = \bigcup_{i=1}^{n} Dom(\pi, g_i)$.
Then:
(Universal Protocol Function of π for $g_1, \ldots, g_n$)

$$f_\pi^\forall : (Con(\pi), Dom(\pi), S_1 \cup \ldots \cup S_n) \longrightarrow \pi Val$$

is a *universal protocol function of* π *for* $g_1, \ldots, g_n$ iff if $C \in Con(\pi, g_i)$ and $x \in Dom(\pi, g_i)$, then $f_\pi^\forall(C, x, s_i) = f_\pi^{g_i}(C, x, s_i)$.

Claim 464 Deepening the connection between protocols and epistemic rationality norms, I show that, given that the content of epistemic rationality norms is captured by norm-kernels [17], it follows that (i) for every norm-kernel (defined over the elements of EDGs), there are "norm functions" that map every pair of condition of application and object of the norm-kernel to the norm-kernel's verdict on that object under that condition, and (ii) with some qualifications, every norm function can be mapped to a protocol whose universal protocol function agrees with it in all verdicts for all objects and conditions of application. This also connects protocols to the analysis of deontic and other intuitive normative concepts which are explicit or implicit in the formulation of norm-kernels, e.g., valences, conditions of application, objects of norms, etc.. Protocols, as was shown in developing PLED, can be used to analyze deontic operators (et al) in a way that is theoretically interesting.

9.3.2.1 The Protocol Function Theorem

Claim 465 Now, the informal idea is that every norm maps conditions of application and objects (inter alia) to a deontic verdict. For every protocol, there are definable protocol functions that map its preconditions to dictated epistemic factors. More precisely, *if all conditions of application are expressible in L^D_{PLEN} and all objects are actions, procedures, processes, transitions, states, or expressible conditions in PLEN, then, for every norm function, f_n, there is a protocol, π, whose universal protocol function, $f^{\forall}_{\pi}$, agrees with f_n for every pair of condition of application and object.* The key thought is that agreement of protocol function, f_{π}, with norm function, f_n, provides a protocol-theoretic translation of the deontic verdict of N in terms of the properties of π. The theorem thus assigns every (deontic) norm kernel a translation in protocol-theoretic terms; the verdicts of the norm-kernel with respect to its objects can thus be read off of the protocol function, which in turn can be read off of the protocol via the execution rules and the structure of EDGs.

Lemma 466 (PF Lemma 1) *Every protocol, π, generates a protocol function, $f^{g_i}_{\pi}$, for every g_i in which π is executable.*

Proof (Proof-Plan) Let g_i be an EDG. Let π be executable in g_i. The key step is noting that $Dom(\pi, g_i)$ is necessarily non-empty given that π is executable (just consider the atomics in any edge executing π). Then, since all of the requisite parts exist for every protocol executable in g_i, $f^{g_i}_{\pi}$ can be constructed. □

Lemma 467 (PF Lemma 2) *Every protocol, π, generates a universal protocol function, $f^{\forall}_{\pi}$.*

Proof This follows by construction from PF1, given the definition of $f^{\forall}_{\pi}$. □
From these, we arrive at the main result binding protocols to protocol functions and finally to norm functions:

Theorem 468 (Protocol Function Theorem (PFT)) *Let $X_{g_1,\ldots,g_n}$ be the union of X_{g_i} for all g_i in which π is executable (e.g., $S_{g_1,\ldots,g_n}$, $T_{g_1,\ldots,g_n}$, $P_{g_1,\ldots,g_n}$). Then: For any f_N, with any $C \in Con(N)$ and any $x \in Dom(N)$, if $C \in L^D_{PLEN}$ and*

$$x \in (P_{PLEN} \cup S_{g_1,\ldots,g_n} \cup T_{g_1,\ldots,g_n} \cup P_{g_1,\ldots,g_n} \cup L^D_{PLEN})$$

then there is a protocol, π, such that the universal protocol function, $f^{\forall}_{\pi}$, for arbitrary $g_1, \ldots, g_n$ in which π is executable is such that $f^{\forall}_{\pi}(C, x, s_i) = f_N(C, x, s_i)$).

Proof Assume $C \in Con(N) \cap L^D_{PLEN}$ and

$$x \in (Dom(N) \cap (P_{PLEN} \cup S_{g1,\dots,gn} \cup T_{g1,\dots,gn} \cup P_{g1,\dots,gn} \cup L^D_{PLEN}))$$

There are five cases: (1) $x \in P_{PLEN}$, (2) $x \in S_{g1,\dots,gn}$, (3) $x \in L^D_{PLEN}$, (4) $x \in T_{g1,\dots,gn}$, (5) $x \in P_{g1,\dots,gn}$. Each case follows by the construction of $f^{\forall}_{\pi}$ from the individual protocol functions of the g_i in which it is executable.

(1) Let $x \in P_{PLEN}$ and $f_N(C, x, s_i) = RO_{\pi}(x)$. Consider any π such that $\pi \Rightarrow x$, $g_1, \dots, g_n$ all execute π, and $\Vdash_{gi} C \equiv \langle\pi\rangle_p \top$ for all g_i. Consider an arbitrary g_i. Every execution of π in g_i contains an execution of x. Thus, for all s, if $v_s(C) = 1$, then $\pi \Rightarrow_{g_i,s} x$. Thus, $f^{gi}_{\pi}(C, x, s_i) = RO_{\pi}(x)$. But g_i was arbitrary. Thus, $f^{\forall}_{\pi}(C, x, s_i) = RO_{\pi}(x) = f_N(C, x, s_i)$. For $f_N(C, x, s_i) = RP_{\pi}(x)$, consider any π such that $\pi \Rightarrow_{g_i,s} x$ for some s in S for each g_i, $g_1, \dots, g_n$ all execute π, and $\Vdash_{gi} C \equiv \langle\pi\rangle_p \top$ for all g_i. Consider an arbitrary g_i. Some execution of π in g_i contains an execution of x. Thus, for some s, $v_s(C) = 1$, and $\pi \Rightarrow_{gi,s} x$. Thus, $f^{gi}_{\pi}(C, x, s_i) = RP_{\pi}(x)$. But g_i was arbitrary. Thus, $f^{\forall}_{\pi}(C, x, s_i) = RP_{\pi}(x) = f_N(C, x, s_i)$. For $f_N(C, x, s_i) = RF_{\pi}(x)$, consider any π such that $\pi \Rightarrow_{g_i,s,x} -x$, $g_1, \dots, g_n$ all execute π, and $\Vdash_{gi} C \equiv \langle\pi\rangle_p \top$ for all g_i. Consider an arbitrary g_i. Every execution of π in g_i contains only executions of $-x$ and no execution of any other protocols (therefore excluding x only). Thus, for all s, if $v_s(C) = 1$, then $\pi \Rightarrow_{g_i,s,x} -x$. Thus, $f^{gi}_{\pi}(C, x, s_i) = RF_{\pi}(x)$. g_i was arbitrary. Thus, $f^{\forall}_{\pi}(C, x, s_i) = RF_{\pi}(x) = f_N(C, x, s_i)$.

(2) Let $x \in S_{g1,\dots,gn}$ and $f_N(C, x, s_i) = RO_{\pi}(x)$. Consider any π such that every execution of π ends with state x, $g_1, \dots, g_n$ all execute π, and $\Vdash_{gi} C \equiv \langle\pi\rangle_p \top$ for all g_i. Consider an arbitrary g_i. Every execution of π in g_i terminates in x. Thus, $f^{gi}_{\pi}(C, x, s_i) = RO_{\pi}(x)$. g_i was arbitrary. Thus, $f^{\forall}_{\pi}(C, x, s_i) = RP_{\pi}(x) = f_N(C, x, s_i)$. For $f_N(C, x, s_i) = RP_{\pi}(x)$, consider any π such that some execution of π ends with state x, $g_1, \dots, g_n$ all execute π, and $\Vdash_{gi} C \equiv \langle\pi\rangle_p \top$ for all g_i. Consider an arbitrary g_i. Some execution of π in g_i terminates in x. Thus, $f^{gi}_{\pi}(C, x, s_i) = RP_{\pi}(x)$. g_i was arbitrary. Thus, $f^{\forall}_{\pi}(C, x, s_i) = RP_{\pi}(x) = f_N(C, x, s_i)$. For $f_N(C, x, s_i) = RF_{\pi}(x)$, consider any π such that such that every execution of π ends with state $-x$, $g_1, \dots, g_n$ all execute π, and $\Vdash_{gi} C \equiv \langle\pi\rangle_p \top$ for all g_i. Consider an arbitrary g_i. Every execution of π in g_i terminates in a state other than x. Thus, $f^{gi}_{\pi}(C, x, s_i) = RF_{\pi}(x)$. g_i was arbitrary. Thus, $f^{\forall}_{\pi}(C, x, s_i) = RF_{\pi}(x) = f_N(C, x, s_i)$.

(3) Let L^D_{PLEN} and $f_N(C, x, s_i) = RO_{\pi}(x)$. Consider any π such that $\Vdash [\pi]_p x$, $g_1, \dots, g_n$ all execute π, and $\Vdash_{gi} C \equiv \langle\pi\rangle_p \top$ for all g_i. Consider an arbitrary g_i. Every execution of π in g_i terminates in a state verifying x. Thus, for all s, if $v_s(C) = 1$, then $v_s([\pi]_p x) = 1$. Thus, $f^{gi}_{\pi}(C, x, s_i) = RO_{\pi}(x)$. g_i was arbitrary. Thus, $f^{\forall}_{\pi}(C, x, s_i) = RP_{\pi}(x) = f_N(C, x, s_i)$. For $f_N(C, x, s_i) = RP_{\pi}(x)$, consider any π such that $\Vdash \langle\pi\rangle_p x$, $g_1, \dots, g_n$ all execute π, and $\Vdash_{gi} C \equiv \langle\pi\rangle_p \top$ for all g_i. Consider an arbitrary g_i. Some execution of

π in g_i terminates in a state verifying x. Thus, for all s, $v_s(C) = 1$, and $v_s(\langle\pi\rangle_p x) = 1$. Thus, $f_\pi^{gi}(C, x, s_i) = RP_\pi(x)$. g_i was arbitrary. Thus, $f_\pi^\forall(C, x, s_i) = RP_\pi(x) = f_N(C, x, s_i)$. For $f_N(C, x, s_i) = RF_\pi(x)$, consider any π such that $\Vdash [\pi]_p - x$, $g_1, \ldots, g_n$ all execute π, and $\Vdash_{gi} C \equiv \langle\pi\rangle_p \top$ for all g_i. Consider an arbitrary g_i. Every execution of π in g_i terminates in a state verifying $-x$. Thus, for all s, if $v_s(C) = 1$, then $v_s([\pi]_p - x) = 1$. Thus, $f_\pi^{gi}(C, x, s_i) = Forb$. g_i was arbitrary. Thus, $f_\pi^\forall(C, x, s_i) = RF_\pi(x) = f_N(C, x, s_i)$.

(4) Let $x \in T_{g_1,\ldots,g_n}$ and $f_N(C, x, s_i) = RO_\pi(x)$. Say that a path, p, is bookended with a transition $t = \langle s, s'\rangle$ iff $s = first(p)$ and $s' = last(p)$. Consider any π such that every execution of π is bookended with transition x, $g_1, \ldots, g_n$ all execute π, and $\Vdash_{gi} C \equiv \langle\pi\rangle_p \top$ for all g_i. Consider an arbitrary g_i. Every execution of π in g_i is bookended with x. Thus, $f_\pi^{gi}(C, x, s_i) = RO_\pi(x)$. g_i was arbitrary. Thus, $f_\pi^\forall(C, x, s_i) = RO_\pi(x) = f_N(C, x, s_i)$. For $f_N(C, x, s_i) = RP_\pi(x)$, consider any π such that some execution of π is bookended with x, $g_1, \ldots, g_n$ all execute π, and $\Vdash_{gi} C \equiv \langle\pi\rangle_p \top$ for all g_i. Consider an arbitrary g_i. Some execution of π in g_i is bookended with x. Thus, $f_\pi^{gi}(C, x, s_i) = RP_\pi(x)$. g_i was arbitrary. Thus, $f_\pi^\forall(C, x, s_i) = RP_\pi(x) = f_N(C, x, s_i)$. For $f_N(C, x, s_i) = RF_\pi(x)$, consider any π such that such that every execution of π is bookended with $-x$, $g_1, \ldots, g_n$ all execute π, and $\Vdash_{gi} C \equiv \langle\pi\rangle_p \top$ for all g_i. Consider an arbitrary g_i. Every execution of π in g_i is bookended with a transition other than x. Thus, $f_\pi^{gi}(C, x, s_i) = RF_\pi(x)$. g_i was arbitrary. Thus, $f_\pi^\forall(C, x, s_i) = RF_\pi(x) = f_N(C, x, s_i)$.

(5) Let $x \in P_{g_1,\ldots,g_n}$ and $f_N(C, x, s_i) = RO_\pi(x)$. Consider any π such that all executions of π contain x as a subpath, $g_1, \ldots, g_n$ all execute π, and $\Vdash_{gi} C \equiv \langle\pi\rangle_p \top$ for all g_i. Consider an arbitrary g_i. Every execution of π in g_i contains x as a subpath. Thus, $f_\pi^{gi}(C, x, s_i) = RO_\pi(x)$. g_i was arbitrary. Thus, $f_\pi^\forall(C, x, s_i) = RO_\pi(x) = f_N(C, x, s_i)$. For $f_N(C, x, s_i) = RP_\pi(x)$, consider any π such that some execution of π contains x as a subpath, $g_1, \ldots, g_n$ all execute π, and $\Vdash_{gi} C \equiv \langle\pi\rangle_p \top$ for all g_i. Consider an arbitrary g_i. Some execution of π in g_i contains x as a subpath Thus, $f_\pi^{gi}(C, x, s_i) = RP_\pi(x)$. g_i was arbitrary. Thus, $f_\pi^\forall(C, x, s_i) = RP_\pi(x) = f_N(C, x, s_i)$. For $f_N(C, x, s_i) = RF_\pi(x)$, consider any π such that such that every execution of π is bookended with $-x$, $g_1, \ldots, g_n$ all execute π, and $\Vdash_{gi} C \equiv \langle\pi\rangle_p \top$ for all g_i. Consider an arbitrary g_i. Every execution of π in g_i is bookended with a transition other than x. Thus, $f_\pi^{gi}(C, x, s_i) = RF_\pi(x)$. g_i was arbitrary. Thus, $f_\pi^\forall(C, x, s_i) = RF_\pi(x) = f_N(C, x, s_i)$.

(6) That there is such a π for any case in the prior reasoning is mechanical given the coherence of the assumptions.

(a) First, for any x where x is a regular or irregular protocol, coherent formula, coherent state, process in any EDG, or transition in any EDG, there is, respectively:
  (i) a graph in which x is executable,
  (ii) a graph in which x is true at some state, s, and there is some set of paths that terminates with s,
  (iii) a graph in which $x \in S$ and there is some set of paths that terminates with x,
  (iv) a graph, g, in which x is in P_g or R_0, or
  (v) a graph, g, in which x is in T_g.
  (vi) That these claims are true for each case is trivial or mechanical. For instance, all regular protocols are executable in some graph (a simple induction suffices: for atomics, edges; for complex protocols, paths built out of edges according to the execution rules). The requisite graphs for irregular protocols can be built by reference to their regular counterparts. For another instance, if x is a formula and it's not inconsistent, then there is some valuation function that maps it to 1.

(b) Second, there is a set of protocols that dictate x—whatever it may be from the foregoing list—for each graph in which the prior conditions hold. Where x is a protocol, there are two immediate ways of showing this. On one path, in each graph in which x is executable, x is dictated by itself and the trivial protocol that describes the entire graph. The choice of all protocols that dictates x across all graphs in which x is executable dictates it universally. Alternately, there is then at least some protocol that dictates x—x itself, an iteration of x itself, any sequential composition of x and a test for x's executability (etc.), any protocol that is implied by x in procedural, ternary, or binary terms, any protocol that is executed only by paths that terminate with the requisite states, etc.
(c) The parallel result for transitions, processes, and states are mechanically elaborated from this.

□

Remark 469 The protocol-theoretic analysis of deontic operators forges logical connections between the properties of protocols and the verdicts of protocol functions. Thus, the representation of norm-kernels by protocols shown in the protocol function theorem is not arbitrary or artifactual; the categorical and deontic verdicts of norm-kernels are mapped to their protocol-theoretic analyses. Given the reasoning in favor of PLED in the prior chapter, it's plausible that these analyses are natural or philosophically plausible. For instance, the conditions under which the protocol function of π assigns π' the value of Ob are exactly the set of conditions under which it's plausible to think that π—as an encoding of the content of some procedural epistemic norm—obligates, dictates, or otherwise requires π' (itself either an epistemic norm or a constitutive or procedural norm for carrying out some action or process).

Claim 470 Given the relation between norm-kernels and protocol functions, it follows that if a norm-kernel, N, differentiates between the (e.g.) rational permissibility of processes and that of transitions, then universal protocol function in perfect agreement to N by PFT also suffices to differentiate processes from transitions. The verdicts of N assigned to processes and transitions will be identical, under the same conditions, to those of the universal protocol function mapped to N by PFT.

9.4 Rule-Following

The constructions bound to protocols by the representation theorems will not serve to make PLEN a plausible account of epistemic norms if those constructions are not themselves plausible means of accounting for the content of epistemic norms. If rules like norm-kernels truly don't determine any courses of action as, for instance, Kripke's Wittgenstein—taken literally—argues,[3] then our result that protocols determine the same restrictions on epistemic dynamics as norm-kernels and the same norm-kernels as restrictions on epistemic dynamics is either trivial or flatly wrong. There will be infinitely many restrictions on dynamics bound to each norm-kernel, there will be no sense in which any set of them is the right one, and vice versa. Indeed, there are numerous constructions in PLEN—execution rules, protocol graphs, and restrictions on epistemic dynamics—that are intended to be formal models of the requirements or dictates of rules—or at least formal models from which the truths about what rules require can be read off. If it is unintelligible that there are truths about what norm-kernels require, then PLEN seems to contradict this by providing an intelligible and (if all works right) rigorous model of it.

Moreover, the core arguments for epistemic proceduralism, which provide the criteria to which the development of PLEN is anchored, require that it is both possible and actual that we follow epistemic norms, rather than merely complying with them. Indeed, the system of norm application, PLEKNA, was developed with the explicit intention of formalizing ways of following norms. How can this project be squared with the apparent unintelligibility of rule-following? If following rules is generally impossible or at least unintelligible, then arguments for the theory rely on (at least) unintelligible premises and the system formalizes (at least) unintelligible ideas. This just won't do. Let me spell out the problems and a putative solution within the toolset of PLEN. The aim of this section is, of course, not to develop a complete, working solution to the rule-following problem, which would likely

[3] Re: §201 "this was our paradox; no course of action could be determined by a rule, because every course of action can be made out to accord with the rule". [18] Naturally, I can't hope to do justice to the task of Wittgenstein exegesis—or even Kripke exegesis. I formulate the problem in the ways that strike me presenting the greatest hurdle to this project.

require a separate volume. Rather, the aim is adding a bit of philosophical breadth to the project by noting potential problems that rule-following might generate for the project and exploring the ideas within the PLEN Framework for potential lines of solution.[4]

9.4.1 The Problems

While it is clear that not all rules are epistemic norms, the formulation of epistemic norms as conditional rules (norm-kernels) or protocols (i.e., program-like instructions), and the general arguments concerning the importance of the regulative function of epistemic norms (i.e., reasoning about what epistemic norms require and then enacting those requirements) suggest that the general problems of rule-following will have force in the special, epistemic case. The epistemic case may, in fact, produce special rule-following considerations.[5]

Kripke's Wittgenstein lays out a challenge and two conditions under which that challenge can be met. The challenge is to provide an account of how an agent can follow rules that dictate indefinitely many actions under indefinitely many circumstances (as distinct from merely complying with them), while also acknowledging that any finite agent could have complied with any given rule only finitely many times. To give a concrete form to the challenge, Kripke introduces the plus and quus case. Consider two functions. On the one hand, there is $+$, which is the usual arithmetic addition function. On the other hand, there is the quaddition function, $\dot{+}$, which is defined:

$$x \dot{+} y = \begin{cases} x + y & \text{if } x < 50 \text{ and } y < 50 \\ 5 & \text{otherwise} \end{cases}$$

If presented with the string of symbols, '$68 + 57 =?$', under the usual conventions for understanding these sorts of strings of symbols, what should be placed on the righthand side of the symbol '$=$' to complete the string? The answer depends on whether the symbol '$+$' denotes addition or quaddition. There is no way to answer this question by appealing to past usage for an agent that hasn't added numbers greater than 50; any of such an agent's sequences of actions that complies with the addition rule complies with the quaddition rule. Of course, there's nothing special about the number 50; the definition provides a recipe for plus-quus type problems for any possible case of finite past usage.

[4] My thanks to an anonymous reviewer for suggesting this.

[5] I won't comment on this further, but below I consider the Intellectualist Regress [11], and the Carroll Regress [4] in connection to this topic. For an epistemic version, see Boghossian's considerations about epistemic rule following [2].

The conditions on solutions to the paradox that give an account of following a rule rather than merely complying with a rule are that:

(Fact) "First, it must give an account of what fact it is (about my mental state) that constitutes my meaning plus, not quus." [8] (p. 11)

(Normativity) "It must, in some sense, show how I am justified in giving the answer '125' to '68 +57'. The 'directions' mentioned in the previous paragraph, that determine what I should do in each instance, must somehow be 'contained' in any candidate for the fact as to what I meant." [8] (p. 11)

This constellation of conditions creates problems for the argument for PLEN that has been built up this point.

The first problem takes some set-up. Addition in some sense defines a rule for inference that dictates infinitely many actions (additions) under infinitely many circumstances (pairs of integers). (Of course, in a human lifetime, the number of additions and integers added will be finite but indefinite.) The same is true for quaddition. There is a suite of questions that this set-up brings in its train. What does it come to to follow the addition rule rather than the quaddition rule?

Presumably, this involves intention rather than mere compliance. After all, given that rules dictate or permit executing some courses of action (those in accordance with them) and dictate or permit avoiding others (those that don't accord), each of those dictated/permitted courses of action complies with the rule. But this can be accidental. Suppose I roll a 6 sided die and write in whatever number I end up rolling. For x and y over 50, there's a 1/6 chance that I comply with the quus rule. For x and y below 3, I will comply with the plus rule. I'm not following either rule in either case, though I'm accidentally complying with them in each.

The question, then, is what does the intention to follow a rule come to? Answering this question might be helped along if we can answer the following questions: How does one know whether one has followed one rule rather than the other in the past? How does one know whether one is following (intends to follow) one rule (rather than the other) in the present or intends to follow one rule (rather than the other) in the future? If we cannot answer the prior questions, then we cannot answer this question: what is required by the rule one intends to follow?

Another way to think of this is to consider π_+ and $\pi_{\dot{+}}$, where these are the protocols for inferring in accordance with the respective functions. It is not the case that, for any given EDG, these will share any set of executions. We can construct EDGs that represent an agent who only ever adds values greater than 50. However, the problem arises for EDGs that represent finite agents who've added finitely many values. There is then a shared graph or set of shared paths for π_+ and $\pi_{\dot{+}}$. That is, there is a countable, non-empty set of states in any such graph, $s_1, \ldots, s_n$, such that $\pi_+ \hat{\curlywedge}_p \pi_{\dot{+}}$ is true. The question then takes the form: what does it mean to follow π_+ rather than $\pi_{\dot{+}}$ even when one's actual history executes a path complying with both?

In the extensive literature, there are primitivist answers, which attempt to provide an account of following a rule as some sort of primitive concept, and reductivist answers, which attempt to answer the question straight-on by providing an account of rule-following that meets the conditions. Kripke expends a great deal of energy setting aside primitivist answers generally as well as a swathe of reductivist answers in favor of a sort of communitarian skeptical solution. Skeptical solutions, here, are those that accept that there is no fact that satisfies the two conditions, but perhaps provide an account of how rule-following can be "legitimate" in some sense. The result is, seemingly, a sort of anti-realism or skepticism about meaning; there is no fact of the matter about whether I mean plus or quus. This is perfectly general, though what matters for our purposes is that it's general for rules. If this is right, then there are no facts that determine whether I am following π_+ or $\pi_{\dot{+}}$. Thus, for a norm-kernel, "If x and y are specified, and you want to know the answer to the equation $'x + y\ =?'$, then add x and y", there is no determinate restriction on epistemic dynamics to be derived, since the norm-kernel could be encoded with either π_+ or $\pi_{\dot{+}}$ (or something in the neighborhood) because there is no determinate meaning for this conditional rule.

That's the first problem. The second problem is that if rule-following is unintelligible, then the core function of epistemic norms in epistemic regulation and planning is unintelligible because it requires rule-following, and so the main case for epistemic proceduralism relies on fog. Moreover, the analyses of norm application and norm adherence in Chap. 8 are attempts to formalize the unintelligible.

I don't intend to litigate the merits of the many rival solutions. Rather, I want to take a short digression to argue for the claim that answers that satisfy Fact and Normativity do not need to posit rules for applying rules that are believed or explicitly represented by would-be rule followers. I then want to explore the formal and conceptual resources in PLEN for associating determinate restrictions on epistemic dynamics with pieces of logico-english like norm-kernels and for making rule-following intelligible.

9.4.2 Rules for Applying Rules

The Carroll Regress [4] famously appears to show that there can be no justification for fundamental rules of deduction. The common modus ponens form of the regress begins by supposing that from propositions, A and B, a third proposition Z appears to follow. But then, what to respond to someone who denies the implication? Well, the conjunction of A and B with the rule, call it C: "From A and B, Z follows." And now, we are off to the regress. For how do we justify the inference from A, B, and C to Z? A further rule, D: "From A, B, and C, Z follows" will be susceptible to the exact same demand for justification, and any further attempts to provide rules of the same form will be further susceptible to parallel demands for justification. The regress is vicious, as the first demand for justification isn't met without meeting the later demands.

The Intellectualist Regress [11] does much the same.[6] If we need to use epistemic norms in order to apply epistemic norms to guide our reasoning, then we are off on a quite similarly structured regress. In order to regulate one's reasoning using N_1, we would need another norm N_2 to guide our reasoning in order to derive the dictates of N_1, and so on.

There is far more to say about these paradoxes than can be said here, but the solution recipe I think best is due to Pollock [11, 13]. The solution relies on the rejection of the assumption that we need to use rules as premises in our justificatory reasoning. Epistemic norms, on this solution, are plans rather than propositions. They justify deductions not by means of further inferences on the basis of additional premises that state the norm, but by functioning as protocols, plans, or programs that (a) dictate patterns of correct inference (which the relevant derivations of conclusions from premises fit), and (b) structure and guide inference directly (though fallibly or interruptibly) by a process of internalization that directly channels internalized into the patterns dictated. Thus, at the first demand for justification, one supplies a description of the norm that one follows, not a premise for which a new norm must be presented. The interlocuter in these cases might need to reason about what the norm requires, but the justification of the inference is given by inferring in the pattern of the norm by executing the internalized plan. In other words, we do not need to use rules to apply rules. All that is required is some account of the ways in which rules can be encoded in cognitive control systems in a way that structures intention and action. This is what plans and programs (and thus protocols), are naturally understood to do [3, 5, 6, 15].

9.4.3 The Solutions

Protocols are constructions that represent epistemic norms (in, arguably, a particularly perspicuous way, with respect to their action-guiding content). This can be understood in two ways. First, each protocol defines a class of restrictions on epistemic dynamics, and epistemic norms can be thought of as other constructions that define classes of restrictions on epistemic dynamics; protocols (or classes of equivalent protocols) are linked (via the representation theorems) by means of identity of the class of their respective restrictions. Second, and the preferred or canonical interpretation as per Chap. 4, protocols directly encode the core procedural content of epistemic norms (a sort of precise set of instructions); each protocol directly encodes the procedural content for a unique epistemic norm (or at least a class of epistemic norms the core content of which is unique up to hyperintensional equivalence). In either case, PLEN is a logical system for reasoning about (classes of) epistemic norms in a variety of ways. Of particular salience are the methods that PLEN provides, by means of its execution rules and the

[6] See also the Rylean Regress [14].

formulae that encode normative verdicts, for rigorously reading off the courses of action required by the epistemic norms that are represented by a given protocol or set of protocols. The logical relations between protocols and their executions, protocol graphs, and normative verdicts are the facts that constitute the determination of correct and incorrect courses of action with respect to those protocols, and thus also with respect to the epistemic norms represented by them (whether by direct encoding or by defining the same restrictions on dynamics).

If we can use norm-kernels to refer to the actual requirements of norms in some way (either their core procedural content or their restrictions on dynamics), the logic of PLEN allows us to rigorously determine the contours of their associated requirements. The determination of requirements on action can be understood as conditional on stipulated associations with protocols. If the norm-kernel, N, is encoded with π (via the intervening protocol and norm functions), then the representation theorems show how to derive restrictions on epistemic dynamics that fully spell out the requirements of N. If the restriction on epistemic dynamics on epistemic factors, X, $R_{\preceq}(X)$, is encoded with π (by means of the intervening reading off rules), then the representation theorems show that there is a norm-kernel, N, that spells out the requirements of that restriction on dynamics (by means of its intervening norm functions and π's universal protocol function). These are the facts that justify carrying on in certain ways. The procedural paths dictated by N are exactly those justified by it. Of course, each protocol dictates courses of action in as many circumstances as can be represented by any EDG (i.e., infinitely many). Thus, an agent capable of using pieces of language (i.e., norm-kernels) to refer to epistemic norms will be able to justify answers about the verdicts of those norms in indefinitely many cases.

This handles Normativity, as long as we can live with the deflated and relational notion of justification that all protocols (hence, norm-kernels) entail merely by definining restrictions on epistemic dynamics. On this account, there won't be a sense in which it is justified (simpliciter) that an agent goes on any particular way; there will only be a sense in which the agent is justified w.r.t. π in going on in a particular way.

How can an agent use linguistic tools (e.g., norm-kernels) to pick out or refer to any particular norm rather than another? In other words, how do we account for this ability to justify determinations of the requirements of the rule in a way that satisfies Fact?

My preferred way is this. Recall the notion, in PLEN (more specifically, PLEKNA), of *knowing adherence*. The key element of knowing adherence comes down to the inclusion of a protocol or set of protocols in one's methodology, Π_M, at a given state, and then complying with it by means of executing a procedural path that one knows lies in accordance with it.[7] What does this sort of "methodological

[7] In other words, the solution is that rule-following is knowing adherence, which is spelled out partially in terms of compliance and knowledge of compliance. There is no skeptical problem of either compliance or knowledge of compliance; it's perfectly clear what compliance with the plus

possession" of π (membership in Π_M) come to? It does not necessarily encode any form of thinking about the protocol by means of linguistic constructions. Rather, it's a formal encoding of any of a number of notions of adoption, endorsement, or acceptance of a norm, including the forms of plan internalization proposed by Pollock to resolve the Intellectualist Regress. Thus, what constitutes my meaning π_+ and not $\pi_{\dotplus}$ is that one of them is internalized in my cognitive control system in a certain way and the other isn't. I can refer to the one that is embodied in my cognitive system directly rather than by means of pointing out instances or past usage. That's the one that my use of norm-kernels or other designative terms (e.g., 'N_+') picks out.

To formulate this in a way more amenable to defending the use of the representation theorems for the philosophical purposes at hand, the idea is that our logico-english items, norm-kernels, can be assigned a determinate class of norm functions (rather than others) as a referent when such a class captures the plan embedded in our control system.[8] The representation theorems determine which norm functions correspond with the protocols (which analyze plans) in our control systems, and also bind them with restrictions on epistemic dynamics. Thus, the proposed solution is this: (epistemic) rule-following is knowing adherence to or deliberate application of epistemic protocols as formalized in PLEKNA, which encodes an intentional state that functions like plan internalization rather than belief formation or linguistic interpretation. This allows us to pick out determinate norm functions by means of norm-kernels, which are then closely bound to restrictions on epistemic dynamics.

Deliberate application of π relies on knowing adherence to π. Knowing adherence to π comes to following one's already (perhaps blindly) adopted norms to come to a state in which π is adopted (perhaps blindly), and in which it is accepted that π dictates (e.g.) p, and then actually embodying or implementing the required path. Thus, what constitutes my following π_+ and not $\pi_{\dotplus}$ is that $\pi_+ \in \Pi_M$ (I've internalized one plan rather than another) rather than $\dotplus$, which settles my referent, allowing the derivation of the knowledge that $p \in P_{\pi_+}$ by my antecedent methodology, followed by my adherence (deliberate or otherwise) to p.

There are three things to note about this putative solution. First, it does not solve the problem of linguistic meaning generally. It is designed only to provide an account of rule-following (as distinct from mere rule-compliance), and, more specifically, to show how logico-english items like norm-kernels might be bound to specific protocols and restrictions on epistemic dynamics. I do not want to wager on how well it solves the problem generally.

Second, as Miller formulates it in the introduction to [9], the Kripkean skeptic is arguing that even if we have perfect epistemic access to certain areas of knowledge, we won't be able to find a fact satisfying Fact and Normativity where "The areas

---

and quus protocols comes to and, presumably, how one could come to know it, especially given the formalization of protocol execution in PLEN and norm compliance in PLEKNA.

[8] A sufficient condition, not a necessary one.

in question are (a) our previous behaviour, linguistic and nonlinguistic; and (b) the entire contents of our previous mental histories." (p. 4) The solution, here, posits that either there is more to (b) than was recognized in the formulation of the paradox or there are third and fourth areas of knowledge to search for answers to the challenge. The contents of our previous mental histories can be understood to contain our internalized plans (i.e., adopted or endorsed norms) or not. If not, then this is a third area to search. Additionally, PLEN formalizes a fourth area of logical knowledge relating restrictions on epistemic dynamics to protocols to norm functions in which we find facts that satisfy Normativity, given that our norm-kernels are mapped to the specified norm functions. When we couple this with facts about plan internalization which we can use to find the referent of our uses of norm-kernels, this generates our answer.

Finally, note that I am not turning my back on conscious, deliberate, ante hoc regulation involving explicit reasoning about epistemic norms and deliberate regulation of epistemic behavior in accordance with them. Rather, I am suggesting that, as noted in Chap. 2, deliberate ante hoc regulation is both real, and not necessary all the way down. It is the possibility, not necessity, of a kind of blind or at least non-linguistically mediated norm adoption at the bottom of some regress that provides a wayout of the Carroll and Intellectualist regresses. The proposal is that the same pattern of solution works here.

Here is another way to provide a means of referring to determinate epistemic norms, restrictions on dynamics, or protocols by means of linguistic items. Pettit [10] provides a so-called "straight" solution to the rule-following problem by means of the notion of exemplification.[9] The paradox, on Pettit's view, is one in which a set of examples (instances of past usage) instantiates an indefinite number of rules (understood as abstract objects—functions in extension or functions in intension), but, as Pettit points out, this does not entail that the same set of examples *exemplifies* the same set of rules to an individual. It is possible that the set of examples in question exemplifies a single, determinate rule for a given individual in much the same way that the same set of perceptual profiles might exemplify a table rather than a congeries of undetached table parts to a given individual. Which rule is picked out by a set of exemplifying examples? According to Pettit, the examples generate a defeasible inclination to carry out certain courses of action, and this inclination picks out a rule: the rule that fits my inclination (i.e., that dictates doing what the inclination inclines me to do) only when the inclination is triggered under a specifiable set of favourable conditions (e.g., conditions that cause the inclination to diverge from the dictates of the rule). See the original article for the requisite clarifications and caveats. Thus, the proposal is that we can interpret $\pi_{+} \in \Pi_M$ and $\pi_{\dotplus} \notin \Pi_M$ representing the facts that (i) π_{+} is exemplified to the agent whose epistemic dynamics is modeled by a given EDG by some sequence of examples, and (ii) $\pi_{\dotplus}$ is not exemplified to that agent by that sequence of examples.

[9] See [7].

Neither of these ways of cashing out the solution is apparently vulnerable to the objections that Kripke raises.[10] Pettit's account seems specifically designed to bypass these problems. I leave it to the reader to confirm for themselves. For my preferred solution, it's closest to the 'algorithm' and 'machine program' solutions. The algorithm solution posits that following the addition rule rather than the quaddition rule comes to adopting an algorithm that matches the addition function rather than the quaddition function (for all relevant inputs). The machine program solution posits that agents embody or implement programs or algorithms for addition rather than quaddition. This, it is argued, is essentially a dispositionalist view and so vulnerable to the objections to dispositionalism.

The core objection to the algorithm solution was that algorithms are simply rules that require further interpretation, and so the paradox applies as well to the algorithm that matches the addition function—how do we know we haven't adopted a qualgorithm matching quaddition for all relevant inputs? Additionally, Kripke argues that most folks just don't possess the conceptual sophistication to adopt algorithms, so when they mean addition by '+', that can't be what accounts for their following the addition rule. The first objection is headed off at the start; it presupposes exactly the sort of intellectualist assumption rejected in solving the Intellectualist Regress. The interpretation of a linguistically presented rule is not necessary to implement an algorithm. The plan internalization account directly bypasses the need for rules for applying rules. One doesn't read the rules and interpret them; they are encoded in one's control system in a way that channels behavior, under favorable conditions, along the program. This also undermines the second objection, as it, too, presupposes an unargued intellectualism about algorithms. One needn't possess any conceptual sophistication to learn to execute an algorithm.

The objections to simple dispositionalism center on error, finitude, and normativity.[11] First, the dispositionalist cannot account for error. I can't read off which function I intend from my dispositions, because I am (perhaps systematically) disposed to give the wrong answers under (at least) some conditions. Second, real agents can only have finite dispositions, which will only ever cover finite stretches of the infinite table specifying the addition function. Finitude also rules out having some dispositions to follow rules; an agent with a finite lifespan or finite computational powers can't add numbers it would take longer than the heat death of the universe to even read. This directly forbids reading off addition rather than quaddition (or a quaddition-like function that covers the same stretches of the addition table). Third, the dispositionalist offers a descriptive solution to a normative problem, mistakenly collapsing what I am disposed to do with what I should do.

The objections to simple dispositionalism misfire when aimed at plan internalization. Pace these objections, plan internalization is both normative and compatible

[10] My nomenclature follows the quite natural taxonomy of Kripkean objections presented in a handout written by Professor Jeff Speaks for teaching purposes [16].

[11] See [8] (pp. 26–33).

with committing error. More, plans entail arbitrarily complex counterfactual dispositions.

Planning is normative, and plans are violable. Plans permit, forbid, and obligate [3, 5, 6, 12]. My plan for taking the elevator to the fifth floor permits any manner of walking toward the elevator, forbids taking the stairs, and obligates pressing at least certain specific buttons. If there is more specific detail in the plan, this may entail further permissions, obligations, or forbiddances (taking the elevator to the fifth floor by 11am may indeed forbid many manners of walkin to the elevator). Plans thus define normative verdicts; an agent has carried out the wrong course of action with respect to their plan when they've done things that either block the completion of the plan or do things that are forbidden by it. More, one can adopt a plan and fail to see it through under unfavorable circumstances (or merely being overridden by other intentions), and so draw criticism for blocking or diverging from the permissions of the plan. One can be disposed to diverge in action from one's own adopted plans, which still hold a place in one's cognitive economy that, under favorable conditions that need not be specifiable in advance, do channel one's actions along the contours of the plan. We don't plan to violate our plans, but we still do so, and it's a commonplace that we can tell when this happens. One can usually sense that one has gone off-plan, even if this doesn't involve the inculcation of any explicit propositional attitude.

Plans can be arbitrarily complex or general, entailing conditional sub-plans for indefinitely—indeed, infinitely—many circumstances [3, 5, 6, 12]. Moreover, we should not conflate the finitude of performance with finitude of disposition. I might be disposed to respond to infinitely many circumstances despite being doomed (or blessed) with only ever facing a finite stream of them. I might even counterfactually be disposed toward actions under conditions that will never be realized in the actual world. For all that has been established, finite systems can, after all, satisfy infinitely many counterfactual conditionals. Consider a vision test in which rational numbers starting with the digits '1.0' or '2.0' are presented on a screen briefly, while one is tasked with raising one finger only if one saw a number starting with '1.0'. It is transparently the case that: one cannot have the disposition to respond to all rational numbers meeting these conditions in sequence, but one can still be disposed to respond to any natural number meeting these conditions (even the ones so long one could not read them in a finite lifetime—all I need to do is see the first digit). For further development, couple this with Blackburn's conception of the usual and perhaps even inherent infinitary nature of dispositions [1]. One important aspect of Blackburn's thinking, here, is that a finite system could be disposed, counterfactually, to do A when no longer finite in some respect. Just the same, the plan I've adopted might require me to do A under C for C that I can, in virtue of some form of finiteness (i.e., with respect to computational power or longevity), never encounter.

Again, these remarks are necessarily too brief to be fully satisfactory, and demand much further development and defense. I do not contend that these remarks constitute an ultimately satisfactory solution to (or even analysis of any formulation of) the rule-following problem(s). It does, however, provide a position from which further, incremental work might be done within the conceptual and formal framework provided by PLEN.

A final note: the solutions here appeal to potentially ontologically puzzling realism about protocols, norms, and rules. These are taken to be quite real abstract or mathematical objects in the sense that there are objective, epistemically accessible truths about how they relate to other abstract objects like epistemic directed graphs and restrictions on epistemic dynamics (indeed, the representation theorems capture some of these truths). If this is objectionable, then there are quite direct reasons to doubt the solutions presented here.

9.5 Representing Epistemic Proceduralism

The representation theorems presented in this chapter, if they work, show that, under the assumption that EDGs provide a model of single-agent epistemic dynamics that is relatively comprehensive, protocols are closely bound to the properties of the regimented and natural language constructions that capture the content of procedural and other epistemic norms. This provides an explicit justification for what, in Chaps. 4, 5, 6, and 7 is stipulated: that protocols are formal devices for encoding procedural epistemic norms. This lays the foundations for the formalization of the clusters of principles of epistemic proceduralism.

The chapter then briefly considered the rule-following paradox, as it presents itself as a pair of potential problems for both the significance of the representation theorems and the central arguments for epistemic proceduralism (the formalization of which motivates the final chapter). The core idea is that PLEN is a system that provides a rigorous framework for thinking about something the rule-following problem suggests is unintelligible: the determination of courses of action in accordance with epistemic rules. This works well enough from the theorists' perspective, where questions of references can be stipulated away or ignored. From the perspective of would be epistemic norm followers, however, this isn't quite as safe. It leaves open room for skepticism about the actual use of language to determine which courses of action accord with a rule and which don't.

A prima facie solution was given, at least in the context of PLEN: the analysis of rule-following in PLEN is knowing adherence which requires protocol or plan internalization, and which does not require the forms of intervening linguistic representation that the paradox turns on. The solution here articulates a sketchy, preliminary account of epistemic rule-following (arguably) without paradox by means of epistemic rules that are not necessarily linguistic items but something closer to abstract objects (not unlike Pettit's functions in extension and functions in extension). It then offers a sketchy, preliminary account of how our linguistic items might be used to pick out those rules.

If all goes right, the stage is set for the final establishment that the PLEN Framework satisfies the critera D1–D3C3.

Appendix 5

Remark 471 This appendix spells out the results discussed in Chap. 9.

Reduction Lemma

Lemma 472 (Reduction Lemma)

(1) For every restriction on the construction of procedures, there is some connected EDG, g, such that there is a corresponding restriction on processes defined over P_g.
(2) For arbitrary connected EDG, g:

(a) Let $R_{\preceq}(P_g)$ be a restriction on processes defined over P_g such that there are no isomorphic paths, p and q, such that p is in the rational set, q is in the irrational set, and, for all pairs of states, s and s′ in p and q, respectively, that are located at the same indices with respect to $s(p)$ and $s(q)$, $v_s = v_{s'}$ and $c_s = c_{s'}$. Then: there is a corresponding restriction on procedural constructions, $R_{\preceq}(\Pi_g)$, defined over Π_g.
(b) For every restriction on transitions defined over T_g, there is a corresponding restriction on processes defined over P_g.
(c) For every restriction on states defined over S_g, there is a corresponding restriction on transitions defined over T_g.
(d) It is not the case that, for every restriction on processes defined over P_g, there is a corresponding restriction on transitions defined over T_g nor a corresponding restriction on states defined over S_g.
(e) It is not the case that, for every restriction on transitions defined over T_g, there is a corresponding restriction on states defined over S_g.

Proof Direct argument for each of clauses 1–6.

(1) For every restriction on the construction of procedures, there is some connected EDG such that there is a corresponding restriction on processes defined over P_g.

(a) Let $R_{\preceq}(\Pi^{\forall}_{v})$ be an arbitrary restriction on procedural construction:

$$R_{\preceq}(\Pi^{\forall}_{v}) = (P(\Pi^{\forall}_{v}), \preceq, T_{\Pi^{\forall}_{v}}, R_{\Pi^{\forall}_{v}}, I_{\Pi^{\forall}_{v}})$$

(b) Let g be an arbitrary connected EDG such that:
(i) If π matches any π_v in $R_{\Pi^{\forall}_{v}}$, then P_{π} is non-empty.
(ii) If π matches any π_v in $I_{\Pi^{\forall}_{v}}$, then P_{π} is non-empty.
(iii) If π matches any π_v in $T_{\Pi^{\forall}_{v}}$, then P_{π} is non-empty.
(c) Define a restriction on P_g:
(i) Let R_{P_g} be the union of P_{π_i} for all π_i that match any π_v in $R_{\Pi^{\forall}_{v}}$.

(ii) Let I_{P_g} be the union of P_{π_i} for all π_i that match any π_v in $I_{\Pi_v^\forall}$.
(iii) Let T_{P_g} be the union of P_{π_i} for all π_i that match any π_v in $T_{\Pi_v^\forall}$.
(iv) Let $R_{P_g} \succ T_{P_g} \succeq I_{P_g}$.
(v) Thus:

$$R_{\preceq}(P_g) = (P(P_g), \preceq, T_{P_g}, R_{P_g}, I_{P_g})$$

is a restriction on epistemic processes.

(d) Thus, for $R_{\preceq}(\Pi_v^\forall)$ and $R_{\preceq}(P_g)$, for any π_v, π that matches π_v, and p such that $p \in P_\pi$, π_v is in $R_{\Pi_v^\forall}$ iff p is in R_{P_g}.
   (i) (LtoR) Let π_v be in $R_{\Pi_v^\forall}$. Then, suppose that π matches it and that p executes π. Because R_{P_g} is the union of all executions of π_i in g for all π_i that match any π_v in $R_{\Pi_v^\forall}$, p is in R_{P_g}.
   (ii) (RtoL) Let arbitrary π and π_v be such that π matches π_v and let p be an arbitrary execution of π and also an element of R_{P_g}. Because R_{P_g} is the union of executions of π_i in g that match π_v in $R_{\Pi_v^\forall}$, π_v is in $R_{\Pi_v^\forall}$.
(e) The arguments for $I_{\Pi_v^\forall}$ and $T_{\Pi_v^\forall}$ follow the same format.
(f) Thus, $R_{\preceq}(\Pi_v^\forall)$ and $R_{\preceq}(P_g)$ correspond.
(g) Thus, because $R_{\preceq}(\Pi_v^\forall)$ was arbitrary, the class of restrictions on procedural construction reduces to the class of restrictions on processes defined over the set of all connected EDGs.

(2) For every restriction on processes defined over P_g such that there are no isomorphic paths, p and q, such that (C1) p is in the rational set, q is in the irrational set, and (C2) for all pairs of states, s and s' in p and q, respectively, that are located at the same indices with respect to $s(p)$ and $s(q)$, $v_s = v_{s'}$ and $c_s = c_{s'}$, there is a corresponding restriction on procedural construction.

(a) Let $R_{\preceq}(P_g)$ be an arbitrary restriction on processes defined over P_g satisfying (C1) and (C2) :

$$R_{\preceq}(P_g) = (P(P_g), \preceq, T_{P_g}, R_{P_g}, I_{P_g})$$

   (i) Let $\Pi_v^\forall$ be the set of all protocol schemata. The (possibly empty) execution sets, P_π, of each protocol π matching any π_v in $\Pi_v^\forall$ are sets of processes and so elements of $P(P_g)$.
(b) Define a restriction on $\Pi_v^\forall$:
   (i) Let $R_{\Pi_v^\forall}$ be the union of all π_v such that every matching π_i is executed by a subset of R_{P_g}.
      (A) If R_{P_g} is non-empty, then there is some set of protocol schemata, $R_{\Pi_v^\forall}$, such that each matching protocol for each schema in $R_{\Pi_v^\forall}$ is executed by some path in R_{P_g}. Let $p \in R_{P_g}$. Then $\{p\} \subseteq R_{P_g}$. Then there is a protocol, π, such that $p \in P_\pi$; $\pi = a_1; \ldots; a_n$ for all a_i such that $p = e_1, \ldots, e_n$ and $e_i = \langle s, a_i, s' \rangle$. Then, immediately, there is a matching schema,

π_v: $f_{\pi v}(\langle a_1, \ldots, a_n\rangle) = \pi$ (indeed, there is a set of matching schemata such that $\langle a_1, \ldots, a_n\rangle$ is in the set of sequences of atomics).

(B) If there are matching schemata, then there are non-matching schemata: simply add an arbitrary choice or parallel execution to π_v, so this reasoning generalizes to the other verdict sets.

(ii) Let $I_{\Pi_v^{\forall}}$ be the union of all π_v such that every matching π_i is executed by a subset of I_{P_g}.

(iii) Let $T_{\Pi_v^{\forall}}$ be the union of all π_v such that every matching π_i is executed by a subset of T_{P_g}.

(iv) Let $R_{\Pi_v^{\forall}} \succ T_{\Pi_v^{\forall}} \succeq I_{\Pi_v^{\forall}}$.

(v) Thus,

$$R_{\preceq}(\Pi_v^{\forall}) = (P(\Pi_v^{\forall}), \preceq, T_{\Pi_v^{\forall}}, R_{\Pi_v^{\forall}}, I_{\Pi_v^{\forall}})$$

is a restriction on epistemic procedural constructions.

(c) Thus, for $R_{\preceq}(P_g)$ and $R_{\preceq}(\Pi_v^{\forall})$, for any π_v, π that matches π_v, and p such that $p \in P_\pi$, π_v is in $R_{\Pi_v^{\forall}}$ iff p is in R_{P_g}.

(i) (LtoR) Let π_v be in $R_{\Pi_v^{\forall}}$. Then, suppose that π matches it and that p executes π. By construction (i.e, $R_{\Pi_v^{\forall}}$ is the union of all schemata such that every matching protocol is executed by a subset of R_{P_g}), it follows from the fact that p executes π that p is in R_{P_g}.

(ii) (RtoL) Let arbitrary π and π_v be such that π matches π_v and let p be an arbitrary execution of π and also an element of R_{P_g}. By construction (i.e., $R_{\Pi_v^{\forall}}$ is the union of all schemata such that every matching protocol is executed by a subset of R_{P_g}), the fact that p is in R_{P_g} implies that π_v is in $R_{\Pi_v^{\forall}}$; $\{p\}$ is a subset of R_{P_g}, so π is executed by a subset of R_{P_g}, and the fact that π matches π_v gives the result.

(d) The arguments for I_{PX} and T_{PX} follow the same format.

(e) Thus, $R_{\preceq}(P_g)$ and $R_{\preceq}(\Pi_v^{\forall})$ correspond.

(f) Thus, because $R_{\preceq}(P_g)$ was an arbitrary restriction over processes satisfying C1 and C2, the class of restrictions on processes satisfying the C1 and C2 reduces to the class of restrictions on procedural construction.

(3) For every restriction on transitions defined over T_g, there is a corresponding restriction on processes defined over P_g.

(a) Let $R_{\preceq}(T_g)$ be an arbitrary restriction on T_g:

$$R_{\preceq}(T_g) = (P(T_g), \preceq, T_{T_g}, R_{T_g}, I_{T_g})$$

(b) Define a restriction on P_g:

(i) For each t in T_g there is some p such that $t = \langle first(p), last(p)\rangle$, and a π such that $p \in P_\pi$. More important, for each p that executes any π, there is some t such that $t = \langle first(p), last(p)\rangle$. The set of all t thusly related to elements of each execution set, P_π, is a set of transitions, and so an element of $P(T_g)$.

(ii) Let R_{Pg} be the union of all p such that $\langle first(p), last(p)\rangle \in R_{T_g}$.

(iii) Let I_{P_g} be the union of all p such that $\langle first(p), last(p)\rangle \in I_{T_g}$.

(iv) Let T_{P_g} be the union of all p such that $\langle first(p), last(p)\rangle \in T_{T_g}$.

(v) Let $R_{P_g} \succ T_{P_g} \succeq I_{P_g}$.

(vi) Thus,

$$R_{\preceq}(P_g) = (P(P_g), \preceq, T_{P_g}, R_{P_g}, I_{P_g})$$

is a restriction on epistemic processes.

(c) Thus, for $R_{\preceq}(P_g)$ and $R_{\preceq}(T_g)$, for any p and any t such that $t = \langle first(p), last(p)\rangle$, p is in R_{P_g} iff t is in R_{T_g}.

(i) (LtoR) Let p be in R_{P_g} and let t bookend p. Trivially, t is in R_{T_g}.

(ii) (RtoL) Let t bookend p and let t be in R_{T_g}. Then p is in R_{P_g}.

(d) The arguments for I_{T_g} and T_{T_g} follow the same format.

(e) Thus, $R_{\preceq}(P_g)$ and $R_{\preceq}(T_g)$ correspond.

(f) Thus, because $R_{\preceq}(T_g)$ was an arbitrary restriction over transitions, the class of restrictions on transitions reduces to the class of restrictions on processes.

(4) For every restriction on states defined over S_g, there is a corresponding restriction on transitions defined over T_g.

(a) Let $R_{\preceq}(S_g)$ be an arbitrary restriction on S_g:

$$R_{\preceq}(S_g) = (P(S_g), \preceq, T_{S_g}, R_{S_g}, I_{S_g})$$

(b) Define a restriction on T_g:

(i) Every subset of S_g is an element of $P(S_g)$. For every $s \in S_g$, there is some $t \in T_g$ such that $s = last(t)$.

(A) Disconnected EDGs–EDGs in which there are states that cannot be reached from any non-identical states by any sequence of edges—present a special case, but do not undermine the argument. Simply consider the t such that, for disconnected state, s, $t = \langle s, s\rangle$ (an execution of $\pi *$ for every π in Π_g).

(ii) Let R_{T_g} be the union of all t such that $s = last(t)$ and $s \in R_{S_g}$.

(iii) Let I_{T_g} be the union of all t such that $s = last(t)$ and $s \in T_{S_g}$.

(iv) Let T_{TX} be the union of all t such that $s = last(t)$ and $s \in I_{S_g}$.

(v) Let $R_{T_g} \succ T_{T_g} \succeq I_{T_g}$.

(vi) Thus,

$$R_{\preceq}(T_g) = (P(T_g), \preceq, T_{T_g}, R_{T_g}, I_{T_g})$$

is a restriction on epistemic transitions.

(c) Thus, for $R_{\preceq}(P_g)$ and $R_{\preceq}(T_g)$, for any t and any $s = last(t)$, t is in R_{T_g} iff s is in R_{S_g}.
   (i) (LtoR) Let t be ended by s and let t be in R_{T_g}. Then trivially s is in R_{S_g}.
   (ii) (RtoL) Let s end t and let s be in R_{S_g}. Trivially, t is in R_{T_g}.
(d) The arguments for I_{S_g} and T_{S_g} follow the same format.
(e) Thus, $R_{\preceq}(T_g)$ and $R_{\preceq}(S_g)$ correspond.
(f) Thus, because $R_{\preceq}(S_g)$ was an arbitrary restriction over states, the class of restrictions on states reduces to the class of restrictions on transitions.

(5) It is not the case that, for every restriction on processes defined over P_g, there is a corresponding restriction on transitions defined over T_g nor a corresponding restriction on states defined over S_g.

(a) Let $R_{\preceq}(P_g)$ be an arbitrary restriction on P_g:

$$R_{\preceq}(P_g) = (P(P_g), \preceq, T_{P_g}, R_{P_g}, I_{P_g})$$

where there are corresponding paths, p and q such that p is in R_{P_g} and q is not in R_{P_g}.
(b) Construct restrictions on transitions and states by procedures parallel to those in steps 3 and 4:
   (i)

$$R_{\preceq}(T_g) = (P(T_g), \preceq, T_{T_g}, R_{T_g}, I_{T_g})$$

   where: for $R_{\preceq}(P_g)$ and $R_{\preceq}(T_g)$, for any p and any t such that $t = \langle first(p), last(p)\rangle$, p is in R_{P_g} iff t is in R_{T_g}.
   (ii)

$$R_{\preceq}(S_g) = (P(S_g), \preceq, T_{S_g}, R_{S_g}, I_{S_g})$$

   where: for $R_{\preceq}(P_g)$ and $R_{\preceq}(S_g)$, for any p and any $s = last(p)$, p is in R_{P_g} iff s is in R_{S_g}.
(c) From the supposition that p and q correspond, it follows—for any t—that, $t = \langle first(p), last(p)\rangle$ iff $t = \langle first(q), last(q)\rangle$ and $s = last(p)$ iff $s = last(q)$.
(d) Thus, for $R_{\preceq}(P_g)$, $R_{\preceq}(T_g)$, and $R_{\preceq}(S_g)$, correspondence must fail.
   (i) Let $p \in P_g$. Suppose that t is such that $t = \langle first(p), last(p)\rangle$. Then $t = \langle first(q), last(q)\rangle$. Thus, by construction, q is also in R_{P_g}. Contradiction.
   (ii) Let $p \in P_g$. Suppose that s is such that $s = last(p)$. Then $s = last(q)$. Thus, by construction, q is also in R_{P_g}. Contradiction.

(6) It is not the case that, for every restriction on transitions defined over T_g, there is a corresponding restriction on states defined over S_g.

(a) Let $R_{\preceq}(T_g)$ be an arbitrary restriction on T_g:

$$R_{\preceq}(T_g) = (P(T_g), \preceq, T_{T_g}, R_{T_g}, I_{T_g})$$

where:

(i) $\langle s, s' \rangle \in R_{T_g}$, $\langle s'', s' \rangle \notin R_{T_g}$, $\langle s'', s' \rangle \in I_{T_g}$, and $\langle s, s' \rangle \notin I_{T_g}$, and
(ii) $\langle s', s''' \rangle \notin R_{T_g}$, $\langle s', s''' \rangle \in I_{T_g}$, $\langle s', s'''' \rangle \in R_{T_g}$, and $\langle s', s'''' \rangle \notin I_{T_g}$.

(A) Trivially, there can be no function from root or terminal states of transitions in $R_{\preceq}(T_g)$ to transitions in $R_{\preceq}(T_g)$ such that t is in R_{T_g} iff s (whether root or terminal) is in R_{S_g}. Suppose there were. Let the function be f. Suppose that s' is in R_{S_g}. Then, whatever $f(s)$, there are transitions with s' as terminal state in both the rational and irrational sets, and there are transitions with s' as the root state in both the rational and irrational sets. f cannot satisfy the correspondence rules.

This completes the proof of the Reduction Lemma. □

Corollary 473 *The set of all arbitrary restrictions on processes contains restrictions that undermine the foregoing reasoning—those restrictions such that there are isomorphic paths that satisfy conditions C1 and C2.*

Proof Let

$$R_{\preceq}(P_g) = (P(P_g), \preceq, T_{P_g}, R_{P_g}, I_{P_g})$$

be such that:

(1) there are isomorphic paths, p and q, where $p \in R_{P_g}$ and $q \in I_{P_g}$,
(2) R_{P_g} and I_{P_g} are exclusive, and
(3) condition C2 holds.
(4) If these conditions hold, then p and q execute the same protocols, by ISOPL. Thus, if we apply the reasoning above to this restriction, we find that some π_v is in $R_{\Pi_v^\vee}$ (in virtue of matching a π executed by p) and also in $I_{\Pi_v^\vee}$ (in virtue of matching the same π executed by q). Thus, the reasoning would get the content of the restriction on processes wrong. It would wrongly assign q to R_{P_g} (in virtue of executing a π that matches a rational π_v) and also wrongly assign p to I_{PX} (in virtue of executing the same π that matches the also irrational π_v), contradicting our assumptions.
(5) If condition C2 fails, a restriction on procedural constructions could be defined that is sensitive to the preconditions and postconditions of the protocols that match them. This would undermine the problem, as would the failure of condition C1, which would enable the restriction on procedural conditions to distinguish schemata that do not have procedurally equivalent matching protocols.

□

Corollary 474 *(RLC1) For each restriction on transitions, there is a restriction on procedures that corresponds to it.*

Proof Restrictions on processes that contain processes satisfying C1 and C2 are not necessary to derive restrictions on transitions by the reasoning for step 2. Thus, exactly the restriction on procedural construction that corresponds to $R_{\preceq}(P_g)$ suffices for this result: simply construct it as an intermediary, and reason as above □

Corollary 475 (RLC2) *It is not the case that for every restriction on procedural construction, there is a corresponding restriction on transitions or states.*

Proof Let $R_{\preceq}(P_g)$ be an arbitrary restriction on a set of processes, P_g:

$$R_{\preceq}(P_g) = (P(P_g), \preceq, T_{P_g}, R_{P_g}, I_{P_g})$$

where there are corresponding paths, p and q such that p is in R_{P_g} and q is not in R_{P_g}. The reasoning in Step 1 works for this graph, and the reasoning in step 5 will show that there is no corresponding restriction on transitions or states. □

Corollaries of Minor Interest

Corollary 476 (RRT2C) *For restrictions on procedural construction, $R^i_{\preceq}(X)$, with finitely large rational, irrational, and threshold sets, there are sets of protocols, Π_R, Π_T, and Π_I, that are singletons, $\{\pi_R\}$, $\{\pi_T\}$, $\{\pi_I\}$, such that for any syntactic schema, π_v, π_v matches a proper subprotocol of π_X iff $\pi_v \in X_{\Pi^\forall_v}$ for all $R^j_{\preceq}(\Pi^\forall_v)$ in $R!$.*

Proof (Proof-Plan) Let $R^i_{\preceq}(\Pi^\forall_v)$ be an arbitrary restriction on procedural constructions. Run the procedure and argument for RRT2 to construct the sets of protocols, Π_R, Π_T, and Π_I, and the set of restrictions $R!$. Construct singleton sets of protocols by the procedure:

(1) (Reading Off Procedure 2) Let $\pi_X = \pi_1 \cup (\ldots(\ldots \cup \pi_n))$ where each π_i is in Π_X according to Reading Off Procedure 1.
(2) (Reading Off Rules) From these sets of protocols, derive verdict sets, $X^i_{\Pi^\forall_v}$, over procedural construction from the syntax of each π_X by the rules:
    (a) For all π_v, $\pi_v \in R^i_{\Pi^\forall_v}$ iff π_v matches π_R.
    (b) For all π_v, $\pi_v \in T^i_{\Pi^\forall_v}$ iff π_v matches π_T.
    (c) For all π_v, $\pi_v \in I^i_{\Pi^\forall_v}$ iff π_v matches π_I.
(3) From π_R, π_T, and π_I, we can derive a set of restrictions, $R!$, containing exactly the restriction with which we started by reasoning parallel to that in the argument for RRT2, replacing arbitrary elements, π_i, of Π_X with subprotocols of π_X.

□

Remark 477 (Qualification) RRT2C only really works for countably large restrictions on epistemic dynamics—or, rather, restrictions such that the (e.g.) rational sets are countably large. The set of all restrictions on procedural construction is uncountably large. This is just a first pass, but, roughly speaking, the set of all restrictions on procedural construction is about the size of the structure,

$$(P(\Pi_v^{\forall}) \times P(P(\Pi_v^{\forall})) \times P(\Pi_v^{\forall}) \times P(\Pi_v^{\forall}) \times P(\Pi_v^{\forall}))$$

since, at the very least, the (e.g.) rational set of schemata is a *set of schemata*, so the space of all possible restrictions must contain every possible rational set of schemata. But this is just the powerset of all syntactic schemata. The set of schemata is countably infinite, so, by Cantor's Theorem, this structure is uncountable. But every protocol is finite; there aren't infinitely many subprotocols for any protocol let alone infinitely many immediate subprotocols. Thus, there can't be protocols of the form of π_X for Π_X that are countably infinite. So, not all restrictions can be assigned triples of protocols by Reading Off Procedure 2; this rule only applies to generate individual, complex protocols to read off finite restrictions on construction. However, if every rational set of protocols (with respect to restrictions on procedural construction) is at most countable, then there is some set of protocols assignable to each element of each restriction on procedural constructions, and there is a triple of countable sets of protocols for every restriction on construction.

Claim 478 The foregoing results show that, for every restriction on epistemic dynamics defined over the elements of EDGs that satisfy the conditions of the Reduction Lemma, there are sets of protocols or, where those sets of elements of EDGs are finite, a set of individual, complex protocols from which to read off that restriction. This implies that, for every restriction on dynamics defined over the elements of EDGs, there is a a set of protocol graphs from which that restriction can be read off. This centers the primary theoretical advantages of protocols on the fidelity of the relations between the syntax of protocols and the central constructions of the formalism, protocol graphs.

Theorem 479 (PGRT) *For every restriction on epistemic dynamics defined over connected graph, g, $R_{\preceq}(X)$ (such that, where $R_{\preceq}(X) = R_{\preceq}(P_g)$, there are no isomorphic paths, p and q, such that p is in the rational set, q is in the irrational set, and, for all pairs of states, s and s' in p and q, respectively, that are located at the same indices with respect to $s(p)$ and $s(q)$, $v_s = v_{s'}$ and $c_s = c_{s'}$), and any triple of sets of protocols, Π_R, Π_T, and Π_I (or complex protocols π_R, π_T, and π_I) identified by RRT2 (or RRT2C, respectively), the extensions of the rational, irrational, and threshold sets of the restriction can be derived from the properties of any protocol graphs of the protocol graphs of elements of Π_R, Π_T, and Π_I, respectively (or the protocol graphs of π_R, π_T, and π_I, respectively).*

Proof (Proof-Plan) First, I define the following procedure for reading off finite restrictions from the protocol graphs of π_R, π_T, and π_I or elements of Π_R, Π_T, and Π_I for arbitrary restriction on epistemic dynamics, $R_{\preceq}(X)$:

(1) (Reading Off Procedure 3)

(a) (Step 1) Define a corresponding restriction on procedural construction, $R_{\preceq}(\Pi_v^{\forall})$, as in the argument for Step 2 of the Reduction Lemma.
(b) (Step 2) Execute the original Reading Off Procedure.
(c) (Step 3) Let $g \in G^{\pi_X}$ for arbitrary $\pi_X \in \Pi_X$ selected by Step 2. Read the verdicts of $R_{\preceq}(X)$ off of g, for each π_X by the following rules:
  (i) For all π_v, there is some $\pi' \in ISub(\pi_X)$, such that, for some p in P_g, π_v matches π', and $p \in maxpath(g) \cap P_{\pi'}$ iff $\pi_v \in R_{\Pi_v^{\forall}}$ (and just so for $T_{\Pi_v^{\forall}}$ and $I_{\Pi_v^{\forall}}$).
  (ii) For all p in P_g and any p' in arbitrary g', $p' \in X_{P_g}$ iff $p \in P_{\pi_X}$ and $p' \sim p$.
  (iii) For all t in T_g and any t' in arbitrary g', $t' \in X_{T_g}$ iff $t \in T_{\pi_X}$ and $t' \sim t$.
  (iv) For all s in S_g and any s' in arbitrary g', $s' \in X_{S_g}$ iff $s = last(p)$, $p \in P_{\pi_X}$, and $s' \sim s$.

(2) The procedure suffices to derive the verdicts of any arbitrary $R_{\preceq}(X)$ on epistemic factors.

(a) Let $R_{\preceq}(X)$ be a restriction on any set of epistemic factors. Then derive a corresponding restriction on procedural constructions, $R_{\preceq}(\Pi_v^{\forall})$, by Step 1.
(b) Derive Π_R by Step 2.
  (i) If $\pi_v \in R_{\Pi_v^{\forall}}$, then from the procedure, π_v matches some protocol, π', in Π_R and converse. That is, if all we knew about π_v is that $R_{\preceq}(\Pi_v^{\forall})$ gives a verdict on it and that π_R was derived by means of Step 1 and Step 2, then we can determine whether $\pi_v \in R_{\Pi_v^{\forall}}$ by checking to see whether it matches any protocols in Π_R.
(c) Derive g by Step 3.
(d) By construction, for any protocol graph, g, of π_R in Π_R, there is a maximal path, p, that executes some $\pi' \in ISub(\pi_R)$, and every such maximal path executes some protocol in Π_R (π_R itself).
(e) Now, suppose that p is a maximal path in g that executes $\pi' \in ISub(\pi_R)$ and π_v matches π'. From Step 3, it follows that $\pi_v \in R_{\Pi_v^{\forall}}$. That is, if all we knew about π_v is that $R_{\preceq}(\Pi_v^{\forall})$ gives a verdict on it, that $\pi_R \in \Pi_R$ was derived by means of Step 1 and Step 2, and that g is a protocol graph of π_R, then we can determine whether $\pi_v \in R_{\Pi_v^{\forall}}$ by checking to see whether there is a maximal path of g that executes a protocol that matches π_v. By construction, for every $\pi_v \in R_{\Pi_v^{\forall}}$, there is a maximal path in any protocol graph of π_R that executes a π_v—matching immediate subprotocol of π_R.

(3) Apply the same procedure and reasoning for $T_{\Pi_v^{\forall}}$ and $I_{\Pi_v^{\forall}}$ by means, respectively, of π_T and π_I in the relevant sets of protocols.
(4) From the Reduction Lemma, $R_{\preceq}(\Pi_v^{\forall})$ corresponds to $R_{\preceq}(X)$, so for any epistemic factor, its verdicts with respect to $R_{\preceq}(X)$ can be read off by the Reading Off Procedure 3 by means of the Reduction Lemma.

□

References

1. Blackburn, S. 2002. The individual strikes back. In *Rule-following and meaning*, eds. Miller and Wright, 28–44 (ch. 3). Mcgill-Queen's University Press.
2. Boghossian, P. 2008. Epistemic rules. *Journal of Philosophy* 105(9):472–500.
3. Bratman, M. 1998. *Intention, plans, and practical reason*. Center for the Study of Language and Information.
4. Carroll, L. 1895. What the tortoise said to achilles. *Mind* 4(14):278–280.
5. Georgeff, M. Ed. 1987. *Reasoning about actions and plans*. Kaufmann.
6. Gibbard, A. 2003. *Thinking how to live*. Harvard University Press.
7. Goodman, N. 1968. *Languages of art: An approach to a theory of symbols*. Bobbs-Merrill.
8. Kripke, S. 1982. *Wittgenstein on rules and private language: An elementary exposition*. Harvard University Press.
9. Miller, A., and C. Wright, Eds. 2002. *Rule-following and meaning*. Mcgill-Queen's University Press.
10. Pettit, P. 1990. The reality of rule-following. *Mind* 99(393):1–21.
11. Pollock, J.L. 1987. Epistemic norms. *Synthese*. 71(1):61–95.
12. Pollock, J.L. 1995. *Cognitive carpentry*. MIT Press.
13. Pollock, J., and J. Cruz. 1999. *Contemporary theories of knowledge*, 2nd edn. Rowman & Littlefield.
14. Ryle, G. 1949. *The concept of mind*. Hutchinson & Co.
15. Schafer, K. 2014. Doxastic planning and epistemic internalism. *Synthese* 191(12):2571–2591.
16. Speaks, J. 2008. Kripke's skeptical paradox. Accessed: https://www3.nd.edu/~jspeaks/courses/2007-8/93914/_HANDOUTS/kripkenstein.pdf.
17. von Wright, G.H. 1960. *Norm and action*. Routledge.
18. Wittgenstein, Ludwig. 1991. *Philosophical Investigations: The German Text, with a Revised English Translation 50th Anniversary Commemorative Edition*. Wiley-Blackwell.

Chapter 10
Philosophical Results III: Formalizing Epistemic Proceduralism

Abstract Prior chapters argued for the intended interpretation (or use) of PLEN as a logic of epistemic norms. On the basis of that interpretation, this chapter uses the PLEN logic of epistemic norms to formalize epistemic proceduralism. More precisely, it shows that, given that protocols can be plausibly interpreted as structures representing or describing the structure of procedural epistemic norms, and protocol-theoretic formulae can be plausibly interpreted as statements expressing properties of epistemic norms, all of the clusters of principles that define epistemic proceduralism (per Chaps. 1–3) can be derived from the logic of protocols in PLEN. Indeed, with few exceptions, each corresponds to a provable theorem, property of protocols, property of EDGs, or suite of inference rules in PLEN. The exceptions can be derived by applying the protocol-theoretic deontic and norm application extensions of PLEN or exploiting the representation theorems.

10.1 Adequate Representation Theses

This chapter provides a sketch of a formal theory of epistemic proceduralism derived from the Revised PLEN framework. This, it is intended, completes the task of proving thesis T3 of the protocol-theoretic account of epistemic norms:

(T3) Procedural epistemic norms can be formally represented and reasoned about by means of a protocol-theoretic framework—a logical framework that deploys protocols to directly encode the procedural structure or content of procedural epistemic norms; more, the protocol-theoretic framework I develop and defend in this book can be shown to be theoretically interesting and useful, in that each of the theoretically central principles characterizing procedural norms (D1–D3 and all sub-clauses and corollaries) can be formally analyzed in a plausible way and then derived from the protocol logic of the framework.

The representation theorems presented in the prior chapter, if they work, show that, under the assumption that EDGs provide a model of single-agent epistemic dynamics that is relatively comprehensive, protocols are closely bound to the properties of the regimented and natural language constructions that capture the content of

R. Jenkins, *A Protocol-theoretic Framework for the Logic of Epistemic Norms*, Synthese Library 467, https://doi.org/10.1007/978-3-031-08597-0_10

procedural and other epistemic norms. This provides an explicit justification for what, in Chaps. 4–7, was stipulated: that protocols are formal devices for encoding procedural epistemic norms.

More than that, they provide the groundwork for formalizing the principles of epistemic proceduralism. The aim of this formalization project, carried out in this chapter, is to satisfy formal adequacy theses similar to those presented in [1]:

> THESIS I. Let s be a social situation involving the intuitive concepts of knowledge, justifiable beliefs and common knowledge among a group of agents. Assume that s is presented in such a way that all the relevant features of s pertaining to knowledge, beliefs and common knowledge are completely determined. Then we may associate to s a mathematical model S. (S is a multi-agent Kripke model; we call these epistemic state models.) The point of the association is that all intuitive judgements concerning s correspond to formal assertions concerning S, and vice-versa.
>
> [...]
>
> THESIS II. Let σ be a social "action" involving and affecting the knowledge (beliefs, common knowledge) of agents. This naturally induces a change of situation; i.e., an operation o taking situations s into situations $o(s)$. Assume that o is presented by assertions concerning knowledge, beliefs and common knowledge facts about s and $o(s)$, and that o is completely determined by these assertions. Then
>
> (a) We may associate to the action σ a mathematical model Σ which we call an epistemic action model. (Σ is also a multi-agent Kripke model.) The point again is that all the intuitive features of, and judgments about, σ correspond to formal properties of Σ.
> (b) There is an operation $\otimes$ taking a state model S and an action model and returning a new state model $S \otimes \Sigma$. So each induces an update operation O on state models: $O(S) = S \otimes \Sigma$.
> (c) The update O is a faithful model of the situation change o, in the sense that for all s: if s corresponds to S as in Thesis I, then again $o(s)$ corresponds to $O(S)$ in the same way; i.e. all intuitive judgements concerning $o(s)$ correspond to formal assertions concerning $O(S)$, and vice-versa. (pp. 166–167)

Clearly, there are differences between their project and mine. For one, PLEN is intended to be a single-agent system rather than a multi-agent system.[1] And so the talk of social situations is not particularly germane for the adequacy of PLEN. Nonetheless, the general thrust of these adequacy theses is exactly what guides the development of PLEN. For the target objects of analysis (analyzanda), we ought to provide formal structures that directly represent them (analyzans). We ought then to provide a plausible or defensible informal account of the core properties of our analyzanda, and the formalization ought to proceed by deriving properties of the analyzans that translate the principles comprising the informal account.

For this project, then, a formalism, F, is theoretically adequate to the analysis of epistemic dynamics, epistemic rationality, and epistemic rationality norms iff it contains elements satisfying the foregoing theses (modified for the single-agent case by replacing state models with states and content functions, update operations and

[1] It is not altogether difficult to see how PLEN could be made into a multi-agent system. This is, however, left to future work.

action models with procedural paths and edge relations) and the following additional adequacy thesis:

THESIS III.: Let N be an epistemic norm concerning the rationality of states, transitions, actions, processes, or procedures of epistemic state change. This naturally induces a restriction on states, transitions, and actions/processes/procedures of epistemic state change; i.e., an operation r taking norms N into sets of states, transitions, and actions/processes/procedures, $r(N)$. Assume that (i) r is presented by assertions concerning procedural operations that construct complex actions and procedures, the procedural content of N, the postconditions, preconditions, and alternatives of complying with N or controlling one's dynamics in accordance with $r(N)$, and about the procedural properties of N and their relation to s and $o(s)$, and that (ii) r is completely determined by these assertions. Then:

(d): We may associate to the norm N a mathematical model π which we call an epistemic protocol. The point is that *all the intuitive features of, and judgments about N correspond to formal properties of* π.

(e): There is an operation $\otimes$ taking an EDG, g, and a norm, N, and returning an EDG $g \otimes N$. So each induces a restriction operation $R_{\preceq}$ on EDGs: $R_{\preceq}(\pi) = g \otimes N$. The operation is the mapping of any protocol, π, that represents N to its execution set in g or, in other words, to a subgraph of g that is bisimilar to a subgraph of any protocol graph of π.

(f): The restriction operation $R_{\preceq}$ is a faithful model of the restrictions on epistemic dynamics, $r(N)$, defined by any norm, N, in the sense that for all N: if N corresponds to π as in Thesis III(d), then $r(N)$ corresponds to $R_{\preceq}(\pi)$ in the same way; i.e. *all intuitive judgements concerning* $r(N)$ *correspond to formal assertions concerning* $R_{\preceq}(\pi)$, *and vice-versa.*

PLEN, of course, is the formalism, protocols are the analyzans of procedural epistemic norms, and epistemic proceduralism—in the form of clusters D1–D3 and their corollaries—comprises the intuitive account of procedural epistemic norms to which formal assertions of PLEN are to correspond. The formalization, then, will be adequate, if there are formal translations of D1–D3 and their corollaries derivable from the protocol logic of PLEN. To be explicit, the correspondence of intuitive judgments and formal assertions is to be spelled out in terms of mapping the intuitive judgment to a string in PLEN's object languages or an assertion in PLEN's meta-theory that, given the modeling intentions of PLEN, plausibly encodes the intuitive judgment, and then deriving the formal string or meta-level assertion from the protocol logic in PLEN. This chapter presents and outlines the derivation of the formal translations of D1–D3, D1C1–D3C3. I present these as pairs of problems and solutions, with minimal discussion, as the formalizations of these principles are all mechanical or immediate.

10.2 Cluster D1 and D1C1

10.2.1 Groundwork

The definitions of the key terms in this chapter are all present in Chaps. 5 and 6. Of special import are the notions of correspondence and reduction among restrictions on epistemic dynamics and the system definitions for PLEDKNA.

The main technical results of PLEN employed in the formalization of D1–D3 and all clauses and corollaries are the protocol logic of Chaps. 4, 5, 6, and 7, the applied logics of Chap. 8, and the representation theorems of Chap. 9.

10.2.2 D1 and D1C1 Elaboration

Problem 480 Formalize the following cluster of principles of epistemic proceduralism:

(D1) Epistemic rationality norms define restrictions on epistemic dynamics distinct from constraints of capacity and feasibility, including:

(1) restrictions on the content of epistemic states,
(2) restrictions on transitions and trajectories of evolution of epistemic state, and
(3) restrictions on procedures and processes of epistemic state change.

Problem 481 Formalize the following cluster of principles of epistemic proceduralism:

(D1C1) Epistemic rationality is dynamic; there are epistemic restrictions on transitions among epistemic states, not just states or contents of states, and there are characteristic relations among restrictions on states and restrictions on transitions. In particular:

(1) There are norms of epistemic rationality that take epistemic transitions as their objects.
    (a) (D1C1.1C) Norms that restrict transitions can be combined to construct norms that restrict trajectories.
(2) There are norms of epistemic rationality that take properties of epistemic transitions as their conditions of application.
(3) (StoT) Norms that restrict epistemic states can be used to generate norms that restrict epistemic transitions in the following sense: there are norms that restrict epistemic states, N_S, such that a corresponding dynamic norm, N_T, can be defined on the basis of N_S's verdicts on states and the relations between states and transitions.

(4) (TtoS) Norms that restrict epistemic transitions can be used to generate norms that restrict epistemic states in the following sense: there are norms that restrict epistemic transitions, N_T, such that a corresponding static norm, N_S, can be defined on the basis of N_T's verdicts on transitions and the relations between states and transitions.
(5) (TtoS Irreducibility) There are dynamic norms, N_T, such that there is no corresponding static norm, N_S, that can be defined on the basis of N_T's verdicts on transitions and the relations between states and transitions.
(6) (StoT Reducibility) If N_S is a norm that restricts epistemic states, then a corresponding dynamic norm, N_T, can be defined on the basis of N_S's verdicts on states and the relations between states and transitions.

10.2.3 D1 and D1C1 Formalization

Solution 482 The dual representation theorems of PLEN, RRT and PFT, suffice to entail each of the following, the derivations of which are immediate or mechanical, and so merely outlined:

(D1F) Protocols define restrictions on epistemic dynamics distinct from constraints of capacity and feasibility, including:

(1) restrictions on the content of epistemic states,

Proof ***(Proof-Plan)*** The Reduction Lemma implies that restrictions on procedures define restrictions on states, and RRT1 entails that every protocol defines restrictions on procedures. Thus, RRT1 and the Reduction Lemma entail that protocols define restrictions on epistemic states and so also on the contents of epistemic states (due to the content function in g) derivable from the restrictions on procedures. □

(2) restrictions on transitions and trajectories of evolution of epistemic state,

Proof ***(Proof-Plan)*** The Reduction Lemma implies that restrictions on procedures define restrictions on transitions and trajectories, and RRT1 entails that every protocol defines restrictions on procedures. Thus, RRT1 and the Reduction Lemma entail that protocols define restrictions on epistemic transitions and trajectories derivable from the restrictions on procedures. □

(3) restrictions on procedures and processes of epistemic state change.

Proof ***(Proof-Plan)*** RRT1 and the first two clauses of the Reduction Lemma suffice immediately. □

(4) There are EDGs, g, such that the foregoing restrictions are distinct from P_g, T_g, S, and C; these restrictions thus diverge from what is possible in g. For $s \in S$, the point holds for $(g', s) \sqsubseteq g$, showing divergence from what is feasible at s in g.

Proof (Proof-Plan) Immediate from the construction of EDGs. □

Solution 483 The dual representation theorems of PLEN, RRT and PFT, suffice to entail each of the following, the derivations of which are immediate or mechanical, and so merely outlined:

(D1C1F) There are protocols that define restrictions on transitions among epistemic states, not just states or contents of states, and there are characteristic relations among restrictions on states and restrictions on transitions. In particular:

(1) (D1C1.1F) There are protocol functions that take epistemic transitions as their objects.

Proof (Proof-Plan) Immediate by construction (i.e., the definitions underlying PFT); there are protocol functions and EDGs, g, such that $Dom(\pi)Tran(g)$ is non-empty. □

(a) (D1C1.1CF) Protocols that define restrictions on transitions can be combined to construct protocols that restrict trajectories or they can be used to define restrictions on trajectories directly.

Proof (Proof-Plan) Immediate by construction. One method is to take the union of state-sequences that link any pairs of states in the rational set for arbitrary $R_{\preceq}(T_g)$ as the rational set (et al). Another might be to take the compositions of rational sets (et al). Yet another might be to simply take the state sequences of a given protocol, which would give the second disjunct. □

(2) (D1C1.2F) There are protocol functions that take properties of epistemic transitions as their conditions.

Proof (Proof-Plan) Immediate by construction (i.e., the definitions underlying PFT); there are protocol functions and EDGs, g, such that $Con(\pi, g)$ contains formulae built with binary operators, $[\ldots]_t$, $\langle\ldots\rangle_t$, $\leftrightarrow_t$, $\curlyvee_t$, and $\curlywedge_t$. □

(3) (D1C1.3F) Protocols that define restrictions on epistemic states can be used to define restrictions on epistemic transitions on the basis of the verdicts on states and the relations between states and transitions. More precisely:

(a) (StoT) There are protocols that define restrictions on epistemic states, π_S, such that a protocol, π_T, defining corresponding restrictions on epistemic transitions can be defined on the basis of π_S's restrictions' verdicts on states and the relations between states and transitions.

Proof (Proof-Plan) Immediate from the Reduction Lemma and RRT. Let $R_{\preceq}(S_g)$ defined over g be derivable from some Π_R, Π_I, Π_I by RRT2. By RRT1, derive a corresponding $R_{\preceq}(\Pi_v^{\forall})$ from Π_R, Π_I, Π_I. By the Reduction Lemma, derive a corresponding $R_{\preceq}(T_g)$. RRT binds these restrictions to protocols on both ends. □

(4) (D1C1.4F) Protocols that define restrictions on epistemic transitions can be used to define restrictions on epistemic states on the basis of the verdicts on transitions and the relations between states and transitions. More precisely:

(a) (TtoS) There are protocols that define restrictions on epistemic transitions, π_T, such that a protocol, π_S, defining corresponding restrictions on epistemic states can be defined on the basis of π_T's restrictions' verdicts on states and the relations between states and transitions.

Proof (Proof-Plan) Immediate from the Reduction Lemma and RRT. Let $R_{\preceq}(T_g)$ defined over g be derivable from some Π_R, Π_I, Π_I by RRT2. By RRT1, derive a corresponding $R_{\preceq}(\Pi_v^{\forall})$ from Π_R, Π_I, Π_I. By the Reduction Lemma, derive a corresponding $R_{\preceq}(S_g)$. RRT binds these restrictions to protocols on both ends. □

(5) (D1C1.5F) There are protocols that define restrictions on epistemic transitions that cannot be reduced to restrictions on epistemic states on the basis of the verdicts on states and the relations between states and transitions. More precisely:

(a) (TtoS Irreducibility) There are protocols that define restrictions on transitions, π_T, such that there is no protocol, π_S, that defines a corresponding restriction on epistemic states that can be defined on the basis of π_T's restrictions' verdicts on transitions and the relations between states and transitions.

Proof (Proof-Plan) Immediate from the Reduction Lemma and RRT. Let $R_{\preceq}(T_g)$ defined over g satisfying the conditions of clauses (d) or (e) of the Reduction Lemma be derivable from some Π_R, Π_I, Π_I by RRT2. By the Reduction Lemma, it can't be reduced to any restriction on states. RRT binds these restrictions to protocols on both ends. □

(6) (D1C1.6F) If π defines a restriction on epistemic states, $R_{\preceq}(S_g)$, that excludes unconnected states, then $R_{\preceq}(S_g)$ reduces to some corresponding restriction on transitions defined on the basis of $R_{\preceq}(S_g)$'s verdicts on states and the relations between states and transitions.

Proof (Proof-Plan) Immediate from the Reduction Lemma. RRT binds these restrictions to protocols on both ends. □

10.3 Cluster D1C2

10.3.1 D1C2 Elaboration

Problem 484 Formalize the following cluster of principles of epistemic proceduralism:

(D1C2) Epistemic rationality is ineliminably procedural; there are epistemic restrictions on epistemic actions, processes, and procedures, not just states or transitions among states, and there are characteristic relations among restrictions on transitions and restrictions on processes. In particular:

(1) There are norms of epistemic rationality that take epistemic actions, processes, and procedures as their objects.
(2) The norms of epistemic rationality take properties of epistemic actions, processes, and procedures as their conditions of application.
(3) Procedural epistemic norms define restrictions on epistemic procedures, which in turn define restrictions on epistemic factors of other kinds.
(4) The procedural norms of epistemic rationality that define restrictions on actions, processes, and procedures are indispensable for the regulative role of epistemic norms.
(5) (TtoA) Norms that restrict epistemic transitions can be used to generate norms in the following sense: there are norms that restrict epistemic transitions, N_T, such that a corresponding norm that restricts actions, N_A, can be defined on the basis of N_T's verdicts on transitions and the relations between actions and transitions.
(6) (TtoP) Norms that restrict epistemic transitions can be used to generate norms in the following sense: there are norms that restrict epistemic transitions, N_T, such that a corresponding norm that restricts processes, N_P, can be defined on the basis of N_T's verdicts on transitions and the relations between processes and transitions.
(7) (TtoPr) Norms that restrict epistemic transitions can be used to generate norms in the following sense: there are norms that restrict epistemic transitions, N_T, such that a corresponding norm that restricts procedures, N_{Pr}, can be defined on the basis of N_T's verdicts on transitions and the relations between procedures and transitions.
(8) (APPrtoT) Norms that restrict epistemic actions, processes, or procedures can be used to generate norms in the following sense: there are norms that restrict epistemic actions, processes, or procedures (respectively), N_A, N_P, or N_{Pr} (respectively), such that a corresponding norm that restricts transitions, N_T, can be defined on the basis of N_A's verdicts (resp. N_P or N_{Pr}) on actions (resp. processes, procedures) and the relations between actions (resp. processes, procedures) and transitions.
(9) (APPrtoS Irreducibility) Not all norms restricting epistemic actions, processes, and procedures can be generated from norms restricting epistemic states on the basis of the verdicts on states and the relations between states

and actions/processes/procedures: There are procedural norms, N_A (resp. N_P, N_{Pr}), such that there is no corresponding static norm, N_S, that can be defined on the basis of N_A's verdicts (resp. N_P or N_{Pr}) on actions (resp. processes, procedures) and the relations between states and actions (resp. processes, procedures).

(10) (APPrtoT Irreducibility) Not all norms restricting epistemic actions, processes, and procedures can be generated from norms restricting epistemic transitions. on the basis of the verdicts on transitions and the relations between actions/processes/procedures and transitions: There are procedural norms, N_A (resp. N_P, N_{Pr}), such that there is no corresponding dynamic norm, N_T, that can be defined on the basis of N_A's verdicts (resp. N_P or N_{Pr}) on actions (resp. processes, procedures) and the relations between transitions and actions (resp. processes, procedures).

10.3.2 D1C2 Formalization

Solution 485 The dual representation theorems of PLEN, RRT and PFT, suffice to entail each of the following, the derivations of which are immediate or mechanical, and so merely outlined:

(1) (D1C2F) There are protocols that define restrictions on epistemic actions, processes, and procedures, not just states or transitions among states, and there are characteristic relations among restrictions on transitions and restrictions on processes.

Proof (Proof-Plan) From RRT1, all protocols define restrictions on procedural construction, and, on the basis of the Reduction Lemma, define restrictions on processes and actions (procedural paths and edges in EDGs), as well as all other epistemic factors, by means of the graph-theoretic properties of EDGs induced by the execution rules for protocols, i.e., path concatenation, iteration, correspondence, and branching. The Reduction Lemma further shows that restrictions on transitions can be reduced to corresponding restrictions on processes and the converse fails. □

(2) (D1C2.1F) There are protocol functions that take epistemic actions, processes, and procedures as their objects.

Proof (Proof-Plan) Let π be such that, for some g, $Dom(\pi) \subseteq (P_g \cup P_{PLEN})$. Define the protocol function, $f_\pi^\forall$, and consider $f_\pi^\forall(C, x, s)$ for some state, s, in g and condition, C, in $\pi Con(g)$. Since $x \in P_g \cap P_{PLEN}$, then the result holds immediately. □

(3) (D1C2.2F) There are protocol functions that take properties of epistemic actions, processes, and procedures as their conditions of application.

Proof (Proof-Plan) Let π be such that, for some g, $\pi Con(g) \subseteq L^R_{PLEN}$. Define the protocol function, $f_\pi^\forall$, and consider $f_\pi^\forall(C, x, s)$ for some s in g and some x in π's domain. Since $C \in \pi Con(g)$, then the result holds immediately, given that protocol-theoretic formulae state properties of epistemic actions, processes, and procedures. □

(4) (D1C2.3F) A restriction on epistemic dynamics of each kind can be derived from each protocol, and every restriction on epistemic dynamics definable over EDGs can be derived from the syntax of some sets of protocols.

Proof (Proof-Plan) The first conjunct is immediate from RRT1 and the Reduction Lemma. The second conjunct is immediate from RRT2 and RRT3. □

(5) (D1C2.4F) All restrictions on states and transitions definable over EDGs and all plausible restrictions on processes definable over EDGs are derivable from restrictions on procedural construction, and the converse fails.

Proof (Proof-Plan) Immediate from the Reduction Lemma of RRT. RRT binds these restrictions to protocols on both ends. □

(6) (D1C2.5F) Protocols that define restrictions on epistemic transitions can be used to define restrictions on epistemic actions on the basis of the verdicts on transitions and the relations between actions and transitions. More precisely:

(a) (TtoA) There are protocols that define restrictions on epistemic transitions, π_T, such that a protocol, π_A, defining corresponding restrictions on epistemic action can be defined on the basis of N_T's restrictions' verdicts on transitions and the relations between actions and transitions.

Proof (Proof-Plan) Immediate from the Reduction Lemma and RRT. Let $R_{\preceq}(T_g)$ defined over g be derivable from some Π_R, Π_I, Π_I by RRT2. By RRT1, derive a corresponding $R_{\preceq}(\Pi_v^{\forall})$ from Π_R, Π_I, Π_I. By the Reduction Lemma, derive a corresponding $R_{\preceq}(T_g)$. RRT binds these restrictions to protocols on both ends. □

(7) (D1C2.6F) Protocols that define restrictions on epistemic transitions can be used to define restrictions on epistemic processes on the basis of the verdicts on transitions and the relations between processes and transitions. More precisely:

(a) (TtoP) There are protocols that restrict epistemic transitions, π_T, such that a corresponding protocol that defines restrictions on epistemic processes, π_P, can be defined on the basis of π_T's restrictions' verdicts on transitions and the relations between processes and transitions.

Proof (Proof-Plan) Immediate from the Reduction Lemma and RRT. Let $R_{\preceq}(T_g)$ defined over g be derivable from some Π_R, Π_I, Π_I by RRT2. By RRT1, derive a corresponding $R_{\preceq}(\Pi_v^{\forall})$ from Π_R, Π_I, Π_I. By the Reduction Lemma, derive a corresponding $R_{\preceq}(P_g)$. RRT binds these restrictions to protocols on both ends. □

(8) (D1C2.7F) Protocols that define restrictions on epistemic transitions can be used to define restrictions on epistemic procedures on the basis of the verdicts

on transitions and the relations between procedures and transitions. More precisely:

(a) (TtoPr) There are protocols that define restrictions on epistemic transitions, π_T, such that a corresponding protocol, π_{Pr}, can be defined on the basis of π_T's restrictions' verdicts on transitions and the relations between procedures and transitions.

Proof (Proof-Plan) Immediate from the Reduction Lemma and RRT. Let $R_{\preceq}(T_g)$ defined over g be derivable from some Π_R, Π_I, Π_I by RRT2. By RRT1, derive some $R_{\preceq}(\Pi_v^{\forall})$ from Π_R, Π_I, Π_I. RRT binds these restrictions to protocols on both ends. □

(9) (D1C2.8F) Protocols that define restrictions on epistemic actions, processes, or procedures can be used to define restrictions on epistemic transitions on the basis of the verdicts on actions, processes, and procedures and the relations between actions/processes/procedures and transitions. More precisely:

(a) (APPrtoT) There are protocols that define restrictions on epistemic actions, processes, or procedures (respectively), π_A, π_P, or π_{Pr} (respectively), such that a protocol defining corresponding restrictions on epistemic, π_T, can be defined on the basis of π_A's restrictions' verdicts (resp. π_P or π_{Pr}) on actions (resp. processes, procedures) and the relations between actions (resp. processes, procedures) and transitions.

Proof (Proof-Plan) Immediate from the Reduction Lemma and RRT. Let $R_{\preceq}(\Pi_v^{\forall})$ defined over g be derivable from some Π_R, Π_I, Π_I by RRT2. By the Reduction Lemma, derive a corresponding $R_{\preceq}(T_g)$. RRT binds these restrictions to protocols on both ends. □

(10) (D1C2.9F) There are protocols that define restrictions on epistemic actions/processes/procedures that cannot be reduced to restrictions on epistemic states on the basis of the verdicts on states and the relations between actions/processes/procedures and states. More precisely:

(a) (APPrtoS Irreducibility) There are protocols defining restrictions on epistemic actions (resp. processes, procedures), π_A (resp. π_P, π_{Pr}), such that there are no protocols, π_S, that define corresponding restrictions on epistemic states that can be defined on the basis of π_A's restrictions' verdicts (resp. π_P or π_{Pr}) on actions and the relations between states and actions (resp. processes, procedures).

Proof (Proof-Plan) Immediate from the Reduction Lemma and RRT. Let $R_{\preceq}(P_g)$ defined over g satisfying the conditions of clauses (d) or (e) of the Reduction Lemma be derivable from some Π_R, Π_I, Π_I by RRT2. By the Reduction Lemma, it can't be reduced to any restriction on states. RRT binds these restrictions to protocols on both ends. □

(11) (D1C2.10F) There are protocols that define restrictions on epistemic actions/processes/procedures that cannot be reduced to restrictions on epistemic transitions on the basis of the verdicts on transitions and

the relations between actions/processes/procedures and transitions. More precisely:

(a) (APPrtoT Irreducibility) There are protocols defining restrictions on epistemic actions (resp. processes, procedures), π_A (resp. π_P, π_{Pr}), such that there are no protocols, π_T, that define corresponding restrictions on transitions that can be defined on the basis of π_A's restrictions' verdicts (resp. π_p or π_{Pr}) on actions and the relations between transitions and actions (resp. processes, procedures).

Proof (Proof-plan) Immediate from the Reduction Lemma and RRT. Let $R_{\preceq}(P_g)$ defined over g satisfying the conditions of clauses (d) or (e) of the Reduction Lemma be derivable from some Π_R, Π_I, Π_I by RRT2. By the Reduction Lemma, it can't be reduced to any restriction on transitions. RRT binds these restrictions to protocols on both ends. □

10.4 Cluster D2

10.4.1 D2 Elaboration

Problem 486 Formalize the following cluster of principles of epistemic proceduralism:

(D2) Epistemic rationality norms define restrictions on the construction of procedures; they determine how basic epistemic actions are to be sequenced, iterated, combined, chosen among, conditioned on tests, avoided, reversed, and cautiously executed. In particular:

(1) Epistemic rationality requires the execution of sequential and iterative procedures under some conditions.
(2) Epistemic rationality requires parallel execution of procedures under some conditions.
(3) Epistemic rationality requires nondeterministic choice among procedures under some conditions.
(4) Sequential and iterative procedures determine sequentially composed paths through epistemic state space.
(5) Parallel execution procedures determine converging and corresponding paths through epistemic state space.
(6) Procedures instructing nondeterministic choice determine branching paths through epistemic state space.
(7) (PrCtoP) Restrictions on procedural construction can be used to generate restrictions on processes in the following sense: there are restrictions on procedural construction, R_{Pr}, such that a corresponding restriction on processes, R_P, matching R_{Pr}'s verdicts can be defined on the basis of R_{Pr}'s verdicts

on procedural constructions and the relations between procedural constructions and processes.

10.4.2 D2 Formalization

Solution 487 The Restriction Representation Theorem suffices to entail each of the following, the derivations of which are immediate or mechanical, and so merely outlined:

(1) (D2F) Protocols define restrictions on the construction of procedures; they determine how basic epistemic actions are to be sequenced, iterated, combined, chosen among, conditioned on tests, avoided, reversed, and cautiously executed.

Proof (Proof-Plan) This is immediate from the interpretation of the semantic structures and from RRT1. First, edges in EDGs are interpreted as token, atomic epistemic actions, so it follows that the induced graph-theoretic structure (path concatenation, iteration, correspondence, and branching) of EDGs represents the sequencing, iteration, parallel execution, and choice sequences of epistemic actions. Second, RRT1 immediately implies that every protocol defines a set of restrictions on procedural construction in the form of restrictions over the space of syntactic schemata. □

(2) (D2.1F) There are protocols that dictate sequential compositions and iterations of subprotocols.

Proof (Proof-Plan) The logic of dictation is immediate from the definition of protocol dictation relations and the R-Derivation System. For instance, protocols of the form $\pi_1; \pi_2$ and $\pi_1 \cap \pi_2$ dictate both π_1 and π_2, so, if either π_i is a sequential composition or iteration, this immediate observation suffices. More complicated instances abound: $\pi_1; (\pi_2 \cap \pi_3)$ dictates $\pi_1; \pi_2$ and $\pi_1; \pi_3$. □

(3) (D2.2F) There are protocols that dictate parallel executions of subprotocols.

Proof (Proof-Plan) The logic of dictation is immediate from the definition of protocol dictation and the R-Derivation System. For instance, protocols of the form $\pi_1; \pi_2$ and $\pi_1 \cap \pi_2$ dictate both π_1 and π_2, so, if either π_i is a parallel execution, this immediate observation suffices. More complicated instances abound: $(\pi_1 \cap \pi_2) \cup (\pi_3 \cap \pi_4)$ dictates $((\pi_1 \cup \pi_3) \cap (\pi_1 \cup \pi_4)) \cap ((\pi_2 \cup \pi_3) \cap (\pi_2 \cup \pi_4))$ and vice versa. □

(4) (D2.3F) There are protocols that dictate nondeterministic choices of subprotocols.

Proof (Proof-Plan) The logic of dictation is immediate from the definition protocol dictation and the R-Derivation System. For instance, protocols of the form $\pi_1; \pi_2$ and $\pi_1 \cap \pi_2$ dictate both π_1 and π_2, so, if either π_i is a nondeterministic choice, this immediate observation suffices. More complicated instances abound: $(\pi_1 \cap \pi_2); (\pi_3 \cap \pi_4)$ dictates $(\pi_1; \pi_3) \cup ((\pi_1; \pi_4) \cup ((\pi_2; \pi_3) \cup (\pi_1; \pi_4)))$.

□

(5) (D2.4F) Sequential composition and iteration are bound to path concatenation and branching by the execution rules and structural display rules.

Proof (Proof-Plan) The graph-theoretic structure of EDGs that execute protocols of the relevant syntactic forms demonstrate this, as does the PGIIRC procedure. For instance:

(a) Every execution of $\pi_1; \pi_2$ must have the form of a concatenation of executions of π_1 and π_2.
(b) PGIIRC shows that the protocol graphs of $\pi_1; \pi_2$ must have the form of a tree of concatenated executions of π_1 and π_2.

□

(6) (D2.5F) Parallel execution is bound to path correspondence by the execution rules and structural display rules.

Proof (Proof-Plan) The graph-theoretic structure of EDGs that execute protocols of the relevant syntactic forms demonstrate this, as does the PGIIRC procedure. For instance:

(a) Every execution of $\pi_1 \cap \pi_2$ must have the form of a corresponding pair of executions of π_1 and π_2.
(b) PGIIRC shows that the protocol graphs of $\pi_1 \cap \pi_2$ must have the form of a tree of corresponding executions of π_1 and π_2.

□

(7) (D2.6F) Nondeterministic choice is bound to path branching by the execution rules and structural display rules.

Proof (Proof-Plan) The graph-theoretic structure of EDGs that execute protocols of the relevant syntactic forms demonstrate this, as does the PGIIRC procedure. For instance:

(a) Every execution of $\pi_1 \cup \pi_2$ must have the form of a branching pair of executions of π_1 and π_2.
(b) PGIIRC shows that the protocol graphs of $\pi_1; \pi_2$ must have the form of a tree of branching executions of π_1 and π_2.

□

(8) (D2.7F) Protocols that define restrictions on procedural construction can be used to generate protocols that define corresponding restrictions on processes in the following sense: there are protocols that define restrictions on procedural construction, π_{Pr}, such that a protocols, π_p, that defines corresponding restrictions on processes can be defined on the basis of π_{Pr}'s restrictions' verdicts on procedural constructions (syntactic schemata) and the relations between procedural constructions and processes.

Proof (Proof-Plan) Immediate from the Reduction Lemma. RRT binds these restrictions to protocols on both ends. □

10.5 Cluster D2C1

10.5.1 D2C1 Elaboration

Problem 488 Formalize the following cluster of principles of epistemic proceduralism:

(D2C1) Procedural norms of epistemic rationality have the internal structure of programs or plans, conditioning actions and procedures on tests, and dictating procedures built with complex operations on action like conditional choice and guarded iteration, which are composed out of more basic procedural operations. In parts:

(1) Epistemic rationality norms dictate or instruct the execution of epistemic actions and procedures.
(2) Epistemic rationality norms specify the conditions under which epistemic actions and procedures are to be executed.
(3) Epistemic rationality norms dictate the construction of complex epistemic procedures formed with relatively basic epistemic actions (and procedures) and the procedural operations of relational algebra, including (inter alia) sequential composition, nondeterministic choice, tests.
(4) The foregoing procedural operations can be combined to form more complex operations like conditional choice and guarded iteration.

10.5.2 D2C1 Formalization

Solution 489 The cluster is formalized by the immediate syntactic and semantic properties of protocols in PLEN, which suffice to entail each of the following, the derivations of which are immediate or mechanical, and so merely outlined:

(1) (D2C1F) Protocols are programs, conditioning actions and procedures on tests, and dictating procedures built with complex operations on action like conditional choice and guarded iteration. This is shown in a piecemeal way:
(2) (D2C1.1F) Protocols dictate atomic and complex protocols and define restrictions on procedural construction.

    Proof (Proof-Plan) The first conjunct is immediate from the logic of dictation. The second conjunct is immediate from RRT. RRT1 and RRT2, in particular, immediately suffice. The definition of restrictions on procedural construction over syntactic schemata gives the result in an even more immediate way. □
(3) (D2C1.2F) Protocols have preconditions, and the syntactic form of protocols determines at which states in an EDG its subprotocols can be executed.

Proof (Proof-Plan) The first conjunct is immediate from the definition of preconditions, and the second conjunct is given by axioms like $[\pi_1; \pi_2]_p A \equiv [\pi_1]_p[\pi_2]_p A$. □

(4) (D2C1.3F) Protocols define restrictions on procedural construction defined over the procedural operations of relational algebra, and each protocol bears dictation relations with some set of protocols.

Proof (Proof-Plan) Immediate from RRT and the definition of restrictions on procedural construction over syntactic schemata, themselves defined over sequential composition, choice, parallel execution, and iteration. The matching schemata are, by definition, those that have the same operator sequences. □

(5) (D2C1.4F) Conditional Choice and Guarded Iteration protocols can be defined over L^R_{PLEN} and P_{PLEN}.

Proof (Proof-Plan) An explicit definition was provided in Chap. 4, reiterated here:

Name	English w/ metavariables	PLEN
Conditional Choice	If A, do π_1; else, do π_2!	$(((A)? \cap -(\neg A)?); \pi_1) * \cup((-(A)? \cap (\neg A)?); \pi_2)*$
Generalization of Conditional Choice	If A, do π_1; if B, do π_2; ...!	$(((A)? \cap -(B)?); \pi_1) * \cup(((B)? \cap -(A)?); \pi_2) * \cup \ldots$
Guarded Iteration	While A, do π	$((A)?; \pi)*; (\neg A)?$
Guarded Repetition	Repeat π until A is true.	$(\pi; (\neg A)?)*; (A)?$

□

10.6 Cluster D3 and D3C1

Claim 490 The distinctive forms of reasoning about and with epistemic norms that are central to formal epistemology and required by the regulative function of epistemic norms can be formalized in PLEN and shown to be sound.

10.6.1 D3 and D3C1 Elaboration

Problem 491 Formalize the following cluster of principles of epistemic proceduralism.

(1) (D3) Epistemic rationality norms can be reasoned about explicitly; terms denoting epistemic rationality norms and statements about the properties of epistemic norms can be embedded in valid arguments.

   (a) (D3.1) Epistemic rationality norms can be equivalent in multiple, distinct ways; reasoning about epistemic rationality must at least sometimes involve reasoning about these equivalence relations.

(b) (D3.2) The essential forms of reasoning about and with epistemic rationality norms include:

   (i) Arguments establishing norm equivalence and deriving facts about norms from premises about norm equivalence.

      (A) Arguments from the features of norms, norm-kernels, or restrictions to norm equivalence.

      (B) Arguments from norm equivalence to propositions about norms, norm-kernels, or restrictions on epistemic dynamics.

      (C) Arguments replacing embedded norms with equivalents in normative, procedural, deontic, modal, and propositional contexts.

   (ii) Arguments deriving normative verdicts (valences, deontic statuses, procedural requirements) from conditions of application, procedural structure, sentential or propositional contexts (including modal and deontic propositions).

   (iii) Arguments establishing dependence, sufficiency, and independence relations among norms.

(2) (D3C1) Where N_i is a term denoting an arbitrary norm, A_i is a formula embedding a norm term (letting $A(N_i, x)$ be a special case encoding N_i's verdict on an arbitrary epistemic factor, x), C_i is an arbitrary formula stating some condition,, and $\equiv_x$ is an arbitrary form of normative equivalence, the following table summarizes schematic inference rules for the foregoing kinds of reasoning:

Basic forms	Equivalence	(In)Dependence, sufficiency
$\dfrac{\begin{array}{c} A_1(N_1) \\ C \end{array}}{A_2(N_2)}$	$\dfrac{\begin{array}{c} N_1 \equiv_x N_2 \\ A_1(N_1) \end{array}}{A_2(N_2)}$	$\dfrac{\begin{array}{c} N_1 \supset_x N_2 \\ A_1(N_1) \end{array}}{A_2(N_2)}$
$\dfrac{\begin{array}{c} A_1(N_1) \\ C_1 \end{array}}{C_2}$	$\dfrac{\begin{array}{c} A_1(N_1) \\ A_2(N_2) \\ \ldots \end{array}}{N_1 \equiv_x N_2}$	$\dfrac{\begin{array}{c} \neg(N_1 \supset_x N_2) \wedge \neg(N_2 \supset_x N_1) \\ A_1(N_1) \end{array}}{A_2(N_2)}$
$\dfrac{\begin{array}{c} C_1 \\ C_2 \\ \ldots \end{array}}{A(N_i)}, \dfrac{\begin{array}{c} C_1 \\ C_2 \\ \ldots \end{array}}{A(N_i, x)}$	$\dfrac{N_1 \equiv_x N_2}{N_1 \equiv_y N_2}$	$\dfrac{\begin{array}{c} A_1(N_1) \\ A_2(N_2) \\ \ldots \end{array}}{N_1 \supset_x N_2}$
$\dfrac{\begin{array}{c} A(N_1, x) \\ C \end{array}}{A(N_2, y)}, \dfrac{\begin{array}{c} A(N_1, x) \\ C \end{array}}{A(N_2, x)}$		$\dfrac{\begin{array}{c} A_1(N_1) \\ A_2(N_2) \\ \ldots \end{array}}{\neg(N_1 \supset_x N_2) \wedge \neg(N_2 \supset_x N_1)}$

Remark 492 Let $N(x)$ encode some verdict of N with respect to x, $A(N)$ encode some statement about a property of N, C encode some non-normative proposition, and $\equiv_x$ encode some form of norm equivalence, and $\supset_x$ encode some form of norm sufficiency and dependence relation depending on left and right direction. These are some of the forms of reasoning about epistemic rationality norms the formalization of which is criterial for a logic of epistemic norms. Or so it's argued in Chaps. 6 and 7. The rules introduced in this section formalize many instances of these argument forms. I don't explicitly present illustrations for each form, but it's not difficult to apply the R-Derivation System to the task.

10.6.2 D3 and D3C1 Formalization

This section adds onto the formalization of the foregoing forms of reasoning that was presented in Sect. 7.3.

Solution 493 The theory of protocol equivalence in PLEN and the R-Derivation system mechanically formalize the forms of reasoning articulated by D3 and D3C1:

(1) (D3F) Protocols can be reasoned about explicitly in PLEN; terms denoting protocols and statements about the properties of protocols can be embedded in valid arguments.

- (a) (D3.1F) Protocols can be equivalent in multiple, distinct ways; reasoning about protocols must at least sometimes involve reasoning about these equivalence relations.
- (b) (D3.2F) The essential forms of reasoning about and with protocols include:
  - (i) Arguments establishing protocol equivalence of protocols and deriving facts about protocols from premises about protocol equivalence.
    - (A) Arguments from the features of protocols, protocol functions, or restrictions on procedural construction and epistemic dynamics defined by protocols to protocol equivalence.
    - (B) Arguments from protocol equivalence to propositions about protocols, protocol functions, and restrictions on procedural construction, processes, transitions, and states.
    - (C) Arguments replacing embedded protocols with equivalents in meta-level, procedural, deontic, modal, and propositional contexts.
  - (ii) Arguments deriving protocol function verdicts (valences, deontic statuses, procedural requirements) from conditions of application, procedural structure, sentential or propositional contexts (including modal and deontic propositions) concerning protocols.
  - (iii) Arguments establishing dependence, sufficiency, and independence relations among protocols.

(2) (D3C1F) The R-Derivation System formalizes most of the schematic forms of reasoning articulated in D3C1, and provides a framework for building complex arguments therefrom. Some of the schematic argument forms are immediate, having been shown sound in Chaps. 5, 6, and 7, and the remaining forms can be immediately derived:

Basic forms, verdicts, and properties	Equivalence	(In)Dependence, sufficiency
$\dfrac{A_1(\pi_1) \quad C}{A_2(\pi_2)}$	$\left[\dfrac{\pi \equiv_G \pi',\ \rho(\pi/\pi')(X_\pi) = X_{\pi'},\ \pi \in ISub(X_\pi)}{X_\pi \equiv_G X_{\pi'}} \right.$ $\dfrac{\pi \equiv_G \pi',\ \rho(\pi/\pi')(X_\pi) = X_{\pi'}}{X_\pi \equiv_G X_{\pi'}}$ $\dfrac{\pi \equiv_G \pi',\ \rho(\pi/\pi')(X_\pi) = X_{\pi'},\ A_{(X_\pi)}}{A_{(X_{\pi'})}}$ $\left. \dfrac{\pi_1 \equiv_G \pi_2}{\pi_1 \supset_G \pi_2},\ \dfrac{\pi_1 \equiv_G \pi_2}{\pi_2 \supset_G \pi_1} \right]$	$\dfrac{\pi_1 \supset_x \pi_2 \quad A_1(\pi_1)}{A_2(\pi_2)}$
$\dfrac{A_1(\pi_1) \quad C_1}{C_2}$	$\left[\dfrac{\pi \equiv_G \pi',\ \rho(\pi/\pi')(X_\pi) = X_{\pi'},\ \pi \in ISub(X_\pi)}{X_\pi \equiv_G X_{\pi'}} \right.$ $\dfrac{\pi \equiv_G \pi',\ \rho(\pi/\pi')(X_\pi) = X_{\pi'}}{X_\pi \equiv_G X_{\pi'}}$ $\dfrac{\pi \equiv_G \pi',\ \pi' \equiv_G \pi''}{\pi \equiv_G \pi''}$ $\dfrac{}{\pi \equiv_G \pi}$ $\left. \dfrac{\pi \equiv_G \pi'}{\pi' \equiv_G \pi} \right]$	$\dfrac{\neg(\pi_1 \supset_x \pi_2) \wedge \neg(\pi_2 \supset_x \pi_1) \quad A_1(\pi_1)}{A_2(\pi_2)}$
$\dfrac{C_1 \quad C_2 \quad \ldots}{A(\pi)},\ \dfrac{C_1 \quad C_2 \quad \ldots}{A(\pi, x)}$	$\left[\dfrac{\pi \equiv_G \pi'}{\pi \equiv_p \pi'} \quad \dfrac{\pi_1 \supset_G \pi_2,\ \pi_2 \supset_G \pi_1}{\pi_1 \equiv_G \pi_2} \right.$ $\dfrac{\pi \equiv_p \pi'}{\pi \equiv_c \pi'}$ $\dfrac{\pi \equiv_c \pi'}{\pi \equiv_t \pi'}$ $\left. \dfrac{\pi \leftrightarrow_r \pi'}{\pi = \pi'} \right]$	$\left[\dfrac{\pi_1 \equiv_G \pi_2}{\pi_1 \supset_G \pi_2},\ \dfrac{\pi_1 \equiv_G \pi_2}{\pi_2 \supset_G \pi_1} \right]$
$\dfrac{A(\pi_1, x) \quad C}{A(\pi_2, x)}$		$\dfrac{A_1(\pi_1) \quad A_2(\pi_2) \quad \ldots}{\neg(\pi_1 \supset_x \pi_2) \wedge \neg(\pi_2 \supset_x \pi_1)}$

Proof (Proof-Plan) Case by case derivation of sound rules of the requisite forms of argument. See, first and foremost, the formalization of the theory of protocol graph isomorphism equivalence and other protocol equivalences by means of the R-Derivation System in Chaps. 5, 6, and 7. From there, we can build further formalizations for complex reasoning about protocols (hence, procedural epistemic norms).

(1) Reasoning from properties of protocol functions and derivable restrictions on epistemic dynamics to protocol equivalence:

   (a) If $f_{\pi}(C, \pi'', s) = f_{\pi'}(C, \pi'', s)$ for all C and all π'', then $\pi \equiv_c \pi'$ or a stricter equivalence holds.

      (i) (Proof-plan) If $f_{\pi}(C, \pi'', s) = f_{\pi'}(C, \pi'', s)$, then, by the rules enumerated in the PFT proof, they deliver identical verdicts on all implementations and subprotocols of π and π' including π and π' themselves. Thus, since π dictates π itself, π' must dictate π, and vice versa. From the definition of dictation, it follows that every execution of π for every g at least corresponds to an execution of π' and vice versa. Thus, $\pi \equiv_c \pi'$. If the stronger condition of the definition of dictation holds—that every π in every g contains an execution of π' and vice versa, then every execution of π in g is an execution of π' and vice versa, and so $\pi \equiv_p \pi'$.

   (b) If every restriction on processes π defines is identical to some restriction on processes π' defines and the converse also holds, then $\pi \equiv_p \pi'$.

      (i) (Proof Sketch) This is immediate from the construction of restrictions on processes in the argument for the Reduction Lemma; let $R_{P_g} = R'_{P_g}$ and just so for the other verdict sets. Then, from construction, $P_{\pi} = P_{\pi'}$.

   (c) If every restriction π defines is identical to some restriction π' defines and the converse also holds, then $\pi \equiv_G \pi'$.

      (i) (Proof-plan) If π and π' define identical restrictions on procedural construction, then they define identical sets of matching protocols. But from the identity of process restrictions, $At(\pi) = At(\pi')$. Thus π and π' are identical. Thus, a fortiori, $\pi \equiv_G \pi'$.

(2) Reasoning from properties of protocols to deontics and back

   (a) Valid basic inferences for protocol-theoretic deontics.

Verdict: Obligation	Verdict: Permission	Verdict: Prohibition/Forbiddance
1. $s \Vdash C \leftrightarrow \langle\pi\rangle_p \top$ 2. $\forall p \in P_\pi(last(p) = s')$ $\therefore$ $f_\pi^\forall(C, s', s) = RO_\pi(s')$	1. $s \Vdash C \leftrightarrow \langle\pi\rangle_p \top$ 2. $\exists p \in P_\pi(last(p) = s')$ $\therefore$ $f_\pi^\forall(C, s', s) = RP_\pi(s')$	1. $s \Vdash C \leftrightarrow \langle\pi\rangle_p \top$ 2. $\neg\exists p \in P_\pi(last(p) = s)'$ $\therefore f_\pi^\forall(C, s', s) = RF_\pi(s')$
1. $s \Vdash C \leftrightarrow \langle\pi\rangle_p \top$ 2. $\forall p \in P_\pi(first(p) =$ $s,\ last(p) = s')$ $\therefore f_\pi^\forall(C, \langle s, s'\rangle, s) =$ $RO_\pi(\langle s, s'\rangle)$	1. $s \Vdash C \leftrightarrow \langle\pi\rangle_p \top$ 2. $\exists p \in P_\pi(first(p) =$ $s,\ last(p) = s')$ $\therefore f_\pi^\forall(C, \langle s, s'\rangle, s) =$ $RP_\pi(\langle s, s'\rangle)$	1. $s \Vdash C \leftrightarrow \langle\pi\rangle_p \top$ 2. $\neg\exists p \in P_\pi(first(p) =$ $s,\ last(p) = s')$ $\therefore f_\pi^\forall(C, \langle s, s'\rangle, s) = RF_\pi(\langle s, s'\rangle)$
1. $s \Vdash C \leftrightarrow \langle\pi\rangle_p \top$ 2. $\{p\} = P_\pi$ $\therefore$ $f_\pi^\forall(C, p, s) = RO_\pi(p)$	1. $s \Vdash C \leftrightarrow \langle\pi\rangle_p \top$ 2. $p \in P_\pi$ $\therefore$ $f_\pi^\forall(C, p, s) = RP_\pi(p)$	1. $s \Vdash C \leftrightarrow \langle\pi\rangle_p \top$ 2. $p \notin P_\pi$ $\therefore f_\pi^\forall(C, p, s) = RF_\pi(p)$
1. $s \Vdash C \leftrightarrow \langle\pi\rangle_p \top$ 2. $\exists q(\forall p \in P_\pi(q \sqsubseteq p))$ $\therefore f_\pi^\forall(C, q) = RO_\pi(q)$	1. $s \Vdash C \leftrightarrow \langle\pi\rangle_p \top$ 2. $\exists q(\exists p \in P_\pi(q \sqsubseteq p))$ $\therefore f_\pi^\forall(C, q, s) = RP_\pi(q)$	1. $s \Vdash C \leftrightarrow \langle\pi\rangle_p \top$ 2. $\exists q(\neg\exists p \in P_\pi(q \sqsubseteq p))$ $\therefore f_\pi^\forall(C, q, s) = RF_\pi(q)$
1. $s \Vdash C \leftrightarrow \langle\pi\rangle_p \top$ 2. $\forall p \in P_\pi(last(p) \Vdash A)$ $\therefore$ $f_\pi^\forall(C, A, s) = RO_\pi(A)$	1. $s \Vdash C \leftrightarrow \langle\pi\rangle_p \top$ 2. $\exists p \in P_\pi(last(p) \Vdash A)$ $\therefore$ $f_\pi^\forall(C, A, s) = RP_\pi(A)$	1. $s \Vdash C \leftrightarrow \langle\pi\rangle_p \top$ 2. $\neg\exists p \in P_\pi(last(p) \Vdash A)$ $\therefore f_\pi^\forall(C, A, s) = RF_\pi(A)$

Verdict: Obligation	Verdict: Permission	Verdict: Prohibition/Forbiddance
1. $C \leftrightarrow$ $\langle a; ((b \cap c); d)\rangle_p \top$ $\therefore f_\pi^\forall(C, (a \cap b), s) =$ $RO_\pi(a \cap b)$ $\therefore f_\pi^\forall(C, (b), s) = Ob$	1. $s \Vdash C \leftrightarrow$ $\langle a; ((b \cup c); d)\rangle_p \top$ $\therefore$ $f_\pi^\forall(C, (b), s) = RP_\pi(b)$	1. $s \Vdash C \leftrightarrow \langle a; ((-b \cap c); d)\rangle_p \top$ $\therefore f_\pi^\forall(C, (b), s) = RF_\pi(b)$
1. $s \Vdash C \leftrightarrow$ $\langle a; ((b \cap c); d)\rangle_p \top$ $\therefore f_\pi^\forall(C, (a \cap b), s) =$ $RO_\pi(a \cap b)$ $\therefore$ $f_\pi^\forall(C, (b), s) = RO_\pi(b)$	1. $s \Vdash C \leftrightarrow$ $\langle a; ((-b \cap c); d)_p\rangle \top$ $\therefore f_\pi^\forall(C, (a; c); d, s) =$ $RP_\pi((a; c); d)$	1. $s \Vdash C \leftrightarrow \langle a; ((-b \cap c); d)_p\rangle \top$ $\therefore$ $f_\pi^\forall(C, (a; b); d, s) = RF_\pi((a; b); d)$

(a) I leave derivation of these rules, which range across the meta and object-level, as exercises. The point is that they are formalizable in PLEN, many of them are immediate, and all can be derived from the protocol logic of PLEN.

(3) Valid basic inferences from conditions of application to properties of norms.

From Conditions of Application and Execution to Propositional Verification 1. $s \Vdash C \leftrightarrow \langle\pi\rangle_p \top$ 2. $p \in P_\pi$ $\therefore first(p) \Vdash C$	From Conditions of Application and Process Structure to Transition Structure 1. $s \Vdash C \leftrightarrow \langle\pi\rangle_p \top$ 2. $\forall p \in P_\pi \exists q (q \sqsubseteq p)$ $\therefore s \Vdash C \rightarrow \exists t =$ $\langle s, s'\rangle (s' = first(q))$	From Syntactic Form (Procedural Structure) to Graph and Process Structure 1. $\pi = \pi_1 \cap \pi_2$ $\therefore \forall p \in P_\pi (\exists q \in P_{\pi 1}, \exists r \in P_{\pi 2} (q \rightleftharpoons r, p = q \vee q = r))$
From Conditions of Application, Syntactic Form, and Propositional Verification to Process Structure, Execution, and Propositional Verification 1. $s \Vdash C \leftrightarrow \langle\pi\rangle_p \top$ 2. $\pi = \pi_1; \pi_2$ 3. $first(p) \Vdash C$ $\therefore p = qr, q \in P_\pi, r \in P_\pi,$ $first(q) \Vdash C$	From Verification to Graph and Process Structure 1. $first(p) \Vdash A \wedge [\pi *]_p (A \rightarrow [\pi]_p A)) \rightarrow [\pi *]_p A$ $\therefore p$ is a loop or $p = q_1 \ldots q_n$ where all $q_i \in P_\pi$	From Syntactic Form (Procedural Structure) and Verdict to Graph and Process Structure 1. $\pi = \pi_1; \pi_2$ 2. $f_\pi^{\forall}(C, \pi_3, s) = RO_\pi$ $\therefore \forall p \in P_\pi (\exists q \in P_{\pi_1}, \exists r \in P_{\pi_2} \exists p' \in P_{\pi_3} (p' \sqsubseteq$ $q \vee p' \sqsubseteq r))$
From Syntactic Form (Procedural Structure) to Graph and Process Structure 1. $\pi = \pi_1; \pi_2$ $\therefore \forall p \in P_\pi (\exists q \in P_{\pi 1}, \exists r \in P_{\pi 2} (qr = p))$	From Syntactic Form (Procedural Structure) and Deontics to Syntactic Form 1. $\pi = \pi_1; \pi_2$ 2. $f_\pi^{\forall}(C, \pi_3, s) = RO_\pi(\pi_3)$ $\therefore$ $\pi_3 \in Sub(\pi_1) \cup Sub(\pi_2) \vee$ $\exists \pi' \mapsto \pi(\pi_3 \in Sub(\pi'))$	

(a) I leave derivation of these rules, which range across the meta and object-level, as exercises. The point is that they are formalizable in PLEN, many of them are immediate, and all can be derived from the protocol logic of PLEN.

(4) (Derived Laws) There are numerous derived laws from the main results of PLEN. Here is a derivation of one object-level law (though, strictly, the derivation could be used to arrive at at least four):

(a) $\langle a; ((b \cap c); d)\rangle_p \top$
(b) $\therefore \langle a\rangle_p \langle((b \cap c); d)\rangle_p \top$
(c) $\therefore \langle a\rangle_p \langle b \cap c\rangle_p \langle d\rangle_p \top$
(d) $\therefore \langle a\rangle_p \langle b\rangle_p \langle d\rangle_p \top \wedge \langle a\rangle_p \langle c\rangle_p \langle d\rangle_p \top$
(e) $\therefore \langle a; (b; d)\rangle_p \top \wedge \langle a; (c; d)\rangle_p \top$
(f)

$$(b(A) \rightarrow \langle a; ((b \cap c); d)\rangle_p \top) \rightarrow (b(A) \rightarrow \langle a\rangle_p \langle c\rangle_p \top)$$

(g) From there, facts about process structure, executability, etc. can be derived, and from there, by the rules above, normative verdicts.

(5) Reasoning from and to Dependence, Independence, and Sufficiency can be formalized at the meta-level. Consider the following definition:

Definition 494 (Verdict Dependence) Protocol π is verdict dependent on π' iff :

$$\forall C \in Con(\pi') \cap Con(\pi) \forall x \in Dom(\pi) \exists y \in Dom(\pi') \\ [f_{\pi}^{\forall}(C, x, s) = X \rightarrow f_{\pi'}^{\forall}(C, y, s) = Y]$$

Remark 495 This is a general schema for dependence and sufficiency relations. Let π represent norm N_1 and π' represent norm N_2. Then verdict dependence tells us that for every condition of application shared by N_1 and N_2, for every object of N_1, x, there is some object, y, of N_2 such that N_1's verdict on x being X under C implies that N_2's verdict on y under C is Y. The idea is that x's being X with respect to N_1 determines that y is Y with respect to N_2. Of course, if y is not Y with respect to N_2, then x is not X with respect to N_1, and so x's being X depends on y's being Y. Reduction strategies, as in Chap. 2, are attempts at establishing this kind of conditional. Whatever our reduction strategy, this is essentially what a claim that N_1 is dependent on N_2 asserts; the content of N_1's verdicts determine and depend on N_2's verdicts.

(a) The protocol logic of PLEN provides a framework in which to reason rigorously about such relations among protocols. In particular, the following rules can be used to reason to verdict dependence and sufficiency from premises concerning protocol equivalence:
   (i) (GVD1) If $\pi \equiv_G \pi'$, then for arbitrary EDG, g:

$$\forall C \in Con(\pi') \cap Con(\pi) \forall y \in Dom(\pi') \cap P_g \forall x \in Dom(\pi) \cap P_g \\ [f_{\pi}^{\forall}(C, x, s) = X \leftrightarrow f_{\pi'}^{\forall}(C, y, s) = X]$$

      (A) (Proof-plan) If two protocols are graph-theoretically equivalent, then, from PRT, they are procedurally equivalent and intersubstitutable in all procedural contexts in PLEN; they will thus bear equivalent relations, with respect to the deontic operators, to the same paths.
   (ii) (GVD2) If $\pi \equiv_G \pi'$, then for any EDG, g:

$$\forall C \in Con(\pi') \cap Con(\pi) \forall y \in Dom(\pi') \cap P_{PLEN} \forall x \in Dom(\pi) \cap P_{PLEN} \\ [f_{\pi}^{\forall}(C, x, s) = X \leftrightarrow f_{\pi'}^{\forall}(C, y, s) = X]$$

      (A) (Proof-plan) If two protocols are graph-theoretically equivalent, then, from PRT, they are procedurally equivalent and intersubstitutable in all procedural contexts in PLEN; they will thus dictate or permit the same protocols.

(b) Protocol implication can also be used as a starting point for reasoning to dependence (et al):

(i) (GVD3) Let $\pi \supset_G \pi'$ iff $\forall g \in G^{\pi} \exists g' \in G^{\pi'} \exists g''[(g'' \sqsubseteq g') \wedge (g \simeq g'')]$. If $\pi \supset_G \pi'$, then for arbitrary EDG, g:

$$\forall C \in Con(\pi') \cap Con(\pi) \forall x \in Dom(\pi) \cap P_{PLEN} \exists y \in Dom(\pi') \cap P_{PLEN}$$
$$[f_{\pi}^{\forall}(C, x, s) = X \rightarrow f_{\pi'}^{\forall}(C, y, s) = X]$$

(A) (Proof-plan) This rule runs on the relations between dictation and the protocol-theoretic deontics. Every maximal execution of π' in any protocol graph of π' is going to contain a (duplicate of a) maximal execution of π, and so if π dictates or permits π'', so does π'.

(c) These rules can also be used to reason from the failure of verdict dependence (i.e., from verdict independence) to the failure of protocol equivalence.

(6) Dependence (et al) can be reasoned about in the object-language, by means of the object-level counterparts of the protocol equivalence relations in PLEN. The object language equivalence relation variants, $\leftrightarrow_X$ or $\equiv_X^s$, in L_{PLEN}^R can be used as stand-ins for establishing more limited kinds of dependence (et al):

(a) (GVD2') If $g, s \models_{PLEN} \pi \equiv_G^s \pi'$, then:

$$\forall C \in Con(\pi') \cap Con(\pi) \forall y \in Dom(\pi') \cap P_{PLEN} \exists x \in Dom(\pi) \cap P_{PLEN}$$
$$[f_{\pi}^{g,s}(C, x, s) = X \leftrightarrow f_{\pi'}^{g,s}(C, y, s) = X]$$

(i) (Proof-plan) This is immediate; the argument for GVD2 applied to arbitrary states in arbitrary graphs, and SUT generalizes the object-level protocol graph isomorphism equivalence to the meta-level protocol graph isomorphism equivalence.

□

10.7 Cluster D3C2

10.7.1 Elaboration of D3C2

Problem 496 Formalize the following cluster of principles of epistemic proceduralism:

(D3C2) Epistemic rationality norms assign deontic properties to epistemic factors including states, transitions, actions, processes, and complex procedures.

(1) Deontic concepts of obligation, permission, forbiddance, and reason can be translated into the procedural concepts of execution, dictation, and compatibility of actions and processes with procedures.
(2) Deontic operators are defined relative to sets of norms.
(3) Epistemic rationality norms assign deontic statuses to epistemic states, transitions, processes, and propositions both directly and indirectly, and these can be conceptualized as normative goal states or postconditions.

10.7.2 Formalization of D3C2

Claim 497 PLED provides a direct translation of D3C2 into PLEN, and these translations can be shown to be immediate consequences of the resulting protocol logic. PLED was given some philosophical grounding in Chap. 8, but all that is relevant to the current project, with respect to our guiding theses of adequate representation, is showing that the Revised PLEN framework can support adaptations that allow us to derive formal translations of the principles in D3C2. And this, the resulting system can do.

Solution 498 The protocol-theoretic logic of deontic operators in PLEN suffices to entail each of the following, the derivations of which are mechanical, and so merely outlined:

(1) (D3C2F) Protocols assign deontic properties to epistemic factors including states, transitions, actions, processes, and complex procedures.

    Proof (Proof-Plan) Immediate from the definition of PLED and less immediate from the representation theorems. First, the protocol-theoretic deontic operators apply to each of these factors and their semantics binds the truth conditions of the resulting deontic statements or normative verdicts to the execution-theoretic and syntactic properties of protocols and their relations to these epistemic factors. Second, restrictions on epistemic dynamics capture deontic verdicts, and RRT shows that every protocol defines a set of restrictions on epistemic dynamics on each class of epistemic factors. Third, PFT directly maps protocols to deontic verdicts on epistemic factors of each kind. □

(2) (D3C2.1F) Deontic concepts of obligation, permission, forbiddance, and reason can be translated into the procedural concepts of execution, dictation, and compatibility of actions and processes with protocol execution conditions.

    Proof (Proof-Plan) The definition of deontic operators in terms of the properties of protocols provides a direct translation of deontic operators into the procedural concepts of execution, dictation, and compatibility (convergence, correspondence, and subpath relation with executions). Every protocol is itself a syntactic object structured by procedural operations, and the properties of each protocol's execution, dictation, and compatibility with procedures (convergence, correspondence, and subpath relation with executions) are determined by the protocol's syntactic (i.e., procedural) structure. The execution rules then directly relate deontics to protocol execution conditions. □

(3) (D3C2.2F) Deontic operators are defined relative to sets of protocols.

Proof ***(Proof-Plan)*** This is direct from construction; the deontics in PLEN are each defined relative to protocols and methodologies, themselves each an analyzans of the commands of some procedural epistemic norm or set thereof. □

(4) (D3C2.3F) Protocols assign deontic statuses to epistemic states, transitions, processes, and propositions both directly and indirectly, and these can be conceptualized as normative goal states or postconditions.

Proof (Proof-plan) The definition of PLED and the derivative definition of protocol functions suffice to show that protocols map the classes of objects of norm-kernels to protocol-theoretic deontic operators by means of the postconditions of protocols. □

Claim 499 These claims are mostly immediate, but the following (similarly immediate) propositions ramify them.

Proposition 500 (D3C2.1) *The definition of the syntax and semantics of L^D_{PLEN} provides an explicit translation of deontic operators into procedural concepts, including execution, dictation, and compatibility of actions and processes with procedures.*

Proposition 501 (D3C2.2) *PFT implies that protocols assign protocol-theoretic analyses of deontic operators to actions and procedures.*

Proposition 502 (D3C2.3) *PFT implies that protocols assign deontic statuses to epistemic states, transitions, processes, and propositional expressions by means of postcondition constructions and the dictation relation.*

10.8 Cluster D3C3

10.8.1 D3C3 Elaboration

Problem 503 Formalize the following cluster of principles of epistemic proceduralism:

(D3C3) Epistemic rationality norms can be deliberately followed/applied/adhered/complied with (as opposed to merely complied with).

(1) Epistemic actions can be procedurally operated upon in knowing compliance with epistemic rationality norms, and this involves determining divergences in actual and possible practice from the rational case.
(2) Natural, heuristic, unconscious epistemic processes can be halted by effortful cognition.

(3) Epistemic processes can be deliberately re-directed in accordance with the restrictions defined by epistemic norms.
(4) Deliberate application of a norm requires knowing adherence to it.
(5) Knowing adherence of a norm is distinct from merely following a norm.
(6) Deliberate application of norms requires reasoning about epistemic rationality norms.
(7) Deliberate construction of procedures out of procedural operations on epistemic action in compliance with norms requires reasoning about procedural content.

10.8.2 D3C3 Formalization

Claim 504 PLEDKNA provides a direct translation of D3C3 into PLEN, and these translations can be shown to be immediate consequences of the resulting protocol logic. It is beyond the scope of this book to fully evaluate the philosophical prospects of PLEDKNA, though it is within the scope of future work on PLEN. All that is relevant to the current project, with respect to our guiding theses of adequate representation, is showing that the Revised PLEN framework can support adaptations that allow us to derive formal translations of the principles in D3C3. And this, the resulting system can do.

Solution 505 The protocol-theoretic logic of procedural knowledge and norm application in PLEN suffices to entail each of the following, the derivations of which are mechanical, and so merely outlined:

(1) (D3C3F) Protocols can be knowingly adhered with and deliberately applied (as opposed to merely complied with).

    Proof (Proof-Plan) Immediate from the definition of PLEDKNA. Both of these properties of protocols are encoded by means of distinct pairs of operators. The feasible and actual knowing adherence and deliberate application operators apply to protocols, and expressions built using these operators are satisfiable in a subclass of EDGs. The feasible variants of these operators can be used to build expressions that encode statements to the effect that a given protocol (or methodology) can be knowingly adhered to or deliberately applied from a given epistemic state in an EDG. The actual variants can be used to build expressions that encode statements to the effect that prior knowing adherence or deliberate application of a protocol brought us to the current state. Knowing adherence and deliberate application have inequivalent truth conditions, and so the system distinguishes these notions. □

(2) (D3C3.1F) Epistemic actions can be procedurally operated upon in knowing compliance with protocols, and this involves determining divergences in actual and possible practice from those paths that are compatible with a given protocol.

    Proof (Proof-Plan) This can be seen by construction given the machinery of PLEDKNA:

(a) There are executable protocols that both dictate complex protocols and can be deliberately applied.
(b) Actual deliberate application of a protocol dictated by another implies that beliefs must be held concerning divergences between feasible paths and execution paths of protocols.
(c) Showing that there are such protocols and deriving the resulting implication is a mechanical construction and evaluation task.

□

(3) (D3C3.2F) Natural, heuristic, unconscious epistemic processes can be halted by effortful cognition.

Proof (Proof-Plan) There are EDGs satisfying the conditions of effortful cognition; g such that g_n is non-empty and there are paths in g_n that prefix paths that verify knowing adherence with protocols at their terminal states . Showing that there are such EDGs is a mechanical construction and evaluation task. □

(4) (D3C3.3F) Epistemic processes can be deliberately re-directed in accordance with the restrictions defined by protocols.

Proof (Proof-Plan) There are EDGs satisfying the conditions of deliberate application; g such that g_n is non-empty and there are paths in g_n that prefix reasoning processes that terminate with states verifying knowing adherence with protocols that verify reasons to follow the protocols knowingly adhered to. Showing that there are such EDGs is a mechanical construction and evaluation task. □

(5) (D3C3.4F) Deliberate application of a protocol requires knowing adherence to it.

Proof (Proof-Plan) Immediate consequence of semantics of deliberate application. □

(6) (D3C3.5F) Knowing adherence of a protocol is distinct from merely following a protocol.

Proof (Proof-Plan) Immediate consequence of semantics of knowing adherence and compliance. □

(7) (D3C3.6F) Deliberate application of a protocol requires reasoning about protocols.

Proof (Proof-Plan) Immediate from construction of deliberate application; the definitions of reasons, reasoning, and deliberate application suffice. One must have defeasible reasons for thinking that π is permissible or obligatory and that it's forbidden to do at least something else. If one has these defeasible reasons, one got to the current state by a process of reasoning. Given that the formulae built into these reasons embedded π in protocol-theoretic contexts, this is reasonably construed as reasoning about π. □

(8) (D3C3.7F) Deliberate construction of procedures out of procedural operations on epistemic action in compliance with protocols requires reasoning about

the execution conditions, procedural structure, equivalence relations, dictation properties, of protocols.

Proof (Proof-Plan) Deliberate application of protocols that dictate other protocols, deliberate application of protocols that implement other protocols, deliberate application of protocols dictated by other protocols, or even knowing restriction of epistemic processes represents the deliberate construction of procedures by procedural operations, as the syntax of dictated or implemented protocols is bound to the graph-theoretic structure of paths—regular control operators correspond to concatenation, correspondence, branching, and iteration. Given this claim, the prior result immediately implies that deliberate procedural construction requires reasoning processes that add reasons about protocols. For instance, a plausible substantive requirement on reasoning about protocols is that, in order to deliberately apply some or another protocol, one must determine whether there is a protocol that one can feasibly execute that implements, dictates, implies, or is equivalent to it. This condition is built into PLEN; to deliberately apply π, one must reason one's way to the knowledge that enables knowing adherence to it, which requires knowing that it's executable. One can acquire this knowledge by paths of reasoning from the implementation, dictation, equivalence, and implication properties of protocols. □

10.9 For Future Work

This short chapter was a no-frills presentation of ways in which to translate the principles that define epistemic proceduralism into the languages of PLEN and then validate them in the logic of PLEN. For every sub-clause and corollary of D1–D3C3, there was a formal assertion in the Revised PLEN Framework that, given the representation theorems of the prior chapter and the epistemological arguments throughout the book, plausibly encode those sub-clauses and corollaries. Each of these then was derived from the protocol logic of the Revised PLEN Framework. This completes the argument for thesis T3. It's now been shown that the Revised PLEN Framework is interesting and theoretically useful for representing and reasoning about epistemic rationality norms, at least in the sense that the core clusters of principles defining epistemic proceduralism can be derived from PLEN.

The objective of the chapter is not to provide a final theory of any of these features of epistemic norms, but to demonstrate PLEN's utility for formally codifying epistemic proceduralism or, in other words, for showing that PLEN satisfies the criteria of theoretical adequacy that epistemic proceduralism imposes on would-be formal frameworks for the logic of epistemic norms. As with much else about PLEN, the main objective is future-directed and open to revision; the goal was demonstrating PLEN's usefulness as a tool by inaugurating projects that apply the system, not completing them. Thoroughly litigating the quality of PLEN's formalization of epistemic proceduralism, especially a fuller exploration of the protocol-theoretic epistemic logic in PLEN, or fully developing a theory choice argument for the PLEN Framework over all plausible rivals, is left, as so much always must be, for future work.

Reference

1. Baltag, A., and L.S. Moss. 2004. Logics for epistemic programs. *Synthese*. 139:165–224.

The manufacturer's authorised representative in the EU is Springer Nature Customer Service Centre GmbH, Europaplatz 3, 69115 Heidelberg, Germany. If you have any concerns regarding our products, please contact ProductSafety@springernature.com

Printed and bound by CPI Group (UK) Ltd, Croydon, CR0 4YY
30/09/2026
02225237-0018